Assumptions (B) for Multiple Linear Regression

(Population) Assumption 1 The study population $\{(Y, X_1, \ldots, X_k)\}$ is a $(k + 1)$-variable Gaussian population.

(Sample) Assumption 2 The sample data are obtained by simple random sampling described in Section 2.3; i.e., a simple random sample of n items is selected from the population and the values of the variables $Y, X_1, \ldots, X_k$ are observed.

(Sample) Assumption 3 The sample values $y_i, x_{i1}, \ldots, x_{ik}$ for $i = 1, \ldots, n$ are measured without error.

Simple Random Sampling

Sample data are obtained by selecting a **simple random sample** of n items from the entire population of N items and recording the values for the response variable Y and the predictor variables $X_1, \ldots, X_k$, for each item in the sample. Refer to Section 1.6.

Random Sampling with Preselected *X* values

Specific values of the predictor variables $X_1, \ldots, X_k$ are preselected by the investigator, and each of these preselected sets of values determines a subpopulation of Y values. A simple random sample of one or more Y values is selected from each of these subpopulations. The number of observations to be sampled from each subpopulation is also predetermined by the investigator.

Regression Analysis:
Concepts and Applications

Regression Analysis: Concepts and Applications

Franklin A. Graybill

Colorado State University

Hariharan K. Iyer

Colorado State University

An Alexander Kugushev Book

Duxbury Press
An Imprint of Wadsworth Publishing Company
Belmont, California

Duxbury Press
An Imprint of Wadsworth Publishing Company
A division of Wadsworth, Inc.

Assistant Editor   *Jennifer Burger*
Editorial Assistant   *Michelle O'Donnell*
Production   *Ruth Cottrell*
Designer   *Cloyce Wall*
Print Buyer   *Randy Hurst*
Copy Editor   *Ruth Cottrell*
Cover: *Stuart Paterson*
Technical Illustrator and Compositor   *Interactive Composition Corporation*
Printer   *Banta Corporation*

This book is printed on acid-free recycled paper.

International Thomson Publishing
The trademark ITP is used under license

Printed in the United States of America
1 2 3 4 5 6 7 8 9 10—98 97 96 95 94

Library of Congress Cataloging-in-Publication Data

Graybill, Franklin A.
    Regression analysis : concepts and applications / Franklin A. Graybill, Hariharan K. Iyer.
        p.   cm.
    Includes index.
    ISBN 0-534-19869-4
    1. Regression analysis. I. Iyer, Hariharan K. II. Title.
QA278.2.G73   1994
519.5'36—dc20   93-44089
                                                                              CIP

Contents

Preface   ix

CHAPTER 1   Review of Basic Statistical Concepts and Matrices   1

1.1   Overview   1

1.2   Basic Ingredients for Statistical Inference   3

1.3   Population   5

1.4   Model   10

1.5   Parameters (Summary Numbers)   14

1.6   Samples and Inferences   20

1.7   Functional Notation   47

1.8   Matrices and Vectors   50

1.9   Multivariate Gaussian Populations   62

1.10   Exercises   70

CHAPTER 2   Regression and Prediction   73

2.1   Overview   73

2.2   Prediction   73

2.3   Regression Analysis   82

2.4   Exercises   97

CHAPTER **3**   Straight Line Regression   99

3.1   Overview   99

3.2   An Example of Straight Line Regression   100

3.3   Straight Line Regression Model—Assumptions (A) and (B)   109

3.4   Point Estimation   112

3.5   Checking Assumptions   132

3.6   Confidence Intervals   161

3.7   Tests   171

3.8   Analysis of Variance   178

3.9   Coefficient of Determination and Coefficient of Correlation   181

3.10   Regression Analysis When There Are Measurement Errors   194

3.11   Regression Through the Origin   210

3.12   Exercises   214

CHAPTER **4**   Multiple Linear Regression   219

4.1   Overview   219

4.2   Notation and Definitions   220

4.3   Assumptions for Multiple Linear Regression   232

4.4   Point Estimation   235

4.5   Residual Analysis   251

4.6   Confidence Intervals   262

4.7   Tests   278

4.8   Analysis of Variance   283

4.9   Comparison of Two Regression Functions (Nested Case) and Coefficients of Determination   291

4.10   Comparing Two Multiple Regressions Models (Nonnested Case)   309

4.11   Lack-of-Fit Analysis   318

4.12   Exercises   335

CHAPTER **5**   Diagnostic Procedures   351

5.1   Overview   351

5.2   Outliers   352

5.3   Leverages or Hat Values   365

5.4   Influential Observations—Cook's Distance and DFFITS   371

5.5   Ill-Conditioning and Multicollinearity   392

5.6   Exercises   399

CHAPTER 6 Applications of Regression I   403

6.1   Overview   403

6.2   Prediction Intervals   403

6.3   Tolerance Intervals   416

6.4   Calibration and Regulation for Straight Line Regression   425

6.5   Comparison of Several Straight Line Regressions—Identical, Parallel, and Intersecting Lines   436

6.6   Intersection of Two Straight Line Regression Functions   450

6.7   Maximum or Minimum of a Quadratic Regression Model   456

6.8   Linear Splines   465

6.9   Exercises   476

CHAPTER 7 Applications of Regression II   501

7.1   Overview   501

7.2   Subset Analysis and Variable Selection   501

7.3   All-Subsets Regression   504

7.4   Alternative Methods for Subset Selection   520

7.5   Growth Curves   551

7.6   Exercises   567

CHAPTER 8 Alternate Assumptions for Regression   571

8.1   Overview   571

8.2   Straight Line Regression with Unequal Subpopulation Standard Deviations   571

8.3   Straight Line Regression—Theil's Method   584

8.4   Exercises   592

CHAPTER 9   Nonlinear Regression   599

9.1   Overview   599

9.2   Some Commonly Used Families of Nonlinear Regression Functions   599

9.3   Statistical Assumptions and Inferences for Nonlinear Regression   605

9.4   Linearizable Models   615

9.5   Exercises   622

Appendix A: Answers to Selected Problems and Exercises   627

Appendix B: Bibliography   645

Appendix D: Data Sets   647

Table D-1: Car Data   649

Table D-2: Car2 Data   657

Table D-3: Grades Data   659

Table D-4: Plastic Data   669

Appendix T: Tables   679

Table T-1: Percentiles of a Standard Gaussian Population   681

Table T-2: Percentiles of a Student's t Population   682

Table T-3: Percentiles of a Chi-Square Population   683

Table T-4: Student's t for m Simultaneous Confidence Intervals for $m = 2, 3, 4, 5, 6$ 684

Table T-5: Percentiles of Snedecor's F Population   689

Table T-6: Table for Obtaining Confidence Bounds for $a_0 \beta_0 + a_1 \beta_1$ Using Theil's Method   692

Table T-7: Charts for Confidence Bounds for the Simple Correlation Coefficient   693

Table T-8: Selected Percentiles of the Noncentral t 695

Index   697

Preface

Our Purpose

This book is intended to be used for a one-semester course in "applied regression" that focuses on concepts and applications to be taught to juniors, seniors, or first-year graduate students. Its prerequisites are: (1) one course in statistical methods, (2) proficiency in high school mathematics, and (3) some familiarity with computers. Of course, it is always useful for students to have more mathematics, statistics, or computing than the minimal prerequisites.

Our Approach

There are two kinds of statistics books—theory books and methods books. Theory books generally start with a discussion of probability and populations (probability distributions) and then discuss sampling, whereas books on statistical methods tend to start directly with data and do not discuss populations thoroughly. We believe that the theoretical approach is sound and that high school mathematics is sufficient to understand populations as they relate to many applications. Consequently we first introduce populations and then discuss sampling, focusing all the while on the concepts underlying statistical inference.

Our Emphasis

We have endeavored to give an accurate and clear account of linear regression with an emphasis on *prediction*. We have deemphasized some traditional concepts such as correlation and tests of hypotheses and significance. We stress standard deviations of prediction errors (when using unbiased predictors), rather than correlations, to judge the adequacy of regression functions, and we use confidence intervals in place of tests. This change in emphasis is more than superficial because it influences the way practical questions are formulated and answered.

Use of Finite Population Models

Although most of the *mathematical theory* of regression is derived using infinite population models, we feel that it is easier for students to understand many concepts—such as random sampling, unbiasedness, sampling distributions, and the relative frequency interpretation of confidence intervals and tests—by using *finite population* models. Moreover, most populations we come across in real problems happen to be finite populations. For this reason, we attempt to explain important concepts using finite populations. On the other hand, the theory based on infinite models is essentially valid for topics covered in this book because real populations under study usually consist of a large number of items, and so infinite population models serve as good approximations.

The Organization

Chapter 1 is a review of what is normally covered in a first course in statistical methods plus some material about matrices, functional notation, and the multivariate Gaussian (normal) population. We suggest that Chapter 1 not be omitted even if students have had one or more courses in statistics because it includes some notation, terminology, and specific prerequisites for the rest of the book. Chapter 2 introduces many of the fundamental concepts underlying regression. Chapter 3 contains a detailed discussion of straight line regression. Chapter 4 gives an in-depth study of multiple linear regression. Chapter 5 deals with diagnostic procedures in regression analysis. Chapters 6 and 7 treat special topics that cover several important practical applications of linear regression. Chapter 8 discusses some procedures, including a distribution-free method, that are valid when traditional assumptions for straight line regression are not appropriate. Finally, Chapter 9 gives a brief discussion of nonlinear regression.

The first four chapters form the core of the book. Thereafter, any or all of the sections in Chapters 5 through 9 can be studied in any order since each section in these chapters depends on only the material in Chapters 1 through 4.

Computing

It is difficult, and practically speaking impossible, to do all of the computing required in multiple regression without the use of some software computing package, yet we did not want details about computing software or hardware to interfere with the flow of the material. Thus we decided to avoid discussion of computing issues in the textbook and to make laboratory manuals available where computing is discussed. The textbook is self-contained and is not dependent on knowledge of any computing package. We include relevant outputs from MINITAB and/or SAS whenever the required computing is, to the say the least, tedious. As a result, students can learn the material without any statistical computing package, although we strongly advise using a suitable one.

We have written two laboratory manuals—one for MINITAB and one for SAS. We chose MINITAB because it is an easy statistical package for students to learn without detracting from the main subject. We also chose SAS because many graduate students know and use this package or want to learn it because it is widely used in government, industry, and business. The laboratory manuals contain instructions for using MINITAB and SAS to solve problems and exercises in the textbook. This arrangement offers students the following choices:

1 For those who wish to use MINITAB, a laboratory manual explaining MINITAB commands needed for regression is available with the textbook.

2 For those who wish to use SAS, a laboratory manual explaining basic SAS commands needed for regression is similarly available with the textbook.

3 For those who do not wish to use computers, the textbook is self-sufficient and they need not concern themselves with the laboratory assignments contained in the laboratory manuals.

The Data Sets

Practically every data set used in the book appears "real," and many are based on real studies. The problem descriptions associated with the data sets in the book do not require knowledge of any particular field, such as physics, chemistry, biology, or engineering. The data sets involve common and well-known topics such as insurance premiums, grade point averages, final and midterm exams, electric bills, grocery costs, strength of plastic containers, professors' salaries, and atmospheric pollutants.

Examples, Problems, and Exercises

In the examples, problems, and exercises we have aimed to ask and answer questions that investigators may encounter in the course of an investigation and not questions that are of interest to statisticians only. In particular, we have tried to explain how to formulate practical questions about population parameters in the statistical model being considered.

There are more than 100 examples (over half of them worked out in detail), and there are 14 tasks and more than 85 data sets throughout the book to illustrate each new procedure as it is presented. Problems at the end of each section help students determine whether they understand the material covered in that section. Exercises at the end of each chapter cover the material in the entire chapter. Answers to selected problems and exercises are given in an appendix.

The Mathematics

We believe that students can learn, understand, and use the techniques of regression correctly and effectively in their applied work without a great deal of mathematics and theoretical statistics. In many statistics courses, students prove theorems without really understanding the statistical concepts the theorems imply, and in some cases this can be an impediment to developing their intuition about statistical methods. Consequently this book requires very little mathematics and no theorems are proven. For many procedures, we give a heuristic explanation to help students understand the underlying ideas. Instructors who wish to teach theoretical (mathematical) regression techniques can supplement this book with appropriate mathematical procedures. This should be much easier than using a mathematical text and supplementing it with appropriate methodological procedures that include data sets, computing techniques, etc.

The Notation

Some students may find the notation difficult and troublesome, but adequate notation is required to understand many of the concepts and procedures. We have simplified the necessary notation as much as possible, and we believe that it will actually help clarify many of the concepts if students study it carefully.

Recommended Coverage

One way to cover the material in this book for a one-semester course that meets three hours per week is as follows:

1 The material in Chapter 1 can be covered rather quickly but deliberately because it is supposed to be a review.

2 Chapter 2, which is an introduction to the concept of regression, can also be covered rather quickly.

3 Chapters 3 and 4 should be covered in depth. However, the instructor may elect to spend very little time on *regression analysis when there are measurement errors* in Section 3.10 and *lack-of-fit analysis* in Section 4.11.

4 After Chapters 1 through 4 have been completed, instructors can teach any section in the remainder of the book. Some instructors may elect to cover the topics in one of the ways described below:
   a Selection of variables, growth curves, and nonlinear regression, along with other topics.
   b Selection of variables, tolerance intervals, and prediction intervals, etc.
   c Selection of variables, nonlinear regression, spline regression, etc.

Noteworthy Features

Some additional noteworthy features of the book are:

1 An exceptional range of applications is presented in Chapters 6 and 7—tolerance intervals, calibration, regulation, intersection of lines, variable selection, growth curves, maximum and minimum of a quadratic, spline regression, and many others.

2 For each and every technique discussed, we stress the assumptions needed for that technique to be valid.

3 We point out that no valid point estimates or confidence intervals exist for certain population parameters unless sampling is carried out by a prescribed method. This relationship between the sampling method and the availability of valid inference procedures is often ignored in many textbooks.

4 We downplay tests of hypotheses and tests of significance but emphasize the use of confidence intervals for making practical decisions from data because confidence intervals are more informative than tests for making these decisions. Tests are used by investigators to assess "statistical significance," but confidence intervals can be used to help assess "practical importance."

5 We use word problems, called *tasks,* to help students see how regression can be used to answer practical questions.

6 We provide an alternative approach for assessing lack-of-fit of prediction models using confidence intervals rather than traditional tests.

7 We stress the fact that correlation must be used with caution in real problems and that it is not an adequate measure of regression functions.

8 We use standard deviations, rather than variances, because standard deviations are easier to interpret in applied problems.

9 All the data sets used in the book are available on a data disk that accompanies each laboratory manual.

10 In the laboratory manuals we show how MINITAB and SAS can be used to analyze the data sets and solve problems and exercises in the textbook. For every topic in regression that is discussed in the book, we discuss a computer program that can be used for computations.

11 For problems that cannot be solved using the built-in commands in MINITAB or SAS, we have written programs for MINITAB and SAS. These programs are called *macros,* and the laboratory manuals explain how to use them. These macros are available on the data disk. They are intended for only the problems discussed in this book, and they are not necessarily valid for other, more complex, problems.

12 We make recommendations as to which procedures should be used in applied problems.

13 We have included several conversations between a statistician and an investigator to help clarify certain concepts.

It is our hope that this book will serve the needs of students of statistics and researchers who wish to have an in-depth understanding of the concepts underlying regression analysis and alternative ways of formulating questions of interest in practical applications.

Acknowledgments

We want to thank the following reviewers for their helpful comments: Roger Chope, University of Oregon; Louis J. Cote, Purdue University; James C. Daly, California Polytechnic State University; Jugal K. Ghorai, University of Wisconsin, Madison; Dallas E. Johnson, Kansas State University; Julia A. Norton, California State University, Hayward; Steven W. Ramsier, Tennessee Technological University; John O. Rawlings, North Carolina State University; Christopher A. Robertson, University of California, Riverside; Diane G. Saphire, Trinity University; and Colin O. Wu, The Johns Hopkins University. We also wish to thank Pat Key, Carolyn Cook, and Jeanie Weitzel for their expert and patient secretarial assistance.

To Jeanne, our children and grandchildren
for the good times.   F. G.

To my grandparents, my parents, all my teachers, and to
Pam, Matthew, Kristin, Kevin, and Geoffrey.   H. I.

Far better an approximate answer to the *right* question, which is often vague, than an exact answer to the *wrong* question, which can always be made precise."

John W. Tukey, *Ann. Math. Stat.,* vol. 33, p. 13, 1962.

1

Review of Basic Statistical Concepts and Matrices

The material in this chapter is generally taught in a first course in statistical methods, which is prerequisite for studying this book. So, depending on the reader's background, this chapter may be treated as a quick review or may be studied in depth.

1.1
Overview

What Is Statistics?

Statistics, in a narrow sense, is a branch of science that deals with making inferences about populations based on samples. In a broader sense, statistics encompasses collection, organization, and summarization of data; presentation of data in tabular and graphical form; developing models for the purpose of understanding random and nonrandom phenomena; use of models for prediction; mathematical approaches to decision making and evaluation of risks; and so forth. **Regression** (and correlation) analysis is an area of statistics that deals with methods for investigating the existence of associations and, if present, the nature of the associations, among various observable quantities. For instance, one might be interested in knowing whether there is any association between age and blood pressure and, if so, what the nature of this association is. One might also be interested in expressing associations in the form of mathematical equations. Regression analysis is one method of investigating the presence of associations if appropriate data are available. The discovery of associations and the ability to express such associations in a precise mathematical form *may* enable one to predict the unobservable value of a variable based on the observed value of one or more associated or related variables. They may also help determine how one might control the values of one variable by manipulating the values of a related variable. For instance, one might be able to manipulate or control one's blood pressure by controlling one's diet if the nature of the association between blood pressure and the dietary components is understood. We do not make any statements regarding that elusive and controversial concept of **cause and effect**. Rather, we simply say that it *may* be possible to control the values of one variable by manipulating the values of a related variable. This can be confirmed only by controlled experimental investigations.

In this book we present the fundamental ideas that form the core of regression (and correlation) analysis, and we give numerous practical applications of the methods developed.

Prerequisites

We assume that you are familiar with the material that is usually taught in an introductory course in statistical methods, which includes concepts of populations, samples, point estimation, confidence interval estimation, tests, and Gaussian distribution (also called normal distribution) models. These are reviewed briefly in Sections 1.2 through 1.6. We further assume that you are acquainted with functions and functional notation and elementary topics in matrix arithmetic. These topics are reviewed in Sections 1.7 and 1.8, respectively.

A brief introduction to the multivariate Gaussian distribution model (also referred to as the multivariate normal distribution) appears in Section 1.9. This topic is not usually taught in an introductory statistics course, but we feel that it is useful background material for studying regression analysis.

Remarks on Computing

Regression analysis requires a great deal of computing, and it will be helpful for you to learn to use computers and statistical computing packages. However, since some of you will not have access to a computer or a suitable computing package, we present most of the basic computations and require you to do only computations that can be done easily on a hand-held calculator. Thus you are not *required* to have knowledge of a computing language or a statistical computing package to read and understand the material in this book. Nevertheless, *we strongly urge that you use a computing package if one is available.*

Practically every statistical computing package contains programs for regression analyses. Some of the popular statistical packages are MINITAB, SAS, SPSS, BMDP, S-PLUS, etc. This textbook is accompanied by a laboratory manual that explains how to use one of the statistical packages—MINITAB or SAS—to perform the calculations needed for regression analysis. The reason that we wrote a laboratory manual using MINITAB is that this system is very easy to learn and does not detract from the main topic, regression. Also, many colleges and universities use this package as a tool for teaching statistics. The reason that we wrote a laboratory manual using SAS is that SAS is widely used by data analysts and researchers in colleges, universities, government, and industry. Also, many graduate students who study regression are already familiar with SAS and would like to analyze their own research using SAS. If computations are needed for the procedures discussed in any section of this book, the corresponding sections of the accompanying laboratory manual explain how to do the calculations in MINITAB or SAS. For example, in Section 3.6 we discuss confidence intervals for simple linear regression, and in Section 3.6 of the laboratory manual we show how to use MINITAB or SAS to perform the calculations needed for confidence intervals. If MINITAB or SAS is not available, you can either ignore the laboratory manual or, if you have another statistical package available, you can work through the assignments in the laboratory manual using that package. Nothing that follows in this book requires knowledge of or the use of computers and statistical computing packages.

1.2
Basic Ingredients for Statistical Inference

One of the aims of science is to relate, describe, and predict events in the world in which we live. These activities are also important in business and our everyday affairs. In almost every aspect of human endeavor, it is useful to be able to predict future events based on present and past information.

To describe situations of interest, we must define them precisely and decide what we want to determine or predict. To illustrate, we first use a very simple example. You are handed a box that contains 10,000 marbles that are indistinguishable except for color—some are white and some are green—and you are asked to determine the proportion of green marbles in the box. You are not allowed to look in the box, but you are allowed to select 20 marbles at random from the box and note their colors. On the basis of this information (the colors of these 20 marbles), you must estimate what proportion of the 10,000 marbles are green. Of course you do not expect that by examining only 20 marbles you will be able to determine exactly the proportion of the entire 10,000 marbles that are green, but you would like to use scientific reasoning to come as close to the true answer as possible. The science of statistics and probability can be useful for making a decision in this situation and for attaching a measure of uncertainty (or certainty) to your result.

The preceding example may seem trivial, but it contains the basic ingredients of the most complicated of problems. Mother Nature has her secrets "locked up in a box" and, in order to make intelligent decisions about everyday activities, we may want to describe or predict the contents of that box. It may be possible to observe a part of Mother Nature's box (observe the colors of the 20 marbles) and make an intelligent decision as to its contents.

The Four Basic Ingredients

From this example we abstract four fundamental concepts that will be useful in examining more complicated situations: **populations**, **models**, **parameters**, and **samples and inferences**. These are summarized in Table 1.2.1.

We now consider a more realistic illustration.

E X A M P L E 1.2.1

Suppose a company that owns a chain of department stores is considering opening a store in city A. There are many things the company wants to know about city A so that it can decide whether it will be profitable to build a store there. One item of interest to the company is last year's average annual income per household in

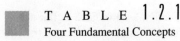

T A B L E 1.2.1
Four Fundamental Concepts

Concept	Description
Population	A **population** of items (sometimes also called a population of units) is a specified set of items or units in which an investigator is interested. The population may be **real** or it may be **conceptual**. It may be a population that existed in the past, or exists now, or will exist in the future, or exists only in one's imagination. A **real** population exists now and the entire population or any part of the population is available now and may be examined. A **conceptual** population is a population that is not available now and cannot be examined. Typically, a conceptual population is one that will exist in the future or exists only in one's imagination. A population may remain constant, or it may change slowly with time, or it may change rapidly with time. In the preceding illustration, the set of 10,000 marbles is the population of items. It is a real population since it is available for study now. Also it is a population that will remain constant, i.e., the colors of the marbles (white or green) will not change with time.
Model	A population **model** is a description of the quantities (numbers or attributes) of interest associated with each item in the population. In the illustration above, the attribute of interest is the color of the marbles and the description is that the marbles are indistinguishable except for color, with some being white and others green.
Parameter	**Parameters** are summary numbers that characterize various aspects of the population and are usually the quantities of interest to the investigator. In the preceding illustration, the proportion p of marbles that are green, characterizes one aspect of the population and is the quantity of interest in the investigation.
Sample	A **sample** is a set of items selected from the population, and observations are made on this set of items. On the basis of this sample a decision is made about the values of the parameters of interest, and this decision is usually accompanied by a measure of the uncertainty in the answer. In the preceding illustration, the set of 20 marbles is a sample from the population of 10,000 marbles. The proportion of green marbles in the sample may be used as an *estimate* of the proportion of all 10,000 marbles in the box that are green. Thus one uses the information contained in the sample to make **inferences** about population characteristics of interest.

the city. The four fundamental concepts outlined in Table 1.2.1 as they apply to this investigation are as follows:

1 The **population** of interest is the set of households in city A last year. A household must be precisely defined (which may not be an easy task!). For all practical purposes, we can regard this population as being constant although families may have moved in or out of the city during the course of the year. Consequently this is a real population that can be studied now.

2 The population **model** may be that the set of annual incomes of the households last year is a Gaussian population with mean μ and standard deviation σ, both of which are unknown.

3 The **parameter** of interest is μ, the average income of the households in the population.

4 A random **sample** of some specified number of households can be selected and last year's annual income of each of the selected households recorded. Based on these data, a decision can be made about the value of μ, the parameter of interest. ■

In the next four sections we discuss in detail each of the four basic ingredients:

- Population
- Model
- Parameters
- Samples and inferences

1.3
Population

A population of items is defined to be any set of items that one wants to study (refer to Table 1.2.1.). In Example 1.3.1 we describe several populations of items.

E X A M P L E   1.3.1

The following are some examples of populations of items.

a The set of all U.S. citizens who paid federal income tax last year.

b The set of all automobiles that will be made by manufacturer A next year.

c The set of all U.S. citizens who were diagnosed as having lung cancer three years ago and are still alive today.

d The set of all trees on a specified tree farm on July 1 two years hence.

e The set of all farms in the states of Iowa and Nebraska last year.

f The set of all grocery stores in the U.S. next year.

g The set of all human beings who will be born in Los Angeles, California, in the year 2000.

h The set of all ball bearings that will be made in a specified production plant next month.

i The set of all plastic containers that a specified manufacturer may make next year, using every possible process temperature between 300° F and 400° F.

j The set of all automobile tires that will be manufactured by a certain company next year using a newly developed tread design.

k The set of all daily record sheets that will be maintained by a particular power plant, containing information including the amount of sulfur-dioxide emitted and the amount of electrical power generated for each day that the power plant will be in operation next year.

l   The set of all daily data records that will be kept by a particular monitoring station, containing information about the total daily flow values at a given monitoring point, and the total daily precipitation recorded at a certain gauge, for a specified river, for next year.

m   The set of all *measured* values of the length of a single bolt.

n   The set of all *measured* values of the breaking strength of a given metal rod.   ■

The set of items to be studied, called the *population of items*, must be precisely defined. The number of items in a population can be finite or infinite; if this number is finite we denote it by N. This number N is generally unknown, but in most practical problems it is large.

Notice that the dates are specified for each of the populations of items described in Example 1.3.1; otherwise the set of items would not be precisely defined. For instance, in Example 1.3.1(e) what was a farm last year may not be a farm this year; in (f) a store that is a grocery store next year may not be a grocery store two years hence, etc.

Note   In Example 1.3.1 the populations in (a)–(h) and (j)–(l) are all finite, whereas in (i), (m), and (n) they may be infinite. The populations described in (a), (c), and (e) are *real populations*, and we can study these populations now. The populations described in (b), (d), (f), (g), (h), (j), (k), and (l) are *conceptual populations* since they are future populations. They do not exist at the present time and cannot be observed *now*. The populations described in (i), (m), and (n) are also conceptual populations. In fact, they are *imagined* populations; the entire population will *never* become available. In the population described in (i), it is impossible for the manufacturer to actually manufacture plastic containers using every possible process temperature between $300°$ F and $400°$ F. But we can certainly *imagine* this population and ask questions about it. For instance, we may want to know what process temperature would lead to production of plastic containers having the required strength. For the population described in (m), it is possible to examine part of the population, viz., the next several measured values, but the entire population can only be *imagined* and will never become available. The population described in (n) is also an *imagined population*. The very act of measurement will destroy the given metal rod and so only *one* value can be observed from this imagined population. Nevertheless, we may want to make inferences about these conceptual populations so that decisions can be made or actions can be taken *now*.

Each item in a population of items possesses one or more characteristics that may be of interest in an investigation, as you can see in Example 1.3.2.

E X A M P L E   **1.3.2**

In Example 1.3.1(a) we may be interested in the age of each person, the I.Q. of each person, the weight of each person, the sex of each person, the political affiliation of each person, the marital status of each person, etc. In Example 1.3.1(b) we may be interested in studying the miles per gallon each automobile will get, the first-year maintenance cost for each automobile, the number of miles each car will be driven the first year after its purchase,,etc. In Example 1.3.1(d) we may be interested in the age of each tree, the diameter of each tree measured 4.5 feet from the ground, the height of each tree, etc. In Example 1.3.1(g) we may be interested in the number of years each person will live, the number of dollars each person will spend on education, the number of siblings each person will have, etc.   ∎

Associated with each item in a population are one or more numbers (age, height, income, etc.) or attributes (sex, marital status, political affiliation, etc.) of interest. In this book we are mainly concerned with numerical quantities associated with each population item. This *set of numbers* is called a *population of numbers* and will be referred to as the **population**. If a single number is associated with each item in the population, then we say that this population of numbers is a **univariate population**. If more than one number is associated with each item in the population, then we say that this population of numbers is a **multivariate population**. A **bivariate population** is a special case of a multivariate population, where each population item has two numbers of interest associated with it. A **trivariate population** is a special case of a multivariate population, where each population item has three numbers of interest associated with it. More generally, if each item has k numbers of interest associated with it, we refer to the population of numbers as a k**-variate** or a k**-variable population**.

In Example 1.3.3 we list specific **sets of numbers (populations)** that can be associated with each population of items listed in Example 1.3.1.

E X A M P L E 1.3.3

The following are populations (of numbers):

a The amount of money in dollars that each U.S. citizen earned as interest income last year.

b The maintenance cost in dollars and the number of miles each car will be driven during its first year.

c The age of each person and the average number of cigarettes each one smoked per day for the five years prior to diagnosis of lung cancer.

d The height, the diameter measured at 4.5 feet above ground level, and the dollar value of each tree, on July 1 two years hence.

e The size of each farm in acres and the profit each farm made.

f The profit, in dollars, that each grocery store will make.

g The number of dollars each person will spend on health care and the number of years each person will live.

h The diameter of each ball bearing.

i The strength of each plastic container and the temperature at which it is made.

j The set of all tread-depths remaining at the end of 50,000 miles of driving of all the automobile tires that will be manufactured by the company next year.

k   The set of all total daily sulfur dioxide emission values and the amount of electrical power generated for each day of next year.

l   The set of all total daily flow values and corresponding daily precipitation values that will be recorded by the monitoring station next year.

m   The set of all measured values of the length of the bolt.

n   The set of all measured values of the breaking strength of the rod.   ∎

Note that in Example 1.3.3 (a), (f), (h), (j), (m), and (n) are examples of univariate populations, (b), (c), (e), (g), (i), (k), and (l) are examples of two-variable (bivariate) populations, and (d) is an example of a 3-variable (trivariate) population.

Notation

At the beginning of an investigation, a target population of items is defined and one or more numbers associated with each item are identified as being of interest. *This resulting population (of numbers) is what an investigator is interested in.*

If a population is a univariate population consisting of N numbers, this population of numbers is written symbolically as $\{Y_1, Y_2, \ldots, Y_N\}$, where Y_I represents the Ith number (the number associated with the Ith item) in the population. Alternatively, we may write $\{Y\}$ without specifying the number of items in the population. If the population is a bivariate population consisting of N pairs of numbers, this population is written symbolically as $\{(Y_1, X_1), (Y_2, X_2), \ldots, (Y_N, X_N)\}$ (or simply as $\{(Y, X)\}$) where Y_I refers to the first quantity and X_I refers to the second quantity associated with the Ith item.

Sometimes a double subscript notation is used, as in $\{(Y_{11}, Y_{12}), (Y_{21}, Y_{22}), \ldots,$ $(Y_{N1}, Y_{N2})\}$. Here Y_{IJ} stands for the Jth quantity of interest associated with the Ith population item. When working with multivariate populations, the double subscript notation is almost always used. For instance, a k-variable population may be denoted by $\{(Y_{11}, Y_{12}, \ldots, Y_{1k}), \ldots, (Y_{N1}, Y_{N2}, \ldots, Y_{Nk})\}$ or simply by $\{(Y_1, Y_2, \ldots, Y_k)\}$. Other symbols such as X or Z may be used instead of Y when convenient.

The population of numbers that is of interest to the investigator is called the **target population**. In some situations, the target population is a real population and is available for study. This is the case for the populations (a), (c), and (e) in Example 1.3.3. But, in other situations, the target population of interest is a conceptual population and is not available for study, as in the case of populations (b), (d), (f), (g), (h), (i), (j), (k), (l), (m), and (n) of Example 1.3.3. Even when the target population is a real population, it may sometimes be impractical or too expensive to study this population, and thus it is unavailable for study. When the target population is unavailable for study, one is sometimes able to find another population that resembles the target population and is a real population that is available for study now. Such a population is called a study population. We discuss these in some detail next.

Target Population and Study Population

Target Population

The target population is the population that is the *target* of a study. Conclusions from an investigation are to be applied to this population. As stated earlier, the target population is often unavailable for study. In many investigations, the numbers in the target population are **future values**, but we want to make decisions about them now. Example 1.3.3(d) is a good illustration. Suppose we are interested in what the dollar value of each tree will be on July 1 two years hence, but we must make this determination now. The target population in this case consists of the dollar values of each of the trees on July 1 two years hence, so this population is unavailable now. Therefore it is not possible to study the target population now. This is also the situation in Example 1.3.3(b), where we want to predict the average first-year maintenance cost of cars that will be produced by manufacturer A *next* year. We must make the prediction at the beginning of the year so we can plan a maintenance budget. But the target population is a future population and is not available for study now.

In fact it is quite common for a target population, which is the population of interest, to be unavailable for study, either because it consists of *future values* or because it is *impractical*, *inconvenient*, or *too expensive* to study this population. Thus we are led to consider another population, the study population.

Study Population

When the target population is unavailable for study, we are sometimes able to find another population that *resembles* the target population and is available for study now. Such a population is called a **study population**. Thus the study population is the population that is actually studied during an investigation. In those situations where the target population itself is available for study, the target population and the study population are one and the same. When the target population is unavailable for study, the study population should be chosen to *resemble* the target population as closely as possible.

To illustrate, consider Example 1.3.3(d). The target population is in the future, so it is not available for study now. The study population can be defined as the set of dollar values of all trees as of the present date. This is a real population that is available for study now. In Example 1.3.3(b) we can use as the study population the first-year maintenance costs for the same make of automobiles last year. In Example 1.3.3(i), no suitable study population may be available. In this case the investigator can produce plastic containers in a pilot plant under several different process temperatures between 300° F and 400° F and measure the strengths of these containers. The data thus generated are all that is available and this collection of pairs of numbers, i.e., the strength of a plastic container and the temperature at which it was made, can be considered the study population.

Methods are available for making *valid statistical inferences* about the study population. When the study population is different from the target population, generalizations from the study population to the target population are subject-matter considerations, and we have to rely on the investigator's judgments. All statistical inference procedures discussed in this book refer to the study population.

1.4
Model

A univariate population to be studied consists of N numbers, where N is generally quite large. In Example 1.3.3(a), for instance, the N numbers are the amounts, in dollars, earned as interest income by each of the N citizens; in Example 1.3.3(f), the N numbers are the profits, in dollars, each grocery store will make next year.

To study a population of N numbers one must have some way of organizing these N numbers. One useful approach is to organize them into a **probability histogram** and find a mathematical function that approximates this histogram. Such a mathematical function is usually called a **probability density function**. Theoretical statisticians use a variety of probability density functions to study different types of populations, the most important among them being **Gaussian** probability density functions. They are the subject of our discussion in this section. The Gaussian population model for multivariate populations is discussed in Section 1.9.

Gaussian Populations

A **theoretical Gaussian population** $\{Y\}$ is completely specified by its mean, denoted by μ_Y, and its standard deviation, denoted by σ_Y. The distribution of this population is described by the probability density function

$$f(Y) = \frac{1}{\sqrt{2\pi\sigma_Y^2}} \exp\left[-\frac{(Y-\mu_Y)^2}{2\sigma_Y^2}\right] \quad \text{for} \quad -\infty < Y < \infty \quad \text{(1.4.1)}$$

The area under the curve, defined by the function in (1.4.1) between the values a and b $(a < b)$, gives the proportion of population values that are greater than a but less than or equal to b (see Figure 1.4.1).

The probability density functions corresponding to two different theoretical Gaussian populations are shown in Figure 1.4.2. These two populations have different values for μ_Y, but they have the same standard deviation σ_Y. It is clear from this figure that changing the value of μ_Y changes only the location of the population distribution and not its shape (curves (a) and (b) in Figure 1.4.2 have the same shape but different locations). On the other hand, changing the value of σ_Y changes the shape of the population distribution but not the location, as exemplified by curves (c) and (d) in Figure 1.4.3. These two curves have the same location (same value of μ_Y) but different shapes (different values of σ_Y).

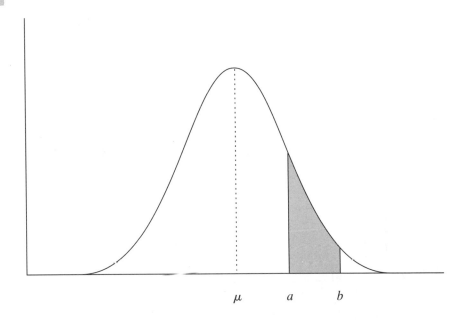

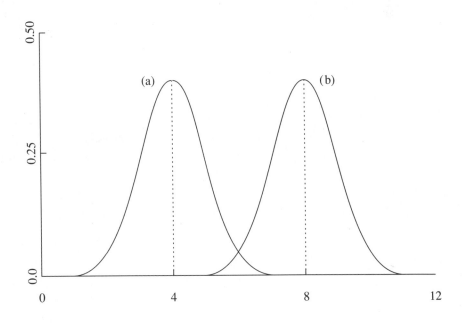

F I G U R E   1.4.3

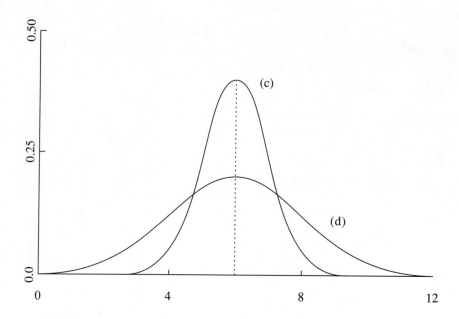

For the sake of completeness we now give the definitions of the **mean**, the **standard deviation**, and the **variance** of a population {Y} consisting of the N numbers $Y_1, Y_2, \ldots, Y_N$.

D E F I N I T I O N   **Mean**

The mean (also called the *average*) of a population {Y} is denoted by μ_Y and is defined by

$$\mu_Y = \frac{1}{N} \sum_{I=1}^{N} Y_I \quad \blacksquare$$
(1.4.2)

D E F I N I T I O N   **Standard Deviation**

The standard deviation of a population {Y} is denoted by σ_Y and is defined by

$$\sigma_Y = \sqrt{\frac{1}{N} \sum_{I=1}^{N} (Y_I - \mu_Y)^2} \quad \blacksquare$$
(1.4.3)

D E F I N I T I O N **Variance**

The variance of a population $\{Y\}$ is the square of its standard deviation. It is denoted by σ_Y^2 and is defined by

$$\sigma_Y^2 = \frac{1}{N} \sum_{I=1}^{N} (Y_I - \mu_Y)^2 \quad \blacksquare \tag{1.4.4}$$

The definitions of mean and standard deviation for infinite populations require concepts from calculus and some knowledge about convergence of infinite series. If you are interested in them you should refer to more advanced books [15], [19], [25], [31].

As stated earlier, in practical problems the N unknown numbers that make up a population could conceptually be used to form a **probability histogram** (a histogram for which the total area equals 1). If this histogram is well approximated by the mathematical function defined in (1.4.1) when the value of the mean and the standard deviation of these N population numbers are substituted for μ_Y and σ_Y, then *we may proceed as if the population under study is Gaussian.* For example, Figure 1.4.4 shows a probability histogram of a population of 4,000 numbers (generated using a computer), with $\mu_Y = 0$ and $\sigma_Y = 1$. The probability density curve of the theoretical Gaussian population with $\mu_Y = 0, \sigma_Y = 1$ in (1.4.1) is also displayed there. From these we can see that the theoretical Gaussian population described by equation (1.4.1) appears to be a good approximation to this finite population of size 4,000. Thus, for most practical purposes, we can regard this finite population consisting of 4,000 numbers as a Gaussian population.

The theoretical Gaussian population is a mathematical abstraction and no such population can exist in real investigations, but it is useful as an approximation to finite populations in many applied problems. It is also used in theoretical statistics to derive point estimates, confidence intervals, and tests for μ_Y and σ_Y. Similar approximations are common in other situations. For example, a circle, which can be defined mathematically, does not exist in the real world, but it is useful as an approximation to the wheel; a rectangle, which can be defined mathematically, does not exist in the real world, but it is useful as an approximation to obtain the area of a tabletop, a farmer's field, the side of a building, etc.

Because we never know the population values in a real problem, it is impossible to be certain that the population under study is actually Gaussian. However, even when the population is not exactly Gaussian, the statistical inference procedures are often accurate enough for making decisions as long as the population is *approximately* Gaussian. In some instances statistical procedures are available for detecting serious violations of the Gaussian assumption.

FIGURE 1.4.4

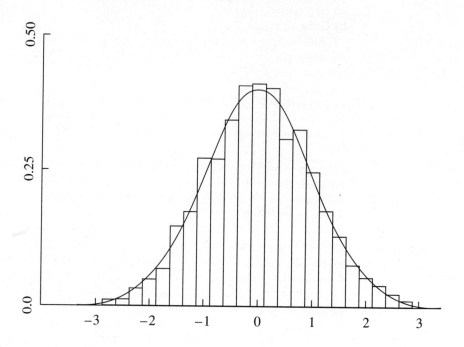

1.5
Parameters (Summary Numbers)

In Table 1.2.1, we described parameters as numbers that summarize certain important characteristics of a population. Even if we had the entire population available (which is rarely the case in a real problem), there would be too many numbers for an investigator to use for making decisions without summarizing them into a smaller set. Thus a judicious summarization of population characteristics is extremely important. Here we discuss this in some detail, first for univariate populations and then for multivariate populations.

Parameters for Univariate Populations

Consider the univariate population $\{Y\}$ and assume for the moment that all the numbers in the population are available. To understand the important characteristics of the population of numbers, it is convenient to summarize them by a smaller set of numbers or perhaps by using suitable graphical techniques. Probability histograms and cumulative frequency curves are two of the commonly used graphical descriptions of populations.

Generally we would like to use one or more (say m) summary numbers, $\theta_1, \ldots, \theta_m$, to describe various characteristics of the entire population. These m

numbers, called (population) **parameters**, would be computed from the entire population if it were known, but in almost all real problems they have to be estimated based on sample data. In this book population parameters are almost always denoted by Greek letters.

Although the m parameters $\theta_1, \ldots, \theta_m$ cannot, in general, tell us everything about the whole population, they often adequately summarize certain important characteristics of the population that are relevant to the study at hand. The **mean** and the **standard deviation** are two particularly useful and important population parameters, especially in the case of Gaussian populations, and we discuss these next.

Mean

Often we would like to use a *single parameter* to *represent* the entire population of numbers. In this book we use the **mean** of the population (also called the **average**), defined in (1.4.2), as the single number that best represents the entire population of numbers. The symbol μ_Y (μ is the Greek letter mu) represents a population mean, and the subscript Y indicates which population is under study. The mean μ_Y is often also used to predict the Y value of any randomly chosen population item. In most real problems the mean of a population is not known because not all of the population elements $\{Y\}$ are known. Nevertheless, it is a number we would like to have available to use to make decisions about a population. So we select a sample from the population and estimate μ_Y. This is discussed later.

In many situations, single numbers such as the mean (or the **median**) are often used to represent the value of each item in a population. For instance, if you plan to retire in the United States, it would be useful to know something about the cost of housing in the various cities where you might choose to live. The mean (average) price of new homes in each city would provide useful information. Also, the average annual cost of living in each city would be helpful, as would the average age of people who live in the area. In many situations the mean is a useful summary of populations: (1) average income of men and women, (2) average number of children per household, (3) average gas mileage for a certain make of cars, (4) average yield per acre of corn in a certain state, (5) the batting average of a certain baseball player, etc.

Standard Deviation

Although the mean μ_Y is often used as the best single number to represent the entire population of numbers $\{Y\}$, no single number can adequately describe or represent an entire population. Therefore an additional summary number is used to tell us how well μ_Y represents the entire population. This number is σ_Y, the standard deviation of the population, which was defined in (1.4.3). Intuitively, the smaller σ_Y is, the more useful μ_Y is as a representative value for the entire population. Likewise, the larger σ_Y is, the less useful μ_Y is as a representative of the whole population. Note that if $\sigma_Y = 0$, then all the numbers in the population $\{Y\}$ are the same and equal to μ_Y, and the mean is a perfect representation of the entire population. But if a substantial proportion of the values of Y are much smaller than the mean and others are much larger, then σ_Y will be large and μ_Y may not adequately represent each population value.

Chebyshev's Theorem

Recall that about 95% of the numbers in a Gaussian population lie between $\mu_Y - 2\sigma_Y$ and $\mu_Y + 2\sigma_Y$, and about 99% of the numbers lie between $\mu_Y - 3\sigma_Y$ and $\mu_Y + 3\sigma_Y$, so if σ_Y is "small," μ_Y does indeed represent the population quite well. In fact, for any population $\{Y\}$, the following fundamental result, due to the mathematician Chebyshev, is valid.

Chebyshev's Theorem

Let $\{Y\}$ be *any* one-variable population with mean μ_Y and standard deviation σ_Y. Then, for any positive number c, the proportion of population values that are greater than $\mu_Y - c\sigma_Y$ but less than $\mu_Y + c\sigma_Y$ (i.e., that are less than c standard deviations away from the mean) is greater than or equal to $1 - 1/c^2$.

For example, if $c = 3$, Chebyshev's theorem says that for any univariate population $\{Y\}$, at least $1 - 1/3^2 = 8/9 = 88.9\%$ of the population values are within 3 standard deviations of the mean. In particular, for a population $\{Y\}$ whose mean is μ_Y and whose standard deviation σ_Y is 2, we can say that at least 88.9% of the population values are less than 3 standard deviations, i.e., 6 units, away from the mean. Similarly, at least $1 - 1/4^2 = 15/16 = 93.75\%$ of the population values are less than 4 standard deviations (8 units) away from the mean. The exact value of such proportions cannot be found without knowing more about the population, but Chebyshev's theorem does give us a bound for such proportions. From the preceding discussion it is clear that μ_Y and σ_Y are two summary numbers that tell us a great deal about a population and that σ_Y can be used to determine how well μ_Y represents the population values (i.e., how close the population values are to μ_Y). We now consider parameters for multivariate populations.

Parameters for Multivariate Populations

Suppose a population of size N is a k-variate population $(k > 1)$. Then there are k quantities associated with each population item, resulting in Nk numbers in all. Each of the k quantities associated with the population items gives rise to a univariate population. Let the k quantities for item I be denoted by $X_{I1}, \ldots, X_{Ik}$. Then the numbers $X_{11}, \ldots, X_{N1}$ form a univariate population with mean μ_1 and standard deviation σ_1, the numbers $X_{12}, \ldots, X_{N2}$ form another univariate population with mean μ_2 and standard deviation σ_2, etc. Thus $\mu_1, \ldots, \mu_k$, and $\sigma_1, \ldots, \sigma_k$ are parameters associated with the k-variate population. The Nk numbers may be schematically represented as in Table 1.5.1.

T A B L E **1.5.1**
Schematic Representation of a *k*-Variate Population of Size *N*

Items	*k* Measurements on Each Item			
	1	2	$\cdots$	*k*
1	X_{11}	X_{12}	$\cdots$	X_{1k}
2	X_{21}	X_{22}	$\cdots$	X_{2k}
$\vdots$	$\vdots$	$\vdots$	$\vdots$	$\vdots$
I	X_{I1}	X_{I2}	$\cdots$	X_{Ik}
$\vdots$	$\vdots$	$\vdots$	$\vdots$	$\vdots$
N	X_{N1}	X_{N2}	$\cdots$	X_{Nk}
Mean	μ_1	μ_2	$\cdots$	μ_k
Standard deviation	σ_1	σ_2	$\cdots$	σ_k

For an illustration, consider Example 1.3.3(d). Let X_{I1}, X_{I2}, and X_{I3} denote the height, the diameter at 4.5 feet above ground level, and the dollar value of the *I*th tree in the population. We thus have a three-variable population, and μ_1, μ_2, and μ_3 represent the mean height, the mean diameter, and the mean dollar value of the trees in the population. Likewise, σ_1, σ_2, and σ_3 represent the standard deviations of the heights, the diameters, and the dollar values, respectively, of the trees in the population.

Coefficient of Correlation

When dealing with multivariate populations there is often a need to summarize associations among various quantities measured on the same item. For instance, how can we summarize the association that may exist between the height of a tree and the diameter of that tree at 4.5 feet above ground level? How can we summarize the association or relationship between the number of miles a car is driven per year and the yearly maintenance cost for the car? One summary measure that is sometimes used for this purpose is the **coefficient of correlation** between two variables (also called the **Pearson correlation** or the **product moment correlation**). This measure of association is denoted by $\rho_{Y,X}$ (ρ is the Greek letter rho) when summarizing the relationship between the variables Y and X. It is defined by

$$\rho_{Y,X} = \frac{\sum_{I=1}^{N}(Y_I - \mu_Y)(X_I - \mu_X)}{\sqrt{\left[\sum_{I=1}^{N}(Y_I - \mu_Y)^2\right]\left[\sum_{I=1}^{N}(X_I - \mu_X)^2\right]}} \tag{1.5.1}$$

for a bivariate population $\{(Y, X)\}$ of N items. Note that (1.5.1) implies $\rho_{Y,X} = \rho_{X,Y}$; i.e., the coefficient of correlation between Y and X is the same as the coefficient of correlation between X and Y. It can be shown that $\rho_{Y,X}$ is a number between -1 and

+1. It is equal to +1 when the population values (Y_I, X_I) all lie on a straight line that has a positive slope, and it is equal to −1 when all the population values lie on a straight line with a negative slope. Figures 1.5.1–1.5.3 are *scatterplots* of bivariate populations $\{(Y, X)\}$ consisting of 500 items, each with a different value of $\rho_{Y,X}$.

F I G U R E   **1.5.1**

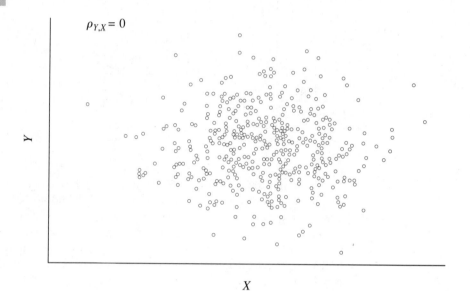

$\rho_{Y,X} = 0$

Y

X

Observe that a positive value of the coefficient of correlation between Y and X indicates that, generally speaking, larger values of Y are associated with larger values of X, and smaller values of Y are associated with smaller values of X. Likewise, when the correlation coefficient is negative, we find that larger values of Y are associated with smaller values of X, and smaller values of Y are associated with larger values of X. Notice that a general lack of *linear* association is indicated when the magnitude of the correlation coefficient is close to zero. *Whereas correlation coefficients may have useful interpretations for some problems (particularly when Y and X are approximately linearly related), they may provide no useful interpretation and, in fact, be a misleading summary quantity in other problems.*

In summary, when dealing with a k-variate population, say $\{(X_1, \ldots, X_k)\}$, the basic summary quantities or parameters that are often used are the means, $\mu_1, \ldots, \mu_k$, and standard deviations, $\sigma_1, \ldots, \sigma_k$, of the k univariate populations, along with the $\binom{k}{2} = k(k-1)/2$ coefficients of correlation, $\rho_{X_1,X_2}, \rho_{X_1,X_3}, \ldots, \rho_{X_{k-1},X_k}$, between the $k(k-1)/2$ pairs of variables $(X_1, X_2), (X_1, X_3), \ldots, (X_{k-1}, X_k)$.

F I G U R E 1.5.2

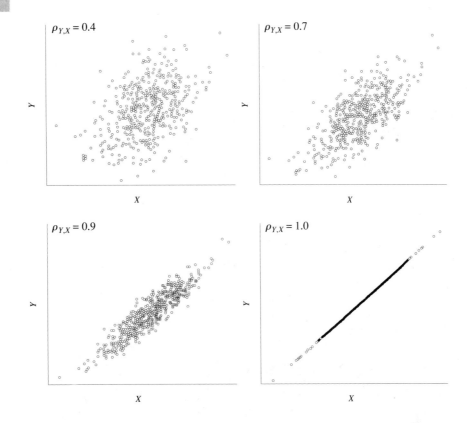

F I G U R E **1.5.3**

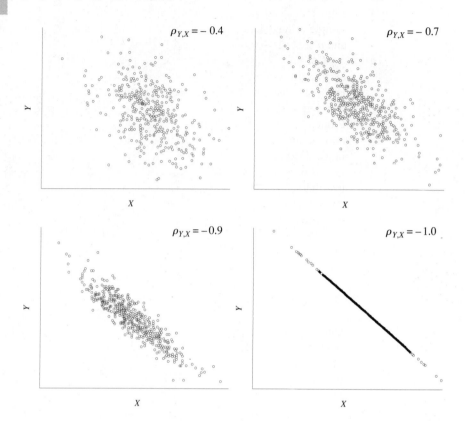

1.6
Samples and Inferences

As stated earlier, an investigator must first define the study population. Whenever possible the target population itself should be the study population. If this is not possible, then the study population should resemble the target population as closely as possible. After the study population is defined, it is described with a model, and parameters that are needed in order to make decisions are identified. The next step is to determine the values of these parameters. Since the population is never completely known in a real situation, an investigator can never know the parameter values exactly. A commonly used procedure is to select a subset of the items, referred to as a **sample**, from the study population and to use the measurements associated with these sample items to *infer* the values of population parameters of interest. If the sample is selected from the population using one of several random sampling procedures, then it is possible to assign a *measure of uncertainty* to the conclusions derived using such a sample. The process of making inferences about the values of population parameters based on random samples is called *statistical inference*.

We use the symbols $y_1, y_2, \ldots, y_n$ to denote the n randomly sampled values from the population $\{Y\}$. Note that lowercase letters (i.e., y) are used to denote sample values, and lowercase n represents the sample size. The values $y_1, y_2, \ldots, y_n$ and n are known numbers.

Consider the population in Example 1.3.3(a). An investigator may be interested in μ_Y, the average interest income earned by the individuals in this population last year. She can select a random sample of n people, say $n = 1,000$, from this population and obtain the amount of interest income earned by each person last year. From these data, she can make inferences on μ_Y. In Example 1.3.3(c) the investigator may be interested in μ_Y, the average age of people who were diagnosed as having lung cancer three years ago and are still alive today, or he may be interested in p, the proportion of persons diagnosed as having lung cancer three years ago who are less than 25 years old now. He can select a random sample of 75 persons (say) from this population of persons and obtain the age of each. From these data, he can make inferences about μ_Y and p. In Example 1.3.3(e) the investigator may be interested in μ_Y, the average profit for all farms in Iowa and Nebraska, so a random sample of n farms, say $n = 80$, can be selected and their profits recorded. From these data, the investigator can make inferences about μ_Y.

In each of these cases, the target population and the study population are one and the same. We now discuss situations where this is not so.

In Example 1.3.3(b), an investigator may be interested in the first-year maintenance costs of the cars in the population. Recall that this population is a conceptual population since it is a future population. It may be reasonable to use the collection of first-year maintenance costs of all cars manufactured by company A *last year* as the study population. A random sample of n cars, say $n = 25$, can be obtained from this study population, and their first-year maintenance costs can be determined by contacting the owners of the cars in the sample. From these data, the investigator can make inferences about the study population. If an investigator feels that the study population resembles the target population sufficiently closely, then he/she can use the conclusions to make decisions about the target population.

In Example 1.3.3(j) an investigator may be interested in the remaining tread-depth (in millimeters) of tires after 50,000 miles of driving. Again the target population is a future population. However, no suitable study population is available because there is no existing information regarding tread-wear for tires manufactured using the new tread design. So the investigator may decide to select the first 100 tires manufactured using the new tread design, mount them on different cars that will then be driven 50,000 miles under conditions similar to those typical customers might experience, and measure the tread-depth remaining. These are the only data available to the investigator. *If it is reasonable to regard these data as a random sample from the target population, then we can use these data to make valid statistical inferences about the target population.*

Simple Random Sample

The size n of the sample to be selected is determined by the investigator based on a careful consideration of costs and the objectives of the study. Samples can be selected from the study population using any one of several random sampling procedures that are discussed in textbooks on sampling methods [32]. A thorough understanding of the advantages and disadvantages of the various sampling methods is required before one of the procedures is selected, and investigators would be wise to consult a professional statistician for advice. The simplest of all sampling procedures is one called *simple random sampling*, and a sample obtained using this procedure is called a **simple random sample**. We assume throughout, unless specifically stated otherwise, that samples are drawn using the simple random sampling procedure. The definition of a simple random sample follows:

D E F I N I T I O N   Simple Random Sample

If a population has N elements, there are

$$H = \binom{N}{n} = \frac{N \times (N-1) \times \cdots \times (N-n+1)}{1 \times 2 \times \cdots \times n}$$

distinct samples of size n that can be obtained. If each of these H samples has an equal chance of being selected, then the sample actually obtained is called a *simple random sample of size n*.   ■

Three Types of Inference Procedures

There are three general types of statistical inference procedures that are commonly used. They are

- Point estimation
- Confidence intervals
- Statistical tests of hypotheses

We discuss these next.

Point Estimation

Suppose θ is an unknown population parameter (θ could be μ_Y, σ_Y, σ_Y^2, $\rho_{Y,X}$, etc.). A **point estimate** of θ is a *number* computed from sample data, to be used by an investigator as the value of θ (because θ is unknown and hence unavailable) in making decisions. Estimation procedures for calculating point estimates are useful when their values are close to the actual values of the unknown population parameters.

Estimates of the Mean and the Standard Deviation of a Univariate Population   If $y_1, y_2, \ldots, y_n$ is a simple random sample from a population $\{Y\}$ whose mean is μ_Y and whose standard deviation is σ_Y, the commonly used estimates of μ_Y and σ_Y, denoted by $\hat{\mu}_Y$ and $\hat{\sigma}_Y$, respectively, are

$$\hat{\mu}_Y = \bar{y} = \frac{1}{n} \sum_{i=1}^{n} y_i \qquad\qquad \text{(1.6.1)}$$

and

$$\hat{\sigma}_Y = \sqrt{\frac{SSY}{(n-1)}} \qquad (1.6.2)$$

where SSY is called the corrected sum of squares for Y, and is defined by

$$SSY = \sum_{i=1}^{n}(y_i - \bar{y})^2$$

The estimate of σ_Y^2 is $\hat{\sigma}_Y^2$, the square of the estimated standard deviation $\hat{\sigma}_Y$. Note that

$$SSY = (n-1)\hat{\sigma}_Y^2$$

Estimate of the Coefficient of Correlation $\rho_{Y,X}$ in a Bivariate Population   If $(y_1, x_1), \ldots,$ (y_n, x_n) is a simple random sample from a bivariate population $\{(Y, X)\}$, the estimate of the coefficient of correlation between Y and X that is widely used is given by

$$\hat{\rho}_{Y,X} = \frac{\sum_{i=1}^{n}(y_i - \bar{y})(x_i - \bar{x})}{\sqrt{\sum_{i=1}^{n}(y_i - \bar{y})^2 \sum_{i=1}^{n}(x_i - \bar{x})^2}} \qquad (1.6.3)$$

Note   *Throughout this book we use a hat symbol ˆ over an unknown parameter to indicate a point estimate of that unknown parameter. For example, $\hat{\sigma}_Y$ represents a point estimate of the unknown population parameter σ_Y.*

Unbiased Estimates   To assess whether a *procedure* for estimating a population parameter is a good procedure, *we must investigate the estimates given by the procedure for every sample that could have been obtained.* Consider the procedure just given of using the sample mean for estimating the population mean. If a population $\{Y\}$ has N items and if the sample size is n, then there are $H = \binom{N}{n}$ possible samples of size n, any one of which could have been selected and, under simple random sampling, each of these possible samples has the same probability of being the actual sample chosen. Conceptually, for each of the H possible samples of size n, we could compute the sample mean. This would result in H sample means $\bar{y}_1, \ldots, \bar{y}_H$, some of which would be close to the population mean and others would not be; some would be larger than the population mean and others would be smaller, but any one of the H possible sample means could end up as the mean of the actual sample selected by the investigator. It can be shown mathematically that the average of all of the H possible sample means is equal to the population mean. We express this fact by saying that the *sample mean is an unbiased estimate of the population mean.* Many investigators and statisticians consider unbiased estimation procedures desirable.

More generally, suppose we want to estimate a parameter θ of a population. We select a random sample of size n from the population and compute an estimate $\hat{\theta}$ of θ using the sample values according to some procedure or formula. In principle, for every possible sample of size n (recall there are H of these), an estimate can be obtained using the same procedure. This would result in H estimates $\hat{\theta}_1, \ldots, \hat{\theta}_H$. If the mean of these H estimates, $\hat{\theta}_1, \ldots, \hat{\theta}_H$, of θ is *equal* to θ, then the procedure used to compute these estimates is said to be an *unbiased estimation procedure*. Of course, in any given problem we compute only one of these H values of $\hat{\theta}$ because we have only one sample of size n available. If we compute this one value $\hat{\theta}$ using an unbiased estimation procedure, then we say that $\hat{\theta}$ is an **unbiased estimate** of θ.

We stated earlier that the sample mean is an unbiased estimate of the population mean. If the sample size n is very large, then $\hat{\sigma}_Y$ as defined in (1.6.2) is approximately an unbiased estimate of σ_Y. For further discussion of the concept of unbiasedness and other desirable properties of estimators, you should consult books on mathematical statistics [15], [19], [25], [31].

To summarize, a point estimate of a parameter θ is a number $\hat{\theta}$, obtained from sample data, that is often used as the value of θ for making various decisions. In addition to a point estimate of θ, we would like to have some way to determine how close $\hat{\theta}$ is to the unknown parameter θ. For most applications in this book, *confidence interval procedures* provide one way of obtaining such information; this is our next topic of discussion.

Confidence Intervals

Suppose an investigator is interested in determining the value of some parameter θ associated with a population under study. As discussed earlier, a point estimate of θ gives us a single value that can be used as the value of θ for making decisions. An alternative approach to estimating the value of θ is to provide, not a single value, but a range of values, say an interval of values, and specify how confident we are that θ is contained in this interval. Such an interval is called a **confidence interval** for θ. This is a reasonable approach because it is often the case that only a range of *possible* or *plausible* values of θ is needed for making decisions and not its *exact* value.

A **two-sided confidence interval** for a population parameter, say θ (where θ could be μ_Y, σ_Y, σ_Y^2, etc.), is an interval whose lower endpoint, say L, and upper endpoint, say U, are computed from sample data. The procedure for computing the confidence interval has the property that the probability of the resulting interval actually containing θ is a prescribed value $1 - \alpha$, which is referred to as the **confidence coefficient** or the **confidence level** associated with the computed confidence interval. We say we are $100(1 - \alpha)\%$ confident that $L \leq \theta \leq U$, and we write $C[L \leq \theta \leq U] = 1 - \alpha$.   **(1.6.4)**

As an illustration, suppose the computations lead to the statement $C[3.86 \leq \theta \leq 6.12] = 0.80$. Then an investigator has 80% confidence that the interval 3.86 to 6.12 contains θ; i.e., we have 80% confidence that θ, the unknown population parameter, is between 3.86 and 6.12.

The confidence level $1 - \alpha$ is selected by the investigator. Values of α that are typically used are $0.01, 0.05$, and 0.10, with corresponding confidence levels $1 - \alpha$

of 0.99, 0.95, and 0.90. The correct value of α to select will depend on the particular problem under consideration, and it need not be restricted to one of the values 0.99, 0.95, 0.90, etc. It may be the case that $1 - \alpha = 0.80$ or 0.85 is appropriate for a particular problem. The value of α (and of $1 - \alpha$) must, of course, lie in the interval from 0 to 1.

The Meaning of a $1 - \alpha$ Confidence Interval

The meaning of a $1 - \alpha$ confidence interval for an unknown population parameter θ, computed using a simple random sample of size n from a population of N items, is as follows:

There are $H = \binom{N}{n}$ possible samples of size n that could be selected from the population. Conceptually, every possible sample of size n could be selected from the population, and a confidence interval for θ could be computed from each sample; there would be H confidence intervals. *The proportion of these H intervals that would include the unknown parameter θ is equal to $1 - \alpha$.* Of course in a real problem an investigator selects only one sample of size n and computes only one confidence interval for θ. Our level of confidence can be quantified by the number $1 - \alpha$ because it is known that a $1 - \alpha$ proportion of all H possible intervals would include the true unknown value of the parameter θ. (1.6.5)

Equal-Tailed Confidence Intervals and One-sided Confidence Bounds

In (1.6.5) the proportion of the H confidence intervals that will fail to contain θ is α. Suppose that in half of these cases the actual value of θ is greater than the corresponding computed upper endpoint U and that in the other half of these cases θ is smaller than the corresponding computed lower endpoint L. Then the confidence interval procedure is said to be **equal-tailed**. Nearly all of the confidence interval procedures discussed in this book are equal-tailed. In this case a two-sided confidence statement

$$C[L \leq \theta \leq U] = 1 - \alpha$$

has the following additional interpretations.

- If we know before looking at the data that only an upper bound for θ is needed for making decisions, then we can be $100(1 - \alpha/2)\%$ confident that θ is less than or equal to U. We will write this as $C[\theta \leq U] = 1 - \alpha/2$ and say that U is an **upper confidence bound** for θ with confidence coefficient equal to $1 - \alpha/2$.

- If we know before looking at the data that only a lower bound for θ is needed for making decisions, then we can be $100(1 - \alpha/2)\%$ confident that θ is greater than or equal to L. We write this as $C[L \leq \theta] = 1 - \alpha/2$ and say that L is a **lower confidence bound** for θ with confidence coefficient equal to $1 - \alpha/2$. **(1.6.6)**

Note   If, before looking at the data, we do not know whether the upper bound or the lower bound for θ is needed for making decisions, but after computing the two-sided $1 - \alpha$ (equal-tailed) confidence interval and looking at L and U we decide to use only L or only U to make a decision, then the appropriate confidence coefficient associated with the decision is $1 - \alpha$ and not $1 - \alpha/2$.

Symmetric Confidence Intervals

In most applications discussed in this book (important exceptions are confidence intervals for standard deviations, variances, and correlation coefficients), a two-sided confidence interval for the unknown parameter θ will be **symmetric** about the best point estimate $\hat{\theta}$ of θ; i.e., the values L and U will be of the form $L = \hat{\theta} - D$ and $U = \hat{\theta} + D$, where $\hat{\theta}$ and D (and hence L and U) are computed from the data, and $D = (U - L)/2$. In these cases the following is a useful way to view a two-sided confidence interval associated with the point estimate $\hat{\theta}$:

If $\hat{\theta}$ is a point estimate of θ and $L \leq \theta \leq U$ is a $1 - \alpha$ **two-sided confidence interval** for θ that is symmetric about $\hat{\theta}$, then we have $100(1 - \alpha)\%$ confidence that $\hat{\theta}$ is within $D = (U - L)/2$ units of the true unknown value θ. **(1.6.7)**

For example, a useful alternative way to describe the confidence statement $C[3.86 \leq \theta \leq 6.12] = 0.80$, symmetric about a point estimate $\hat{\theta}$ of θ, is to say *we have 80% confidence that $\hat{\theta}$ is within $(6.12 - 3.86)/2 = 1.13$ units of θ.*

Note   Some authors use the symbol P instead of C; for example, some authors write $P[\theta \leq 9.67] = 0.80$ and say that the probability is 80% that θ is less than or equal to 9.67. This is an incorrect use of the word *probability* because the statement $\theta \leq 9.67$ is either true or false and hence $P[\theta \leq 9.67]$ is either 0 or 1 (we do not know which), but it is certainly not equal to 0.80. To avoid this incorrect usage of the term *probability*, we use the symbol C instead of P and say confidence instead of probability where the word confidence has the meaning given in (1.6.5).

For many, and in fact most, applications in this book, a $1 - \alpha$ confidence interval is of the form

$$\hat{\theta} - \text{table-value} \times SE(\hat{\theta}) \leq \theta \leq \hat{\theta} + \text{table-value} \times SE(\hat{\theta}) \qquad \textbf{(1.6.8)}$$

where $SE(\hat{\theta})$ denotes the standard error of $\hat{\theta}$, which is a measure of how precisely $\hat{\theta}$ estimates θ. The detailed meaning of $SE(\hat{\theta})$ is as follows:

Suppose the population under study has N items, and a simple random sample of size n is selected. The total number of different possible samples of size n is denoted by H, which equals $\binom{N}{n}$. Conceptually, each possible sample of size n could be selected from the population, and both an estimate $\hat{\theta}$ of θ and the error of estimation $\hat{\theta} - \theta$ could be computed from each sample. The standard deviation of the collection $\{\hat{\theta} - \theta\}$ of estimation errors, which we may write as $SD(\hat{\theta} - \theta)$, tells us how good the estimation procedure is for estimating θ. Calculation of $SD(\hat{\theta} - \theta)$ requires that the entire population of numbers be available. However, a valid estimate of $SD(\hat{\theta} - \theta)$ can often be computed from sample data; it is denoted by $SE(\hat{\theta} - \theta)$, or simply by $SE(\hat{\theta})$ for ease of notation, and it is called the **standard error of $\hat{\theta}$.** **(1.6.9)**

Authors' Recommendation

When the confidence interval for an unknown parameter θ is of the form given in (1.6.8), the investigator should be given the following:

1   The point estimate $\hat{\theta}$ of θ.
2   The standard error of $\hat{\theta}$, viz., $SE(\hat{\theta})$.
3   The degrees of freedom (df) associated with $SE(\hat{\theta})$.

With this information the investigator can easily compute confidence intervals for θ with any confidence coefficient $1 - \alpha$ as needed.

We now illustrate the procedures for computing point estimates and confidence intervals.

E X A M P L E 1.6.1

Company A manufactures compact cars whose gas mileages Y (in miles per gallon) form a Gaussian population with mean μ_Y and standard deviation σ_Y, both of which are unknown. The manager of a car rental company is interested in purchasing a large fleet of these cars from company A next year. In order to adhere to company policy, she will do so only if she is reasonably confident that the average gas mileage of all cars of this type, to be manufactured by company A next year, will be at least 25 miles per gallon when driven over a test route. Thus the target population of numbers is the collection of gas mileages of all cars of this type manufactured by company A next year. This population is unavailable for study now, so the manager decides to study the population of gas mileages of all cars of the type required, manufactured by company A *last year*, because she feels that the study population resembles the target population for all practical purposes. So she obtains a simple random sample of 10 cars manufactured by company A last year and determines their gas mileages. The data are shown in Table 1.6.1 and are also given in the file **table161.dat** on the data disk.

Only one-sided lower confidence bounds are of interest in this problem. The manager wants a 90% lower confidence bound for μ_Y, and she will buy the fleet of cars from company A only if the lower bound exceeds 25. The required lower bound can be obtained from a two-sided 80% confidence interval as explained in (1.6.6). We first compute the following quantities:

$$\hat{\mu}_Y = \bar{y} = 25.302$$

$$SSY = \sum_{i=1}^{10}(y_i - \bar{y})^2 = 3.67796$$

$$\hat{\sigma}_Y = 0.63927$$

$$SE(\hat{\mu}_Y) = \hat{\sigma}_Y/\sqrt{n} = 0.63927/\sqrt{10} = 0.20215$$

A two-sided 80% confidence statement for μ_Y (using $t_{1-\alpha/2:n-1} = t_{0.90:9} = 1.383$ obtained from Table T-2 in the appendix) is

$$C[\hat{\mu}_Y - t_{0.90:9}SE(\hat{\mu}_Y) \leq \mu_Y \leq \hat{\mu}_Y + t_{0.90:9}SE(\hat{\mu}_Y)]$$
$$= C[25.02 \leq \mu_Y \leq 25.58] = 0.80$$

Thus the required one-sided 90% lower confidence bound for μ_Y is 25.02. Observe that the lower endpoint of 25.02 in the two-sided 80% confidence statement can in fact be used as a one-sided 90% lower confidence bound since we knew without looking at the data that it is the lower bound that is needed to make a decision. Thus the manager can be 90% confident that the average gas mileage for cars manufactured last year by company A is greater than or equal to 25.02 mpg. To extrapolate this result to cars that will be manufactured next year by company A is a judgment decision.

Note   In this book, when we demonstrate computations, we may carry several significant digits for all intermediate calculations and report the final results to more

T A B L E 1.6.1

Car number	1	2	3	4	5	6	7	8	9	10
Gas mileage (mpg)	25.72	25.24	25.19	25.88	26.42	24.48	25.11	24.29	25.06	25.63

significant digits than is perhaps necessary. In actual applications we advise that the *final results* be rounded to an appropriate number of significant digits. ∎

Prediction Intervals

In many problems the interest is not in estimating the mean μ_Y of a population $\{Y\}$ but in predicting the Y value of an item that is yet to be chosen from the population. We denote this value by Y_0 and call it a future value to be randomly chosen from the population $\{Y\}$. For instance, consider Example 1.6.1 and suppose you plan to purchase a new car to be made by manufacturer A. You would like to know what you can expect your maintenance cost to be the first year after you purchase it. Thus a future observation Y_0 is to be randomly chosen from a population $\{Y\}$, and you want to predict its value now.

Suppose $y_1, y_2, \ldots, y_n$ is a simple random sample of size n from this population. The predicted value of Y_0 using these sample data is $\bar{y}$, the sample mean. Suppose you also want to obtain an interval, say $L \leq Y_0 \leq U$, such that you are 95% confident this interval will contain the value Y_0 that will be selected from the population. It may be tempting to use a 95% confidence interval for μ_Y, the mean first-year maintenance cost of *all* cars in the population, as the required interval for Y_0. *This is incorrect.* Whereas we can have 95% confidence that the interval for μ_Y will in fact contain μ_Y, our confidence will be lower than 95% that the interval for μ_Y will contain the future value Y_0 because individual values in the population can be, and generally are, different from the population mean. To account for this fact, a 95% confidence interval for a future value Y_0 has to be *wider* than a 95% confidence interval for μ_Y. This interval for Y_0 is called a 95% **prediction interval** and is given by (1.6.8) with θ replaced by Y_0 and $\hat{\theta}$ replaced by $\hat{Y}_0 = \bar{y}$. Also $SE(\hat{\theta})$ is replaced by $SE(\hat{Y}_0) = \hat{\sigma}_Y \sqrt{1 + (1/n)}$. Thus a $1 - \alpha$ confidence statement for Y_0 is

$$C[\hat{Y}_0 - t_{1-\alpha/2:n-1} SE(\hat{Y}_0) \leq Y_0 \leq \hat{Y}_0 + t_{1-\alpha/2:n-1} SE(\hat{Y}_0)] = 1 - \alpha \qquad \text{(1.6.10)}$$

We generally distinguish between a prediction interval and a confidence interval because a confidence interval is for a *fixed unknown parameter* in a population, and a prediction interval is for a *future observation* to be randomly selected from a population.

The Meaning of a $1 - \alpha$ Prediction Interval

The meaning of a $1 - \alpha$ prediction interval for Y_0, a future random observation to be selected from a population $\{Y\}$, computed using a simple random sample of size n from a population of N items, is as follows:

Choose a simple random sample of size n and use this to compute a $1 - \alpha$ two-sided prediction interval for a single future observation. This prediction interval may or may not contain the future observation. Repeat this procedure over and over where each time a new

simple random sample of size n is selected and a prediction inter-   **(1.6.11)**
val is computed for a new future observation. Then in the long run,
$100(1 - \alpha)\%$ of all the prediction intervals will include the corresponding future observation. Our level of confidence that this interval
will include the single future observation can be quantified by the
number $1 - \alpha$.

E X A M P L E   **1.6.2**

Consider Example 1.6.1 and suppose that you are planning to purchase a new car
to be made by company A next year. You want to predict what the miles per gallon
(mpg) will be for the car you will buy. You also want an interval for the miles
per gallon this car will get so you can have 80% confidence that your interval is
correct. The gas mileage for the car you will buy is a future random observation
obtained from the population of gas mileages of all cars that will be manufactured
by company A next year. If we believe that the study population (last year's cars)
closely resembles the target population (next year's cars), we can use last year's data
for inference.

From Example 1.6.1 we get $\hat{\mu}_Y = \bar{y} = 25.302$ and $\hat{\sigma}_Y = 0.63927$. Thus $\hat{Y}_0 =$
25.302 and $SE(\hat{Y}_0) = 0.63927\sqrt{1 + (1/10)} = 0.670$. Hence we get

$$C[24.37 \leq Y_0 \leq 26.23] = 0.80$$

and you have 80% confidence that the new car you will purchase will get between
24.37 and 26.23 miles per gallon of gasoline.   ∎

For convenience, in Table 1.6.2 we have summarized the procedures for computing confidence intervals for the mean μ_Y and the standard deviation σ_Y of a Gaussian
population. In the same table we also describe the procedure for obtaining a confidence interval (prediction interval) for a randomly chosen value Y_0 from a Gaussian
population.

Statistical Tests of Hypotheses

Often an investigator conjectures that a parameter θ of a population is equal to, less
than, or greater than a specified value q. Statistical tests, performed using sample
data, are often used to help decide whether or not the data provide evidence *against*
the conjecture. The investigator formulates an appropriate pair of hypotheses, one
of which is designated as the null hypothesis (NH) and the other as the alternative
hypothesis (AH), and a statistical test is used to determine what evidence the sample
data can provide *against NH in favor of AH*.

A statistical test typically consists of the four steps below.

1   For a population parameter of interest, say θ, and for a specified value q, we
suppose that an investigator is interested in one of the three pairs of hypotheses

T A B L E **1.6.2**

Point Estimates and Confidence Intervals for μ_Y, σ_Y, and Y_0 in a One-Variable Gaussian Population

Notation: $\bar{y} = \dfrac{1}{n}\sum y_i$; $SSY = \sum (y_i - \bar{y})^2$

Inference	Formulas and Procedures
Point estimate of μ_Y, σ_Y	$\hat{\mu}_Y = \bar{y}$ $\hat{\sigma}_Y = \sqrt{SSY/(n-1)}$
Two-sided $1 - \alpha$ confidence intervals for μ_Y	$\hat{\mu}_Y - t_{1-\alpha/2:n-1}SE(\hat{\mu}_Y) \le \mu_Y \le \hat{\mu}_Y + t_{1-\alpha/2:n-1}SE(\hat{\mu}_Y)$ where $SE(\hat{\mu}_Y) = \dfrac{\hat{\sigma}_Y}{\sqrt{n}}$
Two-sided $1 - \alpha$ confidence intervals for σ_Y	$\sqrt{\dfrac{SSY}{\chi^2_{1-\alpha/2:n-1}}} \le \sigma_Y \le \sqrt{\dfrac{SSY}{\chi^2_{\alpha/2:n-1}}}$
Two-sided $1 - \alpha$ confidence intervals for Y_0	$\hat{\mu}_Y - t_{1-\alpha/2:n-1}\hat{\sigma}_Y\sqrt{1+\dfrac{1}{n}} \le Y_0 \le \hat{\mu}_Y + t_{1-\alpha/2:n-1}\hat{\sigma}_Y\sqrt{1+\dfrac{1}{n}}$

(a)–(c) in Table 1.6.3 and wants to determine what evidence the data provide against NH in favor of AH.

2   A random sample of size n is selected from the study population and an appropriate number, say Q_C, called the **test statistic**, is computed using the sample data (the subscript C in Q_C stands for computed value).

3   This number Q_C is referred to an appropriate table of *percentiles* of a theoretical distribution, known as the *reference distribution*, and a number P is determined such that *if the NH is indeed true, then the probability of obtaining a value of Q_C, as unfavorable as or more unfavorable than the value actually obtained,*

TABLE 1.6.3

	NH	AH
(a)	$\theta = q$	$\theta \neq q$
(b)	$\theta \leq q$	$\theta > q$
(c)	$\theta \geq q$	$\theta < q$

is equal to P. This number P is called the **significance probability** or the **P-value** associated with the test of NH versus AH. Since P is a probability, it is a number between 0 and 1. The value of P is a measure of the evidence that the data provide against NH in favor of AH. Small values of P indicate that the data provide evidence against the null hypothesis in favor of the alternative hypothesis, whereas large values of P imply that the data *do not* provide evidence against the null hypothesis (note that large values of P do not necessarily imply that the data provide evidence in support of the null hypothesis).

Note   Computation of exact P-values usually requires detailed statistical tables for the t-distribution, the χ^2-distribution, the F-distribution, etc. Such detailed tables are not easily available, but statistical packages such as MINITAB and SAS may be used to obtain exact P-values in most situations. Alternatively, approximate P-values may be computed by suitably interpolating table values. Appendix T has tables that can be used to obtain bounds for P-values, and these are generally adequate.

4   The investigator decides, based on a detailed knowledge of the problem, what values of P are to be considered small. It seems to be common practice among many investigators to *arbitrarily* select a number α (called the *size* of the test) equal to 0.05 or 0.01 and to consider the value of P to be small if it is less than α. The investigator would then **reject** NH if $P \leq \alpha$ and would *not reject* NH if $P > \alpha$. If this procedure is followed by the investigator, then the probability of rejecting NH when NH is in fact true is guaranteed to be no greater than α. However, the probability of rejecting NH when it is actually false can generally not be determined without further computations.

EXAMPLE 1.6.3

For an illustration consider Example 1.2.1. Suppose the company is interested in building a store in city A only if the average annual income per family is greater than \$20,000. Here $\theta = \mu_Y$, the average annual income per family in city A, and $q = \$20,000$. The null hypothesis and the alternative hypothesis that are appropriate for this problem are

$$\text{NH}: \quad \mu_Y \leq 20,000$$
$$\text{AH}: \quad \mu_Y > 20,000$$

If the investigator chooses $\alpha = .05$, then the probability of rejecting NH (and deciding $\mu_Y > 20,000$) when μ_Y is indeed less than or equal to \$20,000 is at most 5%. By choosing the value of α appropriately and using a statistical test, the investigator can control the probability of incorrectly deciding to build a store if the average annual income is too low (i.e., the probability of rejecting NH if NH is indeed true).

However, it should be pointed out that the probability of the company deciding to build a store when in fact the average annual income is greater than \$20,000 may be unacceptably small unless the sample is sufficiently large. This probability is determined by computing the *power* of the test. ■

The test statistic to compute and the appropriate table to consult for various statistical testing situations have been determined by mathematical theory, and they are described as appropriate. In particular, Boxes 1.6.1 and 1.6.2 summarize statistical tests for the mean μ_Y and the standard deviation σ_Y, respectively, of a univariate Gaussian population.

B O X **1.6.1** **Hypothesis Tests for μ_Y**

Let q be a number specified by the investigator. Compute the statistic

$$t_C = \frac{\hat{\mu}_Y - q}{(\hat{\sigma}_Y/\sqrt{n})}$$

a For testing NH: $\mu_Y = q$ versus AH: $\mu_Y \neq q$, the *P*-value is the value of α such that $|t_C| = t_{1-\alpha/2:n-1}$.

b For testing NH: $\mu_Y \leq q$ versus AH: $\mu_Y > q$, the *P*-value is the value of α such that $t_C = t_{1-\alpha:n-1}$.

c For testing NH: $\mu_Y \geq q$ versus AH: $\mu_Y < q$, the *P*-value is the value of α such that $-t_C = t_{1-\alpha:n-1}$.

B O X **1.6.2** **Hypothesis tests for σ_Y**

Let q be a positive number specified by the investigator. Compute the statistic

$$\chi_C^2 = \frac{(n-1)\hat{\sigma}_Y^2}{q^2}$$

a For testing NH: $\sigma_Y = q$ versus AH: $\sigma_Y \neq q$, the P-value is equal to α, where α is a number between 0 and 1 and satisfies one of the following two conditions (it is not possible for both conditions to be satisfied unless $\alpha = 1$):

$$\chi_C^2 = \chi_{1-\alpha/2:n-1}^2$$

or

$$\chi_C^2 = \chi_{\alpha/2:n-1}^2$$

b For testing NH: $\sigma_Y \leq q$ versus AH: $\sigma_Y > q$, the P-value is the value of α such that $\chi_C^2 = \chi_{1-\alpha:n-1}^2$.

c For testing NH: $\sigma_Y \geq q$ versus AH: $\sigma_Y < q$, the P-value is the value of α such that $\chi_C^2 = \chi_{\alpha:n-1}^2$.

Relationship Between Tests and Confidence Intervals   There is a relationship between size α tests about θ and $1 - \alpha$ equal-tailed confidence intervals for θ, and it is as follows.

1 Suppose we want a size α test of NH: $\theta = q$ against AH: $\theta \neq q$. Then NH is rejected if and only if the two-sided $1 - \alpha$ confidence interval for θ does not contain q.

2 Suppose we want a size α test of NH: $\theta \leq q$ against AH: $\theta > q$. Then NH is rejected if and only if q is less than the one-sided $1 - \alpha$ lower confidence bound L. Recall that this lower confidence bound L is actually the lower endpoint of the $1 - 2\alpha$ two-sided (equal-tailed) confidence interval for θ.

3 Suppose we want a size α test of NH: $\theta \geq q$ against AH: $\theta < q$. Then NH is rejected if and only if q is greater than the $1 - \alpha$ one-sided upper confidence bound U. Recall that this upper bound U is actually the upper endpoint of the $1 - 2\alpha$ two-sided (equal-tailed) confidence interval for θ.

Note   *From a $1 - \alpha$ confidence interval (if it exists), we can obtain the result of a statistical test of size α, but from the result of a statistical test we* **cannot** *obtain the corresponding confidence interval.*

For example, suppose we want to test (with $\alpha = 0.05$) NH: $\mu_Y = 6.0$ against AH: $\mu_Y \neq 6.0$ by using a random sample from a Gaussian population with unknown mean μ_Y and unknown standard deviation σ_Y. If a $1 - \alpha = 0.95$ confidence interval for μ_Y is $13.1 \leq \mu_Y \leq 18.9$, then NH is rejected because 6.0 is not contained in this interval. On the other hand, suppose the $1 - \alpha = 0.95$ confidence interval for μ_Y is $3.5 \leq \mu_Y \leq 10.7$; then NH is not rejected because 6.0 is in this interval.

Authors' Recommendation

We recommend that traditional statistical tests of hypotheses for a parameter, say θ (where one rejects or does not reject NH), never be used if a confidence interval for θ is available because confidence intervals are always more informative than tests, and tests alone (without the accompanying confidence intervals) can be misleading. Since tests are taught and widely used by investigators, we discuss them in this book, but as a general rule we advise against their indiscriminate use.

We now illustrate the procedures just discussed for testing statistical hypotheses.

E X A M P L E **1.6.4**

Consider Example 1.6.1 where the manager of a car rental company is interested in purchasing a large fleet of cars from company A if it can be determined that μ_Y, the average miles per gallon of all automobiles of this type, is at least 25 when the cars are driven over a specified test route. If the manager uses a statistical test to determine this, the appropriate NH and AH are

$$NH: \quad \mu_Y \leq 25$$
$$AH: \quad \mu_Y > 25$$

You may want to refer to Box 1.6.1 for details of the calculations. We obtain

$$t_C = \frac{25.302 - 25}{0.20215} = 1.494$$

From Table T-2 in the appendix we find that the *P*-value corresponding to this value of t_C is between 0.05 and 0.10. Linear interpolation gives a *P*-value equal to 0.0877. The *P*-value obtained from the statistical computing package MINITAB is equal to 0.085 (rounded to 3 decimals).

Suppose the manager chooses $\alpha = 0.10$; i.e., the manager decides to reject NH if $P < 0.10$ (which would correspond to a confidence level of $1 - \alpha = 0.90$ for the lower confidence bound). Then she would reject NH, and decide that the data provide enough evidence to conclude that the gas mileage will exceed 25 mpg. Observe that the significance probability gives us more information than the result of a test using a fixed prechosen value of α. It tells us that NH will be rejected for any value of α greater than 0.0847, but NH will not be rejected for any value of α less than or equal to 0.0847. Notice, however, that much more information is obtained from the two-sided 80% confidence interval for μ_Y. It actually tells us what the value of μ_Y is likely to be. In particular, we recall from Example 1.6.1 that the two-sided 80% confidence interval for μ_Y is given by the confidence statement

$$C[25.02 \leq \mu_Y \leq 25.58] = 0.80$$

On the basis of this confidence interval, the manager may conclude that μ_Y is close enough to 25 to be considered to be equal to 25 mpg for all practical purposes. ∎

**Simultaneous Tests (Tests about Several Parameters)
and Simultaneous Confidence Intervals (Confidence Intervals
for Several Parameters)**

We have discussed how to test a hypothesis about a single parameter θ, how to obtain a confidence interval for a single parameter θ, and we have discussed the relationship between tests and confidence intervals. Researchers often conduct investigations where several parameters $\theta_1, \ldots, \theta_m$ are involved, and they want to examine the relationships among them. We illustrate with two examples.

E X A M P L E **1.6.5**

A company has developed four new chemicals, any one of which can be added to cement in an attempt to increase the strength of cement building blocks. An experiment is conducted to examine the differences among the average strengths, say μ_1, μ_2, μ_3, μ_4, of the cement blocks made with each of the four chemicals; i.e., an experiment is conducted to examine the differences

$$\mu_1 - \mu_2, \ \mu_1 - \mu_3, \ \mu_1 - \mu_4, \ \mu_2 - \mu_3, \ \mu_2 - \mu_4, \ \mu_3 - \mu_4$$

To determine if the average strengths of cement blocks made with the four chemicals are different, many practitioners would conduct a statistical test of

$$\text{NH} : \mu_1 = \mu_2 = \mu_3 = \mu_4 \quad \text{against} \quad \text{AH} : \text{at least one equality does not hold}$$

$$\textbf{(1.6.12)}$$

$\blacksquare$

E X A M P L E **1.6.6**

An experiment is conducted to determine how different water temperatures, at which a new fabric is laundered, affect the strength of the fabric after 100 launderings. Three different water temperatures, $100°$ C, $140°$ C, and $180°$ C, were used, and five different pieces of the fabric were laundered at each of these three temperatures (a total of $5 \times 3 = 15$ different pieces of the fabric were used altogether). After the launderings, the strength of each piece of the fabric was measured. The investigator wants to examine the differences among the average strengths of the various pieces of the fabric laundered at different temperatures. To do this, it is common practice to conduct a statistical test of

$$\text{NH} : \mu_1 = \mu_2 = \mu_3 \quad \text{against} \quad \text{AH} : \text{at least one equality fails} \qquad \textbf{(1.6.13)}$$

where μ_1, μ_2, and μ_3 are the average strengths of the fabric after being laundered 100 times in water at $100°$ C, $140°$ C, and $180°$ C, respectively. $\blacksquare$

Simultaneous Tests

Suppose $\theta_1, \ldots, \theta_m$ are m parameters of interest and that a researcher conjectures that the equalities $\theta_1 = q_1, \ldots, \theta_m = q_m$ are all true ($q_1, \ldots, q_m$ are specified numbers). This is often expressed by a null hypothesis of the form

$$\text{NH}: \left\{ \begin{array}{l} \theta_1 = q_1 \\ \theta_2 = q_2 \\ \vdots \\ \theta_m = q_m \end{array} \right\}$$

The alternative hypothesis AH states that at least one of the equalities in NH is false. A statistical testing procedure will help decide whether or not the NH should be rejected in favor of AH at a specified level α. Such a testing procedure is called a **simultaneous test** if $m \geq 2$.

In Example 1.6.5 let $\theta_1 = \mu_1 - \mu_2$, $\theta_2 = \mu_1 - \mu_3$, and $\theta_3 = \mu_1 - \mu_4$. The null hypothesis in (1.6.12) can be expressed as

$$\text{NH}: \left\{ \begin{array}{l} \theta_1 = 0 \\ \theta_2 = 0 \\ \theta_3 = 0 \end{array} \right\}$$

The alternative hypothesis states that at least one of $\theta_1, \theta_2, \theta_3$ is nonzero. Hence the corresponding statistical test is a simultaneous test (i.e., several hypotheses are tested simultaneously).

In Example 1.6.6 let $\theta_1 = \mu_1 - \mu_2$ and $\theta_2 = \mu_1 - \mu_3$. The NH in (1.6.13) can be expressed as

$$\text{NH}: \left\{ \begin{array}{l} \theta_1 = 0 \\ \theta_2 = 0 \end{array} \right\}$$

The alternative hypothesis states that at least one of θ_1 and θ_2 is nonzero. Again this leads to a simultaneous statistical test.

We have recommended that tests about a single parameter θ not be used without the accompanying confidence interval for θ (if it is available). The same recommendation applies for simultaneous tests about several parameters $\theta_1, \ldots, \theta_m$. However, when $m \geq 2$, there are two types of confidence intervals that can be used: (1) one-at-a-time confidence intervals and (2) simultaneous confidence intervals. We discuss these next.

One-at-a-Time Confidence Intervals

For one-at-a-time confidence intervals with confidence coefficient $1 - \alpha$, we compute confidence intervals for each parameter θ_i, and for any *one* of the statements below,

$$L_1 \leq \theta_1 \leq U_1$$
$$L_2 \leq \theta_2 \leq U_2$$

.
.
.

$$L_m \leq \theta_m \leq U_m$$

we have $1 - \alpha$ confidence that it is correct. However, we do not have $1 - \alpha$ confidence (in fact, in most situations we would have less than $1 - \alpha$ confidence) that *all* preceding m statements are *simultaneously correct*.

Simultaneous Confidence Intervals

For simultaneous confidence intervals with confidence coefficient $1 - \alpha$, we compute confidence intervals for each θ_i such that we have $1 - \alpha$ confidence that all of the confidence intervals below are simultaneously correct.

$$L_1^* \leq \theta_1 \leq U_1^*$$
$$L_2^* \leq \theta_2 \leq U_2^*$$

.

.

.

$$L_m^* \leq \theta_m \leq U_m^*$$

The difference between the interpretation of one-at-a-time confidence statements and simultaneous confidence statements is made clear by the following example.

E X A M P L E 1.6.7

Consider a group of 100 students of whom 80 are females and 20 are males. It is known that 70 of the females and 10 of the males are math majors. Suppose a student is to be randomly selected from this group. We are 80% confident that the chosen student will be a female (because 80 of the 100 students are females). We are also 80% confident that the chosen student will be a math major (because 80 of the 100 students are math majors). However, we cannot be 80% confident that the chosen student will be, *simultaneously*, a female and a math major! In fact we can be only 70% confident that the chosen student is a female math major (because 70 of the 100 students are female math majors).

In the same way, we may have $1 - \alpha$ confidence in the statement $L_1 \leq \theta_1 \leq U_1$ and also have $1 - \alpha$ confidence in the statement $L_2 \leq \theta_2 \leq U_2$, but this does not imply that we have $1 - \alpha$ confidence in the simultaneous statement $L_1 \leq \theta_1 \leq U_1$ *and* $L_2 \leq \theta_2 \leq U_2$. Such simultaneous confidence statements can, however, be made if appropriate table-values and computational procedures are used. ∎

Authors' Recommendation

For each problem, the investigator must decide which type of confidence interval (one-at-a-time or simultaneous) to use. We recommend that simultaneous confidence intervals be used *only* in situations when an investigator must make a decision that depends on knowing all of the values θ_i simultaneously, with a specified level of confidence. That is, an investigator wants to have $1 - \alpha$ confidence that a decision is correct and, for the decision to be correct, *all* of the confidence intervals, $L_i \leq \theta_i \leq U_i$, $i = 1, \ldots, m$, must be simultaneously correct. Thus the investigator wants to have $1 - \alpha$ confidence that all m intervals are correct.

We illustrate with an example.

E X A M P L E **1.6.8**

To be in compliance with environmental regulations, the manager of a coal-fired power-generating plant makes periodic measurements of emissions from the smokestacks. Using these measurements, she can estimate μ_i, for $i = 1, 2, 3, 4$, the average weekly emissions of four different toxic components. These estimates are used to determine whether the plant is in compliance with the regulations. The manager computes upper confidence bounds

$$\mu_1 \leq U_1, \quad \mu_2 \leq U_2, \quad \mu_3 \leq U_3, \quad \mu_4 \leq U_4$$

and she wants to be 99% confident that *all four* of the preceding statements are simultaneously correct. If the value of each U_i is less than the value required by the environmental regulations, she would perhaps conclude that the plant is in compliance. Here, the correctness of the overall decision depends on all of the four confidence statements being simultaneously correct. Thus simultaneous confidence statements are needed. ∎

For many of the problems that we discuss, the formula for confidence intervals for $\theta_1, \theta_2, \ldots, \theta_m$ is given by (this formula does not apply to confidence intervals on standard deviations, variances, and correlation coefficients):

$$\hat{\theta}_i - \text{table-value} \times SE(\hat{\theta}_i) \leq \theta_i \leq \hat{\theta}_i + \text{table-value} \times SE(\hat{\theta}_i) \quad \text{for } i = 1, \ldots, m$$

and the table-value determines whether these m confidence intervals are one-at-a-time or simultaneous. Also, there are several different methods for obtaining simultaneous confidence intervals, each one requiring a different table-value. Which one of these to use depends on the particular problem being studied. We give you the appropriate table-values to use for each situation required in this book.

There is, however, one general method for obtaining simultaneous confidence statements that is easy to apply. This is called the **Bonferroni method**, which is explained in Box 1.6.3.

B O X 1.6.3   **Bonferroni Method for Simultaneous Confidence Statements**

Suppose m one-at-a-time confidence intervals are computed, each with a confidence coefficient equal to $1 - (\alpha/m)$, leading to the following confidence statements.

$$C[L_1 \le \theta_1 \le U_1] = 1 - \frac{\alpha}{m}$$

$$C[L_2 \le \theta_2 \le U_2] = 1 - \frac{\alpha}{m}$$

$$\cdot$$
$$\cdot$$
$$\cdot$$

$$C[L_m \le \theta_m \le U_m] = 1 - \frac{\alpha}{m}$$

Then the following simultaneous confidence statement is valid.

$$C \begin{bmatrix} L_1 \le \theta_1 \le U_1 \\ L_2 \le \theta_2 \le U_2 \\ \cdot \\ \cdot \\ \cdot \\ L_m \le \theta_m \le U_m \end{bmatrix} \ge 1 - \alpha$$

Example 1.6.9 illustrates the use of the Bonferroni method.

E X A M P L E **1.6.9**

Consider the situation described in Example 1.6.1. Suppose we wish to compute a simultaneous 90% two-sided confidence statement for μ_Y and σ_Y of the form

$$C \begin{bmatrix} L_1 \le \mu_Y \le U_1 \\ \text{and} \\ L_2 \le \sigma_Y \le U_2 \end{bmatrix} \ge 0.90$$

Since we want to make *two* statements with simultaneous confidence coefficient greater than or equal to 0.90, we have $m = 2$ and $\alpha = 0.10$. According to the Bonferroni method, we must compute a confidence interval for μ_Y with confidence coefficient $1 - \alpha/m = 0.95$ and a confidence interval for σ_Y with confidence coefficient $1 - \alpha/m = 0.95$. Then the simultaneous confidence coefficient is *at least* $1 - \alpha = 0.90$. You should verify the following confidence interval calculations.

A two-sided 95% confidence statement for μ_Y can be computed using the formula given in Table 1.6.2. For this we need the table value $t_{0.975:9}$, which can be obtained from Table T-2 in the appendix and is equal to 2.262. The required confidence statement is

$$C[\hat{\mu}_Y - t_{0.975:9}SE(\hat{\mu}_Y) \le \mu_Y \le \hat{\mu}_Y + t_{0.975:9}SE(\hat{\mu}_Y)]$$
$$= C[25.302 - (2.262)(0.20215) \le \mu_Y \le 25.302 + (2.262)(0.20215)]$$

$$= C[24.845 \le \mu_Y \le 25.759] = 0.95$$

A two-sided 95% confidence statement for σ_Y, using the table-values $\chi^2_{0.025:9} = 2.699$ and $\chi^2_{0.975:9} = 19.031$ (obtained from Table T-3 in the appendix) in the formula given in Table 1.6.2, is

$$C\left[\sqrt{\frac{SSY}{\chi^2_{0.975:9}}} \le \sigma_Y \le \sqrt{\frac{SSY}{\chi^2_{0.025:9}}}\right]$$

$$= C\left[\sqrt{\frac{3.678}{19.023}} \le \sigma_Y \le \sqrt{\frac{3.678}{2.700}}\right]$$

$$= C[0.4397 \le \sigma_Y \le 1.1674] = 0.95$$

Hence the following simultaneous confidence statement is valid.

$$C\left[\begin{array}{c} 24.845 \le \mu_Y \le 25.759 \\ \text{and} \\ 0.4397 \le \sigma_Y \le 1.1674 \end{array}\right] \ge 0.90 \quad \blacksquare$$

Just as there is a relationship between tests and confidence intervals in the case of single parameters, *there is also a relationship between simultaneous tests and simultaneous confidence intervals when several parameters are of interest.* This relationship is often somewhat complex, and interested readers should consult more advanced books for details [11].

Note   From time to time we present conversations between a professional statistician and an investigator who routinely uses statistical methods to interpret results of experiments. This is intended to emphasize and clarify various topics that are discussed and to show how some of the statistical procedures can be applied to practical problems.

Conversation   1.6

I am a professional statistician with a Ph.D. in statistics. One day I received a telephone call from a person, whom I call the investigator, who wanted to discuss some statistical problems. We set up an appointment and the investigator visited me the following day. Following is an excerpt from our conversation.

Investigator: Good morning, I am an investigator, and I'd like to ask you some questions about statistics.

Statistician: Okay, but first give me some information about your background in statistics, and tell me about the company that employs you.

Investigator: The company I work for makes and sells agricultural products, plastics, soap, over-the-counter medications, canned fruits and vegetables, cosmetics, greeting cards, and many other products. We have a large research and development group and I work in this group. I have a bachelor's degree in computing science, and I have had three courses in statistics—a course in statistical methods, a course in regression, and a course in statistical computer packages.

Statistician: How can I help you?

Investigator: Our agricultural research division has developed a new commercial fertilizer and has conducted a large experiment to determine whether or not the average yield per acre of corn is greater when using the new fertilizer than when using the old fertilizer. The experiment was conducted in five different locations so that we could examine different climatic conditions and soil types. For each location a statistical test of

$$\text{NH} : \theta = 0 \quad against \quad \text{AH} : \theta \neq 0$$

was conducted with $\alpha = .05$, where

$$\theta = \mu_{old} - \mu_{new}$$

is the difference between the average yields of corn in bushels per acre when the two fertilizers, old and new, are used. Here are the results in table form. Will you please examine the entries and see if the computations and conclusions for each location are correct?

Location	*P*-value	Decision (Using $\alpha = 0.05$)
1	0.0474	Reject $\theta = 0$
2	0.0066	Reject $\theta = 0$
3	0.0003	Reject $\theta = 0$
4	0.1280	Do not reject $\theta = 0$
5	0.0772	Do not reject $\theta = 0$

Statistician: Your calculations are correct, but your conclusions would be much more informative if you used 95% confidence intervals instead of 5% tests.

Investigator: What do you mean?

Statistician: A statistical test should never (I repeat, never) be used alone when a confidence interval can be computed, because a confidence interval is always more informative than a statistical test. I can illustrate by using your data and computing a 95% confidence interval for θ for each location. I'll organize the results in a table for each location along with your decision based on a statistical test.

Location	Your Decision Based on a Test	95% Confidence Interval
1	Reject $\theta = 0$	$0.06 \leq \theta \leq 16.67$
2	Reject $\theta = 0$	$10.05 \leq \theta \leq 33.19$
3	Reject $\theta = 0$	$0.101 \leq \theta \leq 0.114$
4	Do not reject $\theta = 0$	$-0.034 \leq \theta \leq 0.021$
5	Do not reject $\theta = 0$	$-0.09 \leq \theta \leq 8.56$

Note The data in this conversation are artificial so that we can make our point in a dramatic fashion. However, results such as these are not uncommon in applied problems.

A test tells you only whether or not you should reject the null hypothesis that $\theta = 0$. In other words, a test tells you what the value of θ is not, but a confidence interval tells you what the plausible values of θ are. And if you know, with a specified confidence (95% in this case), what the plausible values for θ are, then you can make practical decisions based on this knowledge.

For example, for locations 1, 2, and 3, the test indicates that you should reject NH : $\theta = 0$ in each case, but the confidence interval yields the following results. (1) For location 1, farmers who must decide whether or not to use the new fertilizer might conclude that the results are not definitive enough to make a decision, because if $\theta = 0.06$ bu/acre, then the farmer would surely conclude that θ is so small as to be considered negligible in a practical sense. However, if $\theta = 16.67$ bu/acre, then the farmer would surely conclude that θ is not zero and the old fertilizer is better. Since we have 95% confidence that θ is somewhere between 0.06 and 16.67, we need more data to make a definite decision. (2) For location 2, you would surely reject NH that $\theta = 0$ and conclude that $\theta \neq 0$ because we have 95% confidence that $10.05 < \theta \leq 33.19$ bu/acre. (3) For location 3, you would undoubtedly decide that θ is so small that it can be considered negligible for all practical purposes, and it wouldn't make any difference which fertilizer was used.

On the other hand, for locations 4 and 5, the test indicates that NH shouldn't be rejected. But for location 4, you would surely accept NH that $\theta = 0$ because we have 95% confidence that θ is so small as to be practically no different from zero for this problem. And finally, for location 5, we see that you don't have enough data to make a definite decision. Thus you can readily see that a confidence interval gives information that a test does not. If, in fact, the only result of this investigation you were allowed to see was either the conclusion of the tests of size $\alpha = .05$ or the 95% confidence intervals, which would you prefer?

Investigator: I think I see your point. I would definitely choose to see the confidence intervals.

Statistician: It is always true that if the result of an α level test of NH : $\theta = q$ is "do not reject," this implies that the corresponding $1 - \alpha$ confidence interval for θ will contain the value q specified by NH, which is zero in your problem.

Investigator: I remember studying that. And I remember that when the result of an α level test of NH : $\theta = q$ is to "reject," then the corresponding $1 - \alpha$ confidence interval won't contain the value q specified by NH. Is this correct?

Statistician: Yes it is. So you can easily see that confidence intervals give all the information that tests do plus a great deal more.

Investigator: I also remember my statistics instructor telling us that if the result of a test is to not reject NH, this doesn't mean that one can accept NH.

Statistician: That's true. But as you can see from the calculations for locations 4 and 5, even if the result of a test is "do not reject NH," when you examine the corresponding confidence interval the practical result could be (a) accept NH for all practical purposes or (b) there are not enough data to make a decision. Thus a confidence interval in these cases gives considerably more information about the population parameter than a test does.

Investigator: I understand—the test says "reject NH," yet when the corresponding confidence interval is examined, the practical result could be (a) reject NH, (b) accept NH, or (c) there aren't enough data to make a definite decision. Similarly, when the test says "do not reject NH," the corresponding confidence interval may indicate (a) accept NH or (b) there aren't enough data to make a definite decision. Is that correct?

Statistician: That's correct.

Investigator: I see what you are saying, but scientists are constantly formulating scientific hypotheses and they want to determine whether they are correct or incorrect. How can I tell them not to test hypotheses?

Statistician: I am not telling you that scientists should not formulate and test scientific hypotheses. But what I am telling you is that, whenever possible, they should use confidence intervals rather than statistical tests to evaluate scientific hypotheses, particularly when a hypothesis concerns a well-defined population parameter.

Investigator: I do have a question about your table. What exactly do you mean by the statement "not enough data" in locations 1 and 5?

Statistician: By "not enough data" I mean that the confidence interval is too wide to make a definitive statement about the value of the parameter under study, which in this case is θ. To make the case more dramatic, suppose that a confidence statement for a location is

$$C[-16.24 \; bu/acre \; \leq \mu_{old} - \mu_{new} \leq 19.67 \; bu/acre] = 0.95$$

So we have 95% confidence that $\mu_{old} - \mu_{new}$ is some value in this interval. But if $\mu_{old} - \mu_{new} = -16.24$, the lower endpoint of this interval, this means that μ_{new} is 16.24 bu/acre larger than μ_{old}, so the new fertilizer is certainly better. On the other hand, suppose that $\mu_{old} - \mu_{new} = 19.67$, the upper endpoint of this interval.

This means that μ_{old} is 19.67 bu/acre larger than μ_{new}, and so certainly the old fertilizer is better. But since all that the confidence interval tells us is that $\mu_{old} - \mu_{new}$ is somewhere between -16.24 and 19.67 bu/acre, we can arrive at two different conclusions depending on which value in the interval we use. But if the confidence interval were based on enough data, the width of the interval would be small enough so that, with a specified confidence level (say 95%), a single decision would result using any value of $\mu_{old} - \mu_{new}$ in the confidence interval.

It seems to me that in this problem, what your scientists want to know is how much larger (or smaller) μ_{new} is than μ_{old}, and a two-sided confidence interval will give you this information.

Investigator: I think I understand the importance of what you're saying. But since the scientists in our company have always used statistical tests, it won't be easy to get them to change.

Statistician: If they insist on a statistical test to evaluate their scientific hypotheses, then give them a confidence interval too. I think they'll see that confidence intervals are much more informative and are really what they want to help them make decisions.

Investigator: I'll try to convince them. But before I leave, will you clarify something for me about prediction intervals? It has to do with the interpretation of a prediction interval.

Statistician: Certainly.

Investigator: Suppose I compute a 90% prediction interval for Y_0 and I get $C[15.3 \le Y_0 \le 19.4] = 0.90$. Does this mean that I have 90% confidence that all future observations will be between 15.3 and 19.4?

Statistician: *No!* It means that you have 90% confidence that *a single* future observation you obtain will be between 15.3 and 19.4 as explained in (1.6.11).

Investigator: Does it mean that if I select many future observations, 90% of them will be between 15.3 and 19.4?

Statistician: *No!* As I stated above it means that you have 90% confidence that any *one* future observation you obtain will be between 15.3 and 19.4.

Investigator: But I will obtain m future observations, and I want to compute a lower bound, say L, and an upper bound, say U, such that I can be 90% confident that all m future observations are between L and U. Is it possible to compute these bounds L and U?

Statistician: Yes, it is possible. One way to do this is by using the Bonferroni method.

Investigator: You keep referring to future observations. What exactly do you mean by a future observation?

Statistician: I mean any observation that is randomly selected from the population that is being studied and the observation is not known at the time the prediction interval is computed. Generally a future observation is a number to be observed in the future, but it is important to know something about its value before it can be observed for decision-making purposes.

Investigator: One other thing. When you compute a confidence interval for a parameter, say θ, you may *never* know if the interval is correct because θ may never be known. But when you compute a prediction interval for Y_0, a future observation, you can eventually determine whether the interval is correct because, presumably, the future observation will eventually be selected and observed. Is that correct?

Statistician: Yes, you are absolutely right.

Investigator: Thank you. You have certainly helped me understand many fundamental statistical concepts. Can I come to see you again if I have more statistical questions?

Statistician: Certainly. You're welcome anytime.

Problems   1.6

1.6.1 A random sample of size $n = 30$ is obtained from a Gaussian population with unknown mean μ_Y and unknown standard deviation σ_Y. The data are given in Table 1.6.4 and also in the file **table164.dat** on the data disk.

  a Compute $\hat{\mu}_Y$ and a two-sided 80% confidence interval for μ_Y.

  b Write the confidence statement for the confidence interval in (a).

  c Compute a one-sided 95% lower confidence bound for μ_Y.

  d Write the confidence statement corresponding to the confidence bound in (c).

  e State in words the meaning of the confidence statement in (d).

  f Write a short paragraph explaining to an investigator why confidence intervals are more informative than tests.

  g Suppose an investigator wants to test NH : $\sigma_Y \leq 5.0$ against AH : $\sigma_Y > 5.0$ with $\alpha = .05$. Perform this test. What is your conclusion?

T A B L E   **1.6.4**

3.48	7.68	12.96	0.65	12.44	0.16	4.34	3.71	4.67	3.47
5.91	6.73	3.64	10.90	6.37	4.62	9.18	11.67	8.29	5.19
4.30	11.67	8.63	7.84	11.92	11.16	6.13	10.21	8.16	3.59

h Compute a 90% two-sided confidence interval for σ_Y. What is your conclusion about the test in (g) using the confidence interval?

i Compute a 99% two-sided confidence interval for μ_Y. Compare the width of this interval with the width of the 80% confidence interval in (a).

1.7
Functional Notation

The concept and notation of functions are used throughout the book. Here we present a short discussion of this topic.

Functions

Let D be a set of numbers. A function $f(\cdot)$ on the set D is a *rule* that describes how numbers in the set D are *changed, transformed,* or *mapped* to produce other numbers. If x represents a number in D, then the result of applying the rule, i.e., the function $f(\cdot)$, to x results in a number, say z, and this is symbolically denoted by writing $f(x) = z$ or $z = f(x)$. The set D is called the *domain* of the function $f(\cdot)$.

As an example, let D be the set of positive real numbers and $f(\cdot)$ be a function that is defined by the rule "square each number in the set D." Then the result of applying the function $f(\cdot)$ to the number 2 is 4, and this is denoted by writing $f(2) = 4$. Likewise, $f(0.5) = 0.25$, $f(12) = 144$, etc. In general, the result of applying the function $f(\cdot)$ to a number x in D produces the result x^2. This is described by writing $f(x) = x^2$. If the symbol z is used to denote the result of applying the function $f(\cdot)$ to the number x, then we can also write $z = f(x)$ or $z = x^2$. For another illustration, let $f(\cdot)$ be defined by the equation $f(x) = 6x^2 + x - 5$ for any real number x. Then $f(3) = 6(3)^2 + 3 - 5 = 52, f(8) = 387$, etc.

When a function $f(\cdot)$ is specified, its domain D must be specified also. Any letter can be used to represent a function; some examples are $f(\cdot)$, $Y(\cdot)$, and $\mu(\cdot)$. Sometimes a letter with a subscript is used to represent functions; some examples are $f_1(\cdot), f_2(\cdot), \mu_Y(\cdot)$, and $g_t(\cdot)$.

Although, strictly speaking, the symbol $f(\cdot)$ stands for a function and the symbol $f(x)$ stands for the value of the function when applied to the number x in the set D, sometimes people use phrases such as "$f(x)$ is a function" or "let $f(x)$ be a function of x," etc. These phrases are to be interpreted to mean that "$f(\cdot)$ is a function, x is a typical member of the domain D of $f(\cdot)$, and $f(x)$ is the value of the function when applied to the number x." This rarely leads to any confusion because the meaning of the symbol $f(x)$ is usually quite clear from the context.

Independent and Dependent Variables

In the equation $z = f(x)$, x is called the independent variable and z is called the dependent variable. Other letters can be used for dependent and independent variables; for example, we could write the following: $z = f(t), s = v(u), r = g_1(t)$. Some examples of functions appear in (1.7.1), (1.7.2), and (1.7.3).

$$s = f_1(t), \quad 0 \le t \le 3 \quad \text{where} \quad f_1(t) = \frac{2t^2 + 1}{t + 2} \tag{1.7.1}$$

$$z = \mu_Y(x), \quad -5 \le x \le 8 \quad \text{where} \quad \mu_Y(x) = 6x + 4 \tag{1.7.2}$$

$$v = r(u), \quad 0 < u < \infty \quad \text{where} \quad r(u) = \log_e u - 6u + u^2 \tag{1.7.3}$$

The value of s for $t = 3$ in (1.7.1) is $s = f_1(3) = [2(3)^2 + 1]/(3 + 2) = 19/5$. The value of z for $x = 0$ in (1.7.2) is $z = \mu_Y(0) = (6)(0) + 4 = 4$, and for $x = 7$ we get $z = \mu_Y(7) = (6)(7) + 4 = 46$. The value of z in (1.7.2) for $x = 10$ is not defined because $z = \mu_Y(x)$ is defined only for x in the interval $-5 \le x \le 8$.

Functions of Many Variables

The functions just discussed are functions of one (independent) variable, say x. In this book we also need functions of more than one independent variable. A function of three independent variables, say x_1, x_2, x_3, could be denoted by $Y = Y(x_1, x_2, x_3)$, $z = \mu_Y(x_1, x_2, x_3)$, or $z = q(x_1, x_2, x_3)$, etc. Examples are given in (1.7.4) and (1.7.5).

$$z = f(x_1, x_2, x_3), \quad 0 \le x_1 \le 1, \quad 3 \le x_2 \le 8, \quad -4 \le x_3 \le 15 \text{ where} \tag{1.7.4}$$
$$f(x_1, x_2, x_3) = 4x_1^2 + 3x_1 x_2 - 6x_3^2$$

$$z = \mu_Y(x_1, x_2, x_3), \quad 0 \le x_1 < \infty, \quad 0 \le x_2 < \infty, \quad -\infty < x_3 < \infty \tag{1.7.5}$$
$$\text{where } \mu_Y(x_1, x_2, x_3) = \beta_0 + \beta_1 x_1 + \beta_2 x_2 + \beta_3 x_3 \text{ and}$$
$$\beta_0, \beta_1, \beta_2, \beta_3 \text{ represent constants}$$

The value of $\mu_Y(x_1, x_2, x_3)$ in (1.7.5) for $x_1 = 1$, $x_2 = 6$, and $x_3 = -4$ is $\mu_Y(1, 6, -4) = \beta_0 + \beta_1(1) + \beta_2(6) + \beta_3(-4) = \beta_0 + \beta_1 + 6\beta_2 - 4\beta_3$, and this is an unknown number unless values are given for $\beta_0, \beta_1, \beta_2$, and β_3. Nevertheless, the symbolic representations $\mu_Y(x_1, x_2, x_3) = \beta_0 + \beta_1 x_1 + \beta_2 x_2 + \beta_3 x_3$ and $\mu_Y(1, 6, -4) = \beta_0 + \beta_1 + 6\beta_2 - 4\beta_3$ are useful.

Linear Functions

A function $f(x)$ of a single variable x is said to be **linear in** x if it can be written as

$$f(x) = ax + b \tag{1.7.6}$$

where the values of a and b do not depend on the value of x.

A function $f(x_1, \ldots, x_k)$ of k variables $x_1, \ldots, x_k$ is said to be **simultaneously linear in** $x_1, \ldots, x_k$ if it can be written as

$$f(x_1, \ldots, x_k) = a_0 + a_1 x_1 + \cdots + a_k x_k \tag{1.7.7}$$

where the values of $a_0, a_1, \ldots, a_k$ do not depend on the values of $x_1, \ldots, x_k$.

Consider the function $\mu_Y(x) = 6x + 4$ in (1.7.2). It is of the form $ax + b$ with $a = 6$ and $b = 4$, both of which are free of x. So $\mu_Y(x)$ in (1.7.2) is linear in x. The function $f(x_1, x_2, x_3) = 4x_1^2 + 3x_1 x_2 - 6x_3^2$ is not a linear function of the variables x_1, x_2, and x_3 because it cannot be written in the form given in (1.7.7). You should verify that the functions in (1.7.1) and (1.7.3) are not linear functions in the corresponding independent variables.

For further illustration, consider the functions

$$\mu_Y(x_1, x_2, \beta_0, \beta_1, \beta_2) = \beta_0 + \beta_1 x_1 + e^{\beta_2 x_2} \tag{1.7.8}$$

and

$$\mu_Y(x_1, x_2, \beta_0, \beta_1, \beta_2) = \beta_0 + \beta_1 x_1 + \beta_2 x_2 \tag{1.7.9}$$

It can be verified that the function in (1.7.8) is linear in β_0, linear in β_1, and linear in x_1, but it is not linear in β_2, not linear in x_2, and not simultaneously linear in β_1 and x_1. The function in (1.7.9) is linear in each of the variables $x_1, x_2, \beta_0, \beta_1, \beta_2$; is simultaneously linear in x_1, x_2; is simultaneously linear in $\beta_0, \beta_1, \beta_2$; but is not simultaneously linear in $x_1, x_2, \beta_0, \beta_1, \beta_2$.

Problems 1.7

1.7.1 The function $f(x)$ is defined by $f(x) = 6x^2 + 3x^{1/2} - 9x + 4$ for $4 < x < 49$. Find the following.

a $f(4)$

b $f(16)/f(36)$

c $f(3) + 16$

d $f(34) + f(13)$

e $f(64)$

1.7.2 Problems (a)–(c) refer to the function defined by $\mu_Y(x_1, x_2) = \beta_0 + \beta_1 x_1 + \beta_2 x_2^3$ for $-\infty < x_1 < +\infty$, $-\infty < x_2 < +\infty$.

a Compute $\mu_Y(6, 1)$.

b Compute $\mu_Y(15, -4)$.

c Is the function $\mu_Y(x_1, x_2)$ linear in x_1? Is it linear in x_2? Is it simultaneously linear in x_1 and x_2?

1.8
Matrices and Vectors

In this section we introduce some of the basic operations of matrix algebra because numerical computations arising in regression analysis can be presented effectively using the language of matrices. You may want to omit this section initially and read appropriate portions of it when matrices are discussed in later chapters.

What Is a Matrix ?

A **matrix** is defined as a rectangular array of elements. The elements of a matrix are called *scalars*, and they are either numbers (such as 1.3, $-6.0, 0$, etc.) or symbols (such as $x, log(x_5), xy$, etc.) that represent numbers. This array is enclosed in large parentheses or brackets. (We use brackets.) The quantities in (1.8.1)–(1.8.3) are matrices, but (1.8.4) is not because it is not a rectangular array.

$$\begin{bmatrix} x_1 & x_2 \\ x_3 & x_4 \end{bmatrix} \tag{1.8.1}$$

$$\begin{bmatrix} 6 & 2 & 3 \\ 4 & x & y \end{bmatrix} \tag{1.8.2}$$

$$\begin{bmatrix} 5 & x_1 & 4 \\ y_2 & log(x) & 6 \\ 5 & 9 & 6 \end{bmatrix} \tag{1.8.3}$$

$$\begin{bmatrix} 1 & 2 & 3 \\ 4 & 5 \\ 7 & 8 & 9 \end{bmatrix} \tag{1.8.4}$$

The matrix in (1.8.2) has two rows and three columns and is called a "2 by 3 matrix." If a matrix contains r rows and c columns, it is said to be an "r by c matrix" (or a matrix of size "r by c"). The number of rows is always given first when stating the size. The matrix in (1.8.1) is 2 by 2 matrix, and the one in (1.8.3) is a 3 by 3 matrix.

Row Vectors and Column Vectors

If a matrix has only one row, it is usually called a *row vector*. For example, the following 1 by 6 matrix is a row vector.

$$[3 \quad 2 \quad 1 \quad 6 \quad 4 \quad 3] \tag{1.8.5}$$

If a matrix has only one column, it is called a *column vector*. The following 7 by 1 matrix is a column vector.

$$\begin{bmatrix} x_1 \\ x_2 \\ x_3 \\ x_4 \\ x_5 \\ x_6 \\ x_7 \end{bmatrix} \qquad \textbf{(1.8.6)}$$

It is generally clear from the context whether a vector is a row (or a column) vector, and so the word *row* (or *column*) is usually omitted. If a matrix contains only one row and one column, it is a 1 by 1 matrix, say [x]. In this case, the brackets are omitted and the matrix [x] is replaced by the number x. For example, [16] is a 1 by 1 matrix, so we write it simply as 16.

We use boldface italic capital letters to represent matrices and boldface italic lowercase letters to represent column vectors. Scalars arc not boldface. For example, A is a matrix, and it could denote

$$A = \begin{bmatrix} 6 & 3 & 4 \\ 1 & 9 & 3 \\ 2 & 1 & 7 \\ 4 & 2 & 9 \end{bmatrix}$$

In this case, A is a 4 by 3 matrix. If

$$a = \begin{bmatrix} 6 \\ 9 \\ -1 \\ 0 \\ 2 \end{bmatrix}$$

then a is a column vector of length 5 (i.e., a 5 by 1 matrix).

Matrix Elements

It may be desirable to identify the individual elements of matrices or vectors in some systematic manner, so we sometimes write a matrix G as $[g_{ij}]$ or the matrix K as $[k_{ij}]$, etc., where the quantity in brackets denotes the element in the ith row and jth column of the matrix. For a vector, say a, we sometimes write $[a_i]$. A single subscript is used because there is only one row or one column in a vector. Therefore a_i denotes the ith element of a vector a. To illustrate, if we let A denote the following 3 by 2 matrix

$$\begin{bmatrix} 3 & 1 \\ 4 & -4 \\ 6 & 0 \end{bmatrix}$$

we could write

$$A = \begin{bmatrix} 3 & 1 \\ 4 & -4 \\ 6 & 0 \end{bmatrix}$$

or we could write

$$[a_{ij}] = \begin{bmatrix} 3 & 1 \\ 4 & -4 \\ 6 & 0 \end{bmatrix}$$

Notice that a_{12} denotes the element in the first row and the second column, and so in this case $a_{12} = 1$. Verify that $a_{31} = 6$.

As a further illustration, we let a or $[a_i]$ denote the vector

$$a = [a_i] = \begin{bmatrix} 6 \\ 4 \\ -2 \\ 0 \end{bmatrix}$$

Here a_3 is -2, $a_1 = 6$, etc.

Equality of Matrices

Two matrices are said to be equal if and only if the following are true:

1 They are the same size.

2 All corresponding elements are equal.

For example, consider the matrices $A, B, C,$ and D where

$$A = \begin{bmatrix} 6 & 9 \\ 15 & 3 \end{bmatrix} \qquad B = \begin{bmatrix} 6 & 9 \\ 18 & 3 \end{bmatrix} \qquad C = \begin{bmatrix} 6 & 9 & 0 \\ 15 & 3 & 0 \end{bmatrix} \qquad D = \begin{bmatrix} 6 & 9 \\ 18 & 3 \end{bmatrix}$$

Matrix A is not equal to matrix B because element a_{21} is not equal to element b_{21}. Matrix C is not equal to matrix A, B, or D because it is not the same size. Matrix D is equal to B, and we write $D = B$ because all corresponding elements are equal.

In order to effectively use matrices, vectors, and scalars in our discussion, we need to define various operations such as addition, subtraction, multiplication, transposition, etc., for matrices, vectors, and scalars. Because scalars are numbers, the usual rules of arithmetic apply, but matrices and vectors have different definitions.

Transposition of Matrices

For every matrix, say A, there is another matrix derived from it called the *transpose* of A. The transpose of a matrix A is obtained by replacing each row of A by its corresponding column. Thus, if **A** is given by

$$\begin{bmatrix} 3 & -1 \\ 0 & 4 \\ 2 & 6 \end{bmatrix}$$

the transpose of **A** is

$$\begin{bmatrix} 3 & 0 & 2 \\ -1 & 4 & 6 \end{bmatrix}$$

It is easy to see that if **A** is a matrix of size r by c, then the transpose of **A** is a matrix of size c by r. The column vector

$$\begin{bmatrix} a_1 \\ a_2 \\ a_3 \\ a_4 \\ a_5 \end{bmatrix}$$

has as its transpose the row vector

$$[a_1, a_2, a_3, a_4, a_5]$$

which is obtained by arranging the elements a_i in their original order in a row. Thus the transpose of a column vector is a row vector. If the original vector is of size 5 by 1, its transpose is of size 1 by 5. Verify that the transpose of a row vector is a column vector. The transpose of a vector **a** or a matrix **A** is indicated by the notation $\boldsymbol{a}^T$ and $\boldsymbol{A}^T$ (some authors write $\boldsymbol{a}'$ and $\boldsymbol{A}'$). Notice also that $(\boldsymbol{A}^T)^T = \boldsymbol{A}$; that is, the transpose of the transpose of a matrix **A** is equal to the matrix **A**.

An 11 by 2 matrix **D** and its transpose $\boldsymbol{D}^T$ follow:

$$D = \begin{bmatrix} 52 & 0.62 \\ 43 & 0.74 \\ 36 & 0.65 \\ 32 & 0.71 \\ 27 & 0.68 \\ 26 & 0.59 \\ 22 & 0.49 \\ 37 & 0.67 \\ 24 & 0.64 \\ 19 & 0.56 \\ 13 & 0.51 \end{bmatrix}$$

$$D^T = \begin{bmatrix} 52 & 43 & 36 & 32 & 27 & 26 & 22 & 37 & 24 & 19 & 13 \\ 0.62 & 0.74 & 0.65 & 0.71 & 0.68 & 0.59 & 0.49 & 0.67 & 0.64 & 0.56 & 0.51 \end{bmatrix}$$

Addition and Subtraction of Matrices

Addition is defined for two matrices if and only if they are of the same size. For example, consider the matrices A, B, C, and D where

$$A = \begin{bmatrix} 3 & -1 \\ 0 & 4 \\ 2 & 6 \end{bmatrix} \quad B = \begin{bmatrix} 4 & 0 & 2 & 3 \\ 0 & 2 & 6 & 10 \end{bmatrix} \quad C = \begin{bmatrix} 5 & 9 \\ -16 & 4 \\ 0 & 6 \end{bmatrix} \quad D = \begin{bmatrix} 3 & 10 & 4 \\ -1 & 0 & 6 \end{bmatrix}$$

(1.8.7)

Matrices A and C are both 3 by 2 so we can add A and C to form a new matrix, say G. Thus, we write $A + C = G$ or

$$[a_{ij}] + [c_{ij}] = [g_{ij}]$$

and the resultant matrix G is the same size as A (and, of course, the same size as C). We cannot add A and B because they are not the same size, and we cannot add B and C for the same reason. The matrix G that results from adding two matrices A and C is obtained by adding the corresponding elements of A and C; that is, $g_{ij} = a_{ij} + c_{ij}$ for all i and j. We obtain

$$G = A + C = \begin{bmatrix} 3 & -1 \\ 0 & 4 \\ 2 & 6 \end{bmatrix} + \begin{bmatrix} 5 & 9 \\ -16 & 4 \\ 0 & 6 \end{bmatrix} = \begin{bmatrix} 3+5 & -1+9 \\ 0-16 & 4+4 \\ 2+0 & 6+6 \end{bmatrix} = \begin{bmatrix} 8 & 8 \\ -16 & 8 \\ 2 & 12 \end{bmatrix}$$

Notice that $A + C = C + A$.

The matrix S that is the result of subtracting a matrix C from a matrix A, which is written as $S = A - C$, or as $[s_{ij}] = [a_{ij}] - [c_{ij}]$, is obtained by subtracting elements of C from the corresponding elements of A. Hence, subtraction is defined if and only if the two matrices A and C that are involved are the same size; the resulting matrix S is also the same size as A and C. For an illustration, consider the matrices defined in (1.8.7):

$$S = A - C = \begin{bmatrix} 3 & -1 \\ 0 & 4 \\ 2 & 6 \end{bmatrix} - \begin{bmatrix} 5 & 9 \\ -16 & 4 \\ 0 & 6 \end{bmatrix} = \begin{bmatrix} 3-5 & -1-9 \\ 0-(-16) & 4-4 \\ 2-0 & 6-6 \end{bmatrix} = \begin{bmatrix} -2 & -10 \\ 16 & 0 \\ 2 & 0 \end{bmatrix}$$

The transpose of the sum (or difference) of any two matrices is equal to the sum (or difference) of the transpose of each matrix; that is,

$$(A + C)^T = A^T + C^T$$

and

$$(A - C)^T = A^T - C^T$$

You should verify this for the matrices in (1.8.7).

Multiplication of Matrices

The process of adding and subtracting matrices is very similar to that of adding and subtracting numbers. This is not the case for multiplication. For example, with numbers, 6 times 3 and 3 times 6 are both equal to 18, but in multiplying two matrices it is generally not true that A times B and B times A are equal. The quantity A times B, usually written as AB, is defined as multiplying B on the left by A (or A on the right by B), and BA is defined as multiplying A on the left by B (or B on the right by A). The product AB is defined if and only if the number of columns of A (the matrix on the left) is equal to the number of rows of B (the matrix on the right). The matrix, say C, that results from this multiplication has size r by c, where r is the number of rows of A and c is the number of columns of B.

As an illustration, consider the matrices in (1.8.7). Suppose we want to see which of the multiplications AB, AC, and DC are defined. A convenient procedure is to write the size of the two matrices in their corresponding places; that is, we write

$$\begin{array}{cc} A & B \\ 3 \text{ by } 2 & 2 \text{ by } 4 \end{array} \qquad \textbf{(1.8.8)}$$

where the quantity "3 by 2" is the size of A and is placed to the left of the quantity "2 by 4," which is the size of B, because A is to the left of B. Now if the two inner numbers (in this case 2 and 2) are equal, the multiplication is defined, and the matrix that results from this multiplication has size 3 by 4, which is obtained from the two outside numbers in (1.8.8). In general, if the matrix A has size r by s (r rows and s columns) and if B has size t by c (t rows and c columns), then to test whether the multiplication AB is defined, the sizes of A and B are written in the order in which they are to be multiplied; that is, (r by s)(t by c). The multiplication AB is defined if and only if the two inside numbers, s and t, are equal. If AB is defined, the resulting matrix has size r by c, the two outside numbers. Is the multiplication AC in (1.8.7) defined? We get

$$\begin{array}{cc} A & C \\ 3 \text{ by } 2 & 3 \text{ by } 2 \end{array}$$

The two inside numbers 2 and 3 are not equal, and so this multiplication is not defined. The multiplication DC is defined, and the resulting matrix is of size 2 by 2. The same is true for DA. Notice that BA, DB, and BC are not defined; neither are BD and CA.

Before we give the rule for multiplying two matrices, we give the rule for multiplying a row vector a^T on the right by the column vector b, where a and b are both column vectors of length k (so a^T is a row vector of length k). Of course a and b are matrices, because a vector is a special kind of matrix. First we check to see if the multiplication $a^T b$ is defined according to the preceding discussion. We write

$$\begin{array}{cc} a^T & b \\ 1 \text{ by } k & k \text{ by } 1 \end{array}$$

and observe that the middle two numbers are both k so the multiplication is defined; the result is a 1 by 1 matrix that is a scalar; that is, a number. We define $a^T b$ as follows:

$$a^T b = \sum_{i=1}^{k} a_i b_i \qquad \textbf{(1.8.9)}$$

For example, if

$$a^T = [6 \quad -2 \quad 0 \quad 3] \quad \text{and} \quad b = \begin{bmatrix} 4 \\ 9 \\ -8 \\ 7 \end{bmatrix}$$

then $a^T b$ is defined and is equal to

$$a^T b = [6 \quad -2 \quad 0 \quad 3] \begin{bmatrix} 4 \\ 9 \\ -8 \\ 7 \end{bmatrix} = (6)(4) + (-2)(9) + (0)(-8) + (3)(7) = 27$$

This method of multiplying a row vector on the left by a column vector on the right is very important because this is the basic calculation needed for multiplying matrices.

If A is a matrix of size r by t and B is a matrix of size t by c, then AB is defined. Let the resulting matrix of size r by c be denoted by P. We write

$$P = AB$$

The (i, j) element of P is obtained by multiplying the ith row of A (the matrix on the left in the multiplication AB) by the jth column of B (the matrix on the right in the multiplication AB) by the rule in (1.8.9). If this is done for every row in A with every column in B, all the elements in the matrix P are obtained. Another way to state this is

$$p_{ij} = \sum_{m=1}^{t} a_{im} b_{mj}$$

where $[p_{ij}] = P$; $[a_{im}] = A$; and $[b_{mj}] = B$.

As an illustration, we compute P, where

$$P = AB$$

and A and B are as in (1.8.7). We obtain

$$P = AB = \begin{bmatrix} 3 & -1 \\ 0 & 4 \\ 2 & 6 \end{bmatrix} \begin{bmatrix} 4 & 0 & 2 & 3 \\ 0 & 2 & 6 & 10 \end{bmatrix} = \begin{bmatrix} 12 & -2 & 0 & -1 \\ 0 & 8 & 24 & 40 \\ 8 & 12 & 40 & 66 \end{bmatrix}$$

For example, the element in the second row and third column of P—that is p_{23}, which is 24—is obtained by multiplying the second row of A by the third column of B according to the rule in (1.8.9). We obtain

$$p_{23} = [0 \quad 4] \begin{bmatrix} 2 \\ 6 \end{bmatrix} = (0)(2) + (4)(6) = 24$$

Similarly, p_{34}, which is equal to 66, is the product of the third row of A and the fourth column of B. We get

$$p_{34} = [2 \quad 6] \begin{bmatrix} 3 \\ 10 \end{bmatrix} = (2)(3) + (6)(10) = 66$$

You should verify the remaining elements in P.

For another illustration, let us evaluate S, where $S = RQ$ and R and Q are defined as follows:

$$R = \begin{bmatrix} 3 & 5 & -1 \\ 2 & 6 & 0 \end{bmatrix} \qquad Q = \begin{bmatrix} 5 & 4 & 9 & 1 \\ -3 & 0 & 5 & 4 \\ 6 & 1 & 2 & 6 \end{bmatrix}$$

We get

$$S = RQ = \begin{bmatrix} -6 & 11 & 50 & 17 \\ -8 & 8 & 48 & 26 \end{bmatrix}$$

As a final example, let us evaluate Ax, where A is a 2 by 2 matrix (that is, a matrix of size 2 by 2) and x is a 2 by 1 matrix (x is therefore a column vector) defined as follows:

$$A = \begin{bmatrix} 2 & 1 \\ 1 & 3 \end{bmatrix} \qquad x = \begin{bmatrix} x_1 \\ x_2 \end{bmatrix}$$

We get

$$Ax = \begin{bmatrix} 2 & 1 \\ 1 & 3 \end{bmatrix} \begin{bmatrix} x_1 \\ x_2 \end{bmatrix} = \begin{bmatrix} 2x_1 + x_2 \\ x_1 + 3x_2 \end{bmatrix} \qquad \text{(1.8.10)}$$

Clearly, the result of Ax in (1.8.10) is a 2 by 1 matrix, which, of course, is a column vector with two elements; the first element is $2x_1 + x_2$, and the second element is $x_1 + 3x_2$. If we set the product Ax equal to a 2 by 1 vector b where

$$b = \begin{bmatrix} 4 \\ 7 \end{bmatrix}$$

we get

$$Ax = b$$

or

$$\begin{bmatrix} 2 & 1 \\ 1 & 3 \end{bmatrix} \begin{bmatrix} x_1 \\ x_2 \end{bmatrix} = \begin{bmatrix} 4 \\ 7 \end{bmatrix}$$

This becomes (by performing the multiplication Ax)

$$\begin{bmatrix} 2x_1 + x_2 \\ x_1 + 3x_2 \end{bmatrix} = \begin{bmatrix} 4 \\ 7 \end{bmatrix}$$

Because two matrices are equal if and only if the corresponding elements are equal, we get the two equations

$$2x_1 + x_2 = 4$$
$$x_1 + 3x_2 = 7$$

as a result of $Ax = b$. In other words, matrix equations can be used to represent systems of linear equations.

To multiply more than two matrices, we simply multiply any adjoining pair, then multiply this result by any adjoining matrix, and continue until all multiplications are completed. For example, suppose we want to evaluate P, where $P = ABCD$ and where we assume that the sizes are such that all multiplications are defined. One way of calculating P is to first compute AB, then multiply this result on the right by C, and then multiply this result on the right by D.

The Transpose of Products of Matrices

The transpose of the product of two or more matrices is equal to the product in *reverse order* of the transposes; that is, $(AB)^T = B^T A^T$, $(CDE)^T = E^T D^T C^T$, etc.

The Multiplication of a Matrix and a Scalar

For further development of the arithmetic of matrices, we define the multiplication of a scalar and a matrix. This multiplication is defined so that any matrix can be multiplied by any scalar. Also, the result is the same whether the scalar is multiplied on the left or the right of the matrix. The product of a scalar and a matrix is obtained by multiplying every element in the matrix by the scalar. For example, let a scalar c and matrix A be defined as follows:

$$c = 20, \quad A = \begin{bmatrix} 4 & -1 & 0 \\ 3 & 2 & 1 \end{bmatrix}$$

Then

$$cA = Ac = 20 \begin{bmatrix} 4 & -1 & 0 \\ 3 & 2 & 1 \end{bmatrix} = \begin{bmatrix} 4 & -1 & 0 \\ 3 & 2 & 1 \end{bmatrix} 20$$

$$= \begin{bmatrix} 20 \times 4 & 20 \times -1 & 20 \times 0 \\ 20 \times 3 & 20 \times 2 & 20 \times 1 \end{bmatrix} = \begin{bmatrix} 80 & -20 & 0 \\ 60 & 40 & 20 \end{bmatrix}$$

Note that the operation of dividing vectors or matrices by nonzero scalars is equivalent to multiplying the vector or matrix by the reciprocal of the divisor. Thus, if division by 20 had been intended in the previous example, the matrix A would be multiplied by 1/20, or 0.05. This would lead to the result

$$\frac{A}{20} = \frac{1}{20}A = A\frac{1}{20} = 0.05 \begin{bmatrix} 4 & -1 & 0 \\ 3 & 2 & 1 \end{bmatrix} = \begin{bmatrix} 4 & -1 & 0 \\ 3 & 2 & 1 \end{bmatrix} 0.05$$

$$= \begin{bmatrix} 0.05 \times 4 & 0.05 \times -1 & 0.05 \times 0 \\ 0.05 \times 3 & 0.05 \times 2 & 0.05 \times 1 \end{bmatrix} = \begin{bmatrix} 0.20 & -0.05 & 0 \\ 0.15 & 0.10 & 0.05 \end{bmatrix}$$

Special Matrices

Certain matrices arise quite often in applications, and they have been given special names. Some of these special matrices are defined next.

Square Matrix

A matrix having the same number of rows as columns is called a *square matrix*. The matrix

$$A = \begin{bmatrix} 3 & -1 \\ 0 & 4 \\ 2 & 6 \end{bmatrix}$$

in (1.8.7) is not a square matrix but

$$A = \begin{bmatrix} 2 & 1 \\ 1 & 3 \end{bmatrix}$$

in (1.8.10) is a square matrix. In a square matrix A, the elements a_{ii} are called the *diagonal elements*. In the square matrix A in (1.8.10), the numbers 2 and 3 are the diagonal elements of A. The elements not on the diagonal (nondiagonal elements) are called *off-diagonal elements*.

Identity Matrix

A square matrix whose diagonal elements are each equal to 1 and whose off-diagonal elements are equal to 0 is called an *identity matrix* and is denoted by I. The size of I is usually clear from the context. For example,

$$I = \begin{bmatrix} 1 & 0 \\ 0 & 1 \end{bmatrix} \quad \text{and} \quad I = \begin{bmatrix} 1 & 0 & 0 \\ 0 & 1 & 0 \\ 0 & 0 & 1 \end{bmatrix}$$

are both identity matrices. An important feature of an identity matrix is that when it is multiplied by any matrix K for which the multiplication is defined, the result is K. This is written as $KI=IK=K$. Readers can verify this fact for the matrices in (1.8.7) and for identity matrices of appropriate sizes. In many cases, the identity matrix plays a role for matrices similar to that played by the number 1 for numbers. It is also the case that $I^T = I$.

Zero Matrix

A matrix whose elements are all 0 is called a *zero matrix* and is denoted by $\boldsymbol{0}$. The size of a zero matrix is usually clear from the context of the discussion. In many cases, the zero matrix plays a role for matrices similar to that played by the number 0 for numbers.

Diagonal Matrix

A matrix is a diagonal matrix if and only if (1) it is a square matrix and (2) all the off-diagonal elements are equal to zero. For example, I is a diagonal matrix, and any square zero matrix is also a diagonal matrix. For another illustration, the following matrix is also a diagonal matrix.

$$\begin{bmatrix} 0 & 0 & 0 \\ 0 & 4 & 0 \\ 0 & 0 & -1 \end{bmatrix}$$

Symmetric Matrix

A matrix is called a *symmetric matrix* if it is equal to its transpose. Thus, if A is symmetric, $A^T = A$. This requires a symmetric matrix to be a square matrix. Notice that I is a symmetric matrix, and every diagonal matrix is also a symmetric matrix. The following matrix is symmetric.

$$\begin{bmatrix} 6 & -2 & 3 \\ -2 & 5 & 0 \\ 3 & 0 & 4 \end{bmatrix}$$

Matrix Inversion

Although matrices can be added, subtracted, and multiplied (when their sizes allow these operations), *division of matrices* is not defined. However, an operation that is similar to division, called the *inversion* of a matrix, takes the place of division in certain situations. It is a basic operation in matrix algebra, and it is much used in regression calculations.

Let A denote a square matrix. If there exists another matrix B such that $AB = I$, then B is called the inverse of A and is generally denoted by A^{-1}. Thus we have $AA^{-1} = I$. If matrix A has an inverse A^{-1}, then $AA^{-1} = I$ and $A^{-1}A = I$. For example, let A be defined by

$$A = \begin{bmatrix} 7 & 2 \\ 10 & 3 \end{bmatrix}$$

This matrix has an inverse, which is given by

$$A^{-1} = \begin{bmatrix} 3 & -2 \\ -10 & 7 \end{bmatrix}$$

You should verify that for this matrix $A^{-1}A = AA^{-1} = I$.

Not all matrices have inverses; for example, the zero matrix does not have an inverse. One difficult problem in using matrices is finding the inverse of a matrix. A computer is generally used for finding inverses except when we are dealing with small matrices (2 by 2 or perhaps 3 by 3 matrices) .

Matrices are very useful for solving systems of linear equations, a frequent step in regression analysis. To illustrate this let us consider the equations

$$4\hat{\beta}_0 + 8\hat{\beta}_1 = 16 \tag{1.8.11}$$

$$8\hat{\beta}_0 + 20\hat{\beta}_1 = 36 \tag{1.8.12}$$

Using elementary algebra, we can see that the solution to the preceding system of two linear equations is given by $\hat{\beta}_0 = 2$ and $\hat{\beta}_1 = 1$. If we use matrices and vectors, we can write these equations as

$$S\hat{\beta} = g \qquad \text{(1.8.13)}$$

where

$$S = \begin{bmatrix} 4 & 8 \\ 8 & 20 \end{bmatrix}, \qquad \hat{\beta} = \begin{bmatrix} \hat{\beta}_0 \\ \hat{\beta}_1 \end{bmatrix}, \qquad g = \begin{bmatrix} 16 \\ 36 \end{bmatrix}$$

The objective is to find the numbers $\hat{\beta}_0$ and $\hat{\beta}_1$ (in other words, to find the vector $\hat{\beta}$) that satisfy the equations in (1.8.11) and (1.8.12). To find the solution, we use (1.8.13) and multiply both sides of the equation by S^{-1} if it exists. This gives us

$$S^{-1}S\hat{\beta} = S^{-1}g$$

i.e.,

$$I\hat{\beta} = S^{-1}g$$

i.e.,

$$\hat{\beta} = S^{-1}g$$

Therefore, if we find S^{-1} and multiply it on the right by g, we get the desired result. We have not yet explained how to obtain the inverse of a matrix, but you can verify that the following is the inverse of S by showing that $SS^{-1} = I$.

$$S^{-1} = \begin{bmatrix} 1.25 & -0.50 \\ -0.50 & 0.25 \end{bmatrix}$$

Then

$$\hat{\beta} = S^{-1}g = \begin{bmatrix} 1.25 & -0.50 \\ -0.50 & 0.25 \end{bmatrix} \begin{bmatrix} 16 \\ 36 \end{bmatrix} = \begin{bmatrix} 2 \\ 1 \end{bmatrix}$$

or

$$\hat{\beta} = \begin{bmatrix} \hat{\beta}_0 \\ \hat{\beta}_1 \end{bmatrix} = \begin{bmatrix} 2 \\ 1 \end{bmatrix}$$

and so $\hat{\beta}_0 = 2$ and $\hat{\beta}_1 = 1$ as before. Notice that in this illustration, S is a symmetric matrix and so is S^{-1}. It is always the case that the inverse of a symmetric matrix is a symmetric matrix (if the inverse exists).

It is easy to write the inverse of a 2 by 2 matrix when it exists. Let

$$A = \begin{bmatrix} a_{11} & a_{12} \\ a_{21} & a_{22} \end{bmatrix}$$

and $d = a_{11}a_{22} - a_{12}a_{21}$. If $d = 0$ then the matrix A does not have an inverse. If $d \neq 0$, then the inverse of A exists and is given by

$$A^{-1} = \begin{bmatrix} a_{22}/d & -a_{12}/d \\ -a_{21}/d & a_{11}/d \end{bmatrix} \qquad \text{(1.8.14)}$$

By straightforward multiplication we can see that $AA^{-1} = A^{-1}A = I$.

Problems   1.8

1.8.1   Problems (a) and (b) refer to the matrices A, B, and C defined as follows:

$$A = \begin{bmatrix} 9 & 4 & 3 \\ 4 & 16 & 8 \\ 3 & 8 & 12 \end{bmatrix} \qquad B = \begin{bmatrix} 12 & 14 & 3 \\ 4 & 31 & 5 \\ 5 & 13 & 21 \\ 6 & 2 & 31 \end{bmatrix} \qquad C = \begin{bmatrix} 12 & 23 & 17 & 22 \\ 24 & 28 & 19 & 20 \\ 31 & 30 & 41 & 27 \end{bmatrix}$$

a   Compute A^T.

b   Which of the following operations are defined?

$$C + B^T, \quad B + C, \quad B + C^T, \quad AC, \quad CA, \quad B - C^T$$

c   In (b), evaluate the expressions that are defined.

1.8.2   Let

$$A = \begin{bmatrix} 3 & 1 \\ 1 & 2 \end{bmatrix} \qquad y = \begin{bmatrix} 5 \\ 0 \end{bmatrix}$$

Find the 2 by 1 vector $\hat{\beta} = \begin{bmatrix} \hat{\beta}_0 \\ \hat{\beta}_1 \end{bmatrix}$ such that $A\hat{\beta} = y$.

1.8.3   Problems (a) through (d) refer to the matrices X and y defined as follows:

$$X = \begin{bmatrix} 1 & 2 \\ 1 & 3 \\ 1 & 1 \\ 1 & 4 \end{bmatrix} \qquad y = \begin{bmatrix} 1 \\ 6 \\ 3 \\ 2 \end{bmatrix}$$

a   Find $X^T X$.

b   Find $X^T y$.

c   Find $(X^T X)^{-1}$.

d   Find the 2 by 1 vector $\hat{\beta} = \begin{bmatrix} \hat{\beta}_0 \\ \hat{\beta}_1 \end{bmatrix}$ such that $X^T X \hat{\beta} = X^T y$.

1.9
Multivariate Gaussian Populations

In Section 1.4 we discussed univariate Gaussian populations. In this section we discuss multivariate Gaussian populations. This section can be studied now or later when the k-variate Gaussian population is discussed in Chapters 3 and 4.

We consider a k-variable population

$$\{(X_{11}, X_{12}, \dots, X_{1k}), \dots, (X_{N1}, X_{N2}, \dots, X_{Nk})\}$$

written $\{(X_1, X_2, \dots, X_k)\}$ for short.

*We say that this k-variable population is a **k-variate Gaussian population** (also called a k-variate normal population) if the collection of numbers*

$$Z_I = a_1 X_{I1} + a_2 X_{I2} + \cdots + a_k X_{Ik},$$

for $I = 1, 2, \ldots, N$ is a one-variable Gaussian population for every possible choice of the values of the constants $a_1, a_2, \ldots, a_k$.

In reality a population can be Gaussian only if N is infinite as discussed in Section 1.4. So we use Gaussian populations only as approximations. It is easier to determine whether a k-variate population is approximately Gaussian when $k = 1$ than it is when $k > 1$. So to determine if a k-variate ($k > 1$) population is approximately Gaussian, we examine $\{Z\}$ to see if it is approximately a univariate Gaussian population for every choice of the constants $a_1, a_2, \ldots, a_k$. If we find that $\{Z\}$ is approximately Gaussian for every choice of the constants a_i, then we can conclude that the k-variate population $\{(X_1, X_2, \ldots, X_k)\}$ is approximately Gaussian.

In particular, this implies that if $\{(X_1, X_2, \ldots, X_k)\}$ is a k-variate Gaussian population, then each of the k univariate populations $\{X_1\}, \{X_2\}, \ldots, \{X_k\}$ must be a Gaussian population. For instance, by choosing $a_j = 0$ for all $j \neq i$ and taking $a_i = 1$, we can conclude that the population $\{X_i\}$ is a Gaussian population. However, even if each of the k univariate populations $\{X_1\}, \{X_2\}, \ldots, \{X_k\}$ is a Gaussian population, we cannot conclude that $\{(X_1, X_2, \ldots, X_k)\}$ is a k-variate Gaussian population unless we also verify that the collection of numbers $a_1 X_{I1} + a_2 X_{I2} + \cdots + a_k X_{Ik}$ for $I = 1, 2, \ldots, N$ is a one-variable Gaussian population for every possible set of values for the constants $a_1, a_2, \ldots, a_k$. We give two illustrations, one of a bivariate Gaussian population and the other of a bivariate non-Gaussian population.

E X A M P L E 1.9.1

We consider a bivariate population $\{(X_1, X_2)\}$ of size 1,000 stored on the data disk in a file named **bivgauss.dat** (**biv**ariate **Gauss**ian). Of course a theoretical two-variable Gaussian population must contain an infinite number of elements, but for practical applications this is a close enough approximation to a theoretical two-variable Gaussian population. A scatter plot of the data is displayed in Figure 1.9.1.

The mean of the population of X_1 values is $\mu_1 = 2.0573$, and the mean of the population of X_2 values is $\mu_2 = 2.9592$. The standard deviations of these two univariate populations are 3.8535 and 9.5729, respectively. Histograms of these two univariate populations are displayed in Figures 1.9.2 and 1.9.3, respectively.

For illustrations, we examine two different linear combinations of X_1 and X_2. First, we calculate the quantity $Z_I = 2X_{I1} + 4X_{I2}$ for each item in the population. The Z_I form a univariate population with mean equal to 15.951 and standard deviation equal to 38.940. A histogram of the population of numbers $\{Z\}$ is shown in Figure 1.9.4.

Likewise, the linear combination $W_I = 3X_{I1} - 2X_{I2}$ is calculated for each item. The mean of the population $\{W\}$ is 0.2535, and its standard deviation is 22.521. A histogram of the population $\{W\}$ is shown in Figure 1.9.5.

In Figures 1.9.2–1.9.5, we have superimposed the theoretical Gaussian distribution curve over the histograms. Note that $\{X_1\}, \{X_2\}, \{Z\}$, and $\{W\}$ are well

F I G U R E 1.9.1

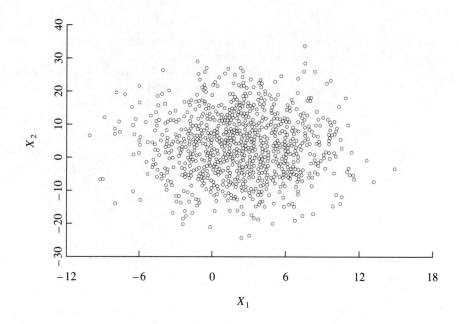

F I G U R E 1.9.2

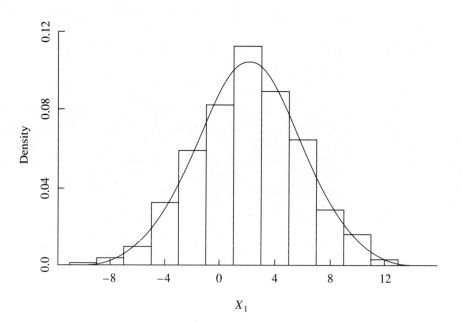

F I G U R E 1.9.3

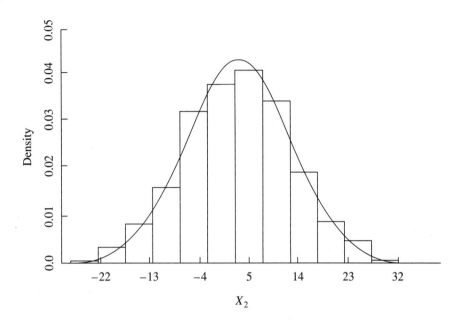

F I G U R E 1.9.4

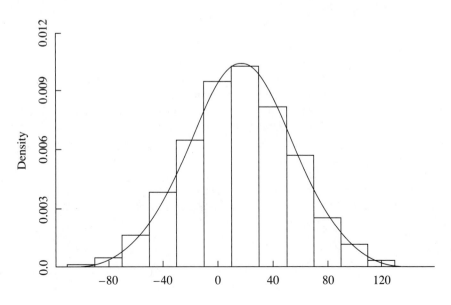

F I G U R E   **1.9.5**

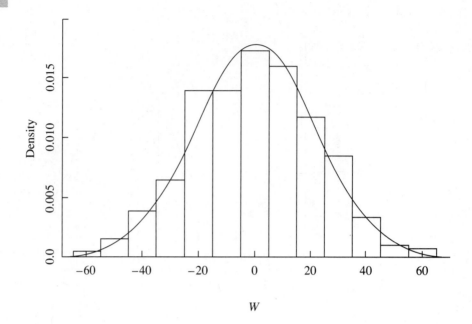

W

approximated by a Gaussian distribution. It may be verified that any other linear function of X_1 and X_2 is also well approximated by a Gaussian distribution. *Thus the bivariate population $\{(X_1, X_2)\}$ is well approximated by a bivariate Gaussian distribution and we say $\{(X_1, X_2)\}$ is a bivariate Gaussian population (approximately).* ■

E X A M P L E   **1.9.2**

Now consider another bivariate population $\{(X_1, X_2)\}$ of size 1,000 stored on the data disk in a file named **bivngaus.dat** (**biv**ariate **n**on-**Gaus**sian). A scatter plot of the data is displayed in Figure 1.9.6.

The mean of the population of X_1 values is $\mu_1 = 2.7958$, and the standard deviation is 3.9842. The mean of the population of X_2 values is $\mu_2 = 1.4921$, and the standard deviation is 11.836. A histogram of the population $\{X_1\}$ is given in Figure 1.9.7 and a histogram of the population $\{X_2\}$ is given in Figure 1.9.8.

Both of these histograms suggest that the two univariate populations $\{X_1\}$ and $\{X_2\}$ are Gaussian populations. But consider the linear function $Z_I = X_{I1} - X_{I2}$. A histogram of the population $\{Z\}$ is given in Figure 1.9.9.

Figure 1.9.10 displays a histogram of the population $\{W\}$, where $W_I = X_{I1} + X_{I2}$, and Figure 1.9.11 shows a histogram of the population $\{V\}$, where $V_I = 8X_{I1} - 3X_{I2}$.

In Figures 1.9.9–1.9.11 we have also displayed the corresponding theoretical Gaussian curve to help us assess the adequacy of the Gaussian distribution as an

F I G U R E 1.9.6

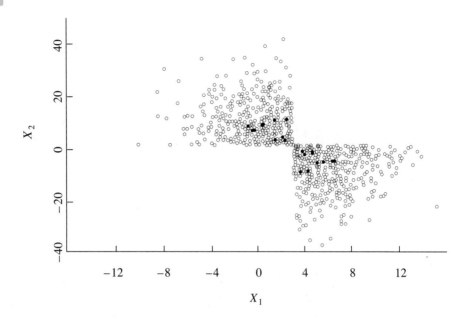

F I G U R E 1.9.6

F I G U R E 1.9.7

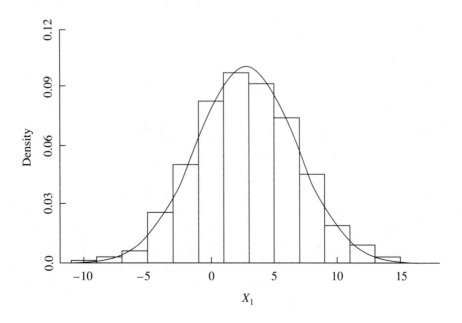

F I G U R E 1.9.7

F I G U R E **1.9.8**

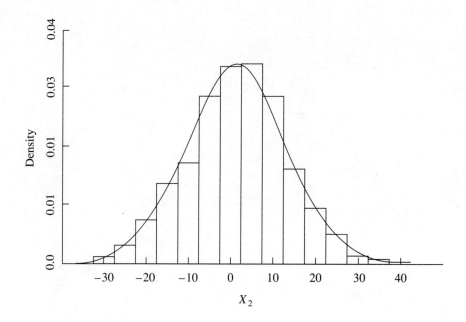

F I G U R E **1.9.9**

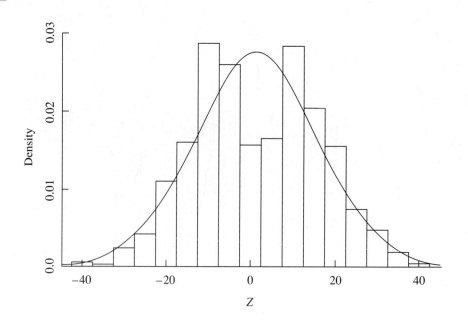

F I G U R E   1.9.10

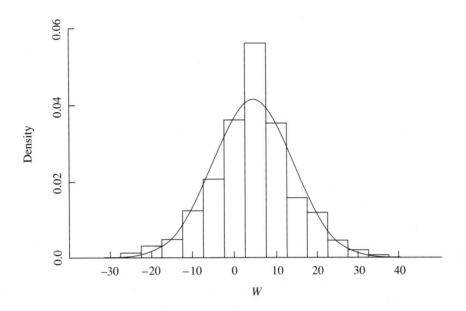

F I G U R E   1.9.11

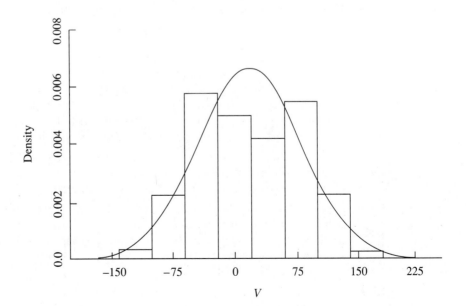

approximation to the populations in question. These figures indicate that none of the three populations $\{Z\}$, $\{W\}$, $\{V\}$ is a Gaussian population. We therefore conclude that the bivariate population $\{(X_1, X_2)\}$ is not a Gaussian population. Thus we have an example of a bivariate population that is not a Gaussian population, and yet the two univariate populations $\{X_1\}$ and $\{X_2\}$ are Gaussian populations (approximately).   ■

You need to refer to more advanced books [11] for further information regarding multivariate Gaussian population models.

1.10
Exercises

1.10.1   A state health agency wants to determine the average number of days of sick leave state employees took last year. Define the target and study populations of items and of numbers. What parameter is to be studied?

1.10.2   In Exercise 1.10.1, the agency wants to predict the average number of days of sick leave state employees will take next year. Define the target and study populations of items and of numbers.

1.10.3   A pharmaceutical company wants to determine what proportion of the bottles of aspirin it will manufacture next month will be damaged before they reach retail stores. Define the target and study populations. What parameter is to be studied?

1.10.4   For an arbitrary population $\{X\}$ of numbers, obtain an upper bound for the proportion of numbers that are more than 3 standard deviations away from the mean (use Chebyshev's theorem). Write down any 20 numbers. Does Chebyshev's theorem apply to this set of numbers (check for many different values of c)?

1.10.5   Consider a population consisting of the following eight numbers (i.e., $N = 8$).

$$3, 7, 5, 2, 9, 8, 5, 6$$

Calculate the mean and the standard deviation for this population. How many distinct samples of size 3 can be obtained from this population? List all possible samples of size 3 that can be obtained from this population. For each sample, compute the sample mean and the sample standard deviation. Is the sample mean an unbiased estimate of the population mean for this population? Is the sample standard deviation an unbiased estimate of the population standard deviation for this population?

1.10.6   Suppose a population is Gaussian with mean 16 and variance 2. Find the proportion of the population that is less than 17.5. Find the proportion of the population that is between 12 and 16. Find the proportion of the population that is greater than 15.

1.10.7   A hospital administrator expects 10,000 patients next year. He wants to determine p, the proportion of patients who will stay less than 7 days. It is assumed that the length of stay Y in days is a Gaussian population (approximately) with mean μ_Y and standard deviation σ_Y. Define the target population of items and of numbers. If $\mu_Y = 8$ and $\sigma_Y = 2$, find p.

1.10.8   A manufacturer of cement blocks runs an experiment to determine the strength Y (in pounds per square inch) of the blocks made by a new method. The target population consists of the strengths of all blocks that will be made by the new process if it is adopted. Twenty blocks are made by the new process, and it is assumed that they are a simple random sample from the target population, which is Gaussian with unknown mean μ_Y and standard deviation σ_Y. From the sample the following are computed: $\sum y_i = 12{,}000$, $\hat{\sigma}_Y = 14$. A 95% confidence statement is to be used to decide if the blocks are strong enough. Which confidence statement gives the appropriate information—a lower confidence bound, an upper confidence bound, or a two-sided confidence interval?

1.10.9 In Exercise 1.10.8, the cement blocks that were made by the old process produced blocks whose average strength was 540 pounds per square inch. To determine if the new process produces stronger blocks, the investigator wants to carry out a hypothesis test. State the appropriate null and alternative hypotheses.

1.10.10 In Exercise 1.10.9, will the null hypothesis be rejected at the 5% level? What is your conclusion about the new process?

1.10.11 In Exercise 1.10.9, compute the *P*-value for the test. Does this *P*-value give you more or less reason to reject the NH than the result of Exercise 1.10.10 ? Why?

1.10.12 In Exercise 1.10.8, compute a 95% lower confidence bound for μ_Y. State your conclusion based on this confidence bound. Which gives you more information—a confidence statement or a test?

1.10.13 The matrices A, B, C are defined as follows:

$$A = \begin{bmatrix} 6 & -1 & 0 \\ 3 & 1 & 5 \\ 2 & 1 & 9 \end{bmatrix} \qquad B = \begin{bmatrix} 3 & 1 \\ 2 & 6 \\ 1 & 0 \end{bmatrix} \qquad C = \begin{bmatrix} 13 & 2 & 1 \\ 0 & -1 & 4 \end{bmatrix}$$

Which of the following operations are defined? (a) $A + B$ (b) $C - B^T$ (c) AB (d) CA^T

1.10.14 In Exercise 1.10.13, evaluate the expressions that are defined.

1.10.15 For the matrix D defined by

$$D = \begin{bmatrix} 52 & 0.62 \\ 43 & 0.74 \\ 36 & 0.65 \\ 32 & 0.71 \\ 27 & 0.68 \\ 26 & 0.59 \\ 22 & 0.49 \\ 37 & 0.67 \\ 24 & 0.64 \\ 19 & 0.56 \\ 13 & 0.51 \end{bmatrix}$$

find $K = D^T D$.

1.10.16 In Exercise 1.10.15, find K^{-1}.

1.10.17 For the function $Y(x) = 6x^3 - 5x + 8$, $-6 \leq x \leq 5$, find $Y(3)$ and $Y(-2)$.

1.10.18 For the function $\mu_Y(x_1, x_2) = 3x_1^2 + 6x_2^2 - 2x_1 x_2 + 4x_1 + 3x_2 - 9$, defined for all real values of x_1, x_2, find $\mu_Y(3, -1)$.

1.10.19 For the function $\mu_Y(x_1, x_2, x_3) = \beta_0 + \beta_1 x_1 + \beta_2 x_2 + \beta_3 x_3$, defined for $-3 \leq x_i \leq 4$ ($i = 1, 2, 3$), find $\mu_Y(1, 3, -2)$.

1.10.20 Which of the following functions are simultaneously linear in all β_i?

a $\mu_Y(x) = \beta_0 + \beta_1 \dfrac{x}{(1+x)}$

b $\mu_Y(x_1, x_2) = \beta_0^2 + \beta_1 x_1 + \beta_2 x_2$

c $\mu_Y(x) = \beta_0 + x \log|\beta_1| + \beta_2 x^2$

d $\mu_Y(x_1, x_2) = \beta_0 + \beta_1 x_1 + \beta_2 x_2 + \beta_3 x_1 x_2$

2

Regression and Prediction

2.1
Overview

In this book we are mainly concerned with the examination and use of associations among variables. Understanding these associations can be useful in many ways, and one of the most important and most common is **prediction**. Knowledge of the associations among variables often leads to methods for *predicting* the value of one quantity by using values of related quantities. Such knowledge may also lead to methods for *controlling* the value of one variable by adjusting the values of related variables, and in some situations it may enhance our understanding of the underlying processes involved. Regression analysis offers us a sensible and sound approach for examining associations among variables and for obtaining good rules for prediction. In Section 2.2 we introduce the subject of prediction as used in various branches of science, in industry, in business, and in everyday affairs. In Section 2.3 we introduce **regression**, which is the subject of the remainder of the book. Section 2.4 contains chapter exercises.

2.2
Prediction

The following example is a simple illustration of the use of prediction.

E X A M P L E 2.2.1

A large university, which receives several thousand applications for admission each year, wants to admit only those students who will successfully complete their first year. To accomplish this, the director of admissions wants to predict the grade point average (GPA) that each applicant would receive at the end of the first year if admitted. Only those students will be admitted whose GPA is predicted to be 3.0

or higher (on a 4-point scale). The GPA is the quantity (usually called the **response variable**) to be predicted.

Suppose the director of admissions decides to use the score on the mathematics part of the Scholastic Aptitude Test (SAT) as the quantity (sometimes called the **predictor variable**, the **predictor factor**, or the **explanatory variable**) on which the prediction is to be based. The predictor variable is often denoted by X and the response variable by Y. It would be convenient to have a rule or a formula that would tell the director of admissions how to compute a predicted GPA of an applicant based on his or her SAT mathematics score. Such a rule or formula is often expressed as a function, say $P_Y(x)$, called a **prediction function**. The notation $P_Y(x)$ means: *predicted Y value of an item whose X value equals x.* If a prediction function $P_Y(x)$ were available, then we could predict that a prospective student who had a SAT score of 480 (i.e., $X = 480$) would obtain a GPA equal to $P_Y(480)$ at the end of one year. For an illustration, suppose that $P_Y(x) = 0.324 + 0.00484x$. Then $P_Y(480) = 0.324 + 0.00484(480) = 2.65$. For a student with an SAT score of 680 (i.e., $X = 680$), the predicted GPA would be $P_Y(680) = 3.62$, etc.

Suppose there are N applicants (N unknown) for admission over, say, the next ten years, and (Y_I, X_I) represents the GPA and SAT, respectively, for the Ith applicant. In general, Y_I and X_I are unknown, but conceptually they could become known if the Ith student took the SAT, was admitted to the university, and completed the first year. We refer to the set of N numbers (Y_I, X_I) for $I = 1, 2, \ldots, N$ as the **target population** under study. Clearly this is a *conceptual* population. Associated with the Ith applicant in the target population there are two numbers of interest, (1) $X_I = $ SAT math score and (2) $Y_I = $ GPA at the end of the first year. If the prediction function used is $P_Y(x)$, then the *predicted* GPA for the Ith applicant is $P_Y(X_I)$ and the *actual* GPA is Y_I. In general, the prediction will not be exact, so Y_I will not be equal to $P_Y(X_I)$, but if the prediction is good, Y_I will be close to $P_Y(X_I)$ for each I. The quantity

$$Y_I - P_Y(X_I)$$

is called the *error of prediction*, i.e., the difference between the predicted GPA and the true GPA for the Ith applicant. If the prediction errors are close to zero for each applicant, then both the predictor factor, viz., SAT, and the prediction function, $P_Y(x)$, are good.   ■

Before we abstract the essential statistical concepts from this example, we give additional examples of cases where prediction is useful.

E X A M P L E 2.2.2

An organization that evaluates performances of automobiles wants to predict the first-year maintenance cost Y of a new car to be made by company A next year using the number of miles X the car will be driven. If $P_Y(x)$ is the prediction function, an individual who knows he will drive 20,000 miles next year can predict his maintenance cost as $P_Y(20,000)$.   ■

E X A M P L E 2.2.3

Suppose a medical research team wants to study how age and weight are related to systolic blood pressure in females between the ages of 25 and 75 currently living in California. The team may want to determine how well the blood pressure Y of a woman can be predicted by using her age X_1 and weight X_2 as explanatory variables.

When more than one explanatory or predictor variable is used, the symbols X_1, X_2, etc., are used to represent these variables and Y_I, X_{I1}, X_{I2}, etc, are used to represent the values of Y, X_1, X_2, etc., for the Ith item in the population.

When there are two predictor variables, we write $P_Y(x_1, x_2)$ for the predicted Y value of an item with $X_1 = x_1$ and $X_2 = x_2$. Note that in this example the research team may not actually be interested in predicting blood pressure, but if a good prediction function $P_Y(x_1, x_2)$ can be found, then an examination of this prediction function may suggest ways of controlling blood pressure, for instance by modifying weight. This information may lead to further investigations, e.g., controlled experiments, in which the team could explicitly study the effect on blood pressure of modifying an individual's weight by diet or exercise. ▪

E X A M P L E 2.2.4

A fertilizer manufacturer wants to predict next year's corn yield Y in bushels per acre, based on the dollars X per acre spent on fertilizer, for farms in the area where the company sells fertilizer. If $P_Y(x)$ represents the prediction function, a customer who plans to spend \$50 per acre for fertilizer next year predicts the yield in bushels per acre to be $P_Y(50)$. ▪

E X A M P L E 2.2.5

An investigator wants to study the pattern of associations among the following variables for U.S.-born individuals who are at least 18 years old now.

$$Y = \text{height of the individual at age 18}$$
$$X_1 = \text{length of the individual at birth}$$
$$X_2 = \text{mother's height at age 18}$$
$$X_3 = \text{father's height at age 18}$$
$$X_4 = \text{paternal grandmother's height at age 18}$$
$$X_5 = \text{paternal grandfather's height at age 18}$$
$$X_6 = \text{maternal grandmother's height at age 18}$$
$$X_7 = \text{maternal grandfather's height at age 18}$$

The investigator may not actually be interested in predicting what an individual's height will be at age 18, but if a good prediction function is found, then this function may yield information regarding what the predominant determinant of an individual's height is—the heights of his maternal ancestors, the heights of his paternal ancestors, both, or neither. ▪

E X A M P L E 2.2.6

A company that owns a large tree farm wants to determine the volume of each tree on the farm, and if the volume of a tree is large enough, the tree will be cut down and sold. The volume of a tree is difficult to determine, and the tree must be cut down to determine the volume accurately. This procedure is not desirable because if the volume is not large enough, it is more profitable to let the tree grow another year. To determine the volume of trees, the owner wants to use an inexpensive method that does not require cutting and hence destroying the trees. One way to accomplish this is to predict the volume Y of a tree by measuring its diameter X_1, say at 4.5 feet from the ground, and its height X_2. The diameter and height of a tree can be measured quickly and inexpensively, and the volume can be predicted without destroying the tree. ■

E X A M P L E 2.2.7

It is known that as the water in a river moves downstream, it carries small rocks (pebbles) along its path and the rocks tend to become smooth and round in shape. The relationship between the roundness Y of the pebbles, called *sphericity*, and the distance X they have been transported may be of interest to a geologist who is trying to determine the source of the rocks. ■

E X A M P L E 2.2.8

A company that produces a certain chemical makes many batches of the chemical each day, and the number of batches is determined by the production superintendent. The quality control section of the company notices an association between the number of batches X made in a day and the percentage of impurities Y in a day's production. A study of this association may help to predict how many batches can be made in a day without allowing the percentage of impurities to get too large. ■

E X A M P L E 2.2.9

A chemical engineer knows that temperature X_1 and pressure X_2 during the production of plastic containers are two factors that determine the strength Y of the final product. He wants to adjust the temperature and pressure to obtain maximum strength, so he studies the relationship between Y, X_1, and X_2. If $P_Y(x_1, x_2)$ is the predicted strength for temperature $X_1 = x_1$ and pressure $X_2 = x_2$, the engineer may want to determine the values of x_1, x_2 that will maximize the predicted strength of the containers, i.e., maximize $P_Y(x_1, x_2)$. ■

E X A M P L E 2.2.10

It is well known that the relationship between the distance S (in feet) traveled by a freely falling object (in a vacuum) starting from rest and the elapsed time T (in seconds) is well described by the relationship $S = (1/2)gT^2$, where g is a physical constant known as the acceleration due to gravity. The constant g also plays a role in explaining many other calculations in physics. It is possible to experimentally determine the value of g by measuring the values of S corresponding to different values of T and appropriately analyzing the resulting pairs of measured values of S and T. Due to errors in measurements as well as uncontrollable fluctuations in experimental conditions, it is almost always the case that the measured values S^* and T^* of distance traveled and time elapsed will not equal the true values S and T, respectively. Consequently, no analysis of the pairs of numbers (S^*, T^*) will yield the exact value of the constant g. An experimenter might be interested in methods of analyzing experimental data that would yield an estimate of g that is as close to the true value of g as possible. The immediate objective is not predicting S from a knowledge of T, but obtaining a good estimate of the value g, the acceleration due to gravity. ■

E X A M P L E 2.2.11

The business manager of a company is preparing a budget for next year, and she must include enough money to purchase a fleet of new cars. She wants to *predict* next year's price of a certain make and model car. She decides that good predictor variables for next year's price Y are the price X_1 of a similar car this year and the rate of inflation X_2. She needs a good prediction function $P_Y(x_1, x_2)$ that will help her predict next year's price of cars that her company is planning to buy. ■

From these examples it is clear that a study of associations or relationships among factors in a system can help in analyzing, predicting, determining, and even controlling the driving forces that affect the system. *Almost every decision that an individual makes is based on prediction, and many of these predictions can be made by systematically studying associations.* Regression analysis deals with the study of these associations.

Populations—A Review

The first step in any investigation where we want to predict a variable Y is the identification of appropriate predictor variables X_1, X_2, etc. Then the *target population* of items (trees, farms, people, etc.) for which the prediction is of interest must be defined. If the target population is available for study, then samples will be selected from this population. If the target population is unavailable for sampling then, typically, we define a *study population* that resembles the target population as closely as possible and from which samples can be obtained, provided that knowledge about the study population will aid the investigator in making predictions in the target population. You should review the material in Section 1.3 concerning target populations and study populations. In particular, keep in mind that valid statistical inferences can be made to the study population only but not to the target population, unless of course the two populations are the same. If the two populations are not the same, it is up to the investigator to *judge* whether or not information about the study population can be used to make decisions about the target population.

Recall that if each item in the population (study population or target population) has one measurement of interest associated with it, that set of numbers will be called a *one-variable* population or a *univariate* population; if each item has a pair of numbers of interest associated with it, that collection of pairs of numbers will be called a *two-variable* population or a *bivariate* population, etc. In the simplest cases where prediction is useful, the target population and the study population are univariate or bivariate populations, whereas in more complex situations where prediction is required the populations are k-variate ($k > 2$) populations.

For an illustration, let us consider Example 2.2.2 where an organization wants to predict the first-year maintenance cost of cars based on the number of miles they will be driven during that year. The target population of items can be the set of automobiles that will be made by manufacturer A next year and driven between 5,000 and 20,000 miles the first year after purchase. The number of automobiles that satisfy this definition will be denoted by N, where N is unknown but presumably quite large. Each automobile in the target population will have many numbers associated with it (price of the car, frequency of repairs, miles driven during the first year after purchase, maintenance cost during the first year, etc.), but only two numbers are of interest in the present investigation—the first-year maintenance cost and the number of miles driven. In fact, the organization is interested in predicting the first-year maintenance cost based on the number of miles the car will be driven during the first year after purchase. So first-year maintenance cost is designated as the response variable Y and miles driven as the predictor variable X. Clearly the target population is unavailable for study because it is a future population. So the investigator decides to use, as the study population, the population of all cars made by manufacturer A last year that were driven between 5,000 and 20,000 miles during the first year after purchase. A simple random sample of n cars, say $n = 50$, is selected, and the first-year maintenance cost and the number of miles driven are recorded. Based on these data the investigator may make valid statistical inferences about the study population. Information about the study population may be useful for making decisions about the target population. This will require subject matter judgment on the part of the investigator. Note that in this example the target population and the study population are *bivariate populations* and are *different* populations.

Now consider Example 2.2.3 where a medical research team wants to predict blood pressure using age and weight as predictor variables. The target population of items is the set of females between the ages of 25 and 75, currently living in California. Each member of this population has many numbers associated with her (blood pressure, age, weight, height, number of years of education, last year's income, etc.), but for this study, only three numbers are of interest—blood pressure, age, and weight. The objective of this study is to understand how systolic blood pressure is related to age and weight. So blood pressure is designated as the response variable Y and age and weight are the two predictor variables X_1 and X_2, respectively. Understanding how blood pressure is related to age and weight may suggest a way of lowering blood pressure by modifying weight. Since the target population is available for study, it is also the study population. In this example the population of interest is a *three-variable population* or a *trivariate population*.

Schematic Representations of Populations

In general a bivariate population of Y and X values is denoted by $\{(Y,X)\}$ and may be schematically exhibited as in Table 2.2.1. In this table the quantities $Y_1, \ldots, Y_N$, represent the values of the response variable Y, and $X_1, \ldots, X_N$ represent the values of the predictor variable X for items $1, 2, \ldots, N$, respectively, in the population.

Likewise, a three-variable population consisting of Y, X_1, and X_2 values is denoted by $\{(Y, X_1, X_2)\}$ and is schematically represented as in Table 2.2.2. Recall that X_{I1} is the value of the first predictor variable X_1 for item I, and X_{I2} is the value of the second predictor variable X_2 for that item.

For an illustration we examine the population data in Table D-1 in Appendix D. These data are assumed to have been obtained last year from the sales-and-maintenance records of automobile dealers in Colorado. The total number of cars included in this data set is 1,242. Thus we have a three-variable population consisting of 1,242 cars. Table D-1 consists of four columns of data as in Table 2.2.2.

T A B L E 2.2.1

A Schematic Representation of a Bivariate Population with Response Variable Y and Predictor Variable X

Item Number I	Response Variable Y	Predictor Variable (Explanatory Variable) X
1	Y_1	X_1
2	Y_2	X_2
$\vdots$	$\vdots$	$\vdots$
I	Y_I	X_I
$\vdots$	$\vdots$	$\vdots$
N	Y_N	X_N

T A B L E 2.2.2

A Schematic Representation of a Trivariate Population with Response Variable Y and Predictor Variables X_1 and X_2

Item Number I	Response Variable Y	Predictor Variable 1 (Explanatory Variable 1) X_1	Predictor Variable 2 (Explanatory Variable 2) X_2
1	Y_1	X_{11}	X_{12}
2	Y_2	X_{21}	X_{22}
$\vdots$	$\vdots$	$\vdots$	$\vdots$
I	Y_I	X_{I1}	X_{I2}
$\vdots$	$\vdots$	$\vdots$	$\vdots$
N	Y_N	X_{N1}	X_{N2}

The first column contains the car numbers I, which range from 1 to 1,242; the second column contains Y, the total amount (in dollars) spent on maintenance during the first year after sale. The third column contains X_1, the price of the car in dollars when it was purchased. Column 4 contains X_2, the number of miles the car was driven the first year after purchase. Thus the three-variable population in Table D-1 is schematically represented in Table 2.2.2. The first three columns of data in Table D-1 represent a bivariate population (column 1 is a column of labels; columns two and three contain Y and X_1, respectively) as in Table 2.2.1. The data in Table D-1 are also stored in the file **car.dat** on the data disk.

For populations such as these, it is of interest to study the relationship of Y with X_1, Y with X_2, and Y with both X_1 and X_2, to determine if either or both X_1, X_2 are useful for predicting Y.

Reasons for Prediction—A Summary

There are at least three reasons why prediction is useful.

1   *The true response values Y are very expensive to obtain,* but the values of the predictor variable X (or $X_1, \ldots, X_k$ in the case of multiple predictor variables) are relatively inexpensive to obtain, so we can use the inexpensive X values to predict the expensive Y values. This is especially useful in cases when an item has to be destroyed to measure the value of the response variable Y, as is the case in Example 2.2.6 where it is very costly to obtain the volume of a tree but relatively inexpensive to measure diameter and height and predict the volume.

2   *The response values are impossible to measure since they are often future values and thus are not available now.* However, for decision-making purposes investigators want to know the values *before* they become available. This is the situation in Examples 2.2.1, 2.2.2, and 2.2.4. Consider, for instance, Example 2.2.4, where we want to predict the yield Y based on the amount X spent on fertilizer. Of course if the yield Y were known, we would not be interested in predicting it. But in many instances where prediction is needed and used, the true value of the response variable is not known because it is a *future* value that we want to know *now*. If X is available now and Y is not, then we can use the value of the prediction function, $P_Y(x)$, to predict the value of Y *now*.

3   *Prediction is not of immediate interest, but the prediction function is the important quantity.* This is illustrated in Example 2.2.3. Certainly blood pressure can be measured very easily and cheaply, and if an individual wants to know her blood pressure she can measure it directly. In this case the *prediction function* is the important quantity because it may give valuable insight into the relationship between blood pressure and weight for individuals of various ages. For example, if a physician knows the prediction function, he may be able to determine how to reduce the blood pressure of a patient to a desirable level by reducing her weight a certain amount through diet or exercise. Example 2.2.10 provides another illustration. In that example an investigator may not be interested in prediction, yet may be interested in knowing the equation relating S and T so as to extract the value of g from it.

You may be able to think of other uses for prediction.

What Is Needed for Prediction?

Two components are needed for prediction:

1. The predictor variables, say $X_1, \ldots, X_k$, and the observed values of these variables.

2. A prediction function or formula, denoted by $P_Y(x_1, \ldots, x_k)$, for predicting the response variable Y using the predictor variables $X_1, \ldots, X_k$.

The Predictor Variables (or Factors) Are Selected by the Investigator

The investigator knows which variable is to be predicted, and her knowledge of the subject suggests factors that may be useful as predictors. She may not know which factors are the best predictors, or how they interrelate, and regression analysis can be helpful in providing answers to these questions.

Prediction Function

The *best* (prediction) function for predicting Y using $X_1, \ldots, X_k$ can be obtained using regression analysis. In the next section, we define the *regression function* and discuss how it is used in prediction.

Problems 2.2

2.2.1 Describe in detail a two-variable population $\{(Y, X)\}$, preferably related to your own field of study, where you want to predict Y using X.

2.2.2 Describe in detail a three-variable population $\{(Y, X_1, X_2)\}$, preferably from your own subject area of interest, where you may want to predict Y using X_1, and X_2.

2.2.3 State why prediction would be useful in your population in Problem 2.2.1.

2.2.4 In Example 2.2.1 suppose that a student has an SAT score of $X = 490$. What is his predicted first-year GPA if the prediction function is $P_Y(x) = 0.324 + 0.00484x$? What is the predicted GPA of a student whose SAT score is 625?

2.2.5 In Problem 2.2.4, assume that student A scored 50 points more than student B on the SAT. What will be the predicted difference between their first-year GPAs?

2.2.6 In Problem 2.2.4 if the director of admissions decides to admit only those students whose first-year GPA is predicted to be 3.0 or higher, what is the lowest a student's SAT score can be if he or she is to be admitted?

2.3

Regression Analysis

Regression analysis is a commonly used method for obtaining a prediction function for predicting the values of a response variable Y using predictor variables $X_1, \ldots, X_k$. We begin by discussing the concept of *subpopulations*, which plays a very important role in defining the *regression function* of Y on $X_1, \ldots, X_k$. We first consider a two-variable population $\{(Y, X)\}$ and suppose that we want to predict the value of Y based on the value of X for any population item.

Subpopulations

For each distinct value of X in the population there is a *subpopulation* of Y values. *The subpopulation corresponding to $X = x$ is the set of all Y values of those items in the population with $X = x$.*

To explain this concept, we use Example 2.2.1 as an illustration. Recall that in that example, the director of admissions wants to predict GPAs using SAT scores. Suppose, for practical purposes, that the director is interested only in the SAT scores (X values) $450, 451, 452, \ldots, 799, 800$. Because the target population is a conceptual population, it is unavailable for study. Consequently she decides to use as the study population the set of applicants who were admitted to the university during the past ten years whose SAT scores X and first-year GPAs Y are available. The individuals in this study population are grouped into subpopulations on the basis of distinct values of X. In other words, all individuals whose SAT score was 450 are placed in a group, all individuals whose SAT score was 451 are placed in a different group, etc. The GPA values of the group of (say N_1) individuals whose SAT score was 450 form a collection of numbers that is denoted by $\{Y(450)\}$, and this group is called the *subpopulation* of Y values determined by the SAT score $X = 450$, i.e., the subpopulation of Y values, all of which have $X = 450$ as their corresponding X value. The mean of this subpopulation of Y values is denoted by $\mu_Y(450)$, and its standard deviation is denoted by $\sigma_Y(450)$. The GPA values of the group of (say N_2) individuals whose SAT score was 451 form a collection of numbers that is denoted by $\{Y(451)\}$, and this group is called the *subpopulation* of Y values with $X = 451$. This subpopulation has mean $\mu_Y(451)$ and standard deviation $\sigma_Y(451)$. More generally, *there is a subpopulation of Y values for each distinct value of X. The subpopulation of Y values, all having the same X value, say $X = x$, is denoted by $\{Y(x)\}$. The mean of this subpopulation is denoted by $\mu_Y(x)$, and its standard deviation is denoted by $\sigma_Y(x)$.*

It is clear why the symbol $\mu_Y(x)$ is used for the mean of a subpopulation— it is the *mean μ* of the *sub*population (*sub*script Y) of Y values corresponding to $X = x$. Similarly, the symbol $\sigma_Y(x)$ is used to denote the standard deviation of the *sub*population of Y values corresponding to $X = x$.

Figure 2.3.1 provides a graphical representation of subpopulations for two different values of $X =$ SAT scores. In this figure, $Y(600)$ represents the Y value of a randomly chosen individual from the subpopulation of all individuals with $X = 600$.

Likewise, $Y(480)$ denotes the Y value of a randomly chosen individual from the subpopulation with $X = 480$.

FIGURE 2.3.1

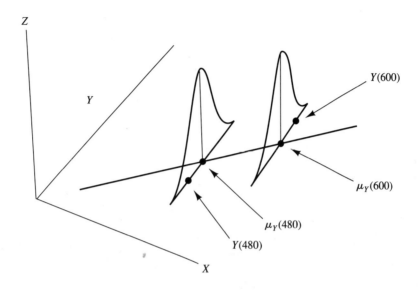

EXAMPLE 2.3.1

To illustrate the concept of subpopulations we consider Example 2.2.2, where the organization wants to predict Y, the first-year maintenance cost of cars, by using X, the distance the car will be driven, as the predictor variable. Table D-1 in Appendix D gives data that represent a three-variable population consisting of $Y = $ first-year maintenance cost of cars, $X_1 = $ sticker price of the cars, and $X_2 = $ miles the cars will be driven the first year after they are purchased. In this example we consider only the two-variable population consisting of Y and X_2 values (i.e., ignore X_1 for the moment). Now if you look at the column of data that represents X_2, you will see the number 14,000 several times. If we look at only the pair of numbers (Y, X_2) for which $X_2 = 14,000$, we get a subpopulation of Y values, all of which have the same X_2 value, namely $X_2 = 14,000$. This subpopulation is shown in Table 2.3.1. The mean Y value for this subpopulation is $\mu_Y(14,000) = \$621.19$, and the standard deviation is $\sigma_Y(14,000) = \$23.18$. You should pick another subpopulation from the population data in Table D-1, say one for which $X_2 = 9,000$, and examine it. ∎

Regression Function

For any given X value, say $X = x$, the *mean* (i.e., average) of the Y values in this subpopulation is denoted by $\mu_Y(x)$, and the standard deviation of these Y values is denoted by $\sigma_Y(x)$. We are now in a position to define the regression function of Y on X.

DEFINITION

The function $\mu_Y(x)$ is called the **regression function** of Y on X and is the *mean* of the subpopulation of Y values for each distinct value of X. ∎

In particular, the mean of a subpopulation of Y values, all of which have $X = x$, is equal to $\mu_Y(x)$. Recall that in this book we use the mean as the best *single* number to represent a population of numbers, and since $\mu_Y(x)$ is the mean of the subpopulation of all Y values that have x for their X value, $\mu_Y(x)$ is the best single number to represent this subpopulation; it is often used as a predictor of any Y value in this subpopulation. For many situations, it can be shown that

BOX 2.3.1   The best prediction function of Y, using X as the predictor variable, is the regression function $\mu_Y(x)$.

Note that although the actual Y values of the items in the subpopulation with $X = x$ are in general not all the same, the predicted value for any of these items will be the same and equal to $\mu_Y(x)$ because they all have the same X value, namely x.

TABLE 2.3.1

Maintenance Cost (Y) and Miles Driven (X_2) for the Subpopulation with $X_2 = 14,000$ Miles

Maintenance Cost Y	Miles Driven X_2
656	14,000
633	14,000
637	14,000
612	14,000
624	14,000
620	14,000
605	14,000
607	14,000
654	14,000
620	14,000
622	14,000
645	14,000
567	14,000
596	14,000
639	14,000
602	14,000

However, if $\sigma_Y(x)$ is small, most of the Y values in this subpopulation will be close to $\mu_Y(x)$, and the probability is high that the Y value to be predicted will be close to the predicted value $\mu_Y(x)$.

The means and standard deviations of subpopulations are of interest in a variety of situations. For instance, consider Example 2.2.2 where the organization wants to predict the first-year maintenance cost of cars. Let μ_Y denote the *average* first-year maintenance cost of *all* cars in the population of cars that will be made by manufacturer A next year. The average maintenance cost of all cars in the subpopulation of cars that will be driven $X = 15{,}000$ miles next year is $\mu_Y(15{,}000)$. If a woman plans to drive 15,000 miles next year, $\mu_Y(15{,}000)$ will be a better representative (predictor) of her maintenance cost than μ_Y will be. Consider another example. Suppose a man in the United States who is 32 years old wants to know if he is overweight. He should *not* compare his weight to μ_Y, the average weight of *all* men, but he should compare his weight with the average weight of all men belonging to the subpopulation of U.S. men who are 32 years old. Because the mean weight of this subpopulation is $\mu_Y(32)$ and the standard deviation is $\sigma_Y(32)$, he knows (using Chebyshev's theorem) that at most 11% of all men in this subpopulation have weights outside the interval $[\mu_Y(32) - 3\sigma_Y(32), \mu_Y(32) + 3\sigma_Y(32)]$. By using such information, he may be able to judge whether or not he is overweight for his age.

Subpopulations, Prediction, and Regression—A Summary of Concepts

We summarize the concepts and ideas just discussed for the case of a single predictor (explanatory) variable.

1   The two-variable population $\{(Y, X)\}$ is partitioned into subpopulations—a subpopulation of Y values for each distinct value of X.

2   The subpopulation of Y values corresponding to any given value of the predictor variable X, say $X = x$, has mean $\mu_Y(x)$ and standard deviation $\sigma_Y(x)$.

3   $\mu_Y(x)$ *is called the regression function of Y on X*, and it is the best single value to represent (predict) any Y value in the subpopulation whose X value is x.

4   If $Y(x)$ denotes the Y value of an item that is to be randomly chosen from the subpopulation with $X = x$, then the best predicted value of $Y(x)$ is the mean, $\mu_Y(x)$, of the subpopulation of all items whose X values equal x.

5   $\sigma_Y(x)$ is the standard deviation of the subpopulation of Y values whose X value is x, and it is used to determine how well $\mu_Y(x)$ represents the entire collection of Y values in the subpopulation whose X values equal x.

6   In most, if not all, applications, $\mu_Y(x)$ and $\sigma_Y(x)$ are unknown and must be estimated from sample data.

7   In theoretical books on statistics, the distribution of the subpopulation $\{Y(x)\}$ is called the *conditional distribution* of Y given $X = x$.

Subpopulations and Regression in the Case of Several Predictor Variables

We now extend the concepts of subpopulations and regression functions to the case where the number of predictor variables is greater than one. When there are k predictor (explanatory) variables, say $X_1, \ldots, X_k$, *each distinct combination of values of $X_1, \ldots, X_k$ in the population determines a subpopulation of Y values.* The subpopulation of Y values determined by $x_1, \ldots, x_k$ is the collection of Y values in the population for which $X_1 = x_1, \ldots, X_k = x_k$. The mean of the Y values belonging to this subpopulation is denoted by $\mu_Y(x_1, \ldots, x_k)$, and the standard deviation is denoted by $\sigma_Y(x_1, \ldots, x_k)$.

B O X 2.3.2   The function $\mu_Y(x_1, \ldots, x_k)$ is called the **regression function of Y on** $X_1, \ldots, X_k$.

Let $Y(x_1, \ldots, x_k)$ represent the Y value of an item to be randomly chosen from the subpopulation with $X_1 = x_1, \ldots, X_k = x_k$. The best value to use to predict $Y(x_1, \ldots, x_k)$ is $\mu_Y(x_1, \ldots, x_k)$, the mean of the subpopulation of Y values with $X_1 = x_1, \ldots, X_k = x_k$. The standard deviation $\sigma_Y(x_1, \ldots, x_k)$ of this subpopulation is a measure of how well $\mu_Y(x_1, \ldots, x_k)$ represents every Y value in this subpopulation (i.e., how good the prediction function is).

When the number of predictor variables is greater than one, we can consider subpopulations determined by any subset of these variables. Consider Example 2.2.3 where a medical research team wants to predict the blood pressure Y of Ms. Smith, a 45-year-old California woman weighing 182 pounds (i.e., $X_1 = 45$ and $X_2 = 182$). There are several subpopulations that the team may consider here. First, there is the entire population of Y values. Then there is a subpopulation of blood pressure values of all women living in California who are 45 years old. There is a subpopulation of blood pressure values of all women living in California who weigh 182 pounds. There is a subpopulation of blood pressures of all women living in California who are 45 years old *and* weigh 182 pounds. In fact, the mean blood pressure, $\mu_Y(45, 182)$, of all women in California who are 45 years old and weigh 182 pounds, is the best predicted value of Ms. Smith's blood pressure (in the absence of any other knowledge about her).

Note   For simplicity of presentation of the *concepts* underlying regression, we have assumed throughout that N is finite, but in actuality the *theory* of regression is typically derived based on N being infinite. In a real application the population size is usually so large that for all practical purposes the results discussed in this book will be valid when all other assumptions are satisfied.

Straight Line Regression

In Chapter 3 we give a detailed presentation of regression with a single predictor X where the population regression function of Y on X is of the form $\mu_Y(x) = \beta_0 + \beta_1 x$. This regression function, *whose graph as a function of X is a straight line*, is not only one of the simplest regression functions but also a very useful one in real

FIGURE 2.3.2

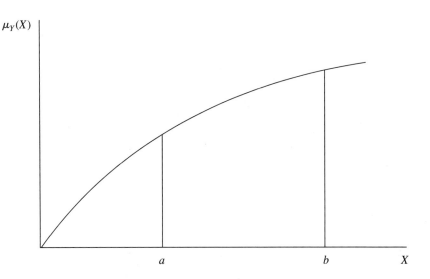

problems. The true relationship between two variables is often linear, but even when it is not, the straight line regression function $\mu_Y(x) = \beta_0 + \beta_1 x$ is sometimes a good approximation to use in the initial stages of an investigation. The regression function may not be a straight line function over the entire range of X values, but it may be an adequate approximation over a limited range that the investigator wants to examine. Consider Figure 2.3.2. A straight line model may be considered adequate if the investigation includes only X values in the range from a to b, but it may not be adequate over the range from 0 to b.

Conversation   To review some of the notions discussed in this section, we present a conversation between an investigator and a statistician.

Conversation 2.3

Investigator: I want to ask you some questions about the notation in your book. Do you have time now?

Statistician: Certainly.

Investigator: Why do you use the notation $\mu_Y(x) = \beta_0 + \beta_1 x$ for a straight line population regression function when most books I've seen use the notation $y = \beta_0 + \beta_1 x$ or $Y_x = \beta_0 + \beta_1 x$?

Statistician: Good question. The reason we use $\mu_Y(x)$ instead of y or Y_x for a straight line population regression function $\mu_Y(x) = \beta_0 + \beta_1 x$ is to help remind you that a regression function is the **mean** of the subpopulation of Y values determined by X. For example, $\mu_Y(6) = \beta_0 + \beta_1(6)$ is the mean of the subpopulation of Y values for which $X = 6$. In statistics Greek letters are often used to denote population parameters, and the Greek letter μ is usually used to denote a population mean. Thus the functional notation $\mu_Y(x)$ denotes a mean, μ, and the subscript Y tells us it is the mean of a population or subpopulation of Y values. The value of X points to that subpopulation of Y values for which the mean is being considered. Similarly, the symbol $\sigma_Y(x)$ denotes the standard deviation of the subpopulation of Y values determined by x. Furthermore, we also use the symbols μ_Y and σ_Y to denote the mean and standard deviation of the *entire* population of Y values when we ignore X.

Investigator: I notice that you also use the symbols $\{Y\}$, $Y(x)$, and $\{Y(x)\}$.

Statistician: Yes, we use the symbol $\{Y\}$ to represent a one-variable population of Y values and the symbol $\{Y(x)\}$ to represent the *subpopulation* of Y values, all of which have x as their common X value. For instance, $\{Y(15.8)\}$ represents a subpopulation of all Y values that have $X = 15.8$ as their common X value. The symbol $Y(x)$ (without the braces around it) is used to represent the Y value of a randomly chosen item from the subpopulation for which $X = x$. Thus $\{Y(x)\}$ denotes the entire subpopulation of Y values having $X = x$ while $Y(x)$ (without the braces, $\{\ \}$) denotes a single randomly chosen Y value from this subpopulation. You might also recall that $\mu_Y(x)$ denotes the mean, and $\sigma_Y(x)$ denotes the standard deviation of the subpopulation $\{Y(x)\}$.

Investigator: Sometimes you use $P_Y(x)$ to represent a prediction function. Why?

Statistician: We use the symbol $P_Y(x)$ to denote a general prediction function of Y. But for a specific problem, we would naturally want to use the best prediction function, and the best prediction function of Y is the regression function $\mu_Y(x)$, the means of subpopulations.

Investigator: I think I understand what you're saying.

Statistician: I might add that notation can often be quite helpful, so it is worth learning.

Investigator: I have one other question. I notice that you don't discuss variance a great deal in your book, yet in the two statistics courses that I took we spent a lot of time studying variance. Why is that?

Statistician: Variance has some mathematical properties that makes it very useful in statistical theory, but for applications the standard deviation is much more important. For example, if a population $\{Y\}$ is Gaussian, then we know what proportion p of Y values are in the interval $\mu_Y - c\sigma_Y$ to $\mu_Y + c\sigma_Y$ for any value c. And for any population $\{Y\}$, Gaussian or not, one can use Chebyshev's theorem to determine a lower bound for this proportion p, for any value of c. Thus, the standard deviation is quite useful in applied problems. You also notice that the standard deviation has the same units as the individual population values and as the mean. Of course either the variance or the standard deviation can be computed from the other.

Investigator: These are all the questions I have for now. Perhaps, I will come to see you again in a few days.

Statistician: Please do so. I am always happy to talk to you.

In many places we ask you to carry out certain **tasks** that may help explain various concepts that have been discussed. These tasks are usually **word problems** that require you to perform appropriate statistical calculations to answer practical questions. Generally, we pose problems that illustrate some aspect of regression. We also supply answers to these questions, and in the problems at the end of this section we ask similar questions about other similar problems. Here we illustrate the concepts in this section with a task.

Task   2.3.1

For illustrative purposes only, suppose that the data in the file **car.dat** on the data disk form a (three-variable) population. These data, which are also given in Table D-1 in Appendix D for convenience, are assumed to have been obtained from last year's sales-and-maintenance records of all automobile dealers who sell a particular make of car in Colorado. The total number of cars included in this data set is 1,242. Thus we have a three-variable population of 1,242 cars. The population data consist of four columns. The first column is the car number I, which ranges from 1 to 1,242; the second column contains Y, the first-year maintenance price (in dollars); the third column contains X_1, the sticker price of the car (in dollars); and the fourth column contains X_2, the number of miles the car was driven during the first year after purchase. In this example we want to study the relationship of Y with X_2, so we are interested only in columns 1, 2, and 4.

The following set of problems refers to this population of 1,242 cars. We give answers to each question, *and the answers and the authors' explanations are in italics.* You will find that using a suitable computer package (SAS, MINITAB, SPSS, BMDP, S-PLUS, etc.) will make it easier to obtain answers to these problems. In the laboratory manual that accompanies this book we present in detail appropriate computer commands that can be used to obtain the answers.

1   First we examine the population data in Table D-1 in Appendix D in detail.

   a   To get an idea of how the Y values in the population are distributed, construct a *frequency histogram* for the maintenance costs Y of all 1,242 cars in the population.

   b   Compute the mean and the standard deviation of Y.

   a   *We construct a frequency histogram for Y = first-year maintenance cost, as shown in Figure 2.3.3. Note that the distribution of Y is not symmetric.*

F I G U R E   **2.3.3**

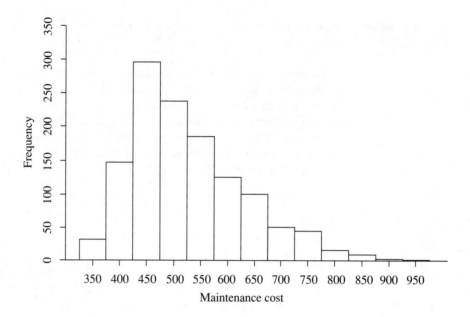

b   *The mean and standard deviation of the population of Y values are*

$$\mu_Y = \$526.14 \text{ and } \sigma_Y = \$105.97$$

respectively. Thus the average first-year maintenance cost of all 1,242 cars in the population is $526.14.

2   Suppose a car is randomly chosen from the preceding population of 1,242 cars.

   a   What is the *actual* first-year maintenance cost for this car?

   b   *Predict* the first-year maintenance cost of this randomly chosen car.

   c   Suppose you plan to purchase a car next year. Predict, if possible, the first-year maintenance cost for *your* car.

   a   *There is no way we can determine exactly the actual first-year maintenance cost of the randomly chosen car unless we know which car was chosen. For instance, if car number 354 is chosen, then we know that the first-year maintenance cost for this car was $483.00. You can check this by examining Table D-1.*

   b   *Because we know that the average first-year maintenance cost of all 1,242 cars is $526.14, our best prediction for the first-year maintenance cost of a randomly chosen car, in the absence of any other relevant information, is $526.14, the mean first-year maintenance cost of all cars in the entire population.*

c   *If we can regard the car you will buy as a randomly chosen car from a population similar to the preceding population of 1,242 cars, then the best predicted value for the first-year maintenance cost of your car is $526.14 as in (b). An appropriate target population here might be the set of all cars, of the same make as the 1,242 cars in the population given in Table D-1, that will be made next year. However, this target population is not available so we might choose to use the population given in Table D-1 as the study population. The value $526.14 is really the best predicted value for the first-year maintenance cost of any randomly chosen car from the study population. Whether or not this is a good predicted value for a randomly chosen car from the target population depends on how closely the study population resembles the target population.*

3   We want to determine whether there is a relationship between Y, the first-year maintenance cost of cars, and X_2, the miles the cars will be driven. Do you think that the first-year maintenance costs of cars in this population are related to the number of miles the cars are driven?

One way to examine the relationship of Y = first-year maintenance cost and X_2 = number of miles the car is driven during the first year after purchase is to plot the values of Y against the values of X_2; so we plot the values of Y on the vertical axis and the values of X_2 on the horizontal axis and study the resulting scatter plot. This plot is given in Figure 2.3.4. (The dashed lines are not part of the computer output.) The scatter plot clearly indicates that the first-year maintenance cost of cars is related to the number of miles they are driven during the first year after purchase because it appears that as the number of miles driven increases, the first-year maintenance cost also tends to increase.

F I G U R E **2.3.4**

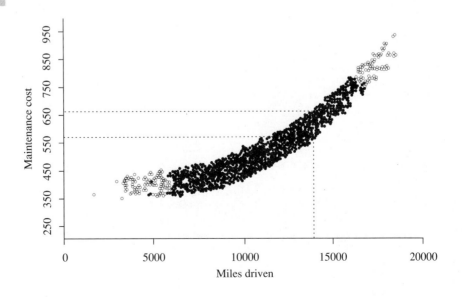

4   In problem 2 (c), suppose you are told that the car you will buy next year will be driven 14,000 miles during the first year after purchase.

   a   Predict what the first-year maintenance cost will be for your car using the scatter plot in Figure 2.3.4.

   b   Examine the subpopulation of all cars that are driven 14,000 miles during the first year. How will this information help you determine the first-year maintenance cost of your car?

   a   *First, suppose that the car you will buy next year can be regarded as a randomly chosen car from a population very similar to the one in Table D-1. In this case the population in Table D-1 serves as the study population. In the scatter plot in Figure 2.3.4 we have drawn a vertical line at miles driven = 14,000 to see where it intersects the plotted points. The Y values of these points of intersection range between $560 and $680. So we see that, in the study population, the first-year maintenance cost for a car driven 14,000 miles during its first year should be somewhere between $560 and $680. Based on this scatter plot you might use the middle value $(560 + 680)/2 = 620$ as the predicted value for the first-year maintenance cost of the car you will buy next year.*

   b   *Because you know you will drive your car 14,000 miles next year, you are not interested in the entire study population of 1,242 cars but only in the subpopulation of cars that were driven 14,000 miles. There are 16 of these cars, and their first-year maintenance costs are as follows:*

Maintenance Cost Y	Miles Driven X
656	14000
633	14000
637	14000
612	14000
624	14000
620	14000
605	14000
607	14000
654	14000
620	14000
622	14000
645	14000
567	14000
596	14000
639	14000
602	14000

The mean of this subpopulation is $\mu_Y(14{,}000) = \$621.19$, and the standard deviation is $\sigma_Y(14{,}000) = \$23.18$. Because the standard deviation of the entire population of Y values is $\sigma_Y = \$105.97$, which is much larger than $\sigma_Y(14{,}000)$, it is clear that $\mu_Y(14{,}000) = \$621.19$, the mean of the subpopulation, is much better for predicting the first-year maintenance cost of a randomly chosen car from this subpopulation than $\mu_Y = \$526.14$, the mean of the entire population. Hence, $\mu_Y(14{,}000) = \$621.19$ is a better value to use to predict the maintenance cost of the car you will buy next year than $\mu_Y = \$526.14$ is (assuming, as before, that the car you will buy belongs to a population of cars which is similar to the population of cars for which the data are given in Table D-1).

5 In problem 4 suppose you know that you will drive your car less than 20,000 miles during the first year, although you are not sure exactly how many miles. What is your predicted first-year maintenance cost?

If all you know is that you will drive your car less than 20,000 miles the first year, then your prediction has to be based on the data you have for all 1,242 cars in the study population, and you cannot just focus attention on any particular group of cars, such as cars that were driven 14,000 miles during their first year. The average first-year maintenance cost of all the cars in the study population is $\mu_Y = \$526.14$. You use this value to predict the first-year maintenance cost of your car if you don't know how many miles you will drive it.

6 In problem 5 suppose you know that you will drive the car 50,000 miles during its first year. Predict the first-year maintenance cost of this car.

Because $X_2 = 50{,}000$ is well outside the range of values of X_2 in the population data at hand, you are uncomfortable making any sort of prediction, and even if you did make a prediction you will have practically no confidence in it for it to be of any real use.

Next we discuss methods for collecting data because carefully planned data collection procedures are an essential part of a successful statistical study. Ideally the collection of data involves sampling from well-defined populations.

Sampling Methods in Regression

In practice, inferences about population parameters are based on the information provided by samples. It is therefore very important to ensure that available resources are used efficiently and that all relevant information is gathered. Ideally the collection of data involves random sampling of well-defined populations. In this section we discuss two commonly used sampling methods, *simple random sampling* and *sampling with preselected X values*. In what follows, Y refers to the response variable and $X_1, \ldots, X_k$ refer to predictor variables.

Simple Random Sampling

Sample data are obtained by selecting a simple random sample of n items from the entire population of N items and recording the values for the response variable Y and the predictor variables $X_1, \ldots, X_k$, for each item in the sample. Refer to Section 1.6.

Random Sampling with Preselected X values

Specific values of the predictor variables $X_1, \ldots, X_k$ are preselected by the investigator, and each of these preselected set of values determines a subpopulation of Y values. A simple random sample of one or more Y values is selected from each of these subpopulations. The number of observations to be sampled from each subpopulation is also predetermined by the investigator.

Note   We assume that the reader is familiar with the concepts of *sampling with replacement* and *sampling without replacement*. In this book, unless otherwise specified, sampling is always assumed to be **without replacement**. However, *if the population size N is very large relative to the sample size n, sampling with and without replacement are equivalent for all practical purposes.*

Whether we collect data using simple random sampling or sampling with preselected X values depends on the objectives of the particular investigation and the availability of the items to be sampled. If simple random sampling is used, a random sample from the population of all items is required. If data are obtained by sampling with preselected X values, we must identify the values of $X_1, \ldots, X_k$ for each item in the population, and we must sample Y values only from those subpopulations corresponding to the values of $X_1, \ldots, X_k$ specified by the investigator. The two methods are illustrated using some of the examples in Section 2.2.

E X A M P L E   **2.3.2**

In Example 2.2.7 where a geologist wants to sample pebbles from a river, suppose he is interested in a specified portion of the river that is 150 miles in length. Conceptually, two numbers are associated with each pebble in this portion of the river, viz., Y_I = sphericity of the Ith pebble and X_I = the distance of the Ith pebble from a given reference point. So the two-variable population consists of N pairs of numbers (Y_I, X_I), one pair for each of the N pebbles.

Suppose the objective is to understand the relationship between sphericity of pebbles and their distances from a reference point. The investigator is thus interested in the regression function of Y on X. If data are obtained from this population by simple random sampling, then the sample of n pebbles must be selected at random from the entire population of pebbles. Note that if this method of sampling is used, there is a chance, albeit small, that the entire sample may be selected from a small portion of the river, say a 5- or 10-mile segment. For this problem the investigator is more likely to obtain data by sampling with preselected X values so that he obtains a sample of pebbles along the full length of the 150-mile portion of the river. He may want to select samples of pebbles every 5 miles; i.e., $X = 0, 5, 10, \ldots, 150$. Thus sampling with preselected X values ensures that he will examine the entire length of the river. Moreover this method is also more convenient, so in this example it is undoubtedly better to preselect the X values and sample the Y values at random from the subpopulations determined by these chosen X values.   ∎

E X A M P L E 2.3.3

Consider Example 2.2.2 where the manager wants to study the relationship between the number of miles a car is driven and its first-year maintenance cost. If she uses simple random sampling in this investigation, a random sample of automobiles of the specified make and year is required, and this may be quite easy to obtain. On the other hand, if she uses preselected X values, she could make certain that the X values *cover* the range of miles driven that is of interest in this study, but this would require the identification of the number of miles that each car in the entire population is driven. This would be an expensive procedure to say the least.   ∎

E X A M P L E 2.3.4

Consider Example 2.2.6 where the diameter X_1 and height X_2 of trees are used to predict volume Y. If sampling with preselected values of X_1, X_2 is used, then an investigator must randomly select Y values (volume) for preselected values of X_1 (diameter) and X_2 (height). This requires the identification of all trees that have the specific diameter and height combinations chosen by the investigator. Suppose that $X_1 = 2$ and $X_2 = 80$ is one combination of X_1, X_2 values chosen by the investigator. Then we must identify all trees that are 2 feet in diameter and 80 feet tall and randomly select one or more trees from this subpopulation of trees. This would be repeated for every combination of values of X_1 and X_2 preselected by the investigator, and it would require the identification and the measurement of the diameter and height of every tree on the farm, an almost impossible task. On the other hand, if simple random sampling is used, a random sample of n trees is selected from the population of all trees. This can be accomplished by giving each tree a number according to its location on a grid and randomly selecting n trees using a set of random numbers generated on a computer. This sampling procedure would be less expensive.   ∎

Remark   Often data are collected under controlled laboratory conditions. If this is the case, then usually only sampling with preselected X values is meaningful. For example, consider the following experiment. The effect of temperature on the growth of soybean seedlings is being studied in a laboratory under controlled conditions. A batch of seedlings is available, and each seedling is to be subjected to a different temperature in the range from 20° C to 35° C. At the end of a week, the growth (in millimeters) of the seedlings is to be observed. In this example, for every value of temperature in the range from 20° C to 35° C, conceptually, there is a subpopulation of heights of soybean seedlings. The experimenter chooses the subpopulations she wishes to sample by choosing the temperature values to be used in the experiment. It would be sensible for her to choose a set of temperature values to cover the range 20° C to 35° C. Thus sampling with preselected X values is a natural choice in this case.

Note that the subpopulations in this remark are all conceptual subpopulations; i.e., they do not currently exist. However, the experimenter can observe a part of each of these subpopulations, viz., those values that are obtained during the experiment. The sample observations are not random samples in a strict sense, but for practical applications, we usually regard the observed values from an experiment as if they are random samples from the specified subpopulations and apply the inference procedures discussed in this book. The results are generally satisfactory.

Summary

Two methods of sampling have been discussed for obtaining data for regression studies. The method to be used in any given situation depends on the problem, the circumstances, and the expenses involved. Generally speaking, the simple random sampling method is less expensive and easier to use in *observational studies,* and sampling with preselected X values is the natural method when data are obtained from controlled experiments. When simple random sampling is used, no control is exerted over the values of the predictor variables in the sample, and consequently there is the undesirable possibility that the values of the predictor variables in the resulting sample will be bunched together. If data are obtained by sampling with preselected X values, investigators can preselect the values of the predictor variables to cover the range that they desire to investigate. It is safe to say that, *in most instances where the investigator is interested in estimating the regression function, a judicious sample obtained by sampling with preselected X values will yield better estimates than one obtained using simple random sampling.* However, investigators may decide to use simple random sampling rather than sampling with preselected X values based on the relative costs associated with the two sampling procedures. They may also decide to use the simple random sampling method if the objectives of the study involve more than just the estimation of the regression function. (We cannot obtain valid estimates of certain parameters if data are obtained by sampling with preselected X values; this is discussed further in Chapter 3.)

Linear and Nonlinear Regression

As stated earlier, we seldom know the true regression function in an applied problem, but we can often postulate a class of functions such that one of the functions in this class will serve as an approximation to the true regression function and is accurate enough for the problem at hand. The simplest classes of functions that are useful in many problems are straight line functions, quadratic functions, etc. This means that we can write an equation for the regression function under study, but it will involve some unknown constants (called parameters). As an example, if an investigator knows that the regression function under study is a straight line (for all practical purposes), but does not know the slope or the intercept of this straight line, then he/she could write down the regression function as $\mu_Y(x) = \beta_0 + \beta_1 x$, where β_0 and β_1 are unknown parameters to be determined or estimated. In this case the regression function is a *linear function of the unknown parameters*. In general, **linear regression** means the regression function is simultaneously linear in the unknown parameters β_i, and **nonlinear regression** means the regression function is not simultaneously linear in the unknown parameters β_i. (Refer to Section 1.7 for a review of the definition of linear functions.)

The theory is much better developed for linear regression than for nonlinear regression. Consequently, most of this book is concerned with linear regression; nonlinear regression is discussed only briefly in Chapter 9.

Some examples of regression functions that are linear are listed in (2.3.1), whereas examples of nonlinear regression functions are listed in (2.3.2). The predictor variables are X_1, X_2, and X_3, and the response variable is Y; $\beta_0, \beta_1, \ldots, \beta_5$, are unknown parameters.

$$
\left.
\begin{aligned}
\mu_Y(x) &= \beta_0 \\
\mu_Y(x) &= \beta_0 + \beta_1 x \\
\mu_Y(x) &= \beta_0 + \beta_1 x + \beta_2 x^2 \\
\mu_Y(x_1, x_2, x_3) &= \beta_0 + \beta_1 x_1 + \beta_2 x_2 + \beta_3 x_3 \\
\mu_Y(x_1) &= \beta_0 + \beta_1 x_1^2 + \beta_2 x_1^{3/2} + \beta_3 / \ln|x_1| \\
\mu_Y(x_1, x_2) &= \beta_0 + \beta_1 e^{x_1} + \beta_2 x_2 + \beta_3 e^{x_1 x_2} \\
\mu_Y(x_1, x_2, x_3) &= \beta_0 + \beta_1 e^{-2x_1} + \beta_2 \sin(x_1 x_2) + \beta_3 x_1 \ln(x_2^2)\tan(x_3) \\
\mu_Y(x_1, x_2, x_3) &= \beta_0 + \beta_1 x_1 + \beta_2 x_2 + \beta_3 x_1 x_2 + \beta_4 x_1^2 + \beta_5 x_1 x_3^2
\end{aligned}
\right\} \quad (2.3.1)
$$

$$
\left.
\begin{aligned}
\mu_Y(x_1) &= \beta_1 e^{\beta_2 x_1} \\
\mu_Y(x_1) &= \beta_0 + \beta_1 e^{\beta_2 x_1} \\
\mu_Y(x_1, x_2) &= \beta_0 + \beta_1 e^{\beta_2 x_1} + \beta_3 e^{\beta_4 x_2} \\
\mu_Y(x_1, x_2, x_3) &= \beta_0 x_1^{\beta_1} x_2^{\beta_2} x_3^{\beta_3} \\
\mu_Y(x_1, x_2) &= \beta_1 x_1 / (\beta_2 e^{\beta_3 x_2})
\end{aligned}
\right\} \quad (2.3.2)
$$

2.4
Exercises

2.4.1 For a two-variable population $\{(Y, X)\}$, define the regression function of Y on X.

2.4.2 What is the relationship of regression to prediction?

2.4.3 Given a two-variable population of values $\{(Y, X)\}$, explain how you would obtain the regression function of Y on X.

There are 151 distinct X_2 values in the population of cars (see Table D-1 in Appendix D) discussed in Task 2.3.1. Hence, the two-variable population of $\{(Y, X_2)\}$ values consists of 151 subpopulations. For each of these 151 subpopulations (i.e., for each distinct value of X_2), we have listed in Table D-2 in Appendix D the subpopulation number (subpop) in the first column, the X_2 value in the second column, the number of Y values (ycount) in the third column, the mean $\mu_Y(x)$ (ymean) in the fourth column, and the standard deviation $\sigma_Y(x)$ (ystdevn) of the Y values in the subpopulation in the fifth column. These data are also stored in the file **car2.dat** on the data disk. Some of the Exercises 2.4.4 through 2.4.10 refer to these subpopulations. The standard deviations are calculated using (1.4.3).

2.4.4 How many items in the population in Table D-1 have $X_2 = 14,300$ (i.e., how many cars were driven 14,300 miles the first year)? How many cars were driven 7,100 miles the first year (i.e., how many items have $X_2 = 7,100$)?

2.4.5 Does the population in Table D-1 have any X_2 values equal to 9,200 (i.e., are there any data for cars that were driven 9,200 miles)?

2.4.6 In the population in Table D-1 in Appendix D, what is the mean first-year maintenance cost of all cars that were driven 8,700 miles the first year; (i.e., what is the value of $\mu_Y(8,700)$)? What is the value of $\sigma_Y(8,700)$?

2.4.7 The mean first-year maintenance cost of *all* cars in the entire population in Table D-1 is denoted by μ_Y, and it is equal to 526.14. If you plan to purchase a car (one that is similar to the cars for which the data appear in Table D-1) and drive it 5,900 miles next year, do you think that 526.14 is a good predictor of your first-year maintenance cost? If you find a better value to predict your first-year maintenance cost, what is it?

2.4.8 For the subpopulation of Y values with $X_2 = 10,000$, what is the value of the mean? What is the value of the standard deviation?

2.4.9 From Exercise 2.4.6, you know the values of $\mu_Y(8,700)$ and $\sigma_Y(8,700)$. Use Chebyshev's theorem to find an upper bound for the probability that the first-year maintenance cost of a car you plan to purchase will be more than $700.00 if you plan to drive it 8,700 miles next year.

2.4.10 Plot the mean $\mu_Y(x)$ against x for

$$x = 7800, 7900, 8000, 8100, 8200, 8300, 8400$$

Do you think that a linear prediction function is adequate if the number of miles driven is between 7,800 and 8,400?

2.4.11 Explain the two sampling methods discussed in this chapter. Describe two studies from your field, one where simple random sampling would be preferred and the other where sampling with preselected X values would be preferred. In each, explain why the particular method would be preferred.

2.4.12 Which of the following regression functions are (simultaneously) linear in the unknown parameters (the symbols $\beta_0, \beta_1, \beta_2, \beta_3, \gamma_0, \gamma_1, \gamma_2, \gamma_3$ refer to unknown parameters)?

a $\mu_Y(x) = \beta_0 + \beta_1 x^4$.

b $\mu_Y(x_1, x_2) = \gamma_0 + \gamma_1 x_1 + \gamma_2 x_2 + \gamma_3 x_1 x_2$.

c $\mu_Y(x) = \beta_0 + \beta_1 x^{\beta_2}$.

d $\mu_Y(x_1, x_2) = \gamma_1 \sqrt{\gamma_2 x_1 + \gamma_3 x_2}$.

e $\mu_Y(x) = \beta_0 + \beta_1 x^{1/2} + \beta_2/x + \beta_3 e^{-2x}$.

f $\mu_Y(x) = \beta_0 + \sin(\beta_1 x)$.

g $\mu_Y(x_1, x_2, x_3) = \beta_0 x_1^{\beta_1} x_2^{\beta_2} x_3^{\beta_3}$.

3

Straight Line Regression

3.1
Overview

In Chapter 2 we defined the regression function $\mu_Y(x_1, \ldots, x_k)$ of a response variable Y on k predictor variables $X_1, \ldots, X_k$ and introduced many of the basic concepts underlying regression. In particular we learned that the best function for predicting the Y value of an item using the values of $X_1, \ldots, X_k$ is the regression function $\mu_Y(x_1, \ldots, x_k)$. In this chapter we focus on the simple but important special case of **straight line regression**. Accordingly, throughout this chapter, we assume that there is only one predictor variable X and that the graph of the regression function of Y on X is a *straight line*, i.e.,

$$\mu_Y(x) = \beta_0 + \beta_1 x \tag{3.1.1}$$

The quantity β_1 is the slope and β_0 is the intercept of the regression line. Thus the mean of the Y values in the subpopulation determined by $X = x$ is given by $\mu_Y(x) = \beta_0 + \beta_1 x$. Recall that $\sigma_Y(x)$ denotes the standard deviation of this subpopulation. If the entire population data are available, then we can calculate exactly the values of β_0, β_1, and $\sigma_Y(x)$ for every allowable x, but since the entire population is almost never available in a real problem, we cannot know the values of β_0, β_1, and $\sigma_Y(x)$ exactly, so we must rely on sample data to *estimate* these and other unknown quantities (parameters). In this chapter we consider point and confidence interval estimation of various quantities of interest, and we also discuss statistical tests. Section 3.3 introduces two sets of assumptions under which the theory of linear regression has been well developed. Section 3.4 discusses point estimation of parameters of interest. Methods for examining the validity of regression assumptions are discussed in Section 3.5. Confidence interval procedures and statistical tests are described in Sections 3.6 and 3.7, respectively. Section 3.8 introduces the analysis of variance. The coefficient of correlation and the coefficient of determination are described in Section 3.9. The effect of measurement errors on inferences about various model parameters is explained in Section 3.10. Section 3.11 considers the special case of the straight line regression model, where the regression line is known to pass through the origin. Chapter exercises appear in Section 3.12. Laboratory as-signments describing the use of a statistical computing package (MINITAB or SAS) for straight line regression are in Chapter 3 of the laboratory manual.

Before proceeding further, we present a detailed illustrative example where the entire population of numbers is assumed to be available, *even though it never is in a real problem*, so that you can get a better grasp of the concepts. This example will also point out how various questions of interest, arising in real problems, can be answered *exactly* when the entire population of numbers is available. Statistical inference procedures, discussed in this chapter and throughout this book, attempt to provide answers to such questions when only sample data, and not the entire population, are available.

3.2
An Example of Straight Line Regression

Table D-3 in Appendix D contains a set of data consisting of 2,600 pairs of numbers (Y, X), where Y is the score (in percent) obtained by a student on a standardized calculus test administered at a certain university, and X is the number of hours (recorded to the nearest hour) that the student spent studying for this test. These data are also stored in the file **grades.dat** on the data disk. For purposes of illustration, we suppose that these data form a *bivariate population* $\{(Y, X)\}$. The size of the population is thus 2,600. An examination of these data shows that there are 13 distinct values of X in the population, and they are $0, 1, 2, \ldots, 12$. The number of observations, the means, and the standard deviations for each of the corresponding 13 subpopulations of Y values are exhibited in Table 3.2.1. A plot of the means of the

T A B L E   3.2.1

Subpopulation Counts, Means, and Standard Deviations for Population Data in Table D-3

Hours X	Number of Items	Subpopulation Mean	Subpopulation Standard Deviation
0	200	45.0	2.881
1	200	49.0	2.881
2	200	53.0	2.881
3	200	57.0	2.881
4	200	61.0	2.881
5	200	65.0	2.881
6	200	69.0	2.881
7	200	73.0	2.881
8	200	77.0	2.881
9	200	81.0	2.881
10	200	85.0	2.881
11	200	89.0	2.881
12	200	93.0	2.881

Y values of these 13 subpopulations against the corresponding X values (i.e., a plot of $\mu_Y(x)$ against x for all allowable x values) is shown in Figure 3.2.1. This plot clearly shows that the regression function of Y on X is of the form $\mu_Y(x) = \beta_0 + \beta_1 x$; i.e., the subpopulation means for Y lie on a straight line when plotted against the corresponding values of X. Furthermore, we can calculate the values of β_0 and β_1 explicitly. In fact, the value of β_0 is 45.0 because the mean value of Y corresponding to $X = 0$ is 45.0 (see Table 3.2.1). Also the value of β_1 is 4.0 because the increase in the mean value of Y for a unit increase in X is easily seen to be 4.0%. Hence the population regression function is

$$\mu_Y(x) = 45.0 + 4.0x$$

Observe also that the subpopulation standard deviations are all equal to 2.881.

F I G U R E 3.2.1

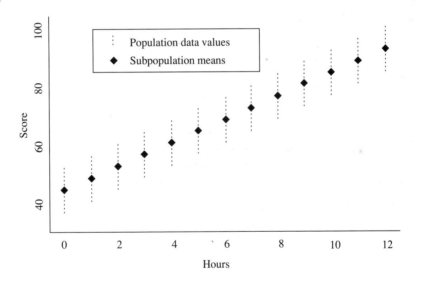

Note   These data are specifically concocted for the purpose of illustration so that the population regression function of Y on X will be *exactly* a straight line and, in addition, the subpopulation standard deviations will all be the same. In most real problems, we cannot expect the population regression function to conform exactly to a straight line model, and the subpopulation standard deviations cannot be expected to all be exactly the same. But in many situations, these idealized conditions may be met *approximately*. You should also be aware that in actual investigations the number of subpopulations of Y values, determined by X, can be quite large, and the sizes of the subpopulations need not all be the same. In this particular example, however, we have deliberately kept the number of subpopulations rather small (13 to be precise) and the sizes of the subpopulations all equal (200 observations in each subpopulation) for ease of discussion.

Thus, because we know the entire population $\{(Y,X)\}$ in this example, we are able to determine *exactly* the values of β_0, β_1 and the subpopulation standard deviations $\sigma_Y(x)$. Any other population summary quantity (parameter) can be calculated exactly as well.

Some Questions of Interest

A student who is considering taking this calculus test may be interested in knowing the answers to the following questions:

1 What is the *average* increase in score per additional hour of studying time?
2 What is the *average* score of students who did not study at all for the test?
3 What is the *best predicted value* of the score of a student who spent 10 hours studying for this test?
4 Of all the students in the population who spent 10 hours studying for the test, what *proportion* obtained a score of 90% or above?

$$(3.2.1)$$

We give answers to these four questions by three methods.

a Answers based on the **entire population data**
b Answers based on **only population parameters**
c Answers based on **only a random sample** from the population

Of course in any real problem we can use only method (c) to obtain answers, but we give the answers to questions (1)–(4) of (3.2.1) by all three methods to help you understand that samples really can help answer questions about the population.

a Answers Based on the Entire Population Data

Answers to the preceding questions based on the entire population data are as follows:

1 The increase in the average score for each additional hour of studying time is equal to β_1, the slope of the regression line of Y on X, which has the value 4.0.
2 The average score of students who did not study at all for the test (i.e., $X = 0$) is $\mu_Y(0)$, which is equal to the intercept β_0 of the regression line, which has the value 45.0.
3 The best predicted value of the score of a student in this population who spent 10 hours studying for this test is $\mu_Y(10) = 45.0 + 4.0(10) = 85.0$.
4 In Table D-3 in Appendix D, an examination of the subpopulation of Y values corresponding to $X = 10$ shows that 11 out of the 200 students in this subpopulation obtained a score of 90% or above. Thus the required proportion is 0.055.

b Answers Based on Only Population Parameters

Clearly, we are able to obtain exact answers to the questions in (3.2.1) when the entire population $\{(Y, X)\}$ is available to us. In many situations, we can answer various questions concerning the population even if we do not know the entire population but know *only certain important summary quantities (parameters)* of the population. To demonstrate this in the present example, we begin by examining the histogram of the subpopulation of Y values determined by $X = 10$. The histogram is in Figure 3.2.2, which suggests that this subpopulation is approximately Gaussian. In fact, we should examine the subpopulation of Y values for each distinct X value to determine if each is approximately Gaussian.

F I G U R E **3.2.2**

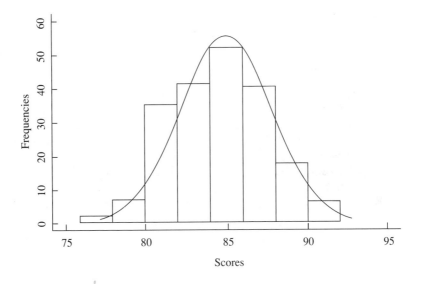

Now suppose that we do not have the entire population $\{(Y, X)\}$ available to us, but suppose we do know that the regression function of Y on X is given by $\mu_Y(x) = 45.0 + 4.0x$ and that each subpopulation of Y values has a standard deviation equal to 2.881. Thus we know the values of β_0 and β_1, which are 45.0 and 4.0, respectively, and we also know that $\sigma_Y(x) = 2.881$ for each allowable x. Furthermore, by plotting the histogram of Y for each distinct value of X, we can demonstrate that each subpopulation of Y values is (approximately) Gaussian. With this information we can answer questions (1)–(4) in (3.2.1). Questions (1)–(3) can be answered knowing only that the regression function of Y on X is $\mu_Y(x) = 45.0 + 4.0x$. To answer question (4) we first observe that the mean Y value for the subpopulation corresponding to an X value of 10 is equal to $45.0 + 4.0(10) = 85$ and that its standard deviation is 2.881. We now use the fact that this subpopulation of Y values is approximately Gaussian. The proportion of values in a Gaussian population, with mean equal to 85.0 and standard deviation equal to 2.881, that equals or exceeds 90 (actually 89.5, to account for the fact that the scores were rounded to the nearest integer) is equal to 0.0594 using Table T-1 in Appendix T. This is close to 0.055, the exact answer to question (4). (The reason for the slight discrepancy between the value 0.055 calculated directly from the population data and the value 0.0594 obtained using a table of Gaussian percentiles is that the theoretical Gaussian distribution is only an approximation to the actual subpopulation distribution.) Thus we can obtain answers to questions (1)–(4) in (3.2.1) if we have the appropriate population parameters β_0, β_1, and $\sigma_Y(x)$ for allowable values of x, even if we do not have the entire population.

To summarize, we can find *exact* answers to questions of interest concerning the population in Table D-3 because we have the entire population data available to us. We also see that we can obtain (nearly) exact answers based only on certain population parameters (summary quantities) because we know that the subpopulations of Y values are (nearly) Gaussian and the subpopulation standard deviations are all equal to 2.881. *It is for this reason that regression analysis focuses its attention on the estimation of various population parameters such as β_0, β_1, $\mu_Y(x)$, $\sigma_Y(x)$, etc.*

c   Answers Based on a Random Sample

We now illustrate how to obtain answers to questions (1)–(4) in (3.2.1) by using a sample rather than the entire population. We do this by calculating, *approximately*, the values of the parameters β_0, β_1, and $\sigma_Y(x)$ and hence obtaining an approximation to the population regression line, using *sample data* from the preceding population. For this purpose we selected a sample of size 26 from the population in Table D-3 by randomly selecting two items from each subpopulation with preselected X values of $0, 1, 2, \ldots, 12$. Thus, the data are obtained by sampling with preselected X values. The sample data are displayed in Table 3.2.2, and they are also stored in the file **grades26.dat** on the data disk.

The 13 subpopulations (one subpopulation of Y values for each value of $X = 0, 1, \ldots, 12$) are displayed in Figure 3.2.3. The sample values are indicated by •. In Figure 3.2.4 the sample values are displayed together with a line that was *visually fitted* to the data. Figure 3.2.5 shows the sample data, the visually fitted line, and the population regression line $\mu_Y(x) = 45.0 + 4.0x$; of course in a real problem only the sample data are available and $\mu_Y(x)$ is not known, but we display $\mu_Y(x)$ to show how the sample data are grouped around it.

If we use the visually fitted line as the *estimate* of the population regression line, then the estimated values of β_0 and β_1 are the values of the intercept and the slope of this line which, from Figure 3.2.4, we judge to be 43.0 and 4.25, respectively (the change in Y as X changes from 0 to 12 is visually approximated as equal to 51 units, and so the slope of the line is estimated to be $51/12 = 4.25$). Based on these sample estimates, we obtain the following approximate answers to questions (1)–(3) of (3.2.1).

1   On the average, the increase in score for each additional hour of studying time is estimated to be 4.25%.

2   The average score of students who did not study at all is estimated to be 43%.

T A B L E 3.2.2

Grades26 Data. Sample of Size 26 from the Population Data in Table D-3 (Sampling with Preselected X Values)

Sample Item Number	Score (in percent) Y	Hours X
1	42	0
2	44	0
3	51	1
4	48	1
5	51	2
6	54	2
7	57	3
8	54	3
9	57	4
10	63	4
11	61	5
12	69	5
13	70	6
14	70	6
15	70	7
16	72	7
17	74	8
18	83	8
19	84	9
20	81	9
21	84	10
22	85	10
23	91	11
24	86	11
25	91	12
26	95	12

3 The predicted value of the score of any student who studies for 10 hours is $43 + 10 \times 4.25 = 85.5\%$.

The answer to question (4) of (3.2.1), based on sample data, depends on procedures to be discussed later (see Problem 3.4.6). Thus we see that even when we have only a *sample* of values from the population, we can obtain useful (though approximate) answers to questions of interest.

■ F I G U R E 3.2.3

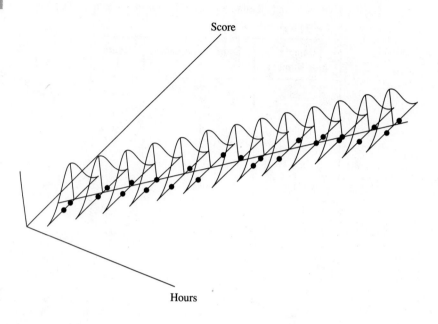

■ F I G U R E 3.2.4

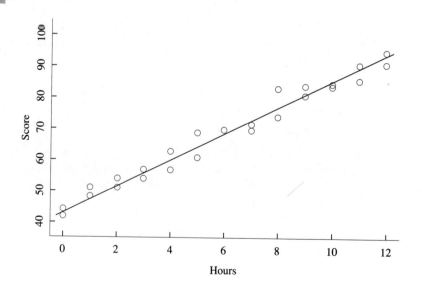

FIGURE 3.2.5

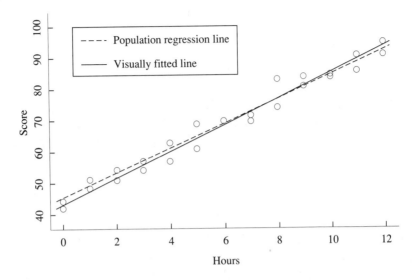

Systematic Methods of Estimation

Although we found an estimate of the population regression line using a straight line that was *visually* judged to provide a good fit to the sample data, we did this for illustration only. It is desirable to obtain an estimate of the population regression line based on a more objective and scientifically sound procedure. Several such methods are available in the literature, and one method that is widely used and has a long history is based on the so-called **method of least squares**. Another method that is becoming increasingly popular is the **method of least absolute deviations**. We use primarily the method of least squares for estimating unknown parameters because it is mathematically simpler than most alternative approaches and because estimates obtained using the method of least squares are best estimates when certain assumptions about the population and the sample are satisfied. If you are interested in circumstances under which other approaches may be desirable, you should consult more advanced books on regression.

In straight line regression, we are usually interested in estimating $\mu_Y(x)$, β_0, β_1, $\sigma_Y(x)$, and $Y(x)$, where $Y(x)$ is the Y value of an item chosen at random from the subpopulation whose X value is x. We may also be interested in estimating $\rho_{Y,X}$, μ_Y, μ_X, σ_Y, and σ_X. The estimation of $\sigma_Y(x)$ for all x is, for all practical purposes, impossible unless we make some simplifying assumptions regarding the population $\{(Y, X)\}$. In the next section we discuss two commonly used sets of assumptions regarding the population $\{(Y, X)\}$ and sampling procedures under which the theory of straight line regression has been extensively studied.

Problems 3.2

3.2.1 Using simple random sampling, a sample of size 26 was selected from the population data in Table D-3 in Appendix D, consisting of scores and hours studied for 2,600 students. The sample data are given in Table 3.2.3 and are also stored in the file **table323.dat** on the data disk. Here X is the number of hours studied, and Y is the percent score obtained on the test for each student in the sample. Examine these data. Plot Y against X, and examine this plot. Does it appear from this plot that the population regression function of Y on X is a straight line?

3.2.2 In Problem 3.2.1, visually fit a straight line to the plotted data. Use this fitted line to obtain approximate values for β_0 and β_1.

3.2.3 Use the data from Problem 3.2.1 to answer questions (1)–(3) of (3.2.1). How do the answers compare with the answers obtained using the entire population?

T A B L E 3.2.3

Sample Item Number	Score (in percent) Y	Hours X
1	44	0
2	86	10
3	87	10
4	58	3
5	85	10
6	55	1
7	63	4
8	48	0
9	57	3
10	54	2
11	82	10
12	90	12
13	56	3
14	67	5
15	81	8
16	57	4
17	47	1
18	47	1
19	44	0
20	48	0
21	54	3
22	45	0
23	51	1
24	91	12
25	58	3
26	100	12

3.2.4 Compare a plot of the data in Table 3.2.3 with a plot of the sample data in Table 3.2.2 (these data are plotted in Figure 3.2.4), and with a plot of the population regression line $\mu_Y(x) = 45.0 + 4.0x$. Which set of sample data do you think best estimates the population regression line? Does this give you any reason to prefer one sampling method over the other?

3.2.5 A simple random sample of size 10 was selected from the population in Table D-3. The data are given in Table 3.2.4 and are also stored in the file **table324.dat** on the data disk. Repeat problems 3.2.1–3.2.3 for these data.

3.2.6 Which sample would you prefer, the one in Problem 3.2.5 or the one in Problem 3.2.1, to estimate the population regression function? Why?

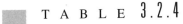

T A B L E   3.2.4

Sample Item Number	Score (in percent) Y	Hours X
1	41	1
2	59	4
3	90	11
4	88	11
5	52	2
6	53	2
7	53	1
8	63	5
9	87	10
10	74	8

3.3
Straight Line Regression Model—Assumptions (A) and (B)

To obtain useful point and confidence interval estimates and tests for parameters associated with a population $\{(Y, X)\}$, we must make some assumptions about the population and about the method used to collect the sample data. One such set of assumptions, which we refer to as assumptions (A), under which the theory for straight line regression has been well developed, is given in Box 3.3.1. Three of these assumptions concern the population and two concern the sample.

B O X 3.3.1   **Assumptions (A) for Straight Line Regression**

A two-variable population $\{(Y, X)\}$ is the study population.

(Population) Assumption 1   The mean of all the Y values in the subpopulation whose X value is x is denoted by $\mu_Y(x)$ and it is given by

$$\mu_Y(x) = \beta_0 + \beta_1 x \qquad \text{for } a \leq x \leq b$$

where β_0 and β_1 are unknown parameters and the allowable values of x lie between a and b.

(Population) Assumption 2   The standard deviations (and hence the variances) of the subpopulations do not depend on the value of X (i.e., they are the same for each subpopulation). This assumption is referred to as the assumption of **homogeneity of standard deviations** or, equivalently, **homogeneity of variances**. This common standard deviation of all the subpopulations is denoted by $\sigma_{Y|X}$. When there is no possibility of confusion, we simply write σ instead of $\sigma_{Y|X}$ to denote this common subpopulation standard deviation.

(Population) Assumption 3   Each subpopulation of Y values, determined by the distinct values of X, is a Gaussian population.

(Sample) Assumption 4   The data are obtained either by simple random sampling or by sampling with preselected X values as discussed in Section 2.3.

(Sample) Assumption 5   The X and Y values of the items in the sample are measured without error (however, see Section 3.10).

Associated with the two-variable population $\{(Y, X)\}$ are several quantities of interest. The most commonly needed quantities are β_0, β_1, $\mu_Y(x)$, $Y(x)$ (the Y value of a randomly chosen item from the subpopulation with $X = x$), and σ. Assumptions (A) in Box 3.3.1 are sufficient for making inferences about these parameters.

In some situations, we may also be interested in μ_Y, σ_Y, μ_X, σ_X, and $\rho_{Y,X}$ and, if data are obtained by sampling with preselected X values, we cannot make valid inferences about these parameters unless every   (3.3.1) subpopulation is sampled and the relative subpopulation sizes are known (which is almost never the case in real problems).

A more restrictive set of assumptions for straight line regression, referred to as assumptions (B), given in Box 3.3.2, is sufficient for making inferences about *all* of the quantities β_0, β_1, $\mu_Y(x)$, $Y(x)$, σ, μ_Y, σ_Y, μ_X, σ_X, and $\rho_{Y,X}$.

BOX 3.3.2   **Assumptions (B) for Straight Line Regression**

(Population) Assumption 1   The two-variable population $\{(Y, X)\}$ that is to be studied is a **bivariate Gaussian population**.

(Sample) Assumption 2   The data are obtained by simple random sampling as discussed in Section 2.3.

(Sample) Assumption 3   The X and Y values of the items in the sample are measured without error (however, see Section 3.10).

Comments

1   For assumptions (B) in Box 3.3.2 to be met, we must obtain sample data by simple random sampling. If, instead, we obtain data by sampling with preselected X values, then no random sample from the $\{X\}$ population or the $\{Y\}$ population is available.

2   If $\{(Y, X)\}$ is a bivariate Gaussian population as in population assumption 1 of Box 3.3.2, then population assumptions 1, 2, 3 in Box 3.3.1 are automatically satisfied. Conversely, if population assumptions 1, 2, and 3 in Box 3.3.1 are satisfied and, additionally, if the one-variable population $\{X\}$ is also a Gaussian population, then population assumption 1 of Box 3.3.2 holds; i.e., the two-variable population $\{(Y, X)\}$ is bivariate Gaussian. Thus, *(population) assumptions (B) for straight line regression imply (population) assumptions (A), but the converse is not generally true.*

3   When the two-variable population $\{(Y, X)\}$ is bivariate Gaussian, then the one-variable populations $\{Y\}$ and $\{X\}$ are both Gaussian populations. However, $\{Y\}$ and $\{X\}$ may both be Gaussian populations and yet $\{(Y, X)\}$ may not be a bivariate Gaussian population. An example of this situation was given in Section 1.9.

4   Let Y_I and X_I be the Y and the X values corresponding to population item I. The predicted Y value for this item is $\mu_Y(X_I) = \beta_0 + \beta_1 X_I$, which is the mean Y value for the subpopulation with $X = X_I$. We write E_I for the difference between the actual value Y_I and the corresponding subpopulation mean $\beta_0 + \beta_1 X_I$. Thus $E_I = Y_I - (\beta_0 + \beta_1 X_I)$. Equivalently,

$$Y_I = \beta_0 + \beta_1 X_I + E_I \tag{3.3.2}$$

which is referred to as the **population regression model**. Under either assumptions (A) or assumptions (B) the population $\{E\}$ is Gaussian with mean zero and standard deviation σ.

    It can never be determined if any of these assumptions are exactly satisfied in a real problem. Investigators may not know for certain that the graph of the population regression function $\mu_Y(x)$ is a straight line, but they may know that this is approximately so. The same can be said for all of the assumptions. The sample assumptions mainly concern collecting the data, and often investigators can make certain they are satisfied. On the other hand, investigators are often restricted by money, time, or other constraints, so the data collection methods may not exactly meet the requirements of randomness, etc. Sometimes the investigators who must analyze the data and draw conclusions from them are not the ones who collected the data. They may know or suspect that errors were made in the sampling procedures or in recording the data. The view we take is that all data contain some information, and the investigators are in the best position to determine whether the assumptions are close enough to being satisfied to allow valid conclusions to be drawn about the population under study. Investigators should always be aware of abnormalities in the data and deal with them.

The next section treats point estimation for the unknown parameters in the straight line regression model. Following this, in Section 3.5, we give methods for examining some of the assumptions in Box 3.3.1 and Box 3.3.2. If they appear not to hold, alternative procedures are sometimes available, and we discuss some of these in Chapter 8.

3.4
Point Estimation

The primary objective in a regression study is to use the sample data to obtain point and confidence interval estimates for the unknown quantities β_0, β_1, σ, $\mu_Y(x)$, and $Y(x)$ and also for various meaningful functions of these quantities. These estimates in turn aid investigators in gaining insight into quite complicated questions about the population under study. In this section we focus our attention on **point estimation**.

Recall that a point estimate of an unknown parameter is a number, computed from observed sample data, that may be used in place of the unknown value of the parameter of interest for making practical decisions. When assumptions (A) or (B) hold, it can be shown mathematically that the best estimates of β_0 and β_1 in (3.1.1) are obtained by the *method of least squares*. Using these estimates we can obtain the best estimates of other quantities of interest. We first describe the method of least squares.

Method of Least Squares

Suppose $(y_1, x_1), \ldots, (y_n, x_n)$ is a sample of size n from the bivariate population $\{(Y, X)\}$, selected using either simple random sampling or sampling with preselected X values, and the population regression function is

$$\mu_Y(x) = \beta_0 + \beta_1 x \tag{3.4.1}$$

The quantity e_i defined by

$$e_i = y_i - \mu_Y(x_i) = y_i - (\beta_0 + \beta_1 x_i) \tag{3.4.2}$$

is the prediction error when we use $\mu_Y(x_i) = \beta_0 + \beta_1 x_i$ to predict y_i for $i = 1, \ldots, n$. The relationship among the observed value y_i, the value of the regression function at x_i, viz., $\beta_0 + \beta_1 x_i$, and the prediction error e_i, is given by

$$y_i = \beta_0 + \beta_1 x_i + e_i \tag{3.4.3}$$

This is referred to as the **sample regression model**.

Since β_0 and β_1 are unknown parameters, we want to use sample data to obtain estimates of them. Under assumptions (A) or (B) it can be shown that the best estimates of β_0 and β_1 are obtained by the method of least squares. The resulting estimates are denoted by $\hat{\beta}_0$ and $\hat{\beta}_1$, respectively. The corresponding estimate of the regression function is denoted by

$$\hat{\mu}_Y(x) = \hat{\beta}_0 + \hat{\beta}_1 x \tag{3.4.4}$$

The prediction error when using $\hat{\mu}_Y(x_i) = \hat{\beta}_0 + \hat{\beta}_1 x_i$ to predict y_i is denoted by $\hat{e}_i$ and is given by

$$\hat{e}_i = y_i - \hat{\beta}_0 - \hat{\beta}_1 x_i \tag{3.4.5}$$

The quantities $\hat{e}_i$ for $i = 1, \ldots, n$ are called **residuals**. They are useful in examining the validity of the assumptions given in Box 3.3.1 as well as those given in Box 3.3.2. This is discussed in Section 3.5.

The *least squares estimates* $\hat{\beta}_0$ and $\hat{\beta}_1$ are chosen in such a way that the quantity $SSE(X)$, called the **sum of squares of prediction errors when X is used to predict Y,** and defined by

$$SSE(X) = \sum_{i=1}^{n} (y_i - \hat{\beta}_0 - \hat{\beta}_1 x_i)^2 = \sum_{i=1}^{n} \hat{e}_i^2 \tag{3.4.6}$$

has the smallest possible value among all the possible choices we could make for $\hat{\beta}_0$ and $\hat{\beta}_1$. When there is no possibility of confusion, we will simply write SSE instead of $SSE(X)$ and refer to it as the **sum of squared errors** or **error sum of squares**.

Note that we are really not interested in predicting the Y values of sample items that were observed because we already know their true values, namely the data values $y_1, \ldots, y_n$. But if the estimated regression function

$$\hat{\mu}_Y(x) = \hat{\beta}_0 + \hat{\beta}_1 x$$

is a good predictor of the *known sample values* y_i for each $X = x_i$ for $i = 1, 2, \ldots, n$, then we have reason to expect that it will be a good prediction function for *all* values of Y in the population corresponding to all allowable values of X. Thus we use the y_i and the x_i values of the items in the sample to assess the performance of the estimate of the population regression function at the sample points. We now enunciate the principle of least squares.

The principle of least squares states that the best estimate of the population regression function $\mu_Y(x) = \beta_0 + \beta_1 x$ is obtained by choosing $\hat{\beta}_0$ and $\hat{\beta}_1$ in (3.4.6) in such a way that the sum of squares of the prediction errors,

$$SSE = \sum_{i=1}^{n} \hat{e}_i^2 = \sum_{i=1}^{n} (y_i - \hat{\beta}_0 - \hat{\beta}_1 x_i)^2 \tag{3.4.7}$$

attains the *least* possible value.

Point Estimates of β_0, β_1, $\mu_Y(x)$ and $Y(x)$

It can be mathematically proven that the values of $\hat{\beta}_1$ and $\hat{\beta}_0$ that minimize $\sum \hat{e}_i^2$ in (3.4.6) are given by

$$\hat{\beta}_1 = \frac{\sum_{i=1}^{n} (x_i - \bar{x})(y_i - \bar{y})}{\sum_{i=1}^{n} (x_i - \bar{x})^2} \tag{3.4.8}$$

and

$$\hat{\beta}_0 = \bar{y} - \hat{\beta}_1 \bar{x} \tag{3.4.9}$$

where

$$\bar{x} = \frac{1}{n} \sum_{i=1}^{n} x_i \quad \text{and} \quad \bar{y} = \frac{1}{n} \sum_{i=1}^{n} y_i$$

The quantities $\hat{\beta}_1$ and $\hat{\beta}_0$, defined in (3.4.8) and (3.4.9), respectively, are known as the *least squares estimates* of the population parameters β_1 and β_0. As mentioned earlier, when assumptions (A) or (B) for straight line regression are satisfied, these are in fact the best estimates of β_1 and β_0, respectively. The best estimate of $\mu_Y(x)$ is

$$\hat{\mu}_Y(x) = \hat{\beta}_0 + \hat{\beta}_1 x \tag{3.4.10}$$

As stated in Section 2.3, *the regression function of Y on X is also the best prediction function for predicting Y using X.* As a result, when only sample data are available, the best predicted value $\hat{Y}(x)$ of a randomly chosen observation $Y(x)$ from the subpopulation with $X = x$ is in fact equal to the estimated subpopulation mean $\hat{\mu}_Y(x)$; i.e.,

$$\hat{Y}(x) = \hat{\beta}_0 + \hat{\beta}_1 x = \hat{\mu}_Y(x) \tag{3.4.11}$$

Point Estimates for Linear Functions of β_0 and β_1

While β_0 and β_1 are important parameters in straight line regression, investigators are quite frequently interested in making inferences about certain linear combinations of β_0 and β_1. Suppose θ denotes the linear combination $a_0\beta_0 + a_1\beta_1$ of β_0 and β_1, where a_0 and a_1 are known numbers. The best point estimate of θ is equal to $\hat{\theta}$ where

$$\hat{\theta} = a_0 \hat{\beta}_0 + a_1 \hat{\beta}_1 \tag{3.4.12}$$

Observe that $\mu_Y(x)$ is a quantity of the form $a_0\beta_0 + a_1\beta_1$ with $a_0 = 1$ and $a_1 = x$; β_1 is also a special case with $a_0 = 0$ and $a_1 = 1$; β_0 is a special case with $a_0 = 1$ and $a_1 = 0$; $\mu_Y(x_1) - \mu_Y(x_2) = (\beta_0 + \beta_1 x_1) - (\beta_0 + \beta_1 x_2) = (x_1 - x_2)\beta_1$ is a special case with $a_0 = 0$ and $a_1 = x_1 - x_2$.

Notation

It is customary to use the notation

$$SXY = \sum_{i=1}^{n} (x_i - \bar{x})(y_i - \bar{y}) \tag{3.4.13}$$

$$SSX = \sum_{i=1}^{n} (x_i - \bar{x})^2 \tag{3.4.14}$$

and

$$SSY = \sum_{i=1}^{n} (y_i - \bar{y})^2 \tag{3.4.15}$$

so that the formula for $\hat{\beta}_1$ in (3.4.8) may be conveniently written as

$$\hat{\beta}_1 = \frac{SXY}{SSX} \tag{3.4.16}$$

Remark

The following alternate (but equivalent) expressions for *SXY*, *SSX*, and *SSY* are sometimes useful.

$$SXY = \sum_{i=1}^{n} x_i y_i - \frac{(\sum_{i=1}^{n} x_i)(\sum_{i=1}^{n} y_i)}{n} = \sum_{i=1}^{n} x_i y_i - n\bar{x}\bar{y} \tag{3.4.17}$$

$$SSX = \sum_{i=1}^{n} x_i^2 - \frac{(\sum_{i=1}^{n} x_i)^2}{n} = \sum_{i=1}^{n} x_i^2 - n\bar{x}^2 \tag{3.4.18}$$

and

$$SSY = \sum_{i=1}^{n} y_i^2 - \frac{(\sum_{i=1}^{n} y_i)^2}{n} = \sum_{i=1}^{n} y_i^2 - n\bar{y}^2 \tag{3.4.19}$$

Point Estimate of σ

Recall that σ is the common standard deviation of the subpopulations of Y values determined by the distinct values of X. The estimate $\hat{\sigma}$ of σ can be calculated using the formula

$$\hat{\sigma} = \sqrt{\frac{SSE}{(n-2)}} \tag{3.4.20}$$

where *SSE* is given by any one of the following equivalent expressions:

$$SSE = \sum_{i=1}^{n} \hat{e}_i^2 = \sum_{i=1}^{n} [y_i - \hat{\mu}_Y(x_i)]^2 = \sum_{i=1}^{n} (y_i - \hat{\beta}_0 - \hat{\beta}_1 x_i)^2$$

The quantity $\frac{SSE}{(n-2)}$, which is under the square root symbol in (3.4.20), is called **mean square error** for predicting Y using X and is denoted by $MSE(X)$, or *MSE* for short.

Thus

$$MSE = \frac{SSE}{(n-2)}$$ (3.4.21)

With this notation the estimate $\hat{\sigma}$ of σ is given by

$$\hat{\sigma} = \sqrt{MSE}$$ (3.4.22)

A convenient formula for calculating SSE using a hand-held calculator is

$$SSE = SSY - \frac{(SXY)^2}{SSX}$$ (3.4.23)

Table 3.4.1 exhibits in detail how the residuals $\hat{e}_i$ enter into the calculation of $\hat{\sigma}$, which is obtained by dividing the sum of the numbers in the last column by $n-2$ and then taking the square root (see 3.4.22).

It can be easily verified that the residuals $\hat{e}_i$ in (3.4.5) must sum to zero. This fact is sometimes used to check the arithmetic involved in the calculation of $\hat{\sigma}$.

All computational formulas for evaluating $\hat{\beta}_0$, $\hat{\beta}_1$, $\hat{\mu}_Y(x)$, $\hat{e}_i$, and $\hat{\sigma}$ are influenced by rounding errors, so it is advisable to carry as many significant digits as possible when performing the required arithmetical operations. The final result may be rounded to the desired number of significant digits.

The calculations required to estimate β_0, β_1, and σ may be conveniently carried out using any standard statistical computing package. We explain the use of a statistical computing package (MINITAB or SAS) for these calculations in Section 3.4 of the laboratory manual.

Terminology

For convenience, Box 3.4.1 summarizes terminology associated with straight line regression, and Box 3.4.2 summarizes the formulas for various parameter estimates of interest.

T A B L E 3.4.1

Sample Item	Observed Y	Observed X	Prediction $\hat{\mu}_Y(x_i) = \hat{\beta}_0 + \hat{\beta}_1 x_i$	Residuals $\hat{e}_i = y_i - \hat{\mu}_Y(x_i)$ $= y_i - \hat{\beta}_0 - \hat{\beta}_1 x_i$	Residuals Squared $\hat{e}_i^2$
1	y_1	x_1	$\hat{\mu}_Y(x_1) = \hat{\beta}_0 + \hat{\beta}_1 x_1$	$\hat{e}_1 = y_1 - \hat{\beta}_0 - \hat{\beta}_1 x_1$	$(y_1 - \hat{\beta}_0 - \hat{\beta}_1 x_1)^2$
2	y_2	x_2	$\hat{\mu}_Y(x_2) = \hat{\beta}_0 + \hat{\beta}_1 x_2$	$\hat{e}_2 = y_2 - \hat{\beta}_0 - \hat{\beta}_1 x_2$	$(y_2 - \hat{\beta}_0 - \hat{\beta}_1 x_2)^2$
$\vdots$	$\vdots$	$\vdots$	$\vdots$	$\vdots$	$\vdots$
n	y_n	x_n	$\hat{\mu}_Y(x_n) = \hat{\beta}_0 + \hat{\beta}_1 x_n$	$\hat{e}_n = y_n - \hat{\beta}_0 - \hat{\beta}_1 x_n$	$(y_n - \hat{\beta}_0 - \hat{\beta}_1 x_n)^2$

B O X **3.4.1**

Population regression function, or simply, the **regression function**:

$$\mu_Y(x) = \beta_0 + \beta_1 x \qquad \text{for } a \le x \le b$$

Sample regression function:

$$\hat{\mu}_Y(x) = \hat{\beta}_0 + \hat{\beta}_1 x$$

Population regression model, or simply, the **regression model**:

$$Y_I = \beta_0 + \beta_1 X_I + E_I \qquad \text{for } I = 1, \ldots, N$$

Sample regression model:

$$y_i = \beta_0 + \beta_1 x_i + e_i \qquad \text{for } i = 1, \ldots, n$$

A randomly chosen Y value from the subpopulation determined by $X = x$:

$$Y(x)$$

Sample prediction function, or simply, **prediction function**:

$$\hat{Y}(x) = \hat{\beta}_0 + \hat{\beta}_1 x$$

Note: $\hat{\mu}_Y(x) = \hat{Y}(x)$.

B O X **3.4.2**

Point Estimates of Various Population Quantities

Suppose a sample $(y_1, x_1), \ldots, (y_n, x_n)$ of size n is obtained from a study population $\{(Y, X)\}$ by simple random sampling or by sampling with preselected X values. Then

$$\bar{x} = \frac{\sum_{i=1}^n x_i}{n} \qquad \bar{y} = \frac{\sum_{i=1}^n y_i}{n}$$

$$SSX = \sum_{i=1}^n (x_i - \bar{x})^2 = \sum_{i=1}^n x_i^2 - \frac{(\sum_{i=1}^n x_i)^2}{n} = \sum_{i=1}^n x_i^2 - n\bar{x}^2$$

$$SSY = \sum_{i=1}^n (y_i - \bar{y})^2 = \sum_{i=1}^n y_i^2 - \frac{(\sum_{i=1}^n y_i)^2}{n} = \sum_{i=1}^n y_i^2 - n\bar{y}^2$$

$$SXY = \sum_{i=1}^n (x_i - \bar{x})(y_i - \bar{y}) = \sum_{i=1}^n x_i y_i - \frac{(\sum_{i=1}^n x_i)(\sum_{i=1}^n y_i)}{n}$$

$$= \sum_{i=1}^n x_i y_i - n\bar{x}\bar{y}$$

$$\hat{\beta}_1 = \frac{SXY}{SSX}$$

$$\hat{\beta}_0 = \bar{y} - \hat{\beta}_1 \bar{x}$$

$$\hat{e}_i = y_i - \hat{\beta}_0 - \hat{\beta}_1 x_i = \text{residual for sample item } i$$

$$SSE = \sum_{i=1}^{n} [y_i - \hat{\mu}_Y(x_i)]^2 = \sum_{i=1}^{n} (y_i - \hat{\beta}_0 - \hat{\beta}_1 x_i)^2 = \sum_{i=1}^{n} \hat{e}_i^2$$

$$= SSY - \frac{(SXY)^2}{SSX}$$

$$MSE = \frac{SSE}{n-2}$$

$$\hat{\sigma} = \sqrt{\frac{\sum_{i=1}^{n} (y_i - \hat{\beta}_0 - \hat{\beta}_1 x_i)^2}{(n-2)}} = \sqrt{\frac{\sum_{i=1}^{n} \hat{e}_i^2}{(n-2)}} = \sqrt{MSE}$$

For $\theta = a_0 \beta_0 + a_1 \beta_1$ the point estimate is

$$\hat{\theta} = a_0 \hat{\beta}_0 + a_1 \hat{\beta}_1$$

Comment

In the sample regression function in (3.4.4), we can substitute $\bar{y} - \hat{\beta}_1 \bar{x}$ for $\hat{\beta}_0$ and write

$$\hat{\mu}_Y(x) = \bar{y} + \hat{\beta}_1 (x - \bar{x}) \tag{3.4.24}$$

Thus $\hat{\mu}_Y(\bar{x}) = \bar{y}$, which demonstrates that the graph of the sample regression function passes through the point $(\bar{x}, \bar{y})$, which is the "center" of the data.

Tasks 3.4.1 and 3.4.2 are intended to provide you with an opportunity to better grasp the concepts discussed so far in straight line regression and also to illustrate the use of the formulas for point estimation of various parameters of interest. The questions in these tasks are posed as word problems and are indicative of the types of questions arising in real applications.

Task   3.4.1

Crystalline forms of certain chemical compounds are used in various electronic devices, and it is often more desirable to have large crystals rather than small ones. Crystals of one particular compound are to be produced by a commercial process, and an investigator wants to examine the relationship between the size of a crystal, as determined by its weight Y in grams, and the number of hours X it takes the crystal to grow to its final size. The following data are from a laboratory study in which 14 crystals of various sizes were obtained by allowing the crystals to grow for different preselected amounts of time. The data are listed in Table 3.4.2 and are also stored in the file **crystal.dat** on the data disk. From the data in Table 3.4.2 we compute

$$\sum_{i=1}^{14} x_i = 210, \qquad \bar{x} = 15 \qquad \sum_{i=1}^{14} y_i = 105.74, \qquad \bar{y} = 7.5529$$

T A B L E **3.4.2**
Crystal Data

Crystal Number	Weight Y (in grams)	Time X (in hours)
1	0.08	2
2	1.12	4
3	4.43	6
4	4.98	8
5	4.92	10
6	7.18	12
7	5.57	14
8	8.40	16
9	8.81	18
10	10.81	20
11	11.16	22
12	10.12	24
13	13.12	26
14	15.04	28

$$\sum_{i=1}^{14}(x_i - \bar{x})^2 = SSX = 910 \qquad \sum_{i=1}^{14}(y_i - \bar{y})^2 = SSY = 244.159$$

$$\sum_{i=1}^{14}(y_i - \bar{y})(x_i - \bar{x}) = SXY = 458.12$$

A plot of the data is displayed in Figure 3.4.1.

Suppose the investigator is reasonably certain that assumptions (A) for straight line regression are (at least approximately) satisfied for x values between 2 hours and 28 hours, i.e.,

$$\mu_Y(x) = \beta_0 + \beta_1 x \qquad \text{for } 2 \le x \le 28$$

She is interested in finding answers to the following questions (our answers appear in italics).

1 Estimate how much the crystals grow per hour on the average.

Because $\mu_Y(x) = \beta_0 + \beta_1 x$ is the regression function of Y on X, the average growth per hour is equal to β_1. The estimate of β_1, calculated using the formula in Box 3.4.2, is

$$\hat{\beta}_1 = \frac{SXY}{SSX} = \frac{458.12}{910} = 0.5034 \qquad \text{(to 4 decimals)}$$

Also from Box 3.4.2 we get

$$\hat{\beta}_0 = \bar{y} - \hat{\beta}_1 \bar{x} = 0.0014 \qquad \text{(to 4 decimals)}$$

FIGURE   3.4.1

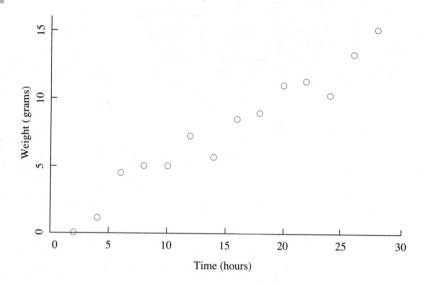

2   If a crystal is allowed to grow for 15 hours, what is its predicted weight?

Here we wish to predict the weight of a single crystal that is allowed to grow for 15 hours. This is considered to be a random observation, Y(15), from the population of all crystals grown for 15 hours. The best predicted value of Y(15) is

$$\hat{Y}(15) = \hat{\beta}_0 + \hat{\beta}_1(15)$$

Using the values $\hat{\beta}_0 = 0.0014$ and $\hat{\beta}_1 = 0.5034$ calculated in (1) above, the best predicted value for the weight of a single crystal at the end of 15 hours is $\hat{Y}(15) = 0.0014 + 0.5034(15) = 7.55$ grams.

3   The crystals are priced depending on the time taken to grow them as well as their actual weight. Crystals that are grown for 8 hours or less are priced at $2 per gram, those that are grown between 8 hours and 16 hours are priced at $10 per gram, and those that are grown for more than 16 hours are priced at $16 per gram. These prices reflect the additional amount of operator intervention necessary to grow crystals for longer periods. Estimate the additional dollars that a crystal will sell for if it is allowed to grow for 24 hours instead of 12 hours.

The weight of a crystal that is grown for 24 hours is estimated to be $\hat{Y}(24) = \hat{\beta}_0 + \hat{\beta}_1(24) = 12.08$ grams, whereas the weight of a crystal grown for 12 hours is estimated to be $\hat{Y}(12) = \hat{\beta}_0 + \hat{\beta}_1(12) = 6.04$ grams. Hence the additional dollars that a crystal grown for 24 hours will fetch compared to a crystal grown for 12 hours, is estimated to be $12.08 \times 16 - 6.04 \times 10 = \$132.88$.

4 An electronic components manufacturer places an order for 100 crystals weighing 12 grams each with a tolerance of ±0.5 gram, i.e., weighing between 11.5 and 12.5 grams. How long should the crystals be allowed to grow? If 100 crystals are grown for this amount of time, how many crystals may be expected to meet the specifications?

Suppose crystals that are allowed to grow for x_0 hours have an average size equal to 12 grams. Then

$$\mu_Y(x_0) = \beta_0 + \beta_1 x_0 = 12$$

which implies that

$$x_0 = \frac{12 - \beta_0}{\beta_1}$$

The estimated value of x_0 is

$$\hat{x}_0 = \frac{12 - \hat{\beta}_0}{\hat{\beta}_1} = 23.84 \ hours$$

Using the formula for $\hat{\sigma}$ in Box 3.4.2, the subpopulation standard deviation is estimated to be

$$\hat{\sigma} = \sqrt{\frac{1}{(n-2)}\left[SSY - \frac{(SXY)^2}{SSX}\right]}$$

$$= \sqrt{\frac{1}{12}\left[244.159 - \frac{(458.12)^2}{910}\right]} = 1.062 \ grams$$

Hence the proportion of crystals, which are grown for 23.84 hours, whose weights lie in the range from 11.5 grams to 12.5 grams, is approximately equal to the proportion of values in a Gaussian population, with mean equal to 12 and standard deviation equal to 1.062, that lie between 11.5 and 12.5. This is equivalent to the proportion of values in a standard Gaussian population that are between

$$\frac{11.5 - 12.0}{1.062} \quad and \quad \frac{12.5 - 12.0}{1.062}$$

i.e., between −0.471 and +0.471. Using Table T-1 in Appendix T, we calculate this proportion to be 0.362. If 100 crystals are grown for 23.84 hours, then we expect $100 \times 0.362 = 36$ (rounded to the nearest integer) crystals to meet the required specifications. This result is approximate because the calculations are based on estimated values of population parameters since their true values are unknown.

Task 3.4.2 further illustrates how regression techniques can be useful in real problems.

Task   3.4.2

An investigator wants to evaluate the performance of a new laboratory method for analyzing the concentration of arsenic (As) in water samples that is much cheaper than the existing method. If the new method is proven to be scientifically acceptable, then it will be adopted by environmental research groups for monitoring the quantity of As in industrial waste water. To investigate the relationship between measured concentrations of As (Y) and actual concentrations (X), the investigator makes several water samples containing *known* (preselected) amounts of As. These water samples are analyzed by a laboratory technician (who is unaware of the actual amounts of As in these solutions) using the new method of analysis. The concentrations are reported in micrograms/milliliter (μg/ml). The data are exhibited in Table 3.4.3 and are also stored in the file **arsenic.dat** on the data disk.

Suppose that, based on experience, the investigator feels assumptions (A) for straight line regression hold at least approximately. In particular the population regression function is

$$\mu_Y(x) = \beta_0 + \beta_1 x$$

and the subpopulation standard deviations for Y are all the same, each equal to σ.

To start an analysis, *the first thing you should always do is to obtain a plot of Y against X.* This plot is shown in Figure 3.4.2. It appears from this plot that the assumption of a straight line model is quite reasonable. Next we compute some basic sums, sums of squares, and sums of crossproducts, the ingredients in the formulas for obtaining estimates of population parameters. We get

$$\sum_{i=1}^{32} x_i = 112 \qquad \bar{x} = 3.5 \qquad \sum_{i=1}^{32} y_i = 113.97 \qquad \bar{y} = 3.5616$$

$$SSX = 168 \qquad SSY = 164.95 \qquad SXY = 165.935$$

From these we compute

$$\hat{\beta}_1 = \frac{SXY}{SSX} = \frac{165.935}{168} = 0.9877$$

$$\hat{\beta}_0 = \bar{y} - \hat{\beta}_1 \bar{x} = 3.5616 - 0.9877(3.5) = 0.1046$$

$$\hat{\sigma} = \sqrt{\frac{1}{(n-2)}\left[SSY - \frac{(SXY)^2}{SSX} \right]} = \sqrt{\frac{1}{30}\left[164.95 - \frac{(165.935)^2}{168} \right]}$$

$$= \sqrt{\frac{1}{30}[164.95 - 163.89538]} = \sqrt{\frac{1.0546}{30}} = 0.1875$$

You should verify these calculations.

Suppose the investigator is interested in obtaining answers to the following questions.

TABLE 3.4.3
Arsenic Data

Sample Item Number	Measured Concentration Y (in μg/ml)	True Concentration X (in μg/ml)
1	0.17	0
2	0.25	0
3	0.01	0
4	0.12	0
5	1.25	1
6	0.86	1
7	1.25	1
8	1.10	1
9	2.01	2
10	2.03	2
11	2.14	2
12	1.74	2
13	3.18	3
14	2.99	3
15	3.23	3
16	3.37	3
17	3.91	4
18	3.90	4
19	3.61	4
20	4.27	4
21	4.88	5
22	5.33	5
23	4.96	5
24	4.98	5
25	6.09	6
26	6.17	6
27	6.07	6
28	5.97	6
29	6.67	7
30	7.02	7
31	7.14	7
32	7.30	7

1 On the average, does the chemical analysis correctly report the absence of As when this is indeed the case? That is, is $\mu_Y(x) = 0$ when $x = 0$? Is $\beta_0 = 0$?

In individual instances, due to the presence of various kinds of disturbances or errors, the analysis may report a nonzero value of As even when the actual concentration of As is zero. The measured concentration of As for water samples containing no As (i.e., x = 0) is equal to β_0 on the average. Thus the investigator wants to know whether or not β_0 is indeed zero.

F I G U R E   *3.4.2*

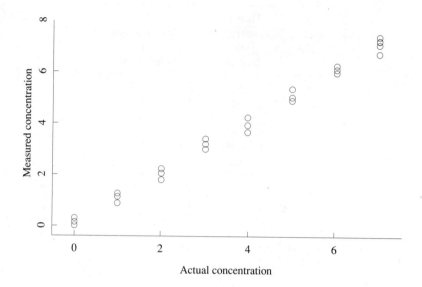

From the preceding computations, the least squares estimate $\hat{\beta}_0$ of β_0 is 0.1046. Before the investigator can decide whether or not β_0 may be regarded as equal to zero for practical purposes, he needs to know how good this estimate is. This is discussed in the section on confidence intervals.

2   On the average, does the chemical analysis result in an underestimate or an overestimate of the true As concentration, or does it provide an *unbiased* estimate of the true As concentration?

In individual instances, the presence of various kinds of errors leads to a measured value that is higher or lower than the true As concentration, and only very rarely will the measured value be exactly equal to the true value. However, if the average of the subpopulation of measured Y values equals the true concentration X for each possible value of X, then the analysis is said to be unbiased. Thus, if the analysis is unbiased, then the regression function of Y on X must be $\mu_Y(x) = x$; i.e., $\beta_0 = 0$ and $\beta_1 = 1$. Here we examine whether or not the value of β_1 is equal to 1.

From the preceding computations, the least squares estimate $\hat{\beta}_1$ of β_1 is equal to 0.9877, which is close to 1.0. Before the investigator is able to decide whether or not β_1 can be regarded as equal to one for practical purposes, he needs to know how good this estimate is. This is discussed in the section on confidence intervals.

3   Suppose, based on the calculations in (1) and (2), the investigator is fairly confident that the new method of chemical analysis for As is *unbiased*; i.e., the regression function of Y on X is $\mu_Y(x) = x$. However, in order for the method to be adopted by water quality monitoring agencies, 99% or more of the reported

concentrations must be accurate to within 1.0 μg/ml. Based on the data at hand, can he conclude that this is indeed the case?

Assuming that $\mu_Y(x) = x$ and that assumptions (A) are valid, 99% of the measured concentrations corresponding to a true concentration x will be between $\mu_Y(x) - 2.576\sigma$ and $\mu_Y(x) + 2.576\sigma$. Hence we want to know whether 2.576σ is less than 1.0. The estimate of σ calculated earlier is 0.1875, and so the estimated value of 2.576σ is equal to 0.483. This is well within the acceptable upper limit of 1.0. However, the investigator realizes that the estimate of σ is itself subject to sampling errors, and so he wants to know if the estimate is sufficiently accurate for decision-making purposes. A confidence interval for σ would be useful for this purpose. This is discussed in the section on confidence intervals.

You should bear in mind that in most instances X and Y are physical quantities that have certain units of measurement associated with them. In this connection the investigator may want to find out what happens if he decides to change the system of units. For instance, the original measurements may have been made in terms of pounds or miles, and it may be necessary to transform the measurements so they are expressed in terms of grams or meters. We discuss this next.

Effect of Change of Units on the Parameters and Their Estimates

The values of the population parameters such as β_0, β_1, and σ depend on the system of units in which Y and X are measured. If X is a measure of distance, then the units of X may be miles, kilometers, inches, feet, millimeters, etc. If X is a measure of weight, then the units of X may be tons, pounds, kilograms, etc. Similarly if Y is a measure of temperature, then the units of Y may be degrees Fahrenheit, degrees Celsius, or degrees Kelvin. Therefore we want to know what effect the choice of units has on the values of population parameters and their point estimates.

Let X_I, Y_I be the values of X and Y for the Ith population item measured using one system of units, and let X_I^*, Y_I^* denote the X and Y values of the same item measured using a second system of units. In most practical applications, different systems of units are *linearly* related. For instance,

$$\text{degrees Fahrenheit} = 32 + \tfrac{9}{5} \text{ (degrees Celsius)}$$

and

$$\text{kilometers} = 1000 \text{ meters}$$

With this in mind, suppose X_I^* and Y_I^* are defined in terms of X_I and Y_I as follows:

$$X_I^* = a + b X_I$$
$$Y_I^* = c + d Y_I$$

Suppose that the population regression function is

$$\mu_Y(x) = \beta_0 + \beta_1 x$$

when X and Y are measured using the first system of units. Also suppose the regression function is

$$\mu_{Y^*}(x^*) = \beta_0^* + \beta_1^* x^*$$

when X^* and Y^* are measured using the second system of units. Then it can be proved mathematically that

$$\beta_1^* = \frac{d}{b}\beta_1$$

and

$$\beta_0^* = c + \frac{d}{b}(b\beta_0 - a\beta_1)$$

Furthermore, if σ is the subpopulation standard deviation when the first system of units is used and σ^* denotes the subpopulation standard deviation when the second system of units is used, then

$$\sigma^* = |d|\sigma$$

The estimated values of the parameters are related in the same manner, viz.,

$$\hat{\beta}_1^* = \frac{d}{b}\hat{\beta}_1$$

$$\hat{\beta}_0^* = c + \frac{d}{b}(b\hat{\beta}_0 - a\hat{\beta}_1)$$

and

$$\hat{\sigma}^* = |d|\hat{\sigma}$$

We use the following example to demonstrate these relationships.

E X A M P L E 3.4.1

The output Y, in kilograms per hour, of a chemical process is related to the temperature X, in degrees Celsius, at which the process is run, and the population regression function has the form $\mu_Y(x) = \beta_0 + \beta_1 x$. The data in Table 3.4.4 were obtained from a pilot plant experiment using preselected X values and are also stored in the file **process1.dat** on the data disk. The estimates of β_0, β_1, and σ are calculated to be

$$\hat{\beta}_0 = -4.959 \qquad \hat{\beta}_1 = 0.0499 \qquad \hat{\sigma} = 0.4095$$

Now suppose the process output is measured in grams per second instead of kilograms per hour, and the process temperature is measured in degrees Fahrenheit instead of degrees Celsius. This gives us

$$Y_I^* = \frac{1000}{3600}Y_I \qquad \text{and} \qquad X_I^* = 32 + \frac{9}{5}X_I \qquad \text{for } I = 1, \dots, N \qquad \text{(3.4.25)}$$

Let the sample data expressed in the new system of units be denoted by y_i^* and x_i^*, respectively, for $i = 1, \dots, n$. Then using (3.4.25) we have the sample data in Table 3.4.5 in terms of the new system of units, and these data are stored in the file **process2.dat** on the data disk. Using the data in Table 3.4.5 we obtain the estimates of the slope and the intercept relative to the new system of units as

$$\hat{\beta}_0^* = -1.624 \qquad \hat{\beta}_1^* = 0.0077 \qquad \hat{\sigma}^* = 0.1137$$

Note that we have $X_I^* = a + bX_I$ and $Y_I^* = c + dY_I$, where $a = 32$, $b = 9/5 = 1.8$, $c = 0.0$, and $d = 1000/3600 = 0.2778$. We may now verify that

$$\frac{d}{b}\hat{\beta}_1 = \frac{0.2778}{1.8}(0.0499) = 0.0077 = \hat{\beta}_1^*$$

$$c + \frac{d}{b}(b\hat{\beta}_0 - a\hat{\beta}_1) = 0 + \frac{0.2778}{1.8}[(1.8)(-4.959) - (32)(0.0499)] = -1.624 = \hat{\beta}_0^*$$

$$|d|\hat{\sigma} = (0.2778)(0.4095) = 0.1137 = \hat{\sigma}^*$$

T A B L E 3.4.4

Chemical Process Data for Example 3.4.1.

Run Number	Process Output (Y) (in kg/hour)	Process Temperature (X) (°C)
1	10.0	300
2	10.5	310
3	11.5	320
4	11.5	330
5	11.9	340
6	12.3	350
7	12.8	360
8	13.1	370
9	14.4	380
10	13.9	390
11	15.7	400

T A B L E 3.4.5

Transformed Data for Table 3.4.4.

Run Number	Process Output (Y*) (g/s)	Process Temperature (X*) (°F)
1	2.77778	572
2	2.91667	590
3	3.19444	608
4	3.19444	626
5	3.30556	644
6	3.41667	662
7	3.55556	680
8	3.63889	698
9	4.00000	716
10	3.86111	734
11	4.36111	752

Thus it is a simple matter to switch from one system of units to another as long as the two measurement systems are *linearly related.* ■

Point estimates for β_0, β_1, linear functions of β_0 and β_1, $\mu_Y(x)$, $Y(x)$, and σ, discussed in this section, are best estimates if assumptions (A) or (B) are satisfied. Valid estimates of μ_Y and σ_Y (respectively, μ_X and σ_X) are obtained using (1.6.1) and (1.6.2) provided the data are obtained by simple random sampling. These estimates are best estimates if assumptions (B) are satisfied.

Before applying any of the inference procedures discussed in this book, we recommend that the investigator carefully examine the data and, combined with his or her own prior experience and knowledge, make a judgment as to whether or not the assumptions underlying the inference procedures are at least approximately met. Statistical procedures are not meant to take the place of good subject matter judgment but to assist in this judgment.

In the next section we present some simple graphical procedures for examining the validity of some of the assumptions underlying regression. But first we explain in the following conversation the difference between fitting straight lines to data using the method of least squares and fitting straight lines **by eye**.

Conversation 3.4

Investigator: Good morning! Do you have time to talk to me now?

Statistician: Certainly. What's on your mind?

Investigator: A scientist I work with has a large data set that includes two variables Y and X. I plotted the data and drew a line through it by eye. It seems to him and to me that the line is a good summarization of the data. The line goes through $\bar{y} = 11.7$ and $\bar{x} = 11.7$, the means of the Y values and X values in the data set. The slope of the line is approximately (as close as we can estimate it by eye) 1.0; i.e., $\hat{\beta}_1^{eye} = 1.0$. The plot of the data and the line are shown in Figure 3.4.3. Figure 3.4.4 shows the plotted data with the line we fitted by eye (solid line) and the least squares line (dashed line). The least squares line has slope $\hat{\beta}_1 = 0.734$, which is considerably different from the slope $\hat{\beta}_1^{eye} = 1$ of the visually fitted line. Isn't this unusual?

Statistician: No, it's not unusual. If the data were obtained from a two-variable Gaussian population by simple random sampling, you would expect the data to be roughly elliptical in shape. The ellipse, the visually fitted line, and the least squares line are shown in Figure 3.4.5. The solid line that you drew by eye is a line such that the data are symmetrical around it: i.e., a line of symmetry. However, the least squares line (the dashed line) is the line that minimizes the sum of squares of *vertical* distances from the line to each point, and *the least squares line will always have a slope that is closer to zero than the line of symmetry (unless, of course, both slopes are zero).*

Both lines go through the point $(\bar{x}, \bar{y})$, where $\bar{x}$ and $\bar{y}$ are the means of the X and Y values in the data set.

Investigator: I see. But it seems to me that the line of symmetry summarizes the data better than the least squares line.

Statistician: I agree that the line of symmetry is more appealing to the eye. However, if assumptions (A) or (B) are satisfied, the least squares line is the *best line for predicting Y using X*. Thus one must be careful in using a visually fitted line.

Investigator: I think I understand. Perhaps I will come to see you again next week.

F I G U R E **3.4.3**

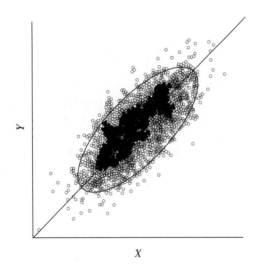

F I G U R E **3.4.4**

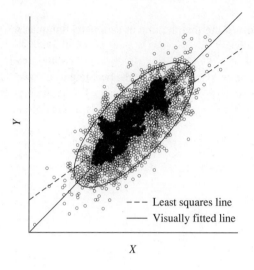

Least squares line
Visually fitted line

F I G U R E **3.4.5**

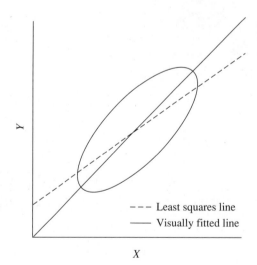

Least squares line
Visually fitted line

Problems   3.4

3.4.1   Consider the crystal data in Task 3.4.1. A different laboratory collects data on crystal growth at 14 preselected times. The data are given in Table 3.4.6 and are also in the file **Table346.dat** on the data disk. Assumptions (A) are presumed to be valid, and the data are obtained by sampling with preselected X values. Calculate the least squares estimates of β_0, β_1, and σ.

3.4.2   Using the results of Problem 3.4.1, answer questions (1)–(3) of Task 3.4.1.

3.4.3   In Problem 3.4.1, what is the estimate of the average weight of crystals that have grown for 19 hours? For 25 hours? For 40 hours?

3.4.4   Twenty-five cars are selected, using simple random sampling, from the car data in the file **car.dat** on the data disk (see also Table D-1 in Appendix D). The first-year maintenance cost Y for these 25 cars and the number of miles X they were driven during the first year after purchase were recorded. The following quantities were computed using these sample data: $\bar{x} = 11{,}364$; $\bar{y} = 532.44$; $SSX = 224{,}617{,}600$; $SSY = 279{,}764$; $SXY = 7{,}391{,}396$.

a   Find $\hat{\beta}_0$, $\hat{\beta}_1$, and $\hat{\mu}_Y(x)$.

b   A prospective new car buyer wants to purchase a car similar to those in the population of Table D-1 and plans to drive it 13,000 miles during the first year. What population quantity is the buyer interested in to predict the first-year maintenance cost for this car, $Y(13{,}000)$ or $\mu_Y(13{,}000)$? Explain. Obtain a point estimate for this quantity.

T A B L E   3.4.6

Crystal Number	Weight Y (in grams)	Time X (in hours)
1	0.10	2
2	1.01	4
3	3.89	6
4	5.14	8
5	5.19	10
6	6.89	12
7	5.29	14
8	8.70	16
9	9.42	18
10	11.38	20
11	11.38	22
12	11.73	24
13	12.95	26
14	15.10	28

c   What population quantity should buyers be interested in if they want to know the average first-year maintenance cost of all cars to be driven 16,000 miles next year? Explain.

3.4.5   Consider the simple random sample of size 26 for the grades example given in Table 3.2.2. These data are in the file **grades26.dat** on the data disk. For these data compute the following quantities: $\bar{x}$; $\bar{y}$; SXY; SSX; SSY; $\hat{\beta}_0$; $\hat{\beta}_1$; SSE; and $\hat{\sigma}$.

3.4.6   In Problem 3.4.5, compute $\hat{\mu}_Y(10)$. Using $\hat{\mu}_Y(10)$ in place of $\mu_Y(10)$ and $\hat{\sigma}$ in place of $\sigma_Y(10)$, answer question (4) of (3.2.1).

3.4.7   Sometimes it is desirable to work with the transformed data

$$x_i^* = x_i - c \qquad y_i^* = y_i - d$$

for suitably chosen constants c and d rather than the original data when the computations are done using a hand-held calculator. Many software packages also transform the data in this manner to combat rounding errors. The quantities SSX, SSY, and SXY can be shown to be the same whether the calculations are done using the untransformed data or the transformed data. Thus the estimates of the slope parameter β_1 and the subpopulation standard deviation σ, calculated using the transformed data, can be shown to be the same as those calculated with the untransformed data. Once $\hat{\beta}_1$ has been calculated, $\hat{\beta}_0$ can be obtained using (3.4.9). To become familiar with this transformation do the following.

a   Compute y_i^* and x_i^* for the data of Problem 3.4.1 using $c = 14$ and $d = 7$.

b   Using the transformed data in (a), compute SSX, SSY, and SXY. Using these compute $\hat{\beta}_1^*$ and $\hat{\sigma}^*$. How do these compare with $\hat{\beta}_1$ and $\hat{\sigma}$ obtained in Problem 3.4.1?

3.5
Checking Assumptions

Inference procedures for regression analyses are strictly valid only when the model assumptions on which the procedures are based are satisfied. However, models are approximations to reality, and so model assumptions will never hold exactly. Nevertheless, if a model is a reasonable approximation of reality, then inferences based on the model may be adequate for real applications.

In an applied problem, an investigator usually knows that some of the assumptions are satisfied, but she cannot be sure if others are. For example, an investigator will generally know whether (sample) assumptions (A) or (B) are satisfied, but she may not know if the (population) assumptions are satisfied. In this section we discuss some procedures for examining the validity of the population assumptions.

For convenience we reproduce here the population regression model given in (3.3.2):

$$Y_I = \beta_0 + \beta_1 X_I + E_I$$

If assumptions (A) or (B) are satisfied, then the population of all E_I is a Gaussian population with mean equal to *zero* and standard deviation equal to σ (i.e., $\sigma_{Y|X}$).

Hence we would like to examine the E_I to determine if they indeed form a Gaussian population with zero mean. However the E_I are population values and are unavailable, so we consider examining the n values of E_I, which we denote by $e_1, \ldots, e_n$, corresponding to the n sample items. However, even these e_i's are not observable, but we can estimate them. An estimate of the e_i corresponding to the ith sample item is the residual $\hat{e}_i$, defined by (see (3.4.5))

$$\hat{e}_i = y_i - \hat{\mu}_Y(x_i) = y_i - \hat{\beta}_0 - \hat{\beta}_1 x_i \qquad \text{for } i = 1, \ldots, n \qquad \textbf{(3.5.1)}$$

These residuals are used to help decide whether the E_I form a Gaussian population with mean zero. If we decide that the E_I do not constitute such a population, then we can conclude that one or more of assumptions (A) or (B) are violated.

We begin by discussing some graphical procedures which—by examining the sample data, the residuals $\hat{e}_i$, and other related quantities—will help us in detecting major violations in model assumptions.

Scatter Plot of *Y* Against *X*

You should routinely plot the sample Y values against the corresponding X values and study this plot carefully before using any of the inference procedures for straight line regression. Failure of the population assumption that $\mu_Y(x)$ is of the form $\beta_0 + \beta_1 x$ is often revealed by such a plot because this assumption states that the graph of the regression function of Y on X is a straight line. If this assumption is correct, then the plotted points should all lie roughly along a straight line.

Figure 3.5.1 is a plot of the sample Y values against the corresponding X values in the crystal data of Task 3.4.1. This plot seems to support the assumption that the regression function of Y on X is of the form $\mu_Y(x) = \beta_0 + \beta_1 x$. On the other hand, if the plots were as shown in Figures 3.5.2–3.5.4, this would tend to suggest that the regression function of Y on X is not of the form $\mu_Y(x) = \beta_0 + \beta_1 x$. The plot in Figure 3.5.4 is particularly interesting because, if we ignore the two points indicated by the symbols +, all remaining points would tend to support the assumption of a straight line regression function. This is not an uncommon situation in practice. Such points are often referred to as **outliers**. We discuss outliers, and how they should be dealt with, in greater detail in Chapter 5.

Standardized Residuals

We have stated that the residuals $\hat{e}_i$ are useful for checking the validity of the model assumptions. Numerous graphical and numerical techniques for checking assumptions using residuals can be found in the regression literature [1], [2]. Most of these

F I G U R E **3.5.1**

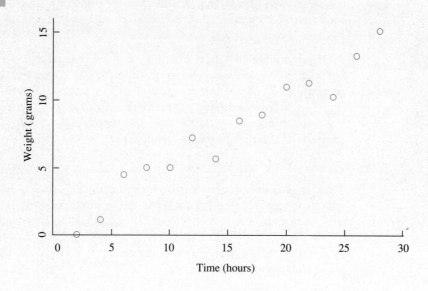

F I G U R E **3.5.2**

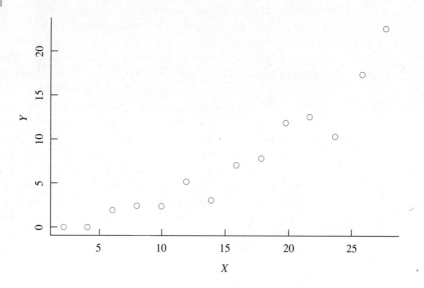

F I G U R E 3.5.3

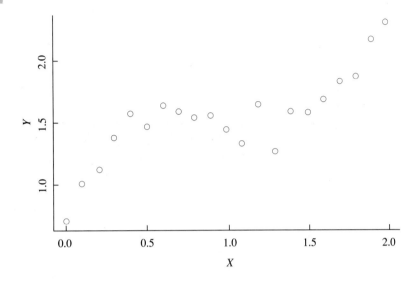

F I G U R E 3.5.4

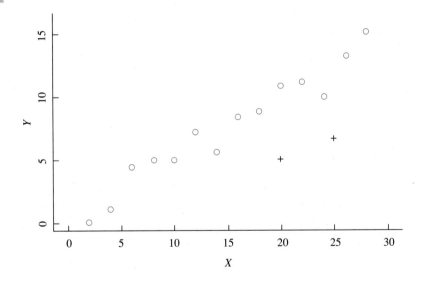

methods *standardize* the residuals $\hat{e}_i$ before using them because the residuals $\hat{e}_i$, corresponding to observations with x values that are far from the center of the range of x values, tend to vary more from sample to sample than those corresponding to observations with x values that are closer to the center. Consequently, they are not directly comparable with one another. The following explanation may help you understand the reason for this.

Recall that the estimated regression line must pass through the 'center' $(\bar{x}, \bar{y})$ of the sample data. A slight change in the estimated value of $\hat{\beta}_1$ will result in a greater change in the value of $\hat{e}_i$ for a point whose x value is far away from the middle (i.e., far from $\bar{x}$) than for a point that is closer to $\bar{x}$ (see Figure 3.5.5). You can think of the line as a see-saw with the point $(\bar{x}, \bar{y})$ (which itself varies from one sample to another) serving as the pivot. Points on the see-saw closer to the pivot move through a smaller vertical distance than points far from the pivot.

F I G U R E   **3.5.5**

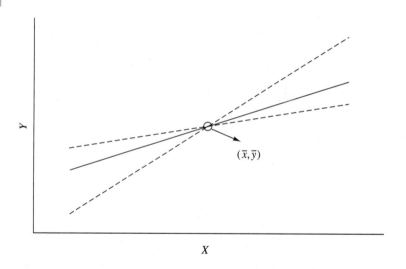

The residuals $\hat{e}_i$ may be made comparable with each other by standardizing them appropriately. It can be shown mathematically that the correct standardization procedure is to divide $\hat{e}_i$ by $\hat{\sigma}\sqrt{1 - h_{i,i}}$, for $i = 1, \ldots, n$, where

$$h_{i,i} = \frac{1}{n} + \frac{(x_i - \bar{x})^2}{SSX} \qquad (3.5.2)$$

and *SSX* is in (3.4.14). The quantities $h_{i,i}$ are usually referred to as **hat values**. More is said about them in Chapter 5.

We are thus led to define the **standardized residual** r_i for observation i as

$$r_i = \frac{\hat{e}_i}{\hat{\sigma}\sqrt{1 - h_{i,i}}} \qquad (3.5.3)$$

If assumptions (A) or (B) are satisfied, then the standardized residuals $r_1, \ldots, r_n$ are (approximately) equivalent to a simple random sample of n observations from a Gaussian population with zero mean and unit standard deviation. **(3.5.4)**

This fact may be used to check the validity of some of the model assumptions.

Plotting Standardized Residuals Against Sample X Values

In some instances a plot of the standardized residuals r_i against x_i is more revealing than a plot of y_i against x_i. When assumptions (A) or (B) are satisfied, the points on the plot of r_i against x_i should be scattered in a random fashion about the horizontal line through the origin (see the dashed line in Figure 3.5.6 for an example), showing no obvious trends or other patterns. If this is found not to be the case, then one or more of these assumptions is likely to be false.

Residual plots corresponding to the scatter plots in Figures 3.5.1–3.5.4 are given in Figures 3.5.6–3.5.9, respectively. Note how the departures from linearity in the plots of Figures 3.5.2 and 3.5.3 appear more prominently in the corresponding residual plots in Figures 3.5.7 and 3.5.8. Also note that the two outliers in the plot shown in Figure 3.5.4 are easily spotted in the residual plot of Figure 3.5.9.

F I G U R E **3.5.6**

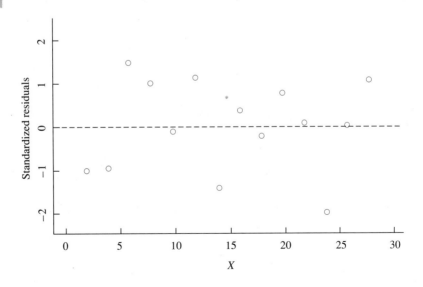

F I G U R E   3.5.7

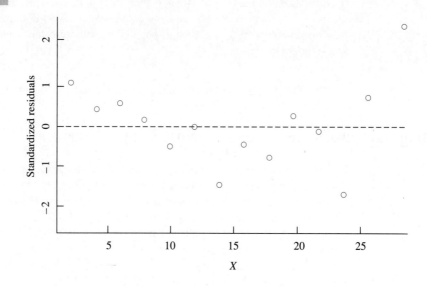

F I G U R E   3.5.8

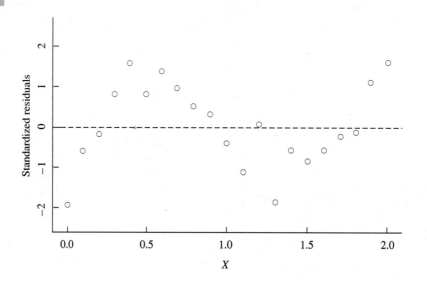

F I G U R E   3.5.9

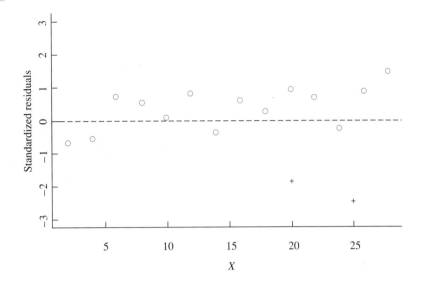

Plotting Standardized Residuals Against Fitted Values $\hat{\mu}_Y(x_i)$

> The estimated subpopulation means $\hat{\mu}_Y(x_i)$ are often referred to as **fitted values** or **predicted values** corresponding to the X values, $x_1, \ldots, x_n$, in the sample.

The fitted values can be shown to be *unrelated* to (independent of) the standardized residuals r_i when assumptions (A) or (B) are satisfied. Hence the points in the plot of r_i versus $\hat{\mu}_Y(x_i)$ should be randomly scattered about the horizontal line through the origin (see the dashed line in Figure 3.5.10 for an example) with no particular pattern. Any systematic pattern observed would indicate that one or more of assumptions (A) or (B) may fail to hold.

In some instances where a pattern is observed instead of a random scatter, it may be possible to diagnose the cause of the observed pattern. In Figures 3.5.10–3.5.15 we show hypothetical plots of standardized residuals versus fitted values for illustration. In each plot the horizontal axis represents the fitted values, and the vertical axis represents standardized residuals.

Figure 3.5.10 is typical of a plot we expect when there are no violations of assumptions. Figures 3.5.11, 3.5.12, and 3.5.13 indicate a possible violation of the assumption of *homogeneity of standard deviations*; i.e., they suggest that the standard deviations of the subpopulations of Y values, determined by the predictor variable X, may not all be the same. Figure 3.5.11 suggests that the standard deviations of the Y values in the various subpopulations increase with increasing values of the subpopulation means, and Figure 3.5.12 indicates that the standard deviations of the Y values in the various subpopulations decrease with increasing values of the

F I G U R E 3.5.10

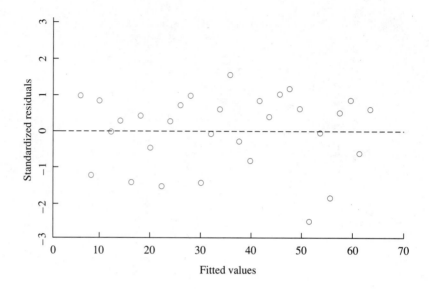

F I G U R E 3.5.11

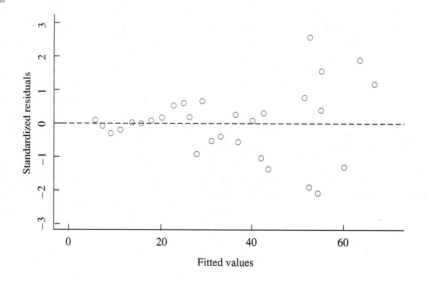

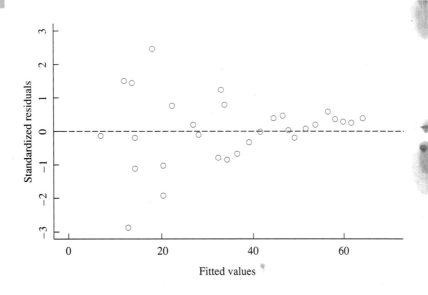

FIGURE 3.5.12

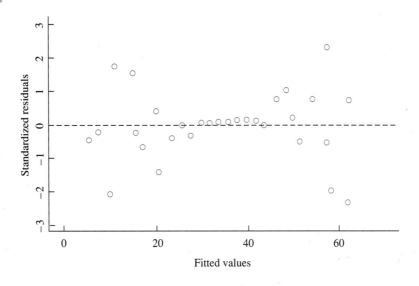

FIGURE 3.5.13

F I G U R E   3.5.14

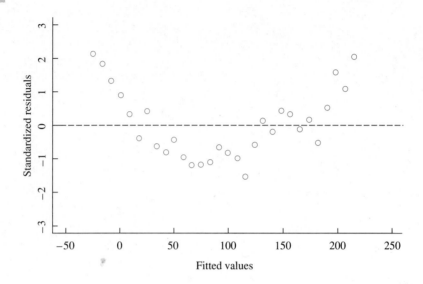

F I G U R E   3.5.15

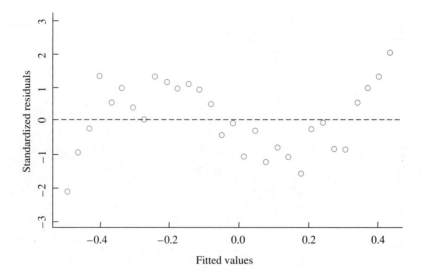

subpopulation means. Figure 3.5.13, on the other hand, suggests that the standard deviations are larger for subpopulations with small means as well as those with large means, whereas the standard deviations are smaller for subpopulations with means near the middle of the possible range of values.

Figures 3.5.14 and 3.5.15 show plots of standardized residuals r_i versus fitted values $\hat{\mu}_Y(x_i)$, which indicate **lack-of-fit** (i.e., they indicate that the graph of the regression function of Y on X is not a straight line). A plot of y_i against x_i, or a plot of r_i against x_i, should also reveal this. When lack of fit is indicated by one of these plots, the investigator may want to formulate an alternate candidate for the regression function, i.e., a quadratic function, a cubic function, a logarithmic function, an exponential function, etc.

Gaussian Rankit-Plot

The **Gaussian rankit-plot** is a graphical tool for checking whether a given set of numbers appears to be a simple random sample from a Gaussian population. Although we primarily use this procedure to assess whether or not the standardized residuals from a regression analysis appear to be a simple random sample from a Gaussian population, the procedure is applicable to any set of numbers. Hereafter, unless otherwise stated, when we say **rankit-plot** we mean a **Gaussian rankit-plot**. The procedure is as follows.

To obtain a rankit-plot of a given set of numbers, say $y_1, \ldots, y_n$, we must first arrange these numbers in increasing order. The ordered data values are denoted by $y_{(1)}, \ldots, y_{(n)}$, with $y_{(1)}$ denoting the smallest of the y_i and $y_{(n)}$ the largest. A rankit-plot of the data values $y_1, \ldots, y_n$ is a plot of the ordered data values $y_{(1)}, \ldots, y_{(n)}$ against the quantities $z_1^{(n)}, \ldots, z_n^{(n)}$, called **Gaussian scores** (or **normal scores**, or **rankits**), which may be thought of as a typical sample of size n (arranged in increasing order) from a *standard* Gaussian population (a Gaussian population with mean zero and standard deviation one). If $y_1, \ldots, y_n$ is a simple random sample from a Gaussian population, then the plot of $y_{(i)}$ against $z_i^{(n)}$ should produce points that more or less all lie on a straight line. If, in addition, the y_i come from a Gaussian population with zero mean and unit standard deviation, then this line should more or less be a line through the origin with unit slope.

Figure 3.5.16 shows the rankit-plot of the data from a simple random sample of size 25 from a Gaussian population with mean 2 and standard deviation 3, while Figure 3.5.17 shows the rankit-plot for a simple random sample of size 30 from a *standard* Gaussian population. In contrast, Figure 3.5.18 is the rankit-plot for a simple random sample of size 28 from a population that is not Gaussian.

Notice how the points in Figure 3.5.16 all lie approximately on a straight line (the dashed line), and the points in Figure 3.5.17 all lie approximately on a straight line through the origin with unit slope (dotted line), but the points in Figure 3.5.18 clearly show a systematic departure from a straight line pattern.

F I G U R E   3.5.16

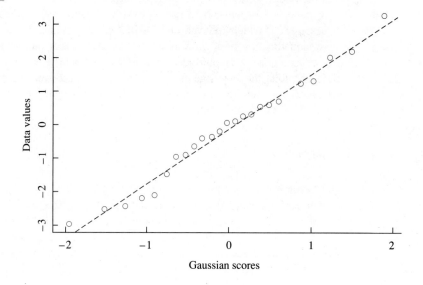

F I G U R E   3.5.17

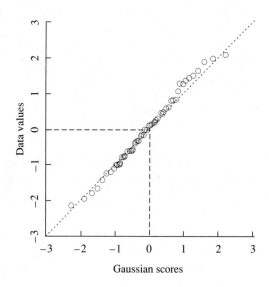

F I G U R E 3.5.18

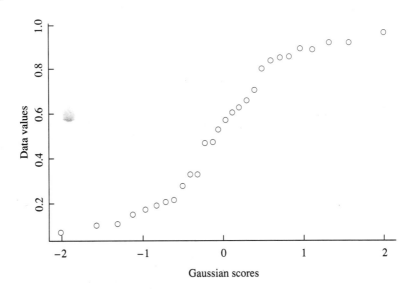

As stated in (3.5.4), when assumptions (A) or (B) are satisfied, the standardized residuals $r_1, \ldots, r_n$ are (approximately) equivalent to a simple random sample of n observations from a standard Gaussian population. So a Gaussian rankit-plot can be used to help determine if assumptions (A) or (B) appear to be violated. We discuss this in some detail, but first we discuss Gaussian scores.

Gaussian Scores (or Normal Scores, or Rankits)

We now explain the **Gaussian scores** $z_i^{(n)}$. Suppose we repeatedly obtain simple random samples of size n from a Gaussian population with mean zero and unit standard deviation (from the population $\{Z\}$, say) and order the sample values from smallest to largest. Suppose the first sample yields the ordered observations

$$z_{1,1} \leq \cdots \leq z_{1,n}$$

and the second sample yields

$$z_{2,1} \leq \cdots \leq z_{2,n}$$

and so forth. If we repeat the process a large number of times, say m times, (i.e., m samples of size n) we will obtain m sets of ordered samples of size n each, which may be organized as in (3.5.5).

$$
\begin{array}{ccccccccc}
z_{1,1} & < & \cdots & < & z_{1,k} & < & \cdots & < & z_{1,n} \\
z_{2,1} & < & \cdots & < & z_{2,k} & < & \cdots & < & z_{2,n} \\
\vdots & < & \vdots & < & \vdots & < & \vdots & < & \vdots \\
z_{j,1} & < & \cdots & < & z_{j,k} & < & \cdots & < & z_{j,n} \\
\vdots & < & \vdots & < & \vdots & < & \vdots & < & \vdots \\
z_{m,1} & < & \cdots & < & z_{m,k} & < & \cdots & < & z_{m,n}
\end{array}
$$

Means	$z_1^{(n)}$	$<$	$\cdots$	$<$	$z_k^{(n)}$	$<$	$\cdots$	$<$	$z_n^{(n)}$

(3.5.5)

Here $z_1^{(n)}$ is the mean of the *smallest* values obtained in each of these m simple random samples of size n, and $z_2^{(n)}$ is the mean of the 2nd smallest values obtained in each of these m random samples of size n, etc. In general, $z_k^{(n)}$ refers to the average of the kth smallest values in simple random samples of size n from a standard Gaussian population; i.e.,

$$
z_k^{(n)} = \frac{1}{m} \sum_{j=1}^{m} z_{j,k}
$$

where $z_{j,k}$ are as in (3.5.5). The values $\{z_k^{(n)}\}$ are called Gaussian scores or normal scores or rankits, and they can be obtained from various computer programs. In Section 3.5 of the MINITAB and SAS laboratory manuals we show how to use computer commands to obtain the numbers

$$
z_1^{(n)} < z_2^{(n)} < \cdots < z_n^{(n)}
$$
(3.5.6)

Rankit-Plot of Standardized Residuals

Recall from (3.5.4) that

if assumptions (A) or (B) are satisfied, then the standardized residuals r_i are (approximately) equivalent to a simple random sample from a standard Gaussian population. This can be checked graphically by examining a rankit-plot of the ordered standardized residuals, which should be (approximately) a straight line through the origin with slope equal to 1 when assumptions (A) or (B) hold.

To obtain a rankit-plot of the standardized residuals, we first order the standardized residuals r_i and denote the ordered values by $r_{(1)}, r_{(2)}, \cdots, r_{(n)}$ so that we have

$$
r_{(1)} < r_{(2)} < \cdots < r_{(n)}
$$

Then we compare these with the Gaussian scores in (3.5.6) by plotting the pairs $(z_i^{(n)}, r_{(i)})$ for $i = 1, \ldots, n$, with $r_{(i)}$ as ordinates and $z_i^{(n)}$ as abscissas. If the plotted points appear to deviate systematically from the line through the origin with slope equal to 1 (which should be plotted on the same graph for ease of reference), this indicates that some or all of the assumptions may be violated. Tables that contain the values $z_i^{(n)}$ for various values of n are available, but the rankit-plot of standardized residuals is more conveniently obtained by computer.

Rankit-Plots of Linear Combinations of y_i and x_i

For inferences on certain parameters (for instance the correlation coefficient $\rho_{Y,X}$ to be discussed in Section 3.9), we need assumptions (B) in Box 3.3.2. In particular, the population $\{(Y, X)\}$ is assumed to be bivariate Gaussian. To check this assumption we must investigate whether or not the sample data, $(y_1, x_1), \ldots, (y_n, x_n)$, appear to be a simple random sample from a bivariate Gaussian population. This can be done by examining the rankit-plots of linear combinations of y_i and x_i. Based on the discussions in Section 1.9, we know that a population $\{(Y, X)\}$ is bivariate Gaussian if and only if the univariate population $\{aY + bX\}$ is Gaussian for every possible choice of values for a and b. Because we have only sample data available, we can examine linear combinations $ay_i + bx_i$ of y_i and x_i in the sample. In practice it is impossible to examine *every* linear combination of y_i and x_i, but we can consider *several* linear combinations and examine the corresponding rankit-plots. For instance, if $v_i = ay_i + bx_i$ is one such linear combination, then we can examine a rankit-plot of the v_i and form an opinion about whether the v_i values appear to be a simple random sample from a Gaussian population. If we consider several such linear combinations (several different values of a and b), and if every linear combination we consider appears to be a simple random sample from a Gaussian population, then we can feel somewhat assured that the bivariate population $\{(Y, X)\}$ under consideration is at least approximately bivariate Gaussian. We give a numerical illustration of this procedure in Example 3.5.2, which is one of two examples we use to illustrate the procedures discussed in this section.

You should study a large number of rankit-plots corresponding to simple random samples from Gaussian populations, as well as non-Gaussian populations, to gain experience in judging the plots and deciding whether or not the plot indicates violations of the assumption of a Gaussian random sample.

For convenience, we list the formulas for computing $\hat{\mu}_Y(x_i)$, $\hat{e}_i$, $h_{i,i}$, and r_i, in Box 3.5.1.

B O X   **3.5.1**   **Formulas for Computing** $\hat{\mu}_Y(x_i)$, $\hat{e}_i$, $h_{i,i}$, **and** r_i

Fitted values $\hat{\mu}_Y(x_i)$:

$$\hat{\mu}_Y(x_i) = \hat{\beta}_0 + \hat{\beta}_1 x_i$$

Residuals $\hat{e}_i$:

$$\hat{e}_i = y_i - \hat{\beta}_0 - \hat{\beta}_1 x_i = y_i - \hat{\mu}_Y(x_i)$$

Hat values $h_{i,i}$:

$$h_{i,i} = \left[\frac{1}{n} + \frac{(x_i - \bar{x})^2}{SSX} \right]$$

Standardized residuals r_i:

$$r_i = \frac{\hat{e}_i}{\hat{\sigma}\sqrt{1 - h_{i,i}}}$$

E X A M P L E   **3.5.1**

We illustrate the preceding discussions using the crystal growth data in Task 3.4.1; the data are in the file **crystal.dat** on the data disk. Recall that for these data $n = 14$, $\hat{\sigma} = 1.062$, $\bar{x} = 15$, $SSX = 910$, and the sample regression function is

$$\hat{\mu}_Y(x) = 0.0014 + 0.5034x$$

Since the X values in this data set are preselected, we know that assumptions (B) are not satisfied, so we examine the data to determine if assumptions (A) appear to be satisfied.

In Figure 3.5.19 we plot y_i, the crystal weights, against x_i, the time it takes to grow to their final sizes. This plot seems to support the assumption that the graph of the regression function of Y on X is a straight line. We calculate the fitted values $\hat{\mu}_Y(x_i)$, the residuals $\hat{e}_i$, the hat values $h_{i,i}$, and the standardized residuals r_i using the formulas in Box 3.5.1, and we list them in Table 3.5.1 along with x_i and y_i.

FIGURE 3.5.19

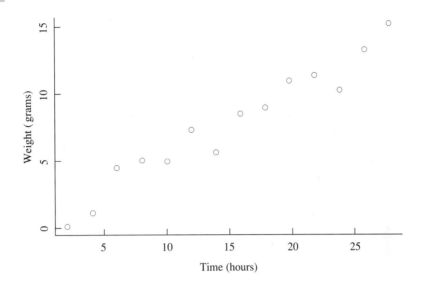

TABLE 3.5.1

Fitted Values, Residuals, Hat Values, and Standardized Residuals for Crystal Data

Weights y_i	Times x_i	Fitted Values $\hat{\mu}_Y(x_i)$	Residuals $\hat{e}_i$	Hat Values $h_{i,i}$	Standardized Residuals r_i
0.08	2	1.0083	−0.92829	0.257143	−1.01438
1.12	4	2.0151	−0.89514	0.204396	−0.94518
4.43	6	3.0220	1.40800	0.160440	1.44726
4.98	8	4.0289	0.95114	0.125275	0.95781
4.92	10	5.0357	−0.11571	0.098901	−0.11481
7.18	12	6.0426	1.13743	0.081319	1.11767
5.57	14	7.0494	−1.47943	0.072527	−1.44682
8.40	16	8.0563	0.34371	0.072527	0.33614
8.81	18	9.0631	−0.25314	0.081319	−0.24874
10.81	20	10.0700	0.74000	0.098901	0.73420
11.16	22	11.0769	0.08314	0.125275	0.08373
10.12	24	12.0837	−1.96371	0.160440	−2.01847
13.12	26	13.0906	0.02943	0.204396	0.03107
15.04	28	14.0974	0.94257	0.257143	1.02999

In Figure 3.5.20 we plot the standardized residuals r_i against the the fitted values $\hat{\mu}_Y(x_i)$.

F I G U R E **3.5.20**

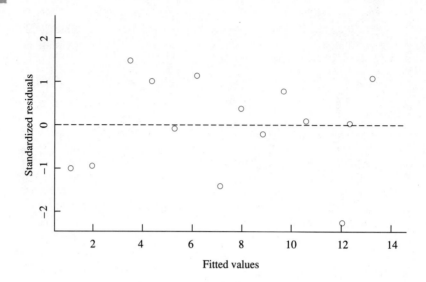

Notice that the points on this plot seem to be randomly scattered about the horizontal dashed line and show no apparent pattern. Based on this plot we do not see any evidence of violations of any of assumptions (A).

Finally we obtain the rankit-plot of r_i shown in Figure 3.5.21.

F I G U R E **3.5.21**

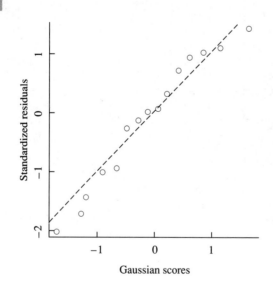

The points on this plot seem to all lie (approximately) on a straight line of unit slope through the origin (the dashed line in Figure 3.5.21), and so this plot is consistent with assumptions (A).

To summarize, for the crystal growth data in Table 3.4.2, the graphical procedures discussed in this section do not point to any violations of assumptions (A). (As stated earlier, we already know that assumptions (B) do *not* hold because the X values in this data set are preselected.) We can feel somewhat confident that the inference procedures discussed in this chapter, which are based on assumptions (A), are valid. ■

E X A M P L E 3.5.2

For another illustration, consider Example 2.2.2 where we are interested in the relationship of Y, the first-year maintenance cost of new cars, and X, the number of miles driven during the first year after purchase. Twenty cars were selected by simple random sampling from the population in Table D-1, in Appendix D, and we want to examine the plausibility of assumptions (A) or (B) for this population, *using only the sample data*. The data are given in Table 3.5.2 and are also stored in the file **car20.dat** on the data disk.

We begin by plotting y_i against x_i (see Figure 3.5.22) to see if it is reasonable to assume that the graph of the regression function of Y on X is a straight line. The plot seems to suggest that the regression function of Y on X may not be a straight line because there is some evidence of curvature. We should be able to better assess this possibility by examining a plot of the standardized residuals r_i against x_i and also of r_i against the fitted values $\hat{\mu}_Y(x_i)$. To do so, we first calculate the estimated regression function $\hat{\mu}_Y(x)$ using (3.4.10) and we get

$$\hat{\mu}_Y(x) = 177.01 + 0.031307x$$

Next we calculate the fitted values $\hat{\mu}_Y(x_i)$, residuals $\hat{e}_i$, hat values $h_{i,i}$, and standardized residuals r_i using formulas in Box 3.5.1. These are listed in Table 3.5.3.

Next we examine a plot of the standardized residuals r_i against x_i. This plot is given in Figure 3.5.23. It strongly suggests that the regression function of Y on X is not a straight line because the points are clearly not randomly scattered about the horizontal line corresponding to $r_i = 0$ (shown as the dotted line in Figure 3.5.23), and there seems to be clear evidence of curvature.

T A B L E   **3.5.2**

Car20 Data (Sample Data from the Population in Table D-1)

Car	Maintenance Cost (in dollars) Y	Miles Driven X
1	456	11200
2	828	17300
3	500	11100
4	489	11000
5	387	6700
6	553	13700
7	531	12400
8	650	15300
9	475	11300
10	474	8200
11	533	12300
12	396	7700
13	618	14300
14	474	8800
15	639	13600
16	457	7100
17	460	8700
18	433	6500
19	621	13100
20	460	9900

F I G U R E   **3.5.22**

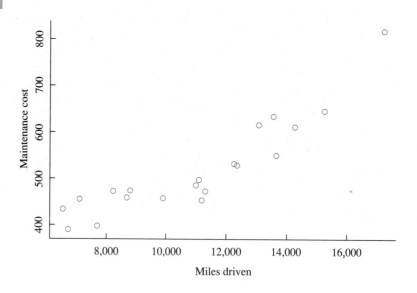

T A B L E 3.5.3
Fitted Values, Residuals, Hat Values, and Standardized Residuals for **car20 data**

Maintenance Costs y_i	Miles Driven x_i	Fitted Values $\hat{\mu}_Y(x_i)$	Residuals e_i	Hat Values $h_{i,i}$	Standardized Residuals r_i
456	11200	527.648	−71.648	0.050206	−1.58529
828	17300	718.623	109.377	0.275645	2.77121
500	11100	524.518	−24.518	0.050046	−0.54243
489	11000	521.387	−32.387	0.050001	−0.71652
387	6700	386.766	0.234	0.155945	0.00550
553	13700	605.917	−52.917	0.091269	−1.19700
531	12400	565.217	−34.217	0.061019	−0.76144
650	15300	656.008	−6.008	0.154964	−0.14094
475	11300	530.779	−55.779	0.050480	−1.23435
474	8200	433.726	40.274	0.095034	0.91290
533	12300	562.086	−29.086	0.059491	−0.64674
396	7700	418.073	−22.073	0.112486	−0.50523
618	14300	624.701	−6.701	0.111733	−0.15332
474	8800	452.511	21.489	0.077855	0.48254
639	13600	602.786	36.214	0.088258	0.81782
457	7100	399.288	57.712	0.137192	1.33975
460	8700	449.380	10.620	0.080433	0.23881
433	6500	380.504	52.496	0.166005	1.23954
621	13100	587.132	33.868	0.074912	0.75930
460	9900	486.949	−26.949	0.057027	−0.59842

F I G U R E 3.5.23

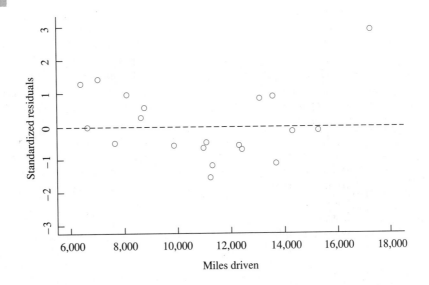

Figure 3.5.24 shows the plot of the standardized residuals r_i against fitted values $\hat{\mu}_Y(x_i)$. The points in this plot are also not randomly scattered about the horizontal dashed line; instead, they suggest the existence of curvature in the regression curve of Y on X.

F I G U R E   **3.5.24**

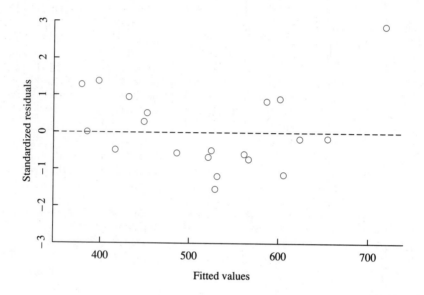

Thus we tentatively conclude that a straight line regression model is not appropriate for the data of this example. You should refer to Figure 2.3.4, which shows a scatter-plot of maintenance cost against miles driven for the population in Task 2.3.1 from which these data are sampled. Notice that the population scatter-plot definitely exhibits curvature. This should give you some confidence that a sample, properly selected, does indeed reflect aspects of the population.

For illustrative purposes, we now examine a rankit-plot of the standardized residuals in Figure 3.5.25 (the dashed line is the line through the origin with unit slope). There does not seem to be anything unusual about this plot that would make us suspect any violations of the assumptions. So based on this plot alone we are unable to detect any departure from assumptions (A) or (B).

Because the sample data in this example are obtained using simple random sampling, it is reasonable to ask if assumptions (B) appear to be satisfied; i.e., it is reasonable to examine whether the data appear to be a simple random sample from a bivariate Gaussian population. This can be done by examining whether or not various linear combinations of y_i and x_i (including y_i and x_i themselves) appear to be a simple random sample from a univariate Gaussian population (see Section 1.9) by using the rankit-plot. We illustrate this by examining rankit-plots of y_i, x_i, and two

F I G U R E   3.5.25

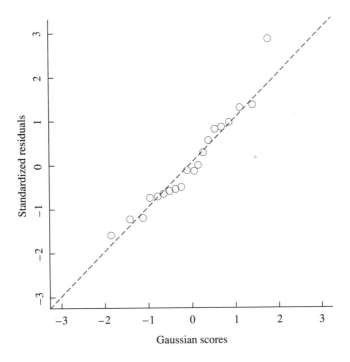

other linear combinations of y_i and x_i, namely $x_i + y_i$ and $x_i - y_i$. These four plots are given in Figure 3.5.26.

Based on these plots we conclude that it is not unreasonable to assume the x_i values are a simple random sample from a Gaussian population. The rankit-plot for Y shows a *hint* of curvature, thus suggesting that the population $\{Y\}$ may not be Gaussian.

The rankit-plots for $y_i + x_i$ and $x_i - y_i$ appear to be consistent with the Gaussian assumption. You are invited to obtain the rankit-plots for several other linear combinations of y_i and x_i for practice in judging them. However, because the sample y_i values do not appear to be a simple random sample from a Gaussian population, the data $(y_1, x_1), \ldots, (y_{20}, x_{20})$ appear not to be a simple random sample from a bivariate Gaussian population. ■

We should point out that if the population $\{(Y, X)\}$ is a bivariate Gaussian population, then the regression function of Y on X must necessarily be of the form $\mu_Y(x) = \beta_0 + \beta_1 x$. Because some of the residual plots have already supplied evidence against this possibility, we should have concluded that it is not very likely for the population $\{(Y, X)\}$ to be bivariate Gaussian. However, we proceeded with the examination of various linear combinations of y_i and x_i to illustrate the procedure.

F I G U R E   **3.5.26**

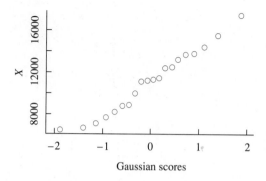

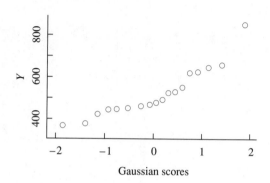

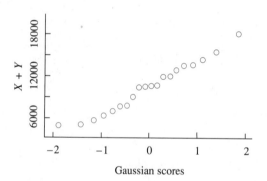

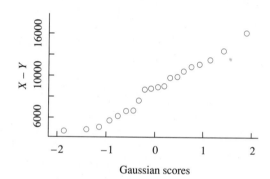

Bear in mind that while some of the procedures for checking assumptions may not detect any violations, other procedures may. It is only with the combined evidence based on many different checks of assumptions that we can arrive at a conclusion regarding the validity (or at least, approximate validity) of the assumptions in any given situation.

Many diagnostic procedures exist for examining the validity of assumptions (A) or (B). Even when these procedures do not suggest the failure of any of the assumptions, there is no guarantee that the assumptions are actually met. This is a necessarily subjective exercise, and at some point the investigator has to either conclude that assumptions (A) or (B) are satisfied (for all practical purposes) and proceed with analyses and inferences, or he/she must conclude that one or more of the assumptions are seriously violated and look for alternative procedures. For certain types of violations of assumptions, we discuss alternative procedures in later chapters.

Authors' Recommendation

When you perform a straight line regression analysis of a set of data $(y_1, x_1), \ldots, (y_n, x_n)$, we suggest you take the following steps:

1 Plot y_i against x_i. Assess the plausibility of a straight line regression model. If the plot indicates that a straight line regression model is not appropriate, then you may want to investigate alternative candidates for a regression function. Chapters 4–9 will be of assistance in this regard.

2 If a straight line regression model appears to be a plausible candidate, then calculate the sample regression function as discussed in Section 3.4. Obtain the **standardized residuals** r_i and the **fitted values** $\hat{\mu}_Y(x_i)$.

3 Plot r_i against x_i and r_i against the **fitted values** $\hat{\mu}_Y(x_i)$. Examine these plots for evidence of unequal subpopulation variances or of lack of fit of the model.

4 Obtain a Gaussian rankit-plot of r_i to evaluate the validity of the assumption that each subpopulation of Y values, determined by X, is a Gaussian population.

5 If you want to determine whether the bivariate population $\{(Y, X)\}$ is Gaussian, and if you obtained data by simple random sampling, then examine the Gaussian rankit-plots of y_i, x_i, and several linear combinations of y_i and x_i to assess whether or not the data appear to be a simple random sample from a bivariate Gaussian population.

6 Make an overall evaluation of the validity (at least approximately) of assumptions (A) or (B) within the context of the particular application in question.

Note: Checking assumptions as recommended here involves a great deal of computing, plotting, etc., which are easily done using one of several statistical computing packages. In Section 3.5 of the laboratory manuals we explain in detail how to do these computations using MINITAB or SAS.

Problems   3.5

Use the following information for Problems 3.5.1–3.5.6.

A coal burning power plant is located a distance of 25 miles from a national park. Emissions from the power plant contain the gas sulfur dioxide (SO_2) (which is linked to acid rain), and consequently a certain fraction of the emitted SO_2 is transported through the atmosphere to the national park. A certain amount of background SO_2 that is not emitted by the power plant is always present at the national park. To assess the power plant's SO_2 contribution to the national park, recordings were made of X, the SO_2 output by the plant in tons/hour, as well as Y, the SO_2 concentrations at the national park in micrograms/cubic meter ($\mu g/m^3$), at various randomly selected times during a particular year. The data are given in Table 3.5.4 and are also stored in the file **so2.dat** on the data disk.

From these data we compute the following quantities.

$$\sum_{i=1}^{14} x_i = 55.760 \qquad \sum_{i=1}^{14} y_i = 123.35$$

$$\sum_{i=1}^{14} x_i^2 = 253.907 \qquad \sum_{i=1}^{14} y_i^2 = 1220.27$$

$$\sum_{i=1}^{14} y_i x_i = 549.355$$

The variations in the Y values for a given X value occur because of changes in the wind direction and wind speed, variations in the rate of dispersion of the gas, etc. We want to find out if higher levels of SO_2 at the national park are associated with higher levels of SO_2 emitted by the power plant. In similar investigations, assumptions (B) have been used, but for these data the experimenter wants to assess the validity of these assumptions.

3.5.1   Plot the values of SO_2 at the national park (y_i) against the values of SO_2 emissions at the power plant (x_i). After examining this plot, do you think a straight line regression function $\mu_Y(x) = \beta_0 + \beta_1 x$ seems reasonable?

3.5.2   Obtain the least squares estimates of β_0 and β_1. Also obtain an estimate of σ.

3.5.3   Table 3.5.5 contains y_i, x_i, the fitted values $\hat{\mu}_Y(x_i)$, the residuals $\hat{e}_i$, the hat values $h_{i,i}$, and the standardized residuals r_i. These quantities are also stored in the file **table355.dat** on the data disk. Plot the standardized residuals r_i against the fitted

T A B L E   **3.5.4**
Power Plant SO_2 Data

Time	Y ($\mu g/m^3$)	X (tons/hour)
1	5.21	1.92
2	7.36	3.92
3	16.26	6.80
4	10.10	6.32
5	5.80	2.00
6	8.06	4.32
7	4.76	2.40
8	6.93	2.96
9	9.36	3.52
10	10.90	4.24
11	12.48	5.12
12	11.70	5.84
13	7.44	3.60
14	6.99	2.80

T A B L E **3.5.5**

Fitted Values, Residuals, Hat Values, and Standardized Residuals for Power Plant SO_2 Data

y_i	x_i	Fitted Values	Residuals	Hat Values	Standardized Residuals
5.21	1.92	5.0465	0.16352	0.205148	0.12115
7.36	3.92	8.6960	−1.33601	0.071553	−0.91583
16.26	6.80	13.9513	2.30865	0.320817	1.85031
10.10	6.32	13.0755	−2.97546	0.243072	−2.25895
5.80	2.00	5.1925	0.60754	0.194978	0.44725
8.06	4.32	9.4259	−1.36592	0.075000	−0.93807
4.76	2.40	5.9224	−1.16237	0.150159	−0.83283
6.93	2.96	6.9442	−0.01424	0.104305	−0.00994
9.36	3.52	7.9661	1.39389	0.078161	0.95892
10.90	4.24	9.2799	1.62006	0.073506	1.11171
12.48	5.12	10.8857	1.59426	0.112062	1.11750
11.70	5.84	12.1996	−0.49957	0.179808	−0.36435
7.44	3.60	8.1121	−0.67209	0.076035	−0.46183
6.99	2.80	6.6523	0.33773	0.115395	0.23718

values $\hat{\mu}_Y(x_i)$. Does this plot suggest any violations of assumptions (B)? If so, what assumption seems to be violated? Round the numbers to two decimals for plotting.

3.5.4 Plot the standardized residuals r_i against x_i. What does this plot suggest regarding assumptions (B)?

3.5.5 The ordered standardized residuals $r_{(i)}$ and the corresponding Gaussian scores $z_i^{(n)}$, where $n = 14$, appear in Table 3.5.6. Obtain a rankit-plot of these standardized residuals. What do you conclude from this plot? For plotting purposes you may suitably round the values of Gaussian scores.

3.5.6 Examine the plausibility of the assumption that the data (y_i, x_i) are a simple random sample from a bivariate Gaussian population by obtaining and examining the **rankit-plots** of y_i, x_i, $x_i + y_i$, and $x_i - y_i$. The Gaussian scores for y_i, x_i, $x_i + y_i$, $x_i - y_i$ appear in Table 3.5.7. Based on these plots, what is your conclusion as to whether (y_i, x_i) is a simple random sample from a bivariate Gaussian population? For plotting purposes you may suitably round the values of Gaussian scores.

TABLE 3.5.6

Ordered Standardized Residuals	Gaussian Scores
−2.25895	−1.70991
−0.93807	−1.20448
−0.91583	−0.89743
−0.83283	−0.65862
−0.46183	−0.45321
−0.36435	−0.26585
−0.00994	−0.08767
0.12115	0.08767
0.23718	0.26585
0.44725	0.45321
0.95892	0.65862
1.11171	0.89743
1.11750	1.20448
1.85031	1.70991

TABLE 3.5.7

Item	y_i	Gaussian Scores for y_i	x_i	Gaussian Scores for x_i	$x_i + y_i$	Gaussian Scores for $(x_i + y_i)$	$x_i - y_i$	Gaussian Scores for $(x_i - y_i)$
1	5.21	−1.20448	1.92	−1.70991	7.13	−1.70991	−3.29	1.20448
2	7.36	−0.26585	3.92	0.08767	11.28	−0.08767	−3.44	0.89743
3	16.26	1.70991	6.80	1.70991	23.06	1.70991	−9.46	−1.70991
4	10.10	0.45321	6.32	1.20448	16.42	0.65862	−3.78	0.45321
5	5.80	−0.89743	2.00	−1.20448	7.80	−0.89743	−3.80	0.26585
6	8.06	0.08767	4.32	0.45321	12.38	0.08767	−3.74	0.65862
7	4.76	−1.70991	2.40	−0.89743	7.16	−1.20448	−2.36	1.70991
8	6.93	−0.65862	2.96	−0.45321	9.89	−0.45321	−3.97	−0.08767
9	9.36	0.26585	3.52	−0.26585	12.88	0.26585	−5.84	−0.45321
10	10.90	0.65862	4.24	0.26585	15.14	0.45321	−6.66	−0.89743
11	12.48	1.20448	5.12	0.65862	17.60	1.20448	−7.36	−1.20448
12	11.70	0.89743	5.84	0.89743	17.54	0.89743	−5.86	−0.65862
13	7.44	−0.08767	3.60	−0.08767	11.04	−0.26585	−3.84	0.08767
14	6.99	−0.45321	2.80	−0.65862	9.79	−0.65862	−4.19	−0.26585

3.6
Confidence Intervals

An interval estimate of a parameter, along with a point estimate, is generally very useful for making decisions because an interval estimate implicitly tells the investigator how well the parameter in question is being estimated, i.e., how close the point estimate is likely to be to the true value. In this connection you should recall the discussions in Section 1.6. In this section we discuss procedures for computing confidence intervals for β_0, β_1, $\mu_Y(x)$, $Y(x)$, σ, and linear functions of the form $a_0\beta_0 + a_1\beta_1$. The results are valid under both assumptions (A) and (B). Note that confidence intervals for μ_Y, σ_Y, μ_X, and σ_X are obtained using the formulas given in Table 1.6.2 and are valid under assumptions (B).

General Form of Confidence Intervals for β_0, β_1, $\mu_Y(x)$, $Y(x)$, and Linear Functions $a_0\beta_0 + a_1\beta_1$

The general form for a $1 - \alpha$ two-sided confidence interval for β_0, β_1, $Y(x)$, $\mu_Y(x)$, and the linear function $a_0\beta_0 + a_1\beta_1$ (let θ denote any one of these quantities and $\hat{\theta}$ denote the point estimate of θ) is

$$\hat{\theta} - (\text{table-value}) \times SE(\hat{\theta}) \leq \theta \leq \hat{\theta} + (\text{table-value}) \times SE(\hat{\theta}) \qquad \text{(3.6.1)}$$

where table-value is the quantity $t_{1-\alpha/2:df}$ obtained from a student's t-table (Table T-2 in Appendix T) with df = degrees of freedom = $n - 2$ in the case of straight line regression. The quantity $SE(\hat{\theta})$, called the standard error of $\hat{\theta}$, is an estimate of the *precision* of $\hat{\theta}$ and is calculated from sample data.

The point estimates for β_1, β_0, $\mu_Y(x)$, $Y(x)$, and $a_0\beta_0 + a_1\beta_1$ are given in Box 3.4.2. Their respective standard errors are given in (3.6.2)–(3.6.6).

$$SE(\hat{\beta}_1) = \hat{\sigma}/\sqrt{SSX} \qquad \text{(3.6.2)}$$

$$SE(\hat{\beta}_0) = \hat{\sigma}\sqrt{\frac{1}{n} + \frac{\bar{x}^2}{SSX}} \qquad \text{(3.6.3)}$$

$$SE(\hat{\mu}_Y(x)) = \hat{\sigma}\sqrt{\frac{1}{n} + \frac{(x - \bar{x})^2}{SSX}} \qquad \text{(3.6.4)}$$

$$SE(\hat{Y}(x)) = \hat{\sigma}\sqrt{1 + \frac{1}{n} + \frac{(x - \bar{x})^2}{SSX}} \qquad \text{(3.6.5)}$$

$$SE(a_0\hat{\beta}_0 + a_1\hat{\beta}_1) = \hat{\sigma}\sqrt{\frac{a_0^2}{n} + \frac{(a_1 - a_0\bar{x})^2}{SSX}} \qquad \text{(3.6.6)}$$

The following relationship between $SE(\hat{\mu}_Y(x))$ and $SE(\hat{Y}(x))$ is sometimes useful.

$$SE(\hat{Y}(x)) = \sqrt{\hat{\sigma}^2 + [SE(\hat{\mu}_Y(x))]^2} \qquad \text{(3.6.7)}$$

Note that $SE(\hat{Y}(x))$, is really an abbreviation for $SE(\hat{Y}(x) - Y(x))$.

Remarks

1 $SE(\hat{\beta}_0)$, $SE(\hat{\beta}_1)$, and $SE(\hat{\mu}_Y(x))$ can be obtained from $SE(a_0\hat{\beta}_0 + a_1\hat{\beta}_1)$ by substituting the appropriate values for a_0 and a_1.

2   By studying (3.6.4) we see that the further that x is from $\bar{x}$, the larger $SE(\hat{\mu}_Y(x))$ becomes. This can be intuitively explained as follows. Recall that the estimated regression line always passes through the point $(\bar{x}, \bar{y})$. Think of the estimated line as a see-saw with its pivot at $(\bar{x}, \bar{y})$. A slight rotation of the line about the pivot causes large deviations far away from the pivot but only small deviations close to the pivot (see Figure 3.5.5 and the discussion pertaining to it). As a result, points on the line (which are the estimated means of subpopulations) close to $(\bar{x}, \bar{y})$ are estimated with greater precision compared to points far away from $(\bar{x}, \bar{y})$. A similar explanation applies to $SE(\hat{Y}(x))$.

3   In any particular application, it is important for you to determine whether a confidence interval for $\mu_Y(x)$ or a confidence interval for $Y(x)$ is required. Some authors call the confidence interval for $Y(x)$ a *prediction interval* because the term confidence interval is traditionally reserved for parameters, but $Y(x)$ is a random variable (it is the Y value of a randomly chosen item from the subpopulation corresponding to $X = x$).

4   Note that even though $\hat{\mu}_Y(x) = \hat{Y}(x)$, i.e.,

$$\hat{\mu}_Y(x) = \hat{Y}(x) = \hat{\beta}_0 + \hat{\beta}_1 x$$

the standard error of $\hat{\mu}_Y(x)$ is not the same as the standard error of $\hat{Y}(x)$. This is so because there is greater uncertainty in predicting a single randomly chosen value from the subpopulation corresponding to $X = x$ than in estimating the mean of the entire subpopulation.

5   For each quantity being estimated, the investigator should be given a point estimate, its standard error, and the degrees of freedom associated with this standard error for obtaining the table t-value. The investigator can substitute these quantities into (3.6.1) and compute the desired confidence intervals by selecting the confidence coefficient $1 - \alpha$ that he wants to use.

Confidence Intervals for σ

A two-sided $1 - \alpha$ confidence interval for σ is given by

$$C\left[\sqrt{\frac{(df)\,\hat{\sigma}^2}{\chi^2_{1-\alpha/2:df}}} \leq \sigma \leq \sqrt{\frac{(df)\,\hat{\sigma}^2}{\chi^2_{\alpha/2:df}}}\right] = 1 - \alpha \qquad \textbf{(3.6.8)}$$

where $df = n - 2$ and $\chi^2_{\alpha/2:df}$ and $\chi^2_{1-\alpha/2:df}$ are obtained from Table T-3 in Appendix T. The formula for a confidence interval for σ is not of the general form given in (3.6.1). In fact the general form for a two-sided $(1 - \alpha)$ confidence interval for a parameter that is a standard deviation (such as σ_Y, σ_X, σ, etc.) is

$$\sqrt{\frac{(df) \, (\text{estimated standard deviation})^2}{\chi^2_{1-\alpha/2:df}}} \leq \sigma$$

$$\leq \sqrt{\frac{(df) \, (\text{estimated standard deviation})^2}{\chi^2_{\alpha/2:df}}}$$

(3.6.9)

where df represents the number of degrees of freedom associated with the estimated standard deviation. Note that the number of degrees of freedom associated with $\hat{\sigma}_X$ or $\hat{\sigma}_Y$ is $n - 1$, whereas that associated with $\hat{\sigma}$ is $n - 2$. It is also worth observing that while the confidence intervals for the quantities, β_0, β_1, $\mu_Y(x)$, $Y(x)$, and $a_0\beta_0 + a_1\beta_1$ are symmetric about the corresponding point estimates, this is not so in the case of σ. However, the confidence interval for σ in (3.6.9) is equal-tailed, contains the point estimate $\hat{\sigma}$, and gives us some indication of how close $\hat{\sigma}$ may be to the true value σ.

One-Sided Confidence Bounds

In this section, the discussion so far has been about two-sided confidence intervals for parameters of interest. However, in some applications one-sided confidence bounds are more useful because an investigator may be interested in only the lower bound or the upper bound for a parameter in a decision-making situation. As discussed in Section 1.6, if a confidence interval is equal-tailed, we can obtain one-sided confidence bounds with confidence coefficient $1 - \alpha$ by first constructing a two-sided confidence interval with confidence coefficient $1 - 2\alpha$ and reading off either the lower or the upper endpoint as appropriate. This is valid for all of the quantities, β_0, β_1, $\mu_Y(x)$, $Y(x)$, and $a_0\beta_0 + a_1\beta_1$, as well as σ, σ_Y, and σ_X, because these confidence intervals are *equal-tailed*.

The following two tasks illustrate the computations required for obtaining confidence intervals in the context of real applications.

Task   3.6.1

Consider the problem discussed in Task 3.4.2 where an investigator wants to evaluate the performance of a new laboratory method, which is less expensive than the current method, for analyzing the concentration of arsenic (As) in water samples. The data appear in Table 3.4.3. For convenience they are reproduced in Table 3.6.1 and are also stored in the file **arsenic.dat** on the data disk.

T A B L E   **3.6.1**

Arsenic Data

Sample Item Number	Measured Concentration Y (μg/ml)	True Concentration X (μg/ml)
1	0.17	0
2	0.25	0
3	0.01	0
4	0.12	0
5	1.25	1
6	0.86	1
7	1.25	1
8	1.10	1
9	2.01	2
10	2.03	2
11	2.14	2
12	1.74	2
13	3.18	3
14	2.99	3
15	3.23	3
16	3.37	3
17	3.91	4
18	3.90	4
19	3.61	4
20	4.27	4
21	4.88	5
22	5.33	5
23	4.96	5
24	4.98	5
25	6.09	6
26	6.17	6
27	6.07	6
28	5.97	6
29	6.67	7
30	7.02	7
31	7.14	7
32	7.30	7

Suppose, based on experience, the investigator feels that population assumptions (A) for straight line regression should hold, at least approximately, where the data are obtained by sampling with preselected X values. In particular, the population regression function is

$$\mu_Y(x) = \beta_0 + \beta_1 x$$

and the subpopulation standard deviations for Y are all the same, each equal to σ.

Some basic sums, sums of squares, and sums of crossproducts—the ingredients in the formulas for obtaining point and confidence interval estimates of population parameters—are as follows:

$$\sum_{i=1}^{n} x_i = 112 \qquad \bar{x} = 3.5 \qquad \sum_{i=1}^{n} y_i = 113.97 \qquad \bar{y} = 3.56156$$

$$SSX = 168 \qquad SSY = 164.95 \qquad SXY = 165.935$$
$$SSE = 1.0544 \qquad MSE = 0.03515$$

From these we compute the point estimates

$$\hat{\beta}_1 = \frac{SXY}{SSX} = \frac{165.935}{168} = 0.98771$$
$$\hat{\beta}_0 = \bar{y} - \hat{\beta}_1 \bar{x} = 3.56156 - 0.98771(3.5) = 0.10458$$
$$\hat{\mu}_Y(x) = \hat{\beta}_0 + \hat{\beta}_1 x = 0.10458 + 0.98771x = \hat{Y}(x)$$
$$\hat{\sigma} = \sqrt{MSE} = \sqrt{0.03515} = 0.1875$$

We now calculate appropriate confidence intervals for β_0, β_1, and σ that will help the investigator obtain practical answers to the following three questions (using $1 - \alpha = 0.95$). In Section 3.6 of the laboratory manual we show how a computer can be used to calculate all quantities needed in this section.

1 On the average, does the chemical analysis correctly report the absence of As when this is indeed the case? (That is, is $\mu_Y(x) = 0$ when $x = 0$? In other words is $\beta_0 = 0$?)

To answer this question we first compute a 95% two-sided confidence interval for β_0. Using the formula in (3.6.3) we compute

$$SE(\hat{\beta}_0) = \hat{\sigma}\sqrt{\frac{1}{n} + \frac{\bar{x}^2}{SSX}} = (0.1875)\sqrt{\frac{1}{32} + \frac{(3.5)^2}{168}} = 0.0605$$

From Table T-2 in Appendix T we obtain $t_{1-\alpha/2:30} = t_{0.975:30} = 2.042$, corresponding to $1 - \alpha = 0.95$ and degrees of freedom $= n - 2 = 30$. Hence a 95% two-sided confidence interval for β_0 is given by

$$C[-0.019 \le \beta_0 \le 0.228] = 0.95$$

Based on this confidence interval, the investigator can decide whether β_0 can be considered close enough to zero to be regarded as negligible for all practical purposes for this problem.

2 Now suppose the investigator wants to examine how close β_1 is to 1 in the model $\mu_Y(x) = \beta_0 + \beta_1 x$. Compute a 95% two-sided confidence interval for β_1 to help determine this.

Using the formula in (3.6.2) we calculate

$$SE(\hat{\beta}_1) = \hat{\sigma}/\sqrt{SSX} = 0.1875/\sqrt{168} = 0.01447$$

From Table T-2 in Appendix T we obtain $t_{1-\alpha/2:30} = t_{0.975:30} = 2.042$. Hence a 95% confidence interval is given by the confidence statement

$$C[0.958 \leq \beta_1 \leq 1.017] = 0.95$$

On the basis of this information, the investigator can decide whether β_1 can be regarded as close enough to 1.0 for all practical purposes in this problem.

3 Suppose that, based on data and previous experience, the investigator is fairly confident that the new method of chemical analysis for As is unbiased; i.e., the regression function of Y on X is $\mu_Y(x) = x$. However, for the method to be adopted by water quality monitoring agencies, they want to know whether a proportion 0.99 or more of the measured concentrations will be accurate to within 1.0 μg/ml. Use the data at hand to help the investigator make a decision.

If $\mu_Y(x) = x$, then 99% of the measured concentrations corresponding to a true concentration x will be between $x - z_{0.995}\sigma$ and $x + z_{0.995}\sigma$, i.e., between $x - 2.575\sigma$ and $x + 2.575\sigma$. Hence we want to know whether 2.575σ is less than 1.0. The estimate of σ from the preceding calculations is equal to 0.1875, and so the estimated value of 2.575σ is equal to 0.483. This is well within the acceptable upper limit of 1.0. However, the investigator realizes that the estimate 0.1875 of σ is subject to sampling errors, so he wants to know if the estimate is sufficiently accurate for decision-making purposes. To investigate this the investigator wants a 95% one-sided upper confidence bound for σ. We use the formula for a 90% two-sided confidence bound for σ given in (3.6.8), and from this we can obtain a one-sided 95% upper confidence bound. The necessary χ^2 table-values, obtained from Table T-3 in Appendix T, are $\chi^2_{0.05:30} = 18.493$ and $\chi^2_{0.95:30} = 43.773$. Thus we get the confidence statement

$$C\left[\sqrt{\frac{30\,(0.1875)^2}{43.773}} \leq \sigma \leq \sqrt{\frac{30\,(0.1875)^2}{18.493}}\right] = 0.90$$

i.e.,

$$C[0.1552 \leq \sigma \leq 0.2388] = 0.90$$

and hence

$$C[0.3996 \leq 2.575\sigma < 0.6149] = 0.90$$

Thus the investigator has 95% confidence that $2.575\sigma \leq 0.6149$. Based on this confidence statement the investigator will perhaps conclude that the variability in the new method of analysis for measuring As is within acceptable limits and may be adopted for routine use in environmental monitoring.

Task 3.6.2

A sample of size 24 was randomly selected using simple random sampling from a population of individuals of a particular ethnic group with ages ranging from 21 to 70 years, and their ages and blood pressures were recorded. The response variable is the average systolic blood pressure Y at 8 A.M. over a two-week period, and the predictor variable is age X of the individual in years. The sample data are listed in Table 3.6.2. They are also stored in the file **agebp.dat** on the data disk.

It is supposed that assumptions (B) in Box 3.3.2 are valid. (You are encouraged to examine the validity of these assumptions using the methods of Section 3.5.) In particular, the regression function of Y on X is of the form $\mu_Y(x) = \beta_0 + \beta_1 x$, where β_0 and β_1 are unknown parameters. Also the subpopulation standard deviations are all equal, and their common value is denoted by σ, which is also an unknown parameter.

TABLE 3.6.2

Age and Blood Pressure Data

Sample Item Number	Blood Pressure (systolic) Y	Age (in years) X
1	116	34
2	112	26
3	151	51
4	161	58
5	122	34
6	129	40
7	119	31
8	158	57
9	144	46
10	150	53
11	111	29
12	148	50
13	135	40
14	126	34
15	172	67
16	100	23
17	139	47
18	135	42
19	163	61
20	128	38
21	159	57
22	177	66
23	135	42
24	149	53

Some of the basic quantities needed to answer the following questions using sample data are

$$\sum_{i=1}^{24} x_i = 1,079 \qquad \sum_{i=1}^{24} y_i = 3,339$$

$$\sum_{i=1}^{24} (x_i - \bar{x})^2 = SSX = 3,608.96 \qquad \sum_{i=1}^{24} (y_i - \bar{y})^2 = SSY = 9,514.63$$

$$\sum_{i=1}^{24} (x_i - \bar{x})(y_i - \bar{y}) = SXY = 5,805.13 \qquad SSE = 176.896 \qquad MSE = 8.04075$$

$$\hat{\beta}_1 = 1.6085 \qquad \hat{\beta}_0 = 66.8081 \qquad \hat{\sigma} = 2.8356 \qquad \hat{\sigma}_Y = 20.3391$$

1 What is the average age of the individuals in the population?

The average age of the individuals in the population is μ_X; it cannot be exactly determined from the sample data but must be estimated. The estimated average age is $\hat{\mu}_X = \bar{x} = 45$ years (rounded to the nearest integer). Note that if data had been obtained by sampling with preselected X values, then, in general, no valid estimate of μ_X would be available (see (3.3.1)).

2 What is the average blood pressure of the individuals in the population?

The average blood pressure of the individuals in the population is μ_Y; it cannot be exactly determined from the sample data. The estimated value of μ_Y is $\hat{\mu}_Y = \bar{y} = 139$ units (to the nearest integer). Note that if data had been obtained by sampling with preselected X values, then, in general, no valid estimate of μ_Y would be available (see (3.3.1).)

3 Without using age as a predictor factor, predict the blood pressure of a randomly chosen individual from this population. Compute a number d such that you can be 95% confident that the predicted value will be within d units of the actual value.

Without using age X as a predictor factor, the best predicted value of the blood pressure Y of a randomly chosen individual from the population is $\hat{Y} = \bar{y} = 139.125$ and the standard error of $\hat{Y}$ is $\hat{\sigma}_Y \sqrt{\frac{1}{n} + 1} = 20.3391\sqrt{1.04167} = 20.7585$. (As we explained earlier, this is actually the standard error of $\hat{Y} - Y$.) From Table T-2 in Appendix T we get $t_{1-\alpha/2:n-1} = t_{0.975:23} = 2.069$. So a 95% confidence statement for Y is

$$C[96.176 \leq Y \leq 182.074] = 0.95$$

i.e.,

$$C[|Y - 139.125| \leq 42.95] = 0.95$$

(You should verify this.) Hence we can be 95% confident that the predicted value of 139.125 will be within $d = 42.95$ units of the actual Y value.

4   What is the average blood pressure of all individuals in the population who are 65 years old? How well can we estimate it based on the sample data (use $1 - \alpha = 0.95$)?

The average blood pressure of all individuals in the population who are 65 years old cannot be determined exactly from sample values. However, based on sample data, we estimate the average blood pressure of all individuals in the population who are 65 years old to be $\hat{\mu}_Y(65) = \hat{\beta}_0 + \hat{\beta}_1(65) = 66.8081 + 1.6085(65) = 171.361$. To find out how good this sample estimate is we calculate its standard error using the formula for $SE(\hat{\mu}_Y(x))$ in (3.6.4). We first calculate the value of $\hat{\sigma}$ as 2.8356. So $SE(\hat{\mu}_Y(65)) = 1.109$. To obtain a 95% confidence interval we use Table T-2 in Appendix T to find that $t_{1-\alpha/2:n-2} = t_{0.975:22} = 2.074$. So using (3.6.1) we obtain a 95% two-sided confidence interval for $\mu_Y(65)$ as

$$[171.361 - 2.074 \times 1.109, \ 171.361 + 2.074 \times 1.109]$$

which is [169.06, 173.66]. Thus we are 95% confident that the average blood pressure of all the individuals in the population who are 65 years old is in the interval [169.06, 173.66].

5   On the average, does blood pressure increase with age? If so, by how much?

The slope of the line relating the average blood pressure $\mu_Y(x)$ of the sub-population with $X = x$ to age X is the parameter β_1. The estimated value of β_1 is 1.6085, which means that the average blood pressure is estimated to increase with age at the rate of 1.6085 units per year. A confidence interval for β_1 will tell us how good our estimate of β_1 is. We first calculate $SE(\hat{\beta}_1)$ using (3.6.2) and obtain the value 0.0472. Hence a 95% two-sided confidence interval for β_1 is $[1.6085 - 2.074 \times 0.0472, \ 1.6085 + 2.074 \times 0.0472]$, i.e., [1.5106, 1.7064]. So we can be 95% confident that the average rate of increase of blood pressure with age is between 1.51 units and 1.71 units per year (rounded to two decimal places).

6   Based on the sample data, can we estimate the average blood pressure of the subpopulation of all newborn babies?

Newborn babies are zero years old, and so the average blood pressure of all newborn babies would be $\mu_Y(0)$, which is equal to β_0. The calculations for the estimate of β_0 using the sample data would yield the value 66.808. However, there is no guarantee that the population regression function $\mu_Y(x) = \beta_0 + \beta_1 x$ is valid for all values of X. Noting that the smallest age in the sample is 23, we immediately know not to extrapolate the estimated regression function to obtain an estimate of the average blood pressure of newborn babies. In short, while we can carry out the calculations for $\hat{\mu}_Y(0)$, the result will not necessarily be meaningful because this involves an excessive amount of extrapolation.

7   Calculate a 95% two-sided prediction interval for the blood pressure of an individual who is 60 years old.

If an individual is randomly chosen from the subpopulation of all individuals who are 60 years old, this individual's blood pressure value would be predicted to be in the interval

$$[\hat{Y}(60) - 2.074 \, SE(\hat{Y}(60)), \ \hat{Y}(60) + 2.074 \, SE(\hat{Y}(60))]$$

with 95% confidence. The value of $\hat{Y}(60)$ is $\hat{\mu}_Y(60) = \hat{\beta}_0 + \hat{\beta}_1(60) = 163.32$, and the value of $SE(\hat{Y}(60))$, using (3.6.5), is 2.98. So this prediction interval is calculated to be [157.139, 169.501], and the confidence statement is

$$C[157.14 \le Y(60) \le 169.50] = 0.95$$

Simultaneous Confidence Intervals for $\mu_Y(x)$ for Several Values of x

There are situations when we want confidence intervals for $\mu_Y(x)$ for several (or all) values of x such that we have confidence $1 - \alpha$ that all the intervals are simultaneously correct (refer to Section 1.6). For example, in evaluating the performance of a rocket during a certain 30-second interval (say, $15 \le X \le 45$), the linear model $\mu_Y(x) = \beta_0 + \beta_1 x$ is assumed to be valid, where Y is the speed of the rocket in feet per second and X is time after the launch in seconds. Data are collected from test firings of small models and estimates are obtained for β_1, β_0, and σ by the formulas in (3.4.8), (3.4.9), and (3.4.22), respectively. A decision must be made about a piece of equipment that is to be installed on the rocket, and the correct decision depends on knowing the average speed at 15 seconds, 20 seconds, 25 seconds, 30 seconds, 35 seconds, 40 seconds, and 45 seconds after the launch. Thus confidence intervals are needed for $\mu_Y(x)$ for $x = 15, 20, 25, 30, 35, 40,$ and 45 seconds such that we have confidence at least $1 - \alpha$ that *all* the confidence intervals are simultaneously correct. Thus we can have confidence $1 - \alpha$ that the decision is correct.

To obtain confidence intervals for $\mu_Y(x)$ for m different values of X so that we can have at least $1 - \alpha$ confidence that they are *all* simultaneously correct, we use the formula in (3.6.1) with the *table-value* given by

$$\text{table-value} = \text{smaller of the two values } t_{1-\alpha/2m:n-2} \text{ and } \sqrt{2F_{1-\alpha:2,n-2}}$$

Table T-4 in Appendix T contains table t-values for m **simultaneous** two-sided confidence intervals with *simultaneous confidence coefficient* greater than or equal to $1 - \alpha$ for $m = 2, 3, \ldots, 6$. Table T-5 contains F values.

Problems   3.6

3.6.1   For the population data in Table D-1 in Appendix D, which is also in the file **car.dat** on the data disk, what parameters must be estimated to obtain the difference between the average first-year maintenance costs of cars driven 15,000 miles and cars driven 10,000 miles?

3.6.2   Parts (a)–(e) refer to the data in Table 3.5.2, which were obtained by simple random sampling from the population data in Table D-1 in Appendix D. These data are also in the file **car20.dat** on the data disk. Assumptions (B) are presumed to hold so the regression function of maintenance cost Y on miles driven X is of the form

$$\mu_Y(x) = \beta_0 + \beta_1 x$$

For these data we have computed the following quantities: $\bar{y} = 521.7$; $\bar{x} = 11,010$; $SSY = 210,568$; $SSX = 175,337,952$; $SXY = 5,489,360$.

a Obtain a point estimate for the quantity of interest in Problem 3.6.1.

b Obtain a 90% confidence interval for the quantity in Problem 3.6.1.

c A company has purchased three cars of the same make as those in Example 3.5.2. One of the cars is to be driven 5,000 miles during its first year, and the other two cars will be driven 12,500 miles each. Predict the total first-year maintenance cost for all three cars together.

d Compute a 90% two-sided confidence interval for the average maintenance cost of all cars driven 12,500 miles during their first year.

e The predictor factor X (miles driven) will be considered to be a useful predictor of Y if σ is less than $50. Compute an 80% confidence statement to help the investigator arrive at a decision.

3.6.3 Consider the arsenic data of Task 3.4.2. Recall that the investigator is interested in determining whether or not the new method of analysis for determining the concentration of As is unbiased. Compute confidence intervals for β_0 and β_1 in Task 3.4.2 such that both confidence intervals are simultaneously correct with 90% confidence. Based on these, help the investigator arrive at an appropriate decision. Give reasons.

3.6.4 Questions (a)–(c) refer to the crystal data of Task 3.4.1.

a Calculate an appropriate 90% confidence statement to help the investigator decide whether the average rate of growth of the crystals is at least 0.4 gram per hour.

b Compute a 95% confidence interval for the weight of a crystal that is allowed to grow for 15 hours.

c Obtain simultaneous two-sided confidence intervals for the weights $Y(10)$ and $Y(25)$ of two individual crystals, one of which is to be grown for 10 hours and the other for 25 hours, such that the investigator can be at least 90% confident that *both* confidence intervals are simultaneously correct. (*Hint*: Use the Bonferroni method described in Box 1.6.3.)

3.7
Tests

Investigators often use statistical tests in an attempt to arrive at answers to practical questions. To do so the investigator is required to formulate a pair of hypotheses, called the null hypothesis (NH) and the alternative hypothesis (AH), and to perform appropriate statistical calculations to obtain a measure, called the **significance probability** or the **P-value** of the evidence contained in the sample data *against* NH. (The results of these calculations do not tell the investigator how much evidence is provided by the sample data in favor of the null hypothesis.) If the evidence against NH is strong, which would be the case if the P-value is small, then it is customary to *reject* the null hypothesis. In the contrary case—i.e., when the P-value is not small—the null hypothesis is not rejected. Neither is it accepted without further analysis.

At this point we refer you back to Section 1.6 (in particular, to the conversation). The discussion in Section 1.6 should convince you that statistical tests are often inappropriate for making practical decisions. Confidence intervals can be much more useful for this purpose.

Formulas and procedures for some statistical tests of size α that pertain to the straight line regression model are summarized in Boxes 3.7.1–3.7.5. The procedures are valid under assumptions (A) or (B).

B O X 3.7.1 **Tests for β_0**

Let q be a specified number. Compute the statistic

$$t_C = \frac{\hat{\beta}_0 - q}{SE(\hat{\beta}_0)}$$

a For testing NH: $\beta_0 = q$ versus AH: $\beta_0 \neq q$, the P-value is the value of α such that $|t_C| = t_{1-\alpha/2:n-2}$.

b For testing NH: $\beta_0 \leq q$ versus AH: $\beta_0 > q$, the P-value is the value of α such that $t_C = t_{1-\alpha:n-2}$.

c For testing NH: $\beta_0 \geq q$ versus AH: $\beta_0 < q$, the P-value is the value of α such that $-t_C = t_{1-\alpha:n-2}$.

B O X 3.7.2 **Tests for β_1**

Let q be a specified number. Compute the statistic

$$t_C = \frac{\hat{\beta}_1 - q}{SE(\hat{\beta}_1)}$$

a For testing NH: $\beta_1 = q$ versus AH: $\beta_1 \neq q$, the P-value is the value of α such that $|t_C| = t_{1-\alpha/2:n-2}$.

b For testing NH: $\beta_1 \leq q$ versus AH: $\beta_1 > q$, the P-value is the value of α such that $t_C = t_{1-\alpha:n-2}$.

c For testing NH: $\beta_1 \geq q$ versus AH: $\beta_1 < q$, the P-value is the value of α such that $-t_C = t_{1-\alpha:n-2}$.

B O X 3.7.3 **Tests for $\mu_Y(x)$**

Let q be a specified number. Compute the statistic

$$t_C = \frac{\hat{\mu}_Y(x) - q}{SE(\hat{\mu}_Y(x))}$$

a   For testing NH: $\mu_Y(x) = q$ versus AH: $\mu_Y(x) \neq q$, the *P*-value is the value of α such that $|t_C| = t_{1-\alpha/2:n-2}$.

b   For testing NH: $\mu_Y(x) \leq q$ versus AH: $\mu_Y(x) > q$, the *P*-value is the value of α such that $t_C = t_{1-\alpha:n-2}$.

c   For testing NH: $\mu_Y(x) \geq q$ versus AH: $\mu_Y(x) < q$, the *P*-value is the value of α such that $-t_C = t_{1-\alpha:n-2}$.

B O X   **3.7.4**   **Tests for $a_0\beta_0 + a_1\beta_1$**

Let q be a specified number. Compute the statistic

$$t_C = \frac{(a_0\hat{\beta}_0 + a_1\hat{\beta}_1) - q}{SE(a_0\hat{\beta}_0 + a_1\hat{\beta}_1)}$$

a   For testing NH: $a_0\beta_0 + a_1\beta_1 = q$ versus AH: $a_0\beta_0 + a_1\beta_1 \neq q$, the *P*-value is the value of α such that $|t_C| = t_{1-\alpha/2:n-2}$.

b   For testing NH: $a_0\beta_0 + a_1\beta_1 \leq q$ versus AH: $a_0\beta_0 + a_1\beta_1 > q$, the *P*-value is the value of α such that $t_C = t_{1-\alpha:n-2}$.

c   For testing NH: $a_0\beta_0 + a_1\beta_1 \geq q$ versus AH: $a_0\beta_0 + a_1\beta_1 < q$, the *P*-value is the value of α such that $-t_C = t_{1-\alpha:n-2}$.

The *P*-values for the tests in Boxes 3.7.1–3.7.4 can be found (at least approximately) by consulting Table T-2 in Appendix T.

B O X   **3.7.5**   **Tests for σ**

Let q be a specified positive number. Compute the statistic

$$\chi_C^2 = \frac{(n-2)\hat{\sigma}^2}{q^2} = \frac{SSE(X)}{q^2}$$

a   For testing NH: $\sigma = q$ versus AH: $\sigma \neq q$, the *P*-value is equal to α where α is a number between 0 and 1 and satisfies

$$\chi_C^2 = \chi_{\alpha/2:n-2}^2 \quad \text{or} \quad \chi_C^2 = \chi_{1-\alpha/2:n-2}^2$$

(only one of these two equalities can be satisfied unless $\alpha = 1$).

b   For testing NH: $\sigma \leq q$ versus AH: $\sigma > q$, the *P*-value is the value of α such that $\chi_C^2 = \chi_{1-\alpha:n-2}^2$.

c   For testing NH: $\sigma \geq q$ versus AH: $\sigma < q$, the *P*-value is the value of α such that $\chi_C^2 = \chi_{\alpha:n-2}^2$.

P-values for the tests in Box 3.7.5 can be found (at least approximately) by consulting Table T-3 in Appendix T. We illustrate these procedures in the following task.

Task   3.7.1

Consider the problem described in Task 3.4.2 in which an investigator wants to evaluate the performance of a new laboratory method for analyzing the concentration of arsenic (As) in water samples. The data appear in Table 3.4.3 and also in the file **arsenic.dat** on the data disk. Recall that the regression function is $\mu_Y(x) = \beta_0 + \beta_1 x$, and the investigator is interested in knowing

- Whether or not β_0 is zero
- Whether or not β_1 is equal to one

Some investigators attempt to answer these questions using statistical tests as follows:

1   Can we conclude that β_0 is zero for all practical purposes?

This question is sometimes translated as follows: How much evidence do the data provide against the hypothesis that $\beta_0 = 0$?

To answer this question, some statisticians and investigators carry out a test of

$$NH: \beta_0 = 0 \qquad against \qquad AH: \beta_0 \neq 0$$

In Task 3.6.1 we obtained $\hat{\beta}_0 = 0.10458$ and $SE(\hat{\beta}_0) = 0.0605$. From Box 3.7.1 we get

$$t_C = \frac{0.10458 - 0}{0.0605} = 1.729$$

Because $n = 32$ for this problem, the degrees of freedom are $n - 2 = 30$ and, using Table T-2 in Appendix T, the P-value for this test is between 0.05 and 0.10. If the investigator uses an α value equal to 0.05, then NH will not be rejected. If an α value equal to 0.10 or greater is used, then NH will be rejected and the investigator will conclude that β_0 is not equal to zero. This hypothesis test does not help the investigator make a practical decision. If NH is not rejected, then the investigator is still not sure whether β_0 is close enough to zero to be considered equal to zero from a practical point of view or whether there are not enough data to arrive at a practical decision. On the other hand, the confidence interval for β_0 given in part (1) of Task 3.6.1 should help the investigator in this regard.

2   Can we conclude that $\beta_1 = 1$ for all practical purposes?

This question is often translated as follows: How much evidence do the data provide against the hypothesis that $\beta_1 = 1$?

To answer this question, some statisticians and investigators carry out a test of

$$NH: \quad \beta_1 = 1 \qquad against \qquad AH: \quad \beta_1 \neq 1$$

In Task 3.6.1 we obtained $\hat{\beta}_1 = 0.98771$ and $SE(\hat{\beta}_1) = 0.01447$. From Box 3.7.2 we get

$$t_C = \frac{0.98771 - 1.0}{0.01447} = -0.85$$

From Table T-2 in Appendix T the P-value for this test is approximately 0.40, so the data do not provide sufficient evidence (at any reasonable value of α such as $0.2, 0.1, 0.05$, etc.) to conclude that $\beta_1 \neq 1.0$.

Again, this hypothesis test does not help the investigator make a practical decision. If NH is not rejected, then the investigator is still not sure whether β_1 is close enough to 1 to be considered equal to 1 from a practical point of view or whether there are not enough data to arrive at a practical decision. On the other hand, the confidence interval for β_1 given in part (2) of Task 3.6.1 should help the investigator in this regard.

Task 3.7.2

Now consider the problem discussed in Task 3.4.1 where an investigator is interested in predicting Y, the weight of crystals used in electronic devices as a function of X, the number of hours the crystals grow. We perform an appropriate statistical test in an attempt to find an answer to the following question.

1 Suppose that crystals are allowed to grow for 24 hours instead of 12 hours. Do the data provide enough evidence to conclude, on the average, that the amount of money the larger crystals can be sold for is more than $50 over what the smaller crystals would fetch?

The average weight of crystals that are allowed to grow for a total of 24 hours is $\mu_Y(24)$ grams, whereas the crystals grown for 12 hours have an average weight of $\mu_Y(12)$ grams. Thus the additional money that the larger crystals would bring (on the average) is equal to $16\mu_Y(24) - 10\mu_Y(12) = 6\beta_0 + 264\beta_1$ dollars. This leads to the pair of hypotheses

$$NH: \quad 6\beta_0 + 264\beta_1 \leq 50 \qquad against \qquad AH: \quad 6\beta_0 + 264\beta_1 > 50$$

So $q = 50$ and let $\theta = 6\beta_0 + 264\beta_1$.

Now recall that in Task 3.4.1 we computed the following quantities: $\bar{x} = 15$; $SSX = 910$; $\hat{\beta}_0 = 0.0014$; $\hat{\beta}_1 = 0.5034$; and $\hat{\sigma} = 1.062$. Thus we have $\hat{\theta} = 6\hat{\beta}_0 + 264\hat{\beta}_1 = 132.906$. Also using (3.6.6) we calculate $SE(\hat{\theta}) = 6.36$. Using the test statistic in Box 3.7.4 for testing θ we get

$$t_C = \frac{\hat{\theta} - 50}{SE(\hat{\theta})} = 13.04$$

Because n = 14 for this problem, the degrees of freedom are 12 and the P-value for this test, obtained from Table T-2 in Appendix T, is less than 0.0005, indicating that the data contain strong evidence against the null hypothesis in favor of the alternative hypothesis.

It is instructive to compute a one-sided 95% lower confidence bound for θ. As usual, this may be obtained as the lower endpoint of a 90% two-sided confidence interval, which is given by

$$C[121.58 \leq \theta \leq 144.240] = 0.90$$

Thus we get

$$C[121.58 \leq \theta] = 0.95$$

Therefore we are 95% confident that the larger crystals can be sold for at least $121 more than the smaller crystals, so we would be led to conclude that they can be sold for at least $50 more than the smaller crystals. Note that the confidence interval supports the result of the test, but it gives considerably more information than the test.

Problems 3.7

3.7.1 A particular brand of cough syrup comes in $\frac{1}{4}$-litre bottles and the manufacturer recommends that after a bottle is unsealed it be kept under cool conditions. The shelf-life of the cough syrup in question is dependent on the temperature at which it is stored. The quality control laboratory of the manufacturing company has obtained the data in Table 3.7.1 on the shelf-life of the cough syrup in question. The data are also in the file **shelflif.dat** on the data disk.

Some basic quantities that are needed for obtaining point estimates and confidence intervals are as follows:

$$\sum_{i=1}^{18} y_i = 11341 \qquad \sum_{i=1}^{18} x_i = 387$$

$$\sum_{i=1}^{18} (y_i - \bar{y})^2 = SSY = 96294.9 \qquad \sum_{i=1}^{18} (x_i - \bar{x})^2 = SSX = 484.5$$

$$\sum_{i=1}^{18} (y_i - \bar{y})(x_i - \bar{x}) = SXY = -6663.49$$

The regression function of shelf-life Y on storage temperature X is assumed to be a straight line $\mu_Y(x) = \beta_0 + \beta_1 x$ for values of x in the range 10°C to 35°C, and assumptions (A) are presumed to be satisfied.

a Define an appropriate target population for this investigation.

b Define an appropriate study population for this investigation.

T A B L E 3.7.1
Shelf-Life Data

Bottle Number	Shelf-Life (Y, in days)	Storage Temperature (X, in °C)
1	727	13
2	760	14
3	730	15
4	716	16
5	683	17
6	665	18
7	641	19
8	663	20
9	653	21
10	615	22
11	585	23
12	614	24
13	592	25
14	564	26
15	537	27
16	537	28
17	552	29
18	507	30

c   Are the data in this investigation obtained by simple random sampling or by sampling with preselected X values?

d   Plot y_i versus x_i. Examine this plot and decide whether a straight line regression model seems reasonable.

e   The director of the laboratory wants to determine whether the data provide evidence (at the 0.05 level) indicating that shelf-life does indeed depend on storage temperature, so he decides to use a statistical test. State an appropriate pair of hypotheses, suitably designating one as the null hypothesis and the other as the alternative hypothesis, and calculate the P-value for this test. What is your conclusion?

f   Estimate, if possible, the average shelf-life for this cough syrup if it is to be stored at $0°C$.

g   Estimate the average shelf-life for this cough syrup if it is to be stored at $15°C$. Also compute a 95% confidence interval for this quantity.

h   Answer part (e) using an appropriate confidence interval instead of a hypothesis test.

i   Do the data provide evidence (at the 0.05 level) indicating that the average shelf-life for bottles of cough syrup stored at $13°C$ is at least 650 days? Carry out an appropriate statistical test and state your conclusions.

j   Construct an appropriate confidence interval to answer part (i).

3.8
Analysis of Variance

For the straight line regression model (and also for more general linear regression models), it is customary to summarize, in a table, certain key numerical quantities that are useful for making inferences. The process of calculating and examining these key numerical quantities is called an **analysis of variance**. The resulting table containing these quantities is called an analysis of variance table. The first key quantity is

$$SSY = \sum_{i=1}^{n}(y_i - \bar{y})^2 \tag{3.8.1}$$

which was defined in (3.4.15). Recall that SSY can also be written as

$$SSY = \sum_{i=1}^{n} y_i^2 - n\bar{y}^2$$

The quantity $\sum_{i=1}^{n} y_i^2$ is called the *uncorrected total sum of squares* for Y, the quantity $n\bar{y}^2$ is called *correction for the mean*, and SSY is called the *corrected total sum of squares for Y*. The adjective *corrected* is often omitted, and SSY is usually simply referred to as the *total sum of squares for Y*.

We know that the best predictor of the Y value of a randomly chosen item from the population, in the absence of any knowledge about its X value, is μ_Y, the mean Y value of all items in the population, and that σ_Y is a measure of how well μ_Y represents the entire population. If data are obtained by simple random sampling, then we can estimate μ_Y by the sample mean $\bar{y}$, and σ_Y by

$$\hat{\sigma}_Y = \sqrt{\frac{\sum_{i=1}^{n}(y_i - \bar{y})^2}{(n-1)}} = \sqrt{\frac{SSY}{(n-1)}}$$

The quantity $SSY/(n-1)$ is sometimes written as MSY and is called the *total mean square for Y*. The divisor, $n-1$, of SSY is called the *degrees of freedom associated with SSY*.

The second key quantity is $SSE(X)$, or SSE for short, the sum of squares of the prediction errors when the sample regression function of Y on X is used to predict the Y values of the sample items. Recall that we defined SSE in (3.4.6), but for convenience we reproduce the definition here.

$$SSE = \sum_{i=1}^{n} \hat{e}_i^2 = \sum_{i=1}^{n}(y_i - \hat{\beta}_0 - \hat{\beta}_1 x_i)^2 \tag{3.8.2}$$

and it has $(n-2)$ degrees of freedom associated with it. The quantity $SSE/(n-2)$ is often denoted by $MSE(X)$, or simply by MSE, and it is referred to as the *mean squared error* or *residual mean square*. Thus

$$MSE(X) = MSE = \frac{SSE(X)}{n-2} \tag{3.8.3}$$

More generally, when a quantity that is a sum of squares of estimated errors of prediction is divided by its associated degrees of freedom, the resulting quantity is referred to as a mean square error. Recall that $\sqrt{MSE} = \hat{\sigma}$ is the estimate of σ.

The third key quantity is the difference $SSY - SSE$, which is the amount by which the total sum of squares SSY is reduced, by using the regression of Y on X, to obtain SSE. This difference is called the *sum of squares due to regression* and is denoted by $SSR(X)$, or simply as SSR. Thus

$$SSR = SSY - SSE \qquad (3.8.4)$$

It can also be easily verified that SSR is in fact equal to the quantity $\hat{\beta}_1^2(SSX)$.

In general, the quantity

$$\frac{SSR}{\text{degrees of freedom associated with } SSR}$$

is referred to as the *mean square due to regression*. For straight line regression, the degrees of freedom associated with SSR is 1, and hence the mean square due to regression, which is denoted by $MSR(X)$, or MSR for short, is the same as the sum of squares due to regression.

The quantities SSY, SSR, and SSE are generally displayed in a table, called an **AN**alysis **O**f **VA**riance (ANOVA) table, as in Table 3.8.1.

T A B L E 3.8.1

ANOVA for Straight Line Regression

Source	Degrees of Freedom df	Sum of Squares SS	Mean square MS	Computed F-Value
Regression	1	SSR	MSR	$F_C = \frac{MSR}{MSE}$
Error	$n-2$	SSE	MSE	
Total	$n-1$	SSY	MSY	

The statistic F_C in the last column of Table 3.8.1 is sometimes used to test NH: $\beta_1 = 0$ against AH: $\beta_1 \neq 0$. The P-value for the test is equal to the value of α for which $F_C = F_{1-\alpha:1,n-2}$. This test is equivalent to the t-test of NH: $\beta_1 = 0$ against AH: $\beta_1 \neq 0$, which is a special case of the test of NH: $\beta_1 = q$ against AH: $\beta_1 \neq q$ described in Box 3.7.2. The reason is the following. It can be shown that the square of the student's t table-value $t_{1-\alpha/2:m}$ is equal to the F table-value $F_{1-\alpha:1,m}$; i.e.,

$$t_{1-\alpha/2:m}^2 = F_{1-\alpha:1,m}$$

Thus to test NH: $\beta_1 = 0$ with AH: $\beta_1 \neq 0$, instead of using the test statistic

$$t_C = \frac{\hat{\beta}_1}{SE(\hat{\beta}_1)} = \frac{\hat{\beta}_1\sqrt{SSX}}{\hat{\sigma}}$$

and comparing it with a student's t table-value, we could use the statistic

$$t_C^2 = F_C = \frac{\hat{\beta}_1^2 SSX}{\hat{\sigma}^2} = \frac{MSR}{MSE}$$

and compare it with an F table-value. The P-value based on the t-test will be identical to the P-value based on the corresponding F-test.

Caution:

Do not forget the fact that the F-test in an ANOVA table for straight line regression can be used only to test NH: $\beta_1 = 0$ against AH: $\beta_1 \neq 0$ and cannot be used to test NH: $\beta_1 = q$ against AH: $\beta_1 \neq q$ for *nonzero* values of q, nor can it be used for *one-sided* tests of hypotheses regarding β_1.

E X A M P L E   **3.8.1**

For the age and blood pressure data in Table 3.6.2 we compute an analysis of variance table and illustrate the F-test for NH: $\beta_1 = 0$ against AH: $\beta_1 \neq 0$. The analysis of variance is displayed in Table 3.8.2. The entries are obtained from the computations in Task 3.6.2. The P-value corresponding to $F_C = 1161.31$ with 1 and 22 degrees of freedom is less than 0.01 using Table T-5 in Appendix T. The hypothesis that $\beta_1 = 0$ can be tested using the t-statistic in Box 3.7.2. From Task 3.6.2 we get

$$t_C = \frac{1.6085 - 0}{0.0472} = 34.078$$

again yielding a P-value less than 0.01 (actually less than 0.0005). Note also that $t_C^2 = 1161.38 = F_C$ (to within rounding error). If assumptions (B) are satisfied, then $\hat{\sigma}_Y = \sqrt{MSY} = \sqrt{9514.6/23} = 20.34$ is a valid estimate of σ_Y. ■

T A B L E   **3.8.2**

ANOVA for Age and Blood Pressure Data in Table 3.6.2

Source	Degrees of Freedom df	Sum of Squares SS	Mean square MS	Computed F-Value
Regression	1	9337.7	9337.7	$F_C = \frac{MSR}{MSE} = 1161.31$
Error	22	176.9	8.0	
Total	23	9514.6	413.68	

Problems 3.8

3.8.1   The following questions refer to the shelf-life data in Table 3.7.1, which are also stored in the file **shelflif.dat** on the data disk.

   a   Present an analysis of variance table.

b Use F_C from the ANOVA table in part (a) to test NH: $\beta_1 = 0$ against AH: $\beta_1 \neq 0$. What is the *P*-value for this test? Interpret the result.

c Calculate t_C for testing NH against AH in part (b). What is the *P*-value for this test? Interpret the result.

d Verify that the square of t_C in part (c) is equal to F_C in (b). Further verify that the *P*-value calculated from the *t* statistic in part (c) is the same as that calculated from the *F* statistic in part (b).

e What conclusion do you draw regarding β_1 based on the test in part (b)?

f Compute a 99% confidence interval for β_1. How will you use this confidence interval to *decide* whether or not β_1 is close enough to zero to be considered negligible for this problem?

g Write a short paragraph outlining your conclusions in parts (b)–(f) and give reasons for your statements.

3.8.2 The following refer to the age and blood pressure data discussed in Task 3.6.2. The data are in Table 3.6.2 and also in the file **agebp.dat** on the data disk.

a Present an analysis of variance table.

b Use F_C from the ANOVA in part (a) to test NH: $\beta_1 = 0$ against AH: $\beta_1 \neq 0$. Compute the *P*-value for this test. Interpret the result.

c Compute t_C for testing NH against AH in part (b). Compute the *P*-value for this test. Interpret the result.

d Verify that the square of t_C in part (c) is equal to F_C in part (b). Further verify that the *P*-value calculated from the *t* statistic in part (c) is the same as that calculated from the *F* statistic in part (b).

e What conclusion do you draw regarding β_1 based on the test in part (b)?

f Compute a 99% confidence interval for β_1. How will you use this confidence interval to *decide* whether or not β_1 is close enough to zero to be considered negligible for this problem?

g Write a paragraph outlining your conclusions in parts (b)–(f) and justify how you reached them.

3.8.3 Consider the grades26 data given in Table 3.2.2, which are also stored in the file **grades26.dat** on the data disk. Repeat parts (a)–(g) of Problem 3.8.2 for these data.

3.9
Coefficient of Determination and Coefficient of Correlation

We have seen that for a two-variable population $\{(Y, X)\}$, the best prediction of the Y value of a randomly chosen item, given that its X value is x is $\mu_Y(x)$, the value of the regression function of Y on X evaluated at $X = x$. If the X value of the item in question is not used, then the best prediction of the Y value of the item is μ_Y. Clearly, investigators have a choice. They can use $\mu_Y(x)$ to predict the Y value of the selected item, or they can use μ_Y to predict its Y value. Of course to use $\mu_Y(x)$ to predict the Y value of the item, we must know its X value, and there may be some costs involved in determining the X value. Also, there is no guarantee that using $\mu_Y(x)$ rather than μ_Y to predict Y will improve the prediction sufficiently to justify the cost associated with measuring or observing X. The following example makes the point clearer.

E X A M P L E   **3.9.1**

Suppose a physician advising a patient with a brain tumor wants to predict the length of time the patient will live if no surgery is performed to remove the tumor. Also suppose that this patient may be regarded as a randomly chosen subject from a population of subjects afflicted with the same type of brain tumor who elected to forego surgery. We assume that the durations between diagnosis and death (referred to as *survival times*) are available for this population of subjects. It is thought that survival time Y is related to tumor severity score X on a scale of 1 to 10, which can be determined from various brain scans. The physician has two options. He can use μ_Y, the average survival time of all the patients in the population, to predict the survival time (i.e., the Y value of his patient, in which case there is no need to know the value of the severity score for this patient's tumor), or he can measure the severity score of the tumor, say its value is $X = 7$, and use $\mu_Y(7)$, the average survival time for all such tumor patients whose tumor severity at the time of diagnosis was equal to 7, to predict the patient's survival time. If $\mu_Y(x)$ is not much better than μ_Y for predicting the Y values (i.e., the survival times), then he has to decide whether the cost of obtaining the X value can be justified.   ▪

Thus the decision to choose between μ_Y and $\mu_Y(x)$ for predicting Y usually depends, at least in part, on (a) the cost of observing the X value, and (b) the improvement in prediction that is made possible by using the X value. In this connection the investigator may be interested in knowing the answers to the following questions:

1 How good is μ_Y as a predictor of the Y value of an item that is to be randomly chosen from the population?

2 How good is $\mu_Y(x)$ as a predictor of the Y value of an item that is to be randomly chosen from the population (note that in order to use $\mu_Y(x)$ to predict the Y value of an item, we must know its X value)?   (3.9.1)

3 How much better is $\mu_Y(x)$ than μ_Y for predicting the Y value of a randomly chosen item?

4 Is μ_Y an adequate predictor of Y?

5 Is $\mu_Y(x)$ an adequate predictor of Y?

We answer these questions using population standard deviations of prediction errors as *summary measures* of how good predictors are.

1 The quantity σ_Y is the measure of how good μ_Y is as a predictor of the value of Y.

2   The quantity $\sigma = \sigma_{Y|X}$ is the measure of how good $\mu_Y(x)$ is as a predictor of the value of Y because, if we know that the X value of the chosen item is x, then we restrict our attention to the subpopulation of all items with $X = x$; $\mu_Y(x)$ is the mean and σ is the standard deviation for this subpopulation. Recall that to use $\mu_Y(x)$, we must know the X value for the item whose Y value is being predicted.

3 σ_Y/σ, or σ_Y^2/σ^2, or $\sigma_Y - \sigma$, or $\sigma_Y^2 - \sigma^2$ (or some other meaningful function of σ_Y and σ) is a measure of how much better $\mu_Y(x)$ is than μ_Y for predicting the value of Y. In this book, we will use σ_Y/σ to describe how much better $\mu_Y(x)$ is than μ_Y for predicting Y.

4   Whether or not μ_Y is adequate for predicting the value of Y depends on the particular problem. An investigator may consider μ_Y to be an adequate predictor of the value of Y if most of the Y values, say at least a proportion p of the population, lie close to μ_Y, say within a distance of d units from μ_Y (p and d are specified by the investigator). It can be shown that when population assumptions (B) for straight line regression are satisfied, a proportion p of the population values lie in the interval $\mu_Y - z_{(1+p)/2}\sigma_Y$ to $\mu_Y + z_{(1+p)/2}\sigma_Y$. So at least a proportion p of the population values will lie in the interval $\mu_Y - d$ to $\mu_Y + d$ provided

$$z_{(1+p)/2}\sigma_Y < d \tag{3.9.2}$$

i.e., if

$$\sigma_Y < d/z_{(1+p)/2} \tag{3.9.3}$$

5   As in item (4), an investigator may consider $\mu_Y(x)$ to be an adequate predictor of the Y value of a randomly chosen item whose X value is known to be equal to x if at least a proportion p of the Y values in the subpopulation with $X = x$ is within d units from the predicted value $\mu_Y(x)$. When population assumptions (A) or (B) for straight line regression are satisfied, this will be true provided that

$$z_{(1+p)/2}\sigma < d \tag{3.9.4}$$

i.e., if

$$\sigma < d/z_{(1+p)/2} \tag{3.9.5}$$

Note   Keep in mind that $\mu_Y(x)$ and μ_Y may *both* be adequate for predicting Y or that *neither one* may be adequate. Also note that we are discussing population prediction functions. In practice we use estimates of these prediction functions based on sample data. The sample prediction functions do not perform as well as the population prediction functions, but if the estimates are based on sufficiently large samples, then we can expect the sample prediction functions to perform almost as well as the population prediction functions.

A further complication arises in judging the adequacy of various prediction functions, viz., in practice we do not know the values of σ_Y or σ. However, sample data may be used to obtain confidence bounds for these quantities that will allow us to determine whether or not the prediction function is adequate. The following task illustrates this point.

Task   3.9.1

To illustrate the preceding ideas, consider the crystal data of Task 3.4.1 where we want to predict the weight Y of a crystal using the amount of time X for which the crystal is grown.

1   Suppose that $\mu_Y(x)$ will be considered an adequate predictor of the Y values if, for each allowable value x of X, a proportion $p = 0.90$ or more of the Y values lie within 0.5 gram of the predicted values. Compute a 95% confidence statement to help decide whether or not $\mu_Y(x)$ is an adequate predictor of Y.

Here $p = 0.90$ and $d = 0.5$. Since assumptions (A) are presumed to be valid, we can conclude using (3.9.5) that $\mu_Y(x)$ is an adequate predictor of Y provided

$$\sigma < d/z_{(1+p)/2}$$

Because $z_{(1+p)/2} = z_{0.95} = 1.645$ from Table T-1 in Appendix T, we can conclude $\mu_Y(x)$ is an adequate predictor of Y if

$$\sigma < 0.5/z_{0.95} = 0.5/1.645 = 0.304$$

Unfortunately we do not know the precise value of σ, and so we do not know whether or not $\sigma \leq 0.304$. In Task 3.4.1 we calculated the point estimate of σ and obtained $\hat{\sigma} = 1.062$. Using (3.6.8) we get the following 95% two-sided confidence interval for σ:

$$C[0.76 \leq \sigma \leq 1.75] = 0.95$$

Thus we can be 95% confident that σ is between 0.76 and 1.75; Using this we would perhaps conclude that $\mu_Y(x)$ is not an adequate predictor of Y.

There are instances where it is difficult, or even impossible, to specify a criterion of adequacy for a prediction function. In such instances, the investigator may be interested in *comparing* the standard deviations of the prediction errors corresponding to each regression function under consideration. In the present situation there are two possible quantities, μ_Y and $\mu_Y(x)$, for predicting Y, and hence the investigator may be interested in comparing σ_Y with σ. She may want either to examine σ_Y and σ individually or to examine some function of σ_Y and σ that may be particularly meaningful in a given problem. One function of σ_Y and σ that has found widespread use in the literature is the coefficient of determination, which we discuss next.

Coefficient of Determination

A commonly used measure that summarizes the performance of $\mu_Y(x)$ as a predictor of Y, *relative* to μ_Y, is the *coefficient of determination* of Y with X, denoted by $\eta_{Y,X}^2$.

This is defined by (η is the Greek letter eta)

$$\eta_{Y,X}^2 = \frac{\sigma_Y^2 - \sigma^2}{\sigma_Y^2} \tag{3.9.6}$$

Recall that σ_Y is the standard deviation of the prediction errors when μ_Y is used to predict Y, whereas σ is the standard deviation of the prediction errors when $\mu_Y(x)$ is used to predict Y. It can be shown that, under assumptions (A) or (B), σ^2 cannot be greater than σ_Y^2, and therefore the quantity on the right-hand side of (3.9.6) cannot be negative. Thus $\eta_{Y,X}^2$ is the *proportional reduction in the variance of prediction errors when using $\mu_Y(x)$ rather than μ_Y to predict Y.*

An alternative measure of relative performance of $\mu_Y(x)$ relative to μ_Y is $\delta_{Y,X}$, the *proportional reduction in the standard deviation of prediction errors*, and it is defined by

$$\delta_{Y,X} = \frac{\sigma_Y - \sigma}{\sigma_Y}$$

These two measures are related by the equation

$$\delta_{Y,X} = 1 - \sqrt{1 - \eta_{Y,X}^2}$$

so that either quantity can be obtained from a knowledge of the other. Because $\eta_{Y,X}^2$ is the measure that is traditionally used by statisticians and practitioners, we consider only this measure although we believe that $\delta_{Y,X}$ is also a meaningful measure.

Relation of $\eta_{Y,X}^2$ to (Pearson's) Coefficient of Correlation $\rho_{Y,X}$

Recall that Pearson's coefficient of correlation (also called the *simple correlation coefficient* or *product moment correlation coefficient*), defined in (1.5.1) and denoted by $\rho_{Y,X}$, is a measure of the *linear association* between Y and X. It can be shown that *when the regression function of Y on X is of the form*

$$\mu_Y(x) = \beta_0 + \beta_1 x$$

(in particular, when population assumptions (A) or (B) are satisfied), the coefficient of determination, $\eta_{Y,X}^2$, of Y with X, is in fact the square of Pearson's correlation coefficient $\rho_{Y,X}$, i.e., $\eta_{Y,X}^2 = \rho_{Y,X}^2$.

This is the reason that the symbol $\rho_{Y,X}^2$ is often used to denote the coefficient of determination of Y with X, but you should be aware that if the regression function of Y on X is not of the form $\mu_Y(x) = \beta_0 + \beta_1 x$ (i.e., it is not linear in x), then $\eta_{Y,X}^2$ is not equal to $\rho_{Y,X}^2$. To avoid any possibility of confusion, we should use the symbol $\eta_{Y,X}^2$ to denote the coefficient of determination of Y with X. However, because this chapter deals with only straight line regression, we are allowed to use $\eta_{Y,X}^2$ and $\rho_{Y,X}^2$ interchangeably in the discussions that follow. Accordingly, *we use the symbol $\rho_{Y,X}^2$ to denote the coefficient of determination in the rest of this chapter.*

Suppose that (population) assumptions (A) or (B) are satisfied. Then $\mu_Y(x) = \beta_0 + \beta_1 x$, and the statements in Box 3.9.1 are true.

B O X **3.9.1**   **Properties of** $\rho^2_{Y,X}$

1 $\sigma \le \sigma_Y$ and consequently $\rho^2_{Y,X} \ge 0$.

2 $\rho^2_{Y,X} = 0$ if and only if $\beta_1 = 0$; i.e., if and only if $\sigma = \sigma_Y$.

3   When $\beta_1 \ne 0$, the sign of $\rho_{Y,X}$ is the same as the sign of β_1.

4   If $\rho^2_{Y,X} = 0$, then $\mu_Y(x) = \beta_0 + \beta_1 x$ is no better for predicting Y than μ_Y is.

5 $\rho^2_{Y,X} = 1$ if and only if $\mu_Y(x)$ is a *perfect predictor* of Y. In this case $\sigma = 0$.

6   The larger the value of $\rho^2_{Y,X}$ is, the better the prediction of Y will be using X; i.e., the predicted values will tend to be closer to the true values.

Remark   As stated in item (4) of Box 3.9.1, the statement $\rho^2_{Y,X} = 0$ means that the regression function of Y on X, namely $\mu_Y(x) = \beta_0 + \beta_1 x$, is no better for predicting Y than the population mean μ_Y is. However, *if $\mu_Y(x)$ is not of the form $\beta_0 + \beta_1 x$, then we cannot draw such a conclusion.*

When the straight line regression model does not hold, it is quite possible that $\rho^2_{Y,X} = 0$ and $\eta^2_{Y,X} \ne 0$, so the function $\mu_Y(x)$ may be a *much better predictor* (in fact, $\mu_Y(x)$ may even be a perfect predictor) of Y than μ_Y is. An illustration of this is given in the conversation later in this section.

Point Estimation for $\rho_{Y,X}$

The quantities $\rho_{Y,X}$ and $\rho^2_{Y,X}$ are population parameters, and the discussion about them has centered around their use and meaning in the population. Valid point estimates of $\rho^2_{Y,X}$ and $\rho_{Y,X}$ can be calculated from sample data according to the formulas given in (3.9.7) and (3.9.8), respectively, provided that assumptions (B) are satisfied. In particular, the data must be obtained by simple random sampling.

$$\hat{\rho}^2_{Y,X} = \frac{SSY - SSE(X)}{SSY} = \frac{SSR(X)}{SSY} = \frac{(SXY)^2}{(SSX)(SSY)} \qquad \text{(3.9.7)}$$

and

$$\hat{\rho}_{Y,X} = \frac{SXY}{\sqrt{(SSX)(SSY)}} \qquad \text{(3.9.8)}$$

If data are obtained by sampling with preselected X values, then no valid estimate of $\rho_{Y,X}$ or $\rho^2_{Y,X}$ is available from the sample data.

As we have stated several times, in the case of straight line regression, investigators can use $\rho^2_{Y,X}$ to decide whether $\mu_Y(x)$ is better than μ_Y for predicting Y. Technically, if population assumptions (A) or (B) are satisfied and $\rho^2_{Y,X} \ne 0$, then we can conclude that $\mu_Y(x)$ is better than μ_Y to predict Y because if $\rho^2_{Y,X} \ne 0$, then it follows that $\sigma < \sigma_Y$. However, in practice, we generally want to know *how much smaller* σ is than σ_Y. To find this out, we can examine the estimated values of σ and σ_Y. An alternate approach is to compute a confidence interval for the *ratio* σ_Y/σ and use this information in making the required decision about how much smaller σ is than σ_Y.

It turns out that to compute a confidence interval for the ratio σ_Y/σ, it is prudent to compute a confidence interval for $\rho_{Y,X}$ as an intermediate step because the table-values that are necessary are readily available in the case of $\rho_{Y,X}$, but that is not the case for σ_Y/σ. Hence we first describe the procedure for obtaining a confidence interval for $\rho_{Y,X}$.

Confidence Interval for $\rho_{Y,X}$

The procedure for computing a two-sided $1 - \alpha$ confidence interval for $\rho_{Y,X}$ is given in Box 3.9.2, and it is valid when assumptions (B) are satisfied.

B O X **3.9.2** **Two-sided $1 - \alpha$ Confidence Interval for $\rho_{Y,X}$**

1 Denote the estimated correlation coefficient $\hat{\rho}_{Y,X}$ by r.

2 Select the chart in Table T-7 in Appendix T corresponding to the desired $1 - \alpha$.

3 Find the number corresponding to the computed correlation coefficient r.

4 Go vertically up the graph along the line at r until you encounter the first curve corresponding to sample size n.

5 Go horizontally from this point on the curve toward the left margin (the ρ margin) until you encounter the vertical axis, say at the point corresponding to $\rho = L$. Then the number L is a $1 - \alpha/2$ lower confidence bound for $\rho_{Y,X}$; i.e., $C[L \leq \rho_{Y,X}] = 1 - \alpha/2$.

6 Go vertically up the graph along the line at r until you encounter the second curve corresponding to sample size n.

7 Go horizontally from this point on the curve toward the left margin (the ρ margin) until you encounter the vertical axis, say at the point corresponding to $\rho = U$. Then the number U is a $1 - \alpha/2$ upper confidence bound for $\rho_{Y,X}$, i.e., $C[\rho_{Y,X} \leq U] = 1 - \alpha/2$.

8 A two-sided $1 - \alpha$ confidence interval for $\rho_{Y,X}$ is $L \leq \rho_{Y,X} \leq U$, and the confidence statement is $C[L \leq \rho_{Y,X} \leq U] = 1 - \alpha$.

Note   When $\hat{\rho}_{Y,X}$ is close to zero or one, the charts in Table T-7 are difficult to read. However, if you are careful in reading the charts, the procedure is adequate for most problems.

Confidence Interval for σ_Y/σ

The procedure for computing a two-sided confidence interval for σ_Y/σ is given in Box 3.9.3, and it is valid under assumptions (B).

B O X **3.9.3** **Two-Sided $1 - \alpha$ Confidence Interval for σ_Y/σ**

1 Obtain L as in Box 3.9.2 corresponding to the desired value of $1 - \alpha$.

2 Obtain U as in Box 3.9.2 corresponding to the desired value of $1 - \alpha$.

3 Let Q_1 denote the larger, and Q_2 the smaller, of the two numbers L^2 and U^2, respectively.

4 Let $U_0 = \dfrac{1}{\sqrt{1-Q_1}}$.

5 If L and U are of the same sign, then let $L_0 = \dfrac{1}{\sqrt{1-Q_2}}$; if they are of opposite signs, then let $L_0 = 1$.

6 A $1 - \alpha$ two-sided confidence statement for σ_Y/σ is given by

$$C[L_0 \leq \sigma_Y/\sigma \leq U_0] = 1 - \alpha$$

L_0 is a $1 - \alpha/2$ lower confidence bound, and U_0 is a $1 - \alpha/2$ upper confidence bound, respectively, for σ_Y/σ.

The following example illustrates the computations discussed in this section.

E X A M P L E **3.9.2**

Consider the power plant SO_2 data given in Table 3.5.4 where we want to study the association between SO_2 concentrations Y at a national park and SO_2 emissions X from a nearby power plant. The data are also in the file **so2.dat** on the data disk. Suppose the investigator wants to know how much improvement in prediction is possible if $\mu_Y(x)$, the regression function of Y on X, is used to predict Y instead of μ_Y; i.e., the investigator wants to know how much smaller σ is than σ_Y. Thus an appropriate population parameter of interest is $\eta_{Y,X}^2$; which is equal to $\rho_{Y,X}^2$ because the regression function of Y on X is assumed to be a straight line. For these data we have

$$SSX = 31.823 \qquad SSY = 133.468 \qquad SXY = 58.0696$$

Using (3.9.8) we obtain

$$\hat{\rho}_{Y,X} = \frac{SXY}{\sqrt{(SSX)(SSY)}} = \frac{58.0696}{\sqrt{(31.823)(133.468)}} = 0.8910$$

Therefore $\hat{\rho}_{Y,X}^2 = \hat{\eta}_{Y,X}^2 = 0.7939$. Thus we estimate that the prediction error *variance* can be reduced 79.39% by using $\mu_Y(x)$ rather than μ_Y to predict Y.

A 95% two-sided confidence interval for $\rho_{Y,X}$ is obtained, using the procedure described in Box 3.9.2, as follows. From the chart in Table T-7 we obtain $L = 0.77$ and $U = 0.95$ (approximately), corresponding to $n = 14$ and $\hat{\rho}_{Y,X} = 0.89$. Hence we get the following confidence statement:

$$C[0.77 \leq \rho_{Y,X} \leq 0.95] = 0.95$$

Now let us use the procedure in Box 3.9.3 to compute a two-sided 95% confidence interval for σ_Y/σ. We have, as previously calculated, $L = 0.77$ and $U = 0.95$ so that $Q_1 = 0.9025$ and $Q_2 = 0.5929$. Because L and U are of the same sign, we get $L_0 = 1/\sqrt{1 - 0.5929} = 1.57$, whereas $U_0 = 1/\sqrt{1 - 0.9025} = 3.20$. Thus we have the confidence statement

$$C[1.57 \leq \sigma_Y/\sigma \leq 3.20] = 0.95$$

In particular if we want only a lower confidence bound for σ_Y/σ, we have

$$C[1.57 \leq \sigma_Y/\sigma] = 0.975$$

so we have 97.5% confidence that the standard deviation of the prediction errors when using μ_Y to predict Y is at least 1.57 times as big as the standard deviation of the prediction errors when using $\mu_Y(x)$ to predict Y. The investigator can use this information to decide whether to use $\mu_Y(x)$ or μ_Y to predict Y. ∎

Remark While the coefficient of determination may be useful for comparing the performance of $\mu_Y(x)$ *relative to* μ_Y for predicting the values of Y, it cannot be used to determine whether or not $\mu_Y(x)$ is an *adequate* prediction function. This is illustrated in the following example.

E X A M P L E **3.9.3**

Consider the problem discussed in Task 3.6.2 concerning the relationship between age and blood pressure. The data are in Table 3.6.2 and also in the file **agebp.dat** on the data disk. The investigator wants to know how much improvement in prediction is possible if the regression function $\mu_Y(x)$ of blood pressure on age is used for predicting blood pressure instead of μ_Y, the mean blood pressure of all individuals in the population (i.e., the investigator wants to know how much smaller σ is than σ_Y). Thus an appropriate population parameter of interest is $\eta_{Y,X}^2$, which is equal to $\rho_{Y,X}^2$ since the regression function of Y on X is a straight line. Using (3.9.8) we obtain

$$\hat{\rho}_{Y,X} = \frac{SXY}{\sqrt{(SSX)(SSY)}} = \frac{5805.13}{\sqrt{(3608.96)(9514.63)}} = 0.9907$$

The fact that the estimated value of $\rho_{Y,X}^2$ is 0.981 means that the improvement in prediction, by using $\mu_Y(x)$ rather than μ_Y to predict Y, appears to be substantial. *However, this does not necessarily imply that age is an adequate predictor of blood pressure.* That depends on the particular application at hand. For instance, suppose the investigator wants to predict blood pressures of individuals accurately to within $d = 5.0$ units for at least a proportion $p = 0.99$ of the individuals in the population. If he uses $\mu_Y(x)$ to predict the Y values, then from (3.9.5) this would be true provided that

$$\sigma < 5/z_{(1+p)/2}$$

i.e., if

$$\sigma < 5/z_{0.995} = 5/2.575 = 1.942$$

because $z_{0.995} = 2.575$ from Table T-1 in Appendix T . Unfortunately, we do not know the exact value of σ based on the sample data, but for these data we have

$\hat{\sigma} = 2.836$, and a 95% confidence interval for σ is given by the confidence statement $C[2.193 \leq \sigma \leq 4.014] = 0.95$. Using this confidence statement the investigator would perhaps conclude that $\mu_Y(x)$ is not an adequate prediction function for this problem.

Observe that $\hat{\sigma}_Y = 20.34$, which is much greater than $\hat{\sigma} = 2.836$. Consequently using age to predict blood pressure does appear to reduce the prediction error considerably, but not quite enough to make it an adequate predictor of blood pressure for this problem.   ■

Authors' Recommendation

The use of $\rho_{Y,X}$ or $\rho_{Y,X}^2$ to determine whether X is an *adequate* predictor of Y is incorrect. Instead the investigator should use σ to judge the adequacy of $\mu_Y(x)$ as a predictor of Y in the context of the study. When an investigator needs to decide whether $\mu_Y(x)$ or μ_Y should be used to predict Y, we recommend that the investigator examine both $\hat{\sigma}$ and $\hat{\sigma}_Y$ and appropriate functions of them. *Do not make this decision based only on the estimated value of $\rho_{Y,X}$ or $\rho_{Y,X}^2$.*

Conversation   3.9

Investigator: I'd like to get your help in interpreting correlation coefficients. One of our scientists says some of his data suggest that the population correlation coefficient between Y and X is about 0.98, and that this means factor X is an adequate predictor of Y. Is that true?

Statistician: No, it is not necessarily true. For example, suppose that $\sigma_Y = 100$ and $\sigma = 20$. If the regression function is $\mu_Y(x) = \beta_0 + \beta_1 x$, then

$$\rho_{Y,X}^2 = \frac{(10{,}000 - 400)}{10{,}000} = 0.96$$

and so the magnitude of $\rho_{Y,X}$ is $\sqrt{0.96} = 0.98$. But if the scientist decides that $\mu_Y(x)$ is an adequate predictor of Y only if σ is less than 10 units, then X is not an adequate predictor of Y for this problem (because $\sigma = 20$ in this example) even though $\rho_{Y,X} = 0.98$. The thing to remember is that $\rho_{Y,X}$ (or $\rho_{Y,X}^2$) by itself is of almost no value in determining whether factor X is an *adequate* predictor of Y. What we do learn from a knowledge of $\rho_{Y,X}^2$, in the case of straight line regression, is *relatively how much better* $\mu_Y(x) = \beta_0 + \beta_1 x$ *is than* μ_Y *as a predictor of* Y.

Investigator: I see what you're saying, but if $\rho_{Y,X} = 0$, doesn't that mean that factor X is of no value in predicting Y?

Statistician: It does if we know *a priori* that the regression function is $\mu_Y(x) = \beta_0 + \beta_1 x$ (i.e., if the regression function is linear in x). Otherwise $\rho_{Y,X} = 0$ cannot be interpreted to mean that X is of no use in predicting Y. I have concocted some data that I will show you and let you make up your own mind as to whether factor X is useful for predicting Y. The data are

Y	X
1.335	-1.9
0.050	5.9
1.230	1.9
0.052	-6.1
0.858	4.0
1.489	0.3
0.861	-4.2

and the estimated correlation coefficient, $\hat{\rho}_{Y,X}$ of Y and X, is 0.008; i.e., $\hat{\rho}_{Y,X}^2 = 0.000064$. So the data suggest that the square of the correlation coefficient between X and Y is essentially equal to zero. We'll plot the data (see Figure 3.9.1) to see what it looks like.

F I G U R E 3.9.1

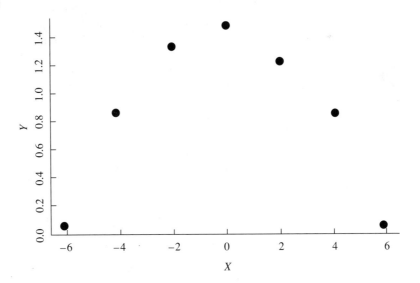

It appears that there is a strong relationship between y_i and x_i. In fact, it appears from the graph that Y can be predicted very well by some function of X, in this case by the square of X. So factor X is clearly very useful in predicting Y even though $\rho_{Y,X}^2$ is essentially zero.

Investigator: I'm getting more discouraged by the minute. First you tell me that a correlation coefficient of 0.98 may *not* indicate that X is an adequate predictor of Y, and then you tell me that even if $\rho^2_{Y,X} = 0$, factor X may be quite useful for predicting Y.

Statistician: That is correct. The correlation coefficient of X with Y is mainly useful when the regression function is $\mu_Y(x) = \beta_0 + \beta_1 x$ (i.e., linear in x), and in other situations it can in fact be misleading. Of course when the population $\{(Y, X)\}$ is bivariate Gaussian, then the regression function of Y on X is of the form $\mu_Y(x) = \beta_0 + \beta_1 x$, and in that case correlation can be a useful summary measure for that population. However, even in that case a correlation coefficient, when used alone, does not tell you how good factor X is for predicting Y. It is the value of σ that tells you how good factor X is for predicting Y. So my advice for the scientists you work with is this: *Do not use the correlation coefficient to decide whether factor X is useful for predicting Y.* On the other hand, if you want to determine how much better $\mu_Y(x) = \beta_0 + \beta_1 x$ is than μ_Y for predicting Y, then compare σ_Y with σ. One way to compare them is by examining their ratio, and we see that

$$\rho^2_{Y,X} = \frac{\sigma^2_Y - \sigma^2}{\sigma^2_Y} = 1 - \left(\frac{\sigma}{\sigma_Y}\right)^2 \tag{3.9.9}$$

From this we get

$$\sqrt{1 - \rho^2_{Y,X}} = \frac{\sigma}{\sigma_Y} \tag{3.9.10}$$

Thus we see that correlation coefficients can be useful in determining how much better $\mu_Y(x) = \beta_0 + \beta_1 x$ is than μ_Y for predicting Y. Of course we've been discussing populations and population parameters, but in a real problem sample values have to be used and the required population quantities have to be estimated.

Investigator: Suppose I have two predictor factors X_1 and X_2, and the regression functions of Y on X_1 and Y on X_2 are both linear. Suppose that ρ^2_{Y,X_1} is two times larger than ρ^2_{Y,X_2}. Does this mean that X_1 is two times better than X_2 for predicting Y?

Statistician: No, it doesn't. Because we are using $\sigma_{Y|X_1}$ and $\sigma_{Y|X_2}$ as the appropriate measures of how good X_1 and X_2, respectively, are for predicting Y, it is better to compare them with one another. From equation (3.9.10) we get

$$\frac{\sqrt{1 - \rho^2_{Y,X_1}}}{\sqrt{1 - \rho^2_{Y,X_2}}} = \frac{\sigma_{Y|X_1}/\sigma_Y}{\sigma_{Y|X_2}/\sigma_Y} = \frac{\sigma_{Y|X_1}}{\sigma_{Y|X_2}}$$

So if $\rho^2_{Y,X_1} = 0.90$ and $\rho^2_{Y,X_2} = 0.45$, then $\rho^2_{Y,X_1} = 2\rho^2_{Y,X_1}$ and

$$\frac{\sigma_{Y|X_1}}{\sigma_{Y|X_2}} = \sqrt{0.10/0.55} = \sqrt{0.18182} = 0.4264$$

On the other hand, if $\rho^2_{Y,X_1} = 0.40$ and $\rho^2_{Y,X_2} = 0.20$, then again $\rho^2_{Y,X_1} = 2\rho^2_{Y,X_2}$ but

$$\frac{\sigma_{Y|X_1}}{\sigma_{Y|X_2}} = \sqrt{0.60/0.80} = \sqrt{0.75} = 0.866$$

So $\sigma_{Y|X_1} = 0.4264\sigma_{Y|X_2}$ when $\rho^2_{Y,X_1} = 2\rho^2_{Y,X_2}$ with $\rho^2_{Y,X_1} = 0.90$ and $\rho^2_{Y,X_2} = 0.45$, but $\sigma_{Y|X_2} = 0.866\sigma_{Y|X_2}$ when $\rho^2_{Y,X_1} = 2\rho^2_{Y,X_2}$ with $\rho^2_{Y,X_1} = 0.40$ and $\rho^2_{Y,X_2} = 0.20$. Thus if one value of ρ^2 is twice as large as another, that does not tell us how much better X_1 is than X_2 for predicting Y.

Investigator: I understand what you're saying, but I'll have to think about all this. Thank you for your time. Perhaps I'll have more questions next week.

Problems 3.9

3.9.1 Consider the sample data in Table 3.2.3, which is also in the file **table323.dat** on the data disk. Assumptions (B) are presumed valid.

  a Plot y_i against x_i to determine whether it appears that the regression function of Y on X is linear in X. Based on this plot, do you think it is?

  b Estimate $\rho_{Y,X}$ the coefficient of correlation between Y and X. Based on this estimate can you tell

   i how good X is for predicting Y?

   ii how much better it is to use $\mu_Y(x)$ rather than μ_y for predicting Y?

   iii *relatively* how much better it is to use $\mu_Y(x)$ for predicting Y than to use μ_Y?

  c Estimate σ and σ_Y. Based on these estimates can you decide whether $\mu_Y(x)$ is an adequate predictor of Y?

3.9.2 Consider the blood pressure data given in Table 3.6.2.

  a–c Repeat (a)–(c) of Problem 3.9.1 for these data.

  d The investigator wants to know whether age is an adequate predictor of blood pressure Y. In other words he wants to know if $\mu_Y(x) = \beta_0 + \beta_1 x$ is an adequate predictor of Y. He will decide that $\mu_Y(x)$ is an adequate predictor of Y if a proportion $p = 0.95$ of the blood pressures of all individuals who are x years old is within 10 units of the predicted value $\mu_Y(x)$. What information about σ is needed for the investigator to determine whether or not $\mu_Y(x) = \beta_0 + \beta_1 x$ is an adequate predictor of Y?

  e In (d) compute an appropriate 90% confidence statement that will help the investigator decide whether $\mu_Y(x) = \beta_0 + \beta_1 x$ is an adequate predictor of Y.

3.9.3 Consider the crystal data in Table 3.4.2.

  a–c Repeat parts (a)–(c) of Problem 3.9.1 for these data.

d   The investigator wants to decide whether X, the number of hours that crystals grow, is a good predictor of Y, the weight of crystals. In particular she wants to decide whether $\mu_Y(x) = \beta_0 + \beta_1 x$ is an adequate predictor of crystal weight Y. She will decide that $\mu_Y(x)$ is indeed an adequate predictor of Y if she determines that a proportion $p = 0.99$ of all crystals that grow x hours will weigh within 1 gram of the predicted value $\mu_Y(x)$. What information about σ is needed for the investigator for her to determine whether or not $\mu_Y(x)$ is an adequate predictor of Y?

e   In (d) compute an appropriate 95% confidence statement that will help the investigator decide whether $\mu_Y(x)$ is an adequate predictor of Y.

3.10
Regression Analysis When There Are Measurement Errors

Thus far in this chapter we have used assumptions (A) or assumptions (B), which among other things require that the response variable Y as well as the predictor variable X can be observed without measurement error. For instance, in Task 3.4.1 it is assumed that the measured (observed) values of crystal weight Y and time X are the true values. The results based on these assumptions are generally satisfactory when measurement errors are present but negligible. However, in many instances the measurement errors are not negligible and it becomes necessary to explicitly account for the presence of these errors in the theoretical development and in the application of linear regression analysis. In this section we consider regression when there are errors in measuring the predictor variable X and/or the response variable Y.

Note   In this section we consider several two-variable populations, so to avoid possible confusion we use the more complete notation $\sigma_{Y|X}$, instead of the simpler notation σ, to denote the standard deviation of the Y values in any subpopulation determined by X.

Measurement Error

Let $(y_1, x_1), \cdots, (y_n, x_n)$ represent sample data from the population $\{(Y, X)\}$ obtained either by simple random sampling or by sampling with preselected X values.

Suppose the true Y value of sample item i is y_i and the *observed* (or recorded or measured) Y value is denoted by v_i. The measurement error in observing the ith sample value of the response variable Y is defined by

D E F I N I T I O N

Y measurement error (for sample item i) = observed Y value (v_i) − true Y value (y_i).   ▪

So we write

$$d_i = v_i - y_i \quad \text{for} \quad i = 1, \ldots, n \tag{3.10.1}$$

Thus d_i is the error in measuring the Y value of the ith sample item.

Suppose the true X value of sample item i is x_i and the *observed* (or recorded or measured) X value is denoted by u_i. The measurement error in observing the ith sample value of the predictor variable X is defined by

DEFINITION

X measurement error (for sample item i) = observed X value (u_i) − true X value (x_i). ∎

So we write

$$e_i = u_i - x_i \quad \text{for} \quad i = 1, \ldots, n \tag{3.10.2}$$

Thus e_i is the error in measuring the X value of the ith sample item.

In this situation it is useful to introduce some additional notation so that we can easily distinguish between true values and measured values. Suppose, as usual, Y_I, X_I denote the true Y and X values for population item I. Now imagine that we measure the Y and X values of each population item. The (conceptual) measured Y value for population item I is denoted by V_I, and the (conceptual) measured X value for this item is denoted by U_I. Thus we have a conceptual two-variable population $\{(V, U)\}$ consisting of the measured values of Y and X for each population item. The true Y and X values for sample item i are denoted by y_i and x_i, respectively. Likewise, the measured Y and X values for sample item i are denoted by v_i and u_i, respectively. Keep in mind that the quantities y_i, x_i will remain unknown due to imprecise measuring techniques, but the quantities v_i, u_i will be known for each sample item.

The assumptions in Box 3.10.1 are made about the measurement errors.

BOX 3.10.1   **Assumptions about Measurement Errors**

1 The errors d_i in measuring the response variable Y are assumed to be a random sample from a Gaussian population with mean zero and standard deviation σ_d.

2 The errors e_i in measuring the predictor variable X are assumed to be a random sample from a Gaussian population with mean zero and standard deviation σ_e.

3 For the *Berkson model* to be discussed, the errors are all assumed to be statistically independent of each other and of Y and U (i.e., the value of any one error is not related to the values of any other errors and is not related to the values of U nor to the values of Y).

4 For the *classical errors in variables model* to be discussed, the errors are all assumed to be statistically independent of each other and of Y and X (i.e., the value of any one error is not related to the values of any other errors and is not related to the values of X nor to the values of Y).

One of the principal reasons for studying the population $\{(Y,X)\}$ is to obtain valid point estimates and confidence intervals *(by **valid** we mean point estimates that are unbiased, or nearly so, and confidence intervals with specified confidence coefficients, or nearly so)* for various population quantities of interest. When assumptions (A) or assumptions (B) are satisfied, the regression function of Y on X is given by

$$\mu_Y(x) = \beta_0 + \beta_1 x \qquad\qquad \textbf{(3.10.3)}$$

with subpopulation standard deviations all equal to $\sigma_{Y|X}$. In this case we are typically interested in the following population quantities:

$$\beta_0, \quad \beta_1, \quad \mu_Y(x) = \beta_0 + \beta_1 x, \quad Y(x), \quad a_0\beta_0 + a_1\beta_1 \qquad \textbf{(3.10.4)}$$

where a_0 and a_1 are specified constants, and x is a specified value of X. Often we are also interested in

$$\sigma_{Y|X} \qquad \rho_{Y,X} \qquad\qquad \textbf{(3.10.5)}$$

If there are measurement errors in X and/or Y, it may *not* be possible to obtain valid estimates for all the quantities in (3.10.4) and (3.10.5).

When the two-variable population $\{(Y,X)\}$ satisfies population assumptions (A) (respectively, population assumptions (B)), and when the measurement errors satisfy the conditions in Box 3.10.1, it can be shown that the two-variable population $\{(V,U)\}$ must also satisfy population assumptions (A) (respectively, population assumptions (B)). In particular, the regression function of V on U is also a straight line function. We denote this function by

$$\mu_V(u) = \beta_0^* + \beta_1^* u \qquad\qquad \textbf{(3.10.6)}$$

where β_0^*, β_1^* may or may not be different from β_0, β_1 in (3.10.3). The standard deviation of the subpopulations of V values determined by U is denoted by $\sigma_{V|U}$. More is said about this later. Because we want to predict Y using the measured value U of X, it is useful to consider another two-variable population, viz., the population $\{(Y,U)\}$. Again we can mathematically prove that this population satisfies population assumptions (A) (respectively, population assumptions (B)) and that the regression function of Y on U is given by

$$\mu_Y(u) = \beta_0^* + \beta_1^* u \qquad\qquad \textbf{(3.10.7)}$$

where the regression coefficients β_0^* and β_1^* in (3.10.7) are *identical* to the regression coefficients in (3.10.6). It is this fact that enables us to use U to predict Y even though we cannot observe the true Y values of the sample items. Since we can observe the values of V and U for each sample item, it is possible to obtain valid estimates for $\mu_V(u)$. But $\mu_V(u)$ is identical to $\mu_Y(u)$. Thus $\hat{\mu}_V(u)$ can be used to predict the true Y value of a randomly chosen item from the study population. Specifically, we can use sample data v_i, u_i to estimate β_0^*, β_1^* and $\mu_V(u)$, and we can use this to predict the true Y values using the measured values of U. The standard deviation of the subpopulation of Y values determined by U is denoted by $\sigma_{Y|U}$.

We now examine the consequences of measurement errors in some detail. First we consider the case when there are errors in measuring the response variable Y only, but the predictor variable X is measured without error.

Measurement Errors in the Response Variable *Y* But Not in the Predictor Variable *X*

Suppose $\{(Y, X)\}$ is the population under study, and assumptions (A) (respectively, assumptions (B)) are satisfied except that the values of the response variable Y are measured with error. Here we assume that the values of the predictor variable X are measured without error. Thus the true X values $x_1, \ldots, x_n$ of the sample items are available; however, the true Y values $y_1, \ldots, y_n$ of these items are not available, but their measured values $v_1, \ldots, v_n$ are. Thus we consider the population $\{(V, X)\}$, which also satisfies population assumptions (A) (respectively, population assumptions (B)) with the regression function given by

$$\mu_V(x) = \beta_0 + \beta_1 x$$

where β_0, β_1 are the same as in (3.10.3).

   In this situation, point estimates for the quantities in (3.10.4) are computed as usual using the formulas in Box 3.4.2, and confidence intervals are computed using the formulas (3.6.1)–(3.6.7) using the data $(v_1, x_1), \ldots, (v_n, x_n)$ in place of $(y_1, x_1), \ldots, (y_n, x_n)$ because $y_1, \ldots, y_n$ are unavailable; the quantity $\hat{\sigma}_{V|X}$ is used in place of $\hat{\sigma}$ (i.e, $\hat{\sigma}_{Y|X}$) in (3.6.2)–(3.6.7). The results are valid if the measurement errors satisfy the assumptions in Box 3.10.1. However, $\sigma_{V|X} \geq \sigma_{Y|X}$, and so these point estimates tend to have larger standard errors, and the confidence intervals tend to be wider than when there are no errors in observing Y.

   On the other hand, there is no valid point estimate or confidence interval for $\sigma_{Y|X}$ or for $\rho_{Y,X}$ in (3.10.5) unless additional information is available about the value of σ_d, the standard deviation of the errors in measuring Y.

   When measurement errors are present in the predictor variable X, the procedures are somewhat more complicated than when there are measurement errors only in the response variable Y. When there are measurement errors in the predictor variable X, we consider two distinct models: (1) the Berkson model, named after Joseph Berkson, who first discussed it in detail, and (2) the classical errors in variables model. These are discussed next.

Berkson Model and Classical Errors in Variables Model

Suppose $\{(Y, X)\}$ is the population under study and assumptions (A) (respectively, assumptions (B)) are satisfied except that measurement errors are present in the predictor variable X. Measurement errors *may* also be present in the response variable Y. Then the *Berkson model* applies when data are obtained by preselecting the *measured* values of X, (i.e., by sampling with preselected U values), for example when the recorded values $u_1, \ldots, u_n$ of the predictor variable X are values on a dial or a gauge, etc., that are set at preselected levels. *Although the dial or gauge settings are the values that are recorded for X, the true X values may be different because the gauge or dial may be in error.* On the other hand, the *classical errors in variables model* applies if data are obtained by simple random sampling. In this case it is impossible to preselect the observed (recorded) value for the predictor variable X because the items are randomly selected. We illustrate these two situations with examples.

E X A M P L E **3.10.1** Berkson Model

Consider the problem where an engineer is interested in studying how the temperature X used in manufacturing aluminum cans is related to the strength Y of the cans. Data are collected by making aluminum cans at various temperatures and measuring their strength. It is known that there are nonnegligible errors in measuring X. The process is run by setting the temperature gauge at a preselected value, say 300°C, and measuring Y, the strength of the can; the temperature gauge is then set at another value, say 325°C, and the Y value is measured; etc. Even though the temperature gauge is set at (say) 300°C, the actual temperature at which the process runs may not be exactly 300°C; i.e., even though 300°C is recorded by observing the gauge, this may not be the true temperature at which the process was run because the gauge may be in error. The true temperature may be 298°C or 304°C, etc. In this problem there may also be errors in measuring the response variable Y. ∎

E X A M P L E **3.10.2** Classical Model

Suppose a biologist is interested in the relationship between the number of damaged blood cells Y and blood sugar levels X in female rats. A random sample of rats is obtained for the study, and a small vial of blood is drawn from each rat. The vials of blood are sent to a laboratory for analysis. Both the counts of damaged cells and the blood sugar levels reported by the laboratory are likely to be different from the unknown true values because of measurement errors. For instance, the laboratory may have recorded the number of damaged cells by examining only a small drop of blood (instead of the entire vial of blood) under a microscope. Similarly, the blood sugar value determined by the laboratory may also be in error. Thus the measured values of X and Y, which we denote by $u_1, \ldots, u_n$ and $v_1, \ldots, v_n$, respectively, are not the true values, and the measurement errors may be nonnegligible. ∎

Consequences of Measurement Errors: Berkson Model

Suppose $\{(Y, X)\}$ is the population under study and assumptions (A) are satisfied, except the values of the predictor variable X, and perhaps the values of the response variable Y, are observed or measured with errors. Further suppose that the data are obtained by preselecting the *measured* values of X, i.e., preselecting the values of $u_1, \ldots, u_n$, using dials or gauges that are subject to measurement errors. If the measurement errors satisfy the assumptions in Box 3.10.1, then it can be shown that the quantities β_0^* and β_1^* in (3.10.6) and in (3.10.7) are the same as β_0 and β_1, respectively, in (3.10.3). *So valid estimates for β_0, β_1, $\mu_Y(x)$, $Y(x)$, and $a_0\beta_0 + a_1\beta_1$ in (3.10.4) can be obtained by the formulas in Box 3.4.2 using the measured data values $(v_1, u_1), \ldots, (v_n, u_n)$ in place of the true values* $(y_1, x_1), \ldots, (y_n, x_n)$. Also, confidence intervals for all the quantities in (3.10.4), except $Y(x)$, can be obtained using the formulas (3.6.1), (3.6.2), (3.6.3), (3.6.4), and (3.6.6) with $\hat{\sigma}_{V|U}$ replacing $\hat{\sigma} = \hat{\sigma}_{Y|X}$.

   On the other hand there are no valid prediction intervals for $Y(x)$ and there are no valid point estimates or valid confidence intervals for $\sigma_{Y|X}$ or $\rho_{Y,X}$ unless additional information is available about the standard deviations σ_d and σ_e of the measurement errors.

Consequences of Measurement Errors: Classical Model

Suppose the study population is $\{(Y, X)\}$ and assumptions (B) hold, except that the X values, and perhaps the Y values, are observed with errors due to imprecise measurement procedures. We suppose that the measurement errors satisfy the assumptions given in Box 3.10.1. Thus data are obtained by simple random sampling; i.e., a simple random sample of n items is chosen from the population and the Y values and the X values are measured for each item. Because the items in the sample are randomly sampled from the population, the X values *cannot be preselected*. The measured Y and X values for the sample items are denoted by $(v_1, u_1), \ldots, (v_n, u_n)$ as usual.

Since population assumptions (B) hold for the study population $\{(Y, X)\}$, the regression function of Y on X is a straight line function. We have denoted this regression function by

$$\mu_Y(x) = \beta_0 + \beta_1 x$$

(see (3.10.3)). However, under the classical errors in variables model, unlike the Berkson model, the quantities β_0^* and β_1^* in (3.10.6) and (3.10.7) are in general different from the quantities β_0 and β_1 in (3.10.3). Therefore, no valid point estimates or confidence interval estimates are available for β_0 or β_1 unless we have information about the size of the measurement errors for X (such as a knowledge of σ_e). Moreover, because the X value of a randomly chosen item cannot be observed (due to errors in measurement), it does not make sense to attempt to use the true X value of the chosen item to predict its Y value. Fortunately, it is possible to obtain valid predictions of the Y value of a randomly chosen item using U, the measured value of X for that item. This is discussed next.

The usual formulas for straight line regression given in Box 3.4.2 are used to compute valid point estimates for β_0^*, β_1^*, $\mu_V(u) = \beta_0^* + \beta_1^* u$, and $a_0 \beta_0^* + a_1 \beta_1^*$, whereas formulas (3.6.1)–(3.6.4) and (3.6.6) are used for confidence intervals with v_i and u_i replacing y_i and x_i, respectively (i.e., regress V on U), and with $\hat{\sigma}_{V|U}$ taking the place of $\hat{\sigma} = \hat{\sigma}_{Y|X}$. The quantity $\hat{\mu}_V(u)$ is also the point estimate for $Y(u)$. Because $\mu_V(u) = \mu_Y(u)$, the computed confidence interval for $\mu_V(u)$ is a valid confidence interval for $\mu_Y(u)$. However, no valid estimate of $\sigma_{Y|X}$ or $\rho_{Y,X}$ is available, and no valid confidence interval is available for $Y(u)$ without additional information regarding the measurement error standard deviations σ_d and σ_e.

Summary

For the classical model where assumptions (B) apply, except that there are measurement errors in the predictor variable X (and possibly in the response variable Y), and where the measurement errors satisfy the assumptions in Box 3.10.1, there are *no* valid estimates or confidence intervals for the regression coefficients β_0 and β_1 in the population regression function of Y on X, given in (3.10.3), without additional information regarding the measurement error standard deviations σ_d and σ_e. However, if the observed sample data v_i, u_i are used in place of the true but unobservable y_i, x_i, respectively, in the formulas in

Box 3.4.2, valid point estimates are obtained for β_0^* and β_1^* (see (3.10.6) and (3.10.7)) and $a_0\beta_0^* + a_1\beta_1^*$, respectively. Valid confidence intervals for β_1^*, β_0^*, and $a_0\beta_0^* + a_1\beta_1^*$ are obtained using (3.6.1) with (3.6.2), (3.6.3), and (3.6.6), respectively, and replacing $\hat{\sigma}_{Y|X}$ in these formulas by $\hat{\sigma}_{V|U}$. Using v_i and u_i, we can also obtain a point estimate and a confidence interval for $\mu_V(u)$, and *these give a valid point estimate and a valid confidence interval for $\mu_Y(u)$*. Also we can compute a point estimate and a confidence interval (prediction interval) for $V(u)$. The point estimate of $V(u)$ is also a valid point estimate of $Y(u)$, but the prediction interval for $V(u)$ is not a valid prediction interval for $Y(u)$. Furthermore, no valid point estimates or confidence intervals exist for

$$\mu_Y(x) = \beta_0 + \beta_1 x, \qquad \rho_{Y,X}, \qquad \sigma_{Y|X}$$

unless information is available about the values of σ_d and σ_e, the standard deviations of the measurement errors associated with Y and X, respectively.

In the following two tasks, we discuss several typical problems encountered when measurement error is present.

Task   3.10.1

In this task we discuss the Berkson Model using the setup of Example 3.10.1. We suppose that an engineer preselected 5 different sets of *measured* values for temperature and ran the process by setting the temperature gauge at each of these preselected values. The *measured values* of the strength of the aluminum cans, along with the *measured values* of the predictor variable, are displayed in Table 3.10.1 and are also stored in the file **cans.dat** on the data disk.

Assumptions (A) are presumed to be valid except that Y and X cannot be observed due to measurement errors. We suppose that these measurement errors satisfy the assumptions in Box 3.10.1. Thus the regression function of the true values of the response variable Y on the true values of the predictor variable X is of the form

$$\mu_Y(x) = \beta_0 + \beta_1 x \tag{3.10.8}$$

Since the U values were obtained by setting the temperature gauge at several preselected values, Berkson's model is applicable, and valid point estimates and confidence intervals for β_0, β_1, and $\mu_Y(x)$ can be calculated by using the measured data values $(v_1, u_1), \ldots, (v_{15}, u_{15})$. We simply use the formulas in Box 3.4.2 for point estimates and the formulas (3.6.1), (3.6.2), (3.6.3), and (3.6.4) for confidence intervals with v_i and u_i replacing y_i and x_i, respectively, in the calculations, and $\hat{\sigma}_{V|U}$ taking the place of $\hat{\sigma}_{Y|X}$.

TABLE 3.10.1

Aluminum Cans Data

Run Number	Observed Strength (newtons) V	Observed Temperature (°C) U
1	18.6	300.0
2	26.3	300.0
3	31.5	300.0
4	20.0	400.0
5	29.2	400.0
6	32.9	400.0
7	29.2	500.0
8	32.5	500.0
9	41.9	500.0
10	31.5	600.0
11	37.6	600.0
12	41.1	600.0
13	34.7	700.0
14	43.2	700.0
15	44.5	700.0

The following quantities can be calculated easily.

$$\bar{v} = 32.9800 \qquad \bar{u} = 500 \qquad SSV = 860.044$$

$$SSU = 300{,}000 \qquad SUV = 12{,}010.0$$

$$SSE(U) = 379.244 \qquad MSE(U) = 29.1726 \qquad \hat{\sigma}_{V|U} = 5.40117$$

$$\hat{\beta}_0 = 12.963 \qquad \hat{\beta}_1 = 0.040033 \qquad SE(\hat{\beta}_0) = 5.124 \qquad SE(\hat{\beta}_1) = 0.009861$$

Now consider the following questions.

1 What is the estimate of the true average strength of aluminum cans that were manufactured with a true temperature of 350° C?

   The true average strength of aluminum cans that were manufactured with a true temperature of 350°C is $\mu_Y(350) = \beta_0 + 350\beta_1$. This is estimated by $\hat{\beta}_0 + 350\hat{\beta}_1 = 26.975$ newtons.

2 What is the estimate of the *true average* strength of aluminum cans that were manufactured with the temperature dial set at 350° C?

   The answer again is $\hat{\mu}_Y(350) = 26.975$ newtons, because $\hat{\mu}_V(350) = \hat{\beta}_0 + \hat{\beta}_1(350) = 12.963 + 0.040033(350) = 26.975$ newtons and $\hat{\mu}_V(u) = \hat{\mu}_Y(u)$ for any allowable value of u, the temperature dial setting.

3   What is a 95% two-sided confidence interval for $\mu_Y(u)$, where $u = 350°$ C is the temperature dial setting during the manufacturing process?

We first compute a 95% two-sided confidence interval for $\mu_V(350)$. We get

$$C[22.58 \leq \mu_V(350) \leq 31.37] = 0.95$$

Because $\mu_V(u) = \mu_Y(u)$, a 95% two-sided confidence interval for $\mu_Y(350)$ is given by

$$C[22.58 \leq \mu_Y(350) \leq 31.37] = 0.95$$

4   What is a 95% two-sided confidence interval for $\mu_Y(x)$, where $x = 350°$ C is the true temperature during the manufacturing process?

The quantity $\mu_Y(x)$ for $x = 350$ is $\beta_0 + 350\beta_1$ and is identical to the quantity $\mu_V(u)$ with $u = 350$. A 95% two-sided confidence interval was computed for $\mu_V(u)$ with $u = 350$ in question (3). So the required confidence interval for $\mu_Y(x)$ with $x = 350$ is given by

$$C[22.58 \leq \mu_Y(350) \leq 31.37] = 0.95$$

5   Compute a point estimate and 95% confidence interval for $\sigma_{Y|X}$.

The answer to this question cannot be obtained without some information about the values of σ_d and σ_e, the standard deviations of the errors in measuring Y and X, respectively.

In the following task we illustrate the procedures for the classical errors in variables model.

Task   3.10.2

A researcher in the exercise science department of a university conducted a study to evaluate the relationship between dietary fat and body fat of competitive runners who ran at least 12 hours per week. A random sample of 18 such runners was obtained, and their body fat Y (in percent) and dietary fat X (in percent), were measured. It is known that there are measurement errors in both Y and X, so the measured dietary fat (the measured value of X) is denoted by U, and the measured body fat (the measured value of Y) by V.

The classical errors in variables model is presumed to apply for this problem. Thus the regression function of Y on X is of the form

$$\mu_Y(x) = \beta_0 + \beta_1 x \tag{3.10.9}$$

But since the true X values are not observable, we cannot obtain valid estimates of β_0 and β_1 without additional information concerning the measurement error standard deviations σ_d and σ_e. However, it is possible to obtain valid predictions of the true values Y using the measured values U. In fact, the regression function of Y on U is given by

$$\mu_Y(u) = \beta_0^* + \beta_1^* u \qquad \text{(3.10.10)}$$

where β_0^* and β_1^* are also the regression coefficients in the regression function of V on U; i.e.,

$$\mu_V(u) = \mu_Y(u) = \beta_0^* + \beta_1^* u \qquad \text{(3.10.11)}$$

Note that β_0^* and β_1^* in (3.10.11) are not the same as the corresponding quantities β_0, β_1 in (3.10.9).

The researcher wants to estimate the regression function $\mu_Y(u)$ for predicting an individual's true body fat Y using u, the measured value of dietary fat. The data are in Table 3.10.2 and also in the file **fat.dat** on the data disk.

The following key quantities can be calculated.

$$\bar{v} = 10.3167 \qquad \bar{u} = 25.8333 \qquad SSV = 36.2650$$

$$SSU = 1010.50 \qquad SUV = 117.450$$

$$SSE(U) = 22.6138 \qquad MSE(U) = 1.4134 \qquad \hat{\sigma}_{V|U} = 1.1889$$

$$\hat{\beta}_1^* = 0.116230 \qquad SE(\hat{\beta}_1^*) = 0.03740 \qquad \hat{\beta}_0^* = 7.31407 \qquad SE(\hat{\beta}_0^*) = 1.006$$

T A B L E 3.10.2
Fat Data

Runner	Measured Body Fat V (%)	Measured Dietary Fat U (%)
1	9.8	22
2	11.7	22
3	8.0	14
4	9.7	21
5	10.9	32
6	7.8	26
7	9.7	30
8	11.6	21
9	8.6	17
10	11.2	35
11	12.3	35
12	10.2	24
13	12.0	24
14	11.6	36
15	10.4	20
16	10.8	37
17	11.5	35
18	7.9	14

A careful study of the following questions and our answers to them will help you understand the concepts related to the classical errors in variables model.

1 What is the estimate of the true average body fat of runners with a true dietary fat of 25%; i.e., what is the estimate of $\mu_Y(x)$ for $x = 25$?

The required true average body fat is $\mu_Y(x)$ with $x = 25$; i.e., $\beta_0 + 25\beta_1$. But we do not have valid estimates for β_0 and β_1 without additional information about the value of σ_e, the standard deviation of the errors in measuring X. So we cannot answer this question with available information.

2 What is the estimate of $\mu_Y(x)$, the true average body fat of runners who have a measured dietary fat of $u = 25$?

The answer is $\hat{\mu}_Y(u) = \hat{\mu}_V(u) = \hat{\beta}_0^* + \hat{\beta}_1^* u$ with $u = 25$, which gives $7.31407 + 0.11623(25) = 10.22$ for the required estimate.

3 What is a 95% two-sided confidence interval for $\mu_Y(u)$, where $u = 25$ is the measured dietary fat?

We first compute a 95% two-sided confidence interval for $\mu_V(u)$ by using formulas (3.6.1) and (3.6.4) with $\hat{\sigma}_{V|U} = 1.1889$ substituted for $\hat{\sigma} = \hat{\sigma}_{Y|X}$. We get

$$C[9.622 \le \mu_V(25) \le 10.818] = 0.95$$

But since $\mu_Y(u) = \mu_V(u)$, the 95% two-sided confidence interval for $\mu_Y(u)$ with $u = 25$ is also [9.622, 10.818]. **Note that for this to be a valid confidence interval, the body fat $u = 25$ must be measured by the same procedure that was used to obtain the sample values u_i.**

4 What is a 95% two-sided confidence interval for $\mu_Y(x)$, where $x = 25$ is the true dietary fat of runners?

The answer to this question cannot be obtained without some information about the value of σ_e, the standard deviation of the errors in measuring X.

5 If an individual's dietary fat is measured to be $u = 25$, estimate this individual's true body fat; i.e., obtain $\hat{Y}(u)$.

The formula in (3.4.11) can be used to obtain $\hat{Y}(u)$, the point estimate for the true body fat, using the measured dietary fat $u = 25$. We get

$$\hat{Y}(u) = \hat{\mu}_Y(u) = \hat{\beta}_0^* + 25\hat{\beta}_1^* = 7.31407 + 0.11623(25) = 10.22$$

6 In question (5), compute a 95% two-sided confidence interval for the true value of this individual's body fat if the measured value of dietary fat is $u = 25$; i.e., compute a 95% two-sided confidence interval for $Y(u)$ with $u = 25$.

A valid 95% confidence interval for $Y(u)$ for any specified value of u cannot be determined based on available information.

7 Compute a point estimate and 95% confidence interval for $\sigma_{Y|X}$ and for $\rho_{Y,X}$.

The required point estimates and confidence intervals cannot be obtained without additional information about the values of σ_d and σ_e, the standard deviations of the errors in measuring Y and X, respectively.

Note   A valid point estimate and confidence interval for $\sigma_{V|U}$ can be obtained using the usual formulas in (3.4.20) and (3.6.8) with y_i, x_i replaced by v_i, u_i. Likewise, a valid point estimate and confidence interval for $\rho_{V,U}$ can be obtained by using the formula in (3.9.8) and the procedure given in Box 3.9.2, respectively, with y_i, x_i replaced by v_i, u_i.

Conversation 3.10

Investigator: Good afternoon. I have some questions about regression when there are measurement errors present. Is this a good time to discuss them with you?

Statistician: Certainly.

Investigator: In Example 3.10.1, suppose the gauge is extremely accurate so there is no error in measuring X, but there is error in measuring Y, the strength of the aluminum cans. (I am presuming that assumptions (A) are satisfied, except that Y cannot be observed due to errors in measurement.) The observed value of Y is denoted by V. Can I ignore the fact that there are measurement errors and proceed as if there are none?

Statistician: You can regress V on X and use all the formulas in Sections 3.4 and 3.6 for point estimates and confidence intervals, *except* there is no valid estimate for $\sigma_{Y|X}$ (and $\rho_{Y,X}$).

Investigator: Does that matter?

Statistician: Only if you need to know $\sigma_{Y|X}$ to make your decisions.

Investigator: Because $\hat{\sigma}_{Y|X}$ (i.e., $\hat{\sigma}$) appears in the formulas for $SE(\hat{\beta}_i)$, and hence in confidence intervals for β_i, don't I need to compute it?

Statistician: For the standard errors and confidence intervals you refer to, the quantity $\hat{\sigma}_{Y|X}$ (i.e., $\hat{\sigma}$) is replaced with $\hat{\sigma}_{V|X}$, and this can be computed. In fact, this is equal to $\sqrt{MSE(X)}$, and it is computed, as usual, by regressing V on X.

Investigator: If assumptions (B) are satisfied except there are errors in measuring X (and perhaps in measuring Y), you state that one cannot estimate the β_i in (3.10.3) and hence one cannot use the estimate of

$$\mu_Y(x) = \beta_0 + \beta_1 x$$

to predict Y using the true value x of X. It appears to me that even if the β_i are known exactly, one cannot use

$$\mu_Y(x) = \beta_0 + \beta_1 x \tag{3.10.12}$$

in (3.10.3) to predict Y because the true value of X cannot be observed so I have no X value to use in (3.10.12). So why are we ever interested in it?

Statistician: You are correct in noting that (3.10.12) cannot be used to predict Y when the true x is unavailable. However, if β_0 and β_1 are known, then you can use the quantity $\beta_0 + \beta_1 u$ to predict Y where u is the measured value corresponding to x. So if β_0 and β_1 are known, they can certainly be useful even if the true x is not known. Furthermore, the investigator might want to examine (3.10.12) to see how the average value of Y is affected by the values of X. To illustrate, consider Example 3.10.2. An investigator might want to know how much the true average number of damaged blood cells $\mu_Y(x)$ changes when the true amount of blood sugar X changes by one unit. The answer is β_1 in (3.10.12). Additionally, the investigator may want to use the estimate of β_0 or β_1 in another formula in another context. For these reasons it is useful to know the values of β_0 and β_1, but no *valid* estimate of β_0 or β_1 is available for this problem without some additional knowledge about the measurement error standard deviation σ_e.

   In practice we can *actually predict* the true average value of Y by using U, the measured value of the predictor, and this can be done as usual by regressing V on U (i.e., use the formulas in Sections 3.4 and 3.6) to obtain point and confidence interval estimates for the β_i and for $\mu_V(u)$. Then use the fact that $\mu_V(u) = \mu_Y(u)$ to obtain a point estimate and confidence interval for $\mu_Y(u)$.

Investigator: I see what you're saying. In general we *can* obtain valid estimates for many of the quantities we want even if there are measurement errors in Y and X.

Statistician: That is correct, but we have to understand the underlying assumptions, limitations, and interpretations.

Investigator: You've said that in some problems the measurement errors are small enough to be considered negligible. How do I determine whether *measurement errors are small enough to be considered negligible* in a given problem?

Statistician: It isn't possible to give a simple answer that will be appropriate for every problem. What you need to do is first specify the objectives of the study and the quantities needed for making decisions, and then examine how the estimates of these quantities will change as a result of errors in measurement. If these changes are small enough so that the decisions aren't affected, then the measurement errors can be considered negligible for that problem. These calculations are best done with the help of a professional statistician.

Investigator: You have stated that some answers aren't available unless σ_e, the standard deviation of the errors in measuring X, is available. Will I ever know it?

Statistician: Probably not, but sometimes an estimate of σ_e is available if the study is planned carefully. This subject is discussed in more advanced books. See reference [7].

Investigator: I see. Are there any other important points I should know in connection with measurement errors?

Statistician: There are many other important considerations you should be aware of in connection with measurement errors, but we will have to discuss them some other time.

However, I'd like to bring to your attention the fact that, in both the Berkson model and the classical model, $\sigma_{Y|X} \leq \sigma_{V|U}$. You can use the measured data to compute a $1 - \alpha$ upper confidence bound for $\sigma_{V|U}$ and get the confidence statement

$$C[\sigma_{V|U} \leq \sqrt{SSE(U)/\chi^2_{\alpha:n-2}}] = 1 - \alpha$$

Then it follows that

$$C[\sigma_{Y|X} \leq \sqrt{SSE(U)/\chi^2_{\alpha:n-2}}] \geq 1 - \alpha$$

is a valid confidence statement. This confidence statement about $\sigma_{Y|X}$ can be useful in practical applications.

Investigator: I think I understand. Thank you. Perhaps I'll come again when I have more questions.

Problems 3.10

3.10.1 A researcher wants to evaluate how different soil temperatures X affect the rate of growth Y of a particular variety of cabbage plants. He conducts an experiment in a greenhouse using seven different soil temperatures (in degrees Fahrenheit). The soil temperatures are controlled by thermostats, but the actual soil temperatures are not necessarily the same as the temperatures at which the thermostats are set. The temperature setting of the thermostats is denoted by U, whereas the true soil temperature is denoted by X. The measurement of growth rate (rate of change of biomass in grams per week) is also subject to errors, and the measured growth rate is denoted by V, whereas the true growth rate is denoted by Y. Due to nonnegligible errors in measurement, only U and V are observable. The data from the experiment are given in Table 3.10.3 and are also stored in the file **cabbage.dat** on the data disk.

It is presumed that (population) assumptions (A) hold for the two-variable population $\{(Y, X)\}$ for $60 \leq X \leq 90$. In particular, the regression function of Y on X is of the form

$$\mu_Y(x) = \beta_0 + \beta_1 x \qquad \text{for} \qquad 60 \leq x \leq 90 \qquad \text{(3.10.13)}$$

We further suppose that the measurement errors satisfy the assumptions given in Box 3.10.1. The results from a regression analysis of V on U are as follows:

$$\bar{v} = 46.7500 \qquad \bar{u} = 75.0000 \qquad SSV = 399.235$$

$$SSU = 1400.00 \qquad SUV = 578.996$$

$$SSE(U) = 159.78 \qquad MSE(U) = 13.315 \qquad \hat{\sigma}_{V|U} = 3.6489$$

$$\hat{\beta}_1^* = 0.4136 \qquad SE(\hat{\beta}_1^*) = 0.09752 \qquad \hat{\beta}_0^* = 15.732 \qquad SE(\hat{\beta}_0^*) = 7.379$$

T A B L E 3.10.3

Cabbage Data

Plant Number	Growth Rate V (grams/week)	Temperature U (°F)
1	35.1	60
2	43.6	60
3	45.3	65
4	37.3	65
5	44.5	70
6	49.5	70
7	46.8	75
8	47.7	75
9	50.2	80
10	51.5	80
11	45.8	85
12	53.8	85
13	54.4	90
14	49.0	90

a   Using the sample data in Table 3.10.3 estimate the parameters β_0 and β_1. Obtain a two-sided 85% confidence interval for the average growth rate of this cabbage plant when the thermostat is set at 72° F. Use $t_{0.925:12} = 1.538$.

b   What is the change in the average growth rate of this particular variety of cabbage plants if the thermostat setting for soil temperature is increased by 5° F? Obtain a two-sided 80% confidence interval for this quantity.

c   If an increase of 5° F in the thermostat setting for soil temperature fails to result in an increase in average growth rate of at least 5 grams per week, then the researcher will conclude that soil temperature is not an important factor relative to his objectives. What will his decision be, based on the confidence interval in part (b)?

d   Suppose the researcher conducts a statistical test to reach a decision in part (c). State the appropriate null and alternative hypotheses for this purpose. Carry out the test and report the P-value. What should his decision be if he uses $\alpha = 0.10$?

e   Predict the growth rate of a cabbage plant if the soil temperature is set at 75° F. Also, if possible, obtain a two-sided 95% confidence interval for the growth rate of this cabbage plant.

f   Obtain a valid estimate, if possible, for the quantity $\sigma_{Y|X}$.

g   Obtain a valid estimate, if possible, for the quantity $\sigma_{V|U}$.

3.10.2   Suppose a researcher wants to predict the first-year maintenance cost Y for minivans purchased next year, using the number of miles X the minivan will be driven during its first year after purchase. The target population is clearly a future population and is unavailable for sampling. Instead, the researcher uses a similar population of minivans from last year as the study population. She selects a simple random sample of

15 minivans from the study population and records the first-year maintenance costs and the miles driven. However, because of inaccurate odometers, the recorded values of miles driven are not the true values but are subject to nonnegligible measurement errors. The maintenance costs may also contain nonnegligible errors becuse of inaccurate record keeping of the minivan owners. Let Y and X represent the true maintenance cost and true miles driven (which are unavailable), respectively, and let V and U represent the corresponding observed, i.e., recorded values. The data are given in Table 3.10.4 and are also stored in the file **minivan.dat** on the data disk.

It is presumed that (population) assumptions (B) hold for the two-variable population $\{(Y, X)\}$. In particular the regression function of Y on X is of the form

$$\mu_Y(x) = \beta_0 + \beta_1 x \tag{3.10.14}$$

We further suppose that the measurement errors satisfy the assumptions given in Box 3.10.1 so that the regression function of V on U is of the form

$$\mu_V(u) = \beta_0^* + \beta_1^* u \tag{3.10.15}$$

and the regression function of Y on U is also of the form

$$\mu_Y(u) = \beta_0^* + \beta_1^* u \tag{3.10.16}$$

where the parameters β_0^* and β_1^* in (3.10.15) are the same as those in (3.10.16).

a   What is the estimated regression function of Y on U?

b   What is the average true first-year maintenance cost for all minivans that will be driven *exactly* 12,500 miles during the first year (give the appropriate population

T A B L E   3.10.4

Minivan Data

Van	Maintenance Costs V (dollars)	Miles Driven U
1	652	16500
2	422	8000
3	724	14200
4	746	18400
5	571	9300
6	644	13900
7	548	11000
8	553	13400
9	792	17200
10	739	16500
11	742	18400
12	763	18700
13	698	17700
14	568	10100
15	663	16300

quantity)? Obtain a valid point estimate of this quantity if possible. If no valid point estimate exists, state why.

c What is the average true first-year maintenance cost of all minivans whose odometer readings indicate that they were driven 12,500 miles during the first year after purchase (give the population quantity)? Obtain a valid point estimate of this quantity if possible. If not, state the reason why it is not possible.

d Is a valid two-sided 90% confidence interval available for the quantity in part (b)? If so, compute it. If not, explain why not.

e Is a valid two-sided 95% confidence interval available for the quantity in part (c)? If so, compute it. If not, explain why not.

3.11
Regression Through the Origin

Recall that when assumptions (A) or (B) for straight line regression are valid, the regression function of Y on X is of the form

$$\mu_Y(x) = \beta_0 + \beta_1 x$$

In some situations the investigator may know, based on subject matter considerations, that the intercept β_0 must equal zero. In such cases the population regression function reduces to

$$\mu_Y(x) = \beta_1 x \tag{3.11.1}$$

and the graph of this function is a straight line that passes through the origin. In this situation, the formulas for computing $\hat{\beta}_1$, $\hat{\mu}_Y(x)$ $\hat{Y}(x)$, $\hat{\sigma}$, $SE(\hat{\beta}_1)$, $SE(\hat{\mu}_Y(x))$, and $SE(\hat{Y}(x))$ must be modified. The appropriate formulas are as follows:

$$\hat{\beta}_1 = \frac{\sum_{i=1}^{n} x_i y_i}{\sum_{i=1}^{n} x_i^2} \tag{3.11.2}$$

$$\hat{\mu}_Y(x) = \hat{\beta}_1 x \tag{3.11.3}$$

$$\hat{Y}(x) = \hat{\beta}_1 x \tag{3.11.4}$$

$$\hat{\sigma} = \sqrt{\frac{1}{n-1} \left[\sum_{i=1}^{n} y_i^2 - \frac{(\sum_{i=1}^{n} x_i y_i)^2}{\sum_{i=1}^{n} x_i^2} \right]} \tag{3.11.5}$$

$$SE(\hat{\beta}_1) = \frac{\hat{\sigma}}{\sqrt{\sum_{i=1}^{n} x_i^2}} \tag{3.11.6}$$

$$SE(\hat{\mu}_Y(x)) = \frac{|x|\hat{\sigma}}{\sqrt{\sum_{i=1}^{n} x_i^2}} \tag{3.11.7}$$

$$SE(\hat{Y}(x)) = \hat{\sigma} \sqrt{1 + \frac{x^2}{\sum_{i=1}^{n} x_i^2}} \tag{3.11.8}$$

$$= \sqrt{\hat{\sigma}^2 + [SE(\hat{\mu}_Y(x))]^2} \tag{3.11.9}$$

The quantity $SSE(X)$ is now given by

$$SSE(X) = \sum_{i=1}^{n}(y_i - \hat{\beta}_1 x_i)^2 \qquad \text{(3.11.10)}$$

which on simplification reduces to

$$SSE(X) = \sum_{i=1}^{n} y_i^2 - \frac{(\sum_{i=1}^{n} x_i y_i)^2}{\sum_{i=1}^{n} x_i^2} \qquad \text{(3.11.11)}$$

The mean square error $MSE(X)$ is now given by

$$MSE(X) = \frac{SSE(X)}{n-1} \qquad \text{(3.11.12)}$$

which is the estimate of σ^2.

Note that the quantity $SSE(X)$ is divided by $n-1$ rather than by $n-2$ as in (3.4.21) to obtain $MSE(X)$ because the number of degrees of freedom associated with $SSE(X)$ is $n-1$ and not $n-2$. More generally, *the number of degrees of freedom associated with a sum of squares of errors is calculated by subtracting from the sample size n, the number of unknown parameters (the β's) in the population regression function*. Thus, when the population regression function is $\mu_Y(x) = \beta_0 + \beta_1 x$, the degrees of freedom associated with the sum of squares of errors

$$SSE(X) = \sum(y_i - \hat{\mu}_Y(x_i))^2 = \sum(y_i - \hat{\beta}_0 - \hat{\beta}_1 x_i)^2$$

is $n-2$, and when the population regression function is $\mu_Y(x) = \beta_0$ (in which case $\beta_0 = \mu_Y$), the degrees of freedom associated with the sum of squares of errors

$$SSY = \sum(y_i - \hat{\mu}_Y(x_i))^2 = \sum(y_i - \bar{y})^2$$

is $n-1$.

Formulas for confidence intervals for β_1, $\mu_Y(x)$, and $Y(x)$ are the same as in (3.6.1), with the understanding that the standard errors to be used are those given in (3.11.6), (3.11.7), and (3.11.8), respectively, and the degrees of freedom to be used for finding table-values is $n-1$. The corresponding procedures for statistical tests for β_1 and $\mu_Y(x)$ are in Boxes 3.7.2 and 3.7.3. The formula for confidence intervals for σ is in (3.6.8), where the degrees of freedom for finding table-values is $n-1$. The procedure for statistical tests for σ is given in Box 3.7.5, where again the degrees of freedom are $n-1$ instead of $n-2$.

The following example illustrates the computations for straight line regression through the origin.

E X A M P L E   3.11.1

It is well known (based on the laws of physics) that when an object is dropped from rest, it will fall because of gravity with continually increasing speed. The speed Y, at the end of X seconds after release, may vary from trial to trial because of air resistance and other random experimental errors. However, the average speed Y (in feet per second) of the object, after a specified amount of elapsed time X in seconds, is given by the regression function

$$\mu_Y(x) = \beta_1 x$$

The quantity β_1 is a constant and is referred to as the acceleration due to gravity. If y_i and x_i are the observed distances and times, respectively, the sample regression model is

$$y_i = \beta_1 x_i + e_i$$

In an experiment conducted to estimate the value of β_1, an object is dropped from rest repeatedly, and each time its speed at the end of a preselected amount of time is recorded. The data appear in Table 3.11.1 and are also stored in the file **gravity.dat** on the data disk.

T A B L E 3.11.1
Gravity Data

Trial Number	Y (ft/sec)	X (sec)
1	63	2
2	128	4
3	194	6
4	257	8
5	322	10
6	387	12
7	451	14

We compute the following quantities:

$$\sum y_i = 1,802 \qquad \sum x_i = 56$$

$$\sum x_i y_i = 18,036 \qquad \sum x_i^2 = 560 \qquad \sum y_i^2 = 580,892$$

$$\hat{\beta}_1 = 32.2071 \qquad \hat{\sigma} = 0.8136 \qquad SE(\hat{\beta}_1) = 0.0344$$

A 95% two-sided confidence interval for β_1 (using $t_{0.975:6} = 2.447$ from Table T-2 as the table-value) is given by the confidence statement

$$C[32.2071 - (2.447)(0.0344) \le \beta_1 \le 32.2071 + (2.447)(0.0344)] = 0.95$$

i.e.,

$$C[32.12 \le \beta_1 \le 32.29] = 0.95$$

A 90% two-sided confidence interval for σ (using $\chi^2_{0.05:6} = 1.635$ and $\chi^2_{0.95:6} = 12.592$ from Table T-3 as the table-values) is

$$\sqrt{\frac{6(0.8136)^2}{12.592}} \le \sigma \le \sqrt{\frac{6(0.8136)^2}{1.635}}$$

i.e.,

$$0.562 \le \sigma \le 1.559$$

Suppose we want to test (using $\alpha = .01$) NH: $\beta_1 = 32$ against AH: $\beta_1 \ne 32$. We compute $t_C = \dfrac{\hat{\beta}_1 - 32}{SE(\hat{\beta}_1)} = 6.02$ and $t_T = t_{1-\alpha/2:6} = t_{0.995:6} = 3.707$, so $t_C > t_T$ and NH is rejected at $\alpha = 0.01$. ∎

Caution

For some situations it may seem that Y must equal zero whenever X is zero and that a straight line regression function with no intercept term is the correct model (i.e., $\mu_Y(x) = \beta_1 x$). This is not always the case. For example, when studying the growth Y of a population of plants as a function of time X, it is true that at time $X = 0$ none of the plants have grown, so the Y value is zero for each plant. Consequently, $\mu_Y(0) = 0$, so the model $\mu_Y(x) = \beta_1 x$ may appear to be correct. If the study period is 30 days to 70 days from germination, the model with no intercept would be very **inadequate** because in these situations the growth model is often of the form given in Figure 3.11.1 (i.e., not a straight line but a nonlinear curve passing through the origin). In the range from $X = 30$ to $X = 70$, a model that is linear in X may fit quite well (dotted line in Figure 3.11.1), but if it is forced through the origin (i.e., if β_0 is required to be zero), the fit will be bad. So if we use a straight line model for this situation for $30 \le x \le 70$, we do not require β_0 to be zero even though we know that $\mu_Y(0) = 0$.

F I G U R E 3.11.1

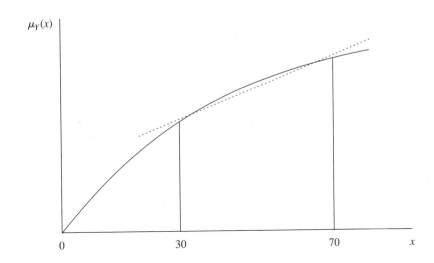

Problems 3.11

3.11.1 Consider the crystal data of Task 3.4.1. These data are given in Table 3.4.2 and are also stored in the file **crystal.dat** on the data disk. You should refer to Task 3.4.1 for a description of these data. Suppose that assumptions (A) hold with the regression function $\mu_Y(x)$ given by

$$\mu_Y(x) = \beta_1 x \quad \text{for} \quad 0 \le x \le 30 \tag{3.11.13}$$

i.e., a straight line through the origin.

a Plot Y against X and visually evaluate whether or not a straight line through the origin appears to be a reasonable model for these data.

b Obtain point estimates for β_1 and σ.

c On the average, what is the increase in weight of crystals for each additional hour of growth? Give the population parameter that answers the question.

d Compute a two-sided 80% confidence interval for the quantity of interest in part (c).

e Compute a two-sided 90% confidence interval for σ.

f The regression function $\mu_Y(x)$ in (3.11.13) will be considered adequate for predicting the weights of crystals provided the predicted weights will be within 3 grams of the true weights for at least a proportion $p = 0.9$ of the crystals in the population. Compute an appropriate 95% confidence interval to help the investigator determine whether or not $\mu_Y(x)$ in (3.11.13) is adequate for predicting values of Y.

3.11.2 Consider the arsenic data of Task 3.4.2. These data are given in Table 3.4.3 and are also stored in the file **arsenic.dat** on the data disk. Refer back to Task 3.4.2 for a description of these data. Suppose assumptions (A) hold with the regression function of Y on X given by

$$\mu_Y(x) = \beta_1 x \quad \text{for} \quad 0 \le x \le 7, \tag{3.11.14}$$

i.e., a straight line through the origin. The chemical analysis will be deemed to provide unbiased estimates of the true As concentrations if the slope β_1 is between 0.96 and 1.04. Compute an appropriate 99% confidence interval to help the investigator decide whether or not the chemical analysis provides unbiased estimates of As concentrations.

3.12
Exercises

3.12.1 Give two examples (preferably from your own field) in which you want to study the relationship between a response variable Y and one or more factor variables for the purpose of prediction or other reasons. Describe carefully the target population, the study population, and convenient sampling methods for each example. If possible describe the form of $\mu_Y(x)$.

3.12.2 The SAT scores X and the the corresponding GPAs (Y) at the end of the first term for a simple random sample of ten students from a major university are as follows:

SAT Scores	321	358	640	270	443	669	582	451	791	594
GPA	1.97	2.19	2.98	1.89	2.66	3.14	2.20	2.02	3.07	3.11

a If assumptions (B) are valid, for which of the following parameters can a valid estimate be obtained?

$$\beta_0, \quad \beta_1, \quad \sigma, \quad \mu_Y(x), \quad \mu_Y, \quad \mu_X, \quad \sigma_X, \quad \sigma_Y, \quad \rho_{Y,X}$$

b Find point estimates of all parameters in part (a) for which valid estimates are available. Show all your calculations.

c Find 90% two-sided confidence intervals for all parameters in part (a), except $\rho_{Y,X}$, for which valid confidence intervals are available. Compute a 95% two-sided confidence interval for $\rho_{Y,X}$.

d Plot y_i versus x_i. From the plot do you have reason to believe that the regression function of Y on X is a straight line function?

e Estimate how much better $\mu_Y(x)$ is than μ_Y for predicting GPA.

f Using population parameters, write an expression for the difference between the average GPA of all students whose SAT score is 550 and the average GPA of all students whose SAT score is 500.

g Obtain a point estimate and a 95% two-sided confidence interval for the quantity in part (f).

h Repeat parts (f) and (g) for SAT scores of 700 and 600 (instead of 550 and 500), respectively.

3.12.3 A simple random sample of twenty U.S. males was selected, and the following information was recorded for each individual.

X = Number of grams of fat consumed per day. (This is usually calculated from a detailed record of food intake over a period of several days and then averaged to obtain fat intake per day).
Y = Total cholesterol in blood in milligrams per deciliter. (This is obtained from a blood test).

Suppose assumptions (B) for straight line regression are satisfied so the model is $\mu_Y(x) = \beta_0 + \beta_1 x$. The data are given in Table 3.12.1 and are also stored in the file **chol.dat** on the data disk.

T A B L E   3.12.1

Cholesterol Data

Sample Item Number	Total Cholesterol Y, (in mg/dl)	Daily Fat Intake X, (in g)
1	130	21
2	163	29
3	169	43
4	136	52
5	187	56
6	193	64
7	170	77
8	115	81
9	196	84
10	237	93
11	214	98
12	239	101
13	258	107
14	283	109
15	242	113
16	289	120
17	298	127
18	271	134
19	297	148
20	316	157

Some basic numerical summaries that you will need are: $\sum x_i = 1{,}814$

$$\sum y_i = 4{,}403 \quad SSX = 27{,}674.2 \quad SSY = 72{,}098.6 \quad SXY = 39{,}495.9$$

a   Plot y_i against x_i.

b   Estimate the regression function of total cholesterol on daily fat intake.

c   Plot this regression line on the same graph as in part (a).

d   Compute the standard errors for the intercept and slope estimates (i.e., compute $SE(\hat{\beta}_0)$ and $SE(\hat{\beta}_1)$).

e   Estimate σ.

f   Construct a 95% upper confidence bound for the population intercept β_0.

g   Construct a 99% two-sided confidence interval for the population slope β_1. Does it appear that $\beta_1 = 0$? Why?

h   Construct a 95% two-sided confidence interval for σ.

i   Construct a 99.5% lower confidence bound for the population slope β_1. Interpret this confidence bound (see part (g)).

j   Find the estimate of the average value of total cholesterol for those people whose daily fat intake is 50 grams (i.e., find $\hat{\mu}_Y(50)$). Find the estimate of the cholesterol of an individual chosen at random from this subpopulation (i.e., $\hat{Y}(50)$).

Construct a two-sided 95% confidence interval for $Y(50)$ and $\mu_Y(50)$. Obtain a 95% two-sided confidence interval for the cholesterol level of an individual whose daily fat intake is 60 grams.

k Construct a 97.5% upper confidence bound for the mean cholesterol level of the subpopulation for which the daily fat intake is 25 grams.

l Estimate the correlation coefficient between total cholesterol and fat intake (i.e., find $\hat{\rho}_{Y,X}$). Construct a 95% two-sided confidence interval for $\rho_{Y,X}$, the population correlation coefficient between Y and X. What does this say about the linear relationship between cholesterol and fat intake?

m Perform the following test of hypothesis:

$$\text{NH: } \beta_1 \leq 2 \text{ against AH: } \beta_1 > 2$$

Compute the P-value. What is your interpretation of the result of this test if you decide to reject NH when $P < 0.005$?

n Use the confidence interval in part (g) to decide between NH and AH in part (m). What is your interpretation?

3.12.4 This is a continuation of Exercise 3.12.3.

a Estimate σ_Y and σ.

b Exhibit an analysis of variance table.

c From the analysis of variance table, obtain the computed F statistic F_C for the test of NH: $\beta_1 = 0$ versus AH: $\beta_1 \neq 0$.

d What is the P-value for this test?

e Obtain a 95% confidence interval for σ_Y/σ.

f What do you conclude about the relationship between cholesterol and fat intake?

3.12.5 In Exercise 3.12.4 suppose that a company manufacturing margarine makes the following claim: The difference between the average blood cholesterol level for the subpopulation of individuals consuming 100 grams of fat per day and the average cholesterol level of the subpopulation of individuals consuming 50 grams of fat per day does not exceed 40 milligrams per deciliter (mg/dl). If the manufacturer's claim is true, then perhaps some people would be willing to include extra fat in their diets, thinking that the resulting increase in cholesterol is small enough so that there is no need for concern.

a State the appropriate null hypothesis and alternative hypothesis to test the manufacturer's claim.

b Use the data of Exercise 3.12.3 and calculate the P-value for this test.

c What do you conclude about the manufacturer's claim?

d Compute an appropriate confidence interval and make a decision about the manufacturer's claim.

4

Multiple Linear Regression

4.1
Overview

In Chapter 2 we pointed out that the *population regression function of Y on* $X_1, \ldots, X_k$ *is the best function for predicting a response variable Y, using the predictor variables* $X_1, \ldots, X_k$. In applied problems, the population prediction function is seldom available, and sample data are used to estimate it.

In many situations we either know or assume the *form* of the population regression function so that the regression function is known except for some unknown parameters (regression coefficients). In Chapter 3 we discussed point estimation, confidence interval estimation, and tests for unknown parameters when the population regression function of the response variable Y on the predictor variable X is a straight line function of X; i.e., $\mu_Y(x) = \beta_0 + \beta_1 x$. In particular, the number of predictor variables is one. In this chapter we consider the situation where there are several predictor (or explanatory) variables, say $X_1, \ldots, X_k$, and the population regression function Y on $X_1, \ldots, X_k$ is of the form

$$\mu_Y(x_1, \ldots, x_k) = \beta_0 + \beta_1 x_1 + \cdots + \beta_k x_k \qquad \text{(4.1.1)}$$

When data for the entire population are available, the constants $\beta_0, \beta_1, \ldots, \beta_k$ can be calculated exactly. However, as we pointed out earlier, this is seldom the case in applied problems, and so the constants $\beta_0, \beta_1, \ldots, \beta_k$ are unknown parameters that have to be estimated using sample data. This chapter is primarily concerned with point and confidence interval estimation of the parameters in the **multiple linear regression function** of (4.1.1). The formulas and procedures for point and confidence interval estimation, discussed in Chapter 3, are special cases (with $k = 1$) of the formulas and procedures that are discussed in this chapter.

The chapter is organized as follows. Section 4.2 includes definitions and notation that are used throughout this chapter and elsewhere in the book. Section 4.3 contains a discussion of the assumptions required for valid statistical inference procedures for multiple linear regression. Procedures for point estimation are given in Section 4.4. Section 4.5 discusses some methods for investigating the validity of the assumptions required for multiple linear regression. Confidence interval procedures for multiple linear regression are given in Section 4.6. A discussion of statistical tests is provided in Section 4.7. Analysis of variance in the case of multiple linear regression is discussed in Section 4.8. Procedures for comparing two regression functions are given in Sections 4.9 and 4.10. Section 4.9 also discusses multiple correlation, the coefficient of determination, multiple-partial correlation, and partial coefficient of determination. Procedures for evaluating the lack-of-fit of a straight line regression model are provided in Section 4.11, and Section 4.12 contains exercises.

For a clear and accurate explanation of the concepts developed in this chapter, we need to introduce some notation that may initially appear complicated. However, if you take the time to understand the notation, it will help you to understand the concepts as well as the procedures.

4.2
Notation and Definitions

The word *multiple* in multiple linear regression means there is more than one predictor variable. Throughout this chapter, it is assumed that there are k predictor variables that are denoted by $X_1, \ldots, X_k$. Recall from Section 2.3 that the word *linear* means the regression function, denoted by $\mu_Y(x_1, \ldots, x_k)$, is *linear in the unknown parameters* $\beta_0, \beta_1, \ldots, \beta_k$ (see (4.1.1)). The word *regression* means that the study is concerned with the prediction of the response variable Y using the relationship between Y and the k predictor variables $X_1, X_2, \ldots, X_k$.

We begin with two examples illustrating how regression functions can be useful in practical problems.

E X A M P L E   4 . 2 . 1

Consider the population of high school graduates who were admitted to a particular university during the past ten years and who completed at least the first year of coursework after being admitted. Suppose the director of admissions at this university is interested in investigating how well Y, the first year grade point average (GPA) of a student, can be predicted by using the following quantities:

$X_1 =$ the score on the mathematics part of the Scholastic Aptitude Test (SATmath)

$X_2 =$ the score on the verbal part of the Scholastic Aptitude Test (SATverbal)

$X_3 =$ the grade point average of all high school mathematics courses (HSmath)

$X_4 =$ the grade point average of all high school English courses (HSenglish)

If a relationship does exist between Y and X_1, X_2, X_3, X_4, then it may be possible to predict the performance of new applicants during their first year in college based on their SAT scores and high school grades; this would provide the director of admissions with a very valuable tool for making decisions about whether or not an applicant should be recommended for financial assistance.

Suppose, in this example, that the population regression function of Y on X_1, X_2, X_3, X_4 is of the form

$$\mu_Y(x_1, x_2, x_3, x_4) = \beta_0 + \beta_1 x_1 + \beta_2 x_2 + \beta_3 x_3 + \beta_4 x_4$$

This is a special case of the multiple linear regression function in (4.1.1) with $k = 4$. Methods discussed in this chapter can therefore be used to investigate the relationship between Y and X_1, X_2, X_3, X_4. ∎

E X A M P L E 4.2.2
Consider Example 2.2.9. Suppose that the regression function of Y, the strength of plastic containers, on X_1 and X_2, the temperature and pressure, respectively, during the production process, is of the form

$$\mu_Y(x_1, x_2) = \beta_0 + \beta_1 x_1 + \beta_2 x_2$$

This is a special case of the multiple linear regression function in (4.1.1) with $k = 2$. ∎

Other situations where multiple linear regression may be applicable are described in Examples 2.2.3, 2.2.5, and 2.2.6.

Basic Observable Variables and Derived Variables

It is useful to make a distinction between *basic observable variables* and *derived variables*. If Z represents the height of an individual, then Z is a basic observable variable and $\sqrt{Z}$ or $\log Z$ are derived variables. Thus derived variables are known functions of observable variables. The quantities x_i in the multiple linear regression function in (4.1.1) may be values of basic observable variables or derived variables. For instance, if the variables Z_1 and Z_2 are basic observable variables and the regression function is

$$\mu_Y(z_1, z_2) = \beta_0 + \beta_1 z_1 + \beta_2 z_1^2 + \beta_3 \sqrt{z_2}$$

then we can define the variables X_1, X_2, X_3 by

$$X_1 = Z_1 \qquad X_2 = Z_1^2 \qquad X_3 = \sqrt{Z_2}$$

and their values x_1, x_2, x_3 by

$$x_1 = z_1 \qquad x_2 = z_1^2 \qquad x_3 = \sqrt{z_2}$$

so that the regression function can be written as

$$\mu_Y(x_1, x_2, x_3) = \beta_0 + \beta_1 x_1 + \beta_2 x_2 + \beta_3 x_3$$

This demonstrates the versatility of the regression function in (4.1.1). More examples are given in (4.2.1).

$$\mu_Y(z_1, z_2) = \beta_0 + \beta_1 z_1 + \beta_2 z_1 z_2 + \beta_3 z_1^2 \tag{4.2.1a}$$

$$\mu_Y(z_1, z_2) = \beta_0 + \beta_1 e^{z_1} + \beta_2(\cos z_1 + \cos z_2) + \beta_3 \sqrt{z_1} \tag{4.2.1b}$$

$$\mu_Y(z_1, z_2, z_3) = \beta_0 + \beta_1 z_1 + \beta_2 z_1^2 + \beta_3 \frac{z_1}{z_1 + z_2 + z_3} + \beta_4 \log |z_3| \quad \textbf{(4.2.1c)}$$

Each of these regression functions may be written in the form given in (4.1.1) by suitably defining variables $X_1, X_2, \ldots$, etc. For instance, in (4.2.1a) we define $X_1 = Z_1, X_2 = Z_1 Z_2$, and $X_3 = Z_1^2$ so that the regression function can be rewritten as

$$\mu_Y(x_1, x_2, x_3) = \beta_0 + \beta_1 x_1 + \beta_2 x_2 + \beta_3 x_3$$

Here each X_i is a *known* function of the Z_i and involves no unknown parameters. In (4.2.1b) we use $X_1 = e^{Z_1}, X_2 = (\cos Z_1 + \cos Z_2)$, and $X_3 = \sqrt{Z_1}$, which allow us to rewrite the regression function as

$$\mu_Y(x_1, x_2, x_3) = \beta_0 + \beta_1 x_1 + \beta_2 x_2 + \beta_3 x_3$$

Similarly, in (4.2.1c) we use $X_1 = Z_1, X_2 = Z_1^2, X_3 = Z_1/(Z_1 + Z_2 + Z_3)$, and $X_4 = \log |Z_3|$ and rewrite the regression function as

$$\mu_Y(x_1, x_2, x_3, x_4) = \beta_0 + \beta_1 x_1 + \beta_2 x_2 + \beta_3 x_3 + \beta_4 x_4$$

Polynomial Regression

A population regression model that is quite useful in many applied problems uses the polynomial function given by

$$\mu_Y(x) = \beta_0 + \beta_1 x + \beta_2 x^2 + \cdots + \beta_k x^k$$

which is a kth degree polynomial in the predictor X. We can write this as

$$\mu_Y(x_1, x_2, \ldots, x_k) = \beta_0 + \beta_1 x_1 + \beta_2 x_2 + \cdots + \beta_k x_k$$

where $X_1 = X, X_2 = X^2, \ldots, X_k = X^k$. This is a multiple regression model using the derived variables $X_1, \ldots, X_k$, where X is the basic observable variable.

Notation

The starting point of any investigation involving multiple linear regression is the precise definition of the $(k + 1)$ variable population $\{(Y, X_1, \ldots, X_k)\}$. To identify the elements of this population, we use a double subscript notation for the values of the predictor variables. In this notation, $X_{I,J}$ refers to the value of the Ith population item corresponding to the Jth predictor variable; i.e., the first subscript I is the *reference number* or *label* of the population item under consideration, and the second subscript J refers to the Jth predictor variable. For example, $X_{4,8}$ is the value of the 8th predictor variable of the 4th population item; $X_{24,6}$ is the value of the 6th predictor variable of the 24th population item, etc. Of course, Y_I denotes the value of the response variable for the Ith population item.

Table 4.2.1 gives a schematic display of the five-variable population $\{(Y, X_1, X_2, X_3, X_4)\}$ discussed in Example 4.2.1. We suppose that this population has N items. The items in this population are *all high school graduates who were admitted to the university during the past ten years and completed at least the first year of coursework after being admitted.* Of course, all of the population values Y_I and $X_{I,J}$ may not be known in a real problem, but conceptually they exist.

In Table 4.2.1, $X_{6,3}$ represents the high school math GPA of the 6th individual, $X_{32,2}$ represents the SAT verbal score of the 32nd individual, etc. The theory for deriving point estimates, confidence intervals, and tests assumes that N, the population size, is infinite, but in most practical applications N is indeed finite. However, N is generally very large (but unknown) so the theory is approximately valid for all practical purposes.

T A B L E 4.2.1
Schematic Display of the Population in Example 4.2.1

Item (Individual Student) I	GPA at the End of One Year Y	SAT Math Score X_1	SAT Verbal Score X_2	High School Math GPA X_3	High School English GPA X_4
1	Y_1	$X_{1,1}$	$X_{1,2}$	$X_{1,3}$	$X_{1,4}$
2	Y_2	$X_{2,1}$	$X_{2,2}$	$X_{2,3}$	$X_{2,4}$
⋮	⋮	⋮	⋮	⋮	⋮
I	Y_I	$X_{I,1}$	$X_{I,2}$	$X_{I,3}$	$X_{I,4}$
⋮	⋮	⋮	⋮	⋮	⋮
N	Y_N	$X_{N,1}$	$X_{N,2}$	$X_{N,3}$	$X_{N,4}$

Parameters of a $(k+1)$-Variable Population

A $(k+1)$-variable population $\{(Y, X_1, X_2, \ldots, X_k)\}$ has associated with it several parameters that are useful in describing and summarizing it. The parameters

$$\mu_Y, \mu_{X_1}, \ldots, \mu_{X_k}$$

are the means of each one-variable population, and

$$\sigma_Y, \sigma_{X_1}, \ldots, \sigma_{X_k}$$

are the standard deviations of each one-variable population. They are useful for summarizing the one-variable populations $\{Y\}, \{X_1\}, \ldots, \{X_k\}$. The parameters

$$\rho_{Y,X_1}, \ldots, \rho_{Y,X_k}, \rho_{X_1,X_2}, \ldots, \rho_{X_{k-1},X_k}$$

are the coefficients of correlation of all pairs of variables, which may be useful for investigating the relationships between pairs of variables.

The regression function $\mu_Y(x_1, \ldots, x_k)$ *can be used for predicting Y using the k predictor variables* $X_1, \ldots, X_k$. *It is in fact the best prediction function for predicting Y using* $X_1, \ldots, X_k$. When the regression function $\mu_Y(x_1, \ldots, x_k)$ is of the form given in (4.1.1), the parameters $\beta_0, \beta_1, \ldots, \beta_k$ completely determine this regression function. Therefore we are interested in knowing the values of these (and possibly other) parameters to help make decisions about the population. Of course the population values are unknown, so the parameters that summarize the population are unknown. As in simple linear regression, a sample is selected from the population, and inferences about population parameters are made by using the sample data.

The Population Regression Model

For the Ith item in the population, the true value of the response variable is Y_I, and the predicted value using the regression function in (4.1.1), which is the best prediction function, is

$$\mu_Y(X_{I,1}, X_{I,2}, \ldots, X_{I,k}) = \beta_0 + \beta_1 X_{I,1} + \beta_2 X_{I,2} + \cdots + \beta_k X_{I,k}$$

The error of prediction for the Ith element of the population is denoted by E_I; it is the difference between the true value Y_I and the predicted value $\mu_Y(X_{I,1}, X_{I,2}, \ldots, X_{I,k})$ and is given by

$$E_I = Y_I - \mu_Y(X_{I,1}, \ldots, X_{I,k})$$

or

$$E_I = Y_I - (\beta_0 + \beta_1 X_{I,1} + \cdots + \beta_k X_{I,k})$$

This equation is generally written as

$$Y_I = \beta_0 + \beta_1 X_{I,1} + \cdots + \beta_k X_{I,k} + E_I \qquad \text{(4.2.2)}$$

for $I = 1, \ldots, N$ and is called the **population regression model**. We use the symbol $Y(x_1, \ldots, x_k)$ to denote the Y value of a randomly chosen item with $X_1 = x_1, \ldots, X_k = x_k$.

Sample

To make statistical inferences about the unknown parameters in the population regression function in (4.1.1), we must use sample data to compute point estimates and interval estimates of these unknown parameters, so we select a sample of size n from the population. A schematic display of the sample data corresponding to the population in Table 4.2.1 is given in Table 4.2.2. Note that we use lower-case letters y, $x_{i,j}$, n, etc., to emphasize that these quantities refer to the sample and not to the population. For example, $(y_i, x_{i,1}, x_{i,2}, x_{i,3}, x_{i,4})$ are the five measurements on the ith sample item selected from the population. The sample size n and the sample observations are, of course, known quantities.

You are encouraged to carry out the following task to develop familiarity with the terminology as well as some of the concepts underlying multiple regression.

T A B L E   4.2.2

Schematic Display of a Sample from the Population in Table 4.2.1

Item (Individual Student) i	GPA at the End of One Year Y	SAT Math Score X_1	SAT Verbal Score X_2	High School Math GPA X_3	High School English GPA X_4
1	y_1	$x_{1,1}$	$x_{1,2}$	$x_{1,3}$	$x_{1,4}$
2	y_2	$x_{2,1}$	$x_{2,2}$	$x_{2,3}$	$x_{2,4}$
$\vdots$	$\vdots$	$\vdots$	$\vdots$	$\vdots$	$\vdots$
i	y_i	$x_{i,1}$	$x_{i,2}$	$x_{i,3}$	$x_{i,4}$
$\vdots$	$\vdots$	$\vdots$	$\vdots$	$\vdots$	$\vdots$
n	y_n	$x_{n,1}$	$x_{n,2}$	$x_{n,3}$	$x_{n,4}$

Task   4.2.1

In this task we illustrate the concept of subpopulations when the number of predictor variables is greater than one.

Table D-4 in Appendix D contains a three-variable population $\{(Y, X_1, X_2)\}$ where Y is the strength of plastic containers, and X_1 and X_2 are, respectively, the temperature (in degrees Celsius) and the pressure (in pounds per square inch) used during the production process. These data are also in file **plastic.dat** on the data disk. There are 1,650 items in this (artificial) population. Column 1 contains the item numbers for the items in the population. Column 2 contains the strengths of the plastic containers, and columns 3 and 4 contain the temperatures and pressures, respectively, used during the production of each item. These population data may be thought of as having been obtained from the records, for the past two years, of the research and development division of a company that manufactures plastic containers.

The means of Y, X_1, and X_2 are

$$\mu_Y = 30 \qquad \mu_{X_1} = 250 \qquad \mu_{X_2} = 15$$

and the standard deviations are

$$\sigma_Y = 7.39 \qquad \sigma_{X_1} = 31.62 \qquad \sigma_{X_2} = 3.42$$

An examination of the population reveals that there are 66 distinct subpopulations of Y values determined by the distinct pairs of values of (X_1, X_2). Note that this is a very special population in which the standard deviation of each subpopulation is the same. We say more about this in Section 4.3. The means and the standard deviations of these 66 subpopulations of Y values are listed in Table 4.2.3. They are also stored in the file **table423.dat** on the data disk.

T A B L E **4.2.3**

The Sixty-Six Subpopulation Means and Standard Deviations for the Population Data in Table D-4

Subpopulation	Temperature	Pressure	Y Mean	Y Standard Deviation
1	200	10	25	1.7076
2	200	12	23	1.7076
3	200	14	21	1.7076
4	200	16	19	1.7076
5	200	18	17	1.7076
6	200	20	15	1.7076
7	210	10	27	1.7076
8	210	12	25	1.7076
9	210	14	23	1.7076
10	210	16	21	1.7076
11	210	18	19	1.7076
12	210	20	17	1.7076
13	220	10	29	1.7076
14	220	12	27	1.7076
15	220	14	25	1.7076
16	220	16	23	1.7076
17	220	18	21	1.7076
18	220	20	19	1.7076
19	230	10	31	1.7076
20	230	12	29	1.7076
21	230	14	27	1.7076
22	230	16	25	1.7076
23	230	18	23	1.7076
24	230	20	21	1.7076
25	240	10	33	1.7076
26	240	12	31	1.7076
27	240	14	29	1.7076
28	240	16	27	1.7076
29	240	18	25	1.7076
30	240	20	23	1.7076
31	250	10	35	1.7076
32	250	12	33	1.7076
33	250	14	31	1.7076
34	250	16	29	1.7076
35	250	18	27	1.7076
36	250	20	25	1.7076
37	260	10	37	1.7076
38	260	12	35	1.7076
39	260	14	33	1.7076
40	260	16	31	1.7076

(Continued)

T A B L E   4.2.3
(Continued)

Subpopulation	Temperature	Pressure	Y Mean	Y Standard Deviation
41	260	18	29	1.7076
42	260	20	27	1.7076
43	270	10	39	1.7076
44	270	12	37	1.7076
45	270	14	35	1.7076
46	270	16	33	1.7076
47	270	18	31	1.7076
48	270	20	29	1.7076
49	280	10	41	1.7076
50	280	12	39	1.7076
51	280	14	37	1.7076
52	280	16	35	1.7076
53	280	18	33	1.7076
54	280	20	31	1.7076
55	290	10	43	1.7076
56	290	12	41	1.7076
57	290	14	39	1.7076
58	290	16	37	1.7076
59	290	18	35	1.7076
60	290	20	33	1.7076
61	300	10	45	1.7076
62	300	12	43	1.7076
63	300	14	41	1.7076
64	300	16	39	1.7076
65	300	18	37	1.7076
66	300	20	35	1.7076

1   Verify that the population regression function of Y on X_1 and X_2 is of the form
(4.1.1) with $k = 2$, i.e., linear in X_1 and X_2 (for those values of the pair (X_1, X_2)
occurring in the population). In particular, what are the values of β_0, β_1, and β_2?

Suppose that the regression function of Y on X_1 and X_2 is of the form
$\mu_Y(x_1, x_2) = \beta_0 + \beta_1 x_1 + \beta_2 x_2$. *Then, in particular, the average Y value of each
subpopulation must be equal to the value obtained when the corresponding val-
ues of temperature X_1 and pressure X_2 are substituted into the regression function. For instance, each one of the 66 equations given in Table 4.2.4 (obtained by
examining the subpopulation values in Table 4.2.3) must be satisfied. Consider
the first two equations in Table 4.2.4, viz.,*

$$\beta_0 + \beta_1(200) + \beta_2(10) = 25$$
$$\beta_0 + \beta_1(200) + \beta_2(12) = 23$$

T A B L E   4.2.4

Subpopulation Number	X_1	X_2	Value of the Regression Function = $\mu_Y(X_{i1}, X_{i2}) = \beta_0 + \beta_1 X_{i1} + \beta_2 X_{i2}$ = Y mean
1	200	10	$\mu_Y(200, 10) = \beta_0 + \beta_1(200) + \beta_2(10) = 25$
2	200	12	$\mu_Y(200, 12) = \beta_0 + \beta_1(200) + \beta_2(12) = 23$
⋮	⋮	⋮	⋮
7	210	10	$\mu_Y(210, 10) = \beta_0 + \beta_1(210) + \beta_2(10) = 27$
⋮	⋮	⋮	⋮
66	300	20	$\mu_Y(300, 20) = \beta_0 + \beta_1(300) + \beta_2(20) = 35$

We subtract the second equation from the first and obtain $-2\beta_2 = 2$ *or* $\beta_2 = -1$. *Next, consider the first and the last equation in Table 4.2.4, viz.,*

$$\beta_0 + \beta_1(200) + \beta_2(10) = 25$$
$$\beta_0 + \beta_1(300) + \beta_2(20) = 35$$

Subtract the first equation from the second equation above and obtain $100\beta_1 + 10\beta_2 = 10$. *Substituting* $\beta_2 = -1$, *we get* $\beta_1 = 0.2$. *Finally, substituting the value* $\beta_1 = 0.2$ *and* $\beta_2 = -1.0$ *into the first equation in Table 4.2.4, viz., into*

$$\beta_0 + \beta_1(200) + \beta_2(10) = 25$$

we get $\beta_0 = -5$. *Hence, if the regression function of Y on* X_1, X_2 *is indeed of the form* $\mu_Y(x_1, x_2) = \beta_0 + \beta_1 x_1 + \beta_2 x_2$, *then it must be given by*

$$\mu_Y(x_1, x_2) = -5 + 0.2x_1 - x_2 \tag{4.2.3}$$

You should verify that all of the subpopulation means for Y are obtained by substituting the appropriate temperature and pressure values into (4.2.3). Thus we obtain

$$\beta_0 = -5 \qquad \beta_1 = 0.2 \qquad \beta_2 = -1 \tag{4.2.4}$$

2 Examine the Y values in the subpopulation with $X_1 = 220$ and $X_2 = 16$.
     Table 4.2.5 contains these subpopulation values, and they are also stored in the file **table425.dat** *on the data disk.*

3 Compute the mean and the standard deviation of the Y values in the subpopulation in (2) where $X_1 = 220$, and $X_2 = 16$.
     There are three ways to obtain the Y mean $\mu_Y(220, 16)$.

i   *Compute the mean of the Y values in the subpopulation in (2) by direct calculation; i.e.,* $\mu_Y(220, 16) = \overline{Y} = (1/25) \sum Y_i$.

ii *Look up the Y mean corresponding to subpopulation number 16 in Table 4.2.3 where temperature* $X_1 = 220$ *and pressure* $X_2 = 16$.

iii *Use (4.2.3) and plug in* $X_1 = 220$ *and* $X_2 = 16$.

T A B L E 4.2.5

Row	Item Number I	Y	X_1	X_2
1	76	22.8	220	16
2	183	22.2	220	16
3	332	24.2	220	16
4	473	25.8	220	16
5	551	24.5	220	16
6	592	22.2	220	16
7	623	23.3	220	16
8	683	26.5	220	16
9	858	25.0	220	16
10	883	20.4	220	16
11	888	21.7	220	16
12	900	20.7	220	16
13	931	22.3	220	16
14	951	26.7	220	16
15	952	22.6	220	16
16	978	22.6	220	16
17	989	21.5	220	16
18	993	22.7	220	16
19	1165	21.7	220	16
20	1228	24.1	220	16
21	1316	23.9	220	16
22	1463	20.7	220	16
23	1526	21.2	220	16
24	1557	22.2	220	16
25	1572	23.5	220	16

To apply method (i) we compute the mean of the Y values in Table 4.2.5 and get $\mu_Y(220, 16) = \overline{Y} = 23.0$. To apply method (ii) we look up subpopulation number 16 and observe that $\mu_Y(220, 16) = 23.0$. To apply method (iii) we use (4.2.3) and get

$$\mu_Y(220, 16) = -5.0 + 0.2(220) - 16 = 23.0$$

There are two ways to get the standard deviation of a subpopulation. For instance, the standard deviation of the subpopulation with $X_1 = 220$ and $X_2 = 16$ can be obtained using one of the following two ways.

iv *By directly computing the standard deviation of the Y values in the subpopulation in (2) using the definition of standard deviation.*

v *By looking up the Y standard deviation for subpopulation 16 in Table 4.2.3.*

By both of these procedures we get $\sigma_Y(220, 16) = 1.7076$.

4   The population regression model is

$$Y_I = \beta_0 + \beta_1 X_{I,1} + \beta_2 X_{I,2} + E_I$$

For the subpopulation with $X_1 = 220$ and $X_2 = 16$ we get

$$E_I = Y_I - \mu_Y(220, 16) = Y_I - \beta_0 - \beta_1(220) - \beta_2(16)$$
$$= Y_I + 5 - 0.2(220) + 1(16) = Y_I - 23$$

The E_I's satisfy the following conditions:

i   The mean of the E_I values is zero.

ii   The standard deviation of the E_I values in any subpopulation is the same as the standard deviation of the Y_I values in that subpopulation.

To see that this is the case in this subpopulation, calculate the values of E_I for the items in the subpopulation referred to in (2). Verify that the mean of these E_I values is zero and that their standard deviation is the same as the standard deviation of the Y values in this subpopulation, namely 1.7076.

By subtracting the subpopulation mean Y value of 23 from each of the Y values in the subpopulation in Table 4.2.5, we obtain the E_I values, which are

```
-0.20000  -0.80000   1.20000   2.80000   1.50000  -0.80000   0.30000
 3.50000   2.00000  -2.60000  -1.30000  -2.30000  -0.70000   3.70000
-0.40000  -0.40000  -1.50000  -0.30000  -1.30000   1.10000   0.90000
-2.30000  -1.80000  -0.80000   0.50000
```

You should check these and also check that the mean of these numbers is 0 (to within rounding error) and that the standard deviation is 1.7076, the same as the standard deviation of the Y values in Table 4.2.5, as it should be.

In Task 4.2.1 we examined a population and its subpopulations whose values were all *known*. However, in applied problems the entire population is seldom known and so the population parameters must be estimated using sample data. For these estimates to be valid, certain assumptions must be satisfied, as in the case of straight line regression. These assumptions are stated and discussed in the next section.

Problems   4.2

Problems 4.2.1–4.2.5 refer to the population data given in Table D-4 in Appendix D and also in the file **plastic.dat** on the data disk. Consider the subpopulation of Y values with $X_1 = 240$ and $X_2 = 18$. This subpopulation has 25 values and they are given in Table 4.2.6 along with the corresponding E_I values. These subpopulation

T A B L E 4.2.6

Row	Population Item Number	Y	X_1	X_2	E_I
1	19	24.6	240	18	−0.400
2	44	24.2	240	18	−0.800
3	71	23.7	240	18	−1.300
4	174	26.5	240	18	1.500
5	175	24.3	240	18	−0.700
6	212	24.2	240	18	−0.800
7	216	23.7	240	18	−1.300
8	218	24.6	240	18	−0.400
9	235	22.7	240	18	−2.300
10	378	28.7	240	18	3.700
11	407	22.4	240	18	−2.600
12	584	26.1	240	18	1.100
13	588	27.8	240	18	2.800
14	948	26.2	240	18	1.200
15	953	27.0	240	18	2.000
16	962	25.3	240	18	*
17	1118	22.7	240	18	−2.300
18	1135	24.8	240	18	−0.200
19	1209	25.5	240	18	0.500
20	1285	24.2	240	18	−0.800
21	1302	28.5	240	18	3.500
22	1376	25.9	240	18	*
23	1406	24.7	240	18	−0.300
24	1447	23.5	240	18	−1.500
25	1459	23.2	240	18	−1.800

* These values were omitted and you will be asked to supply them.

values are also stored in **table426.dat** on the data disk. For this subpopulation, use the following:

$$\sum Y_I = 625.0 \qquad \sum Y_I^2 = 15697.90$$

4.2.1 **a** Show that the mean of this subpopulation is 25.0, i.e. $\mu_Y(240, 18) = 25.0$.

b The subpopulation of E_I values can be obtained by computing

$$E_I = Y_I - \mu_Y(240, 18) = Y_I - \beta_0 - \beta_1(240) - \beta_2(18)$$

Substituting the values for $\beta_0, \beta_1, \beta_2$ from (4.2.4) we get

$$E_I = Y_I - [-5 + 0.2X_{I1} - X_{I2}] = Y_I - [-5 + 0.2(240) - 18] = Y_I - 25$$

In Table 4.2.6, the 25 values of E_I have been computed except for population items 962 and 1376 (where the asterisk * appears). Compute the E_I values for items 962 and 1376.

c   Verify that the sum of the E_I values in Table 4.2.6 (including items 962 and 1376) is zero.

d   Find the standard deviation of the Y_I values in this subpopulation.

e   Find the standard deviation of all 25 E_I values in this subpopulation.

f   In Problem 4.2.1(e), is this standard deviation what you should get? (Examine Table 4.2.3.)

4.2.2   Predict the strength of a single plastic container that was produced with a temperature of 280° and a pressure of 10 units.

4.2.3   Predict the strength of a single plastic container if it is produced with a temperature of 205° and a pressure of 11 units.

4.2.4   What is the *average* strength of all plastic containers in the study population produced using a temperature of 280° and a pressure of 18 units? Do you know this exactly or do you have to estimate it?

4.2.5   For the subpopulation with $X_1 = 280$ and $X_2 = 10$, find a number L such that 80% of the plastic containers produced would have strength greater than or equal to L.

4.3
Assumptions for Multiple Linear Regression

To obtain valid point and confidence interval estimates for parameters in the multiple linear regression function

$$\mu_Y(x_1, \ldots, x_k) = \beta_0 + \beta_1 x_1 + \cdots + \beta_k x_k$$

we must make some assumptions about the $(k+1)$-variable population $\{(Y, X_1, \ldots, X_k)\}$ and about the method used to obtain a sample from this population. One set of assumptions under which the theory for multiple linear regression has been extensively developed is given in Box 4.3.1 and is referred to as assumptions (A) throughout this book. Three of the assumptions concern the population and two concern the sample.

BOX **4.3.1**   **Assumptions (A) for Multiple Linear Regression**

Notation: The $(k+1)$-variable population $\{(Y, X_1, \ldots, X_k)\}$ is the study population.

(Population) Assumption 1.   The mean $\mu_Y(x_1, \ldots, x_k)$ of the subpopulation of Y values with $X_1 = x_1, \ldots, X_k = x_k$ is

$$\mu_Y(x_1, \ldots, x_k) = \beta_0 + \beta_1 x_1 + \cdots + \beta_k x_k$$

where $\beta_0, \beta_1, \ldots, \beta_k$ are unknown parameters and $x_1, \cdots, x_k$ belong to the set of allowable values (sometimes called the *domain*) of the predictor variables.

(Population) Assumption 2.   The standard deviation of the Y values in the subpopulations with $X_1 = x_1, \ldots, X_k = x_k$ does not depend on the values $x_1, \ldots, x_k$ (i.e., the standard deviations are the same for each subpopulation determined by specified values of the predictor variables $X_1, \ldots, X_k$). This common standard deviation of all the subpopulations is denoted by $\sigma_{Y|X_1,\ldots,X_k}$. When there is no possibility of confusion, we use the simpler notation σ instead of the more complete notation $\sigma_{Y|X_1,\ldots,X_k}$.

(Population) Assumption 3.   Each subpopulation of Y values, determined by specified values of $X_1, \ldots, X_k$, is a Gaussian population.

(Sample) Assumption 4.   The sample data are obtained by simple random sampling or by sampling with preselected values of $X_1, \ldots, X_k$, discussed in Section 2.3. The number of items in the sample is n.

(Sample) Assumption 5.   All sample values $y_i, x_{i,1}, \ldots, x_{i,k}$ for $i = 1, \ldots, n$ are observed without error (but read Section 3.10).

Several quantities of interest are associated with the $(k + 1)$-variable population $\{(Y, X_1, \ldots, X_k)\}$. They include

a   Parameters and randomly chosen Y values associated with subpopulations determined by $X_1, \ldots, X_k$, viz.,

$$\beta_0, \beta_1, \ldots, \beta_k, Y(x_1, x_2, \ldots, x_k), \mu_Y(x_1, \ldots, x_k), \sigma = \sigma_{Y|X_1,\ldots,X_k}$$

b   Parameters of individual populations $\{Y\}, \{X_1\}, \ldots, \{X_k\}$, viz.,

$$\mu_Y, \sigma_Y, \mu_{X_1}, \sigma_{X_1}, \mu_{X_2}, \sigma_{X_2}, \ldots, \mu_{X_k}, \sigma_{X_k}$$

c   Correlation coefficients of all two-variable populations, viz.,

$$\rho_{Y,X_1}, \ldots, \rho_{Y,X_k}, \rho_{X_1,X_2}, \ldots, \rho_{X_{k-1},X_k}$$

The assumptions in Box 4.3.1 are sufficient for making valid inferences about the quantities in (a), but in some situations the investigator may also be interested in making inferences about the quantities in (b) and (c). If data are obtained by sampling with preselected values of $X_1, \ldots, X_k$, we cannot make valid inferences about the parameters in (b) and (c) unless every subpopulation is sampled and the relative subpopulation sizes are known, which is almost never the case in real applications. If data are obtained by simple random sampling, then valid point estimates of the quantities in (b) and (c) are available. See (1.6.1), (1.6.2), and (1.6.3). A more restrictive set of assumptions, referred to as assumptions (B) and given in Box 4.3.2, is sufficient for making inferences about *all* of the quantities in (a), (b), and (c).

B O X   **4.3.2**   **Assumptions (B) for Multiple Linear Regression**

(Population) Assumption 1.   The study population $\{(Y, X_1, \ldots, X_k)\}$ is a $(k + 1)$-variable Gaussian population.

(Sample) Assumption 2.   The sample data are obtained by simple random sampling described in Section 2.3; i.e., a simple random sample of n items is selected from the population and the values of the variables $Y, X_1, \ldots, X_k$, are observed.

(Sample) Assumption 3.   The sample values $y_i, x_{i,1}, \ldots, x_{i,k}$, for $i = 1, \ldots, n$ are measured without error.

We make several comments about the assumptions.

1   For assumptions (B) to be met, sample data must be obtained by simple random sampling.

2   If $\{(Y, X_1, \ldots, X_k)\}$ is a $(k+1)$-variable Gaussian population as in (population) assumption 1 of Box 4.3.2, then (population) assumptions 1, 2, 3 in Box 4.3.1 are automatically satisfied, and $\mu_Y(x_1, \ldots, x_k)$ is indeed of the form

$$\mu_Y(x_1, \ldots, x_k) = \beta_0 + \beta_1 x_1 + \cdots + \beta_k x_k$$

Thus, *assumptions* (B) *for multiple linear regression imply assumptions* (A); *however, the converse is true only when the k-variable population* $\{(X_1, \ldots, X_k)\}$ *is Gaussian and data are obtained by simple random sampling.*

3   Strictly speaking, in (population) assumption 1 of Box 4.3.1, we should clearly state what the allowable values of the predictor variables are. This can be done by stating a lower limit and an upper limit for each predictor variable: for instance, $a_1 \leq X_1 \leq b_1, \ldots, a_k \leq X_k \leq b_k$. For simplicity of presentation, we often omit specification of these limits, but investigators and statisticians should know what these limits are for each given problem.

4   If we are interested only in point estimation and not in confidence intervals or tests, then (population) assumptions 1 and 2 and (sample) assumptions 4 and 5 of regression assumptions (A) in Box 4.3.1 are sufficient, and the Gaussian assumption is not needed.

Some Comments on Measurement Errors

In many applied problems, the values of $X_1, \ldots, X_k$ and Y of the sample items are not known exactly but are subject to measurement errors. For many applications of regression, measurement errors can be ignored provided that they are small. For such situations the inference procedures developed in Chapters 3 through 7 should be adequate. However, there are situations where measurement errors should not be ignored, and certain modifications may be required for the inference procedures to be valid. For the straight line regression model, these are discussed in Section 3.10. For the multiple linear regression model involving measurement errors in the predictor variables $X_1, \ldots, X_k$, and possibly in the response variable Y, results similar to those in Section 3.10 are applicable. You should consult more advanced textbooks [7].

Often, when discussing statistical procedures, we make no mention of measurement errors. While we want you to be aware of the additional complications that may arise when measurement errors are not small, we want to assure you that the methods developed in Chapters 3 through 7, under the assumption of no measurement errors, are adequate for most applications.

Unless specifically stated otherwise, we assume that the values of $X_1, \ldots, X_k$ and Y of the sample items are measured without appreciable error and that (sample) assumption 5 in Box 4.3.1 and/or (sample) assumption 3 in Box 4.3.2 are satisfied.

Note that the (artificial) population of Task 4.2.1 satisfies (population) assumptions 1 and 2 in Box 4.3.1. It can never be determined if all of the assumptions in Box 4.3.1 or 4.3.2 are exactly satisfied in a real problem. Investigators may not know for certain that $\mu_Y(x_1, \ldots, x_k)$ is of the form in (4.1.1), but they can generally determine the form of the regression function that fits the problem so that the theory will be approximately valid. The same can be said for all of the assumptions. The sample assumptions mainly concern collecting the data, and often the investigator is restricted by money, time, or other constraints so the data collection methods may not exactly meet the requirements of randomness, etc. Sometimes the investigator who must analyze the data and draw conclusions from them is not the one who collected the data. We may know or suspect that some errors were made in the sampling procedures or in recording the data. The view we take is that all data contain some information, and the investigator is in the best position to determine whether the assumptions are close enough to being satisfied to allow valid conclusions to be drawn about the populations under study. Investigators should always be aware of abnormalities in the data and deal with them. In this chapter, we give inference procedures that are valid if assumptions (A) or (B) for multiple linear regression hold. However, the results are generally reliable and useful when the assumptions are approximately satisfied or if the sample size is large.

The next section discusses point estimation for the unknown parameters in the multiple linear regression model. In Section 4.5, we consider some methods for examining the validity of assumptions (A) and (B) and, if they appear not to be satisfied, alternative procedures are sometimes available. These are discussed in Chapter 8.

4.4
Point Estimation

One of the main objectives of multiple regression analysis is to use sample data to obtain point and confidence interval estimates for the unknown quantities $\beta_0, \beta_1, \ldots, \beta_k, \mu_Y(x_1, \ldots, x_k), Y(x_1, \ldots, x_k)$, and σ, and also for selected functions of these quantities. In Section 3.4 we discussed point estimation for straight line regression. In this section we discuss point estimation for multiple linear regression with k predictor factors when assumptions (A) and (B) are satisfied. Consequently, the population regression function is of the form given in (4.1.1). A sample of size n is selected using simple random sampling or by sampling with preselected values of $X_1, \ldots, X_k$ from the population $\{(Y, X_1, \ldots, X_k)\}$ of N items. The sample data may be organized as shown in Table 4.4.1.

The *method of least squares* is used to obtain point estimates of $\beta_0, \beta_1, \ldots, \beta_k$, and we discuss this next.

T A B L E **4.4.1**

Schematic Representation of Sample Data Values

Y	X_1	X_2	$\cdots$	X_k
y_1	$x_{1,1}$	$x_{1,2}$	$\cdots$	$x_{1,k}$
y_2	$x_{2,1}$	$x_{2,2}$	$\cdots$	$x_{2,k}$
$\vdots$	$\vdots$	$\vdots$	$\vdots$	$\vdots$
y_i	$x_{i,1}$	$x_{i,2}$	$\cdots$	$x_{i,k}$
$\vdots$	$\vdots$	$\vdots$	$\vdots$	$\vdots$
y_n	$x_{n,1}$	$x_{n,2}$	$\cdots$	$x_{n,k}$

Least Squares Estimates of $\beta_0, \beta_1, \ldots, \beta_k$

Consider the population regression function

$$\mu_Y(x_1, \ldots, x_k) = \beta_0 + \beta_1 x_1 + \cdots + \beta_k x_k \qquad (4.4.1)$$

which we would like to use to predict Y with $X_1, \ldots, X_k$ as predictor factors. However, since the β_i are not known, we cannot use this function. Thus we must use sample data given in Table 4.4.1 to obtain estimates of the β_i and an estimate of $\mu_Y(x_1, \ldots, x_k)$ in (4.4.1). To obtain estimates of β_i, we use the **principle of least squares**. The resulting estimates are denoted by $\hat{\beta}_0, \hat{\beta}_1, \ldots, \hat{\beta}_k$, respectively. The corresponding estimate of the regression function is denoted by

$$\hat{\mu}_Y(x_1, \ldots, x_k) = \hat{\beta}_0 + \hat{\beta}_1 x_1 + \cdots + \hat{\beta}_k x_k \qquad (4.4.2)$$

Consider the n sample observations in Table 4.4.1. The predicted Y value of the ith sample item, with $x_{i,1}, \ldots, x_{i,k}$ as the values of the predictor variables, is given by $\hat{\mu}_Y(x_{i,1}, \ldots, x_{i,k}) = \hat{\beta}_0 + \hat{\beta}_1 x_{i,1} + \cdots + \hat{\beta}_k x_{i,k}$. The corresponding prediction error is denoted by $\hat{e}_i$ and is given by

$$\hat{e}_i = y_i - [\hat{\beta}_0 + \hat{\beta}_1 x_{i,1} + \cdots + \hat{\beta}_k x_{i,k}] \qquad (4.4.3)$$

The quantities $\hat{e}_i, i = 1, \ldots, n$ are called *residuals*. They are useful in examining the validity of the assumptions given in Box 4.3.1 as well as those given in Box 4.3.2. This is discussed in Section 4.5.

The least squares estimates $\hat{\beta}_0, \hat{\beta}_1, \ldots, \hat{\beta}_k$ of $\beta_0, \beta_1, \ldots, \beta_k$ are chosen in such a way that the quantity $SSE(X_1, \ldots, X_k)$, which is defined by

$$SSE(X_1, \ldots, X_k) = \sum_{i=1}^{n} (y_i - \hat{\beta}_0 - \hat{\beta}_1 x_{i,1} - \cdots - \hat{\beta}_k x_{i,k})^2 = \sum_{i=1}^{n} \hat{e}_i^2 \qquad (4.4.4)$$

attains its smallest possible value among all the possible choices we could make for estimating $\beta_0, \beta_1, \ldots, \beta_k$. The corresponding minimum value of $SSE(X_1, \ldots, X_k)$ is called the **sum of squared errors** for predicting Y using the estimated regression function

$$\hat{\mu}_Y(x_1, \ldots, x_k) = \hat{\beta}_0 + \hat{\beta}_1 x_1 + \cdots + \hat{\beta}_k x_k$$

Note that we are really not interested in predicting the Y values of the sample items because we already know their true values, namely $y_1, \ldots, y_n$. But if the estimated regression function

$$\hat{\mu}_Y(x_1, \ldots, x_k) = \hat{\beta}_0 + \hat{\beta}_1 x_1 + \cdots + \hat{\beta}_k x_k$$

is a good predictor of the *known sample values* y_i corresponding to the n sample items, then we have reason to expect that it will be a good prediction function for *all* values of Y in the population. Thus we use the y_i and the $x_{i,j}, j = 1, \ldots, k$, values of the items in the sample to assess the performance of the estimated regression function.

For the sake of completeness, we now enunciate the **principle of least squares**.

The principle of least squares states that the best estimate of the population regression function $\mu_Y(x_1, \ldots, x_k) = \beta_0 + \beta_1 x_1 + \cdots + \beta_k x_k$, using sample data, is obtained by choosing $\hat{\beta}_0, \hat{\beta}_1, \ldots, \hat{\beta}_k$ in (4.4.2) in such a way that the sum of squares of the prediction errors

$$SSE = \sum_{i=1}^{n} \hat{e}_i^2 = \sum_{i=1}^{n} (y_i - \hat{\beta}_0 - \hat{\beta}_1 x_{i,1} - \cdots - \hat{\beta}_k x_{i,k})^2 \qquad \text{(4.4.5)}$$

is the minimum possible.

Normal Equations

It is convenient to use matrices to compute the least squares estimates of $\beta_0, \beta_1, \ldots, \beta_k$. It can be shown mathematically that the least squares estimates $\hat{\beta}_0, \hat{\beta}_1, \ldots, \hat{\beta}_k$ of the parameters $\beta_0, \beta_1, \ldots, \beta_k$ are obtained by solving the following matrix equation for $\hat{\beta}$:

$$X^T X \hat{\beta} = X^T y \qquad \text{(4.4.6)}$$

where y is an $n \times 1$ vector, X is an $n \times (k+1)$ matrix, and $\hat{\beta}$ is a $(k+1) \times 1$ vector, given by

$$y = \begin{bmatrix} y_1 \\ y_2 \\ \vdots \\ y_n \end{bmatrix} \qquad X = \begin{bmatrix} 1 & x_{1,1} & x_{1,2} & \cdots & x_{1,k} \\ 1 & x_{2,1} & x_{2,2} & \cdots & x_{2,k} \\ \vdots & \vdots & \vdots & \vdots & \vdots \\ 1 & x_{i,1} & x_{i,2} & \cdots & x_{i,k} \\ \vdots & \vdots & \vdots & \vdots & \vdots \\ 1 & x_{n,1} & x_{n,2} & \cdots & x_{n,k} \end{bmatrix} \qquad \hat{\beta} = \begin{bmatrix} \hat{\beta}_0 \\ \hat{\beta}_1 \\ \vdots \\ \hat{\beta}_i \\ \vdots \\ \hat{\beta}_k \end{bmatrix} \qquad \text{(4.4.7)}$$

respectively. Note that the y vector and X matrix are known since the entries are the data values, except that X has a column of 1's as its first column. The equations in (4.4.6) are known as the **normal equations**. The solution of these normal equations is

$$\hat{\beta} = (X^T X)^{-1} X^T y \qquad (4.4.8)$$

provided that the inverse of $X^T X$ exists (which is always the case in this book). The estimate of the regression function $\mu_Y(x_1, \ldots, x_k)$ is

$$\hat{\mu}_Y(x_1, \ldots, x_k) = \hat{\beta}_0 + \hat{\beta}_1 x_1 + \cdots + \hat{\beta}_k x_k \qquad (4.4.9)$$

and the *best* predicted Y value for a randomly chosen item from the subpopulation with $X_1 = x_1, \ldots, X_k = x_k$ is

$$\hat{Y}(x_1, \ldots, x_k) = \hat{\beta}_0 + \hat{\beta}_1 x_1 + \cdots + \hat{\beta}_k x_k \qquad (4.4.10)$$

Note that, as in the case of straight line regression, the regression function is the best function for predicting Y using $X_1, \ldots, X_k$.

E X A M P L E 4.4.1

In this example we use a small (artificial) data set with three predictors (i.e., $k = 3$) to illustrate how to form the y vector and the X matrix from the data. We also show how to use matrix calculations to solve for $\hat{\beta}$. The data are given below in Table 4.4.2 and are also in the file **table442.dat** on the data disk. The regression function of Y on $X_1, \ldots, X_k$ is assumed to be of the form

$$\mu_Y(x_1, x_2, x_3) = \beta_0 + \beta_1 x_1 + \beta_2 x_2 + \beta_3 x_3$$

T A B L E 4.4.2

Y	X_1	X_2	X_3
6	3	9	16
9	6	13	13
12	4	3	17
5	8	2	10
13	3	4	9
2	2	4	7

and assumptions (A) are presumed to hold. Thus the y vector, the X matrix, and the $\hat{\beta}$ vector are

$$
y = \begin{bmatrix} 6 \\ 9 \\ 12 \\ 5 \\ 13 \\ 2 \end{bmatrix}
\qquad
X = \begin{bmatrix} 1 & 3 & 9 & 16 \\ 1 & 6 & 13 & 13 \\ 1 & 4 & 3 & 17 \\ 1 & 8 & 2 & 10 \\ 1 & 3 & 4 & 9 \\ 1 & 2 & 4 & 7 \end{bmatrix}
\qquad
\hat{\beta} = \begin{bmatrix} \hat{\beta}_0 \\ \hat{\beta}_1 \\ \hat{\beta}_2 \\ \hat{\beta}_3 \end{bmatrix}
$$

From these we get

$$
X^T = \begin{bmatrix} 1 & 1 & 1 & 1 & 1 & 1 \\ 3 & 6 & 4 & 8 & 3 & 2 \\ 9 & 13 & 3 & 2 & 4 & 4 \\ 16 & 13 & 17 & 10 & 9 & 7 \end{bmatrix}
\quad
X^T X = \begin{bmatrix} 6 & 26 & 35 & 72 \\ 26 & 138 & 153 & 315 \\ 35 & 153 & 295 & 448 \\ 72 & 315 & 448 & 944 \end{bmatrix}
\quad
X^T y = \begin{bmatrix} 47 \\ 203 \\ 277 \\ 598 \end{bmatrix}
$$

So

$$
\hat{\beta} = (X^T X)^{-1} X^T y = \begin{bmatrix} 2.59578 & -0.15375 & -0.01962 & -0.13737 \\ -0.15375 & 0.03965 & -0.00014 & -0.00144 \\ -0.01962 & -0.00014 & 0.01234 & -0.00431 \\ -0.13737 & -0.00144 & -0.00431 & 0.01406 \end{bmatrix} \begin{bmatrix} 47 \\ 203 \\ 277 \\ 598 \end{bmatrix}
$$

$$
= \begin{bmatrix} 3.20975 \\ -0.07573 \\ -0.11162 \\ 0.46691 \end{bmatrix}
$$

Thus

$$
\hat{\beta}_0 = 3.20975 \qquad \hat{\beta}_1 = -0.07573 \qquad \hat{\beta}_2 = -0.11162 \qquad \hat{\beta}_3 = 0.46691
$$

$$
\hat{\mu}_Y(x_1, x_2, x_3) = 3.20975 - 0.07573 x_1 - 0.11162 x_2 + 0.46691 x_3
$$

and

$$
\hat{Y}(x_1, x_2, x_3) = \hat{\mu}_Y(x_1, x_2, x_3) = 3.20975 - 0.07573 x_1 - 0.11162 x_2 + 0.46691 x_3
$$

∎

Point Estimates for Linear Functions of $\beta_0, \beta_1, \ldots, \beta_k$

A question of interest to an investigator can often be formulated in terms of a question involving a linear combination of the parameters $\beta_0, \beta_1, \ldots, \beta_k$, given by

$$
\theta = a_0 \beta_0 + a_1 \beta_1 + \cdots + a_k \beta_k = a^T \beta
$$

where the components a_i in the vector $a^T = [a_0, a_1, \ldots, a_k]$ are specified by the investigator. For example, we may want to estimate $\beta_1 - \beta_2$, in which case $a_1 = 1$, $a_2 = -1$, and all other $a_i = 0$; or we may want to estimate $2\beta_0 - \beta_1 + 3\beta_k$, in which case $a_0 = 2$, $a_1 = -1$, $a_k = 3$, and all other a_i equal zero.

The point estimate of $\theta = \boldsymbol{a}^T \boldsymbol{\beta}$ is

$$\hat{\theta} = \boldsymbol{a}^T \hat{\boldsymbol{\beta}} = a_0 \hat{\beta}_0 + a_1 \hat{\beta}_1 + \cdots + a_k \hat{\beta}_k \qquad \text{(4.4.11)}$$

where the $\hat{\beta}_i$ are computed by the formula in (4.4.8).

Observe that $\mu_Y(x_1, \ldots, x_k)$ itself is a quantity of the form $\boldsymbol{a}^T \boldsymbol{\beta}$ with $a_0 = 1$, $a_1 = x_1, \ldots, a_k = x_k$. Likewise, every β_i is a special case of $\boldsymbol{a}^T \boldsymbol{\beta}$ with $a_i = 1$ and all remaining elements of $\boldsymbol{a}$ equal to zero.

Residuals

If we use the estimated regression function $\hat{\mu}(x_1, \ldots, x_k)$ to predict the Y value of sample item i, which has $x_{i,1}, \ldots, x_{i,k}$ as the values for the predictor variables, the predicted Y value is $\hat{\mu}_Y(x_{i,1}, \ldots, x_{i,k})$, and the error of prediction is $\hat{e}_i = y_i - \hat{\mu}_Y(x_{i,1}, \ldots, x_{i,k})$. See (4.4.3). Thus

$$\hat{e}_i = y_i - \hat{\mu}_Y(x_{i,1}, \ldots, x_{i,k}) = y_i - \hat{\beta}_0 - \hat{\beta}_1 x_{i,1} - \cdots - \hat{\beta}_k x_{i,k} \quad \text{for} \quad i = 1, \ldots, n$$

$$\text{(4.4.12)}$$

The quantities $\hat{e}_i$ are called **residuals** and they can be computed from sample data because $y_i, x_{i,1}, \ldots, x_{i,k}$ and $\hat{\beta}_i$ are all known. As discussed in Chapter 3, residuals are useful in examining the validity of assumptions (A) and (B) given in Boxes 4.3.1 and 4.3.2, respectively. This is discussed in Section 4.5.

Point Estimate of σ (i.e., $\sigma_{Y|X_1,\ldots,X_k}$)

The estimate of σ is given by

$$\hat{\sigma} = \sqrt{\frac{\sum_{i=1}^{n} \hat{e}_i^2}{(n - k - 1)}} = \sqrt{\frac{\sum_{i=1}^{n} (y_i - \hat{\beta}_0 - \hat{\beta}_1 x_{i,1} - \cdots - \hat{\beta}_k x_{i,k})^2}{(n - k - 1)}} \qquad \text{(4.4.13)}$$

The quantity

$$SSE(X_1, \ldots, X_k) = \sum_{i=1}^{n} (y_i - \hat{\beta}_0 - \hat{\beta}_1 x_{i,1} - \cdots - \hat{\beta}_k x_{i,k})^2 = \sum_{i=1}^{n} \hat{e}_i^2 \qquad \text{(4.4.14)}$$

is referred to as the **sum of squared errors** (sometimes also called **sum of squares due to error** or **error sum of squares** or **residual sum of squares**) for the estimated regression of Y on $X_1, \ldots, X_k$. The quantity

$$MSE(X_1, \ldots, X_k) = \frac{SSE(X_1, \ldots, X_k)}{(n - k - 1)} \qquad \text{(4.4.15)}$$

is called the **mean squared error** (or **error mean square**, or **residual mean square**) for the estimated regression of Y on $X_1, \ldots, X_k$. When there is no possibility of confusion, we write SSE for $SSE(X_1, \ldots, X_k)$ and MSE for $MSE(X_1, \ldots, X_k)$. With this notation, the estimate of σ can be written as

$$\hat{\sigma} = \sqrt{MSE} \qquad \text{(4.4.16)}$$

Equivalently, the estimate of σ^2 is

$$\hat{\sigma}^2 = MSE = \frac{SSE}{n-k-1} \qquad \text{(4.4.17)}$$

If we write

$$\hat{e} = \begin{bmatrix} \hat{e}_1 \\ \hat{e}_2 \\ \vdots \\ \hat{e}_n \end{bmatrix} \qquad \text{(4.4.18)}$$

for the vector of residuals, then by using (4.4.12) for $\hat{e}_i$, we get

$$\hat{e} = y - X\hat{\beta}$$

The sum of squared errors *SSE* can be expressed in various equivalent ways using matrix notation. For instance, we have

$$SSE = \hat{e}^T \hat{e} \qquad \text{(4.4.19)}$$
$$SSE = (y - X\hat{\beta})^T (y - X\hat{\beta}) \qquad \text{(4.4.20)}$$
$$SSE = y^T y - \hat{\beta}^T X^T y \qquad \text{(4.4.21)}$$
$$SSE = y^T [I - X(X^T X)^{-1} X^T] y \qquad \text{(4.4.22)}$$

Degrees of Freedom Associated with $\hat{\sigma}^2$

Note that to obtain $\hat{\sigma}^2$ we divide the sum of squared errors *SSE* by $(n-k-1)$ (see (4.4.17)). This number $(n-k-1)$ is called the **degrees of freedom for error** or the **degrees of freedom associated with** $\hat{\sigma}^2$. Observe that *the degrees of freedom for error equals n, the sample size, minus the number of β's in the model*. This is a general rule. In Chapter 1 we saw that the estimated variance for a one-variable population $\{Y\}$ is

$$\hat{\sigma}_Y^2 = \frac{\sum_{i=1}^n (y_i - \hat{\mu}_Y)^2}{(n-1)}$$

the divisor is $(n-1)$ because the model is $y_i = \mu_Y + e_i$, and there is one β (say β_0, represented by μ_Y) in the model. In Chapter 3 the estimated variance $\hat{\sigma}^2$ was seen to be

$$\hat{\sigma}^2 = \frac{\sum_{i=1}^n (y_i - \hat{\beta}_0 - \hat{\beta}_1 x_i)^2}{(n-2)}$$

The divisor is $(n-2)$ because the straight line regression function $\mu_Y(x) = \beta_0 + \beta_1 x$ contains two β's (β_0 and β_1).

Straight Line Regression

Observe that the formula in (4.4.8) for $\hat{\beta}$ is also valid for straight line regression. In that case it is a reexpression of formulas (3.4.8) and (3.4.9) using matrix notation. Note that for straight line regression, the X matrix is of size n by 2, the vectors β and $\hat{\beta}$ are 2 by 1, and the vector y is n by 1. We display them here.

$$X = \begin{bmatrix} 1 & x_1 \\ \vdots & \vdots \\ 1 & x_i \\ \vdots & \vdots \\ 1 & x_n \end{bmatrix} \quad \beta = \begin{bmatrix} \beta_0 \\ \beta_1 \end{bmatrix} \quad \hat{\beta} = \begin{bmatrix} \hat{\beta}_0 \\ \hat{\beta}_1 \end{bmatrix} \quad y = \begin{bmatrix} y_1 \\ \vdots \\ y_i \\ \vdots \\ y_n \end{bmatrix} \qquad \text{(4.4.23)}$$

Whereas matrix notation is not really necessary for expressing the formulas for parameter estimates in straight line regression, it is invaluable for multiple linear regression; otherwise the formulas for parameter estimates would be too cumbersome, if not impossible, to write down.

Estimates of population parameters in a regression problem may be calculated using formulas (4.4.8), (4.4.9), (4.4.10), (4.4.14), (4.4.15), (4.4.16), etc., but the computations are tedious. It is often convenient to use any one of several readily available statistical packages such as SAS, SPSS, SPLUS, BMDP, MINITAB, and so forth, to obtain the estimates. The printouts from these packages are quite similar in content, but differences do exist in the style of presentation of the results. We will present computer output from MINITAB or SAS whenever appropriate. The following example illustrates a typical computer output (obtained using MINITAB) for regression analysis that can be used for obtaining estimates of the unknown parameters in the multiple linear regression model.

E X A M P L E **4.4.2**

Consider the situation described in Example 4.2.1. Suppose that the population regression function of Y (GPA at the end of one year) on X_1 (SATmath), X_2 (SATverbal), X_3 (HSmath), and X_4 (HSenglish) is of the form

$$\mu_Y(x_1, x_2, x_3, x_4) = \beta_0 + \beta_1 x_1 + \beta_2 x_2 + \beta_3 x_3 + \beta_4 x_4 \qquad \text{(4.4.24)}$$

and that (population) assumptions (B) for multiple linear regression hold. The director of admissions decides to obtain an estimate of this regression function based on a sample of 20 students, obtained using simple random sampling, from all students who were admitted over the past four years and who completed their first year. The data for the 20 students in the sample are given in Table 4.4.3. For convenience these data are also stored in the file **gpa.dat** on the data disk. The MINITAB output from a regression analysis of Y on X_1, X_2, X_3, X_4 for the data in Table 4.4.3 is shown in Exhibit 4.4.1. As mentioned earlier, other computer packages produce similar output but may differ slightly in the style of presentation. Only portions of the MINITAB output that are relevant to the present discussion are shown in Exhibit 4.4.1. Other aspects of the MINITAB output will be discussed as the need arises.

T A B L E   **4.4.3**
GPA Data

Student	First-Year GPA	SATmath	SATverbal	HSmath	HSenglish
i	Y	X_1	X_2	X_3	X_4
1	1.97	321	247	2.30	2.63
2	2.74	718	436	3.80	3.57
3	2.19	358	578	2.98	2.57
4	2.60	403	447	3.58	2.21
5	2.98	640	563	3.38	3.48
6	1.65	237	342	1.48	2.14
7	1.89	270	472	1.67	2.64
8	2.38	418	356	3.73	2.52
9	2.66	443	327	3.09	3.20
10	1.96	359	385	1.54	3.46
11	3.14	669	664	3.21	3.37
12	1.96	409	518	2.77	2.60
13	2.20	582	364	1.47	2.90
14	3.90	750	632	3.14	3.49
15	2.02	451	435	1.54	3.20
16	3.61	645	704	3.50	3.74
17	3.07	791	341	3.20	2.93
18	2.63	521	483	3.59	3.32
19	3.11	594	665	3.42	2.70
20	3.20	653	606	3.69	3.52

E X H I B I T   **4.4.1**
MINITAB Output for Regression Analysis of Data in Table 4.4.3

```
The regression equation is
GPA = 0.162 + 0.00201 SATmath + 0.00125 SATverb +
      0.189 HSmath + 0.088 HSengl                        (4.4.25)


Predictor       Coef       Stdev     t-ratio       p
Constant      0.1615      0.4375        0.37    0.717   (4.4.26)
SATmath    0.0020102   0.0005844        3.44    0.004   (4.4.27)
SATverb    0.0012522   0.0005515        2.27    0.038   (4.4.28)
HSmath       0.18944     0.09187        2.06    0.057   (4.4.29)
HSengl        0.0876      0.1765        0.50    0.627   (4.4.30)

s = 0.2685      R-sq = 85.3%    R-sq(adj) = 81.4%       (4.4.31)
```

The estimated regression equation is in (4.4.25), and the values of $\hat{\beta}_0, \hat{\beta}_1, \hat{\beta}_2, \hat{\beta}_3,$ and $\hat{\beta}_4$ can be obtained from this. They are also listed under the heading `Coef` in (4.4.26), (4.4.27), (4.4.28), (4.4.29), and (4.4.30), respectively. The value of $\hat{\sigma}$ is the quantity labeled s in (4.4.31). Thus

$$\hat{\beta}_0 = 0.1615 \qquad \hat{\beta}_1 = 0.0020102 \qquad \hat{\beta}_2 = 0.0012522 \qquad \hat{\beta}_3 = 0.18944$$

$$\hat{\beta}_4 = 0.0876 \qquad \hat{\sigma} = 0.2685$$

Note that the $\hat{\beta}_i$ given in the estimated regression equation are often *rounded* values of the $\hat{\beta}_i$ given under `Coef`. The estimated regression function of Y on X_1, X_2, X_3, X_4 is

$$\hat{\mu}_Y(x_1, x_2, x_3, x_4) = 0.162 + 0.00201x_1 + 0.00125x_2 + 0.189x_3 + 0.088x_4$$

$$(4.4.32)$$

and the best prediction $\hat{Y}(x_1, x_2, x_3, x_4)$ of the Y value of an item from the subpopulation with $X_1 = x_1, X_2 = x_2, X_3 = x_3, X_4 = x_4$ is

$$\hat{Y}(x_1, x_2, x_3, x_4) = 0.162 + 0.00201x_1 + 0.00125x_2 + 0.189x_3 + 0.088x_4 \quad (4.4.33)$$

For example, the predicted first-year GPA for an applicant with SATmath = 730, SATverbal = 570, HSmath = 3.2, and HSenglish = 2.7, is

$$\hat{Y}(730, 570, 3.2, 2.7) = 0.162 + 0.00201(730) + 0.00125(570)$$

$$+ 0.189(3.2) + 0.088(2.7)$$

$$= 3.2 \text{ (rounded to one decimal)}$$

The director of admissions can use this information to make a decision regarding whether or not this student should be recommended for financial assistance.

We must not lose sight of the fact that such predictions may not be reliable. For instance, it is hard to believe that the applicant just referred to would get a GPA of exactly 3.2 at the end of the first year. To get an idea of how good the prediction is, we must also compute appropriate confidence intervals. This is discussed in Section 4.6.

Suppose for a moment that the true subpopulation standard deviation σ is indeed equal to 0.2685. Then, based on the assumption that the subpopulations of Y values are Gaussian, a proportion $p = 0.8$ of the first-year GPAs are within $z_{0.9} \times (0.2685) = (1.28) \times (0.2685) = 0.3437$ unit of $\mu_Y(x_1, x_2, x_3, x_4)$, the mean of the subpopulation to which the individual belongs. These calculations suggest that, if $\mu_Y(x_1, x_2, x_3, x_4)$ is used to predict the Y values of applicants, then 80% of the GPAs will be within 0.3437 unit of $\mu_Y(x_1, x_2, x_3, x_4)$. However, to account for the fact that the number 0.2685 is the *estimated value and not the true value* of σ, we must use a confidence interval for σ. This is discussed in Section 4.6.

In Exhibit 4.4.2 we give a SAS output for the GPA data of Example 4.4.2 analogous to the MINITAB output in Exhibit 4.4.1. Compare the SAS output with the MINITAB output and note the similarities and differences. Only those portions

E X H I B I T **4.4.2**

SAS Output for Regression Analysis of Data in Table 4.4.3

The SAS System   0:00 Saturday,   Jan 1, 1994

Model: MODEL1
Dependent Variable: GPA

Analysis of Variance

Source	DF	Sum of Squares	Mean Square	F Value	Prob>F	
Model	4	6.26432	1.56608	21.721	0.0001	
Error	15	1.08150	0.07210			
C Total	19	7.34582				

Root MSE	0.26851	R-square	0.8528			**(4.4.34)**
Dep Mean	2.59300	Adj R-sq	0.8135			
C.V.	10.35535					

Parameter Estimates

Variable	DF	Parameter Estimate	Standard Error	T for H0: Parameter=0	Prob > |T|	
INTERCEP	1	0.161550	0.43753205	0.369	0.7171	**(4.4.35)**
SATMATH	1	0.002010	0.00058444	3.439	0.0036	**(4.4.36)**
SATVERB	1	0.001252	0.00055152	2.270	0.0383	**(4.4.37)**
HSMATH	1	0.189440	0.09186804	2.062	0.0570	**(4.4.38)**
HSENGL	1	0.087564	0.17649628	0.496	0.6270	**(4.4.39)**

of the SAS output that are relevant to the present discussion are shown in this exhibit. Other parts of the SAS output will be discussed as the need arises.

The estimates of $\beta_0, \beta_1, \beta_2, \beta_3$, and β_4 are given in (4.4.35)–(4.4.39), respectively, under the column labeled Parameter Estimate. The estimate of σ is in (4.4.34) corresponding to the label Root MSE. Compare these results with the results obtained from MINITAB. ∎

In Task 4.4.1 we show how regression can be useful for solving practical problems. Use the computer output and perform all the subsequent calculations needed to solve the problems.

Task   4.4.1

The questions in this task refer to the (artificial) population given in Table D-4 in Appendix D and also in the file **plastic.dat** on the data disk. Refer to Task 4.2.1 at this point.

Suppose we undertake a study to determine the relationship of Y, the strength of plastic containers with temperature X_1 and pressure X_2 for the population in the file **plastic.dat**, but the population data are unavailable. However, a simple random sample of size 16 from this population is available. They are given in Table 4.4.4 and are also stored in the file **table444.dat** on the data disk.

Even though we have the complete population in this example, we use the sample data to illustrate how they can be used to make inferences about population quantities. This gives us an opportunity to compare the estimated parameter values with the true population parameter values.

T A B L E 4.4.4
A Simple Random Sample of Size 16 from the Plastic Data Population

Sample Item Number	Population Item Number	Strength Y	Temperature (°C) X_1	Pressure (psi) X_2
1	1150	36.6	260	10
2	1186	20.7	230	18
3	200	36.5	290	18
4	1305	16.4	200	16
5	783	23.2	200	10
6	1066	26.6	230	14
7	1023	22.5	210	16
8	448	17.0	200	20
9	945	32.7	290	18
10	508	34.4	260	10
11	704	32.4	260	12
12	1135	24.8	240	18
13	107	26.8	220	12
14	742	37.7	280	12
15	749	26.7	260	20
16	1585	24.6	250	20

Suppose that assumptions (A) for multiple linear regression are satisfied and that the regression function is $\mu_Y(x_1, x_2) = \beta_0 + \beta_1 x_1 + \beta_2 x_2$. For questions (1) through (7), we express the answer in three parts. They are

a   In terms of the *symbols* for the **population parameters**, namely β_0, β_1, etc.

b   In terms of the *values* of the **population parameters** from Task 4.2.1.

c In terms of *estimates* of these **population parameters** based on the sample data in Table 4.4.4.

Even though part (c) is the answer we seek, we deliberately provide parts (a) and (b) so you can get accustomed to thinking about the population, the population parameters, and finally the estimates of these parameters.

A SAS output from a regression analysis of the sample data in Table 4.4.4 appears in Exhibit 4.4.3.

E X H I B I T 4.4.3
SAS Output for Regression Analysis of Data in Table 4.4.4

```
                    The SAS System          0:00 Saturday, Jan 1, 1994
```

Model: MODEL1
Dependent Variable: STRENGTH

Analysis of Variance

Source	DF	Sum of Squares	Mean Square	F Value	Prob>F
Model	2	678.56980	339.28490	153.361	0.0001
Error	13	28.76020	2.21232		
C Total	15	707.33000			

Root MSE	1.48739	R-square	0.9593	**(4.4.40)**
Dep Mean	27.47500	Adj R-sq	0.9531	
C.V.	5.41361			

Parameter Estimates

Variable	DF	Parameter Estimate	Standard Error	T for H0: Parameter=0	Prob > |T|	
INTERCEP	1	-6.522361	3.36888546	-1.936	.0749	**(4.4.41)**
TEMP	1	0.192693	0.01235387	15.598	0.0001	**(4.4.42)**
PRESSURE	1	-0.834794	0.10145367	-8.228	0.0001	**(4.4.43)**

1 Estimate the regression function $\mu_Y(x_1, x_2)$, the mean strength of the plastic containers corresponding to a production temperature of x_1 (degrees Celsius), and pressure of x_2 (pounds per square inch); i.e., find $\hat{\beta}_0$, $\hat{\beta}_1$, and $\hat{\mu}_Y(x_1, x_2)$.

a   *The population regression function has the form*

$$\mu_Y(x_1, x_2) = \beta_0 + \beta_1 x_1 + \beta_2 x_2$$

b   *In Task 4.2.1 we found the population regression function to be*

$$\mu_Y(x_1, x_2) = -5 + 0.2x_1 - x_2$$

c   *From (4.4.41)–(4.4.43) the sample estimate of the regression function is*

$$\hat{\mu}_Y(x_1, x_2) = -6.52 + 0.193x_1 - 0.835x_2$$

2   Estimate the difference between the average strengths of two batches of plastic containers if they were produced using the same value for pressure, but their production temperatures differed by 1°C.

Suppose the production temperature and pressure for one batch of plastic containers are x_1 (°C) and x_2 (psi), whereas for the second batch they are $x_1 + 1$ (°C) and x_2 (psi), respectively.

a   *The difference between the average strengths of these two batches is* $\mu_Y(x_1 + 1, x_2) - \mu_Y(x_1, x_2) = [\beta_0 + \beta_1(x_1 + 1) + \beta_2 x_2] - [\beta_0 + \beta_1 x_1 + \beta_2 x_2] = \beta_1.$

b   *The population value of β_1 is 0.2.*

c   *The estimate of β_1 based on the sample is $\hat{\beta}_1 = 0.193$ from (4.4.42).*

3   Estimate the difference between the average strengths of two batches of products if they were produced using the same value for temperature but their production pressures differed by 1 psi.

Suppose the production temperature and pressure for one batch of plastic containers are x_1 (°C) and x_2 (psi), whereas for the second batch they are x_1 (°C) and $x_2 + 1$ (psi), respectively.

a   *The difference between the average strengths of these two batches is* $\mu_Y(x_1, x_2 + 1) - \mu_Y(x_1, x_2) = [\beta_0 + \beta_1 x_1 + \beta_2(x_2 + 1)] - [\beta_0 + \beta_1 x_1 + \beta_2 x_2] = \beta_2.$

b   *The population value of β_2 is -1.0.*

c   *The estimated value of β_2 is $\hat{\beta}_2 = -0.835$ from (4.4.43).*

4   Estimate the difference between the average strengths that would result if a temperature of 260°C is used during production instead of 250°C, while the pressure is kept fixed at some value, say x_2.

a   *The required quantity is* $\mu_Y(260, x_2) - \mu_Y(250, x_2) = [\beta_0 + \beta_1(260) + \beta_2 x_2] - [\beta_0 + \beta_1(250) + \beta_2 x_2] = 10\beta_1.$

b   *The population value of this quantity is $10\beta_1 = 10(0.2) = 2.0$.*

c   *The estimated value of this quantity is $10\hat{\beta}_1 = 10(0.193) = 1.93$.*

Thus, based on the sample, we estimate that on the average the strength of the containers will increase by 1.93 units if a temperature of 260°C is used during production instead of a temperature of 250°C, while the pressure is kept fixed at some value.

5 If a pressure of 16 psi is used during production instead of 14 psi, while the temperature is kept fixed at some value, say x_1, estimate the resulting difference in the average strength of the plastic containers.

   a *The quantity of interest is $\mu_Y(x_1, 16) - \mu_Y(x_1, 14) = [\beta_0 + \beta_1 x_1 + \beta_2(16)] - [\beta_0 + \beta_1 x_1 + \beta_2(14)] = 2\beta_2$.*

   b *The population value of this quantity is $2\beta_2 = 2(-1) = -2.0$.*

   c *The estimated value of this quantity is $2\hat{\beta}_2 = 2(-0.835) = -1.67$.*

Thus the sample data suggest that using a pressure of 16 psi instead of 14 psi, while keeping the temperature fixed at some value will decrease the average strength of the plastic containers by 1.67 units.

6 Estimate the difference in the average strengths between containers manufactured at a temperature of 260°C and pressure of 16 psi and those manufactured at a temperature of 240°C and pressure of 14 psi.

   a *The quantity of interest is $\mu_Y(260, 16) - \mu_Y(240, 14) = [\beta_0 + 260\beta_1 + 16\beta_2] - [\beta_0 + 240\beta_1 + 14\beta_2] = 20\beta_1 + 2\beta_2$.*

   b *The population value of this quantity is $20(0.2) + 2(-1) = 2$.*

   c *The estimated value of this quantity is $20(0.193) + 2(-0.835) = 2.19$.*

7 Estimate the standard deviation of the strength of plastic containers manufactured under identical temperature and pressure conditions.

   First note that plastic containers, manufactured under identical temperature and pressure conditions, all belong to the same subpopulation.

   a *Recall from the results of Task 4.2.1 that the subpopulation standard deviations for strength are all equal. This common subpopulation standard deviation is denoted by σ.*

   b *The population value of the standard deviation is $\sigma = 1.7076$ (from Task 4.2.1).*

   c *The sample estimate of this standard deviation is $\hat{\sigma} = 1.487$ from (4.4.40).*

8 What is the interpretation of the parameter β_0 in this problem?

   The quantity β_0 is the value of the regression function when X_1 is zero and X_2 is zero. Do not, however, be tempted to conclude that this is the average strength of the plastic containers when the process temperature is set at 0°C and pressure at 0 psi. We know the regression function to be valid only for the range of values 200–300°C for temperature, and 10–20 psi for pressure. The pair of values, 0°C and 0 psi, are far removed from the allowable range of values, and no meaningful conclusions can be derived from the value of β_0 (or $\hat{\beta}_0$) regarding the average strength of the containers manufactured at 0°C and 0 psi. In all likelihood, it may not even make any sense to run the process at these values of temperature and pressure.

Problems   4.4

Use the following information for Problems 4.4.1–4.4.12. A simple random sample of size 10 is selected from the population discussed in Task 4.2.1, where an investigator wants to study the relationship of strength (Y) of plastic containers to the predictor factors temperature (X_1) and pressure (X_2). The sample data are given in Table 4.4.5, and are also stored in the file **table445.dat** on the data disk. We suppose that assumptions (A) for multiple regression are satisfied. A MINITAB output from a regression analysis of Y on X_1 and X_2 is given in Exhibit 4.4.4.

T A B L E **4.4.5**

Sample Item Number	Population Item Number	Strength Y	Temperature X_1	Pressure X_2
1	1001	40.2	290	12
2	260	38.2	270	12
3	1085	30.2	240	12
4	1267	18.5	210	20
5	733	28.2	250	16
6	1173	35.3	260	12
7	438	27.3	220	12
8	129	28.1	210	10
9	1072	35.0	250	12
10	1381	16.7	210	20

E X H I B I T **4.4.4**
MINITAB Output for Regression of Plastic Data in Table 4.4.5

```
The regression equation is
 strength = - 0.65 + 0.187 temp - 1.07 pressure

Predictor        Coef        Stdev        t-ratio         p
Constant        -0.646       4.509         -0.14       0.890
temp             0.18721     0.01477       12.68       0.000
pressure        -1.0653      0.1157        -9.21       0.000

s = 1.117       R-sq = 98.4%      R-sq(adj) = 97.9%
```

4.4.1 Exhibit the X matrix and the y vector for these data (see (4.4.7)).

4.4.2 Compute $X^T X$ and $X^T y$.

4.4.3 The matrix $(X^T X)^{-1}$ is as follows:

$$(X^T X)^{-1} = \begin{bmatrix} 16.300423 & -0.050442 & -0.293043 \\ -0.050442 & 0.000175 & 0.000602 \\ -0.293043 & 0.000602 & 0.010723 \end{bmatrix}$$

Calculate $\hat{\beta}$ using (4.4.8) and compare the results with the $\hat{\beta}_i$ given in the computer output in Exhibit 4.4.4.

4.4.4 Compute SSY, $SSX(1)$, and $SSX(2)$ where $SSX(i)$ is SSX for X_i.

4.4.5 Are valid estimates available for the following population quantities?

$$\mu_Y \qquad \mu_{X_1} \qquad \mu_{X_2} \qquad \sigma_Y \qquad \sigma_{X_1} \qquad \sigma_{X_2}$$

Explain.

4.4.6 What is the value of $\hat{\sigma}$ (i.e., $\hat{\sigma}_{Y|X_1,X_2}$)? Compare this with the population value.

4.4.7 Estimate $\mu_Y(300, 16)$ using the sample data in Table 4.4.5. Compare this estimate with the population value.

4.4.8 Write the symbol for the population quantity that represents the best predicted value for the strength of a plastic container that was produced using a temperature of $280°C$ and a pressure of 19 psi.

4.4.9 In Problem 4.4.8, is that value available from the population? If not, why not?

4.4.10 What is the estimate of $\mu_Y(280, 18)$ using the sample data in Table 4.4.5?

4.4.11 Use the *population values of the parameters* and determine what proportion of the plastic containers has a strength greater than 31 units if they are produced with a temperature of $250°C$ and a pressure of 16 psi.

4.4.12 Estimate the quantity of interest in Problem 4.4.11 using the data in Table 4.4.5.

4.5
Residual Analysis

Residuals in multiple linear regression were defined in (4.4.12). They are useful for checking the validity of assumptions (A) and (B). For reasons similar to those discussed in Section 3.5, it is customary to examine the residuals after appropriately *standardizing* them.

Standardized Residuals

The **standardized residuals** for multiple linear regression are defined by

D E F I N I T I O N

$$r_i = \frac{\hat{e}_i}{\hat{\sigma}\sqrt{1 - h_{i,i}}} \qquad (4.5.1)$$

where $h_{i,i}$ is the ith diagonal element of the matrix

$$H = X(X^T X)^{-1} X^T \quad \blacksquare \qquad (4.5.2)$$

The matrix H is sometimes called the **hat matrix** and the quantities $h_{i,i}$ are called the **hat values**. If assumptions (A) or (B) for multiple linear regression are satisfied, then the standardized residuals $r_1, \ldots, r_n$ will be (approximately) a simple random sample of n observations from a Gaussian population with mean zero and unit standard deviation. This fact is useful in examining the validity of some of the model assumptions.

Plotting Standardized Residuals Against the Predictors

In some instances a plot of the standardized residuals r_i against the sample values $x_{i,j}$ of the predictor variable X_j ($j = 1, \ldots, k$) can be useful in detecting inadequacies in the assumed regression function. When assumptions (A) or (B) are satisfied, the points on this plot should be scattered about the X_j axis in a random fashion, showing no obvious trends or other patterns. If this is found not to be the case, then one or more of assumptions (A) or (B) are likely to be violated.

Plotting Standardized Residuals Against Fitted Values

The estimated subpopulation means $\hat{\mu}_Y(x_{i,1}, \ldots, x_{i,k})$ corresponding to the sample values $(x_{i,1}, \ldots, x_{i,k})$ are often referred to as **fitted values** or simply **fits**. When assumptions (A) or (B) are satisfied, the fitted values are *independent* of the standardized residuals r_i. Hence the points in the plot of r_i versus $\hat{\mu}_Y(x_{i,1}, \ldots, x_{i,k})$ for $i = 1, \ldots, n$ should be randomly scattered about the horizontal axis (axis of fitted values), with no specific pattern. If any systematic pattern is observed, then this may indicate violation of one or more of assumptions (A) in Box 4.3.1 or assumptions (B) in Box 4.3.2. In some instances where a pattern is observed instead of a random scatter, it may be possible to diagnose the cause of the observed pattern. Refer to Section 3.5 for details.

Rankit-Plots of Standardized Residuals

If assumptions (A) or (B) are satisfied, then the standardized residuals r_i will be (approximately) a simple random sample from a *standard* Gaussian population (Gaussian with mean zero and standard deviation one). This can be checked graphically by examining a (Gaussian) rankit-plot of the standardized residuals (i.e., a plot of the standardized residuals against the Gaussian scores) as discussed in Section 3.5.

Rankit-Plots of Linear Combinations of $X_1, \ldots, X_k$ and Y

As discussed in Section 4.3, for inferences on certain parameters we need assumptions (B) in Box 4.3.2 and, in particular, we need the assumption that the population $\{(Y, X_1, \ldots, X_k)\}$ is Gaussian. To check this assumption, we investigate whether or not the sample data $(y_1, x_{1,1}, \ldots, x_{1,k}), \ldots, (y_n, x_{n,1}, \ldots, x_{n,k})$ appear to be a simple random sample from a $(k+1)$-variable Gaussian population. We can do this by examining the rankit-plots of linear combinations of $x_{i,1}, \ldots, x_{i,k}$ and y_i. Recall that a theoretical population $\{(Y, X_1, \ldots, X_k)\}$ is Gaussian if and only if every linear combination $b_0 Y + b_1 X_1 + b_2 X_2 + \cdots + b_k X_k$ is Gaussian. Since only sample data are available, we examine linear combinations of y_i and $x_{i,1}, \ldots, x_{i,k}$ in the sample. In practice, it is impossible to examine *every* linear combination of y_i and $x_{i,1}, \ldots, x_{i,k}$, but we can consider several linear combinations and examine the corresponding rankit-plots. We illustrate in the following example.

E X A M P L E 4.5.1

A utility company is interested in investigating how Y, the electricity consumption by each household, is related to monthly income (X_1), number of persons (X_2) in the household, and the living area (X_3) of the house (or apartment). A simple random sample of 34 households served by this utility company is surveyed, and the following information is obtained for each household:

Y = (total) electric bill (in dollars) for the past year

X_1 = monthly income for the household (in dollars)

X_2 = number of persons in the household

X_3 = living area (in square feet) of the house or apartment

The data are displayed in Table 4.5.1 and are also stored in the file **electric.dat** on the data disk. The investigator assumes that the regression function of Y on X_1, X_2, and X_3 is of the form (at least approximately)

$$\mu_Y(x_1, x_2, x_3) = \beta_0 + \beta_1 x_1 + \beta_2 x_2 + \beta_3 x_3 \qquad \text{(4.5.3)}$$

for $2{,}000 \le x_1 \le 6{,}000$, $x_2 = 1, 2, 3, 4, 5, 6$, and $500 \le x_3 \le 4{,}000$. We are interested in checking the validity of assumptions (A) for multiple linear regression. For this purpose we carry out a residual analysis of the data. The computations, plots, etc., were done using MINITAB although any regression package can be used. The computer output containing these results is given in Exhibit 4.5.1 where we have also printed the fitted values $\hat{\mu}_Y(x_{i,1}, x_{i,2}, x_{i,3})$ (labeled `fits`), the residuals $\hat{e}_i$ (labeled `residual`), the standardized residuals r_i (labeled `stdresid`), and the Gaussian scores $z_i^{(n)}$ (labeled `nscores`) of the standardized residuals.

T A B L E **4.5.1**

Electric Bill Data

Household	Bill Y	Income X_1	Persons X_2	Area X_3
1	228	3220	2	1160
2	156	2750	1	1080
3	648	3620	2	1720
4	528	3940	1	1840
5	552	4510	3	2240
6	636	3990	4	2190
7	444	2430	1	830
8	144	3070	1	1150
9	744	3750	2	1570
10	1104	4790	5	2660
11	204	2490	1	900
12	420	3600	3	1680
13	876	5370	1	2550
14	840	3180	7	1770
15	876	5910	2	2960
16	276	3020	2	1190
17	1236	5920	3	3130
18	372	3520	2	1560
19	276	3720	1	1510
20	540	4840	1	2190
21	1044	4700	6	2620
22	552	3270	2	1350
23	756	4420	2	1990
24	636	4480	2	2070
25	708	3820	4	1850
26	960	5740	2	2700
27	1080	5600	3	3030
28	480	3950	2	1700
29	96	2290	3	890
30	1272	5580	5	3270
31	1056	5820	2	2660
32	156	3160	2	1330
33	396	2880	4	1280
34	768	3780	3	1950

E X H I B I T 4.5.1

MINITAB Output for Regression Analysis of Electric Bill Data

```
The regression equation is
Bill = - 358 + 0.075 Income + 55.1 Persons + 0.281 Area
```

Predictor	Coef	Stdev	t-ratio	p
Constant	-358.4	198.7	-1.80	0.081
Income	0.0751	0.1361	0.55	0.585
Persons	55.09	29.05	1.90	0.068
Area	0.2811	0.2261	1.24	0.223

```
s = 135.4      R-sq = 85.1%     R-sq(adj) = 83.7%
```

ROW	Bill	Income	Persons	Area	fits	residual	stdresid	nscores
1	228	3220	2	1160	319.75	-91.755	-0.73740	-0.57777
2	156	2750	1	1080	206.86	-50.865	-0.40260	-0.41240
3	648	3620	2	1720	507.23	140.772	1.08468	1.24834
4	528	3940	1	1840	509.92	18.084	0.14376	0.18319
5	552	4510	3	2240	775.36	-223.361	-1.68098	-2.09417
6	636	3990	4	2190	777.32	-141.322	-1.11948	-1.24834
7	444	2430	1	830	112.54	331.455	2.61553	2.09417
8	144	3070	1	1150	250.59	-106.586	-0.82029	-0.66662
9	744	3750	2	1570	474.83	269.170	2.05367	1.67229
10	1104	4790	5	2660	1024.64	79.362	0.62575	0.57777
11	204	2490	1	900	136.73	67.270	0.53344	0.49336
12	420	3600	3	1680	549.57	-129.568	-0.97566	-0.86326
13	876	5370	1	2550	816.95	59.054	0.46565	0.41240
14	840	3180	7	1770	763.66	76.339	0.70903	0.66662
15	876	5910	2	2960	1027.86	-151.860	-1.20401	-1.42855
16	276	3020	2	1190	313.16	-37.161	-0.28362	-0.18319
17	1236	5920	3	3130	1131.49	104.514	0.82864	0.76119
18	372	3520	2	1560	454.74	-82.737	-0.62388	-0.49336
19	276	3720	1	1510	400.62	-124.622	-0.95231	-0.76119
20	540	4840	1	2190	675.93	-135.926	-1.05102	-1.10154
21	1044	4700	6	2620	1061.72	-17.719	-0.14601	-0.03644
22	552	3270	2	1350	376.92	175.079	1.32665	1.42855
23	756	4420	2	1990	643.24	112.765	0.85709	0.86326
24	636	4480	2	2070	670.23	-34.232	-0.25848	-0.10951
25	708	3820	4	1850	668.97	39.026	0.30005	0.33413
26	960	5740	2	2700	942.00	18.000	0.14687	0.25790
27	1080	5600	3	3030	1079.33	0.668	0.00538	0.03644
28	480	3950	2	1700	526.40	-46.401	-0.35265	-0.33413
29	96	2290	3	890	229.07	-133.067	-1.04615	-0.97539
30	1272	5580	5	3270	1255.47	16.530	0.14132	0.10951
31	1056	5820	2	2660	936.77	119.234	1.04082	1.10154
32	156	3160	2	1330	363.03	-207.034	-1.56974	-1.67229
33	396	2880	4	1280	438.12	-42.116	-0.33233	-0.25790
34	768	3780	3	1950	638.99	129.009	1.00438	0.97539

First we plot the standardized residuals r_i against monthly income X_1.

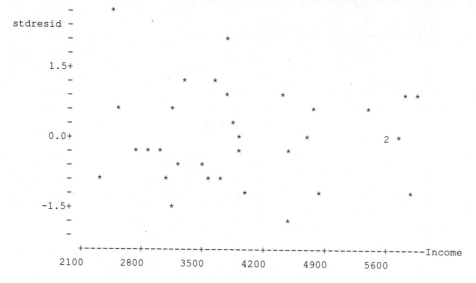

The standardized residuals exhibit only a random scatter about the horizontal line through the origin with stdresid = 0 (which is not shown in the plot, but you should draw this line for ease of interpretation) and no definite pattern is seen. So this plot does not point to any violations of assumptions (A).

Next we plot the standardized residuals r_i against X_2, the number of persons in the household.

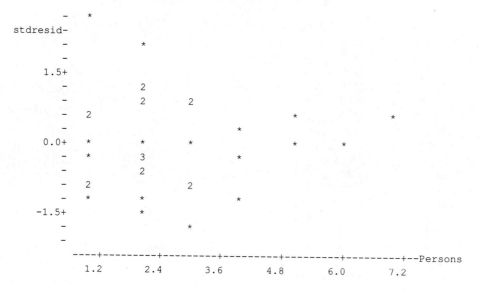

This plot suggests that perhaps the assumption of homogeneity of standard deviations may *not* be satisfied. The standardized residuals seem to vary over a wider range for small values of X_2 than for large values of X_2. However, this apparent pattern may be due to the fact that there are only seven observations with X_2 values equal to 4 or more. Nevertheless, further examination may be warranted.

Next we plot the standardized residuals r_i against living area X_3 and also against fitted values $\hat{\mu}_Y(x_{i,1}, x_{i,2}, x_{i,3})$.

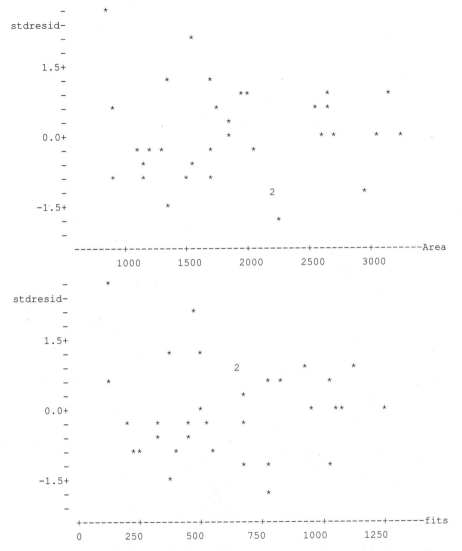

These plots might suggest the possibility that the standard deviations of subpopulations may not all be the same. There seems to be a greater variability in the values of the standardized residuals corresponding to small values of `fits` (and area) than for large values of `fits` (and area). It is possible that this apparent pattern arises due to the two data values giving rise to the two largest positive standardized residuals (check the data values for sample items 7 and 9). Further investigation may be warranted.

To examine whether or not the subpopulations determined by X_1, X_2, and X_3 are Gaussian, we obtain a rankit-plot of the standardized residuals.

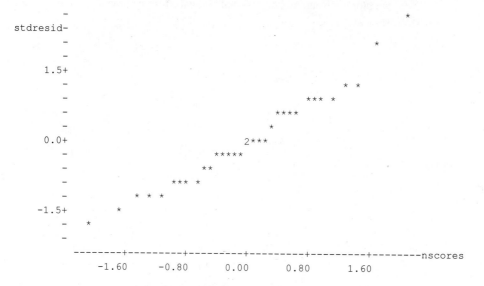

This plot appears to be consistent with the assumption that the residuals are a simple random sample from a standard Gaussian population and there is no indication of violations of the assumption that subpopulations determined by X_1, X_2, X_3 are Gaussian.

To summarize, the only part of assumptions (A) whose validity is in some doubt is the homogeneity of standard deviations. In a real study, the investigator should probably consult a professional statistician for advice. ■

In Section 4.5 of the laboratory manuals we show how to carry out the computations and plots to examine the plausibility of assumptions (A) or (B) using **MINITAB** and/or **SAS**.

There may exist other diagnostic procedures which might indicate that the assumptions for multiple linear regression may be violated. Even when these procedures do not suggest the failure of any of the assumptions, there is no guarantee that the assumptions are actually met. This is necessarily a subjective exercise, and at some point the investigator must either conclude that the assumptions required for a valid analysis are satisfied (for all practical purposes) and proceed with the analyses and inferences, or he/she must conclude that one or more of the assumptions is seriously violated and look for alternative procedures. For certain types of violations of assumptions, we discuss alternative procedures in Chapter 8.

Authors' Recommendation

When performing a multiple linear regression analysis of a set of data $(y_1, x_{1,1}, \ldots, x_{1,k}), \ldots, (y_n, x_{n,1}, \ldots, x_{n,k})$, we suggest that you include the following steps.

1   Obtain the standardized residuals r_i and the fitted values $\hat{\mu}_Y(x_{i,1}, \ldots, x_{i,k})$, denoted here by $\hat{\mu}_i$ for ease of notation.

2   Plot r_i against $\hat{\mu}_i$ and also r_i against $x_{i,j}$ for $j = 1, \cdots, k$. Examine these plots for evidence of unequal subpopulation variances or an incorrect model.

3   Obtain a rankit-plot of r_i to evaluate the validity of the assumption that each subpopulation of Y values is a Gaussian population.

4   If you wish to examine the validity of assumptions (B) and the data are obtained by simple random sampling, then examine the Gaussian rankit-plots of $y_i, x_{i,1}, \ldots$, and $x_{i,k}$, and several linear combinations of these, to assess whether or not the data appear to be a simple random sample from a $(k + 1)$-variable Gaussian population.

5   Make an overall evaluation of the validity (at least approximately) of assumptions (A) or (B) within the context of the particular application in question.

Problems   4.5

4.5.1   Consider Task 4.4.1 where an investigator wants to study the relationship of the strength (Y) of plastic containers to the predictor factors, temperature (X_1) and pressure (X_2). A simple random sample of size 16 is selected from the population and assumptions (A) are presumed to be satisfied. The data are given in Table 4.4.4 and are also stored in the file **table444.dat** on the data disk. We want to perform a residual analysis to determine whether assumptions (A) seem to be valid for this problem. A SAS output for regression appears in Exhibit 4.5.2, along with a printout of y_i (STRENGTH), $x_{i,1}$ (TEMP), $x_{i,2}$ (PRESSURE), fitted values $\hat{\mu}_Y(x_{i,1}, x_{i,2})$ (FITS), residuals $\hat{e}_i$ (RESIDUAL), standardized residuals r_i (STDRESID), and Gaussian scores $z_i^{(n)}$ (NSCORES) with $n = 16$.

a   Plot the standardized residuals r_i against fits $\hat{\mu}_Y(x_{i,1}, x_{i,2})$ (for convenience round all numbers to one decimal).

b   Plot the standardized residuals r_i against $X_1 = $ temperature.

c   Plot the standardized residuals r_i against $X_2 = $ pressure.

d   Plot the standardized residuals r_i against the Gaussian scores (i.e., rankits) $z_i^{(n)}$.

What do you conclude regarding the plausibility of assumptions (A) for this problem?

E X H I B I T 4.5.2
SAS Output for Problem 4.5.1

```
                            The SAS System   0:00 Saturday,   Jan 1, 1994
Model: MODEL1
Dependent Variable: STRENGTH
                          Analysis of Variance
                             Sum of        Mean
       Source        DF    Squares       Square     F Value      Prob>F
       Model          2    678.56980    339.28490    153.361     0.0001
       Error         13     28.76020      2.21232
       C Total       15    707.33000

            Root MSE        1.48739    R-square      0.9593
            Dep Mean       27.47500    Adj R-sq      0.9531
            C.V.            5.41361
                          Parameter Estimates
                        Parameter      Standard    T for H0:
       Variable   DF    Estimate         Error    Parameter=0    Prob > |T|
       INTERCEP    1    -6.522361     3.36888546     -1.936        0.0749
       TEMP        1     0.192693     0.01235387     15.598        0.0001
       PRESSURE    1    -0.834794     0.10145367     -8.228        0.0001
```

S		S		P		R		S	
A	P	T		R		E		T	N
M	O	R		E		S		D	S
P	P	E		S		I		R	C
I	I	N	T	S	F	D		E	O
T	T	G	E	S	I	U		S	R
E	E	T	M	U	T	A		I	E
M	M	H	P	R	S	L		D	S
				E					

```
 1   1150   36.6   260   10   35.2298    1.37021    1.03884    0.76184
 2   1186   20.7   230   18   22.7707   -2.07066   -1.47494   -1.28155
 3    200   36.5   290   18   34.3322    2.16778    1.68383    1.76883
 4   1305   16.4   200   16   18.6595   -2.25947   -1.68822   -1.76883
 5    783   23.2   200   10   23.6682   -0.46823   -0.37926   -0.39573
 6   1066   26.6   230   14   26.1098    0.49017    0.34362    0.39573
 7   1023   22.5   210   16   20.5864    1.91361    1.38608    1.28155
 8    448   17.0   200   20   15.3203    1.67971    1.34590    0.98815
 9    945   32.7   290   18   34.3322   -1.63222   -1.26783   -0.98815
10    508   34.4   260   10   35.2298   -0.82979   -0.62912   -0.56918
11    704   32.4   260   12   33.5602   -1.16020   -0.83814   -0.76184
12   1135   24.8   240   18   24.6976    0.10242    0.07251    0.07720
13    107   26.8   220   12   25.8525    0.94750    0.68899    0.56918
14    742   37.7   280   12   37.4141    0.28594    0.21643    0.23349
15    749   26.7   260   20   26.8818   -0.18185   -0.13559   -0.07720
16   1585   24.6   250   20   24.9549   -0.35492   -0.26203   -0.23349
```

4.5.2 Consider the GPA data of Example 4.4.2, which is also stored in the file **gpa.dat** on the data disk. We want to perform a residual analysis to determine if assumptions (B) seem to be valid for this problem. A MINITAB output for regression appears in Exhibit 4.5.3, along with a printout of the data, the residuals $\hat{e}_i$, fits $\hat{\mu}_Y(x_{i,1}, x_{i,2}, x_{i,3}, x_{i,4})$, standardized residuals r_i, and Gaussian scores $z_i^{(n)}$ with $n = 20$.

    a Plot the standardized residuals r_i against the fits $\hat{\mu}_Y(x_{i,1}, x_{i,2}, x_{i,3}, x_{i,4})$ (for convenience round all numbers to one decimal).

    b Plot the standardized residuals r_i against $X_1 = $ SATmath.

    c Plot the standardized residuals r_i against $X_2 = $ SATverbal.

    d Plot the standardized residuals r_i against $X_3 = $ HSmath.

    e Plot the standardized residuals r_i against $X_4 = $ HSenglish.

    f Obtain Gaussian rankit-plots for several linear combinations of $Y, X_1, X_2, X_3,$ and X_4.

Based on parts (a)–(f), judge whether or not the five-variable population $\{(Y, X_1, X_2, X_3, X_4)\}$ may be assumed to be multivariate Gaussian.

4.5.3 In Problem 4.5.2, plot the standardized residuals r_i against the Gaussian scores (i.e., rankits) $z_i^{(n)}$. Assess whether or not the standardized residuals appear to be a simple random sample from a standard Gaussian population.

4.5.4 Based on the results of Problems 4.5.2 and 4.5.3, decide whether or not assumptions (B) appear to hold for this problem.

E X H I B I T   **4.5.3**
MINITAB Output for Problem 4.5.3

```
The regression equation is
GPA = 0.162 + 0.00201 SATmath + 0.00125 SATverb +
      0.189 HSmath + 0.088 HSengl


Predictor        Coef        Stdev     t-ratio         p
Constant       0.1615       0.4375        0.37     0.717
SATmath     0.0020102    0.0005844        3.44     0.004
SATverb     0.0012522    0.0005515        2.27     0.038
HSmath        0.18944      0.09187        2.06     0.057
HSengl         0.0876       0.1765        0.50     0.627


s = 0.2685      R-sq = 85.3%      R-sq(adj) = 81.4%
```

E X H I B I T   4.5.3
(Continued)

student	SAT math	SAT verb	HS math	HS engl	fits	residual	stdresid	nscores
1	1.97	321	247	2.30	2.63	1.7821	0.1879	0.7966	0.7420
2	2.74	718	436	3.80	3.57	3.1833	-0.4433	-1.9039	-1.8713
3	2.19	358	578	2.98	2.57	2.3945	-0.2045	-0.8595	-1.1269
4	2.60	403	447	3.58	2.21	2.4031	0.1969	0.8708	0.9172
5	2.98	640	563	3.38	3.48	3.0981	-0.1181	-0.4655	-0.4460
6	1.65	237	342	1.48	2.14	1.5340	0.1160	0.5137	0.5874
7	1.89	270	472	1.67	2.64	1.8429	0.0471	0.1998	0.4460
8	2.38	418	356	3.73	2.52	2.3749	0.0051	0.0226	0.0617
9	2.66	443	327	3.09	3.20	2.3271	0.3329	1.4335	1.4038
10	1.96	359	385	1.54	3.46	1.9600	-0.0000	-0.0000	-0.0617
11	3.14	669	664	3.21	3.37	3.2410	-0.1010	-0.4127	-0.3132
12	1.96	409	518	2.77	2.60	2.3848	-0.4248	-1.6851	-1.4038
13	2.20	582	364	1.47	2.90	2.3197	-0.1197	-0.5717	-0.5874
14	3.90	750	632	3.14	3.49	3.3610	0.5390	2.2510	1.8713
15	2.02	451	435	1.54	3.20	2.1848	-0.1648	-0.6919	-0.9172
16	3.61	645	704	3.50	3.74	3.3302	0.2798	1.2155	1.1269
17	3.07	791	341	3.20	2.93	3.0414	0.0286	0.1549	0.1859
18	2.63	521	483	3.59	3.32	2.7845	-0.1545	-0.6295	-0.7420
19	3.11	594	665	3.42	2.70	3.0726	0.0374	0.1645	0.3132
20	3.20	653	606	3.69	3.52	3.2403	-0.0403	-0.1622	-0.1859

4.6
Confidence Intervals

The practical importance of confidence intervals cannot be overemphasized, and you should recall the discussions of Section 1.6 in this regard. In this section we give formulas for computing confidence intervals for the unknown parameters in multiple linear regression.

Confidence Intervals for β_i, $\mu_Y(x_1, \ldots, x_k)$, $Y(x_1, \ldots, x_k)$, and $\sum_{i=0}^{k} a_i \beta_i = a^T \beta$

The general form for $1 - \alpha$ two-sided confidence intervals for

$$\beta_0, \beta_1, \ldots, \beta_k, \mu_Y(x_1, \ldots, x_k), Y(x_1, \ldots, x_k)$$

and linear functions

$$a^T \beta = \sum_{i=0}^{k} a_i \beta_i = a_0 \beta_0 + a_1 \beta_1 + \cdots + a_k \beta_k$$

for a specified vector a is

$$\hat{\theta} - \text{table-value} \times SE(\hat{\theta}) \leq \theta \leq \hat{\theta} + \text{table-value} \times SE(\hat{\theta}) \tag{4.6.1}$$

where θ refers to any of the quantities

$$\beta_0, \beta_1, \dots, \beta_k, \mu_Y(x_1, \dots, x_k), Y(x_1, \dots, x_k), \text{ or } a^T \beta$$

and $\hat{\theta}$ refers to the corresponding estimate. The quantity to be used as the table-value is $t_{1-\alpha/2:df}$ obtained from a student's t-table (Table T-2 in Appendix T) with df = degrees of freedom = $n - k - 1 = n - (k + 1)$, where $k + 1$ is the number of parameters β_i in the regression function in (4.1.1). As usual, the quantity $SE(\hat{\theta})$ is the standard error of $\hat{\theta}$ and is an estimate of the *precision* of $\hat{\theta}$. These confidence intervals are valid under assumptions (A) or (B).

From the two-sided $1 - \alpha$ confidence interval in (4.6.1) a $1 - \alpha/2$ lower confidence bound or a $1 - \alpha/2$ upper confidence bound can be obtained as explained in Section 1.6. (4.6.2)

Point estimates for $\beta = [\beta_0, \beta_1, \dots, \beta_k]^T$, $\mu_Y(x_1, \dots, x_k)$, $Y(x_1, \dots, x_k)$, and $a^T \beta$ are given in (4.4.8), (4.4.9), (4.4.10), and (4.4.11), respectively. The formulas for standard errors require elements from the matrix C where C is defined by

$$C = (X^T X)^{-1} \tag{4.6.3}$$

and the matrix X is defined in (4.4.7). We write $c_{i,j}$ for the (i, j) element of the matrix C. Formulas for standard errors of various estimated quantities of interest are listed in (4.6.4)–(4.6.8).

$$SE(\hat{\beta}_{i-1}) = \hat{\sigma} \sqrt{c_{i,i}} \qquad \text{for } i = 1, \dots, (k + 1) \tag{4.6.4}$$

$$SE(\hat{\mu}_Y(x_1, \dots, x_k)) = \hat{\sigma} \sqrt{x^T C x} \tag{4.6.5}$$

$$SE(\hat{Y}(x_1, \dots, x_k)) = \hat{\sigma} \sqrt{1 + (x^T C x)} \tag{4.6.6}$$

$$= \sqrt{\hat{\sigma}^2 + [SE(\hat{\mu}_Y(x_1, \dots, x_k))]^2} \tag{4.6.7}$$

$$SE(a^T \hat{\beta}) = \hat{\sigma} \sqrt{a^T C a} \tag{4.6.8}$$

The vector x in (4.6.5) and (4.6.6) is given by

$$x = [1, x_1, x_2, \dots, x_k]^T \tag{4.6.9}$$

and the vector a in (4.6.8) is given by

$$a = [a_0, a_1, \dots, a_k]^T \tag{4.6.10}$$

where the linear combination $a^T\beta$ of interest is $a_0\beta_0 + a_1\beta_1 + \cdots + a_k\beta_k$. For instance, in Example 4.4.2 if we wish to compute a confidence interval for $\mu_Y(720, 570, 3.2, 2.7)$, then we would compute $SE(\hat{\mu}_Y(720, 570, 3.2, 2.7))$ using (4.6.5) with $x = [1, 720, 570, 3.2, 2.7]^T$.

Remarks

1 Note that $SE(\hat{\beta}_i)$ for $i = 0, \ldots, k$ and $SE(\hat{\mu}_Y(x_1, \ldots, x_k))$ can be obtained from (4.6.8) by using appropriate vectors a. For instance, $SE(\hat{\beta}_0)$ is obtained from (4.6.8) by taking $a = [1, 0, \ldots, 0]^T$, $SE(\hat{\beta}_1)$ is obtained by taking $a = [0, 1, 0, \ldots, 0]^T$, $SE(\mu_Y(x_1, \ldots, x_k))$ is obtained by taking $a = [1, x_1, \ldots, x_k]^T$, etc.

2 In any particular application, it is important for the user to determine whether a confidence interval for $\mu_Y(x_1, \ldots, x_k)$ or a confidence interval for $Y(x_1, \ldots, x_k)$ is required. Some authors call the confidence interval for $Y(x_1, \ldots, x_k)$ a **prediction interval** because the term confidence interval is traditionally reserved for *parameters*, but $Y(x_1, \ldots, x_k)$ is a *random variable*. (It is the Y value of a randomly chosen item with $X_1 = x_1, \ldots, X_k = x_k$.)

3 Note that even though

$$\hat{\mu}_Y(x_1, \ldots, x_k) = \hat{Y}(x_1, \ldots, x_k) = \hat{\beta}_0 + \hat{\beta}_1 x_1 + \cdots + \hat{\beta}_k x_k \qquad \textbf{(4.6.11)}$$

their standard errors are not the same. This is so because there is greater uncertainty in predicting $Y(x_1, \ldots, x_k)$, a single randomly chosen value from a subpopulation with $X_1 = x_1, \ldots, X_k = x_k$ than in estimating the mean $\mu_Y(x_1, \ldots, x_k)$ of the entire subpopulation.

Confidence Interval for σ

When regression assumptions (A) or (B) hold, a two-sided $1 - \alpha$ confidence interval for σ is given by

$$\sqrt{\frac{(n-k-1)\hat{\sigma}^2}{\chi^2_{1-\alpha/2:n-k-1}}} \leq \sigma \leq \sqrt{\frac{(n-k-1)\hat{\sigma}^2}{\chi^2_{\alpha/2:n-k-1}}} \qquad \textbf{(4.6.12)}$$

where $\chi^2_{\alpha/2:n-k-1}$ and $\chi^2_{1-\alpha/2:n-k-1}$ may be obtained from Table T-3 in Appendix T. It is worth noting that the quantity $(n-k-1)\hat{\sigma}^2$ in (4.6.12) is in fact equal to $SSE = SSE(X_1, \ldots, X_k)$ (refer to (4.4.17)), and so we can rewrite (4.6.12) in terms of SSE. Thus the $(1 - \alpha)$ two-sided confidence interval for σ in (4.6.12) may be reexpressed in the form

$$\sqrt{\frac{SSE}{\chi^2_{1-\alpha/2:n-k-1}}} \leq \sigma \leq \sqrt{\frac{SSE}{\chi^2_{\alpha/2:n-k-1}}} \qquad \textbf{(4.6.13)}$$

The formula for a confidence interval for σ in (4.6.12) is not of the general form given in (4.6.1). In fact, the general form for a two-sided $(1 - \alpha)$ confidence interval for a parameter that is a standard deviation (when a valid confidence interval is available) is

$$\sqrt{\frac{(df)\,(\text{estimated standard deviation})^2}{\chi^2_{1-\alpha/2:df}}} \le \sigma \qquad\qquad (4.6.14)$$

$$\le \sqrt{\frac{(df)\,(\text{estimated standard deviation})^2}{\chi^2_{\alpha/2:df}}}$$

where df represents the number of degrees of freedom associated with the estimate $\hat{\sigma}$. Note that the number of degrees of freedom associated with $\hat{\sigma}$ is $n - (k + 1)$ when the regression function of Y on $X_1, \ldots, X_k$ is the one given in (4.1.1); i.e., it has $(k + 1)$ regression coefficients $\beta_0, \beta_1, \ldots, \beta_k$. It is also worth observing that while the confidence intervals for the quantities, $\beta_0, \beta_1, \ldots, \beta_k, \mu_Y(x_1, \ldots, x_k)$, $Y(x_1, \ldots, x_k)$, and $a^T\beta$ are symmetric about the corresponding point estimates, this is not so in the case of σ. However, the confidence interval for σ is equal-tailed and contains the point estimate $\hat{\sigma}$, and hence it gives us some indication of how close $\hat{\sigma}$ might be to the population value σ.

One-Sided Confidence Bounds

In this section, most of the discussion has been about two-sided confidence intervals for a parameter of interest. However, in some applications an investigator may be interested in only the lower bound or the upper bound for a parameter in a decision-making situation, and hence one-sided confidence bounds are useful. As discussed in Section 1.6, we can obtain one-sided confidence bounds with confidence coefficient $1 - \alpha/2$ by first constructing a two-sided confidence interval with confidence coefficient $1 - \alpha$ and reading off either the lower or the upper endpoint as appropriate. This is valid for all of the quantities, $\beta_0, \beta_1, \ldots, \beta_k$, $\mu_Y(x_1, \ldots, x_k)$, $Y(x_1, \ldots, x_k)$, and $a^T\beta$, as well as σ, because all of these confidence intervals discussed so far are *equal-tailed*.

We illustrate the use of the preceeding formulas in the following task.

Task 4.6.1

Here we consider some practical questions that may arise in the context of Example 4.4.2, where a director of admissions is studying how $X_1 = $ SATmath, $X_2 = $ SATverbal, $X_3 = $ HSmath, and $X_4 = $ HSenglish can be used to predict $Y = $ GPA at the end of the first year after admission to a certain university. The population regression function is assumed to be of the form

$$\mu_Y(x_1, x_2, x_3, x_4) = \beta_0 + \beta_1 x_1 + \beta_2 x_2 + \beta_3 x_3 + \beta_4 x_4$$

and population assumptions (B) are presumed valid. A sample of 20 students was selected using simple random sampling from all students who were admitted over the past four years and completed the first year. The data are given in Table 4.4.3 and are also in the file **gpa.dat** on the data disk.

In Exhibit 4.6.1 we give a MINITAB output from a regression analysis of GPA on SATmath, SATverbal, HSmath, and HSenglish along with the $C = (X^T X)^{-1}$ matrix. The matrix C is needed to obtain standard errors and hence confidence intervals for $a^T \beta$. A similar output can be obtained from most statistical computing packages.

E X H I B I T **4.6.1**
MINITAB Output for Task 4.6.1

```
The regression equation is
GPA = 0.162 + 0.00201 SATmath + 0.00125 SATverb +
      0.189 HSmath + 0.088 HSengl
```
(4.6.15)

Predictor	Coef	Stdev	t-ratio	p	
Constant	0.1615	0.4375	0.37	0.717	(4.6.16)
SATmath	0.0020102	0.0005844	3.44	0.004	(4.6.17)
SATverb	0.0012522	0.0005515	2.27	0.038	(4.6.18)
HSmath	0.18944	0.09187	2.06	0.057	(4.6.19)
HSengl	0.0876	0.1765	0.50	0.627	(4.6.20)

```
s = 0.2685     R-sq = 85.3%     R-sq(adj) = 81.4%
```
(4.6.21)

```
The C matrix is
```

$$C = \begin{bmatrix} 2.6551249569 & 0.0013784337 & -0.0003619435 & -0.2006841873 & -0.8521280877 \\ 0.0013784337 & 0.0000047375 & -0.0000004016 & -0.0003559377 & -0.0008620172 \\ -0.0003619435 & -0.0000004016 & 0.0000042188 & -0.0001953018 & -0.0002966849 \\ -0.2006841873 & -0.0003559377 & -0.0001953018 & 0.1170561208 & 0.0472194123 \\ -0.8521280877 & -0.0008620172 & -0.0002966849 & 0.0472194123 & 0.4320523096 \end{bmatrix}$$

(4.6.22)

In the computer output in Exhibit 4.6.1, the quantities in (4.6.16)–(4.6.20) under the heading `coef` are the $\hat{\beta}_i$, the point estimates of the corresponding β's. Also in (4.6.16)–(4.6.20) under `Stdev` are the $SE(\hat{\beta}_i)$, the standard errors of the $\hat{\beta}_i$'s. The value of $\hat{\sigma}$ is the quantity labeled s in (4.6.21), and the matrix $C = (X^T X)^{-1}$ is given in (4.6.22).

Regression calculations are notoriously prone to rounding errors. To minimize this problem, it is advisable to carry as many significant digits as possible for all intermediate calculations. For this reason, we have used ten-decimal accuracy in the matrix C given in (4.6.22). Some computer programs will round results to four or five decimals unless you specifically ask for more. Final results may, of course, be appropriately rounded.

We now consider several practical questions concerning the GPA data.

1 Suppose that the current requirement to qualify as a candidate for financial assistance at this university is that applicants must have a grade point average of 1.5 or better in high school mathematics courses. The director of admissions is considering a recommendation by a faculty committee to increase this requirement to 2.5 or better, so it is of interest to know what the difference will be between the *average* first-year GPA of students who received a grade point average of 2.5 in high school mathematics courses and that of students who received a grade point average of 1.5 in high school mathematics courses, other variables being equal. The director of admissions would like to have an estimate of this difference along with an indication of how good this estimate is.

Let the values of X_1, X_2, and X_4 for the two groups of students being compared, be x_1, x_2, and x_4, respectively. Then the required difference is in fact equal to

$$\mu_Y(x_1, x_2, 2.5, x_4) - \mu_Y(x_1, x_2, 1.5, x_4)$$
$$= [\beta_0 + \beta_1 x_1 + \beta_2 x_2 + \beta_3(2.5) + \beta_4 x_4] - [\beta_0 + \beta_1 x_1 + \beta_2 x_2 + \beta_3(1.5) + \beta_4 x_4]$$
$$= \beta_3[2.5 - 1.5] = (1)\beta_3 = \beta_3.$$

From (4.6.19) above under `coef` we obtain $\hat{\beta}_3 = 0.18944$. A confidence interval for β_3 tells us how well we can estimate β_3 using the sample data. For illustration, we compute a 90% two-sided confidence interval for β_3 using the formula in (4.6.1). Here $1 - \alpha = 0.90$, so we have $\alpha = 0.10$. Also $n - k - 1 = 20 - 4 - 1 = 15$. From Table T-2 in Appendix T we obtain $t_{1-\alpha/2:n-k-1} = t_{.95:15} = 1.753$. From (4.6.21) we get $\hat{\sigma} = 0.2685$. To compute the standard error of $\hat{\beta}_3$ we use the formula in (4.6.4) with $i = 4$. For this we need $c_{4,4}$, the $(4, 4)$ element of the matrix C. This is equal to 0.11706 (rounded to five decimals) and is obtained from (4.6.22). So $SE(\hat{\beta}_3) = 0.2685\sqrt{0.11706}$, which equals 0.09186. We can also obtain $SE(\hat{\beta}_3)$ from (4.6.19) under `Stdev` in the MINITAB output in Exhibit 4.6.1. Putting together the various quantities, we get

$$C[0.18944 - (1.753)(0.09186) \leq \beta_3 \leq 0.18944 + (1.753)(0.09186)] = 0.90$$

i.e.,

$$C[0.0284 \leq \beta_3 \leq 0.3505] = 0.90$$

This means that if X_1, X_2, and X_4 remain fixed, we can state with 90% confidence that an increase in HSmath of 1 grade point unit (from 1.5 units to 2.5 units) is associated with an increase in the average first-year GPA of at least 0.0284 grade point unit and not more than 0.3504 grade point unit. The director of admissions has to decide whether the information provided by this confidence statement is adequate for the purpose at hand, or if the interval is too wide for decision-making purposes; in the latter case a larger sample would be necessary to obtain the required information.

2 One of the applicants has furnished the following information in support of her application for financial aid. She has a score of 594 in SATmath, a score of 665 in SATverbal, a grade point average of 3.42 in high school mathematics courses, and a grade point average of 2.70 in high school English courses. The financial aid committee will grant her financial aid if there is evidence that she will obtain a GPA of 2.5 or higher at the end of the first year. The committee will use a 95% lower confidence bound to help make this decision.

So we are interested in a lower bound for $Y(594, 665, 3.42, 2.70)$. We compute a 90% two-sided confidence interval for $Y(594, 665, 3.42, 2.70)$, the lower endpoint of which gives us the required 95% lower confidence bound. The values of $\hat{\beta}_i$ can be obtained from (4.6.16)–(4.6.20), and using these we obtain $\hat{Y}(594, 665, 3.42, 2.70) = 3.0726$. Next we need the standard error of $\hat{Y}(594, 665, 3.42, 2.70)$. This is computed using the formula in (4.6.6) or (4.6.7). The vector x is equal to $[1, 594, 665, 3.42, 2.70]^T$, and the matrix C is in (4.6.22). So we get $x^T C x = 0.2831$. From (4.6.21) we obtain $\hat{\sigma} = 0.2685$. So $SE(\hat{Y}(x_1, x_2, x_3, x_4)) = (0.2685)\sqrt{1 + 0.2831} = 0.3041$. From Table T-2 we obtain $t_{1-\alpha/2:n-k-1} = t_{0.95:15} = 1.753$. Using (4.6.1) we get

$$C[3.0726 - (1.753)(0.3041) \leq Y(594, 665, 3.42, 2.7)$$
$$\leq 3.0726 + (1.753)(0.3041)] = 0.90$$

i.e.,

$$C[2.54 \leq Y(594, 665, 3.42, 2.7) \leq 3.61] = 0.90$$

In particular we have 95% confidence that this student will have a GPA no smaller than 2.54 at the end of the first year. The financial aid committee will use this information to help decide whether or not to grant assistance to this student.

3 The director of admissions wants to determine how good the population regression function $\mu_Y(x_1, x_2, x_3, x_4)$ is for predicting Y = first-year GPA (it is assumed that the model $\mu_Y(x_1, x_2, x_3, x_4) = \beta_0 + \beta_1 x_1 + \beta_2 x_2 + \beta_3 x_3 + \beta_4 x_4$ holds). For this reason the director wants a point and an interval estimate of the subpopulation standard deviation σ.

The quantity σ is a measure of how good the four predictor variables X_1, X_2, X_3, X_4 together are for predicting the value of Y. The point estimate of σ is given in (4.6.21) and is equal to 0.2685. We now compute a 90% two-sided confidence interval for σ using (4.6.12). We have $n - k - 1 = 15$,

$1 - \alpha/2 = .95,\ \alpha/2 = 0.05,$ and, from Table T-3, $\chi^2_{0.05:15} = 7.261,$ and $\chi^2_{0.95:15} = 24.996.$ Hence the confidence statement is

$$C\left[\sqrt{\frac{15(0.2685)^2}{24.996}} \leq \sigma \leq \sqrt{\frac{15(0.2685)^2}{7.261}}\right] = 0.90$$

i.e.,

$$C[0.208 \leq \sigma \leq 0.386] = 0.90$$

So we have 90% confidence that the subpopulation standard deviation σ is between 0.208 grade point unit and 0.386 grade point unit.

We can directly read $\hat{\sigma}$, $\hat{\beta}_i$, and the standard errors of $\hat{\beta}_i$ from the regression output of most statistical computer packages. These are the ingredients required to compute confidence intervals for each β_i. However, the point estimate and the standard error of $\boldsymbol{a}^T \boldsymbol{\beta}$ are not given directly in all statistical packages but must be computed by using the $\boldsymbol{C}$ matrix in (4.6.3). In Section 4.6 of the laboratory manuals we discuss how confidence intervals can be obtained using computer commands.

Conversation   4.6.1

Investigator: I want to ask you some questions about confidence intervals and in particular about confidence coefficients. Can you discuss these with me now?

Statistician: Certainly.

Investigator: One of the scientists with whom I work says that 90%, 95%, and 99% confidence coefficients are part of statistical theory and that we should always use $1 - \alpha = 0.90, 0.95,$ or $0.99.$ Is that correct?

Statistician: No, it is not correct. You can use any value between zero and one for the confidence coefficient $1 - \alpha.$

Investigator: That's what I thought. But most statistics books almost always use a confidence coefficient of 90%, 95%, or 99%. Why is that?

Statistician: I guess this is partly due to tradition or habit. In the 1930s, when modern statistical inference was beginning to include confidence intervals, some books used confidence intervals with 90%, 95%, or 99% confidence coefficients, and it appears that other books continued to use these values. As people read and studied these and subsequent books, these values became quite standard although there is nothing sacred about these or any other values for confidence coefficients.

Investigator: What does it really mean to have a confidence of, say, 95% that an interval contains a parameter θ?

Statistician: I will try to explain it this way. Suppose that you are handed a box containing 95 white balls and 5 red balls, and the balls are indistinguishable except for color. You are going to select a ball at random from this box, and we say the probability is 95% that the ball you choose will be white, or we have 95% confidence that the ball drawn will be white. Now, theory and experience tell us that if you draw a ball repeatedly (with replacement) from the box many times, say 100,000 times, the percentage of white balls drawn will be very close to 95%.

Investigator: But in practice I'll draw a ball only once.

Statistician: That's correct, and it isn't completely clear what the 95% probability means for this one draw. But suppose you are handed two boxes. Box 1 contains 99 white balls and 1 red ball, and Box 2 contains 50 white balls and 50 red balls. You can randomly choose one ball from either box and, if the ball is white, you get a valuable prize. From which box would you choose this one ball?

Investigator: I'd choose a ball from Box 1—the box that has 99 white balls and one red ball.

Statistician: Why would you select Box 1?

Investigator: Because there are more white balls in Box 1 than in Box 2.

Statistician: Well, let's change the problem slightly. Suppose Box 1 still contains 99 white balls and 1 red ball, but Box 2 now contains 1,000 white balls and 1,000 red balls. Box 2 now has more white balls in it than Box 1 has. Which box would you choose a ball from now?

Investigator: I'd still choose Box 1.

Statistician: Why?

Investigator: I just *feel* that if I select a ball from Box 1, I have a better chance of getting a white ball than if I select a ball from Box 2. In fact, if I draw a ball from Box 1 100,000 times (with replacement), and if you draw a ball from Box 2 100,000 times (with replacement), I'll get a white ball about 95% of the time, but you'll get a white ball only about 50% of the time.

Statistician: That's right, but you are going to draw only one ball. From which box will you draw it?

Investigator: Box 1.

Statistician: Why?

Investigator: Since the ratio of the number of white balls to the number of red balls in Box 1 is much greater than their ratio in Box 2, I feel that I'm more apt to get a white ball if I draw it from Box 1 rather than Box 2.

Statistician: More apt? What does that mean?

Investigator: I have a better chance, or a higher probability. In fact, I'll have 95% probability of getting a white ball if I draw it from Box 1, but I'll have only a 50% probability of getting a white ball if I draw it from Box 2.

Statistician: I guess you've answered your original question: "What does 95% confidence mean?" When an investigator selects a simple random sample of size n from a population of size N and computes a confidence interval for a parameter θ, he/she is using one of $H = \binom{N}{n}$ possible intervals. We can view this process as follows. Suppose that a box contains H balls, and each ball has written on it a confidence interval for θ. Now it is known (from theory) that for 95% of these balls, the confidence intervals written on them actually cover the unknown parameter θ. Color these balls white. Color the balls red if the confidence intervals written on them do not cover θ. Thus selecting a sample of size n and computing a confidence interval for θ with confidence coefficient 95% is equivalent to choosing a ball at random and noting its color. Since 95% of the balls are white, we have 95% confidence that the chosen ball will be white and the confidence interval will cover θ.

Investigator: I think I see your point, but since scientists want to be certain that their confidence interval covers the unknown parameter θ, why should they use a 95% or even a 99% confidence interval? Why shouldn't I advise them to use a 99.9999% confidence interval? Then they can be practically certain that their interval includes θ.

Statistician: Because, in general, the larger the confidence coefficient is, the wider the confidence interval will be, and so to reach a decision a scientist must also consider the width of the interval. Let me give you a simple illustration. Suppose a company is considering moving to city A, and it needs to know μ, the average annual income of wage earners in the city. Suppose a sample is selected and a 99.9999% confidence interval for μ, is computed as

$$\$6.59 \leq \mu \leq \$750,000$$

The investigator is quite certain that the confidence interval is correct, but it is so wide that it is useless for making a decision. Now suppose that a 90% confidence interval for μ is

$$\$21,000 \leq \mu \leq \$23,981$$

(Of course these numbers were chosen to make our point in a dramatic manner.) The investigator is less certain that μ is in this interval, but the interval is useful. So you can see why a very large confidence coefficient may not be the thing to use.

Investigator: Are you saying that a confidence interval with confidence coefficient of either 90%, 95%, or 99% will always have a desirable width? Is that the reason these values are generally used?

Statistician: No, I'm not saying that. That isn't the case. What I am saying is that the larger the confidence coefficient is, the wider the confidence interval will be, so using confidence coefficients very close to 1 may result in very wide confidence intervals that will be of no help in making decisions.

Investigator: So what should I do? In one situation the confidence interval is practically certain (99.9999% confidence) to cover θ, but the confidence interval is so wide that it is not useful. In the other situation the result is not so certain (90% confidence), but if it is not too wide it may be useful.

Statistician: That's correct, but in many situations we can have a desirable confidence coefficient (say 95%, 99%, etc.) and still have the width such that the result is useful.

Investigator: How do I do this?

Statistician: By designing the study carefully before any sample values are selected and then choosing the **sample size** judiciously! In general, in statistical inference problems for a fixed confidence coefficient $1 - \alpha$, one can reduce the width of the interval by increasing n, the size of the sample. So in many problems if you choose the proper value for the sample size n, you may be able to obtain a confidence interval of desired width and desired confidence coefficient $1 - \alpha$. You must remember, however, that the confidence coefficient you specify may be 0.95, but due to the fact that the assumptions are generally not exactly satisfied, the actual confidence coefficient will typically differ from the specified nominal value. It may be 0.94 or 0.96, or even 0.92 or 0.98. But if you choose a confidence coefficient equal to $1 - \alpha$, and if the assumptions are approximately satisfied, then the actual confidence coefficient will be close to $1 - \alpha$.

Investigator: That sounds good to me, but in almost all of the problems that I've been associated with, the data have already been collected. Thus the sample size n has already been fixed.

Statistician: Then you may be much more restricted in what conclusions you can draw from the data. That is why it's important to design the study carefully before samples are selected. Perhaps we can discuss this later.

Investigator: I'd like to do that. But let me ask you one final question. What value should I recommend that the scientists use for the confidence coefficient $1 - \alpha$?

Statistician: I can't answer that question definitively. You might explain to the scientists you work with about confidence coefficients as we have discussed them using balls in boxes and ask them to think about the risk they are willing to take in obtaining an incorrect conclusion from their experiment. Then let them make their own decision about what confidence coefficient they want to use.

Investigator: I'll try that. Can I come to see you again after I discuss this with them?

Statistician: Certainly.

Conversation 4.6.2

Investigator: Good morning. I know you didn't expect to see me again so soon! Do you have time to talk with me now? This won't take very long.

Statistician: Certainly. How can I help?

Investigator: We're working on a problem where we must make a census projection of the resident population of the United States in the year 2010. This will help my company determine whether or not it should start thinking about plans for expanding its operations over the next several years. I have obtained the following U.S. Bureau of Census data of the resident population for every 10 years from 1790 through 1990.

Resident Population (in millions) Y	Year X
3.929	1790
5.308	1800
7.240	1810
9.638	1820
12.866	1830
17.069	1840
23.192	1850
31.443	1860
39.818	1870
50.156	1880
62.948	1890
75.995	1900
91.972	1910
105.711	1920
122.775	1930
131.669	1940
150.697	1950
179.323	1960
203.302	1970
226.546	1980
248.710	1990

I plotted Y against X for the data, and the result is

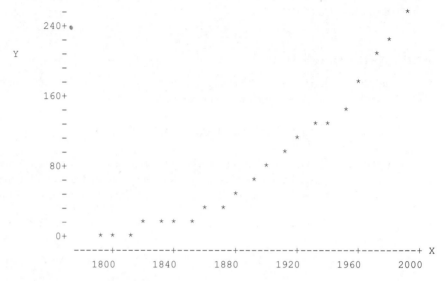

I used least squares and fitted a second-degree polynomial

$$f(x) = \beta_0 + \beta_1 x + \beta_2 x^2$$

to the data. The fitted least squares polynomial is denoted by $\hat{f}(x)$ where

$$\hat{f}(x) = 21010.7 - 23.3819x + 0.0065068x^2$$

I substituted $x = 2{,}010$ and the result is $\hat{f}(2010) = 301.201$. So based on these calculations the resident population in the United States in the year 2010 is estimated to be 301.201 million. Is this a valid estimate? Can I use regression techniques to obtain a valid confidence interval for $f(2010)$?

Statistician: Your estimate of 301.2 million may be a *reasonable* estimate, but you don't have a regression setup that satisfies assumptions (A) or (B). You also do not have a *random* sample from a well-defined population. What you have is a set of 21 pairs of points, and I don't see how you can define a target or study population from which you collected a random sample. The curve that you computed is a way to summarize the pattern exhibited by these 21 data points. I'm not saying that the data do not contain a great deal of information—they do. But they don't seem to fit into the framework of regression assumptions.

Investigator: We want to publish these results along with others in a professional journal, and if we don't include the results of statistical tests or confidence intervals, it won't be accepted for publication. What would you advise?

Statistician: As I stated earlier, your data contain a great deal of information about the U.S. resident population for the past 21 decades (in fact, you have the entire set of available data for these decades), and you don't need confidence intervals, tests, etc. to make useful statements for these decades. However, you want to extrapolate to the year 2010, and extrapolation is always risky unless you can define a population that includes the extrapolated points. Perhaps you can look at your model as a *process* rather than a population and use a *stochastic process model*. You might want to talk to a professional statistician who works with such models.

Investigator: That's a possibility. I'll look into it. Thanks. I'll see you again next week.

Problems 4.6

The following questions refer to the problem discussed in Example 4.5.1 and the data in Table 4.5.1, which are also in the file **electric.dat** on the data disk. Recall that

$Y = $ (total) electric bill (in dollars) for the past year

$X_1 = $ monthly income for the household (in dollars)

$X_2 = $ number of persons in the household

$X_3 = $ living area (in square feet) of the house or apartment

Suppose that assumptions (A) are met and the data are obtained by simple random sampling. In particular, the regression function of Y on X_1, X_2, and X_3 is of the form

$$\mu_Y(x_1, x_2, x_3) = \beta_0 + \beta_1 x_1 + \beta_2 x_2 + \beta_3 x_3 \qquad \textbf{(4.6.23)}$$

The utility company is interested in predicting the electricity usage patterns by various households for next year so the target population is a *future* population. However, the study population is assumed to be very similar to the target population, and so the sample data from the study population may be used to make inferences about the target population. A SAS output from a regression analysis of these data appears in Exhibit 4.6.2.

Note that the C matrix is obtained from the first four rows and columns of the matrix given in (4.6.24). The rows and columns of the C matrix are labeled by the names of the predictors corresponding to $\beta_0, \beta_1, \beta_2$, and β_3, respectively. Thus the $(2, 2)$-element of the C matrix is 0.0000010099689, etc. The first four numbers in the last row (and column) of the matrix in (4.6.24), labeled BILL, are the regression coefficients, and the final number in the last row (and column) is $SSE(X_1, X_2, X_3, X_4)$. Use this information to solve the following problems.

4.6.1   Are valid point estimates available for any of the following parameters?

$$\mu_{X_1}, \mu_{X_2}, \mu_{X_3}, \mu_Y, \sigma_{X_1}, \sigma_{X_2}, \sigma_{X_3}, \sigma_Y$$

Explain.

4.6.2   In Problem 4.6.1, find the point estimates of the parameters for which valid estimates are available.

E X H I B I T 4.6.2
SAS Output for Problem 4.6.1

```
                      The SAS System      0:00 Saturday, Jan 1, 1994
Model: MODEL1

            X'X Inverse, Parameter Estimates, and SSE


              INTERCEP        INCOME       PERSONS         AREA          BILL


INTERCEP   2.153683547   -0.001377673   -0.25570104   0.0021517892   -358.4415686
INCOME    -0.001377673   1.0099689E-6    0.000175464  -1.655901E-6    0.075136905
PERSONS   -0.25570104    0.000175464     0.046002015  -0.00029998    55.087632718
AREA       0.0021517892  -1.655901E-6    -0.00029998   2.7878446E-6   0.2811036938
BILL      -358.4415686   0.075136905    55.087632718   0.2811036938   550163.42009
```

(4.6.24)

```
Dependent Variable: BILL


                    Analysis of Variance

                      Sum of         Mean
Source         DF    Squares        Square       F Value      Prob>F

Model           3 3151504.8152  1050501.6051     57.283       0.0001
Error          30 550163.42009   18338.78067
C Total        33 3701668.2353


     Root MSE      135.42075     R-square       0.8514
     Dep Mean      619.41176     Adj R-sq       0.8365
     C.V.           21.86280


                    Parameter Estimates

               Parameter      Standard     T for H0:
  Variable  DF  Estimate        Error    Parameter=0    Prob > |T|

  INTERCEP   1  -358.441569  198.73583019    -1.804        0.0813
  INCOME     1     0.075137    0.13609408     0.552        0.5850
  PERSONS    1    55.087633   29.04515215     1.897        0.0675
  AREA       1     0.281104    0.22610987     1.243        0.2234
```

4.6.3   Exhibit the values of

$$\hat{\beta}_0, \hat{\beta}_1, \hat{\beta}_2, \hat{\beta}_3, SE(\hat{\beta}_0), SE(\hat{\beta}_1), SE(\hat{\beta}_2), SE(\hat{\beta}_3), \hat{\mu}_Y(x_1, x_2, x_3)$$

4.6.4   How well can Y, the annual electric bill for a randomly chosen household, be predicted using X_1, X_2, and X_3?

4.6.5   How well can Y, the annual electric bill for a randomly chosen household, be predicted if no predictor factors are used?

4.6.6   What is the difference between the average annual electric bill of households consisting of five individuals and households with four individuals if they have the same monthly income and the same living area? Obtain a point estimate and a 95% upper confidence bound for this quantity.

4.6.7   What is the difference between the average annual electric bill of households with a monthly income of $4,000.00 and households with a monthly income of $3,000.00 if they have the same number of individuals in the household and the same living area? Obtain a point estimate and a 95% upper confidence bound for this quantity.

4.6.8   What is the difference between the average annual electric bill of households with a living area of 2,000 square feet and households with a living area of 1,500 square feet if they have the same monthly income and the same number of individuals in the household? Obtain a point estimate and a 95% upper confidence bound for this quantity.

4.6.9   The monthly income of a *particular* household is $3,200. There are six individuals in the household, and the living area equals 2,800 square feet. Predict the annual monthly electric bill for this household.

4.6.10   In Problem 4.6.9, compute a 90% upper confidence bound for the annual electric bill for this household.

4.6.11   Repeat Problems 4.6.9 and 4.6.10 for the *average* annual electric bill of all households with this monthly income, number of individuals in the household, and living area.

4.6.12   What is the difference between the average annual electric bill for households consisting of seven individuals with a living area of 3,500 square feet and a monthly income of $5,400.00 and that for households consisting of four individuals with a living area of 2,400 square feet and a monthly income of $5,000.00? Estimate this difference and also compute a 95% two-sided confidence interval for this quantity.

4.6.13   The model in (4.6.23) would be considered adequate for predicting Y if the standard deviation of the prediction errors—i.e., σ—is less than $50.00. If this is not so, then the investigator will look for additional explanatory variables to include in the regression model. To assist the investigator in making a decision in this connection, compute a two-sided 95% confidence interval for σ. Do you believe that the investigator needs to look for additional explanatory variables? Or do you believe that the model in (4.6.23) is adequate? Explain.

4.7
Tests

It has been stated several times that *statistical tests alone should never be used in situations where appropriate confidence intervals for parameters of interest are available.* Refer to Sections 1.6 and 3.6 for a more thorough discussion in this regard. However, as we said earlier, statistical tests are quite popular among many practitioners. We feel that you should be familiar with some of the commonly used testing procedures because of their widespread use so that you can properly interpret published results of investigations. Formulas and procedures for some statistical tests that pertain to the multiple linear regression model in (4.1.1) are summarized in Boxes 4.7.1 and 4.7.2. Here θ denotes any one of the quantities

$$\beta_0, \beta_1, \ldots, \beta_k, \mu_Y(x_1, \ldots, x_k)$$

or

$$a^T \beta = a_0\beta_0 + a_1\beta_1 + \cdots + a_k\beta_k$$

The quantity $\hat{\theta}$ denotes the point estimate of θ.

B O X **4.7.1** **Statistical Tests for $\beta_i, i = 0, \ldots, k, \mu_Y(x_1, \ldots, x_k)$ or $a^T\beta$ (θ Stands for Any One of These)**

Let q be a specified number (the investigator specifies q). Compute the statistic

$$t_C = \frac{\hat{\theta} - q}{SE(\hat{\theta})}$$

1 For testing NH: $\theta = q$ versus AH: $\theta \neq q$, the *P*-value is the value of α such that $|t_C| = t_{1-\alpha/2:n-k-1}$.

2 For testing NH: $\theta \leq q$ versus AH: $\theta > q$, the *P*-value is the value of α such that $t_C = t_{1-\alpha:n-k-1}$.

3 For testing NH: $\theta \geq q$ versus AH: $\theta < q$, the *P*-value is the value of α such that $-t_C = t_{1-\alpha:n-k-1}$.

B O X **4.7.2** **Statistical Tests for σ**

Let q be a positive number specified by the investigator. Compute the statistic

$$\chi_C^2 = \frac{(n-k-1)\hat{\sigma}^2}{q^2} = \frac{SSE}{q^2}$$

1 For testing NH: $\sigma = q$ versus AH: $\sigma \neq q$, the *P*-value is equal to α where α is a number between 0 and 1 and satisfies

$$\chi_C^2 = \chi_{\alpha/2:n-k-1}^2 \quad \text{or} \quad \chi_C^2 = \chi_{1-\alpha/2:n-k-1}^2$$

(only one of these two equalities can be satisfied unless $\alpha = 1$).

2 For testing NH: $\sigma \le q$ versus AH: $\sigma > q$, the *P*-value is the value of α such that $\chi_C^2 = \chi_{1-\alpha:n-k-1}^2$.

3 For testing NH: $\sigma \ge q$ versus AH: $\sigma < q$, the *P*-value is the value of α such that $\chi_C^2 = \chi_{\alpha:n-k-1}^2$.

We illustrate the use of these procedures with Example 4.7.1.

E X A M P L E 4.7.1

(a) Consider the GPA problem discussed in Task 4.6.1, and suppose the financial aid committee is interested in knowing whether the average GPA at the end of the first year of applicants with SATmath = 594, SATverbal = 665, HSmath = 3.42, and HSenglish = 2.70 would equal or exceed 2.5. We might consider the following pair of hypotheses:

$$NH : \mu_Y(594, 665, 3.42, 2.70) \le 2.5$$

versus

$$AH : \mu_Y(594, 665, 3.42, 2.70) > 2.5$$

Here we have $\theta = \mu_Y(594, 665, 3.42, 2.70)$ and $q = 2.5$. The appropriate test procedure is in part 2 in Box 4.7.1. From (4.6.16)–(4.6.20) we know that

$$\hat{\mu}_Y(x_1, x_2, x_3, x_4) = 0.1615 + 0.0020102x_1 + 0.0012522x_2 + 0.18944x_3 + 0.0876x_4$$

so the value of $\hat{\mu}_Y(594, 665, 3.42, 2.70)$ is 3.0726. To calculate $SE(\hat{\mu}_Y(594, 665, 3.42, 2.70))$ we use the formula in (4.6.5). From (4.6.21) we have $\hat{\sigma} = 0.2685$. Also from (4.6.22) we have the matrix C, and using it we get $x^T C x = 0.2831$ where $x^T = [1, 594, 665, 3.42, 2.70]$. So $SE(\hat{\mu}_Y(594, 665, 3.42, 2.70)) = 0.2685\sqrt{0.2831} = 0.1429$. Also, $n - k - 1 = 20 - 4 - 1 = 15$. Hence

$$t_C = \frac{3.0726 - 2.5}{0.1429} = 4.007$$

The value of α for which $t_C = 4.007 = t_T = t_{1-\alpha:15}$ is less than 0.005 by Table T-2 in Appendix T, so the *P*-value is less than 0.005. So if NH were indeed true, then the probability of obtaining a value for t_C as large as, or larger than, 4.007 is less than 0.005. Since this probability is so small, the committee will very likely conclude that the average first-year GPA of applicants with SATmath = 594, SATverbal = 665, HSmath = 3.42, and HSenglish = 2.70, is greater than 2.5 (i.e., reject NH). A confidence interval for $\theta = \mu_Y(594, 665, 3.42, 2.70)$ would provide the committee with additional information.

(b) Test procedures given in Box 4.7.2 can be used to help decide whether σ is equal to, greater than or equal to, or less than or equal to a specified value q. To illustrate these procedures, we consider the following problem.

Suppose the director of admissions wants to know how well he can predict the first-year GPAs of applicants using X_1 = SATmath, X_2 = SATverbal, X_3 = HSmath, and X_4 = HSenglish. The prediction would be considered *adequate for this problem* if σ is less than 0.2. To help determine whether σ is less than 0.2, the director might consider the following statistical test.

$$\text{NH: } \sigma \geq 0.2 \qquad \text{versus} \qquad \text{AH: } \sigma < 0.2 \qquad\qquad \textbf{(4.7.1)}$$

If NH is rejected, then the predictors X_1, X_2, X_3, and X_4 together will be considered to be adequate for predicting Y.

Using the procedure given in part 3 of Box 4.7.2 with $q = 0.2$, we obtain

$$\chi_C^2 = \frac{(n - k - 1)\hat{\sigma}^2}{(0.2)^2} = \frac{SSE}{(0.2)^2} = \frac{1.0815}{0.04} = 27.04$$

From Table T-3 in Appendix T we find that $\chi_C^2 = 27.04$ is between $\chi_{0.95:15}^2 = 24.996$ and $\chi_{0.975,15}^2 = 27.488$ so the value of α for which χ_C^2 is equal to $\chi_{\alpha:15}^2$ is between 0.95 and 0.975. Hence the P-value for the preceding hypothesis test is between 0.95 and 0.975. So NH will not be rejected at any of the usual α levels. The director of admissions would perhaps conclude that the predictors X_1, X_2, X_3, X_4 together may not be adequate for predicting Y. ∎

Most computer programs for regression routinely output t_C values for testing whether or not the various β_i are equal to zero (i.e., $q = 0$). In Box 4.7.1 note that when $q = 0$, the expression for t_C becomes

$$t_C = \frac{\hat{\theta}}{SE(\hat{\theta})}$$

In the computer output in Exhibit 4.6.1, the quantities in (4.6.16)–(4.6.20) under 't-ratio' are the quantities t_C in Box 4.7.1 when $q = 0$. These quantities cannot be used for values of q other than zero.

Problems   4.7

The following problems refer to Example 4.5.1. The data are given in Table 4.5.1 and are also stored in the file **electric.dat** on the data disk. These data are repeated in Table 4.7.1 for convenience.

Recall that

$$Y = \text{(total) annual electric bill (in dollars) for the past year}$$
$$X_1 = \text{monthly income for the household (in dollars)}$$
$$X_2 = \text{number of individuals in the household}$$
$$X_3 = \text{living area (in square feet) of the house or apartment}$$

Suppose that assumptions (A) are met and the data are obtained by simple random sampling. In particular, the regression function of Y on X_1, X_2, and X_3 is of the form

$$\mu_Y(x_1, x_2, x_3) = \beta_0 + \beta_1 x_1 + \beta_2 x_2 + \beta_3 x_3 \qquad\qquad \textbf{(4.7.2)}$$

T A B L E **4.7.1**

Electric Bill Data

Sample Item Number	Bill Y	Income X_1	Persons X_2	Area X_3
1	228	3220	2	1160
2	156	2750	1	1080
3	648	3620	2	1720
4	528	3940	1	1840
5	552	4510	3	2240
6	636	3990	4	2190
7	444	2430	1	830
8	144	3070	1	1150
9	744	3750	2	1570
10	1104	4790	5	2660
11	204	2490	1	900
12	420	3600	3	1680
13	876	5370	1	2550
14	840	3180	7	1770
15	876	5910	2	2960
16	276	3020	2	1190
17	1236	5920	3	3130
18	372	3520	2	1560
19	276	3720	1	1510
20	540	4840	1	2190
21	1044	4700	6	2620
22	552	3270	2	1350
23	756	4420	2	1990
24	636	4480	2	2070
25	708	3820	4	1850
26	960	5740	2	2700
27	1080	5600	3	3030
28	480	3950	2	1700
29	96	2290	3	890
30	1272	5580	5	3270
31	1056	5820	2	2660
32	156	3160	2	1330
33	396	2880	4	1280
34	768	3780	3	1950

A SAS output containing the results of a regression analysis for this problem is given in Exhibit 4.7.1. Answer the following.

4.7.1 The utility company wants to know whether the annual electric bill for a household is dependent on the monthly income in each subpopulation of households having a specified number of individuals and a specified living area.

E X H I B I T 4.7.1
SAS Output for Regression Analysis of Electric Data

```
                        The SAS System   0:00 Saturday,   Jan 1, 1994

Dependent Variable: BILL

                        Analysis of Variance

                     Sum of          Mean
Source          DF   Squares         Square      F Value      Prob>F

Model            3 3151504.8152 1050501.6051      57.283      0.0001
Error           30 550163.42009  18338.78067
C Total         33 3701668.2353


     Root MSE     135.42075    R-square      0.8514
     Dep Mean     619.41176    Adj R-sq      0.8365
     C.V.          21.86280


                     Parameter Estimates

               Parameter     Standard    T for H0:
Variable  DF   Estimate      Error       Parameter=0     Prob > |T|

INTERCEP   1  -358.441569  198.73583019     -1.804         0.0813
INCOME     1     0.075137    0.13609408      0.552         0.5850
PERSONS    1    55.087633   29.04515215      1.897         0.0675
AREA       1     0.281104    0.22610987      1.243         0.2234
```

a Show that, for the model in (4.7.2), the average annual electric bill is *not* dependent on the monthly income in each subpopulation of households having a specified number of individuals and a specified living area if and only if $\beta_1 = 0$.

b Formulate an appropriate pair of hypotheses to test this and calculate the *P*-value. Use $\alpha = 0.05$ and state your conclusion.

4.7.2 In Problem 4.7.1, suppose, for company purposes, the manager is willing to conclude that the annual electric bill does not depend on income if the difference between the average electric bills of households whose monthly incomes differ by $1,000 (the other factors being the same) is less than $120. Compute a 95% confidence interval for the parameter of interest and make a decision. Which is more informative, the confidence interval of this problem or the test of Problem 4.7.1?

4.7.3 The regression function in (4.7.2) is considered adequate for predicting Y if the standard deviation of the prediction errors—i.e., σ—is less than $50.00. If this is not so, then the investigator will look for additional explanatory variables to include in the regression model. With this in mind, the investigator wants to test (using $\alpha = 0.05$)

$$\text{NH:}\ \sigma \geq 50 \qquad \text{against} \qquad \text{AH:}\ \sigma < 50$$

Carry out this test and state your conclusions. Which is more informative, the confidence statement obtained in Problem 4.6.13 or this test?

4.8
Analysis of Variance

Recall that for the straight line regression model we used a table to summarize certain key numerical quantities that are useful for computing confidence intervals, standard errors, and tests. The process of calculating, tabulating, and examining these key numerical quantities was termed analysis of variance, which is often abbreviated to ANOVA. We now discuss analysis of variance for the multiple linear regression model.

As in the case of straight line regression, the first key quantity is

$$SSY = \sum_{i=1}^{n}(y_i - \bar{y})^2 \tag{4.8.1}$$

This is called the **total sum of squares** (of deviations from the sample mean) for Y. This is the sum of squares of prediction errors when $\hat{\mu}_Y = \bar{y}$ is used to predict the Y values of the sample items. (Note that we are really not interested in predicting the Y values of the sample items because we already know them, but we do this to assess how good the predictions are likely to be when predicting unknown Y values.)

We know that the best predictor of the Y value of any item in the population without using any of the predictor variables $X_1, \ldots, X_k$, is μ_Y, the mean Y value of all items in the population. We also know that σ_Y is a measure of how well μ_Y represents the entire population of Y values. If data are obtained by simple random sampling, then we can estimate μ_Y by the sample mean $\bar{y}$ and estimate σ_Y^2 by $\sum_{i=1}^{n}(y_i - \bar{y})^2/(n-1) = SSY/(n-1)$. The quantity $SSY/(n-1)$ is sometimes written as MSY and is called the *total mean square for Y*. The divisor $n-1$ of SSY is called the *degrees of freedom associated with SSY* (or with $\hat{\sigma}_Y$).

The second key quantity is the sum of squares of the prediction errors when the sample regression function $\hat{\mu}_Y(x_1, \ldots, x_k)$ is used to predict the Y values of the sample items (in order to assess the performance of the sample regression function as a prediction function for Y). This quantity is called the **sum of squared errors** and is denoted by $SSE(X_1, \ldots, X_k)$, or SSE for short. It was defined in (4.4.14) to be

$$SSE = \sum_{i=1}^{n}\hat{e}_i^2 = \sum_{i=1}^{n}(y_i - \hat{\beta}_0 - \hat{\beta}_1 x_{i,1} - \cdots - \hat{\beta}_k x_{i,k})^2 \tag{4.8.2}$$

and it has $(n-k-1)$ degrees of freedom associated with it. The quantity $MSE(X_1, \ldots, X_k)$ (or MSE for short), called the *mean squared error*, was defined in

(4.4.15) to be

$$MSE = \frac{SSE}{(n - k - 1)}$$ (4.8.3)

Note that $MSE(X_1, \ldots, X_k) = \hat{\sigma}^2$ is the estimate of σ^2, regardless of whether the data are obtained by simple random sampling or by sampling with preselected X values.

The third key quantity is called the **sum of squares due to regression** and is denoted by $SSR(X_1, \ldots, X_k)$ (*SSR* for short). This quantity is the difference

$$SSR = SSY - SSE$$ (4.8.4)

and it has k degrees of freedom associated with it. Note that the degrees of freedom associated with *SSR* is the difference between the degrees of freedom for *SSY* and the degrees of freedom for *SSE*, i.e., $k = (n - 1) - (n - k - 1)$. The quantity

$$MSR = \frac{SSR}{k}$$ (4.8.5)

is called the *mean square due to regression*. All of these quantities are generally displayed in an ANOVA table such as Table 4.8.1.

When assumptions (A) or (B) hold, the statistic F_C in the last column of the analysis of variance table can be used to test the null hypothesis that μ_Y is equal to $\mu_Y(x_1, \ldots, x_k)$ against the alternative hypothesis that $\mu_Y(x_1, \ldots, x_k)$ is not equal to μ_Y. An equivalent way of stating this hypothesis is as follows:

$$\text{NH: } \beta_1 = \beta_2 = \cdots = \beta_k = 0 \qquad \text{against}$$
$$\text{AH: at least one of } \beta_1, \beta_2, \ldots, \beta_k \text{ is not zero} \qquad (4.8.6)$$

A test of the NH in (4.8.6) can be carried out using the numerical quantities exhibited in the analysis of variance table. Specifically, if a statistical test is used, the test statistic can be calculated as

$$F_C = \frac{MSR}{MSE}$$ (4.8.7)

If NH in (4.8.6) is true, then the quantity F_C in (4.8.7) has an F-distribution with k degrees of freedom for the numerator and $n - k - 1$ degrees of freedom for the denominator. The P-value for this test is the value of α such that $F_C = F_{1-\alpha:k,n-k-1}$.

Example 4.8.1 explains how to use an analysis of variance table for multiple linear regression obtained from a computer output.

T A B L E **4.8.1**

ANOVA for Multiple Linear Regression

Source	Degrees of Freedom (df)	Sum of Squares (SS)	Mean Square (MS)	Computed F-Value
Regression	k	SSR	MSR	$F_C = \dfrac{MSR}{MSE}$
Error	$n - k - 1$	SSE	MSE	
Total	$n - 1$	SSY	MSY	

E X A M P L E 4.8.1

Table 4.8.2 is an ANOVA table for the GPA data of Example 4.4.2 (see Table 4.4.3). The quantity *SSY* is computed by using (4.8.1), *SSE* by using (4.8.2), and *SSR* by using (4.8.4). You will notice that this computing can be quite an arduous task if it is not done on a computer. The test statistic F_C for the test of

$$\text{NH: } \beta_1 = \beta_2 = \beta_3 = \beta_4 = 0$$

against

$$\text{AH: at least one of } \beta_1, \beta_2, \beta_3, \beta_4 \text{ is not zero}$$

is 21.72 with 4 degrees of freedom for the numerator and 15 degrees of freedom for the denominator. If we use $\alpha = 0.01$, then we would reject NH since the *P*-value for the test is seen to be less than 0.01 using Table T-5 in Appendix T. Hence we conclude that at least one of β_1, β_2, β_3, β_4 is nonzero so the regression function $\mu_Y(x_1, x_2, x_3, x_4)$ is not equal to μ_Y.

An examination of $\hat{\sigma}$ and $\hat{\sigma}_Y$ gives us an idea of how well $\mu_Y(x_1, x_2, x_3, x_4)$ and μ_Y, respectively, predict the values of Y. From Table 4.8.2 we find $\hat{\sigma}_Y = \sqrt{0.3866} = 0.6218$. In comparison, $\hat{\sigma} = \sqrt{0.0721} = 0.2685$, which is considerably lower than 0.6218. In Section 4.9 we discuss appropriate confidence interval procedures (valid under assumptions (B)) for comparing σ_Y and σ in any multiple linear regression problem. We note that for this problem, $\hat{\sigma}_Y$ is a valid estimate of σ_Y because data are obtained by simple random sampling. Do not take it for granted that a valid estimate of σ_Y exists. Check to see what the assumptions are. ∎

Exhibit 4.8.1 shows a computer output for regression analysis of the GPA data (obtained using MINITAB) that includes an ANOVA table. Of course the ANOVA in the computer output of Exhibit 4.8.1 is the same as the one we computed in Table 4.8.2, except that the computer output includes a *P*-value (rounded to three decimal places in this case) and does not include *MSY*.

Reminder The *F* test in an ANOVA table can be used to test only

$$\text{NH: } \beta_1 = \beta_2 \ldots = \beta_k = 0 \qquad \text{against} \qquad \text{AH: at least one } \beta_i \neq 0$$

For one-sided tests or tests of $\beta_1 = \beta_2 = \ldots = \beta_k = q$ where $q \neq 0$, an ANOVA table cannot be used directly.

T A B L E 4.8.2

ANOVA for GPA Data

Source of Variation	Degrees of Freedom (DF)	Sum of Squares (SS)	Mean square MS	Computed F-Value
Regression	4	6.2643	1.5661	$F_C = \dfrac{1.5661}{0.0721} = 21.72$
Error	15	1.0815	0.0721	
Total	19	7.3458	0.3866	

E X H I B I T 4.8.1

MINITAB Output for Regression Analysis of GPA Data

```
The regression equation is
GPA = 0.162 + 0.00201 SATmath + 0.00125 SATverb +
      0.189 HSmath + 0.088 HSengl
```

Predictor	Coef	Stdev	t-ratio	p
Constant	0.1615	0.4375	0.37	0.717
SATmath	0.0020102	0.0005844	3.44	0.004
SATverb	0.0012522	0.0005515	2.27	0.038
HSmath	0.18944	0.09187	2.06	0.057
HSengl	0.0876	0.1765	0.50	0.627

s = 0.2685   R-sq = 85.3%   R-sq(adj) = 81.4%

Analysis of Variance

SOURCE	DF	SS	MS	F	p	
Regression	4	6.2643	1.5661	21.72	0.000	**(4.8.8)**
Error	15	1.0815	0.0721			
Total	19	7.3458				

Problems 4.8

4.8.1   A MINITAB output from a regression analysis is given in Exhibit 4.8.2 for the electric bill data of Table 4.5.1. Calculate the P-value for a test of

NH: $\beta_1 = \beta_2 = \beta_3 = 0$ against   AH: at least one of β_1, β_2, and β_3 is not zero

What is your conclusion if you use $\alpha = 0.01$?

4.8.2   The manager of the marketing division of a grocery store chain wants to conduct a study in a particular city where the company wants to open a store to understand the relationship between the number of dollars Y a household spends in grocery stores each month and the following variables—monthly income X_1 for the household,

E X H I B I T 4.8.2

MINITAB Output for Regression Analysis of Electric Data

```
The regression equation is
Bill = - 358 + 0.075 Income + 55.1 Persons + 0.281 Area
```

Predictor	Coef	Stdev	t-ratio	p
Constant	-358.4	198.7	-1.80	0.081
Income	0.0751	0.1361	0.55	0.585
Persons	55.09	29.05	1.90	0.068
Area	0.2811	0.2261	1.24	0.223

```
s = 135.4        R-sq = 85.1%      R-sq(adj) = 83.7%
```

Analysis of Variance

SOURCE	DF	SS	MS	F	p
Regression	3	3151505	1050502	57.28	0.000
Error	30	550163	18339		
Total	33	3701668			

number of children X_2 in the household, and the number of adults X_3 in the household. A group of 27 grocery shoppers are selected by simple random sampling from a study population and are requested to provide the needed information. The data for these 27 shoppers are given in Table 4.8.3 and are also stored in the file **grocery.dat** on the data disk.

T A B L E 4.8.3
Grocery Shoppers Data

Sample Item Number	Amount Spent in Grocery Stores Each Month Y (in dollars)	Monthly Income X_1 (in dollars)	Number of Children X_2	Number of Adults X_3
1	486	3800	2	2
2	164	1200	1	2
3	245	5000	0	1
4	714	5700	2	2
5	565	5600	1	2
6	728	4500	3	2
7	221	4400	0	1
8	209	2200	1	1
9	299	2300	3	1
10	477	4700	2	1
11	711	4300	4	2
12	379	3100	2	2
13	738	4900	4	1
14	325	3000	2	1
15	517	2000	4	2
16	441	2700	2	2
17	168	3400	1	1
18	525	2200	4	2
19	201	3800	0	1
20	358	4700	0	2
21	202	1400	3	2
22	272	1200	3	2
23	257	1700	2	2
24	376	1900	3	2
25	697	4900	3	2
26	248	2600	0	2
27	507	5100	2	1

Suppose that assumptions (A) for multiple linear regression are satisfied; thus the regression function of Y on X_1, X_2, and X_3 is of the form

$$\mu_Y(x_1, x_2, x_3) = \beta_0 + \beta_1 x_1 + \beta_2 x_2 + \beta_3 x_3 \qquad \textbf{(4.8.9)}$$

A SAS output for a regression of Y on X_1, X_2, X_3 follows.

E X H I B I T **4.8.3**
SAS Output for Regression Analysis of Grocery Data

The SAS System   0:00 Saturday,   Jan 1, 1994

Model: MODEL1 Dependent Variable: AMOUNT

Analysis of Variance

Source	DF	Sum of Squares	Mean Square	F Value	Prob>F
Model	3	881100.31876	293700.10625	104.399	0.0001
Error	23	64704.42198	2813.23574		
C Total	26	945804.74074			

Root MSE	53.03994	R-square	0.9316	
Dep Mean	408.51852	Adj R-sq	0.9227	
C.V.	12.98349			

Parameter Estimates

Variable	DF	Parameter Estimate	Standard Error	T for H0: Parameter=0	Prob > |T|
INTERCEP	1	-324.608237	51.16247889	-6.345	0.0001
INCOME	1	0.105141	0.00760990	13.816	0.0001
CHILDREN	1	96.601256	8.27176593	11.678	0.0001
ADULTS	1	110.760869	22.58900750	4.903	0.0001

Calculate the *P*-value for a test of

$$\text{NH:} \quad \beta_1 = \beta_2 = \beta_3 = 0$$

$$\text{AH:} \quad \text{at least one of } \beta_1, \beta_2, \text{ and } \beta_3 \text{ is not zero}$$

What is your conclusion if you use $\alpha = 0.01$?

4.8.3 An investigator is studying a population of males who have lived in mountain isolation for several generations, and she is interested in investigating the relationship between the heights Y of these males at age 18 years and the following variables.

$X_1 = $ Length at birth

$X_2 = $ Mother's height at age 18

$X_3 = $ Father's height at age 18

$X_4 = $ Maternal grandmother's height at age 18

$X_5 = $ Maternal grandfather's height at age 18

$$X_6 = \text{Paternal grandmother's height at age 18}$$
$$X_7 = \text{Paternal grandfather's height at age 18}$$

All heights and lengths are in inches. A simple random sample of 20 males of age 18 or more was drawn from the study population, and all the preceding information was recorded. The data are given in Table 4.8.4 and are also stored in the file **age18.dat** on the data disk.

Assumptions (B) are presumed to hold. In particular, the regression function is of the form

$$\mu_Y(x_1, x_2, x_3, x_4, x_5, x_6, x_7) = \beta_0 + \beta_1 x_1 + \beta_2 x_2 + \beta_3 x_3 + \beta_4 x_4 + \beta_5 x_5 + \beta_6 x_6 + \beta_7 x_7$$

A MINITAB output from a regression analysis of Y on $X_1, X_2, X_3, X_4, X_5, X_6, X_7$ follows in Exhibit 4.8.4.

Calculate the P-value for a test of

$$\text{NH:} \quad \beta_1 = \beta_2 = \beta_3 = \beta_4 = \beta_5 = \beta_6 = \beta_7 = 0$$

against

$$\text{AH:} \quad \text{at least one of } \beta_1, \beta_2, \beta_3, \beta_4, \beta_5, \beta_6, \text{ and } \beta_7 \text{ is not zero}$$

What is your conclusion if you use $\alpha = 0.05$?

T A B L E **4.8.4**
Heights at Age 18 of a Random Sample of Mountain People

Sample Item Number	Y	X_1	X_2	X_3	X_4	X_5	X_6	X_7
1	67.2	19.7	60.5	70.3	65.7	69.3	65.7	67.3
2	69.1	19.6	64.9	70.4	62.6	69.6	64.6	66.4
3	67.0	19.4	65.4	65.8	66.2	68.8	64.0	69.4
4	72.4	19.4	63.4	71.9	60.7	68.0	64.9	67.1
5	63.6	19.7	65.1	65.1	65.5	65.5	61.8	70.9
6	72.7	19.6	65.2	71.1	63.5	66.2	67.3	68.6
7	68.5	19.8	64.3	67.9	62.4	71.4	63.4	69.4
8	69.7	19.7	65.3	68.8	61.5	66.0	62.4	67.7
9	68.4	19.7	64.5	68.7	63.9	68.8	62.3	68.8
10	70.4	19.9	63.4	70.3	65.9	69.0	63.7	65.1
11	67.5	18.9	63.3	70.4	63.7	68.2	66.2	68.5
12	73.3	20.8	66.2	70.2	65.4	66.6	61.7	64.0
13	70.0	20.3	64.9	68.8	65.2	70.2	62.4	67.0
14	69.8	19.7	63.5	70.3	63.1	64.4	65.1	67.0
15	63.6	19.9	62.0	65.5	64.1	67.7	62.1	66.5
16	64.3	19.6	63.5	65.2	63.9	70.0	64.2	64.5
17	68.5	21.3	66.1	65.4	64.8	68.4	66.4	70.8
18	70.5	20.1	64.8	70.2	65.3	65.5	63.7	66.9
19	68.1	20.2	62.6	68.6	63.7	69.8	66.7	68.0
20	66.1	19.2	62.2	67.3	63.6	70.9	63.6	66.7

EXHIBIT 4.8.4

MINITAB Output for Regression Analysis of Data in Table 4.8.4

```
The regression equation is
Y = - 78.3 + 1.37 X1 + 0.782 X2 + 1.05 X3 - 0.120 X4 + 0.091 X5
    + 0.088 X6 - 0.102 X7
```

Predictor	Coef	Stdev	t-ratio	p
Constant	-78.27	26.96	-2.90	0.013
X1	1.3718	0.5207	2.63	0.022
X2	0.7824	0.1992	3.93	0.002
X3	1.0514	0.1358	7.74	0.000
X4	-0.1199	0.1717	-0.70	0.498
X5	0.0914	0.1301	0.70	0.496
X6	0.0883	0.1613	0.55	0.594
X7	-0.1017	0.1549	-0.66	0.524

s = 1.004   R-sq = 91.7%   R-sq(adj) = 86.9%

Analysis of Variance

SOURCE	DF	SS	MS	F	p
Regression	7	133.657	19.094	18.95	0.000
Error	12	12.088	1.007		
Total	19	145.746			

4.9
Comparison of Two Regression Functions (Nested Case) and Coefficients of Determination

The notation is quite heavy in this section and the next, but if you study it carefully, it will help you to learn and understand the material.

Consider a $(k + 1)$-variable population $\{(Y, X_1, \ldots, X_k)\}$. We know that the best function for predicting Y using the k predictor variables $X_1, \ldots, X_k$ is the regression function $\mu_Y(x_1, \ldots, x_k)$. However, to predict Y, investigators may want to use a smaller number of variables, which for ease of notation we take to be the first m variables $X_1, \ldots, X_m (m < k)$. They may want to know how much better the regression function of Y on $X_1, \ldots, X_k$ is for predicting Y than the regression function of Y on $X_1, \ldots, X_m$. It can be shown that when population assumptions (B) hold for the $(k + 1)$-variable population $\{(Y, X_1, \ldots, X_k)\}$, the regression function of Y on $X_1, \ldots, X_k$ cannot be worse than the regression function of Y on $X_1, \ldots, X_m$ for predicting Y.

Notation

In this section, we work with more than one regression function. It therefore becomes necessary to introduce some notational conventions to help avoid confusion. When dealing with two or more regression functions based on different sets of predictor variables, we distinguish between them by using appropriate superscripts. For instance, in the discussion in the previous paragraph, we would use the symbol $\mu_Y^{(A)}(x_1, \ldots, x_k)$ to denote the regression function of Y on $X_1, \ldots, X_k$, which we will refer to as model A, and the symbol $\mu_Y^{(B)}(x_1, \ldots, x_m)$ to denote the regression function of Y on $X_1, \ldots, X_m$, which we will refer to as model B. When discussing subpopulation standard deviations, we must again distinguish between the subpopulation standard deviation for model A and the subpopulation standard deviation for model B. We do this either by using the more complete notation $\sigma_{Y|X_1,\ldots,X_k}$ and $\sigma_{Y|X_1,\ldots,X_m}$ or, when there is no possibility of confusion, by abbreviating these to σ_A and σ_B, respectively. Recall that to judge how good a regression function is for predicting Y, the measure we use is the corresponding subpopulation standard deviation.

We illustrate with an example.

E X A M P L E 4.9.1

For an illustration, consider Example 2.2.3 and suppose an investigator wants to determine if age X_1 and weight X_2 together are better than age X_1 alone for predicting blood pressure Y and, if so, how much better. This can be stated as follows: How much better is the regression function $\mu_Y^{(A)}(x_1, x_2)$ than the regression function $\mu_Y^{(B)}(x_1)$ for predicting Y? So the investigator will be interested in comparing the subpopulation standard deviation $\sigma_{Y|X_1,X_2} = \sigma_A$ with $\sigma_{Y|X_1} = \sigma_B$. ∎

The decision of whether to use the full set $X_1, \ldots, X_k$ of predictor variables, or the subset $X_1, \ldots, X_m$, usually depends, at least in part, on

1   The cost of observing the values of the additional variables $X_{m+1}, \ldots, X_k$

2   The improvement in prediction that is made possible by using the full set of predictors rather than the subset of predictors under consideration

As explained earlier, we let $\mu_Y^{(A)}(x_1, \ldots, x_k)$ denote the regression function of Y on the k predictor variables $X_1, \ldots, X_k$ and call this model A. We let $\mu_Y^{(B)}(x_1, \ldots, x_m)$ denote the regression function of Y on the subset of m predictor variables $X_1, \ldots, X_m$ and call this model B.

The investigator is likely to be interested in obtaining the answers to the following questions.

1   How well does model A predict Y?

2   Is model A an adequate predictor of Y?

3   How well does model B predict Y?

4   Is model B an adequate predictor of Y?

5   How much better is model A than model B for predicting Y?

In this section we discuss methods for answering these and other questions. An important difference between assumptions (A) and assumptions (B) should be noted, especially when discussing several regression models. *Even though population assumptions* (A) *may hold for* $\{(Y, X_1, \ldots, X_k)\}$, *this does not imply that population assumptions* (A) *will hold for* $\{(Y, X_1, \ldots, X_m)\}$, *where* $\{X_1, \ldots, X_m\}$ *is a subset of the variables in* $\{X_1, \ldots, X_k\}$. *But if population assumptions* (B) *hold for* $\{(Y, X_1, \ldots, X_k)\}$, *then population assumptions* (B) *must hold for* $\{(Y, X_1, \ldots, X_m)\}$ *for every subset* $\{X_1, \ldots, X_m\}$ *of predictors from the full set* $\{X_1, \ldots, X_k\}$. This leads us to make the following assumption.

> Throughout this section, we presume that assumptions (B) are valid. This means that the $(k+1)$-variable population $\{(Y, X_1, \ldots, X_k)\}$ and the $(m+1)$-variable population $\{(Y, X_1, \ldots, X_m)\}$ (where $m <$ k) both satisfy (population) assumptions (B) in Box 4.3.2. This also means that the data are obtained by simple random sampling. **(4.9.1)**

Remarks

a   By virtue of (4.9.1), the regression functions $\mu_Y^{(A)}(x_1, \ldots, x_k)$—i.e., model A— and $\mu_Y^{(B)}(x_1, \ldots, x_m)$—i.e., model B—are of the form

$$\mu_Y^{(A)}(x_1, \ldots, x_k) = \beta_0^A + \beta_1^A x_1 + \cdots + \beta_k^A x_k \tag{4.9.2}$$

and

$$\mu_Y^{(B)}(x_1, \ldots, x_m) = \beta_0^B + \beta_1^B x_1 + \cdots + \beta_m^B x_m \tag{4.9.3}$$

respectively.

b   The regression function for model B is said to be **nested** in the regression function for model A because the set of predictor variables in $\mu_Y^{(B)}(x_1, \ldots, x_m)$ is a *subset* of the set of predictor variables in $\mu_Y^{(A)}(x_1, \ldots, x_k)$.

c   We use the superscripts A and B to distinguish between the β coefficients in the two regression functions $\mu_Y^{(A)}(x_1, \ldots, x_k)$ and $\mu_Y^{(B)}(x_1, \ldots, x_m)$, respectively. The regression coefficients β_i^A and β_i^B for $i = 0, 1, 2, \ldots, m$ are different unless $\beta_{m+1}^A = \cdots = \beta_k^A = 0$, in which case the two regression functions are identical and $\beta_i^A = \beta_i^B$ for $i = 1, \ldots, m$.

d   As mentioned earlier, we sometimes use σ_A to denote the subpopulation standard deviation for model A and σ_B to denote the subpopulation standard deviation for model B.

In Example 4.9.1 the set of factors in model B (i.e., $\{X_1\}$) is a subset of the set of factors in model A (i.e., $\{X_1, X_2\}$), and so model B is nested in model A. However, if model A is $\mu_Y^{(A)}(x_1) = \beta_0^A + \beta_1^A x_1$ and model B is $\mu_Y^{(B)}(x_2) = \beta_0^B + \beta_2^B x_2$, then the set of factors in model B (i.e., $\{X_2\}$) is not a subset of the set of factors in model A (i.e., $\{X_1\}$), so for this case the discussion in this section does not apply, but the discussion in Section 4.10 does.

Subpopulation Standard Deviations as Measures of Goodness of Prediction

As usual, we use the quantity σ_A as a summary measure of how well model A in (4.9.2) predicts Y. Likewise, the quantity σ_B is a summary measure of how well model B in (4.9.3) predicts Y.

Relationship Between Subpopulation Standard Deviations in Nested Models

Under the assumption in (4.9.1), it can be shown that

$$\sigma_A \leq \sigma_B \tag{4.9.4}$$

i.e., the subpopulation standard deviation for the larger model (model A) is smaller than or equal to the subpopulation standard deviation for the smaller nested model (model B). To elaborate on this, consider the k predictor factors $X_1, \ldots, X_k$, and let S_1 be any subset of these k factors and S_2 be any subset of the factors in S_1. Then the following inequalities are true.

$$0 \leq \sigma_A \leq \sigma_{S_1} \leq \sigma_{S_2} \leq \sigma_Y \tag{4.9.5}$$

For instance consider the five-variable population of Example 4.2.1. Let $S_1 = \{X_1, X_2, X_4\}$ and $S_2 = \{X_1, X_2\}$. Then σ_{S_1} stands for $\sigma_{Y|X_1,X_2,X_4}$ and σ_{S_2} stands for $\sigma_{Y|X_1,X_2}$. By (4.9.5) the following is true:

$$0 \leq \sigma_{Y|X_1,X_2,X_3,X_4} \leq \sigma_{Y|X_1,X_2,X_4} \leq \sigma_{Y|X_1,X_2} \leq \sigma_Y$$

On the other hand, if $S_1 = \{X_1, X_2, X_3\}$ and $S_2 = \{X_1, X_4\}$, then S_2 is not a subset of S_1 (and S_1 is not a subset of S_2), so it is not known whether $\sigma_{Y|X_1,X_2,X_3}$ is larger or smaller than $\sigma_{Y|X_1,X_4}$. However, from (4.9.5) we get

$$0 \leq \sigma_{Y|X_1,X_2,X_3,X_4} \leq \sigma_{Y|X_1,X_2,X_3} \leq \sigma_Y$$

and

$$0 \leq \sigma_{Y|X_1,X_2,X_3,X_4} \leq \sigma_{Y|X_1,X_4} \leq \sigma_Y$$

To illustrate inequalities such as (4.9.5), we consider the simple case described in Example 2.2.2, where Y = first-year maintenance cost of a new car and X = number of miles the car is driven the first year. The target population $\{(Y, X)\}$ is the set of all cars that will be made by company A next year and driven between 5,000 and 20,000 miles. The study population is a set of similar cars made by company A last year. The quantity σ_Y is the standard deviation of the first-year maintenance costs of *all* the cars in the study population. If only the subpopulation of cars that were driven 10,000 miles is considered, then the standard deviation of the first-year maintenance costs of all cars driven 10,000 miles the first-year is equal to $\sigma_{Y|X}$.

We would typically expect the first-year maintenance costs of these cars to be more similar than for the entire population of cars, and hence we would expect $\sigma_{Y|X} \leq \sigma_Y$.

Adequacy of Prediction Functions

Whether or not a prediction function is adequate for predicting Y depends on the particular problem.

We consider a prediction function to be adequate for predicting values of Y if a proportion p (p is taken to be greater than 0.5) of the Y values are within d units of the corresponding predicted values. The values of p and d are specified by the investigator.

Let us examine the conditions under which the regression function $\mu_Y(x_1, \ldots, x_k)$ may be regarded as an adequate prediction function for predicting values of Y. When population assumptions (A) or (B) hold for $\{(Y, X_1, \ldots X_k)\}$, a proportion p of the Y values with $X_1 = x_1, \ldots, X_k = x_k$ will lie in the interval

$$[\mu_Y(x_1, \ldots, x_k) - z_{(1+p)/2}\sigma_{Y|X_1,\ldots,X_k}, \mu_Y(x_1, \ldots, x_k) + z_{(1+p)/2}\sigma_{Y|X_1,\ldots,X_k}]$$

So if $\mu_Y(x_1, \ldots, x_k)$ is to be within d units of a proportion p of the Y values in the corresponding subpopulation, we must have

$$z_{(1+p)/2}\sigma_{Y|X_1,\ldots,X_k} \leq d \qquad \text{i.e.,} \qquad \sigma_{Y|X_1,\ldots,X_k} \leq d/z_{(1+p)/2}$$

Sometimes an investigator may say "$\mu_Y(x_1, \ldots, x_k)$ is adequate for predicting Y values if $\sigma_{Y|X_1,\ldots,X_k}$ is less than a specified value d^*." In this case d and d^* are related by the equation

$$d^* = d/z_{(1+p)/2} \qquad \text{i.e.,} \qquad d = d^* z_{(1+p)/2}$$

For example, in Problem 4.7.3, the investigator considers the regression function in (4.7.2) to be adequate for predicting the annual electric bill if $\sigma_{Y|X_1,X_2,X_3} \leq \$50$. So here $d^* = 50$. The numbers d and p (or d^* and p) specified by an investigator are based on various practical considerations in the context of the particular problem.

We now apply the preceding criterion of adequacy to the regression functions $\mu_Y^{(A)}(x_1, \ldots, x_k)$ and $\mu_Y^{(B)}(x_1, \ldots, x_m)$ given in (4.9.2) and (4.9.3), respectively. By (4.9.1), assumptions (B) for multiple regression hold for both models. So we can conclude that $\mu_Y^{(A)}(x_1, \ldots, x_k)$ is an adequate predictor of Y if $z_{(1+p)/2}\sigma_A \leq d$, i.e., if $\sigma_A \leq d/z_{(1+p)/2}$. Similarly, $\mu_Y^{(B)}(x_1, \ldots, x_m)$ is an adequate predictor of Y if $\sigma_B \leq d/z_{(1+p)/2}$. It is possible that $\mu_Y^{(A)}(x_1, \ldots, x_k)$ and $\mu_Y^{(B)}(x_1, \ldots, x_m)$ are *both* adequate for predicting Y or that *neither* $\mu_Y^{(A)}(x_1, \ldots, x_k)$ nor $\mu_Y^{(B)}(x_1, \ldots, x_m)$ is adequate for predicting Y. Of course, in a real problem the regression functions $\mu_Y^{(A)}(x_1, \ldots, x_k)$ and $\mu_Y^{(B)}(x_1, \ldots, x_m)$ are not known and must be estimated. This adds another source of uncertainty to the problem. Nevertheless, $\hat{\sigma}_A$ and $\hat{\sigma}_B$, the estimates of σ_A and σ_B, respectively, and the confidence intervals for these, are useful for determining the adequacy of prediction functions.

Performance of Model A Relative to Model B

There are instances where it is difficult to specify a number d to determine whether a prediction function is adequate. In these instances and in other situations, we may be interested in the relative performances of competing prediction functions. In the present situation, two prediction functions, $\mu_Y^{(A)}(x_1, \ldots, x_k)$ and $\mu_Y^{(B)}(x_1, \ldots, x_m)$ (i.e., two models A and B), are being considered, and hence we may want to compare σ_A and σ_B. We can either examine σ_A and σ_B individually or examine some function of σ_A and σ_B that may be particularly meaningful in a given problem. For instance, we may want to examine the following functions of σ_A and σ_B to determine how much better $\mu_Y^{(A)}(x_1, \ldots, x_k)$ is than $\mu_Y^{(B)}(x_1, \ldots, x_m)$ for predicting Y.

1 $\sigma_B - \sigma_A$

2 σ_B/σ_A

3 $\sigma_B^2 - \sigma_A^2$

4 σ_B^2/σ_A^2

In this book we use the ratio σ_B/σ_A for a comparison of σ_A relative to σ_B. A function of σ_A and σ_B that is related to the ratio σ_B/σ_A and has found widespread use in the literature is called the *multiple coefficient of determination*, and we discuss this next.

Multiple Coefficient of Determination

A commonly used measure that summarizes the performance of $\mu_Y^{(A)}(x_1, \ldots, x_k)$ (model A) as a predictor of Y relative to $\mu_Y^{(B)}(x_1, \ldots, x_m)$ (model B) is the **multiple-partial coefficient of determination** of Y with $X_{m+1}, \ldots, X_k$ when $X_1, \ldots, X_m$ are held fixed. It is denoted by $\rho^2_{Y(X_{m+1}, \ldots, X_k)|X_1, \ldots, X_m}$ and is defined by

$$\rho^2_{Y(X_{m+1}, \ldots, X_k)|X_1, \ldots, X_m} = \frac{\sigma^2_{Y|X_1, \ldots, X_m} - \sigma^2_{Y|X_1, \ldots, X_k}}{\sigma^2_{Y|X_1, \ldots, X_m}} = \frac{\sigma_B^2 - \sigma_A^2}{\sigma_B^2} = 1 - \frac{1}{(\sigma_B/\sigma_A)^2}$$

$$(4.9.6)$$

Thus $\rho^2_{Y(X_{m+1}, \ldots, X_k)|X_1, \ldots, X_m}$ is the *proportional reduction in the variance of prediction errors by using* $\mu_Y^{(A)}(x_1, \ldots, x_k)$ *to predict* Y, *relative to using* $\mu_Y^{(B)}(x_1, \ldots, x_m)$ *to predict* Y.

The positive square root of $\rho^2_{Y(X_{m+1}, \ldots, X_k)|X_1, \ldots, X_m}$ is generally called the *multiple-partial correlation coefficient* of Y with $X_{m+1}, X_{m+2}, \ldots, X_k$ when $X_1, \ldots, X_m$ are held fixed. It is denoted by $\rho_{Y(X_{m+1}, \ldots, X_k)|X_1, \ldots, X_m}$. The word *multiple* in multiple-partial coefficient of determination means there is more than one predictor variable; i.e., $X_1, \ldots, X_k, k > 1$. The word *partial* means that some of the predictor variables are held fixed.

To examine (4.9.6) in more detail, consider only the numerator

$$\sigma_B^2 - \sigma_A^2 \qquad \text{(4.9.7)}$$

There are two equivalent ways of looking at (4.9.7).

1   As a measure of how much better the k factors $X_1, \ldots, X_k$ together are than the m factors $X_1, \ldots, X_m$ for predicting values of Y.

2   As a measure of how much factors $X_{m+1}, X_{m+2}, \ldots, X_k$ contribute to predicting Y in addition to what factors $X_1, X_2, \ldots, X_m$ contribute.

Consider the special case in Example 4.9.1 where the relationship of blood pressure Y with age X_1 and weight X_2 is studied. Let $A = \{X_1, X_2\} = \{\text{age, weight}\}$, $B = \{X_1\} = \{\text{age}\}$, and $C = \{X_2\} = \{\text{weight}\}$.

1   σ_A is a measure of how good age and weight together are for predicting blood pressure Y.

2   σ_B is a measure of how good age alone is for predicting blood pressure Y.

3   σ_C is a measure of how good weight alone is for predicting blood pressure Y.

4   The quantity $\sigma_B^2 - \sigma_A^2$ is often used in the following two equivalent ways.
   a   As a measure of how much better age and weight, X_1 and X_2, together are for predicting blood pressure Y than age X_1 alone is.
   b   As a measure of how much weight contributes to predicting Y in addition to what age contributes.

When model B in (4.9.3) uses no predictors, i.e., when $m = 0$, then $\mu_Y^{(B)} = \beta_0^B = \mu_Y$ is the quantity used to predict Y under model B and so σ_B is simply σ_Y. In this case there are no predictor variables to the right of the '|' symbol in $\rho_{Y(X_{m+1}, \ldots, X_k)|X_1, \ldots, X_m}^2$ (since $m = 0$), and therefore it is simply written as $\rho_{Y(X_1, \ldots, X_k)}^2$ and is called the **multiple coefficient of determination** of Y with $X_1, \ldots, X_k$. The word *partial* is omitted since no predictors are held fixed. Thus $\rho_{Y(X_1, \ldots, X_k)}^2$ measures relatively how much better $\mu_Y^{(A)}(x_1, \ldots, x_k)$ is for predicting Y than μ_Y is. For completeness, we give the definition of $\rho_{Y(X_1, \ldots, X_k)}^2$ in (4.9.8).

D E F I N I T I O N

$$\rho_{Y(X_1, \ldots, X_k)}^2 = \frac{\sigma_Y^2 - \sigma_{Y|X_1, \ldots, X_k}^2}{\sigma_Y^2} = \frac{\sigma_Y^2 - \sigma_A^2}{\sigma_Y^2} = 1 - \frac{1}{(\sigma_Y/\sigma_A)^2} \qquad \blacksquare \qquad \text{(4.9.8)}$$

The case when $k = 1$ and $m = 0$ was discussed in Section 3.9, and we write ρ_{Y,X_1}^2 instead of $\rho_{Y(X_1)}^2$. Also when $m = k - 1$, the word *multiple* is often omitted; for instance, the quantity $\rho_{Y,X_1|X_2}^2$ is simply called the *partial coefficient of determination* of Y with X_1 where X_2 is held fixed.

To summarize:

1   $\rho_{Y(X_{m+1}, X_{m+2}, \ldots, X_k)}^2$ is the multiple coefficient of determination of Y with the predictor variables in parentheses, namely $X_{m+1}, X_{m+2}, \ldots, X_k$.

2 $\rho^2_{Y(X_{m+1},X_{m+2},...,X_k)|X_1,X_2,...,X_m}$ is the multiple-partial coefficient of determination of Y with $X_{m+1}, X_{m+2}, \ldots, X_k$ (the variables in parentheses) when the variables $X_1, X_2, \ldots, X_m$ (the variables following the vertical line '|') are held fixed.

Properties of $\rho^2_{Y(X_{m+1},...,X_k)|X_1,...,X_m}$

Under population assumptions (B), the statements in Box 4.9.1 hold.

B O X 4.9.1

1 $0 \le \rho^2_{Y(X_{m+1},...,X_k)|X_1,...,X_m} \le 1$.

2 $\rho^2_{Y(X_{m+1},...,X_k)|X_1,...,X_m} = 0$ if and only if $\beta^A_{m+1} = \ldots = \beta^A_k = 0$ in (4.9.2). In this case the two regression functions, $\mu^{(A)}_Y(x_1,\ldots,x_k)$ in (4.9.2) and $\mu^{(B)}_Y(x_1,\ldots,x_m)$ in (4.9.3), are identical. Moreover, $\sigma_A = \sigma_B$.

3 If $\rho^2_{Y(X_{m+1},...,X_k)|X_1,...,X_m} > 0$, then at least one of the parameters $\beta^A_{m+1},\ldots,\beta^A_k$ in (4.9.2) is nonzero. Also $\sigma_A < \sigma_B$.

4 $\rho^2_{Y(X_{m+1},...,X_k)|X_1,...,X_m} = 1$ if and only if $\sigma_A = 0$. In this case $\mu^{(A)}_Y(x_1,\ldots,x_k)$ is a *perfect predictor* of Y.

For the case when $m = 0$ (i.e., model B is $\mu^{(B)}_Y = \beta^B_0 = \mu_Y$ and it contains none of the predictors $X_1,\ldots,X_k$), the statements in Box 4.9.1 specialize to those in Box 4.9.2.

B O X 4.9.2

1 $0 \le \rho^2_{Y(X_1,...,X_k)} \le 1$.

2 $\rho^2_{Y(X_1,...,X_k)} = 0$ if and only if $\beta^A_1 = \ldots = \beta^A_k = 0$ in (4.9.2). In this case the regression function $\mu^{(A)}_Y(x_1,\ldots,x_k)$ is equal to $\mu^{(B)}_Y = \mu_Y$. Moreover, $\sigma_A = \sigma_Y$.

3 If $\rho^2_{Y(X_1,...,X_k)} > 0$, then at least one of the parameters $\beta^A_1,\ldots,\beta^A_k$ in (4.9.2) is nonzero. Also $\sigma_A < \sigma_Y$.

4 $\rho^2_{Y(X_1,...,X_k)} = 1$ if and only if $\sigma_A = 0$. In this case $\mu^{(A)}_Y(x_1,\ldots,x_k)$ is a *perfect predictor* of Y.

E X A M P L E 4.9.2

To illustrate the preceding concepts, let us consider Example 2.2.3 where Y is (systolic) blood pressure, X_1 is age, and X_2 is weight. Suppose we want to examine the relationship among Y, X_1, and X_2 for the population of all females in California between the ages of 25 and 75. Suppose the three-variable population $\{(Y, X_1, X_2)\}$ is Gaussian.

- The two-variable populations $\{(Y, X_1)\}$ and $\{(Y, X_2)\}$ and the one-variable population $\{Y\}$ are also Gaussian.
- The average blood pressure of the entire population $\{Y\}$ is μ_Y, and the standard deviation is σ_Y.
- The average blood pressure for the subpopulation of all females who are x_1 years old is

$$\mu_Y^{(B)}(x_1) = \beta_0^B + \beta_1^B x_1 \qquad \textbf{(4.9.9)}$$

with standard deviation σ_B.

- The average blood pressure for the subpopulation of all females who weigh x_2 pounds is

$$\mu_Y^{(C)}(x_2) = \beta_0^C + \beta_2^C x_2 \qquad \textbf{(4.9.10)}$$

with standard deviation σ_C.

- The average blood pressure for the subpopulation of all females whose age is x_1 and weight is x_2 is

$$\mu_Y^{(A)}(x_1, x_2) = \beta_0^A + \beta_1^A x_1 + \beta_2^A x_2 \qquad \textbf{(4.9.11)}$$

with standard deviation σ_A.

Also

$$\sigma_Y, \sigma_B, \sigma_C, \sigma_A$$

are the quantities that are used as summary measures of how good μ_Y, $\mu_Y^{(B)}(x_1)$, $\mu_Y^{(C)}(x_2)$, and $\mu_Y^{(A)}(x_1, x_2)$, respectively, are as predictors of Y. From (4.9.5) we have

$$0 \leq \sigma_A \leq \sigma_B \leq \sigma_Y \qquad \text{and} \qquad 0 \leq \sigma_A \leq \sigma_C \leq \sigma_Y \qquad \textbf{(4.9.12)}$$

We may want to determine the answer to several questions about X_1 and X_2 as predictors of Y. Some of these questions follow.

1. How good are age and weight together as predictors of blood pressure Y?
2. How well can Y be predicted if no predictor variables are used?
3. How good is age alone as a predictor of Y (if weight is ignored)?
4. How good is weight alone as a predictor of Y (if age is ignored)?
5. How good is weight as a predictor of Y in the subpopulation of all females who are 35 years old?
6. How much better are age and weight together for predicting blood pressure than is age alone?
7. How good is weight as a predictor of blood pressure after age has been accounted for?

8 *Relatively*, how much better are age and weight together for predicting blood pressure than age alone?

9 *Relatively*, how good is weight for predicting blood pressure after age has been taken into account?

The answers to these questions are given next in terms of the population parameters.

1 A measure of how good age and weight together are for predicting blood pressure—i.e., how good $\mu_Y^{(A)}(x_1, x_2)$ is for predicting Y—is σ_A because this is the standard deviation of Y in the subpopulation determined by fixed values of X_1 and X_2.

2 When no predictor variables are used to predict Y, the best value to use to predict Y is μ_Y, the mean of the Y values in the entire population. A measure of how good μ_Y is for predicting Y is provided by σ_Y, the standard deviation of the population $\{Y\}$.

3 The best prediction function for predicting Y using age alone is $\mu_Y^{(B)}(x_1)$, the regression function of Y on X_1. A measure of how good $\mu_Y^{(B)}(x_1)$ is for predicting Y is provided by σ_B, the standard deviation of the Y values in the subpopulation determined by X_1 (age) when X_2 (weight) is not considered.

4 The best prediction function for predicting Y using weight alone is $\mu_Y^{(C)}(x_2)$, the regression function of Y on X_2. A measure of how good $\mu_Y^{(C)}(x_2)$ is for predicting Y is provided by σ_C, the standard deviation of the Y values in the subpopulation determined by X_2 (weight) when X_1 (age) is not considered.

5 σ_A, the same as the answer to question (1), because assumptions (B) in Box 4.3.2 imply that the standard deviations of the subpopulations determined by X_1 and X_2 are the same for all values of X_1 and X_2.

6 $\sigma_B - \sigma_A$, or σ_B/σ_A, or $\sigma_B^2 - \sigma_A^2$, or σ_B^2/σ_A^2, depending on which measure you want to use.

7 Same answer as (6).

8 $\dfrac{\sigma_B^2 - \sigma_A^2}{\sigma_B^2} = \rho_{Y,X_2|X_1}^2$ if we use variances as the summary measures.

9 Same answer as (8).

We illustrate other important points using this example.

1 In the multiple regression model in (4.1.1), the coefficient β_i (for $i = 1, \ldots, k$) is the change in $\mu_Y(x_1, \ldots, x_k)$ per unit change in X_i when the other factors are held fixed. For example, in the model in (4.9.11), β_2^A is the change in the average blood pressure per unit change in weight for all females of the same age (say, 35 years old), i.e., when X_1 (age) is held fixed.

Note *Sometimes it is not meaningful to change one variable, say X_i, and hold other variables fixed. For example, consider the regression model in (4.2.1a). If we let $X_1 = Z_1$, $X_2 = Z_1 Z_2$, and $X_3 = Z_1^2$, we get the multiple linear regression* *model*

$$\mu_Y(x_1, x_2, x_3) = \beta_0 + \beta_1 x_1 + \beta_2 x_2 + \beta_3 x_3$$

and clearly we cannot hold X_1 and X_2 fixed and let X_3 change.

2 Some predictor factors are controllable factors and some are noncontrollable factors. In the preceding example, suppose β_1^A and β_2^A in (4.9.11) are positive, which means that $\mu_Y^{(A)}(x_1, x_2)$ decreases as x_1 and/or x_2 decreases. If a physician recognizes that a patient's blood pressure should be reduced, the model indicates that it *may be possible* to reduce it by lowering the patient's age or weight. Of course it is not possible to reduce the patient's age (X_1 is a noncontrollable factor). But weight is a controllable factor (at least controllable to some extent), and it *may be possible* to reduce the patient's weight by diet and hence possibly reduce the blood pressure. Whether changing the value of X_2 will actually result in a change in the value of Y cannot be known based on observational studies, and it has to be studied using controlled experiments. ■

The quantity $\rho^2_{Y(X_{m+1},\dots,X_k)|X_1,\dots,X_m}$ is a population parameter, and so it is unknown in any real problem. A valid point estimate of $\rho^2_{Y(X_{m+1},\dots,X_k)|X_1,\dots,X_m}$ can be calculated from sample data according to the formula given in (4.9.13) if assumptions (B) are satisfied; in particular, *the data must be obtained by simple random sampling. If data are obtained by sampling with preselected values of $X_1, \dots, X_k$, then no valid estimate of $\rho^2_{(X_{m+1},\dots,X_k)|X_1,\dots,X_m}$ is available.*

Point Estimate of $\rho^2_{Y(X_{m+1},\dots,X_k)|X_1,\dots,X_m}$

The point estimate of $\rho^2_{Y(X_{m+1},\dots,X_k)|X_1,\dots,X_m}$, the multiple-partial coefficient of determination of Y with $X_{m+1}, \dots, X_k$ when $X_1, \dots, X_m$ are held fixed, is given by

$$\hat{\rho}^2_{Y(X_{m+1},\dots,X_k)|X_1,\dots,X_m} = \frac{SSE(X_1, \dots, X_m) - SSE(X_1, \dots, X_k)}{SSE(X_1, \dots, X_m)} \tag{4.9.13}$$

The sum of squares $SSE(X_1, \dots, X_k)$ was defined in (4.4.14), and the sum of squares $SSE(X_1, \dots, X_m)$ is defined similarly but using only the variables $X_1, \dots, X_m$. For notational convenience we write $SSE(A)$ for $SSE(X_1, \dots, X_k)$ and $SSE(B)$ for $SSE(X_1, \dots, X_m)$ because these correspond to models A and B in (4.9.2) and (4.9.3), respectively. Thus we have

$$\hat{\rho}^2_{Y(X_{m+1},\dots,X_k)|X_1,\dots,X_m} = \frac{SSE(B) - SSE(A)}{SSE(B)} \tag{4.9.14}$$

An Alternate Point Estimate of $\rho^2_{Y(X_{m+1},\ldots,X_k)|X_1,\ldots,X_m}$

The definition of $\rho^2_{Y(X_{m+1},\ldots,X_k)|X_1,\ldots,X_m}$ in (4.9.6) might suggest that the point estimate of $\rho^2_{Y(X_{m+1},\ldots,X_k)|X_1,\ldots,X_m}$ would be given by

$$\frac{\hat{\sigma}^2_{Y|X_1,\ldots,X_m} - \hat{\sigma}^2_{Y|X_1,\ldots,X_k}}{\hat{\sigma}^2_{Y|X_1,\ldots,X_m}}$$

and indeed this estimate is sometimes used. It is called *the estimate of the multiple-partial coefficient of determination (adjusted for degrees of freedom) of Y with* $X_{m+1},\ldots,X_k$ *when* $X_1,\ldots,X_m$ *are held fixed* and is written as $Adj[\hat{\rho}^2_{Y(X_{m+1},\ldots,X_k)|X_1,\ldots,X_m}]$. Thus,

$$Adj[\hat{\rho}^2_{Y(X_{m+1},\ldots,X_k)|X_1,\ldots,X_m}] = \frac{\hat{\sigma}^2_{Y|X_1,\ldots,X_m} - \hat{\sigma}^2_{Y|X_1,\ldots,X_k}}{\hat{\sigma}^2_{YX_1,\ldots,X_m}} \tag{4.9.15}$$

$$= \frac{MSE(X_1,\ldots,X_m) - MSE(X_1,\ldots,X_k)}{MSE(X_1,\ldots,X_m)} \tag{4.9.16}$$

The mean square $MSE(X_1,\ldots,X_k)$ was defined in (4.4.15), and the mean square $MSE(X_1,\ldots,X_m)$ is defined similarly but using only the variables $X_1,\ldots,X_m$. For notational convenience we write $MSE(A)$ for $MSE(X_1,\ldots,X_k)$ and $MSE(B)$ for $MSE(X_1,\ldots,X_m)$ because these correspond to models A and B in (4.9.2) and (4.9.3), respectively. Thus we have

$$Adj[\hat{\rho}^2_{Y(X_{m+1},\ldots,X_k)|X_1,\ldots,X_m}] = \frac{MSE(B) - MSE(A)}{MSE(B)} \tag{4.9.17}$$

Relationship Between $Adj[\hat{\rho}^2_{Y(X_{m+1},\ldots,X_k)|X_1,\ldots,X_m}]$ and $\hat{\rho}^2_{Y(X_{m+1},\ldots,X_k)|X_1,\ldots,X_m}$

The following equations relate $Adj[\hat{\rho}^2_{Y(X_{m+1},\ldots,X_k)|X_1,\ldots,X_m}]$ and $\hat{\rho}^2_{Y(X_{m+1},\ldots,X_k)|X_1,\ldots,X_m}$.

$$1 - Adj[\hat{\rho}^2_{Y(X_{m+1},\ldots,X_k)|X_1,\ldots,X_m}] = \frac{n-m-1}{n-k-1}(1 - \hat{\rho}^2_{Y(X_{m+1},\ldots,X_k)|X_1,\ldots,X_m}) \tag{4.9.18}$$

$$1 - \hat{\rho}^2_{Y(X_{m+1},\ldots,X_k)|X_1,\ldots,X_m} = \frac{n-k-1}{n-m-1}(1 - Adj[\hat{\rho}^2_{Y(X_{m+1},\ldots,X_k)|X_1,\ldots,X_m}]) \tag{4.9.19}$$

It follows that $Adj[\hat{\rho}^2_{Y(X_{m+1},\ldots,X_k)|X_1,\ldots,X_m}]$ is always less than or equal to $\hat{\rho}^2_{Y(X_{m+1},\ldots,X_k)|X_1,\ldots,X_m}$. Note that $Adj[\hat{\rho}^2_{Y(X_{m+1},\ldots,X_k)|X_1,\ldots,X_m}]$ can take negative values, and if it does, we replace the negative value with zero because it is estimating $\rho^2_{Y(X_{m+1},\ldots,X_k)|X_1,\ldots,X_m}$, which is always greater than or equal to zero. On the other hand, the estimate $\hat{\rho}^2_{Y(X_{m+1},\ldots,X_k)|X_1,\ldots,X_m}$ is never negative. Unless we specifically state otherwise, we always use the estimate in (4.9.13).

Point Estimates for $\rho^2_{Y(X_1,...,X_k)}$

When $m = 0$, i.e., when model B uses none of the predictors $X_1, \ldots, X_k$, the formulas in (4.9.13)–(4.9.16) reduce to

$$\hat{\rho}^2_{Y(X_1,...,X_k)} = \frac{SSY - SSE(X_1, \ldots, X_k)}{SSY} \tag{4.9.20}$$

and

$$Adj[\hat{\rho}^2_{Y(X_1,...,X_k)}] = \frac{\hat{\sigma}^2_Y - \hat{\sigma}^2_{Y|X_1,...,X_k}}{\hat{\sigma}^2_Y} \tag{4.9.21}$$

$$= \frac{MSY - MSE(X_1, \ldots, X_k)}{MSY} \tag{4.9.22}$$

respectively.

Investigators often use $\rho^2_{Y(X_{m+1},...,X_k)|X_1,...,X_m}$ to decide whether $\mu_Y^{(A)}(x_1, \ldots, x_k)$ is a better predictor of Y than $\mu_Y^{(B)}(x_1, \ldots, x_m)$. Technically, if $\rho^2_{Y(X_{m+1},...,X_k)|X_1,...,X_m} > 0$, then we can conclude that σ_A (i.e., $\sigma_{Y|X_1,...,X_k}$) is smaller than σ_B (i.e., $\sigma_{Y|X_1,...,X_m}$) and hence $\mu_Y^{(A)}(x_1, \ldots, x_k)$ is better than $\mu_Y^{(B)}(x_1, \ldots, x_m)$ for predicting Y. However, in practice, we generally want to know *how much smaller σ_A is than σ_B.* (Recall that, as stated in (4.9.4), σ_A is never greater than σ_B.) An obvious procedure is to examine the estimated values and confidence intervals for σ_A and σ_B. Additionally, we could use a confidence interval for σ_B/σ_A for this purpose.

Confidence Interval for $\sigma_{Y|X_1,...,X_m}/\sigma_{Y|X_1,...,X_k}$ (i.e., σ_B/σ_A)

A two-sided confidence interval for $\sigma_{Y|X_1,...,X_m}/\sigma_{Y|X_1,...,X_k}$ with confidence coefficient equal to $1 - \alpha$, can be obtained by first obtaining a confidence interval for $\rho^2_{Y(X_{m+1},...,X_k)|X_2,...,X_m}$. Confidence intervals for $\sigma_Y/\sigma_{Y|X_1,...,X_k}$ can be obtained as a special case of this procedure with $m = 0$. We do not discuss this here and you are referred to [10]. However, a procedure using the Bonferroni method, given in Box 4.10.2 in the next section, can be used here to obtain confidence intervals for σ_B/σ_A with confidence coefficient greater than or equal to $1 - \alpha$, but the resulting confidence intervals are wide and thus cannot be recommended for routine use in applications. We discuss them in Box 4.10.2 for illustrative purposes only.

We illustrate the procedures discussed in this section in Examples 4.9.3 and 4.9.4.

E X A M P L E 4.9.3

For the GPA data in Example 4.4.2, we show how to compute point estimates for $\rho^2_{Y(X_1,X_2,X_3,X_4)}$ and for σ_B/σ_A. We suppose that assumptions (B) in Box 4.3.2 are satisfied. The quantities needed in the formulas in (4.9.20) and (4.9.21) can be obtained from the ANOVA in Table 4.8.2. These quantities are

$$SSY = 7.3458 \qquad SSE(X_1, X_2, X_3, X_4) = 1.0815 \qquad MSY = 0.3866$$

$$MSE(X_1, X_2, X_3, X_4) = 0.0721$$

The point estimate of $\rho^2_{Y(X_1,X_2,X_3,X_4)}$, computed using (4.9.20), is

$$\hat{\rho}^2_{Y(X_1,X_2,X_3,X_4)} = \frac{7.3458 - 1.0815}{7.3458} = \frac{6.2643}{7.3458} = 0.853, \text{ i.e., } 85.3\%$$

This is the quantity labeled R-sq in the computer output in Exhibit 4.8.1. It is an estimate of relatively how much better the regression function

$$\mu_Y(x_1, x_2, x_3, x_4) = \beta_0 + \beta_1 x_1 + \beta_2 x_2 + \beta_3 x_3 + \beta_4 x_4 \qquad \textbf{(4.9.23)}$$

is for predicting GPA than μ_Y is. Note that the quantity labeled R-sq(adj) in Exhibit 4.8.1 is the alternate estimate $adj[\hat{\rho}^2_{Y(X_1,X_2,X_3,X_4)}]$ of $\rho^2_{Y(X_1,X_2,X_3,X_4)}$ given in (4.9.21), and it is equal to

$$adj[\hat{\rho}^2_{Y(X_1,X_2,X_3,X_4)}] = \frac{MSY - MSE(X_1, X_2, X_3, X_4)}{MSY} = \frac{0.3866 - 0.0721}{0.3866}$$

$$= 0.814 \text{ (i.e., } 81.4\%)$$

A point estimate of σ_B/σ_A is given by $\sigma_B/\sigma_A = \sqrt{0.3866/0.0721} = 2.32$. ∎

E X A M P L E 4.9.4

In Example 4.9.3, suppose the director of admissions wants to compare the performance of the model

$$\mu_Y^{(B)}(x_3, x_4) = \beta_0^B + \beta_3^B x_3 + \beta_4^B x_4 \qquad \textbf{(4.9.24)}$$

which uses only X_3 (HSmath) and X_4 (HSenglish), with the model

$$\mu_Y^{(A)}(x_1, x_2, x_3, x_4) = \beta_0^A + \beta_1^A x_1 + \beta_2^A x_2 + \beta_3^A x_3 + \beta_4^A x_4 \qquad \textbf{(4.9.25)}$$

for predicting Y. If model A is not much better than model B, then she may decide that applicants need not take the SAT.

Also, suppose the director decides that the prediction function $\mu_Y^{(A)}(x_1, x_2, x_3, x_4)$ would be considered adequate for predicting values of Y if a proportion $p = 0.95$ of the Y values being predicted are within 0.8 grade point unit of $\mu_Y^{(A)}(x_1, x_2, x_3, x_4)$, i.e., if $z_{0.975}\sigma_A \le 0.8$, or equivalently, $\sigma_A \le 0.8/1.96 = 0.41$. Likewise, the prediction function $\mu_Y^{(B)}(x_3, x_4)$ would be adequate for predicting values of Y if a proportion $p = 0.95$ of the Y values being predicted is within 0.8 grade point unit of $\mu_Y^{(B)}(x_3, x_4)$, i.e., if $\sigma_B \le 0.41$. We now examine the adequacy of model A and model B and also compare them relative to one another. Assumptions (B) for regression are presumed valid.

To obtain the required point estimates and confidence intervals we need $SSE(X_3, X_4)$ and $SSE(X_1, X_2, X_3, X_4)$. We can get these from the ANOVA tables for the models in (4.9.24) and (4.9.25) given in Exhibit 4.9.1.

The standard deviation σ_B for model B is estimated to be 0.3771 (see (4.9.26)). A 90% confidence interval for σ_B is given by the statement

$$C[0.296 \le \sigma_B \le 0.528] = 0.90$$

E X H I B I T 4.9.1
MINITAB Output for Example 4.9.4

```
Regression Analysis for Model B

The regression equation is
gpa = - 0.340 + 0.417 hsmath + 0.579 hsengl

Predictor        Coef        Stdev      t-ratio         p
Constant      -0.3400       0.5644        -0.60     0.555
hsmath         0.4171       0.1054         3.96     0.001
hsengl         0.5790       0.1857         3.12     0.006
```

$s = 0.3771$ R-sq = 67.1%   R-sq(adj) = 63.2%   **(4.9.26)**

```
Analysis of Variance

SOURCE        DF           SS           MS          F         p
Regression     2       4.9278       2.4639      17.32     0.000
Error         17       2.4180       0.1422
Total         19       7.3458
```

```
Regression Analysis for Model A

The regression equation is
gpa = 0.162 + 0.00201 satmath + 0.00125 satverb +
      0.189 hsmath + 0.088 hsengl

Predictor        Coef        Stdev      t-ratio         p
Constant       0.1615       0.4375         0.37     0.717
satmath      0.0020102    0.0005844         3.44     0.004
satverb      0.0012522    0.0005515         2.27     0.038
hsmath        0.18944      0.09187          2.06     0.057
hsengl         0.0876       0.1765          0.50     0.627
```

$s = 0.2685$ R-sq = 85.3%   R-sq(adj) = 81.4%   **(4.9.27)**

```
Analysis of Variance

SOURCE        DF           SS           MS          F         p
Regression     4       6.2643       1.5661      21.72     0.000
Error         15       1.0815       0.0721
Total         19       7.3458
```

From this the director of admissions can conclude, with 90% confidence, that σ_B is between 0.296 and 0.528 grade point unit. Thus she is led to conclude that more information is necessary to decide whether or not model B is *adequate* for predicting Y.

The standard deviation σ_A for model A is estimated to be 0.2685 (see (4.9.27)), and a two-sided 90% confidence interval for σ_A is given by

$$C[0.208 \le \sigma_A \le 0.386] = 0.90$$

From this interval the director can conclude, with 90% confidence, that σ_A is between 0.208 and 0.386 grade point unit. Thus, according to the criterion of adequacy stated earlier, she might conclude that model A is *adequate* for predicting Y.

A point estimate of σ_B/σ_A is $\hat{\sigma}_B/\hat{\sigma}_A = 0.3771/0.2685 = 1.404$, i.e., σ_B is estimated to be about 1.404 times as large as σ_A. ∎

Tests for $\rho^2_{Y(X_{m+1},\ldots,X_k)|X_1,\ldots,X_m}$ and $\rho^2_{Y(X_1,\ldots,X_k)}$

When comparing the performances of the two regression functions, $\mu_Y^{(A)}(x_1,\ldots,x_k)$ in (4.9.2) and $\mu_Y^{(B)}(x_1,\ldots,x_m)$ in (4.9.3), it is common to formulate this as a hypothesis testing problem. The null and the alternative hypotheses considered are

NH: $\rho^2_{Y(X_{m+1},\ldots,X_k)|X_1,\ldots,X_m} = 0$

or, equivalently, $\sigma_{Y|X_1,\ldots,X_k} = \sigma_{Y|X_1,\ldots,X_m}$

or, equivalently, $\beta_{m+1} = \cdots = \beta_k = 0$

versus   (4.9.28)

AH: $\rho^2_{Y(X_{m+1},\ldots,X_k)|X_1,\ldots,X_m} > 0$

or, equivalently, $\sigma_{Y|X_1,\ldots,X_k} < \sigma_{Y|X_1,\ldots,X_m}$

or, equivalently, at least one of $\beta_{m+1},\ldots,\beta_k$ is nonzero

This test procedure is given in Box 4.9.3 and is valid if assumptions (A) or (B) hold.

BOX 4.9.3 **Hypothesis Test for $\rho^2_{Y(X_{m+1},\ldots,X_k)|X_1,\ldots,X_m}$**

For a size α test of the NH versus AH in (4.9.28), compute

$$F_C = \frac{[SSE(X_1,\ldots,X_m) - SSE(X_1,\ldots,X_k)]/(k-m)}{MSE(X_1,\ldots,X_k)}$$

The *P*-value for the test is the value of α for which $F_C = F_{1-\alpha:k-m,n-k-1}$. The quantities $SSE(X_1,\ldots,X_m)$, $SSE(X_1,\ldots,X_k)$, and $MSE(X_1,\ldots,X_k)$ can be obtained by regressing Y on $X_1,\ldots,X_m$ and Y on $X_1,\ldots,X_k$.

Authors' Recommendation

To determine whether the regression function $\mu_Y(x_1, \ldots, x_k)$ is adequate for predicting Y, we recommend that a confidence interval for $\sigma_{Y|X_1,\ldots,X_k}$ be examined. To decide whether the regression function $\mu_Y^{(A)}(x_1, \ldots, x_k)$ or the function $\mu_Y^{(B)}(x_1, \ldots, x_m)$ should be used to predict Y, we recommend that confidence intervals for each of the two standard deviations, $\sigma_{Y|X_1,\ldots,X_k}$ and $\sigma_{Y|X_1,\ldots,X_m}$, be examined. Additionally, the investigator may want to examine a confidence interval for $\sigma_{Y|X_1,\ldots,X_m}/\sigma_{Y|X_1,\ldots,X_k}$. Equipped with such information, the investigator is better able to make a practical decision, taking into account such factors as cost of obtaining the observations for the variables under consideration, the desired level of accuracy of the predictions, etc. Do not settle for hypothesis tests only and thus waste valuable information contained in the data.

Problems   4.9

4.9.1   Consider the data of Problem 4.8.3 given in Table 4.8.4. An investigator is studying a population of males who have lived in mountain isolation for several generations and wants to investigate the relationship between the heights Y of these males at age 18 years and the following variables.

$$X_1 = \text{length at birth}$$
$$X_2 = \text{mother's height at age 18}$$
$$X_3 = \text{father's height at age 18}$$
$$X_4 = \text{maternal grandmother's height at age 18}$$
$$X_5 = \text{maternal grandfather's height at age 18}$$
$$X_6 = \text{paternal grandmother's height at age 18}$$
$$X_7 = \text{paternal grandfather's height at age 18}$$

All heights and lengths are in inches. A simple random sample of 20 males of age 18 or more was drawn from the study population, and all the preceding information was recorded. The data are also stored in the file **age18.dat** on the data disk. Assumptions (B) are presumed to hold. The investigator will consider a prediction function to be adequate for predicting values of Y if a proportion $p = 0.90$ of the Y values in the population are within $d = 2.0$ inches of the corresponding predicted value $\mu_Y(x_1, \ldots, x_7)$. You can use the computer output (obtained using SAS) in Exhibit 4.9.2 to answer questions (a)–(d). Model 1 is the regression of Y on $X_1, X_2, X_3, X_4, X_5, X_6, X_7$. Model 2 is the regression of Y on X_1, X_2, X_3.

E X H I B I T   4.9.2
SAS Output for Problem 4.9.1

```
                        The SAS System    0:00 Saturday,  Jan 1, 1994

Model: MODEL1    Dependent Variable: Y

                         Analysis of Variance
                        Sum of        Mean
Source          DF      Squares       Square      F Value     Prob>F

Model           7      133.65740     19.09391      18.955     0.0001
Error          12       12.08810      1.00734
C Total        19      145.74550


     Root MSE        1.00366      R-square        0.9171
     Dep Mean       68.53500      Adj R-sq        0.8687
     C.V.            1.46445


                         Parameter Estimates
                  Parameter       Standard     T for H0:
Variable   DF     Estimate          Error    Parameter=0    Prob > |T|
INTERCEP    1    -78.268378     26.96236510      -2.903        0.0133
X1          1      1.371816      0.52067159       2.635        0.0218
X2          1      0.782423      0.19923906       3.927        0.0020
X3          1      1.051413      0.13581316       7.742        0.0001
X4          1     -0.119914      0.17173269      -0.698        0.4983
X5          1      0.091436      0.13011662       0.703        0.4956
X6          1      0.088343      0.16132682       0.548        0.5940
X7          1     -0.101743      0.15489810      -0.657        0.5237
```
___

```
Model: MODEL2
Dependent Variable: Y


                         Analysis of Variance

                        Sum of        Mean
Source          DF      Squares       Square      F Value     Prob>F

Model           3      131.89367     43.96456      50.783     0.0001
Error          16       13.85183      0.86574
C Total        19      145.74550


     Root MSE        0.93045      R-square        0.9050
     Dep Mean       68.53500      Adj R-sq        0.8871
     C.V.            1.35763
```

Parameter Estimates

Variable	DF	Parameter Estimate	Standard Error	T for H0: Parameter=0	Prob > |T|
INTERCEP	1	-78.232762	13.23928437	-5.909	0.0001
X1	1	1.350302	0.44744522	3.018	0.0082
X2	1	0.692465	0.16017457	4.323	0.0005
X3	1	1.102495	0.09907801	11.128	0.0001

Consider the following two regression functions:

$$\text{model A: } \mu_Y^{(A)}(x_1, x_2, x_3, x_4, x_5, x_6, x_7) = \beta_0^A + \beta_1^A x_1 + \beta_2^A x_2 + \beta_3^A x_3$$
$$+ \beta_4^A x_4 + \beta_5^A x_5 + \beta_6^A x_6 + \beta_7^A x_7$$
$$\text{model B: } \mu_Y^{(B)}(x_1, x_2, x_3) = \beta_0^B + \beta_1^B x_1 + \beta_2^B x_2 + \beta_3^B x_3$$

a Compute an appropriate 80% confidence statement to help the investigator decide whether model A is adequate for predicting Y.

b Compute an appropriate 80% confidence statement to help the investigator decide whether model B is adequate for predicting Y.

c Explain the meaning of

$$\rho_{Y(X_4,X_5,X_6,X_7)}^2 \quad \rho_{Y(X_1,X_2,X_3)}^2 \quad \rho_{Y(X_4,X_5,X_6,X_7)|X_1,X_2,X_3}^2 \quad \rho_{Y(X_1,X_2,X_3,X_4,X_5,X_6,X_7)}^2$$

Obtain point estimates for

$$\rho_{Y(X_1,X_2,X_3)}^2 \quad \rho_{Y(X_4,X_5,X_6,X_7)|X_1,X_2,X_3}^2 \quad \rho_{Y(X_1,X_2,X_3,X_4,X_5,X_6,X_7)}^2$$

d Estimate the ratio σ_B/σ_A.

4.10
Comparing Two Multiple Regression Models (Nonnested Case)

The problem of comparing the performances of two regression functions for predicting Y was considered in Section 4.9 for the situation where the set of predictor variables $X_1, \ldots, X_m$ used in one regression function $\mu_Y^{(B)}(x_1, \ldots, x_m)$ is a subset of the predictor variables $X_1, \ldots, X_k$ used in the other regression function $\mu_Y^{(A)}(x_1, \ldots, x_k)$. For brevity we referred to these two regression functions as model B and model A, respectively, and the fact that the set of predictor variables in model B is a subset of the set of predictor variables in model A was expressed by saying that *model B is nested in model A*.

However, in practice, situations arise where an investigator is faced with choosing between two models when neither model is nested in the other; i.e., there are predictor variables in one set that are not in the other set and vice versa, although some of the predictor variables may belong to both sets. To avoid any possible confusion in the notation, we use superscripts A and B to represent the predictor variables in the two sets, respectively. In this section, model A may not contain the full set of k predictors but contains only $r(r < k)$ of the k predictors $X_1, \ldots, X_k$. Thus the r predictor variables in model A will be denoted by $X_1^A, \ldots, X_r^A$ and the m predictor variables in model B will be denoted by $X_1^B, \ldots, X_m^B$. To reiterate, *there are predictor variables in model A that are not in model B, and there are predictor variables in model B that are not in model A. Additionally, there **may** be some predictor variables that occur in both model A and model B.* We denote the union of the two collections of predictor variables, $X_1^A, \ldots, X_r^A$ and $X_1^B, \ldots, X_m^B$, by the collection $X_1, \ldots, X_k$; i.e., $X_1, \ldots, X_k$ include all the predictor variables that are in model A or model B. The choice of a model (model A or model B) for predicting Y depends not only on how good one prediction function is relative to the other but also on many other things, including costs involved in using one set of predictor variables relative to the other set, etc. We discuss this problem of comparing two nonnested models under the assumption given in Box 4.10.1 below.

B O X **4.10.1**   **Assumptions for Comparing Two Nonnested Models**

Throughout this section we suppose that assumptions (B) for regression hold for $\{(Y, X_1, \ldots, X_k)\}$. Consequently, assumptions (B) also hold for $\{(Y, X_1^A, \ldots, X_r^A)\}$ and $\{(Y, X_1^B, \ldots, X_m^B)\}$. In particular, the sample data are obtained by simple random sampling.

As a consequence of this assumption, the regression function corresponding to model A is of the form

$$\mu_Y^{(A)}(x_1^A, \ldots, x_r^A) = \beta_0^A + \beta_1^A x_1^A + \cdots + \beta_r^A x_r^A \qquad \text{(4.10.1)}$$

while the regression function corresponding to model B is of the form

$$\mu_Y^{(B)}(x_1^B, \ldots, x_m^B) = \beta_0^B + \beta_1^B x_1^B + \cdots + \beta_m^B x_m^B \qquad \text{(4.10.2)}$$

E X A M P L E **4.10.1**

To illustrate the notation of this section, consider the setup in Example 4.4.2. Suppose the director of admissions wishes to compare the performances of the regression function of Y on X_3 (HSmath) and X_4 (HSenglish) (call this model A) and the regression function of Y on X_1 (SATmath) and X_2 (SATverbal) (call this model B) as predictors of Y. Since the assumption in Box 4.10.1 implies that both three-variable populations, $\{(Y, X_3, X_4)\}$ and $\{(Y, X_1, X_2)\}$, satisfy (population) assumptions (B), the regression function of Y on X_3, X_4 is of the form

$$\mu_Y^{(A)}(x_3, x_4) = \beta_0^A + \beta_3^A x_3 + \beta_4^A x_4$$

and the regression function of Y on X_1, X_2 is of the form

$$\mu_Y^{(B)}(x_1, x_2) = \beta_0^B + \beta_1^B x_1 + \beta_2^B x_2$$

Here the value of r is 2 and the value of m is also 2. The predictor variables in model A are $\{X_1^A, \ldots, X_r^A\} = \{X_3, X_4\}$, while the predictor variables in model B are $\{X_1^B, \ldots, X_m^B\} = \{X_1, X_2\}$. Since neither model is *nested* in the other, the discussions of Section 4.9 do not apply. You must use the discussions in this section. Observe that the union of the two sets of predictor variables has $k = 4$ predictors and is the set $\{X_1, X_2, X_3, X_4\}$.

If on the other hand the director of admissions were interested in comparing the model

$$\mu_Y^{(A)}(x_1, x_3, x_4) = \beta_0^A + \beta_1^A x_1 + \beta_3^A x_3 + \beta_4^A x_4$$

with the model

$$\mu_Y^{(B)}(x_1, x_2) = \beta_0^B + \beta_1^B x_1 + \beta_2^B x_2$$

then we would have $r = 3$ and $m = 2$. The predictor variables in model A would be $\{X_1^A, \ldots, X_r^A\} = \{X_1, X_3, X_4\}$, while those in model B would be $\{X_1^B, \ldots, X_m^B\} = \{X_1, X_2\}$. The union of the two sets of predictor variables is the set $\{X_1, X_2, X_3, X_4\}$ so that $k = 4$. Note that X_1 appears in both sets of variables. ∎

For notational convenience, let σ_A denote $\sigma_{Y|X_1^A, \ldots, X_r^A}$, the subpopulation standard deviation for model A, and let σ_B denote $\sigma_{Y|X_1^B, \ldots, X_m^B}$, the subpopulation standard deviation for model B. If $\sigma_A < \sigma_B$, then the regression function $\mu_Y^{(A)}(x_1^A, \ldots, x_r^A)$ is a better predictor of Y than the regression function $\mu_Y^{(B)}(x_1^B, \ldots, x_m^B)$. On the other hand, if $\sigma_A > \sigma_B$, then the regression function $\mu_Y^{(B)}(x_1^B, \ldots, X_m^B)$ is a better predictor of Y than the regression function $\mu_Y^{(A)}(x_1^A, \ldots, x_r^A)$. In practice it is highly unlikely that the two standard deviations will be exactly equal. So the investigator is actually interested in knowing *how much bigger or smaller* σ_A is than σ_B. The obvious approach is to examine the point estimates and confidence intervals for both σ_A and σ_B. However, one could additionally calculate a confidence interval for the *ratio* σ_B/σ_A and make a practical decision based on these results.

A procedure for computing a two-sided confidence interval for σ_B/σ_A with confidence coefficient greater than or equal to $1 - \alpha$ is given in Box 4.10.2. This procedure uses the Bonferroni method.

B O X **4.10.2** **Two-Sided Confidence Intervals for σ_B/σ_A Using the Bonferroni Method**

1   Let r and m denote the number of predictors in models A and B, respectively, and let n be the number of sample observations.

2   Regress Y on the predictors in model A and obtain the sum of squared errors $SSE(A)$.

3   Regress Y on the predictors in model B and obtain the sum of squared errors $SSE(B)$.

4 Compute a $1 - \alpha/2$ two-sided confidence interval for σ_A, which is given by

$$C[L_A \leq \sigma_A \leq U_A] = 1 - \alpha/2$$

where

$$L_A = \sqrt{\frac{SSE(A)}{\chi^2_{1-\alpha/4:n-r-1}}} \quad \text{and} \quad U_A = \sqrt{\frac{SSE(A)}{\chi^2_{\alpha/4:n-r-1}}}$$

5 Compute a $1 - \alpha/2$ two-sided confidence interval for σ_B, which is given by

$$C[L_B \leq \sigma_B \leq U_B] = 1 - \alpha/2$$

where

$$L_B = \sqrt{\frac{SSE(B)}{\chi^2_{1-\alpha/4:n-m-1}}} \quad \text{and} \quad U_B = \sqrt{\frac{SSE(B)}{\chi^2_{\alpha/4:n-m-1}}}$$

6 We have the following confidence statement for σ_B/σ_A:

$$C[L_B/U_A \leq \sigma_B/\sigma_A \leq U_B/L_A] \geq 1 - \alpha$$

7 Equivalently, a confidence statement for σ_A/σ_B is given by

$$C[L_A/U_B \leq \sigma_A/\sigma_B \leq U_A/L_B] \geq 1 - \alpha$$

8 If only a one-sided confidence bound is needed, then either

$$C[L_B/U_A \leq \sigma_B/\sigma_A] \geq 1 - \alpha/2$$

or

$$C[\sigma_B/\sigma_A \leq U_B/L_A] \geq 1 - \alpha/2$$

may be used.

We illustrate the procedure described in Box 4.10.2 in Example 4.10.2.

E X A M P L E   4.10.2

In Example 4.4.2, suppose the director of admissions wants to determine how much better, or worse, the set of predictors $\{X_3, X_4\}$ is for predicting $Y =$ GPA than the set of predictors $\{X_1, X_2\}$. As required in Box 4.10.1, we suppose that (population) assumptions (B) hold for the five-variable population $\{(Y, X_1, X_2, X_3, X_4)\}$. Hence the two regression functions being compared are of the form

$$\mu_Y^{(A)}(x_3, x_4) = \beta_0^A + \beta_3^A x_3 + \beta_4^A x_4 \tag{4.10.3}$$

and

$$\mu_Y^{(B)}(x_1, x_2) = \beta_0^B + \beta_1^B x_1 + \beta_2^B x_2 \tag{4.10.4}$$

For this purpose we compute a confidence interval for the ratio σ_B/σ_A. If model A is judged to be better than model B, then the director of admissions will *consider* the possibility of not requiring applicants to submit their SAT scores (although such a decision would perhaps be based on more elaborate studies).

By examining Box 4.10.2, we see that to compute confidence intervals for σ_B/σ_A we need $SSE(X_3, X_4)$ and $SSE(X_1, X_2)$, the sum of squared errors for the models in (4.10.3) and (4.10.4). These sums of squares can be obtained from the ANOVA tables for these models. We have obtained them using SAS and they are given in Exhibit 4.10.1 and Exhibit 4.10.2, respectively.

We obtain $SSE(A) = SSE(X_3, X_4) = 2.41803$, $SSE(B) = SSE(X_1, X_2) = 1.38838$, $\hat{\sigma}_A = 0.37714$, and $\hat{\sigma}_B = 0.28578$. The table-values needed to calculate L_A, U_A, L_B, and U_B in Box 4.10.2 can be obtained from Table T-3 in Appendix T, although interpolation may be required for some values of α and/or degrees of freedom.

In this example, for a 95% two-sided confidence interval for σ_B/σ_A, we need the values of $\chi^2_{\alpha/4:n-r-1} = \chi^2_{0.0125:17}$ and $\chi^2_{1-\alpha/4:n-r-1} = \chi^2_{0.9875:17}$ for computing L_A and U_A, respectively. These are also the table-values for computing L_B and U_B in this example because $n - m - 1$ is also equal to 17. These table-values (obtained from SAS) are

$$\chi^2_{0.9875:17} = 32.644 \qquad \text{and} \qquad \chi^2_{0.0125:17} = 6.664$$

so the value of L_A is 0.272, and the value of U_A is 0.602. We also have $L_B = 0.206$ and $U_B = 0.456$. From these we obtain the following confidence statement:

$$C[0.342 \leq \sigma_B/\sigma_A \leq 1.678] \geq 0.95 \qquad \text{(4.10.5)}$$

Thus there is no clear-cut evidence indicating the superiority of one model over the other. The director of admissions will have to decide, based on the confidence interval in (4.10.5), either

- that the ratio σ_B/σ_A is close enough to 1 that for this problem the two models can be considered to be equally good for predicting GPA, or

- that the sample size is not large enough to determine, sufficiently precisely, the amount by which σ_A is larger or smaller than σ_B, and additional data are required for making a decision. ■

E X H I B I T 4.10.1
SAS Output for Model A in Example 4.10.2

The SAS System   0:00 Saturday,   Jan 1, 1994

Dependent Variable: GPA

Analysis of Variance

Source	DF	Sum of Squares	Mean Square	F Value	Prob>F
Model	2	4.92779	2.46389	17.322	0.0001
Error	17	2.41803	0.14224		
C Total	19	7.34582			

Root MSE	0.37714	R-square	0.6708
Dep Mean	2.59300	Adj R-sq	0.6321
C.V.	14.54467		

Parameter Estimates

Variable	DF	Parameter Estimate	Standard Error	T for H0: Parameter=0	Prob > |T|
INTERCEP	1	-0.340013	0.56441815	-0.602	0.5548
HSMATH	1	0.417116	0.10544213	3.956	0.0010
HSENGL	1	0.579021	0.18573410	3.117	0.0063

E X H I B I T **4.10.2**
SAS Output for Model B in Example 4.10.2

The SAS System 0:00 Saturday, Jan 1, 1994

Dependent Variable: GPA

Analysis of Variance

Source	DF	Sum of Squares	Mean Square	F Value	Prob>F
Model	2	5.95744	2.97872	36.473	0.0001
Error	17	1.38838	0.08167		
C Total	19	7.34582			

Root MSE	0.28578	R-square	0.8110	
Dep Mean	2.59300	Adj R-sq	0.7888	
C.V.	11.02117			

Parameter Estimates

Variable	DF	Parameter Estimate	Standard Error	T for H0: Parameter=0	Prob > |T|
INTERCEP	1	0.507142	0.26672665	1.901	0.0743
SATMATH	1	0.002606	0.00044323	5.879	0.0001
SATVERB	1	0.001574	0.00055547	2.834	0.0115

Problems 4.10

4.10.1   Consider the GPA data of Example 4.4.2 where we presume that assumptions (B) are valid. Exhibits 4.10.3–4.10.6 give MINITAB computer outputs for the following regression functions.

1   Model A: $\mu_Y^{(A)}(x_1) = \beta_0^A + \beta_1^A x_1$.

2   Model B: $\mu_Y^{(B)}(x_2) = \beta_0^B + \beta_2^B x_2$.

3   Model C: $\mu_Y^{(C)}(x_3) = \beta_0^C + \beta_3^C x_3$.

4   Model D: $\mu_Y^{(D)}(x_4) = \beta_0^D + \beta_4^D x_4$.

a   The director of admissions asks you to determine how much better (or worse) X_1 (SATmath) is as a predictor of Y than X_3 (HSmath) is as a predictor of Y.

   i   Obtain 95% confidence intervals for $\sigma_{Y|X_1}$ and $\sigma_{Y|X_3}$.

   ii   Obtain a two-sided 90% confidence interval for $\sigma_{Y|X_1}/\sigma_{Y|X_3}$.

b   The director of admissions asks you to determine how much better (or worse) X_2 (SATverbal) is as a predictor of Y than X_4 (HSenglish) is as a predictor of Y.

   i   Obtain 95% confidence intervals for $\sigma_{Y|X_2}$ and $\sigma_{Y|X_4}$.

   ii   Obtain a two-sided 90% confidence interval for $\sigma_{Y|X_2}/\sigma_{Y|X_4}$.

c   Write a short report to the director of admissions summarizing the results in (a) and (b).

E X H I B I T   4.10.3
MINITAB Output for Problem 4.10.1
Regression of $Y = $ GPA on $X_1 = $ SATmath

```
The regression equation is
GPA = 0.967 + 0.00318 SATmath


Predictor          Coef        Stdev     t-ratio          p
Constant         0.9670       0.2496        3.87      0.001
SATmath       0.0031783    0.0004652        6.83      0.000


s = 0.3370       R-sq = 72.2%       R-sq(adj) = 70.6%


Analysis of Variance

SOURCE           DF           SS          MS          F          p
Regression        1       5.3015      5.3015      46.68      0.000
Error            18       2.0443      0.1136
Total            19       7.3458
```

E X H I B I T 4.10.4

MINITAB Output for Problem 4.10.1

Regression of $Y = $ GPA on $X_2 = $ SATverb

```
The regression equation is
GPA = 1.13 + 0.00306 SATverb
```

Predictor	Coef	Stdev	t-ratio	p
Constant	1.1281	0.4145	2.72	0.014
SATverb	0.0030630	0.0008367	3.66	0.002

```
s = 0.4837     R-sq = 42.7%     R-sq(adj) = 39.5%
```

Analysis of Variance

SOURCE	DF	SS	MS	F	p
Regression	1	3.1350	3.1350	13.40	0.002
Error	18	4.2108	0.2339		
Total	19	7.3458			

E X H I B I T 4.10.5

MINITAB Output for Problem 4.10.1

Regression of $Y = $ GPA on $X_3 = $ HSmath

```
The regression equation is
GPA = 1.15 + 0.507 HSmath
```

Predictor	Coef	Stdev	t-ratio	p
Constant	1.1473	0.3675	3.12	0.006
HSmath	0.5066	0.1236	4.10	0.001

```
s = 0.4595     R-sq = 48.3%     R-sq(adj) = 45.4%
```

Analysis of Variance

SOURCE	DF	SS	MS	F	p
Regression	1	3.5454	3.5454	16.79	0.001
Error	18	3.8004	0.2111		
Total	19	7.3458			

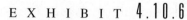

E X H I B I T   4.10.6

MINITAB Output for Problem 4.10.1

Regression of $Y = $ GPA on $X_4 = $ HSenglish

```
The regression equation is
GPA = 0.249 + 0.779 HSengl


Predictor        Coef        Stdev      t-ratio          p
Constant       0.2487       0.7332         0.34      0.738
HSengl         0.7790       0.2407         3.24      0.005


s = 0.5079        R-sq = 36.8%       R-sq(adj) = 33.3%


Analysis of Variance


SOURCE         DF          SS           MS         F         p
Regression      1      2.7019       2.7019     10.47     0.005
Error          18      4.6439       0.2580
Total          19      7.3458
```

4.11
Lack-of-Fit Analysis

In Sections 4.9 and 4.10 we discussed the important problem of determining which of *two* regression functions is better for predicting Y. The two regression functions being compared used different sets of predictor variables, say A and B. Both model A and model B were multiple linear regression models. For instance, in Example 4.9.4 we compared the regression function

$$\mu_Y^{(A)}(x_1, x_2, x_3, x_4) = \beta_0^A + \beta_1^A x_1 + \beta_2^A x_2 + \beta_3^A x_3 + \beta_4^A x_4$$

with the regression function

$$\mu_Y^{(B)}(x_3, x_4) = \beta_0^B + \beta_3^B x_3 + \beta_4^B x_4$$

Here A is the set $\{X_1, X_2, X_3, X_4\}$, and B is the set $\{X_3, X_4\}$ (B is nested in A). In Example 4.10.2 we compared the regression function

$$\mu_Y^{(A)}(x_3, x_4) = \beta_0^A + \beta_3^A x_3 + \beta_4^A x_4$$

with the regression function

$$\mu_Y^{(B)}(x_1, x_2) = \beta_0^B + \beta_1^B x_1 + \beta_2^B x_2$$

Here A is the set $\{X_3, X_4\}$ and B is the set $\{X_1, X_2\}$ (neither set is nested in the other). In both examples the regression functions being compared have known forms; in fact, they are both multiple *linear* regression functions.

In many real problems investigators do not know the population *regression* function $\mu_Y(x)$ or even its *form*. In such instances the investigator may want to find a *prediction* function whose mathematical form is reasonably simple and which approximates the true, but unknown, regression function adequately for the problem at hand. It is very unlikely that population regression functions are simple functions of the form $\beta_0 + \beta_1 x$ or $\beta_0 + \beta_1 x^2$, etc., but it is often the case that they can be well approximated by simple functions such as these. Figures 4.11.1–4.11.3 illustrate the point.

F I G U R E 4.11.1

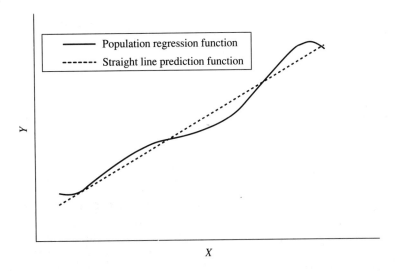

F I G U R E 4.11.2

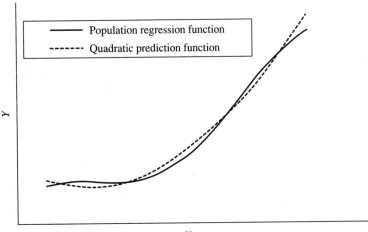

F I G U R E   4.11.3

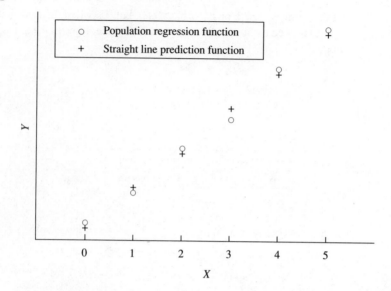

In Figure 4.11.1 a *linear* prediction function may be an adequate approximation to the regression function, and in Figure 4.11.2 a *quadratic* prediction model may be a satisfactory approximation to the regression function. In Figure 4.11.3 the population regression function is defined only at isolated values of the predictor variable X (for example when X is the number of children in a household), so the graph of the population regression function is not a continuous curve. In any event, since the population regression function $\mu_Y(x)$ can perhaps never be known exactly, we replace it with a function $P_Y(x)$ whose *form* is known and which is an adequate approximation to the population regression function. In these cases we are interested in the difference

$$\mu_Y(x) - P_Y(x)$$

to determine if indeed the function $P_Y(x)$ is an adequate approximation to the true unknown population regression function $\mu_Y(x)$, at least for the X values of interest in the investigation. When sample data are available, plots of the sample data are often useful in developing suitable classes of functions to consider in an attempt to find such an approximation to the regression function. We illustrate with two examples.

E X A M P L E   4.11.1

In using a regression function for predicting blood pressure (Y) as a function of age (X) for men between the ages of 20 and 50 of a certain ethnic background, an investigator is not sure what the regression function is. The conjecture is that the function $P_Y(x) = \beta_0 + \beta_1 x$ will provide an adequate approximation, and data are collected to check whether this is indeed the case.   ■

EXAMPLE 4.11.2

An investigator wants to find a prediction function for predicting Y, the first-year maintenance cost of a new car, using the predictor variable X, the miles the car is driven the first year. He has reason to believe that the regression function $\mu_Y(x)$ of Y on X can be adequately approximated by a quadratic function of X of the form

$$P_Y(x) = \beta_0 + \beta_1 x + \beta_2 x^2$$

Data are collected to see whether this is in fact an adequate model to use for predicting Y, the first-year maintenance cost. ∎

When there is only one predictor variable, straight line functions of the form

$$P_Y(x) = \beta_0 + \beta_1 x$$

are useful in many situations. If the plot of the sample data suggests that a straight line function may serve as a good approximation to the regression function, then an obvious question of interest to the investigator is: What is the difference between the true unknown *regression* function $\mu_Y(x)$ and the proposed straight line function $P_Y(x) = \beta_0 + \beta_1 x$?

In the rest of this section we explain a procedure for answering this question.

Lack-of-Fit Analysis for Straight Line Prediction Functions

Suppose that an investigator who is interested in studying the relationship between a response variable Y and a predictor variable X for $a \leq X \leq b$ obtains a sample of size n by simple random sampling or by sampling with preselected X. Let $x_1, \ldots, x_m$ denote the distinct x values in the chosen sample. Suppose (conceptually) that the entire subpopulation corresponding to each x_i in the sample is available. We can then calculate the corresponding subpopulation means $\mu_Y(x_1), \ldots, \mu_Y(x_m)$, which we denote by $\mu_1, \ldots, \mu_m$, respectively, for ease of notation. Figure 4.11.1 shows one example of what the graph for $\mu_Y(x)$ may look like; it is obtained by plotting the mean of Y for each subpopulation corresponding to each distinct value of X in the sample. These subpopulation means μ_i need not lie exactly on a straight line, but they may lie approximately on a straight line (e.g., Figures 4.11.1 and 4.11.3). If this is so, it seems reasonable to use the least squares straight line fitted to the points $(x_1, \mu_1), \ldots, (x_m, \mu_m)$ as an approximation to the true regression function $\mu_Y(x)$. We denote this least squares straight line function by

$$P_Y(x) = \beta_0 + \beta_1 x \tag{4.11.1}$$

where

$$\beta_1 = \frac{\sum_{i=1}^m (x_i - \bar{x})(\mu_i - \bar{\mu})}{\sum_{i=1}^m (x_i - \bar{x})^2} \quad \text{and} \quad \beta_0 = \bar{\mu} - \beta_1 \bar{x} \tag{4.11.2}$$

*The function in (4.11.1), in general, is not the population **regression** function of Y on X (unless it so happens that the population regression function is truly a straight line), but it is the least squares straight line approximation to the regression function* at the points $x_1, \ldots, x_m$. We refer to the function $P_Y(x)$ as the *proposed prediction function*. Because the μ_i are not known, β_0 and β_1 are also unknown, but we can estimate them by collecting appropriate sample data and using the fact that $\bar{y}_i$ is an estimate of μ_i for $i = 1, \ldots, m$.

Let $\theta_i, i = 1, \ldots, m$, denote the differences between the unknown population regression function $\mu_Y(x)$ and the proposed prediction function $P_Y(x)$ at the sample points $x_1, \ldots, x_m$, respectively. Thus

$$\theta_i = \mu_Y(x_i) - P_Y(x_i) = \mu_Y(x_i) - [\beta_0 + \beta_1 x_i] \qquad \textbf{(4.11.3)}$$

The quantities $\theta_1, \ldots, \theta_m$ are called lack-of-fit constants. If the θ_i are small enough so that the investigator can regard them as being negligible for the problem under consideration, then the proposed straight line function will be an adequate approximation to the unknown population regression function (which is the *best* prediction function), at least for the values of X in the sample. For this reason we want to investigate the lack-of-fit constants $\theta_1, \ldots, \theta_m$.

Remarks   Traditionally, statisticians and practitioners have examined the differences $\mu_Y(x) - P_Y(x)$ by performing the following hypothesis test:

$$\text{NH:} \quad \mu_Y(x) = \beta_0 + \beta_1 x \qquad \text{against} \qquad \text{AH:} \quad \mu_Y(x) \neq \beta_0 + \beta_1 x \qquad \textbf{(4.11.4)}$$

If NH is rejected at level α, then the investigator might conclude that the *proposed* straight line prediction function $P_Y(x) = \beta_0 + \beta_1 x$ is not the *true* regression function. The test in (4.11.4) is often called a **lack-of-fit test**. The difficulty with this test is that the result of the test does not shed any light on the actual magnitude of the difference between the proposed prediction function and the unknown population regression function. The differences (the values of the θ_i), even if detected by the test (i.e., if NH is rejected), may be negligible for the practical problem being investigated. On the other hand, if the NH is not rejected, it does not imply that the proposed prediction function is indeed the true regression function, or even that it is a good prediction function for the problem. Thus a hypothesis test for *lack-of-fit* is of very little value when practical decisions are to be made. We therefore proceed to describe methods for obtaining point and confidence interval estimates of the lack-of-fit constants $\theta_1, \ldots, \theta_m$. This information can be used by the investigator to decide whether the differences between the proposed prediction function $P_Y(x) = \beta_0 + \beta_1 x$ and the true, but unknown, regression function $\mu_Y(x)$ are negligible for all practical purposes for the problem under study.

Estimation and Confidence Intervals for Lack-of-Fit Constants for Straight Line Prediction Functions

To obtain point and confidence interval estimates for the lack-of-fit constants $\theta_1, \ldots, \theta_m$, we must obtain sample data from the two-variable population $\{(Y, X)\}$ and make some assumptions. We make the assumptions given in Box 4.11.1.

B O X 4.11.1 **(Straight Line) Lack-of-Fit Assumptions**

1 The form of the regression function $\mu_Y(x)$ of Y on X is unknown.

2 The investigator is interested in determining whether the straight line function $P_Y(x) = \beta_0 + \beta_1 x$, which is the least squares approximation to the true regression function at m ($m > 2$) distinct points $x_1, \ldots, x_m$ between a and b (see Figures 4.11.1 and 4.11.3), provides a good approximation to $\mu_Y(x)$. The points $(x_1, \ldots, x_m)$ are preselected by the investigator.

3 The subpopulation of Y values determined by $X = x_i$ is a Gaussian population with mean $\mu_Y(x_i)$ (written μ_i for short) and standard deviation $\sigma_Y(x_i)$ (written simply as σ_i), both of which are unknown.

4 $\sigma_1 = \sigma_2 = \cdots = \sigma_m$ and their common value is denoted by σ.

5 From each subpopulation in (3), determined by $X = x_i$, a simple random sample of n_i items is selected. The Y values of these sample items are denoted by $y_{i,1}, y_{i,2}, \ldots, y_{i,n_i}$. Furthermore, $n_i \geq 1$ for all i, and $n_i > 1$ for at least one value of i. The sample sizes $n_1, \ldots, n_m$ are selected by the investigator. The total number of observations in the sample is $n = n_1 + \cdots + n_m$.

Remark Although part (5) requires only that $n_i > 1$ for *at least* one value of i, it is desirable to choose the values of n_i so that the quantity $\sum_{i=1}^{m}(n_i - 1) = (n - m)$ is not too small because, as we see later, the estimate of σ is based on $(n - m)$ degrees of freedom.

T A B L E 4.11.1

X Values	Y Values	Mean of Y Values	Estimate of μ_i	Estimate of σ_i
x_1	$y_{1,1}, y_{1,2}, \ldots, y_{1,n_1}$	$\bar{y}_1 = \frac{1}{n_1}\sum_{j=1}^{n_1} y_{1,j}$	$\hat{\mu}_1 = \bar{y}_1$	$\hat{\sigma}_1 = \sqrt{\frac{\sum_{j=1}^{n_1}(y_{1,j}-\bar{y}_1)^2}{(n_1-1)}}$
x_2	$y_{2,1}, y_{2,2}, \ldots, y_{2,n_2}$	$\bar{y}_2 = \frac{1}{n_2}\sum_{j=1}^{n_2} y_{2,j}$	$\hat{\mu}_2 = \bar{y}_2$	$\hat{\sigma}_2 = \sqrt{\frac{\sum_{j=1}^{n_2}(y_{2,j}-\bar{y}_2)^2}{(n_2-1)}}$
$\vdots$	$\vdots$	$\vdots$	$\vdots$	$\vdots$
x_m	$y_{m,1}, y_{m,2}, \ldots, y_{m,n_m}$	$\bar{y}_m = \frac{1}{n_m}\sum_{j=1}^{n_m} y_{m,j}$	$\hat{\mu}_m = \bar{y}_m$	$\hat{\sigma}_m = \sqrt{\frac{\sum_{j=1}^{n_m}(y_{m,j}-\bar{y}_m)^2}{(n_m-1)}}$

A schematic representation of the sample data is given in Table 4.11.1. This table also displays *estimates* of the subpopulation means μ_i and standard deviations σ_i.

The point estimates of β_0 and β_1 we use are (substituting y_i for μ_i in (4.11.2))

$$\hat{\beta}_1 = \frac{\sum_{i=1}^{m}(x_i - \bar{x})(\bar{y}_i - \bar{y})}{\sum_{i=1}^{m}(x_i - \bar{x})^2} \qquad \text{and} \qquad \hat{\beta}_0 = \bar{y} - \hat{\beta}_1\bar{x} \qquad \text{(4.11.5)}$$

where

$$\bar{y}_i = \frac{1}{n_i}\sum_{j=1}^{n_i} y_{i,j} \qquad \bar{y} = \frac{\bar{y}_1 + \cdots + \bar{y}_m}{m} \qquad \bar{x} = \frac{x_1 + \cdots + x_m}{m} \qquad \text{(4.11.6)}$$

The point estimate of

$$P_Y(x) = \beta_0 + \beta_1 x$$

is

$$\hat{P}_Y(x) = \hat{\beta}_0 + \hat{\beta}_1 x$$

and the point estimate of θ_i is

$$\hat{\theta}_i = \bar{y}_i - \hat{P}_Y(x_i) = \bar{y}_i - \hat{\beta}_0 - \hat{\beta}_1 x_i \quad \text{for} \quad i = 1, 2, \ldots, m \qquad \text{(4.11.7)}$$

where $\hat{\beta}_0$ and $\hat{\beta}_1$ are computed by (4.11.5); i.e., by regressing the means $\bar{y}_1, \ldots, \bar{y}_m$ on $x_1, \ldots, x_m$, the distinct x values. Thus

$$\hat{\beta} = (A^T A)^{-1} A^T \bar{y} \qquad \text{(4.11.8)}$$

where

$$\hat{\beta} = \begin{bmatrix} \hat{\beta}_0 \\ \hat{\beta}_1 \end{bmatrix} \qquad A = \begin{bmatrix} 1 & x_1 \\ 1 & x_2 \\ \vdots & \vdots \\ 1 & x_m \end{bmatrix} \qquad \bar{y} = \begin{bmatrix} \bar{y}_1 \\ \bar{y}_2 \\ \vdots \\ \bar{y}_m \end{bmatrix} \qquad \text{(4.11.9)}$$

Note If we regress the means $\bar{y}_i$ on the distinct x_i values for $i = 1, \ldots, m$, then the $\hat{\theta}_i$ in (4.11.7) are obtained by the same formulas by which the residuals $\hat{e}_i$ were computed in (4.4.12).

We can obtain confidence intervals for $\theta_1, \ldots, \theta_m$ such that *all m intervals are simultaneously correct with at least* $(1 - \alpha)$ *confidence*. The confidence interval for θ_i is of the same form as in (4.6.1), viz.,

$$\hat{\theta}_i - \text{(table-value)}\, SE(\hat{\theta}_i) \leq \theta_i \leq \hat{\theta}_i + \text{(table-value)}\, SE(\hat{\theta}_i) \qquad \text{(4.11.10)}$$

where

$$SE(\hat{\theta}_i) = \hat{\sigma}\sqrt{v_{ii}} \qquad \text{(4.11.11)}$$

The quantity v_{ii} in (4.11.11) is the ith diagonal element of the matrix V, which is given by

$$V = QDQ^T \qquad \text{(4.11.12)}$$

with

$$Q = I - A(A^T A)^{-1} A^T \qquad \text{(4.11.13)}$$

where I is an $m \times m$ identity matrix and

$$D = \begin{bmatrix} 1/n_1 & 0 & 0\ldots & 0 \\ 0 & 1/n_2 & 0\ldots & 0 \\ \vdots & \vdots & \vdots & \vdots \\ 0 & 0 & 0 & 1/n_m \end{bmatrix} \qquad \text{(4.11.14)}$$

Thus D is a diagonal matrix with the ith diagonal element equal to $1/n_i$. Also,

$$\hat{\sigma}^2 = \frac{\sum (n_i - 1)\hat{\sigma}_i^2}{\sum (n_i - 1)} \qquad \text{(4.11.15)}$$

where

$$\hat{\sigma}_i^2 = \frac{1}{(n_i - 1)} \sum_{i=1}^{n_i} (y_{i,j} - \bar{y}_i)^2 \qquad \text{(4.11.16)}$$

for those x_i with $n_i > 1$ (we take $\hat{\sigma}_i^2$ to be zero if $n_i = 1$). The table-value is the smaller of the two values

$$t_{1-\alpha/2m:dfd} \qquad \text{and} \qquad \sqrt{(dfn)F_{1-\alpha:dfn,dfd}}$$

with $dfn = m - 2 =$ degrees of freedom for the numerator, and $dfd = n - m =$ degrees of freedom for the denominator. The table-value $t_{1-\alpha/2m:dfd}$ can be obtained from Table T-4 in Appendix T, and $F_{1-\alpha:dfn,dfd}$ can be obtained from Table T-5, also in Appendix T.

The quantity $\sum (n_i - 1)\hat{\sigma}_i^2$ in the numerator of (4.11.15) is usually referred to as the **sum of squares for pure error** (denoted by $SS(Pure\ error)$), and the quantity $\sum (n_i - 1)$ in the denominator of (4.11.15) is called the **degrees of freedom for pure error** denoted by $df(Pure\ error)$. The estimate of σ^2 is called the **mean square for pure error** and is denoted by $MS(Pure\ error)$. Thus we have

$$SS(Pure\ error) = \frac{\sum_{i=1}^{m} \sum_{j=1}^{n_i} (y_{i,j} - \bar{y}_i)^2}{\sum_{i=1}^{m} (n_i - 1)} = (n - m)\hat{\sigma}^2 \qquad \text{(4.11.17)}$$

$$df(Pure\ error) = n - m \qquad \text{(4.11.18)}$$

and

$$MS(Pure\ error) = \hat{\sigma}^2 \qquad \text{(4.11.19)}$$

We illustrate these computations in Example 4.11.3.

E X A M P L E **4.11.3**

Consider Example 4.11.1 where an investigator is studying the relationship between age and blood pressure of men of certain ethnic background. Suppose the investigator preselects five distinct values of age, chooses several men using simple random sampling from each of the five selected age groups, and records their ages (x_i values) and blood pressures (y_i values). Using these sample data, we want to check whether $P_Y(x) = \beta_0 + \beta_1 x$ is close enough to the unknown regression function $\mu_Y(x)$ so that $P_Y(x)$ can be used to predict blood pressure using age. The data appear in Table 4.11.2 and in the file **bp.dat** on the data disk. There are five subpopulations represented in the sample, so $m = 5$. Also

$$n = 25 \qquad n_1 = 4 \qquad n_2 = 5 \qquad n_3 = 6 \qquad n_4 = 6 \qquad n_5 = 4$$

The distinct values of x are 25, 30, 35, 40, and 45. The estimated values of the subpopulation means corresponding to the distinct values of x are

$$\bar{y}_1 = 108.250 \quad \bar{y}_2 = 114.400 \quad \bar{y}_3 = 121.833 \quad \bar{y}_4 = 134.167 \quad \bar{y}_5 = 142.000$$

respectively.

The estimated values for the corresponding subpopulation standard deviations are $\hat{\sigma}_1 = 3.775$, $\hat{\sigma}_2 = 2.191$, $\hat{\sigma}_3 = 5.456$, $\hat{\sigma}_4 = 2.401$, and $\hat{\sigma}_5 = 3.830$. So the estimate of σ is

$$
\hat{\sigma} = \sqrt{\frac{\sum_{i=1}^{5}(n_i - 1)\hat{\sigma}_i^2}{\sum_{i=1}^{5}(n_i - 1)}}
$$

$$
= \sqrt{\frac{3(3.775)^2 + 4(2.191)^2 + 5(5.456)^2 + 5(2.401)^2 + 3(3.830)^2}{3 + 4 + 5 + 5 + 3}} = 3.766
$$

T A B L E **4.11.2**
Blood Pressure Data

Item Y	Blood Pressure X	Age	Item Y	Blood Pressure X	Age
1	104	25	14	114	35
2	107	25	15	121	35
3	113	25	16	132	40
4	109	25	17	132	40
5	114	30	18	133	40
6	114	30	19	134	40
7	114	30	20	136	40
8	118	30	21	138	40
9	112	30	22	141	45
10	127	35	23	145	45
11	125	35	24	145	45
12	127	35	25	137	45
13	117	35			

The A matrix and the $\bar{y}$ vector in (4.11.9) are

$$A = \begin{bmatrix} 1 & 25 \\ 1 & 30 \\ 1 & 35 \\ 1 & 40 \\ 1 & 45 \end{bmatrix} \qquad \bar{y} = \begin{bmatrix} 108.250 \\ 114.400 \\ 121.833 \\ 134.167 \\ 142.000 \end{bmatrix} \qquad \textbf{(4.11.20)}$$

A MINITAB output for calculating the regression of $\bar{y}$ on the distinct values of X appears in Exhibit 4.11.1.

From the residuals in (4.11.21) we note that $\hat{\theta}_i$ for $i = 1, \ldots, 5$ are

$$\hat{\theta}_1 = 1.57340 \quad \hat{\theta}_2 = -1.00330 \quad \hat{\theta}_3 = -2.29700 \quad \hat{\theta}_4 = 1.31030$$
$$\hat{\theta}_5 = 0.41660$$

We obtain v_{ii} for $i = 1, \ldots, 5$ from the diagonal elements of the matrix V in Exhibit 4.11.1, which was computed using the formulas in (4.11.12)–(4.11.14). We get

$$v_{11} = 0.088667 \quad v_{22} = 0.146333 \quad v_{33} = 0.141333 \quad v_{44} = 0.130333$$
$$v_{55} = 0.083333$$

To calculate the standard errors of the $\hat{\theta}_i$ we use the formula in (4.11.11) and get

$$SE(\hat{\theta}_1) = 1.12140$$
$$SE(\hat{\theta}_2) = 1.44063$$
$$SE(\hat{\theta}_3) = 1.41580$$
$$SE(\hat{\theta}_4) = 1.35959$$
$$SE(\hat{\theta}_5) = 1.08715$$

The table-value needed for computing 95% confidence intervals for θ_i is the smaller of the two values

$$\sqrt{(m-2)F_{1-\alpha:m-2,n-m}} = \sqrt{3F_{0.95:3,20}} = \sqrt{3(3.10)} = 3.05$$

and

$$t_{1-\alpha/2m:20} = t_{0.995:20} = 2.845$$

obtained from Tables T-5 and T-4 in Appendix T, respectively. Hence the required table-value is 2.845. The confidence intervals for θ_i are

$$-1.617 \le \theta_1 \le 4.764$$
$$-5.102 \le \theta_2 \le 3.095$$
$$-6.325 \le \theta_3 \le 1.731$$
$$-2.558 \le \theta_4 \le 5.178$$
$$-2.676 \le \theta_5 \le 3.509$$

E X H I B I T 4.11.1
MINITAB Output for Example 4.11.3

```
The regression equation is
ymeans = 63.0 + 1.75 x
```

Predictor	Coef	Stdev	t-ratio	p
Constant	63.043	4.255	14.82	0.001
x	1.7453	0.1192	14.65	0.001

```
s = 1.884      R-sq = 98.6%     R-sq(adj) = 98.2%
```

Analysis of Variance

SOURCE	DF	SS	MS	F	p
Regression	1	761.55	761.55	214.54	0.001
Error	3	10.65	3.55		
Total	4	772.20			

means of y	distinct x values	residuals =estimate of theta(i)	
108.250	25	1.57340	**(4.11.21)**
114.400	30	-1.00330	
121.833	35	-2.29700	
134.167	40	1.31030	
142.000	45	0.41660	

MATRIX $(A^T A)^{-1}$:

```
 5.100   -0.140
-0.140    0.004
```

```
MATRIX D
0.25000000   0.00000000   0.00000000   0.00000000   0.00000000
0.00000000   0.20000000   0.00000000   0.00000000   0.00000000
0.00000000   0.00000000   0.16666667   0.00000000   0.00000000
0.00000000   0.00000000   0.00000000   0.16666667   0.00000000
0.00000000   0.00000000   0.00000000   0.00000000   0.25000000
```

```
MATRIX V
 0.088667  -0.089333  -0.040667  -0.005333   0.046667
-0.089333   0.146333  -0.031333  -0.019000  -0.006667
-0.040667  -0.031333   0.141333  -0.026000  -0.043333
-0.005333  -0.019000  -0.026000   0.130333  -0.080000
 0.046667  -0.006667  -0.043333  -0.080000   0.083333
```

We can be at least 95% confident that all five of the confidence intervals are simultaneously correct. As a consequence, we can be at least 95% confident that the proposed straight line function $P_Y(x) = \beta_0 + \beta_1 x$ does not deviate from the true unknown regression function $\mu_Y(x)$ by more than 6.325 blood pressure units (absolute value of the lower bound for θ_3) *for any of the subpopulations represented in the sample.* ∎

Cautionary Remark   Remember that the lack-of-fit of the proposed straight line regression model has been investigated only for the subpopulations represented in the sample; i.e., for $X = x_1, X = x_2, \ldots, X = x_m$. Even if the proposed straight line function $P_Y(x)$ agrees *exactly* with the true unknown regression function $\mu_Y(x)$ for these selected subpopulations, we still cannot conclude from these data that the two functions are identical for all values of X. Keep this point in mind when making any practical conclusions based on the results of a lack-of-fit analysis.

In Section 4.11 in the laboratory manuals we show you how to use a program that we have supplied on the data disk to do the computations necessary to obtain point estimates and confidence intervals for the lack-of-fit constants θ_i.

The Traditional Lack-of-Fit Test for Straight Line Prediction Functions

The statistical test of (4.11.4) is usually referred to as a *lack-of-fit test*. Although we recommend against using only a statistical test to examine the lack-of-fit of a proposed function, it seems to be common practice among statisticians and investigators to use such a test. We therefore describe this test procedure for the sake of completeness.

The true regression function $\mu_Y(x)$ is unknown, but the investigator postulates that it is

$$P_Y(x) = \beta_0 + \beta_1 x$$

for some unknown constants β_0 and β_1. Assumptions for lack-of-fit analysis, given in Box 4.11.1, are presumed to be valid. Thus m values of X are preselected (which are denoted by $x_1, x_2, \ldots, x_m$) and for each i, n_i values of Y are obtained by simple random sampling from the subpopulation corresponding to $X = x_i$. A schematic representation of the sample is in Table 4.11.1. The test of

$$\text{NH:} \quad \mu_Y(x) = \beta_0 + \beta_1 x \text{ for some } \beta_0, \beta_1 \tag{4.11.22}$$

against

$$\text{AH:} \quad \mu_Y(x) \neq \beta_0 + \beta_1 x \text{ for any } \beta_0, \beta_1$$

is conducted as follows.

1   Compute the estimate of σ^2 as in (4.11.15). Recall that this is called the *mean square for pure error* and is denoted by *MS(Pure error)*. Also calculate *SS(Pure error)*, the *sum of squares for pure error* given in (4.11.17).

2 Obtain *SSE* from the regression of Y on X using *all* of the sample observations.

3 Compute the *sum of squares for lack-of-fit*, *SS(Lack-of-fit)*, by the formula

$$SS(Lack\text{-}of\text{-}fit) = SSE - SS(Pure\ error)$$

4 Compute the *mean square for lack-of-fit*, *MS(Lack-of-fit)*, by the formula

$$MS(Lack\text{-}of\text{-}fit) = SS(\text{Lack-of-fit})/(m-2)$$

5 Compute the test statistic F_C by

$$F_C = \frac{MS(Lack\text{-}of\text{-}fit)}{MS(Pure\ error)}$$

6 The *P*-value for the test is the value of α for which $F_C = F_{1-\alpha:m-2,n-m}$.

We illustrate the procedure for conducting the traditional lack-of-fit test in Example 4.11.4.

E X A M P L E 4.11.4

For the age and blood pressure problem discussed in Example 4.11.3, we carry out the traditional lack-of-fit test. To compute *SSE*, we used SAS and regressed Y on X. The relevant part of the computer output appears in Exhibit 4.11.2.

From (4.11.23) we get *SSE* = 340.5 (rounded to one decimal). The estimate of σ is $\hat{\sigma} = 3.766$ (from Example 4.11.3), so the sum of squares for pure error, using (4.11.17), is calculated to be *SS(Pure error)* $= 20(3.766)^2 = 283.7$ with *df(Pure error)* $= n - m = 20$. Also *MS(Pure error)* $= 14.18$. Hence the sum of squares for lack-of-fit is $340.5 - 283.7 = 56.8$ with degrees of freedom $m - 2 = 3$. The mean square for lack-of-fit is $56.8/3 = 18.9$. The test statistic is $F_C = 18.9/14.18 = 1.33$ with 3 and 20 degrees of freedom. So from Table T-5 in Appendix T the *P*-value is between 0.1 and 0.5. Hence we do not reject NH at any of the commonly used α levels.

Although the null hypothesis is not rejected, we *cannot* conclude that the proposed straight line function $P_Y(x) = \beta_0 + \beta_1 x$ is correct. However, an examination of the confidence intervals for θ_i for $i = 1, \ldots, 5$ will help the investigator determine whether or not the deviations between the true unknown regression function and the proposed straight line function are small enough to be ignored for the problem under study. ∎

E X H I B I T **4.11.2**

SAS Output for Example 4.11.4

The SAS System   0.00 Saturday Jan 1, 1994

Dependent Variable: BP

Analysis of Variance

Source	DF	Sum of Squares	Mean Square	F Value	Prob>F
Model	1	3337.24976	3337.24976	225.417	0.0001
Error	23	340.51024	14.80479		
C Total	24	3677.76000			

(4.11.23)

Root MSE	3.84770	R-square	0.9074	
Dep Mean	124.36000	Adj R-sq	0.9034	
C.V.	3.09400			

Parameter Estimates

Variable	DF	Parameter Estimate	Standard Error	T for H0: Parameter=0	Prob > |T|
INTERCEP	1	62.310987	4.20380966	14.823	0.0001
AGE	1	1.762756	0.11740836	15.014	0.0001

Problems 4.11

4.11.1   The bowl-life Y (in seconds) of a breakfast cereal, which is usually eaten with milk, is defined to be the amount of time that the cereal will retain its crunchiness. This bowl-life depends on the temperature X (in degrees Celsius) of the milk. The regression function $\mu_Y(x)$ of Y on X is unknown, but the investigator postulates that the regression function of Y on X is adequately approximated by a straight line prediction function of the form

$$P_Y(x) = \beta_0 + \beta_1 x \qquad \textbf{(4.11.24)}$$

which is the least squares approximation to $\mu_Y(x)$ for $x = 40, 45, 50$, and 55. To analyze the lack-of-fit of the proposed model, data were collected at these preselected values of X by adding the milk to 5 ounces of the cereal under each test condition. The data are displayed in Table 4.11.3 and are also stored in the file **cereal.dat** on the data disk.

Because there are multiple observations in at least one of the four subpopulations represented in the sample, it is possible to perform a lack-of-fit analysis using the methods of Section 4.11.

Several quantities that are needed for a lack-of-fit analysis in this problem follow.

$$\bar{y} = 9.06667$$
$$\bar{x} = 47.5000$$
$$SSY = 195.653$$
$$SSX = 750.000$$
$$SXY = -257.500$$

T A B L E **4.11.3**
Cereal Bowl-Life Data

Sample Item Number	Bowl-Life Y (minutes)	Temperature X (°C)
1	13.8	40.0
2	14.8	40.0
3	11.1	40.0
4	11.3	40.0
5	9.7	40.0
6	8.7	40.0
7	12.5	45.0
8	12.7	45.0
9	10.9	45.0
10	10.5	45.0
11	6.5	45.0
12	7.1	45.0
13	10.5	50.0
14	10.2	50.0
15	8.7	50.0
16	8.2	50.0
17	6.5	50.0
18	5.2	50.0
19	8.7	55.0
20	8.7	55.0
21	6.8	55.0
22	6.6	55.0
23	3.6	55.0
24	4.3	55.0

$$SSE = 107.245$$

$$SS(Pure\ error) = 107.130$$

The V matrix is

$$\begin{bmatrix} 0.050000 & -0.066667 & -0.016667 & 0.033333 \\ -0.066667 & 0.116667 & -0.033333 & -0.016667 \\ -0.016667 & -0.033333 & 0.116667 & -0.066667 \\ 0.033333 & -0.016667 & -0.066667 & 0.050000 \end{bmatrix}$$

a   What is the value of m, the number of distinct subpopulations represented in the sample?

b   What is n? What are the values of $n_1, \ldots, n_m$?

c   Calculate the estimates of the subpopulation means $\mu_i = \mu_Y(x_i)$ and the estimates of the subpopulation standard deviations $\sigma_i = \sigma_Y(x_i)$ for each subpopulation represented in the sample.

d   Calculate the pure error estimate of σ, and show that it is equal to $\hat{\sigma} = 2.31441$.

e   Estimate the lack-of-fit constants $\theta_1, \ldots, \theta_m$.

f   Show that $SE(\hat{\theta}_i)$ are

$$0.517518, \qquad 0.790522, \qquad 0.790522, \qquad 0.517518$$

respectively.

g   Show that the two-sided confidence intervals for $\theta_1, \ldots, \theta_m$, such that we have at least 90% confidence that all of them are simultaneously correct, are

Lower	Upper
−1.25265	1.10272
−1.69058	1.90730
−1.79058	1.80730
−1.21933	1.13604

Use $t_{0.9875:20} = 2.423$ and $F_{0.9:2,20} = 2.59$.

h   Find a number d such that we can say with at least 90% confidence that the population regression function and the proposed linear prediction function will not differ by more than d units in any of the m subpopulations included in the sample.

i   Suppose an investigator decides that the proposed prediction function in (4.11.24) is an adequate approximation of the true regression function if the differences $\theta_1, \ldots, \theta_m$ between the two functions can be shown to be less than or equal to 2 minutes. Based on the lack-of-fit analysis in (g), can the prediction function in (4.11.24) be regarded as close enough to the true regression function for the problem under study?

j   Perform a traditional lack-of-fit test of the model in (4.11.24). What is your conclusion using $\alpha = .10$?

Conversation   4.11

Investigator: Good morning. Do you have time to talk to me?

Statistician: Certainly. How can I help you?

Investigator: I want to discuss lack-of-fit. Can you explain the difference between the following?

1   Checking the lack-of-fit of a straight line model and

2   Checking whether model A is better than model B for predicting Y where the two models are

$$\text{Model A: } \mu_Y^{(A)}(x) = \beta_0^A + \beta_1^A x$$
$$\text{Model B: } \mu_Y^{(B)} = \beta_0^B$$

Statistician: In the simplest terms, in (1) you are checking to determine whether a proposed model $P_Y(x) = \beta_0 + \beta_1 x$ is close enough to the true *unknown* regression function $\mu_Y(x)$ so that $P_Y(x)$ can be used in place of $\mu_Y(x)$. In (2), you are assuming model A is correct and checking to see whether it is better than model B for predicting Y.

For case (2), both models are specified (they of course both contain unknown parameters), but in case (1) you do not specify the *true* regression function $\mu_Y(x)$ because you do not know what it is. You just want to know if the model you propose, namely $P_Y(x) = \beta_0 + \beta_1 x$, can be used in place of the true unknown regression model $\mu_Y(x)$.

Investigator: One of the people I work for says that he would rather use the *test* for lack-of-fit than the simultaneous confidence intervals you propose.

Statistician: Why is that?

Investigator: He says the test is easier. All he has to do to test for lack-of-fit is compute a P-value, and on the basis of that he can reject or not reject the hypothesis that the proposed model $P_Y(x) = \beta_0 + \beta_1 x$ is equal to the unknown true model $\mu_Y(x)$. On the other hand, after the confidence intervals are obtained for the lack-of-fit constants, he has to spend a considerable amount of time determining whether any of the differences are important in his problem.

Statistician: That's exactly right. It seems to me that an investigator would be much more comfortable with the decision if he examined the confidence intervals than if he just used the P-value.

To perform a test, he would examine two hypotheses:

$$\text{NH:} \quad \beta_0 + \beta_1 x \quad \text{*is the true model*}$$

against

$$\text{AH:} \quad \beta_0 + \beta_1 x \quad \text{*is not the true model*}$$

However, if he looked at the confidence intervals, they would help him decide whether the proposed model $P_Y(x) = \beta_0 + \beta_1 x$ is adequate for his problem. It is unlikely that $P_Y(x) = \beta_0 + \beta_1 x$ is exactly the true model, but it may be close enough to be useful in a specified problem.

Investigator: I see your point, but the investigator says if NH is rejected, he will assume that the model $\beta_0 + \beta_1 x$ is not adequate for his problem, but if NH is not rejected, he will assume the model is adequate.

Statistician: You recall our previous conversation in Chapter 1 where we decided that results can be quite different if confidence intervals are used instead of tests. So if the investigator insists on using a *P*-value to make the decision, why don't you give him the *P*-value **and** the confidence intervals and ask him to examine them also.

Investigator: I will do that, but he may say that a test for lack-of-fit is easier to compute than simultaneous confidence intervals.

Statistician: Tell him that isn't the case if he has the MINITAB or SAS macro provided on the data disk and discussed in Section 4.11 in the laboratory manuals. This macro computes the simultaneous confidence intervals he needs.

Investigator: I'll explain that to him.

4.12
Exercises

4.12.1 While performing experiments to study the absorption of a certain drug in mice, an investigator administered a specified dose of the drug to a laboratory mouse and determined the drug concentration in blood samples drawn from the mouse at times ranging from 20 to 420 minutes. The drug concentrations C in the blood and the times T when blood was drawn are given in Exhibit 4.12.1 along with a MINITAB printout from a regression analysis of these data. The data also appear in the file **mouse.dat** on the data disk.

If we let

$$Y = log_{10}(C) \qquad X_1 = log_{10}(T) \qquad X_2 = X_1^2 = [log_{10}(T)]^2$$

the regression function is given by

$$\mu_Y(t) = \beta_0 + \beta_1 log_{10}(t) + \beta_2 [log_{10}(t)]^2$$

i.e., by

$$\mu_Y(x_1, x_2) = \beta_0 + \beta_1 x_1 + \beta_2 x_2 \qquad\qquad \textbf{(4.12.1)}$$

Population assumptions (A) are presumed to hold for $\{(Y, X_1, X_2)\}$.

E X H I B I T 4.12.1

MINITAB Output for Mouse Data

```
                         RAW AND TRANSFORMED DATA


            C          T          Y          X1              X2
          conc       time
          1.02        20      0.008600    1.30103         1.69268
          1.08        40      0.033424    1.60206         2.56660
          1.10        60      0.041393    1.77815         3.16182
          1.06        90      0.025306    1.95424         3.81906
          0.95       150     -0.022276    2.17609         4.73537
          0.77       210     -0.113509    2.32222         5.39270
          0.60       300     -0.221849    2.47712         6.13613
          0.42       420     -0.376751    2.62325         6.88144


The regression equation is


   Y  = - 1.31 + 1.62 X1 - 0.479 X2


Predictor         Coef       Stdev      t-ratio         p
Constant        -1.3101      0.2005      -6.53       0.001
X1               1.6222      0.2091       7.76       0.001
X2              -0.47926     0.05266     -9.10       0.000


s = 0.02441       R-sq = 98.1%      R-sq(adj) = 97.4%


Analysis of Variance


SOURCE         DF          SS          MS          F         p
Regression      2      0.156172    0.078086     131.03    0.000
Error           5      0.002980    0.000596
Total           7      0.159152
```

a   Plot C against T.

b   Plot Y against X_1.

c   The estimated regression equation can be written as

$$\hat{\mu}_{log_{10}(C)}(t) = \hat{\mu}_Y(t) = -1.31 + 1.62\, log_{10}(t) - 0.479[log_{10}(t)]^2$$

Compute $\hat{\mu}_Y(30)$. On the graph in (b), superimpose the graph of $\hat{\mu}_Y(t)$.

d   If β_2 is zero in the regression function in (4.12.1), it would mean that this regression function is a straight line in X_1. On the other hand, if β_2 is nonzero, then the regression function in (4.12.1) is a quadratic function of X_1. Suppose the investigator will consider the quadratic term to be negligible for this problem if β_2 is less than 0.0002 in magnitude. In that case a linear regression function would be used for this problem. Compute an appropriate 95% confidence interval for β_2 and state what the investigator's conclusion will be.

   e Suppose the investigator wants to use a statistical test to help determine whether the data provide evidence (at $\alpha = 0.05$) suggesting that the regression function in (4.12.1) is a quadratic function and not a straight line function of X_1 (i.e., $\beta_2 = 0$). Formulate an appropriate statistical test to help the investigator determine this and carry out the test. What is the P-value for this test? What is your conclusion based on this test? Compare this with your answer for part (d).

4.12.2 An organization that evaluates the performance of automobiles wants to predict the first-year maintenance cost Y of a new car as a function of the number of miles X the car will be driven. With this in mind a sample of 17 cars was selected and the owners were asked to report maintenance costs after the cars were driven a specified number of miles; i.e., the data were obtained by sampling with preselected X values to cover a wide range of miles driven. The values of X (in miles) and Y (in dollars) are recorded. The data and computer output appear in Exhibit 4.12.2 and are also stored in **car17.dat** on the data disk. Assumptions (A) are presumed valid for the population $\{(Y, X)\}$, with the regression function of Y on X given by

$$\mu_Y(x) = \beta_0 + \beta_1 x + \beta_2 x^2 \qquad\qquad \textbf{(4.12.2)}$$

which can be written as $\mu_Y(x_1, x_2) = \beta_0 + \beta_1 x_1 + \beta_2 x_2$ where $x_1 = x$ and $x_2 = x^2$. If $\beta_2 = 0$ then the regression function reduces to

$$\mu_Y(x) = \beta_0 + \beta_1 x \qquad\qquad \textbf{(4.12.3)}$$

   a Plot Y against X.

   b Assume that the regression function of Y on X is given in (4.12.3).

     i What are the estimates of β_0, β_1, $\sigma_{Y|X}$, $\mu_Y(x)$, and $Y(x)$?

       You will notice that there are nine different subpopulations represented in the sample data, with two or more observations from six of the nine subpopulations, so it is possible to carry out a lack-of-fit analysis. The investigator wants to know if the model

$$P_Y(x) = \beta_0 + \beta_1 x$$

       is an adequate approximation for the regression function. Questions (ii)–(vi) pertain to this.

     ii Find *SS(Pure error)*, *degrees of freedom(Pure error)*, and *MS(Pure error)*.

E X H I B I T 4.12.2
MINITAB Output for Problem 4.12.2

carno	Y mtcost	X1 miles	X2 [miles]^2
1	272	3000	9000000
2	300	5000	25000000
3	287	7000	49000000
4	327	9000	81000000
5	330	10000	100000000
6	386	14000	196000000
7	442	18000	324000000
8	522	22000	484000000
9	604	25000	625000000
10	266	3000	9000000
11	313	7000	49000000
12	336	10000	100000000
13	328	10000	100000000
14	367	14000	196000000
15	397	14000	196000000
16	483	18000	324000000
17	537	22000	484000000

Regression of Y on $X_1 = X$ and $X_2 = X^2$

The regression equation is
Y = 259 + 0.00331 X1 +0.00000042 X2.

Predictor	Coef	Stdev	t-ratio	p
Constant	258.72	12.36	20.93	0.000
X1	0.003309	0.002084	1.59	0.135
X2	0.00000042	0.00000007	5.58	0.000

s = 12.85   R-sq = 98.6%   R-sq(adj) = 98.4%

Analysis of Variance

SOURCE	DF	SS	MS	F	p
Regression	2	161689	80845	489.28	0.000
Error	14	2313	165		
Total	16	164002			

EXHIBIT 4.12.2

(Continued)

Regression of Y on $X_1 = X$

```
The regression equation is
Y = 201 + 0.0146 X1
```

Predictor	Coef	Stdev	t-ratio	p
Constant	200.68	11.57	17.35	0.000
X1	0.0146230	0.0008238	17.75	0.000

```
s = 22.29        R-sq = 95.5%      R-sq(adj) = 95.2%
```

Analysis of Variance

SOURCE	DF	SS	MS	F	p
Regression	1	156551	156551	315.12	0.000
Error	15	7452	497		
Total	16	164002			

iii The matrix V defined in (4.11.12), the estimates of the lack-of-fit constants $\hat{\theta}_i$ given in (4.11.7), and the standard errors of the $\hat{\theta}_i$ given in (4.11.11) follow:

$$V = \begin{bmatrix} 0.403773 & -0.216777 & -0.073971 & -0.154221 & -0.030178 & -0.021355 & -0.012766 & 0.009490 & 0.096004 \\ -0.216777 & 0.692293 & -0.167147 & -0.226212 & -0.104619 & -0.065611 & -0.030664 & 0.018966 & 0.099769 \\ -0.073971 & -0.167147 & 0.439303 & -0.130442 & -0.027233 & -0.021770 & -0.024194 & -0.010920 & 0.016374 \\ -0.154220 & -0.226211 & -0.130443 & 0.812562 & -0.090897 & -0.072117 & -0.065053 & -0.041275 & -0.032345 \\ -0.030178 & -0.104619 & -0.027233 & -0.090897 & 0.329122 & -0.004859 & -0.019136 & -0.016192 & -0.036008 \\ -0.021355 & -0.065611 & -0.021770 & -0.072117 & -0.004859 & 0.330090 & -0.022910 & -0.023325 & -0.098143 \\ -0.012766 & -0.030664 & -0.024194 & -0.065053 & -0.019136 & -0.022910 & 0.444380 & -0.067048 & -0.202610 \\ 0.009490 & 0.018966 & -0.010920 & -0.041275 & -0.016192 & -0.023324 & -0.067048 & 0.412542 & -0.282240 \\ 0.096004 & 0.099769 & 0.016374 & -0.032345 & -0.036008 & -0.098143 & -0.202610 & -0.282240 & 0.439199 \end{bmatrix}$$

The estimates of the lack-of-fit constants θ_i are

$$\hat{\theta}_1 = 21.9485 \qquad \hat{\theta}_2 = 23.1066 \qquad \hat{\theta}_3 = -6.7354$$
$$\hat{\theta}_4 = -9.5773 \qquad \hat{\theta}_5 = -20.1649 \qquad \hat{\theta}_6 = -27.8488$$
$$\hat{\theta}_7 = -8.3661 \qquad \hat{\theta}_8 = -1.0499 \qquad \hat{\theta}_9 = 28.6871$$

Show that the standard errors of the estimates $\hat{\theta}_i$ are

$$SE(\hat{\theta}_1) = 9.5429 \qquad SE(\hat{\theta}_2) = 12.4956 \qquad SE(\hat{\theta}_3) = 9.9540$$
$$SE(\hat{\theta}_4) = 13.5376 \qquad SE(\hat{\theta}_5) = 8.6157 \qquad SE(\hat{\theta}_6) = 8.6284$$
$$SE(\hat{\theta}_7) = 10.0113 \qquad SE(\hat{\theta}_8) = 9.6460 \qquad SE(\hat{\theta}_9) = 9.9528$$

iv Show that simultaneous confidence intervals for $\theta_1, \ldots, \theta_9$, with confidence coefficients greater than or equal to 95%, are given by (use $t_{0.9972:8} = 3.7856$ and $F_{0.95:7,8} = 3.501$)

θ	Lower	Upper
1	−13.9194	57.8164
2	−23.8593	70.0725
3	−44.1485	30.6777
4	−60.4597	41.3051
5	−52.5478	12.2180
6	−60.2795	4.5819
7	−45.9945	29.2623
8	−37.3053	35.2055
9	−8.7214	66.0956

v An investigator will use a straight line model for this problem if none of the lack-of-fit constants θ_i, $i = 1, \ldots, 9$ exceed \$75 in magnitude. Using the results from (iv), would the investigator use a straight line model?

vi Perform the traditional lack-of-fit test for a straight line model using $\alpha = 0.05$. What is your conclusion? Compare this with your conclusion in part (v).

c Assume that the regression function of Y on X is given by (4.12.2). What are the estimates of β_0, β_1, β_2, $\sigma_{Y|X_1,X_2}$, $\mu_Y(x_1, x_2)$, and $Y(x_1, x_2)$?

d Plot the estimates of the regression functions in (4.12.2) and (4.12.3) on the same graph. Also show the observed data points on this graph.

4.12.3 The height (Y) in inches of a plant during the first few days after germination is related to the temperature (X_1) in degrees Fahrenheit at which it is grown and the time (X_2) in days after germination. Assume that the regression function of Y on X_1 and X_2 is of the form

$$\mu_Y(x_1, x_2) = \beta_0 + \beta_1 x_1 + \beta_2 x_2 + \beta_3 x_1 x_2 \tag{4.12.4}$$

which can be written as

$$\mu_Y(x_1, x_2, x_3) = \beta_0 + \beta_1 x_1 + \beta_2 x_2 + \beta_3 x_3$$

where $x_3 = x_1 x_2$.

Twenty plants were included in an experiment. Each plant was grown at a prechosen temperature $(60°F, 70°F, 80°F, 90°F, \text{ or } 100°F)$ for a preselected number of days (6 days or 12 days), at the end of which its height was recorded. The data and computer output are given in Exhibit 4.12.3 and are also stored in the file **plant.dat** on the data disk. Assumptions (A) are presumed to hold, and data were obtained by preselecting the values of X_1 and X_2.

The computer output in Exhibit 4.12.4 (obtained using MINITAB) lists the values of $X_1, X_2, X_1 X_2, Y, \hat{\mu}_Y(x_1, x_2) = $ fits, $\hat{e}_i = $ residuals, $r_i = $ standardized residuals, and Gaussian scores (nscores) $= z_i^{(n)}$ for the model in (4.12.4).

E X H I B I T 4.12.3
MINITAB Output for Plant Growth Data

Y	X1	X2	X1X2
3.11	60	6	360
2.04	60	6	360
4.36	60	12	720
4.60	60	12	720
2.98	70	6	420
3.65	70	6	420
6.31	70	12	840
7.05	70	12	840
4.21	80	6	480
4.31	80	6	480
7.86	80	12	960
8.45	80	12	960
4.86	90	6	540
4.25	90	6	540
9.63	90	12	1080
9.59	90	12	1080
5.66	100	6	600
5.28	100	6	600
10.89	100	12	1200
11.23	100	12	1200

```
The regression equation is
Y = 1.70 - 0.0203 X1 - 0.548 X2 + 0.0151 X1X2
```

Predictor	Coef	Stdev	t-ratio	p
Constant	1.697	1.513	1.12	0.279
X1	-0.02030	0.01862	-1.09	0.292
X2	-0.5477	0.1595	-3.43	0.003
X1X2	0.015100	0.001963	7.69	0.000

```
s = 0.3724     R-sq = 98.4%     R-sq(adj) = 98.1%
```

Analysis of Variance

SOURCE	DF	SS	MS	F	p
Regression	3	140.149	46.716	336.84	0.000
Error	16	2.219	0.139		
Total	19	142.368			

E X H I B I T   **4.12.4**

Diagnostics for the Model in (4.12.4) for the Plant Growth Data

X1 temp	X2 time	X1X2 (temp)(time)	Y height	fits	residuals	stdresid	nscores
60	6	360	3.11	2.629	0.481000	1.54373	1.40377
60	6	360	2.04	2.629	-0.589000	-1.89035	-1.87129
60	12	720	4.36	4.779	-0.419000	-1.34475	-1.12690
60	12	720	4.60	4.779	-0.179000	-0.57449	-0.58740
70	6	420	2.98	3.332	-0.352000	-1.02520	-0.91718
70	6	420	3.65	3.332	0.318000	0.92618	0.91718
70	12	840	6.31	6.388	-0.078000	-0.22718	-0.18593
70	12	840	7.05	6.388	0.662000	1.92808	1.87129
80	6	480	4.21	4.035	0.175000	0.49533	0.44602
80	6	480	4.31	4.035	0.275000	0.77837	0.74198
80	12	960	7.86	7.997	-0.137000	-0.38777	-0.31325
80	12	960	8.45	7.997	0.453000	1.28219	1.12690
90	6	540	4.86	4.738	0.122000	0.35533	0.31325
90	6	540	4.25	4.738	-0.488000	-1.42130	-1.40377
90	12	1080	9.63	9.606	0.024000	0.06990	0.18593
90	12	1080	9.59	9.606	-0.016000	-0.04660	-0.06165
100	6	600	5.66	5.441	0.219000	0.70286	0.58740
100	6	600	5.28	5.441	-0.161000	-0.51672	-0.44602
100	12	1200	10.89	11.215	-0.325000	-1.04306	-0.74198
100	12	1200	11.23	11.215	0.014999	0.04814	0.06165

a   Plot height versus temperature using different symbols for points corresponding to different times while using the same symbol for points corresponding to the same times.

b   Plot height versus time using different symbols for points corresponding to different temperatures while using the same symbol for points corresponding to the same temperatures.

c   How were the data obtained for this study—by simple random sampling or by sampling with preselected X_1, X_2 values ?

d   Carry out a residual analysis for the model in (4.12.4) to examine the validity of assumptions (A) for this problem. What are your conclusions?

e   Estimate $\beta_0, \beta_1, \beta_2, \beta_3, \sigma_{Y|X_1,X_2,X_3}, \mu_Y(x_1, x_2, x_3)$, and $Y(x_1, x_2, x_3)$.

f   For the model in (4.12.4) test NH: $\beta_2 = \beta_3 = 0$ against AH: at least one of β_2, β_3 is nonzero. Use $\alpha = 0.05$. You may need the results from Exhibit 4.12.5. State your conclusion.

g   Using the model in (4.12.4) estimate the average height in inches, at the end of 10 days, of plants that are grown at 65° F. Construct a 95% two-sided confidence interval for this mean height. Describe in words the meaning of this confidence interval.

E X H I B I T 4.12.5

MINITAB Output for Regression of Y on X_1 for the Plant Growth Data

```
The regression equation is
Y = - 3.23 + 0.116 X1


Predictor         Coef       Stdev     t-ratio        p
Constant        -3.232       2.855       -1.13    0.272
X1              0.11560     0.03514       3.29    0.004


s = 2.223      R-sq = 37.5%      R-sq(adj) = 34.1%


Analysis of Variance


SOURCE         DF           SS          MS         F        p
Regression      1       53.453      53.453     10.82    0.004
Error          18       88.915       4.940
Total          19      142.368
```

4.12.4 An investigator is interested in studying how the height Y at age 18 years of a group of people who have lived in mountain isolation for several generations is related to the following variables.

$X_1 = $ Length at birth
$X_2 = $ Mother's height at age 18
$X_3 = $ Father's height at age 18
$X_4 = $ Maternal grandmother's height at age 18
$X_5 = $ Maternal grandfather's height at age 18
$X_6 = $ Paternal grandmother's height at age 18
$X_7 = $ Paternal grandfather's height at age 18

All heights and lengths are in inches. A simple random sample of 20 males of age 18 or more was drawn, and all the above information was recorded. The data and computer output are given in Exhibit 4.12.6 and are also stored in the file **age18.dat** on the data disk. Assumptions (B) are presumed to hold.

E X H I B I T **4.12.6**
Data and MINITAB Output for Problem 4.12.4

Sample Item Number	Y	X_1	X_2	X_3	X_4	X_5	X_6	X_7
1	67.2	19.7	60.5	70.3	65.7	69.3	65.7	67.3
2	69.1	19.6	64.9	70.4	62.6	69.6	64.6	66.4
3	67.0	19.4	65.4	65.8	66.2	68.8	64.0	69.4
4	72.4	19.4	63.4	71.9	60.7	68.0	64.9	67.1
5	63.6	19.7	65.1	65.1	65.5	65.5	61.8	70.9
6	72.7	19.6	65.2	71.1	63.5	66.2	67.3	68.6
7	68.5	19.8	64.3	67.9	62.4	71.4	63.4	69.4
8	69.7	19.7	65.3	68.8	61.5	66.0	62.4	67.7
9	68.4	19.7	64.5	68.7	63.9	68.8	62.3	68.8
10	70.4	19.9	63.4	70.3	65.9	69.0	63.7	65.1
11	67.5	18.9	63.3	70.4	63.7	68.2	66.2	68.5
12	73.3	20.8	66.2	70.2	65.4	66.6	61.7	64.0
13	70.0	20.3	64.9	68.8	65.2	70.2	62.4	67.0
14	69.8	19.7	63.5	70.3	63.1	64.4	65.1	67.0
15	63.6	19.9	62.0	65.5	64.1	67.7	62.1	66.5
16	64.3	19.6	63.5	65.2	63.9	70.0	64.2	64.5
17	68.5	21.3	66.1	65.4	64.8	68.4	66.4	70.8
18	70.5	20.1	64.8	70.2	65.3	65.5	63.7	66.9
19	68.1	20.2	62.6	68.6	63.7	69.8	66.7	68.0
20	66.1	19.2	62.2	67.3	63.6	70.9	63.6	66.7

The regression of Y on $X_1, X_2, X_3, X_4, X_5, X_6, X_7$

```
The regression equation is
Y = - 78.3 + 1.37 X1 + 0.782 X2 + 1.05 X3 - 0.120 X4 +
    0.091 X5 + 0.088 X6 - 0.102 X7


Predictor        Coef        Stdev       t-ratio         p

Constant       -78.27        26.96        -2.90       0.013
X1              1.3718       0.5207         2.63       0.022
X2              0.7824       0.1992         3.93       0.002
X3              1.0514       0.1358         7.74       0.000
X4             -0.1199       0.1717        -0.70       0.498
X5              0.0914       0.1301         0.70       0.496
X6              0.0883       0.1613         0.55       0.594
X7             -0.1017       0.1549        -0.66       0.524

s = 1.004       R-sq = 91.7%     R-sq(adj) = 86.9%
```

Analysis of Variance

SOURCE	DF	SS	MS	F	p
Regression	7	133.657	19.094	18.95	0.000
Error	12	12.088	1.007		
Total	19	145.746			

THE C MATRIX IS

721.671	-1.494	-2.020	-2.062	-2.279	-2.030	-0.557	-1.496
-1.494	0.269	-0.047	0.011	-0.027	-0.001	-0.014	0.016
-2.020	-0.047	0.039	-0.003	0.004	0.007	0.010	-0.010
-2.062	0.011	-0.003	0.018	0.006	0.005	-0.009	0.010
-2.279	-0.027	0.004	0.006	0.029	0.002	0.001	0.001
-2.030	-0.001	0.007	0.005	0.002	0.017	-0.002	0.002
-0.557	-0.014	0.010	-0.009	0.001	-0.002	0.026	-0.011
-1.496	0.016	-0.010	0.010	0.001	0.002	-0.011	0.024

The regression of Y on X_2 and X_3

The regression equation is
Y = - 61.2 + 0.895 X2 + 1.06 X3

Predictor	Coef	Stdev	t-ratio	p
Constant	-61.20	14.55	-4.21	0.001
X2	0.8947	0.1768	5.06	0.000
X3	1.0556	0.1189	8.88	0.000

s = 1.131   R-sq = 85.1%   R-sq(adj) = 83.3%

Analysis of Variance

SOURCE	DF	SS	MS	F	p
Regression	2	124.009	62.005	48.49	0.000
Error	17	21.736	1.279		
Total	19	145.746			

E X H I B I T   4.12.6

Data and MINITAB Output for Problem 4.12.4

```
THE C MATRIX IS

    165.675      -1.669      -0.855
     -1.669       0.024       0.002
     -0.855       0.002       0.011
```

The regression of Y on X_1, X_2, and X_3

```
The regression equation is
Y = - 78.2 + 1.35 X1 + 0.692 X2 + 1.10 X3


Predictor        Coef        Stdev      t-ratio         p
Constant       -78.23        13.24        -5.91     0.000
X1             1.3503       0.4474         3.02     0.008
X2             0.6925       0.1602         4.32     0.001
X3            1.10250      0.09908        11.13     0.000


s = 0.9305      R-sq = 90.5%      R-sq(adj) = 88.7%



Analysis of Variance


SOURCE         DF           SS          MS          F          p
Regression      3      131.894      43.965      50.78      0.000
Error          16       13.852       0.866
Total          19      145.746


THE C MATRIX IS

    202.461      -2.917      -1.233      -0.957
     -2.917       0.231      -0.035       0.008
     -1.233      -0.035       0.030       0.000
     -0.957       0.008       0.000       0.011
```

For parts (a) and (b), suppose that assumptions (B) hold and that the model is

model A: $\mu_Y^{(A)}(x_1, x_2, x_3, x_4, x_5, x_6, x_7) =$

$$\beta_0^A + \beta_1^A x_1 + \beta_2^A x_2 + \beta_3^A x_3 + \beta_4^A x_4 + \beta_5^A x_5 + \beta_6^A x_6 + \beta_7^A x_7 \quad \textbf{(4.12.5)}$$

Answer the following questions. For ease of notation let σ_A denote $\sigma_{Y|X_1,X_2,X_3,X_4,X_5,X_6,X_7}$ and $\rho_{Y(A)}^2$ denote $\rho_{Y(X_1,X_2,X_3,X_4,X_5,X_6,X_7)}^2$.

a   What might be an appropriate target population of interest? Is the target population identical with the study population? Explain.

b   i   Estimate β_i^A ($i = 0, 1, 2, 3, 4, 5, 6, 7$), and $\mu_Y^{(A)}(x_1, x_2, x_3, x_4, x_5, x_6, x_7)$.

ii Estimate σ_A.

iii Compute 95% two-sided confidence intervals for β_i^A, $i = 0, 1, 2, 3, 4, 5, 6, 7$. Explain in words the meaning of the computed confidence interval for β_1^A.

iv The investigator wants to examine the difference in the average heights at age 18 between two groups of subjects whose lengths at birth differed by 1 inch, but the subjects have the same set of values for $X_2, X_3, X_4, X_4, X_5, X_6, X_7$. Express this difference in terms of the *parameters* of the multiple regression function in (4.12.5) and obtain a 95% confidence interval for it.

v To evaluate how good the regression function in (4.12.5) is for predicting Y, the investigator wants to compare σ_A with σ_Y. Calculate a two-sided confidence interval for σ_Y / σ_A with confidence coefficient greater than or equal to 90%. Explain in words the meaning of this confidence statement.

vi To assist the investigator in evaluating how good the regression function in (4.12.5) is for predicting Y, *relative to* μ_Y (when no predictors are used), compute a point estimate for $\rho_{Y(A)}^2$.

vii Predict the height at age 18 of an *individual* who belongs to the subpopulation with

$$X_1 = \text{length at birth} = 20 \text{ inches}$$
$$X_2 = \text{mother's height at age } 18 = 60 \text{ inches}$$
$$X_3 = \text{father's height at age } 18 = 72 \text{ inches}$$
$$X_4 = \text{maternal grandmother's height at age } 18 = 61 \text{ inches}$$
$$X_5 = \text{maternal grandfather's height at age } 18 = 71 \text{ inches}$$
$$X_6 = \text{paternal grandmother's height at age } 18 = 62 \text{ inches}$$
$$X_7 = \text{paternal grandfather's height at age } 18 = 70 \text{ inches}$$

viii In (vii) estimate the *average* height at age 18 of all individuals in the subpopulation.

ix Compute a 95% lower confidence bound for the height at age 18 of an individual randomly chosen from the subpopulation in (vii). The value of $x^T C x$, should you need it, is given to be 2.5321.

x In (vii) compute a 95% lower confidence bound for the average height at age 18 of all individuals in the subpopulation.

xi Explain why the answers to (vii) and (viii) are the same but the lower bounds in (ix) and (x) are not equal to each other.

c Since assumptions (B) are presumed to hold for the eight-variable population $\{(Y, X_1, \ldots, X_7)\}$, it follows that assumptions (B) hold for the three-variable population $\{(Y, X_2, X_3)\}$ and also for the four-variable population $\{(Y, X_1, X_2, X_3)\}$. Exhibit 4.12.6 gives the computer outputs for model C and model D defined in (4.12.6) and (4.12.7), respectively.

i Consider model C defined by

$$\mu_Y^{(C)}(x_2, x_3) = \beta_0^C + \beta_2^C x_2 + \beta_3^C x_3 \qquad \textbf{(4.12.6)}$$

Note that this model uses only mother's height at age 18 and father's height at age 18 to predict son's height at age 18. Estimate β_0^C, β_2^C, β_3^C, and σ_C.

ii   Consider model D defined by

$$\mu_Y^{(D)}(x_1, x_2, x_3) = \beta_0^D + \beta_1^D x_1 + \beta_2^D x_2 + \beta_3^D x_3 \tag{4.12.7}$$

Note that this model uses length at birth, mother's height at age 18, and father's height at age 18 to predict son's height at age 18. Estimate β_0^D, β_1^D, β_2^D, β_3^D, and σ_D.

iii   To evaluate how much better model D is than model C for predicting Y, the investigator wants to compare σ_D with σ_C. Calculate an approximate 90% two-sided confidence interval for σ_C/σ_D. Explain in words the meaning of this confidence statement.

For (iv)–(xiii), use model C in (4.12.6).

iv   Express the difference (in terms of population parameters) in the average heights at age 18 between two subpopulations of individuals if every mother of the first group of individuals was 1 inch taller (at age 18) than every mother of the individuals in the second group, but fathers' heights are the same for both subpopulations. Estimate this difference and compute a 95% lower confidence bound for it.

v   An investigator is interested in determining whether the average height at age 18 of a subpopulation of individuals is at least 1/4 inch greater than the average height at age 18 of another subpopulation of individuals if every mother of the first group of individuals was 1 inch taller (at age 18) than every mother of the individuals in the second group, but fathers' heights are the same for both subpopulations. Formulate an appropriate pair of hypotheses and carry out the test. Calculate the P-value for this test. State your conclusions using $\alpha = 0.05$.

vi   Which is more informative—the confidence bound in (iv) or the hypothesis test in (v)? Why?

vii   What is the difference between the average heights at age 18 of two subpopulations if every father of the first group of individuals was 1 inch taller (at age 18) than every father of the individuals in the second group, given mothers' heights are the same for both subpopulations? Estimate this difference and obtain a 95% upper confidence bound for it.

viii   Do the data provide evidence (at $\alpha = 0.05$) indicating that the average height at age 18 of a subpopulation of individuals is at most 1 inch greater than the average height at age 18 of another subpopulation of individuals if every father of the first group of individuals was 1 inch taller (at age 18) than every father of the individuals in the second group, if the mothers' heights are the same for both subpopulations? Formulate an appropriate pair of hypotheses and carry out the test. Calculate the P-value for this test. State your conclusion using $\alpha = 0.05$.

ix Which is more informative—the confidence bound in (vii) or the hypothesis test in (viii)? Why?

x Consider the subpopulation of all individuals with $X_2 = 58$. For this subpopulation, estimate the coefficient of determination of Y with X_3. The quantity $SSE(X_2)$ is equal to 122.491.

xi Compute a two-sided 90% confidence interval for $\sigma_{Y|X_2,X_3}$.

xii Predict the height at age 18 of a child who is now 2 years old if it is known that his mother was 60 inches tall at age 18 and his father was 72 inches tall at age 18. Also compute a two-sided 95% confidence interval for the height at age 18 of *this* child.

xiii Consider all individuals in the subpopulation determined by
  mother's height at age 18 $= X_2 = 60$
  father's height at age 18 $= X_3 = 72$
Estimate the mean height at age 18; i.e., $\mu_Y^{(C)}(60, 72)$, of the individuals in this subpopulation. Also compute a 90% two-sided confidence interval for this mean height.

d For (i)–(iv) use model D in (4.12.7).

i Estimate the height at age 18 of an individual in the population whose length at birth is 20 inches, whose mother's height at age 18 was 60 inches, and whose father's height at age 18 was 72 inches.

ii Compute a 95% two-sided confidence interval for the height at age 18 of a randomly chosen baby belonging to the subpopulation in (i).

iii Compute a 95% two-sided confidence interval for the average height at age 18 of all individuals belonging to the subpopulation in (i).

iv Estimate $\rho^2_{Y(X_1)|X_2,X_3}$. Explain in words the meaning of $\rho^2_{Y(X_1)|X_2,X_3}$.

5

Diagnostic Procedures

5.1
Overview

In Chapters 3 and 4 we discussed procedures for point and confidence interval estimation of parameters of interest in a linear regression model. While most of these procedures are valid under assumptions (A), some procedures relied on assumptions (B). In actual applications of these procedures, the investigator needs to decide whether or not the assumptions are reasonable for the given situation. To assist the investigator in making this judgment, we suggested that a **residual analysis** be carried out. Several graphical methods of examining the residuals to assess the validity of the model assumptions were suggested. These checks of model assumptions are by no means the only ones, and there are many other graphical and numerical *diagnostic procedures* that can be of assistance in the understanding of the sample data. Some of these procedures are discussed in this chapter.

Section 5.2 introduces **studentized deleted residuals** that are useful in identifying *outliers*, i.e., sample observations that have unusual values for the *response* variable Y. Section 5.3 contains a discussion of **hat values** or **leverages** that are useful in identifying sample observations that have unusual values for the *predictor* variables X_i. Section 5.4 deals with the identification of **influential observations**, i.e., observations whose inclusion or exclusion from the analyses may give very different results. Problems associated with **ill-conditioned** data matrices, such as *numerical instability* of regression computations, and **multicollinearity** are discussed in Section 5.5. Section 5.6 contains chapter exercises. In the laboratory manuals we discuss some MINITAB and SAS commands that can be used to carry out the computations required for the procedures in this chapter.

5.2
Outliers

Often we find that there are a few (perhaps one or two) sample values that do not seem to be consistent with the regression model being fitted, but the remaining values do agree with the model. If we can identify these few and exclude them, the remaining sample may satisfy assumptions (A) or (B). Such situations are often revealed by residual plots discussed in Chapters 3 and 4. If the response y_i corresponding to sample item i deviates considerably more from its fitted value $\hat{\mu}_Y(x_{i,1}, \ldots, x_{i,k})$ than other sample values deviate from their fitted values, we say that sample observation i (case i) is an **outlier**. Such sample values need to be examined further to find out why they do not agree more closely with the fitted model. Following are some of the possible reasons.

1   Even when the model holds exactly, there are occasionally some observations that fall far from the regression curve, giving rise to an apparent outlier. (Recall that when sampling from a Gaussian population with mean zero and standard deviation one we do expect values outside the interval $(-3, 3)$ about 1% of the time). Such data values are typically included in the analysis.

2   An examination of the circumstances under which an observed value was obtained reveals that an error of some sort was made and that the sample value is incorrect. Such instances occur due to instrument malfunction during an investigation, human errors, or even errors occurring during data transcription or entry of data into a computer. If such an explanation is available, then the sample value in question should be excluded from the analysis.

3   The model under consideration does describe most of the observations adequately, but there are some combinations of the predictor factors for which the model does not give an adequate approximation, resulting in apparent outliers. Here the model should be reexamined to determine the range of values of predictor factors for which it is adequate. Observed values outside this range should be analyzed separately, or a better model should be constructed.

Occasionally we can find no explanation for an apparent outlier, and this poses a dilemma. Should this sample value be included in the analysis or excluded from the analysis? An often-used tactic is to analyze the data with the value in question included and then analyze it again with it excluded. If the results of the two analyses do not differ much, or if decisions reached are the same with either of the two sets of results, then at least our conclusions will be unaffected by the presence or absence of the value in question. If the two analyses result in different conclusions, we are forced to present both situations. Until an explanation is found for the outlying point, we have no way of verifying which conclusion is correct. In experimental situations, the factor values corresponding to the outlier may perhaps be repeated, one or more additional observations may be obtained, and the data reanalyzed. The advice of a professional statistician might be useful here.

While the standardized residuals r_i defined in (4.5.1) may be used to identify unusual observations (called *outliers*), keep in mind that the ith fitted value

$\hat{\mu}_Y(x_{i,1}, \ldots, x_{i,k})$, and hence the ith residual $\hat{e}_i$, will already be affected by including the outlying observations. For this reason, many authors suggest the use of **studentized deleted residuals**, which we discuss next.

Studentized Deleted Residuals

Studentized deleted residuals are defined as follows. Delete case i (the ith sample value) from the data and carry out the regression analysis with the remaining data. Then use the estimated regression coefficients to predict the response for sample item i. More specifically, omit the ith sample point from the data and use the remaining data to compute a regression of Y on the k predictor variables $X_1, \ldots, X_k$. Let

$$\hat{\beta}_{(-i)0}, \; \hat{\beta}_{(-i)1}, \; \hat{\beta}_{(-i)2}, \ldots, \; \hat{\beta}_{(-i)k}$$

denote the estimates of the regression coefficients and

$$\hat{\sigma}_{(-i)Y|X_1,\ldots,X_k}$$

which we abbreviate as

$$\hat{\sigma}_{(-i)}$$

denote the estimate of the subpopulation standard deviation σ with the ith sample point omitted from the data. Likewise, let

$$\hat{Y}_{(-i)}(x_{i,1}, \ldots, x_{i,k})$$

be the estimate of $Y(x_{i,1}, \ldots, x_{i,k})$ when the ith sample value is not used. Thus

$$\hat{Y}_{(-i)}(x_{i,1}, \ldots, x_{i,k}) = \hat{\beta}_{(-i)0} + \hat{\beta}_{(-i)1}x_{i,1} + \cdots + \hat{\beta}_{(-i)k}x_{i,k}$$

is the predicted value corresponding to the observed y_i when the ith sample value is excluded from the anaylsis. The estimates $\hat{\beta}_{(-i)0}, \ldots, \hat{\beta}_{(-i)k}$ are not influenced by sample item i because it is not included in the analysis. Thus

$$y_i - \hat{Y}_{(-i)}(x_{i,1}, \ldots, x_{i,k}) \tag{5.2.1}$$

is the difference between the observed value y_i and predicted value of y_i when sample item i is not used to estimate $\beta_0, \ldots, \beta_k$. This difference between the actual value and the predicted value is then divided by its standard error to obtain the *studentized deleted residual* for sample item i. It can be shown that

$$SE(y_i - \hat{Y}_{(-i)}(x_{i,1}, \ldots, x_{i,k})) = \frac{\hat{\sigma}_{(-i)}}{\sqrt{1 - h_{i,i}}} \tag{5.2.2}$$

where $h_{i,i}$ is the ith diagonal element of the hat-matrix $H = X(X^TX)^{-1}X^T$ defined in (4.5.2) where all observations are included in the X matrix.

D E F I N I T I O N

We define the *studentized deleted residual* for case i, denoted by T_i, as

$$T_i = \frac{y_i - \hat{Y}_{(-i)}(x_{i,1}, \ldots, x_{i,k})}{\hat{\sigma}_{(-i)} / \sqrt{1 - h_{i,i}}} \qquad \blacksquare \qquad (5.2.3)$$

It can be shown that T_i has a student's-t distribution with $(n - 1) - k - 1 = n - k - 2$ degrees of freedom. For this reason studentized deleted residuals are sometimes called t-residuals.

Thus T_i is computed and examined for each $i = 1, \ldots, n$. Roughly speaking, if the absolute value of any T_i is greater than 2 (some investigators use 3), this indicates that the ith sample observation should be carefully scrutinized as a possible outlier. Outliers stand out more clearly when we examine studentized deleted residuals T_i than when we examine the standardized residuals r_i. We illustrate the use of studentized deleted residuals in Example 5.2.1.

E X A M P L E 5.2.1

In a study to investigate how automobile insurance premiums Y (in dollars) for collision coverage are related to the age X_1 (in months) and the purchase price X_2 (in dollars) of a car, a simple random sample of 36 car owners was selected from a study population, and each was asked to provide the above information. The data are displayed in Table 5.2.1 and are also stored in the file **premiums.dat** on the data disk.

T A B L E 5.2.1

Premiums Data

Observation Number	Premium Y (dollars)	Age of Car X_1 (months)	Price of Car X_2 (dollars)
1	221	57	11804
2	448	8	12926
3	515	6	14054
4	632	12	17486
5	48	47	8700
6	189	30	8570
7	581	34	18982
8	102	39	9198
9	404	33	14986
10	83	59	8473
11	280	56	13891
12	565	13	16127
13	1105	10	29480
14	388	46	15868
15	435	2	10782

T A B L E 5.2.1

(Continued)

Observation Number	Premium Y (dollars)	Age of Car X_1 (months)	Price of Car X_2 (dollars)
16	309	11	8645
17	322	17	9086
18	741	32	22559
19	500	34	14969
20	626	1	14861
21	1051	34	29733
22	845	4	22893
23	278	59	15198
24	333	56	16696
25	650	34	20411
26	772	27	23128
27	477	19	16507
28	443	37	13704
29	692	3	16472
30	618	36	18422
31	1050	7	27110
32	643	45	22968
33	116	46	9177
34	269	9	8977
35	259	38	10514
36	491	16	13739

Exhibit 5.2.1 contains a MINITAB output for the regression of Y on X_1, X_2. The following are included: observation number, age X_1, price X_2, premium Y, the standardized residuals r_i (labeled stdresid), the fitted values $\hat{\mu}_Y(x_{i,1}, x_{i,2})$ (labeled fits), the residuals $\hat{e}_i$ (labeled residual), and the studentized deleted residuals T_i (labeled tresid). ■

E X H I B I T 5.2.1

MINITAB Output for Example 5.2.1

```
The regression equation is
premium  = 6.9 - 5.10 age + 0.0395 price


Predictor        Coef        Stdev      t-ratio          p
Constant         6.90        24.52         0.28      0.780
Age           -5.0996       0.3894       -13.10      0.000
Price        0.039533     0.001209        32.69      0.000


s = 41.94        R-sq = 97.7%      R-sq(adj) = 97.6%
```

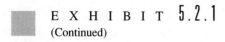

E X H I B I T **5.2.1**

(Continued)

Analysis of Variance

SOURCE	DF	SS	MS	F	p
Regression	2	2492087	1246043	708.49	0.000
Error	33	58038	1759		
Total	35	2550125			

Unusual Observations

Obs.	age	premium	Fit	Stdev.Fit	Residual	St.Resid	
27	19.0	477.00	562.58	7.85	-85.58	-2.08R	**(5.2.4)**
28	37.0	443.00	359.97	7.99	83.03	2.02R	**(5.2.5)**

R denotes an obs. with a large st. resid.

obsno	premium	age	price	fits	residual	stdresid	tresid
1	221	57	11804	182.87	38.1338	0.95962	0.95844
2	448	8	12926	477.10	-29.1040	-0.72149	-0.71614
3	515	6	14054	531.90	-16.8965	-0.41918	-0.41388
4	632	12	17486	636.98	-4.9763	-0.12178	-0.11995
5	48	47	8700	111.15	-63.1519	-1.57678	-1.61473
6	189	30	8570	192.71	-3.7062	-0.09163	-0.09025
7	581	34	18982	583.93	-2.9259	-0.07124	-0.07016
8	102	39	9198	171.64	-69.6364	-1.71939	-1.77448
9	404	33	14986	431.05	-27.0514	-0.65491	-0.64914
10	83	59	8473	40.98	42.0176	1.07656	1.07924
11	280	56	13891	270.47	9.5287	0.23852	0.23508
12	565	13	16127	578.15	-13.1512	-0.32131	-0.31690
13	1105	10	29480	1121.33	-16.3350	-0.43316	-0.42776
14	388	46	15868	399.62	-11.6244	-0.28516	-0.28115
15	435	2	10782	422.94	12.0571	0.30633	0.30208
16	309	11	8645	292.56	16.4359	0.41454	0.40927
17	322	17	9086	279.40	42.5995	1.06031	1.06237
18	741	32	22559	735.53	5.4651	0.13511	0.13309
19	500	34	14969	425.28	74.7202	1.80976	1.87775
20	626	1	14861	589.30	36.7021	0.91975	0.91754
21	1051	34	29733	1008.95	42.0543	1.12130	1.12584
22	845	4	22893	891.53	-46.5284	-1.17327	-1.18023
23	278	59	15198	306.84	-28.8421	-0.72822	-0.72294
24	333	56	16696	381.36	-48.3615	-1.21368	-1.22275
25	650	34	20411	640.42	9.5814	0.23453	0.23114
26	772	27	23128	783.53	-11.5273	-0.28539	-0.28138
27	477	19	16507	562.58	-85.5760	-2.07728	-2.19403
28	443	37	13704	359.97	83.0284	2.01678	2.12100

E X H I B I T 5.2.1

(Continued)

29	692	3	16472	642.79	49.2136	1.22453	1.23420
30	618	36	18422	551.59	66.4119	1.61684	1.65923
31	1050	7	27110	1042.94	7.0596	0.18290	0.18020
32	643	45	22968	685.41	-42.4088	-1.06941	-1.07182
33	116	46	9177	135.11	-19.1088	-0.47524	-0.46959
34	269	9	8977	315.89	-46.8884	-1.18467	-1.19221
35	259	38	10514	228.76	30.2385	0.74147	0.73631
36	491	16	13739	468.45	22.5526	0.55066	0.54476

In many regression computer packages, the output will indicate when a sample value should be examined as a possible outlier by tagging a standardized (or studentized deleted) residual if its absolute value is larger than a certain number, generally larger than 2 (some packages use 3). As you can see in (5.2.4) and (5.2.5), sample observations 27 and 28 have been tagged with an R to indicate that these observations have standardized residuals greater than 2 in magnitude and should be examined.

We plot the studentized deleted residuals against the corresponding observation numbers to see if any stand out as being unduly large. We could examine the magnitudes of studentized deleted residuals in the preceding exhibit, but with a plot of them against observation numbers we can view all of them together.

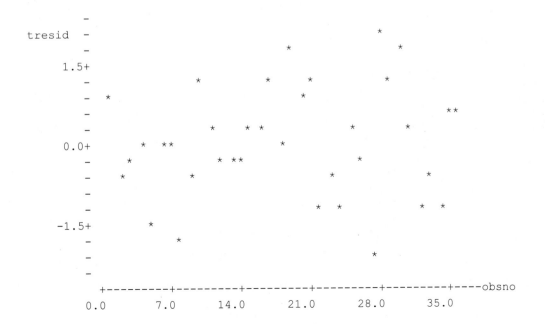

This plot does not indicate the presence of any outliers because none of the studentized deleted residuals are exceptionally large in magnitude. For instance, none of them exceeds 3 units in magnitude, and the studentized deleted residuals for observations 27 and 28 are just slightly greater than 2 in magnitude.

Suppose, for illustration, that the Y value for sample number 36 in Table 5.2.1 had been incorrectly entered as 1491 instead of 491. We carry out the calculation of studentized deleted residuals along with other diagnostic statistics for this modified data set. The results are given in Exhibit 5.2.2.

E X H I B I T 5.2.2
MINITAB Output for Example 5.2.1—Modified Data

```
The regression equation is
premium = 102 - 6.24 age + 0.0373 price
```

Predictor	Coef	Stdev	t-ratio	p
Constant	101.8	104.6	0.97	0.338
Age	-6.244	1.662	-3.76	0.001
Price	0.037324	0.005161	7.23	0.000

s = 179.0   R-sq = 70.1%   R-sq(adj) = 68.3%

Analysis of Variance

SOURCE	DF	SS	MS	F	p
Regression	2	2476278	1238139	38.66	0.000
Error	33	1056902	32027		
Total	35	3533181			

Unusual Observations

Obs.	Age	Premium	Fit	Stdev.Fit	Residual	St.Resid	
36	16.0	1491.0	514.7	38.5	976.3	5.59R	**(5.2.6)**

R denotes an obs. with a large st. resid.

E X H I B I T 5.2.2
(Continued)

obsno	premium	age	price	fits	residual	stdresid	tresid
1	221	57	11804	186.48	34.523	0.20358	0.2006
2	448	8	12926	534.29	-86.293	-0.50129	-0.4955
3	515	6	14054	588.88	-73.881	-0.42951	-0.4241
4	632	12	17486	679.51	-47.515	-0.27248	-0.2686
5	48	47	8700	133.06	-85.061	-0.49768	-0.4919
6	189	30	8570	234.35	-45.350	-0.26275	-0.2590
7	581	34	18982	597.99	-16.991	-0.09695	-0.0955
8	102	39	9198	201.60	-99.597	-0.57626	-0.5703
9	404	33	14986	455.09	-51.089	-0.28984	-0.2858
10	83	59	8473	49.66	33.336	0.20015	0.1972
11	280	56	13891	270.62	9.384	0.05505	0.0542
12	565	13	16127	622.55	-57.548	-0.32948	-0.3250
13	1105	10	29480	1139.66	-34.664	-0.21540	-0.2123
14	388	46	15868	406.84	-18.841	-0.10831	-0.1067
15	435	2	10782	491.73	-56.733	-0.33776	-0.3332
16	309	11	8645	355.78	-46.779	-0.27648	-0.2726
17	322	17	9086	334.78	-12.777	-0.07452	-0.0734
18	741	32	22559	743.99	-2.986	-0.01730	-0.0170
19	500	34	14969	448.21	51.789	0.29394	0.2898
20	626	1	14861	650.22	-24.220	-0.14223	-0.1401
21	1051	34	29733	999.26	51.741	0.32328	0.3189
22	845	4	22893	931.27	-86.274	-0.50979	-0.5040
23	278	59	15198	300.67	-22.667	-0.13411	-0.1321
24	333	56	16696	375.31	-42.309	-0.24881	-0.2452
25	650	34	20411	651.33	-1.327	-0.00761	-0.0075
26	772	27	23128	796.44	-24.441	-0.14180	-0.1397
27	477	19	16507	599.27	-122.269	-0.69550	-0.6900
28	443	37	13704	382.27	60.735	0.34571	0.3410
29	692	3	16472	697.86	-5.861	-0.03418	-0.0337
30	618	36	18422	564.60	53.397	0.30464	0.3004
31	1050	7	27110	1069.94	-19.937	-0.12105	-0.1192
32	643	45	22968	678.08	-35.084	-0.20732	-0.2043
33	116	46	9177	157.11	-41.108	-0.23957	-0.2361
34	269	9	8977	380.66	-111.658	-0.66109	-0.6553
35	259	38	10514	256.96	2.041	0.01173	0.0116
36	1491	16	13739	514.69	976.312	5.58610	23.5827

In (5.2.6) we note that observation number 36 is tagged as having a standardized residual greater than 2. We now plot the standardized residuals r_i against observation numbers, and also plot the studentized deleted residuals T_i against the observation numbers, to see if any of these residuals stand out as being unduly large.

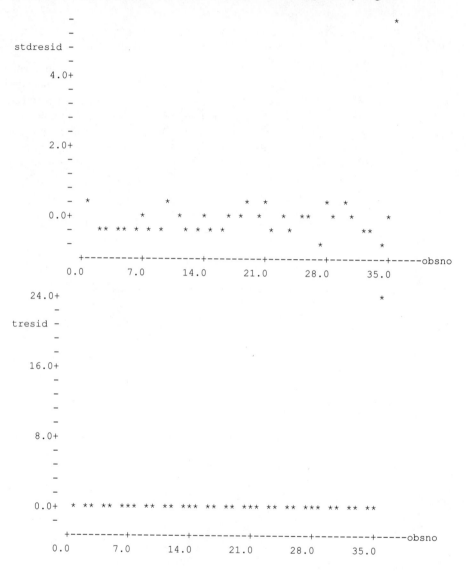

Both plots clearly indicate that observation 36 is an outlier, although the studentized deleted residual for case 36 is considerably bigger than its standardized residual. This demonstrates the fact that it is often easier to identify outlying observations based on their studentized deleted residuals T_i than on their standardized residuals r_i.

In a real problem we would have to look into why observation 36 is an outlier and make a decision as to whether or not it should be included in any further analyses. As mentioned earlier, if no explanation can be found, we should carry out the analysis both ways, once with the questionable point included and once with it excluded.

Problems   5.2

5.2.1 Consider Task 3.4.1 where crystalline forms of certain chemical compounds are used in various electronic devices and where it is often more desirable to have large crystals than small ones. Crystals of one particular compound are to be produced by a commercial process, and an investigator wants to examine the relationship between Y, the weight of a crystal in grams and X, the time taken (in hours) for the crystal to grow to its final size. The following data are from a laboratory study in which 14 crystals of various sizes were obtained by allowing the crystals to grow for different preselected amounts of time. The data are reproduced in Table 5.2.2, and are also stored in the file **crystal.dat** on the data disk. A SAS output from a regression analysis of these data is in Exhibit 5.2.3.

T A B L E   5.2.2

Crystal Number	Weight Y (in grams)	Time X (in hours)
1	0.08	2
2	1.12	4
3	4.43	6
4	4.98	8
5	4.92	10
6	7.18	12
7	5.57	14
8	8.40	16
9	8.81	18
10	10.81	20
11	11.16	22
12	10.12	24
13	13.12	26
14	15.04	28

E X H I B I T 5.2.3
SAS Output for Problem 5.2.1

The SAS System

0:00 Saturday, Jan 1, 1994

Model: MODEL1
Dependent Variable: WEIGHT

Analysis of Variance

Source	DF	Sum of Squares	Mean Square	F Value	Prob>F
Model	1	230.63070	230.63070	204.578	0.0001
Error	12	13.52819	1.12735		
C Total	13	244.15889			

Root MSE	1.06177	R-square	0.9446	
Dep Mean	7.55286	Adj R-sq	0.9400	
C.V.	14.05782			

Parameter Estimates

Variable	DF	Parameter Estimate	Standard Error	T for H0: Parameter=0	Prob > |T|
INTERCEP	1	0.001429	0.59938725	0.002	0.9981
TIME	1	0.503429	0.03519723	14.303	0.0001

OBS	WEIGHT	TIME	FITS	RESIDUAL	STDRESID	TRESID
1	0.08	2	1.0083	-0.92829	-1.01438	-1.01572
2	1.12	4	2.0151	-0.89514	-0.94518	-0.94063
3	4.43	6	3.0220	1.40800	1.44726	1.52513
4	4.98	8	4.0289	0.95114	0.95781	0.95424
5	4.92	10	5.0357	-0.11571	-0.11481	-0.10998
6	7.18	12	6.0426	1.13743	1.11767	1.13055
7	5.57	14	7.0494	-1.47943	-1.44682	-1.52456
8	8.40	16	8.0563	0.34371	0.33614	0.32335
9	8.81	18	9.0631	-0.25314	-0.24874	-0.23877
10	10.81	20	10.0700	0.74000	0.73420	0.71929
11	11.16	22	11.0769	0.08314	0.08373	0.08018
12	10.12	24	12.0837	-1.96371	-2.01847	-2.37793
13	13.12	26	13.0906	0.02943	0.03107	0.02975
14	15.04	28	14.0974	0.94257	1.02999	1.03285

a Examine a plot of Y versus X and decide if there appear to be any outliers in this data set. If so, state which observations you regard as outliers. Why?

b Examine the standardized residuals and studentized deleted residuals given in the computer output in Exhibit 5.2.3 and decide whether there appear to be any outliers in this data set. If so, state which observations you regard as outliers. Why?

5.2.2 For this problem the data are the same as in Table 5.2.1 except one Y value has been changed (suppose it was incorrectly recorded). A SAS output containing various diagnostic statistics for the incorrect data is given in Exhibit 5.2.4. Identify the sample item that has been recorded incorrectly by first looking at the studentized deleted residuals and then by checking the standardized residuals.

E X H I B I T 5.2.4
SAS Output for Problem 5.2.2

The SAS System

0:00 Saturday, Jan 1, 1994

Model: MODEL1
Dependent Variable: PREMIUM

Analysis of Variance

Source	DF	Sum of Squares	Mean Square	F Value	Prob>F
Model	2	2214896.6332	1107448.3166	40.958	0.0001
Error	33	892283.67239	27038.89916		
C Total	35	3107180.3056			

Root MSE	164.43509	R-square	0.7128	
Dep Mean	513.36111	Adj R-sq	0.6954	
C.V.	32.03108			

Parameter Estimates

Variable	DF	Parameter Estimate	Standard Error	T for H0: Parameter=0	Prob > |T|
INTERCEP	1	-53.402539	96.13561819	-0.555	0.5823
AGE	1	-2.472629	1.52683741	-1.619	0.1149
PRICE	1	0.040413	0.00474230	8.522	0.0001

E X H I B I T 5.2.4
(Continued)

OBS	PREMIUM	AGE	PRICE	FITS	RESIDUAL	STDRESID	TRESID
1	221	57	11804	282.69	-61.695	-0.39595	-0.3908
2	448	8	12926	449.20	-1.197	-0.00757	-0.0075
3	515	6	14054	499.73	15.271	0.09662	0.0952
4	632	12	17486	623.59	8.409	0.05248	0.0517
5	48	47	8700	181.98	-133.979	-0.85315	-0.8495
6	189	30	8570	218.76	-29.760	-0.18765	-0.1849
7	581	34	18982	629.65	-48.651	-0.30211	-0.2979
8	102	39	9198	221.89	-119.885	-0.75493	-0.7499
9	404	33	14986	470.63	-66.633	-0.41142	-0.4062
10	83	59	8473	143.13	-60.133	-0.39294	-0.3878
11	280	56	13891	369.51	-89.510	-0.57143	-0.5655
12	565	13	16127	566.20	-1.197	-0.00746	-0.0073
13	1105	10	29480	1113.25	-8.252	-0.05581	-0.0550
14	388	46	15868	474.13	-86.133	-0.53888	-0.5330
15	435	2	10782	377.39	57.613	0.37331	0.3684
16	309	11	8645	268.77	40.229	0.25877	0.2551
17	322	17	9086	271.76	50.243	0.31894	0.3146
18	741	32	22559	779.15	-38.154	-0.24057	-0.2371
19	500	34	14969	467.47	32.527	0.20092	0.1980
20	626	1	14861	544.71	81.295	0.51957	0.5137
21	1051	34	29733	1064.13	-13.133	-0.08931	-0.0880
22	845	4	22893	861.89	-16.886	-0.10859	-0.1070
23	1278	59	15198	414.91	863.088	5.55770	21.6335
24	333	56	16696	482.87	-149.869	-0.95922	-0.9580
25	650	34	20411	687.40	-37.402	-0.23349	-0.2301
26	772	27	23128	814.51	-42.513	-0.26843	-0.2646
27	477	19	16507	566.72	-89.718	-0.55543	-0.5495
28	443	37	13704	408.93	34.068	0.21105	0.2080
29	692	3	16472	604.87	87.134	0.55294	0.5470
30	618	36	18422	602.07	15.926	0.09888	0.0974
31	1050	7	27110	1024.89	25.110	0.16592	0.1635
32	643	45	22968	763.54	-120.539	-0.77521	-0.7704
33	116	46	9177	203.73	-87.728	-0.55644	-0.5505
34	269	9	8977	287.13	-18.133	-0.11684	-0.1151
35	259	38	10514	277.54	-18.542	-0.11596	-0.1142
36	491	16	13739	462.27	28.728	0.17889	0.1762

5.3
Leverages or Hat Values

The **hat matrix** H was defined in (4.5.2), and the diagonal elements $h_{i,i}$ of the hat matrix were used in the calculation of standardized residuals (see (4.5.1)) and also the studentized deleted residuals (see (5.2.2) and (5.2.3)). The numbers $h_{i,i}$ are called **leverages** or **hat values**.

Hat values (leverages) are determined entirely by the sample values of the predictor variables, $X_1, \ldots, X_k$, and are not affected by the values of the response variable Y.

Hat values have a practical interpretation. Loosely speaking, the value $h_{i,i}$ is a measure of how *typical* or *atypical* the values of the predictor variables are for observation i. Recall that hat values are used in calculating $SE(\hat{\mu}_Y(x_{i,1}, \ldots, x_{i,k}))$ and $SE(\hat{Y}(x_{i,1}, \ldots, x_{i,k}))$. In fact, from (4.6.5), using $x^T = (x_{i,1}, \ldots, x_{i,k})$ we have

$$SE(\hat{\mu}_Y(x_{i,1}, \ldots, x_{i,k})) = \hat{\sigma}\sqrt{h_{i,i}}$$

and from (4.6.6) we have

$$SE(\hat{Y}(x_{i,1}, \ldots, x_{i,k})) = \hat{\sigma}\sqrt{1 + h_{i,i}}$$

To better understand the practical meaning of the leverages, it is useful to explicitly examine what they are in straight line regression. Recall that for straight line regression ($k = 1$), the observations satisfy the model

$$y = X\beta + e$$

in matrix form where

$$y = \begin{bmatrix} y_1 \\ \vdots \\ y_n \end{bmatrix} \qquad X\beta + e = \begin{bmatrix} 1 & x_1 \\ \vdots & \vdots \\ 1 & x_n \end{bmatrix} \begin{bmatrix} \beta_0 \\ \beta_1 \end{bmatrix} + \begin{bmatrix} e_1 \\ \vdots \\ e_n \end{bmatrix} \qquad \text{(5.3.1)}$$

The hat matrix is

$$H = X(X^T X)^{-1} X^T \qquad \text{(5.3.2)}$$

$$= \begin{bmatrix} 1 & x_1 \\ \vdots & \vdots \\ 1 & x_n \end{bmatrix} \begin{bmatrix} n & \sum_{i=1}^n x_i \\ \sum_{i=1}^n x_i & \sum_{i=1}^n x_i^2 \end{bmatrix}^{-1} \begin{bmatrix} 1 & \cdots & 1 \\ x_1 & \cdots & x_n \end{bmatrix}$$

From this we find that

$$h_{i,i} = \frac{1}{n} + \frac{(x_i - \bar{x})^2}{SSX} \qquad \text{for} \qquad i = 1, \ldots, n \qquad \text{(5.3.3)}$$

From (5.3.3) we see that the values of x_i that are far away from $\bar{x}$ will result in large values of $h_{i,i}$ relative to the others. It can be shown mathematically that $\sum_{i=1}^n h_{i,i} = 2$

for the straight line regression model and, more generally, that

$$\sum_{i=1}^{n} h_{i,i} = p \qquad \text{(5.3.4)}$$

for the multiple regression model, where p is the number of β parameters in the regression function. The *average* of $h_{1,1}, \ldots, h_{n,n}$ is thus equal to p/n. In a set of data, if any $h_{i,i}$ value is large relative to p/n, say greater than $2p/n$ or $3p/n$, then the corresponding sample item may be considered to have an unusual set of values for the predictor variables, and the corresponding data point is called a *high leverage point*. Some of the possible reasons for such unusual values are

- The values of one or more of the predictor variables are recorded incorrectly.

- The corresponding sample item may not actually belong to the population under investigation.

- The investigator may have designed the study to include population items that have extreme values for the predictor variables. If the investigator is certain that the regression function being used is correct, then such a design may result in more precise estimates of the parameters. In this case, however, the investigator is already aware of such extreme cases and the hat values (leverages) provide a confirmation of this.

Roughly speaking, if $h_{i,i}$ is large, then for multiple regression with k predictors, this implies that the ith observation $x_{i,1}, \ldots, x_{i,k}$ is far removed from the means $\bar{x}_1, \ldots, \bar{x}_k$ of all observations. This unusual set of hat values might have an influence on point estimates and confidence intervals for the population regression quantities. This influence may be useful or it may be detrimental. An examination of the hat values is thus advisable, particularly in observational studies where it is not uncommon for the sample to be contaminated with observations that really do not belong to the study population. They are especially useful in conjunction with the procedures discussed in the next section for examining *influential observations*. Example 5.3.1 illustrates the use of hat values to find unusual predictor (X) values.

E X A M P L E   **5.3.1**

A wildlife biologist interested in a certain species of mammals has collected the data in Table 5.3.1 on 25 offspring of that species under the age of 12 months. The data were obtained by simple random sampling from a target population. The quantities observed are

$$Y = \text{weight of the offspring (in pounds)}$$
$$X_1 = \text{age of the offspring (in months)}$$
$$X_2 = \text{length of the offspring (in inches)}$$

The investigator is interested in the relationship among Y, X_1, and X_2. It is supposed that assumptions (A) hold and the regression function of Y on X_1 and X_2 is given by

$$\mu_Y(x_1, x_2) = \beta_0 + \beta_1 x_1 + \beta_2 x_2$$

The data are also stored in the file **mammalwt.dat** on the data disks.

Using MINITAB we have computed several relevant quantities that are given in Exhibit 5.3.1. This exhibit also contains a plot of the hat values (`hatvals`) against observation numbers and a plot of length X_2 against age X_1.

Observe that the hat value corresponding to sample observation 18 is high relative to the other hat values. The value of p in this problem is 3 because there are three β parameters in the regression function. Hence $p/n = 3/25 = 0.12$ and $3p/n = 0.36$. The value $h_{18,18}$ is 0.618561, which is much greater than $3p/n = 0.36$. We therefore conclude that the values of X_1 and X_2 for offspring 18 are unusual relative to the X_1 and X_2 values for the other offspring in the sample. From Table 5.3.1

T A B L E 5.3.1

Mammal Offspring Data

Observation Number	Weight Y	Age X_1	Length X_2
1	16.4	10	23.1
2	16.3	7	22.8
3	18.2	12	24.0
4	14.8	6	22.1
5	14.7	5	21.5
6	16.4	11	23.2
7	17.3	10	22.7
8	13.0	1	20.2
9	18.8	12	24.2
10	16.3	11	23.5
11	15.0	8	22.5
12	16.8	9	22.6
13	16.0	8	23.2
14	18.4	11	23.4
15	13.5	5	21.2
16	12.4	2	20.6
17	15.6	9	22.2
18	16.6	4	23.4
19	16.4	10	23.1
20	14.1	3	21.0
21	17.9	10	23.1
22	17.7	11	22.8
23	16.3	10	23.5
24	16.6	8	22.3
25	15.2	6	22.5

E X H I B I T **5.3.1**
MINITAB Output for Example 5.3.1

```
The regression equation is
weight = - 5.94 + 0.200 age + 0.902 length

Predictor       Coef       Stdev     t-ratio        p
Constant      -5.939       5.636      -1.05      0.303
Age           0.19969     0.08829      2.26      0.034
Length        0.9021      0.2753       3.28      0.003

s = 0.7261     R-sq = 82.3%     R-sq(adj) = 80.7%


Unusual Observations
Obs.     age     length      Fit  Stdev.Fit  Residual   St.Resid
 18      4.0     16.600    15.970    0.571      0.630      1.41 X
```
                                                                      (5.3.5)
```
hatvals

   0.057404    0.076032    0.121121    0.056115    0.087465    0.084756    0.085440
   0.274175    0.144781    0.079048    0.041413    0.055036    0.091957    0.078075
   0.124700    0.205291    0.109248    0.618561    0.057404    0.149319    0.057404
   0.132624    0.075377    0.052852    0.084405
```

```
        -
  0.60+                                                    *
        -
hatvals -
        -
        -
        -
  0.40+
        -
        -
        -             *
        -
  0.20+                                           *
        -                              *
        -                   *                *      *
        -        *      * * *        *    * *       * *
        -      *     *              * *          *    *  *
        -    *      *                    * *      *  *     *
  0.00+
         +---------+---------+---------+---------+---------+-----obsno
        0.0       5.0      10.0      15.0      20.0      25.0
```

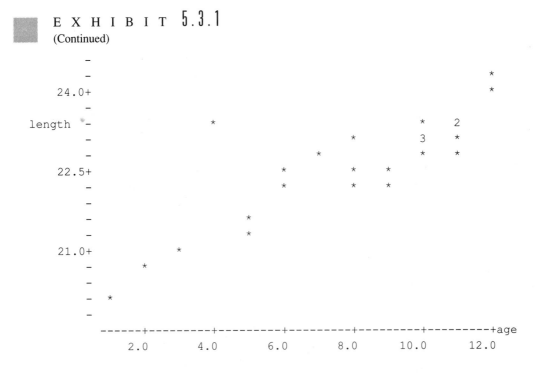

we find that the age and length for offspring 18 are 4 months and 23.4 inches, respectively. Neither of these values in itself is unusual; what is unusual is the combination $X_1 = 4$, $X_2 = 23.4$. Offspring 18 is rather long for its age. An examination of the plot of X_2 against X_1 given in Exhibit 5.3.1 makes this clearer.

Aside from the possibility that this information may be of interest to the investigator in its own right, it is important to be aware of sample observations with high hat values. Such observations may have a strong influence on the estimates of the regression coefficients and, unless the assumed form of the regression function is known to be valid at all of the (x_1, x_2) values occurring in the sample, the estimates and confidence intervals for various quantities in regression may strongly depend on whether or not the observation under question is included in the analysis. Some computer programs warn about this possibility by tagging cases with high hat values. In (5.3.5) of Exhibit 5.3.1 note that observation 18 has been tagged by MINITAB using the symbol "x", indicating that the hat value for observation 18 is greater than $3p/n$. Further examination of such cases is warranted, and we discuss this in the next section. ∎

Problems   5.3

Problems 5.3.1–5.3.3 refer to the insurance premium data in Table 5.2.1. The MINITAB output from a regression analysis of these data is given in Exhibit 5.3.2. The exhibit also contains the hat values and a plot of the hat values against observation numbers.

E X H I B I T   5.3.2
MINITAB Output for Problems 5.3.1–5.3.3

```
The regression equation is
premium = 6.9 - 5.10 age + 0.0395 price


Predictor         Coef        Stdev      t-ratio          p
Constant          6.90        24.52         0.28      0.780
Age            -5.0996       0.3894       -13.10      0.000
Price         0.039533     0.001209        32.69      0.000


s = 41.94      R-sq = 97.7%      R-sq(adj) = 97.6%



Analysis of Variance

SOURCE         DF           SS           MS          F         p
Regression      2      2492087      1246043     708.49     0.000
Error          33        58038         1759
Total          35      2550125



  hatvals

  0.102106    0.074772    0.076159    0.050586    0.087922    0.069855    0.040922
  0.067328    0.029894    0.133848    0.092546    0.047447    0.191368    0.055131
  0.119122    0.106149    0.082200    0.069749    0.030746    0.094581    0.200206
  0.105777    0.108070    0.097195    0.051014    0.072351    0.035019    0.036306
  0.081593    0.040688    0.152939    0.105813    0.080722    0.109292    0.054343
  0.046241
```

E X H I B I T 5.3.2
(Continued)

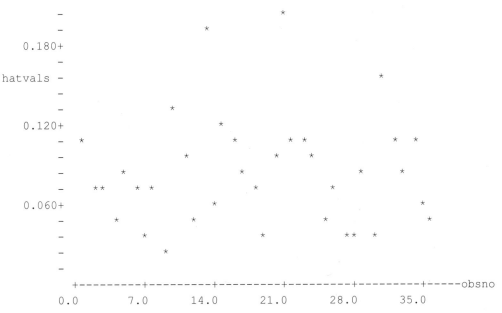

5.3.1 **a** What is the value of n? What is the value of p? What is the value of $2p/n$?

b Are there any hat values that are larger than $2p/n$? If so, what are they?

5.3.2 Carry out an analysis of hat values for these data and explain your findings in a short report.

5.3.3 If the Y value (premium) for car 8 in the sample is changed from its value 102 to an incorrect value 302, how will this change the hat value $h_{8,8}$? How will it change the hat values $h_{i,i}$ for $i \neq 8$?

5.4
Influential Observations—Cook's Distance and DFFITS

There are instances when we find that the conclusions derived from a regression analysis are highly influenced by one or more specific sample observations. A sample observation is said to be an **influential observation** if the exclusion of this observation from the analysis results in conclusions that are very different from the conclusions reached when this observation is included in the analysis. Figure 5.4.1 illustrates one such situation.

Observe that while most of the points in Figure 5.4.1 suggest a straight line regression function with a positive slope, the point with the largest X value (this point is labeled as P in the figure) will unduly influence the estimation of the slope

F I G U R E   5.4.1

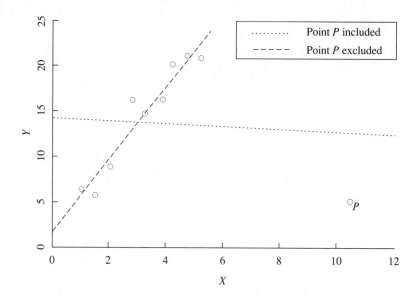

and the intercept, and its inclusion in the analysis will actually result in a negative estimate for the slope. In this section we discuss two commonly used diagnostic measures for identifying such influential observations. They are:

1   Cook's distance

2   DFFITS

Cook's Distance

The *influence* of any data point may be assessed by examining the amount by which the estimates of the regression coefficients (i.e., β parameters) change if this data point is deleted from the analysis. This approach leads to the measure known as *Cook's distance* [1], [2].

Cook's distance for sample observation i is defined by

$$c_i = \frac{1}{p}\left(\frac{h_{i,i}}{1 - h_{i,i}}\right) r_i^2 \qquad \text{(5.4.1)}$$

where $h_{i,i}$ is the hat value for sample item i, r_i is the ith standardized residual defined in (4.5.1), and p is the number of β's in the regression function. If $\hat{\beta}_j$ is the estimate of β_j using all of the data points and $\hat{\beta}_{j(-i)}$ is the estimate of β_j using all of the data points *but excluding observation i*, then c_i *is a measure of the difference between*

$$(\hat{\beta}_0, \hat{\beta}_1, \ldots, \hat{\beta}_k)$$

and

$$(\hat{\beta}_{0(-i)}, \hat{\beta}_{1(-i)}, \ldots, \hat{\beta}_{k(-i)})$$

A large value of c_i indicates that the estimates of some or all of the β's will change substantially if observation i is omitted from the analysis. Some authors recommend examining those cases for which c_i is greater than the tabled F-value $F_{0.50:p,n-p}$ in Table T-5 of Appendix T. The interpretation of Cook's distance is illustrated in Example 5.4.1.

DFFITS

The *influence* of an individual sample observation may also be assessed by examining, for each i, the amount by which the predicted Y value for sample item i changes when this item is excluded from the analysis; i.e., by examining the quantity

$$\hat{Y}(x_{i,1}, \ldots, x_{i,k}) - \hat{Y}_{(-i)}(x_{i,1}, \ldots, x_{i,k})$$

We sometimes write this quantity as $\hat{Y}_i - \hat{Y}_{(-i)i}$ for ease of notation. We first standardize this quantity by dividing by its standard error based on data with the ith sample value removed. This standard error is $\hat{\sigma}_{(-i)}\sqrt{h_{i,i}}$, where $h_{i,i}$ is the hat value for the ith observation using all the data and the standardized value is called DFFITS (*dif*ference in the *fit*ted value–*s*tandardized) for the ith observation. Thus we define

$$\text{DFFITS}_i = \frac{\hat{Y}_i - \hat{Y}_{(-i)i}}{\hat{\sigma}_{(-i)}\sqrt{h_{i,i}}} \tag{5.4.2}$$

If the magnitude of $\hat{Y}_i - \hat{Y}_{(-i)i}$ is large relative to its standard error, this tells us that it makes a difference whether or not we include the ith observation in predicting $Y(x_{i,1}, \ldots, x_{i,k})$. In other words, if the absolute value of DFFITS_i is large, the ith sample observation is said to influence the estimate of $Y(x_{i,1}, \ldots, x_{i,k})$ and hence is called an *influential observation*. An alternate formula for computing DFFITS_i is

$$\text{DFFITS}_i = T_i\sqrt{\frac{h_{i,i}}{1 - h_{i,i}}} \tag{5.4.3}$$

where T_i is the studentized deleted residual for sample item i defined in (5.2.3). Some authors have suggested that observations for which the absolute value of DFFITS_i is greater than $2\sqrt{p/n}$ may be considered *influential* and should be examined.

Example 5.4.1 illustrates the use and interpretation of Cook's distance and DFFITS.

E X A M P L E 5.4.1

Consider the artificial data in Table 5.4.1, which consists of 10 observations of a predictor variable X and a response variable Y. The data are also stored in the file **table 541.dat** on the data disk.

■ T A B L E **5.4.1**

Observation Number	Y	X
1	6.8	1.0
2	6.1	1.5
3	9.5	2.0
4	16.2	2.5
5	15.2	3.0
6	16.4	3.5
7	21.5	4.0
8	22.5	4.5
9	22.3	5.0
10	4.7	11.0

The plot of these data is shown in Figure 5.4.1. Note that if we were to carry out a straight line regression analysis for these data, the results using all the data values would be very different from the results obtained when sample observation 10 is excluded. The fitted straight lines with and without sample observation 10 are also shown in Figure 5.4.1. Thus it is clear that observation 10 is an *influential* data point. An examination of the Cook's distances or DFFITS should also reveal this. First we plot Y against X.

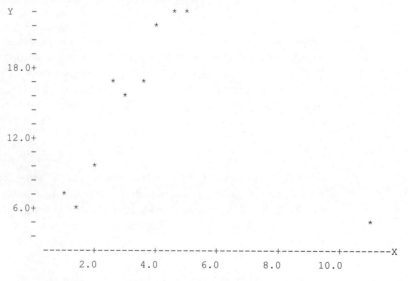

Next we regress Y on X and compute the standardized residuals (stdresid), Cook's distances (cooksd), DFFITS (dffits), and hat values (hatvals). The results are in Exhibit 5.4.1.

E X H I B I T 5.4.1

MINITAB Output for Example 5.4.1

```
The regression equation is
Y = 14.5 - 0.106 X

Predictor        Coef        Stdev      t-ratio          p
Constant       14.521        4.006         3.63      0.007
X             -0.1055       0.8599        -0.12      0.905

s = 7.327      R-sq = 0.2%      R-sq(adj) = 0.0%

Analysis of Variance

SOURCE         DF          SS          MS        F         p
Regression      1        0.81        0.81     0.02     0.905
Error           8      429.47       53.68
Total           9      430.28

Unusual Observations
Obs.       X           Y        Fit  Stdev.Fit   Residual    St.Resid
  10     11.0        4.70      13.36      6.61      -8.66     -2.74RX


R denotes an obs. with a large st. resid.
X denotes an obs. whose X value gives it large influence.

obsno       Y        X     stdresid      hatvals      cooksd      dffits

    1      6.8      1.0    -1.16791     0.207989     0.1791     -0.6147
    2      6.1      1.5    -1.23997     0.172865     0.1607     -0.5900
    3      9.5      2.0    -0.70981     0.144628     0.0426     -0.2820
    4     16.2      2.5     0.28319     0.123278     0.0056      0.0998
    5     15.2      3.0     0.14394     0.108815     0.0013      0.0471
    6     16.4      3.5     0.32368     0.101240     0.0059      0.1023
    7     21.5      4.0     1.06509     0.100551     0.0634      0.3596
    8     22.5      4.5     1.22081     0.106749     0.0891      0.4376
    9     22.3      5.0     1.20843     0.119835     0.0994      0.4613
   10      4.7     11.0    -2.74104     0.814050    16.4458    -21.7503
```

An examination of the diagnostic statistics in Exhibit 5.4.1 reveals that observation 10 is an influential observation. Various plots may also be examined.

First we plot Cook's distances against the corresponding observation numbers.

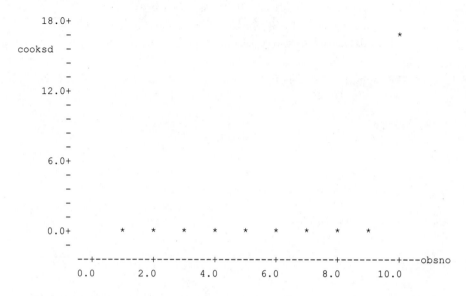

The value of $F_{0.5:2,8}$ is 0.76, as given in Table T-5 in Appendix T and so sample observations with Cook's distances that are greater than 0.76 are candidates for further scrutiny. In this problem the only such case is observation 10. Next we plot DFFITS against sample item numbers.

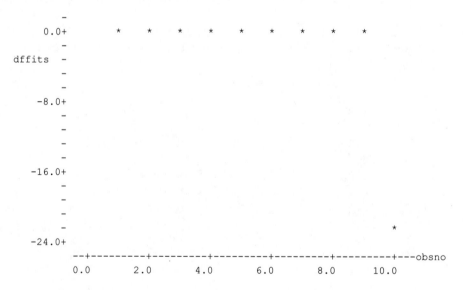

We observe that the value of DFFITS for case 10 is very large in absolute value relative to the remaining cases. The value of $2\sqrt{p/n}$ is $2\sqrt{2/10} = 0.894427$, and the value of DFFITS for case 10 is -21.7503, which is larger than 0.894427 in absolute value. Thus case 10 is identified as an influential observation.

Next we examine a plot of the hat values against observation numbers.

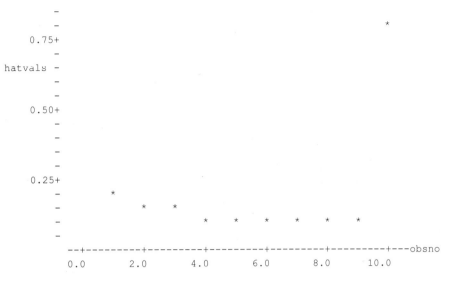

This plot indicates that case 10 is a *high leverage point*. It has a hat value of 0.81405, which is larger than $2p/n = 0.4$. Its X value is unusual relative to other X values. We thus have a possible explanation of why case 10 is an *influential* observation. ■

Although in this example an examination of a plot of Y against X would have revealed the fact that case 10 has an unusual X value, and that its inclusion or exclusion from the regression analysis may unduly affect the values of the estimated parameters, this may not be possible in general in problems involving several predictor variables. In such cases, diagnostic statistics such as hat values (leverages), Cook's distances, and DFFITS are of great assistance in identifying and interpreting influential observations.

To see what effect observation 10 has on the conclusions, we now reanalyze the data with case 10 removed. A computer output from a regression analysis for the data in Table 5.4.1, with sample item number 10 removed, is given in Exhibit 5.4.2. Note that the parameter estimates have changed substantially as a result of deleting case 10. When all the data are included in the analysis we get (see Exhibit 5.4.1)

$$\hat{\beta}_0 = 14.521 \qquad \hat{\beta}_1 = -0.1055 \qquad \hat{\sigma} = 7.327 \tag{5.4.4}$$

When case 10 is omitted from the analysis we get (see Exhibit 5.4.2)

$$\hat{\beta}_{(-10),0} = 1.627 \qquad \hat{\beta}_{(-10),1} = 4.5133 \qquad \hat{\sigma}_{(-10)} = 1.932 \tag{5.4.5}$$

The predicted Y value for case 10 using $\hat{\beta}_0$ and $\hat{\beta}_1$ in (5.4.4) is

$$\hat{Y}(11) = 14.521 + 11(-0.1055) = 13.3605$$

but if the estimates in (5.4.5) are used (i.e., if sample number 10 is omitted) we get

$$\hat{Y}_{(-10)}(11) = 1.627 + 11(4.5133) = 51.2733$$

E X H I B I T **5.4.2**

MINITAB Output for Example 5.4.1—Observation 10 Removed

```
Observation        Y       X
  Number
     1            6.8     1.0
     2            6.1     1.5
     3            9.5     2.0
     4           16.2     2.5
     5           15.2     3.0
     6           16.4     3.5
     7           21.5     4.0
     8           22.5     4.5
     9           22.3     5.0



The regression equation is
Y = 1.63 + 4.51 X



Predictor        Coef      Stdev     t-ratio        p
Constant        1.627      1.629        1.00    0.351
X               4.5133     0.4988       9.05    0.000


s = 1.932        R-sq = 92.1%      R-sq(adj) = 91.0%



Analysis of Variance

SOURCE          DF          SS          MS         F        p
Regression       1       305.55      305.55     81.86    0.000
Error            7        26.13        3.73
Total            8       331.68
```

Clearly we obtain substantially different values for $\hat{\beta}_i$ and for $\hat{Y}(11)$ depending on whether or not we include case 10 in the analysis. In a real study the investigator would now look for possible explanations for observation 10 being unusual. It may be that the values for this item are incorrectly recorded, or it may be the case that a straight line regression model is not appropriate over the entire range of data values. The true population regression function may be a quadratic function or some other type of function over the range of the X values in the sample data. The investigator has to make a decision, using all available information and having considered various possible explanations, about how to treat influential observations. When no explanations are available concerning influential observations, it may be best to present results from both analyses, one with the sample item in question included and one with it excluded.

E X A M P L E **5.4.2**

The data for this example were obtained by a slight modification of the data for Example 5.4.1. Specifically, suppose that the Y value for item 10 had been wrongly recorded as 4.7 instead of the correct value, which equals 52.2. Table 5.4.2 gives the corrected data set. These data are also stored in the file **table542.dat** on the data disk. The results of a regression analysis of these data, performed using MINITAB, appear in Exhibit 5.4.3.

T A B L E **5.4.2**

Observation Number	Y	X
1	6.8	1.0
2	6.1	1.5
3	9.5	2.0
4	16.2	2.5
5	15.2	3.0
6	16.4	3.5
7	21.5	4.0
8	22.5	4.5
9	22.3	5.0
10	52.2	11.0

E X H I B I T 5.4.3

MINITAB Output for Example 5.4.2

```
The regression equation is
Y = 1.37 + 4.61 X


Predictor        Coef        Stdev      t-ratio         p
Constant       1.3701       0.9910         1.38     0.204
X              4.6052       0.2127        21.65     0.000


s = 1.813       R-sq = 98.3%       R-sq(adj) = 98.1%


Analysis of Variance


SOURCE          DF          SS          MS          F         p
Regression       1       1539.7      1539.7     468.59     0.000
Error            8         26.3         3.3
Total            9       1566.0


Unusual Observations


Obs.       X          Y        Fit  Stdev.Fit   Residual    St.Resid
 10      11.0     52.200     52.028      1.636      0.172      0.22 X
X denotes an obs. whose X value gives it large influence.


obsno     Y        X   stdresid       hatvals       cooksd       dffits

   1     6.8      1.0    0.51119      0.207989     0.034312     0.249145
   2     6.1      1.5   -1.32110      0.172865     0.182378    -0.638922
   3     9.5      2.0   -0.64455      0.144628     0.035122    -0.254616
   4    16.2      2.5    1.95417      0.123278     0.268486     0.948144
   5    15.2      3.0    0.00829      0.108815     0.000004     0.002710
   6    16.4      3.5   -0.63336      0.101240     0.022593    -0.204024
   7    21.5      4.0    0.99407      0.100551     0.055235     0.332089
   8    22.5      4.5    0.23718      0.106749     0.003361     0.076967
   9    22.3      5.0   -1.23266      0.119835     0.103436    -0.472709
  10    52.2     11.0    0.22044      0.814050     0.106369     0.432762
```

(5.4.6)

We plot Y against X. This plot shows that case 10 has an unusual X value relative to the X values of the other cases, but it appears that the estimated regression line will not change substantially if this point is omitted from the analysis.

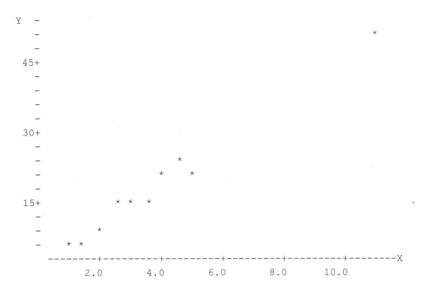

Next we plot Cook's distances against observation numbers.

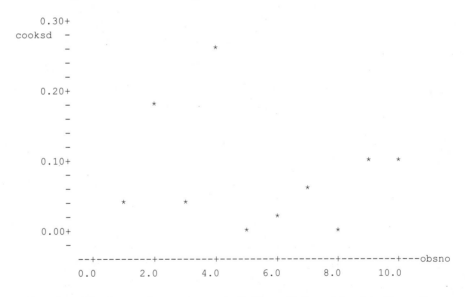

Based on this plot, we do not see any indication of influential points. To confirm this we calculate $F_{0.5:p,n-p} = F_{0.5:2,8} = 0.76$. Because all of the Cook's distances are less than this value, we can conclude that none of the observations is influential.

Next we plot the DFFITS against the observation numbers.

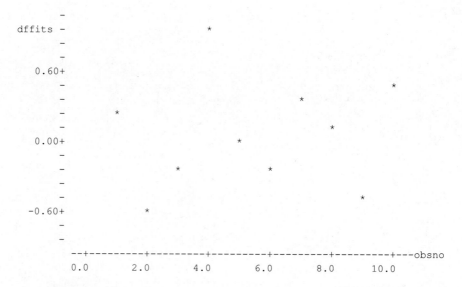

Again, based on the plot of DFFITS against observation numbers, one might conclude that none of the observations is influential since none is greater than $2\sqrt{p/n} = 0.8944$ in absolute value.

Finally, we plot the hat values against the observation numbers.

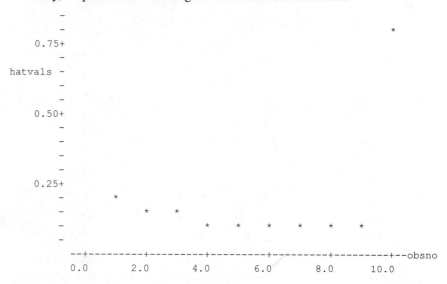

Note that, although case 10 has an unusual X value relative to the remaining data points, its inclusion or exclusion in the analysis does not substantially affect the estimated regression function. You should confirm this statement by verifying that the estimated regression function when case 10 is excluded is

$$\hat{Y}_{(-10)}(x) = 1.63 + 4.51x$$

which is not very different from the estimated regression function

$$\hat{Y}(x) = 1.37 + 4.61x$$

calculated with case 10 included.

However, you should note that although inclusion or exclusion of observations with large hat values may or may not substantially affect the estimated regression function, it *will* affect the standard errors of predicted values and the standard errors of the estimated regression coefficients, thus affecting the widths of confidence intervals. In the present example note that $SE(\hat{\beta}_i) = 0.4988$ when observation 10 is excluded, but $SE(\hat{\beta}_i) = 0.2127$ when observation 10 is included in the analysis. ▪

In Box 5.4.1 we summarize the formulas for computing various diagnostic measures discussed in Sections 5.2–5.4.

B O X 5.4.1 **Summary of Formulas for Diagnostic Statistics**

Notation:

Y is the response variable.

$X_1, \ldots, X_k$ are predictor variables.

The regression function of Y on $X_1, \ldots, X_k$ is

$$\mu_Y(x_1, \ldots, x_k) = \beta_0 + \beta_1 x_1 + \cdots + \beta_k x_k$$

y_i is the observed Y value for sample item i for $i = 1, \ldots, n$.

$x_{i,1}, \ldots, x_{i,k}$ are values of the predictor variables $X_1, \ldots X_k$ for sample item i for $i = 1, \ldots, n$.

$\hat{\beta}_0, \hat{\beta}_1, \ldots, \hat{\beta}_k$ are the least squares estimates of $\beta_0, \beta_1, \ldots, \beta_k$ using all n sample observations.

$\hat{\beta}_{(-i)0}, \hat{\beta}_{(-i)1}, \ldots, \hat{\beta}_{(-i)k}$ are the least squares estimates of $\beta_0, \beta_1, \ldots, \beta_k$ calculated after *leaving out the ith observation* but using the remaining $n - 1$ sample observations.

$\hat{\sigma}$ = the estimated subpopulation standard deviation using all n sample observations.

$\hat{\sigma}_{(-i)}$ = the estimated subpopulation standard deviation calculated after *leaving out the ith observation* but using the remaining $n - 1$ sample observations.

$\hat{Y}(x_{i,1}, \ldots, x_{i,k})$ is the estimated Y value of an item with $X_1 = x_{i,1}, \ldots, X_k = x_{i,k}$ calculated using all n sample observations and is given by

$$\hat{Y}(x_{i,1}, \ldots, x_{i,k}) = \hat{\beta}_0 + \hat{\beta}_1 x_{i,1} + \cdots + \hat{\beta}_k x_{i,k}$$

$\hat{Y}_{(-i)}(x_{i,1}, \ldots, x_{i,k})$ is the estimated Y value of an item with $X_1 = x_{i,1}, \ldots, X_k = x_{i,k}$ calculated after *leaving out the ith observation* but using the remaining $n - 1$ sample observations and is given by

$$\hat{Y}_{(-i)}(x_{i,1}, \ldots, x_{i,k}) = \hat{\beta}_{(-i)0} + \hat{\beta}_{(-i)1}x_{i,1} + \cdots + \hat{\beta}_{(-i)k}x_{i,k}$$

The **residual** corresponding to observation i is $\hat{e}_i$ and is given by

$$\hat{e}_i = y_i - (\hat{\beta}_0 + \hat{\beta}_1 x_{i,1} + \cdots + \hat{\beta}_k x_{i,k})$$

The **hat matrix H** is defined by

$$H = X(X^T X)^{-1} X^T$$

The **hat value** corresponding to observation i is $h_{i,i}$, which is given by

$$h_{i,i} = \text{the } i\text{th diagonal element of the hat matrix } H$$

Observations with $h_{i,i} > 2p/n$ should be examined as potentially influential observations.

The **standardized residual** corresponding to observation i is r_i and is given by

$$r_i = \frac{\hat{e}_i}{\hat{\sigma}\sqrt{1 - h_{i,i}}}$$

Observations with $|r_i| > 2$ should be examined as possible outliers.

Studentized deleted residual for observation i is given by

$$T_i = \frac{y_i - \hat{Y}_{(-i)}(x_{i,1}, \ldots, x_{i,k})}{\hat{\sigma}_{(-i)} \Big/ \sqrt{1 - h_{i,i}}}.$$

Observations with $|T_i| > 2$ should be examined as possible outliers.

Cook's distance for observation i is denoted by c_i and is defined by

$$c_i = \frac{1}{p}\left(\frac{h_{i,i}}{1 - h_{i,i}}\right)r_i^2$$

where p is the number of β parameters in the model. Thus $p = k + 1$ for the model $\mu_Y(x_1, \ldots, x_k) = \beta_0 + \beta_1 x_1 + \cdots + \beta_k x_k$. Observations with $c_i > F_{0.5:p,n-p}$ should be examined as possible influential observations.

DFFITS for observation i is denoted by DFFITS_i and is given by

$$\text{DFFITS}_i = T_i \sqrt{\frac{h_{i,i}}{1 - h_{i,i}}}$$

where T_i is the studentized deleted residual for observation i defined above. Observations with $|\text{DFFITS}_i| > 2\sqrt{p/n}$ should be examined as possible influential observations.

Problems   5.4

Problems 5.4.1 and 5.4.2 refer to the data in Table 5.4.2. In Exhibit 5.4.4 is a SAS output for regression analysis of these data. This exhibit also includes several diagnostic measures.

E X H I B I T 5.4.4
SAS Output for Problems 5.4.1–5.4.2

```
                          The SAS System
                                      0:00  Saturday, Jan  1, 1994
Model: MODEL1
Dependent Variable: Y
```

Analysis of Variance

Source	DF	Sum of Squares	Mean Square	F Value	Prob>F
Model	1	1539.71399	1539.71399	468.585	0.0001
Error	8	26.28701	3.28588		
C Total	9	1566.00100			

Root MSE	1.81270	R-square	0.9832	
Dep Mean	18.87000	Adj R-sq	0.9811	
C.V.	9.60625			

Parameter Estimates

Variable	DF	Parameter Estimate	Standard Error	T for H0: Parameter=0	Prob > |T|
INTERCEP	1	1.370110	0.99103083	1.383	0.2042
X	1	4.605234	0.21274399	21.647	0.0001

E X H I B I T 5.4.4
(Continued)

obs	Y	X	fits	residual	stdresid	tresid	hatvals	cooksd	dffits
1	6.8	1.0	5.97534	0.82466	0.51119	0.48618	0.20799	0.03431	0.24914
2	6.1	1.5	8.27796	-2.17796	-1.32110	-1.39760	0.17287	0.18238	-0.63892
3	9.5	2.0	10.58058	-1.08058	-0.64454	-0.61921	0.14463	0.03512	-0.25462
4	16.2	2.5	12.88320	3.31680	1.95417	2.52849	0.12328	0.26849	0.94814
5	15.2	3.0	15.18581	0.01419	0.00829	0.00776	0.10882	0.00000	0.00271
6	16.4	3.5	17.48843	-1.08843	-0.63336	-0.60789	0.10124	0.02259	-0.20402
7	21.5	4.0	19.79105	1.70895	0.99407	0.99323	0.10055	0.05523	0.33209
8	22.5	4.5	22.09366	0.40634	0.23718	0.22264	0.10675	0.00336	0.07697
9	22.3	5.0	24.39628	-2.09628	-1.23266	-1.28110	0.11983	0.10344	-0.47271
10	52.2	11.0	52.02769	0.17231	0.22044	0.20683	0.81405	0.10637	0.43276

5.4.1 From Exhibit 5.4.4 obtain the following:

a $h_{3,3}$ c Cook's distance c_4 e r_3

b DFFITS$_4$.   d $\hat{Y}(3.5)$ f $\hat{e}_3$

g Studentized deleted residual T_6

5.4.2

a What is the largest (in absolute value) studentized deleted residual? Is this large enough to investigate the corresponding observation as a possible outlier?

b What is the largest hat value? Is this large enough to investigate the corresponding observation as a possible influential observation (high leverage point)?

c What is the largest Cook's distance? Is this large enough to conclude that the corresponding observation may be an influential observation?

d What is the largest (in absolute value) DFFITS value? Is this large enough to conclude that the corresponding observation may be an influential observation?

e Write a brief report summarizing your findings in parts (a)–(d).

Problems 5.4.3 through 5.4.6 refer to the data in Table 5.4.3. These data are also stored in the file **table543.dat** on the data disk. SAS output for regression of Y on X is given in Exhibit 5.4.5.

T A B L E 5.4.3

Observation Number	Y	X	Observation Number	Y	X
1	16.3	11.0	4	27.3	14.0
2	16.8	12.0	5	28.3	15.0
3	20.1	13.0	6	50.3	32.0

E X H I B I T 5.4.5

SAS Output for Problems 5.4.3–5.4.6 (Regression of *Y* on *X* with Observation Number 6 Included)

The SAS System

0:00 Saturday, Jan 1, 1994

Model: MODEL1
Dependent Variable: Y

Analysis of Variance

Source	DF	Sum of Squares	Mean Square	F Value	Prob>F
Model	1	761.07908	761.07908	62.991	0.0014
Error	4	48.32925	12.08231		
C Total	5	809.40833			

Root MSE	3.47596	R-square	0.9403	
Dep Mean	26.51667	Adj R-sq	0.9254	
C.V.	13.10859			

Parameter Estimates

Variable	DF	Parameter Estimate	Standard Error	T for H0: Parameter=0	Prob > |T|
INTERCEP	1	1.219517	3.48898438	0.350	0.7443
X	1	1.564772	0.19715657	7.937	0.0014

obs	Y	X	fits	residual	stdresid	tresid	hatvals	cooksd	dffits
1	16.3	11	18.43201	-2.13201	-0.70945	-0.65714	0.25255	0.08503	-0.38197
2	16.8	12	19.99678	-3.19678	-1.04302	-1.05865	0.22252	0.15568	-0.56636
3	20.1	13	21.56155	-1.46155	-0.46979	-0.41856	0.19893	0.02740	-0.20858
4	27.3	14	23.12633	4.17367	1.32741	1.53687	0.18177	0.19572	0.72437
5	28.3	15	24.69110	3.60890	1.14034	1.20211	0.17105	0.13416	0.54605
6	50.3	32	51.29222	-0.99223	-1.74337	-3.08082	0.97319	55.16410	-18.56178

E X H I B I T 5. 4. 5 (Continued)
(Regression of *Y* on *X* with Observation Number 6 Excluded)

The SAS System

0:00 Saturday, Jan 1, 1994

Model: MODEL1
Dependent Variable: Y

Analysis of Variance

Source	DF	Sum of Squares	Mean Square	F Value	Prob>F
Model	1	119.02500	119.02500	30.764	0.0116
Error	3	11.60700	3.86900		
C Total	4	130.63200			

Root MSE	1.96698	R-square	0.9111	
Dep Mean	21.76000	Adj R-sq	0.8815	
C.V.	9.03942			

Parameter Estimates

Variable	DF	Parameter Estimate	Standard Error	T for H0: Parameter=0	Prob > |T|
INTERCEP	1	-23.090000	8.13387362	-2.839	0.0657
X	1	3.450000	0.62201286	5.547	0.0116

obs	Y	X	fits	residual	stdresid	tresid	hatvals	cooksd	dffits
1	16.3	11	14.86	1.44	1.15753	1.27051	0.6	1.00491	1.55605
2	16.8	12	18.31	-1.51	-0.91755	-0.88330	0.3	0.18041	-0.57825
3	20.1	13	21.76	-1.66	-0.94355	-0.91868	0.2	0.11129	-0.45934
4	27.3	14	25.21	2.09	1.26998	1.52494	0.3	0.34561	0.99831
5	28.3	15	28.66	-0.36	-0.28938	-0.23965	0.6	0.06281	-0.29351

5.4.3 Exhibit the following:

a $\hat{Y}(13)$

b $\hat{Y}_{(-6)}(13)$

c $\hat{\sigma}$

d $\hat{\sigma}_{(-6)}$

5.4.4 Use the formula in (5.2.3) and compute T_6. Compare this value with the "tresidual" value for observation 6 in Exhibit 5.4.5.

5.4.5 Use the formula in (5.3.3) and compute $h_{6,6}$. Compare this value with the $h_{i,i}$ value for observation 6 in Exhibit 5.4.5.

5.4.6 Study the diagnostic statistics given in Exhibit 5.4.5 and write a short report summarizing your findings.

Conversation 5.4

Investigator: Good afternoon. I'm somewhat confused about all the diagnostic procedures. Do you have some time to help me sort these out?

Statistician: Certainly. I'll be glad to do what I can. What seems to be the difficulty?

Investigator: I've read about fits, residuals, standardized residuals, studentized deleted residuals, Gaussian scores, hat values, leverages, DFFITS, Cook's distances, and yet I haven't seen confidence intervals or tests for these quantities. It's been said that if some of these quantities are large, then this means something, etc. I really don't understand how to interpret them.

Statistician: All of these quantities have been developed to help determine whether certain assumptions are satisfied. And to check these assumptions, you must make other assumptions. First, let's discuss **residuals**. For simplicity we'll consider straight line regression even though the concepts apply to multiple regression as well.

Residuals are "estimates" of the random errors e_i in the model

$$y_i = \beta_0 + \beta_1 x_i + e_i$$

Assumptions (A) or (B) underlie the mathematical theory of regression as we have been describing it and imply the following about the residuals e_i.

The e_i for $i = 1, \ldots, n$ form a simple random sample from a population of errors E, and this population is Gaussian with mean zero and standard deviation σ (which is unknown). **(5.4.7)**

If these assumptions are not satisfied at least approximately, then many of the results (such as confidence intervals on β_i, etc.) may not be correct. So it's important to determine whether these assumptions are at least approximately satisfied. We'd like to examine the population of errors $\{E\}$, but they aren't available to us because we only have a sample of size n, viz., e_i, from this population. But even e_i aren't observable. However, we can estimate the e_i, and these estimates are the residuals. So we use the residuals to check assumptions. It may be helpful to view the residuals $\hat{e}_i$ in two ways.

1 As estimates of e_i for $i = 1, \ldots, n$ where e_i is a simple random sample from the population of errors $\{E\}$. From this point of view, residuals are used to help determine whether the population of errors $\{E\}$ is, at least approximately, Gaussian with mean zero.

2   The ith residual $\hat{e}_i = y_i - \hat{Y}(x_i)$ is the difference between the observed y_i and the estimated value, namely $\hat{Y}(x_i)$; i.e., the ith residual is a measure of how far y_i is from its estimated value. From this point of view, residuals can help determine whether any of the sample Y values is an outlier; i.e., if any of the Y values do not seem to agree with the estimated regression function. So you can see why residuals are important!

Investigator: I think I see what you're saying. But we not only use residuals, we also use *standardized residuals*.

Statistician: That's correct. The residuals $\hat{e}_i$ typically have different standard deviations that depend on the x_i values and are therefore not directly comparable with one another. Thus, to eliminate this dependence on x_i, we standardize them so that the standardized residuals approximate a simple random sample of size n (approximate because they are correlated) from a Gaussian population with mean zero and standard deviation *one* if the statement in (5.4.7) is correct. Thus we can use Gaussian *scores* to see whether the standardized residuals appear to be a simple random sample of size n from a Gaussian population with mean zero and standard deviation one. If they do not appear to be a simple random sample from a Gaussian population with mean zero and standard deviation one, then we suspect that (5.4.7) is not correct. To help evaluate whether the population of errors $\{E\}$ is Gaussian with mean zero, we plot the standardized residuals against Gaussian scores and examine how close the points on this plot are to a line through the origin with slope one. If they are close, we conclude that (5.4.7) is close enough to being correct so that our statistical procedures are approximately valid. This was discussed in Section 4.5.

Investigator: I think I see why standardized residuals are useful, but can you give me some additional insight about studentized *deleted* residuals? What is so important about deleted?

Statistician: First, let's discuss **fits**, viz., the fitted values, which are the estimates of the sample y_i values based on the estimated regression function. Of course we know the value of y_i, and so we are not interested in estimating it, but if the ith fit, namely $\hat{Y}(x_i)$, is close to y_i for each $i = 1, 2, \ldots, n$, then we feel that $\hat{Y}(x)$ will be close to $Y(x)$ for other x values. In fact, as you know, residuals $\hat{e}_i$ are measures of how good the fits are because

$$\hat{e}_i = y_i - \hat{Y}(x_i)$$

If one of the residuals, say the ith one, has a magnitude that is substantially different from the others, this means that something could be wrong with either y_i or $\hat{Y}(x_i)$. To decide which is wrong, we compute the regression of Y on X after omitting (y_i, x_i), the ith sample observation. Then $\hat{Y}_{(-i)}(x_i)$ is another estimate of y_i, but this estimate is not influenced by y_i because it is not used in computing $\hat{Y}_{(-i)}(x_i)$. So

$$y_i - \hat{Y}_{(-i)}(x_i)$$

is a measure of how far the ith sample value y_i is from its estimate. We standardize this quantity by dividing it by its standard error, which is

$$SE(y_i - \hat{Y}_{(-i)}(x_i)) = \hat{\sigma}_{(-i)}/\sqrt{1 - h_{i,i}}$$

and the standardized quantity is T_i, the studentized deleted residual given in (5.2.3). So the studentized deleted residual is similar to the standardized residual, except the studentized deleted residual is computed without using y_i.

Investigator: This helps me see why studentized deleted residuals are useful.

Statistician: In addition, it is true that if (5.4.7) obtains, then T_i is distributed as student's t. Hence if T_i is larger in absolute value than, say 3, then it seems unlikely that (5.4.7) is correct. There are other statistics that can be used to check these assumptions, and one of them is Cook's distance. This is similar to studentized deleted residuals. Cook's distance for sample observation i, denoted by c_i, is a measure of how much the two sets of estimated regression coefficients differ when they are obtained by the following two methods: (1) by using all the sample data or (2) by using all the sample data except the ith observation. Another useful diagnostic measure is DFFITS. DFFITS$_i$ is a measure of how much the two estimates of $Y(x_{i1}, \ldots, x_{ik})$, the Y value of sample item i, differ when one estimate is obtained by using all the sample data and the other is obtained by using all the sample data except y_i.

Investigator: Can you summarize the usefulness of hat values for me?

Statistician: Hat values are useful because they identify sample observations whose X values are substantially further from the center (mean) of the data than the remainder of the X values. Observations with large hat values are called *high leverage points*, and each such sample value should be investigated to see if point estimates and confidence intervals are changed substantially when it is included or excluded from the analysis.

Investigator: You use terms like "close enough," "approximately," "similar to," "previous judgment," and "knowledge," etc. I thought that statistical inference was a very objective procedure and that personal judgment shouldn't enter into making decisions.

Statistician: Scientists would, perhaps, like to make decisions that do not depend on their personal judgments, but that is not possible. Investigators make decisions based on assumptions, and if they want to check to see whether these assumptions are correct, then they must make other assumptions, etc. So someplace in this chain of events, they must make some assumptions. In regression we make certain assumptions (A), or (B), but we ought to use all available information, including our judgments and previous experiences, to examine these assumptions. The evaluation of assumptions is necessarily a subjective exercise using descriptive statistics. Thus, even though we don't have exact procedures available to check assumptions, the approximate procedures are useful and important. If, based on these procedures, we believe that assumptions (A) or (B) are approximately satisfied, then point estimates, confidence intervals, etc. are valid enough to be useful.

Investigator: Thanks for your time. Perhaps I'll come again soon.

5.5
Ill-Conditioning and Multicollinearity

The computation of parameters in a multiple linear regression model often involves a large number of arithmetical operations, and the results may be affected by rounding errors. Typically the computations are done using a calculator or a computer. This usually means that only a finite number of significant digits are kept in the computer's memory. For most desktop computers, this is about seven digits when using single precision arithmetic and about twelve digits when using double precision. As a simple illustration that dramatically shows the effect of rounding, consider the problem of calculating the quantity c where

$$c = 10^{10} \left(\sqrt{2} - \frac{2 \times 29 \times 37 \times 659}{1000000} \right)$$

If we carry out the calculations, rounding to the nearest seven significant digits at each step, we get

$$\sqrt{2} = 1.414214$$
$$2 \times 29 = 58$$
$$58 \times 37 = 2146$$
$$2146 \times 659 = 1414214$$
$$\frac{2 \times 29 \times 37 \times 659}{1000000} = 1.414214$$
$$c = 10^{10}(1.414214 - 1.414214) = 0$$

If we carry out the calculations, rounding to the nearest twelve significant digits at each step, we get

$$\sqrt{2} = 1.41421356237$$
$$2 \times 29 = 58$$
$$58 \times 37 = 2146$$
$$2146 \times 659 = 1414214$$
$$\frac{2 \times 29 \times 37 \times 659}{1000000} = 1.414214$$
$$c = 10^{10}(1.41421356237 - 1.414214) = 10^{10}(-0.00000043763) = -4376.3$$

Certainly in this problem there is a considerable loss of accuracy in the result when only seven significant digits are kept. Admittedly this will not be the case in every problem, but regression calculations are particularly prone to rounding errors. Example 5.5.1 illustrates this point.

E X A M P L E 5.5.1

Recall that the estimate of β in a multiple linear regression model may be written as

$$\hat{\beta} = (X^T X)^{-1} X^T y$$

where X is the matrix of predictors (with the first column being all 1's in the case of a regression model with an intercept), and y is the column vector consisting of the corresponding values of the response variable. Suppose, for illustration, that

$$X = \begin{bmatrix} 1 & 2.001 \\ 1 & 1.998 \\ 1 & 2.003 \\ 1 & 1.999 \end{bmatrix} \qquad y = \begin{bmatrix} 3.01 \\ 2.99 \\ 2.94 \\ 2.99 \end{bmatrix}$$

We calculate $\hat{\beta}$ in two ways. First, we use exact arithmetic with no loss of accuracy at any stage. In parallel to this we compute $\hat{\beta}$ again, but this time we *round* all intermediate calculations to seven significant digits. The actual calculations are given in Table 5.5.1. Most statistical packages compute $\hat{\beta}$ by using special numerical techniques; they do not apply the formula $\hat{\beta} = (X^T X)^{-1} X^T y$ directly. However, we use this formula to illustrate how rounding errors can be troublesome, and even the special numerical techniques are not immune to rounding error problems.

T A B L E 5.5.1

Illustration of the Effect of Rounding in Regression Calculations

Exact Calculations	**Rounding to Seven Significant Digits**
$X^T X = \begin{bmatrix} 4 & 8001/1000 \\ 8001/1000 & 16004015/1000000 \end{bmatrix}$	$X^T X = \begin{bmatrix} 4.0000 & 8.0010 \\ 8.0010 & 16.00402 \end{bmatrix}$
$X^T y = \begin{bmatrix} 1193/100 \\ 2386286/100000 \end{bmatrix}$	$X^T y = \begin{bmatrix} 11.930 \\ 23.86286 \end{bmatrix}$
$(X^T X)^{-1} = \begin{bmatrix} 16004015/59 & -8001000/59 \\ -8001000/59 & 4000000/59 \end{bmatrix}$	$(X^T X)^{-1} = \begin{bmatrix} 202582.5 & -101278.5 \\ -101278.5 & 50632.91 \end{bmatrix}$
$\hat{\beta} = (X^T X)^{-1} X^T y = \begin{bmatrix} 115609/5900 \\ -490/59 \end{bmatrix}$	$\hat{\beta} = (X^T X)^{-1} X^T y = \begin{bmatrix} 14.00000 \\ -7.000000 \end{bmatrix}$
$= \begin{bmatrix} 19.59475 \\ -8.305085 \end{bmatrix}$ (final result rounded to seven significant digits)	

Thus rounding to seven significant digits has resulted in inaccuracy in the results, even in the first significant digit for β_1. ∎

Sometimes the errors due to inexact arithmetic are not very serious and can be safely ignored. However, this is not always so, and it is possible for rounding errors in regression analysis to result in serious errors in the parameter estimates, which can lead to erroneous conclusions or decisions. The seriousness of the errors due to rounding usually depends on a property of the X matrix called the **condition of X**. Some X matrices are **well-conditioned**, while others, like the one in Example 5.5.1, are said to be **ill-conditioned**. We say that the X matrix is ill-conditioned if small errors in the sample data values translate into large errors in the final results. *This may be so even if all the calculations are carried out exactly.* When the X matrix in a regression problem is ill-conditioned, rounding is likely to lead to substantial errors in the results. Many of the software packages for regression use double precision arithmetic and also employ numerical techniques that minimize the effects of rounding. If the X matrix is **ill-conditioned**, then some computer packages inform the user that this is the case.

We illustrate these concepts in Example 5.5.2.

E X A M P L E   5.5.2

Suppose we want to develop a function for predicting the weight Y using age X_1 and length X_2 for babies with ages ranging from 1 month to 12 months, and that a sample of size 12 was selected by first preselecting the ages and then randomly choosing one baby from each preselected age group. The length and weight of each chosen baby are recorded along with age. The data are displayed in Table 5.5.2. We

T A B L E   5.5.2

Observation Number	Weight Y (pounds)	Age X_1 (months)	Length X_2 (inches)
1	9.2	1	20.4
2	9.8	2	20.9
3	9.1	3	22.1
4	9.6	4	21.7
5	11.7	5	22.9
6	10.7	6	24.2
7	12.7	7	24.9
8	13.0	8	26.1
9	13.4	9	26.9
10	14.7	10	27.6
11	14.4	11	28.1
12	15.2	12	29.2

suppose that assumptions (A) for regression are satisfied with

$$\mu_Y(x_1, x_2) = \beta_0 + \beta_1 x_1 + \beta_2 x_2 \tag{5.5.1}$$

We estimate the parameters β_i using the formula (4.4.8) and carrying out all calculations exactly (by hand). The results are

$$X^T X = \begin{bmatrix} 12 & 78 & 295 \\ 78 & 650 & 2035.7 \\ 295 & 2035.7 & 7351.16 \end{bmatrix}$$

$$X^T y = \begin{bmatrix} 143.5 \\ 1018.5 \\ 3598.99 \end{bmatrix}$$

$$(X^T X)^{-1} = \begin{bmatrix} 63417951/236068 & 1357051/118034 & -824135/59017 \\ 1357051/118034 & 29723/59017 & -35460/59017 \\ -824135/59017 & -35460/59017 & 42900/59017 \end{bmatrix}$$

$$\hat{\beta} = (X^T X)^{-1} X^T y = \begin{bmatrix} 5743609/2360680 \\ 421987/1180340 \\ 34577/118034 \end{bmatrix} = \begin{bmatrix} 2.433031584 \\ 0.357513089 \\ 0.292941017 \end{bmatrix} \quad \begin{array}{l} \text{(final result} \\ \text{rounded to} \\ \text{nine} \\ \text{decimals)} \end{array}$$

Now suppose that X_2 and Y values are measured slightly inaccurately, resulting in the data in Table 5.5.3. Note that the data values in Table 5.5.3 differ from the corresponding values in Table 5.5.2 by at most plus or minus 0.1.

T A B L E   5.5.3

Observation Number	Weight Y (pounds)	Age X_1 (months)	Length X_2 (inches)
1	9.3	1	20.5
2	9.7	2	20.8
3	9.2	3	22.2
4	9.5	4	21.6
5	11.8	5	23.0
6	10.6	6	24.1
7	12.8	7	25.0
8	12.9	8	26.0
9	13.5	9	27.0
10	14.6	10	27.5
11	14.5	11	28.2
12	15.1	12	29.1

We again estimate the parameters β_i using formula (4.4.8) and carrying out all calculations exactly (by hand). The results are

$$X^TX = \begin{bmatrix} 12 & 78 & 295 \\ 78 & 650 & 2035.1 \\ 295 & 2035.1 & 7350.40 \end{bmatrix}$$

$$X^Ty = \begin{bmatrix} 143.5 \\ 1017.9 \\ 3598.42 \end{bmatrix}$$

$$(X^TX)^{-1} = \begin{bmatrix} 63612799/275428 & 1351165/137714 & -825305/68857 \\ 1351165/137714 & 29495/68857 & -35280/68857 \\ -825305/68857 & -35280/68857 & 42900/68857 \end{bmatrix}$$

$$\hat{\beta} = (X^TX)^{-1}X^Ty = \begin{bmatrix} -377089/2754280 \\ 335833/1377140 \\ 58877/137714 \end{bmatrix}$$

$$= \begin{bmatrix} -0.13691019 \\ 0.243862643 \\ 0.427530970 \end{bmatrix} \quad \text{(final result rounded to nine decimals)}$$

We see that small perturbations (changes or errors) in the sample values have resulted in substantial changes in the estimated parameter values, even though the calculations are exact. ∎

The matrix X may be ill-conditioned because of one or both of the following reasons:

1   One or more columns of X consist of elements all of which are *very nearly* equal to zero.

2   One or more columns of X are *very nearly* obtainable as linear combinations of the remaining columns. In this case we say that **multicollinearity** exists among the columns of X. This is what happens in Example 5.5.2. You may verify that, in Example 5.5.2, X_1 and X_2 are nearly linearly related and that in fact

$$x_{i,2} \approx 19.2106 + 0.82657x_{i,1}$$

for each $i = 1, \ldots, 12$, causing the X matrix to be ill-conditioned.

Multicollinearity among the columns of X can occur due to one or more of the following reasons:

a   In the population, one or more of the predictor variables $X_1, \ldots, X_k$ is nearly an exact linear combination of some or all of the remaining predictor variables. For instance, variable X_j may be very nearly an exact linear combination of the remaining predictors so that we have

$$X_{I,j} \approx c_0 + c_1 X_{I,1} + \cdots + c_{j-1} X_{I,j-1} + c_{j+1} X_{I,j+1} + \cdots + c_k X_{I,k} \qquad \text{(5.5.2)}$$

for every $I = 1, \ldots, N$. In this situation we say that **multicollinearity** exists among the predictor variables in the population. If sample data are obtained by simple random sampling, then the sample values of the predictor variables also tend to exhibit a relation such as (5.5.2), resulting in multicollinearity among the columns of the X matrix, making it ill-conditioned. When the predictor variables exhibit multicollinearity in the population, then even if the data are obtained by sampling with preselected X values, we are unable to avoid an ill-conditioned X matrix because a relation such as (5.5.2) holds for every set of values $(X_{I,1}, \ldots, X_{I,k})$ that occurs in the population.

Consider, for instance, a study in which the predictor variables X_1, X_2, X_3, X_4, and X_5 are heights at ages 4, 5, 6, 7, and 8, respectively, of a population of children, and the response variable Y is height at age 9. In all likelihood, the height at age 8 can be predicted very well using the heights at ages 4, 5, 6, and 7 in a linear prediction function. This means that X_5 is very nearly a linear function of X_1, X_2, X_3, and X_4 in the population, and no matter how the sample is selected, the sample values of X_5 will also be very nearly a linear function of the sample values of X_1, X_2, X_3, and X_4.

b Data were obtained by sampling with preselected X values, but practical constraints such as cost, infeasibility of obtaining samples of the response variable at certain combinations of the predictors, etc., may have resulted in a choice of preselected values for the predictors leading to an ill-conditioned X matrix.

c The design of the study is bad. Here investigators could have selected values of the predictor variables in such a way that the X matrix would not be ill-conditioned, but they failed to take advantage of this opportunity.

Presence of *multicollinearity* among the columns of the X matrix has the following implications:

a Computations are very sensitive to rounding, and even if several significant digits are retained during various steps of the calculations, they often yield incorrect values for estimates of various parameters. This can perhaps be overcome by using double precision or multiple precision calculations.

b The results are highly sensitive to errors in the sample data. Even seemingly negligible errors in the measurements can lead to results that have no resemblence to the results that would be obtained if there were no errors in the data. Because practically all measurements are subject to errors, the resulting statistics cannot be taken seriously when the columns of the X matrix exhibit multicollinearity. The standard errors of the parameter estimates may reflect this situation by taking on values that are extremely large relative to the magnitude of the estimates.

c Based on the sample at hand, it is not possible to separate the influences of each of the predictors on the response. This is again related to the fact that the estimated regression coefficients tend to have large standard errors relative to their magnitudes. Whereas we may be able to find good prediction functions, we have to choose arbitrarily from among several sets of nearly equally good prediction functions. Knowledge related to the field of application can often guide us in making a rational selection.

Variance Inflation Factors (VIF)

Several diagnostic procedures have been proposed in the literature for detecting the presence of approximate linear relationships among the columns of the X matrix. Associated with each predictor variable X_j is a number denoted by VIF_j, called the **variance inflation factor** for X_j, which is defined as

$$\text{VIF}_j = \frac{1}{1 - \hat{\rho}^2_{X_j(X_1,\ldots,X_{j-1},X_{j+1},\ldots,X_k)}} \tag{5.5.3}$$

where $\hat{\rho}^2_{X_j(X_1,\ldots,X_{j-1},X_{j+1},\ldots,X_k)}$ is the sample coefficient of determination of X_j on the remaining predictor variables (see (4.9.20)). If one or more of these variance inflation factors is large, we can conclude that *nearly linear relationships* exist among the columns of the X matrix. It has been suggested, as a rule of thumb, that values of VIF_j greater than 10.0 may be considered large enough for us to suspect serious multicollinearity. Note that $\hat{\rho}^2_{X_j(X_1,\ldots,X_{j-1},X_{j+1},\ldots,X_k)}$ in (5.5.3) is not in general a valid estimate of $\rho^2_{X_j(X_1,\ldots,X_{j-1},X_{j+1},\ldots,X_k)}$ (since no valid estimate is available if the X values are preselected), but it can be computed using the formula in (4.9.20).

What can or should an investigator do if the diagnostics reveal multicollinearity among the columns of X? Two cases need to be distinguished.

1 *The main objective of the study is to obtain a good prediction function for the response variable.* If the data were obtained by simple random sampling, then the existence of multicollinearity among the columns of X indicates that, even in the population, some of the predictor variables are nearly linear functions of other predictors and thus are redundant. Prediction functions based on only a subset of the predictors can be found that are nearly as good as the one based on *all* the predictors.

   If the data were obtained by sampling with preselected X values, then the preselected values of the predictor variables were not chosen appropriately; i.e., the sampling design was bad. In this case any prediction function based on the sample cannot be expected to predict the Y values very well for values of the predictor variables that are not similar to the ones in the sample. More data have to be collected using a better sampling design.

2 *The main objective is to estimate the parameters in the regression function or to assess the importance of each predictor in predicting the response variable Y.* There are two ways in which this situation can be handled:

   a If the multicollinearity among the columns of X is due to a bad sampling design (not preselecting the values of the predictors judiciously), then the investigator may be able to compensate for this mistake by obtaining an additional sample of suitable size by judiciously preselecting the values of the predictor variables at which to sample the response variable. The pre-selected values of the predictors should be chosen to make the X matrix well-conditioned. A professional statistician's assistance can be invaluable here.

b If the multicollinearity among the columns of X is due to the existence of exact (or nearly exact) linear relationships among the predictor variables in the population, then realistically the only way to obtain useful unbiased or nearly unbiased estimates of the parameters is to collect additional data. In cases of serious multicollinearity, the sample size needed to obtain useful unbiased or nearly unbiased estimates of the parameters may be extremely large. Some authors have suggested the use of alternate approaches such as **ridge regression**, but we do not discuss them here. To find out more about these approaches, consult other texts [6].

When multicollinearity among the predictors is detected, it is often useful to understand the nature of the multicollinearity—i.e., understand which predictor variables are approximately linearly related. If you are interested in this topic, you may refer to other texts [1], [30].

5.6
Exercises

5.6.1 Consider Problem 4.12.4 where an investigator is studying a population of people who have lived in mountain isolation for several generations. She is interested in studying the relationship of Y, the height of males at age 18, to the following variables.

$$X_1 = \text{length at birth}$$
$$X_2 = \text{mother's height at age 18}$$
$$X_3 = \text{father's height at age 18}$$
$$X_4 = \text{maternal grandmother's height at age 18}$$
$$X_5 = \text{maternal grandfather's height at age 18}$$
$$X_6 = \text{paternal grandmother's height at age 18}$$
$$X_7 = \text{paternal grandfather's height at age 18}$$

All heights and lengths are in inches. A random sample of 20 males of age 18 or more was obtained from the study population, and the preceding information was recorded. The data for this problem are a modification (for illustrative purposes) of the data in Problem 4.12.4. The data appear in Table 5.6.1. For convenience, they are also stored in the file **table561.dat** on the data disk.

T A B L E   **5.6.1**

Observation Number	Y	X_1	X_2	X_3	X_4	X_5	X_6	X_7
1	67.2	19.7	60.5	70.3	65.7	69.3	65.7	67.3
2	69.1	19.6	64.9	70.4	62.6	69.6	64.6	66.4
3	67.0	19.4	65.4	65.8	66.2	68.8	64.0	69.4
4	72.4	19.4	63.4	71.9	60.7	68.0	64.9	67.1
5	63.6	19.7	65.1	65.1	65.5	65.5	61.8	70.9
6	72.7	19.6	65.2	71.1	63.5	66.2	67.3	68.6
7	68.5	19.8	64.3	67.9	62.4	71.4	63.4	69.4
8	69.7	19.7	65.3	68.8	61.5	66.0	62.4	67.7
9	68.4	19.7	64.5	68.7	63.9	68.8	62.3	68.8
10	70.4	19.9	63.4	70.3	65.9	69.0	63.7	65.1
11	67.5	18.9	63.3	70.4	63.7	68.2	66.2	68.5
12	73.3	18.3	63.1	65.2	65.4	66.6	61.7	64.0
13	70.0	20.3	64.9	68.8	65.2	70.2	62.4	67.0
14	69.8	19.7	63.5	70.3	63.1	64.4	65.1	67.0
15	63.6	19.9	62.0	65.5	64.1	67.7	62.1	66.5
16	64.3	19.6	63.5	65.2	63.9	70.0	64.2	64.5
17	68.5	21.3	66.1	65.4	64.8	68.4	66.4	70.8
18	70.5	20.1	64.8	70.2	65.3	65.5	63.7	66.9
19	68.1	20.2	62.6	68.6	63.7	69.8	66.7	68.0
20	66.1	19.2	62.2	67.3	63.6	70.9	63.6	66.7

Exhibit 5.6.1 contains a computer output from a regression analysis for this problem. Observe that variance inflation factors are given as part of the output. Other diagnostic statistics such as residuals, standardized residuals, fitted values, studentized deleted residuals, DFFITS, etc. are also given in the computer output.

a Are there any sample items that have high leverage values? If so, what are the sample item numbers?

b Are there any sample items that are outliers? If so, what are the sample item numbers?

c Are there any sample items that are influential observations? If so, what are the sample item numbers?

d Write a short summary discussing unusual sample items. Explain what an investigator should do about them.

e Is there any indication of multicollinearity among the predictor variables? If so, explain what an investigator should do about it.

EXHIBIT 5.6.1

MINITAB Output for Exercise 5.6.1

The regression equation is
$$Y = 5.2 - 0.77\ X1 + 1.01\ X2 + 0.635\ X3 + 0.093\ X4 - 0.134\ X5$$
$$+ 0.210\ X6 - 0.589\ X7$$

Predictor	Coef	Stdev	t-ratio	p	VIF
Constant	5.15	56.80	0.09	0.929	
X1	-0.769	1.082	-0.71	0.491	1.4
X2	1.0113	0.4744	2.13	0.054	1.6
X3	0.6355	0.2934	2.17	0.051	1.6
X4	0.0926	0.3953	0.23	0.819	1.2
X5	-0.1343	0.2915	-0.46	0.653	1.2
X6	0.2104	0.3799	0.55	0.590	1.5
X7	-0.5891	0.3616	-1.63	0.129	1.6

s = 2.312   R-sq = 56.0%   R-sq(adj) = 30.3%

Analysis of Variance

SOURCE	DF	SS	MS	F	p
Regression	7	81.597	11.657	2.18	0.113
Error	12	64.148	5.346		
Total	19	145.746			

Unusual Observations

Obs.	x1	y	Fit	Stdev.Fit	Residual	St.Resid
12	18.3	73.300	68.711	1.786	4.589	3.12R
16	19.6	64.300	67.751	1.623	-3.451	-2.10R

R denotes an obs. with a large st. resid.

EXHIBIT **5.6.1**
(Continued)

obsno	fits	residual	stdresid	tresid	hatvals	cooksd	dffits
1	66.8079	0.39214	0.25649	0.24625	0.562734	0.01058	0.27935
2	71.3694	-2.26945	-1.14638	-1.16311	0.266875	0.05980	-0.70176
3	67.6532	-0.65315	-0.36642	-0.35280	0.405627	0.01145	-0.29145
4	70.6491	1.75085	0.94053	0.93564	0.351738	0.06000	0.68919
5	65.7059	-2.10593	-1.31866	-1.36532	0.522893	0.23822	-1.42933
6	71.9299	0.77010	0.42613	0.41111	0.389044	0.01445	0.32806
7	66.7398	1.76019	0.96058	0.95722	0.371873	0.06828	0.73652
8	69.8332	-0.13318	-0.07290	-0.06981	0.375660	0.00040	-0.05415
9	68.1376	0.26237	0.12876	0.12337	0.223320	0.00060	0.06615
10	70.5209	-0.12094	-0.06682	-0.06399	0.387266	0.00035	-0.05087
11	69.6793	-2.17931	-1.17245	-1.19294	0.353678	0.09403	-0.88247
12	68.7108	4.58917	3.12471	6.93025	0.596496	1.80421	8.42614
13	69.1580	0.84199	0.46716	0.45140	0.392318	0.01761	0.36269
14	70.3097	-0.50974	-0.27241	-0.26163	0.345025	0.00489	-0.18989
15	64.9012	-1.30122	-0.74929	-0.73479	0.435849	0.05422	-0.64585
16	67.7511	-3.45107	-2.09591	-2.52033	0.492823	0.53356	-2.48441
17	66.2495	2.25047	1.67526	1.83247	0.662419	0.68838	2.56694
18	71.0736	-0.57355	-0.30545	-0.29359	0.340418	0.00602	-0.21092
19	67.0123	1.08770	0.55363	0.53696	0.277929	0.01475	0.33313
20	66.5074	-0.40744	-0.20295	-0.19464	0.246015	0.00168	-0.11118

6

Applications of Regression I

6.1
Overview

In this chapter we discuss several applications of linear regression. Section 6.2 deals with prediction intervals. Tolerance intervals are discussed in Section 6.3, and the problem of estimating X from a knowledge of Y, commonly known as *the inverse prediction problem* or *calibration problem*, is considered in Section 6.4. Section 6.5 discusses the comparison of two or more regression lines. The intersection of two regression lines is discussed in Section 6.6, maximum and minimum of a quadratic regression function in Section 6.7, and spline regression in Section 6.8. In the laboratory manuals we present and explain programs we have written that can be used to perform the calculations required in this chapter.

Each of Sections 6.2 through 6.8 is self-contained, so sections of interest can be studied without a knowledge of the material in other sections. Only the material in Chapters 3 and 4 is prerequisite.

6.2
Prediction Intervals

Consider Example 2.2.2. In that example a car agency predicts Y, the first-year maintenance cost of new cars, based on X, the number of miles the cars will be driven. Suppose the target population is the set of all cars to be made by manufacturer A and driven between 5,000 and 20,000 miles, *next* year. A reasonable study population might be the set of similar cars made by manufacturer A, and driven between 5,000 and 20,000 miles, *last* year.

The questions asked about *next* year's cars will be answered by the agency using a sample from the study population of *last* year's cars. As usual, the *statistical* inference (point estimates, confidence intervals, etc.) from the sample to the study population is valid if the assumptions used are valid. The inference from the *study* population to the *target* population is not statistical inference but is *judgment infer-ence.*

Suppose that the (study) population regression function is given by $\mu_Y(x) = \beta_0 + \beta_1 x$. Then $\mu_Y(x)$ is the average first-year maintenance cost of all cars in the study population that were driven x miles the first year. Suppose, however, that you are interested not only in $\mu_Y(x)$, the *average* first-year maintenance cost of *all* cars that were driven x miles, but you are also interested in the first-year maintenance cost of a car you will purchase, which is considered to be *randomly chosen* from all cars that were driven x miles. The first-year maintenance cost of this randomly chosen car is denoted by $Y(x)$ and is the quantity that you want to determine. We call $Y(x)$ a random observation to be chosen from the subpopulation with $X = x$. We do not know $Y(x)$, but by using sample data we can obtain a point and interval estimate of it. The point estimate of this randomly chosen car is denoted by $\hat{Y}(x) = \hat{\beta}_0 + \hat{\beta}_1 x$, which is the same as $\hat{\mu}_Y(x)$, the point estimate of the average maintenance cost of all cars driven x miles. You should not be surprised that the same quantity is used for the point estimate of $Y(x)$ and of $\mu_Y(x)$, because, as we pointed out several times, the mean of the subpopulation is the best predictor for any individual element from that subpopulation.

Obviously we don't expect the point estimate $\hat{Y}(x)$ to be exactly equal to $Y(x)$, and therefore we want to know how good the estimate is. To this end it would be very useful to have an interval

$$[L, U]$$

computed from sample data such that, with a specified degree of confidence (say, $1 - \alpha$), a future randomly chosen Y value from the subpopulation corresponding to $X = x$, i.e., $Y(x)$, will lie in this interval. We write this as

$$C[L \leq Y(x) \leq U] = 1 - \alpha$$

Such an interval is called a two-sided *prediction interval* for $Y(x)$.

There is a close resemblance between *prediction intervals* and confidence in-tervals. The interval is called a *prediction interval* if the quantity of interest is an observation that will be chosen at random; in this particular case it is $Y(x)$. We refer to this random observation as a *future observation* because it is randomly chosen and not observed until *after the interval is computed.* The interval is called a *confidence interval* if the quantity of interest is a fixed, unknown parameter such as the mean $\mu_Y(x)$.

The procedure for computing prediction intervals for straight line regression was discussed in Section 3.6 and for multiple regression in Section 4.6. In this section we generalize the formulas for prediction intervals to situations involving the *average* and the *sum* of h future observations. These prediction intervals follow the general form given in (4.6.1), which is

$$\hat{\theta} - (\text{table-value}) \times SE(\hat{\theta}) \leq \theta \leq \hat{\theta} + (\text{table-value}) \times SE(\hat{\theta}) \qquad \textbf{(6.2.1)}$$

where θ and $\hat{\theta}$ are $Y(x_1, \ldots, x_k)$ and $\hat{Y}(x_1, \ldots, x_k)$, respectively. In particular, sup-pose we have a $(k+1)$-variable study population $\{(Y, X_1, \ldots, X_k)\}$ and we want to predict a single future value of Y to be selected at random from the subpopulation determined by $X_1 = x_1, \ldots, X_k = x_k$; i.e., we want to **estimate** the value $Y(x_1, \ldots, x_k)$. The estimate $\hat{Y}(x_1, \ldots, x_k)$ and the appropriate standard error are in (4.4.10) and (4.6.6), respectively. They are repeated here.

$$\hat{Y}(x_1, \ldots, x_k) = \hat{\mu}_Y(x_1, \ldots, x_k) = \hat{\beta}_0 + \hat{\beta}_1 x_1 + \cdots + \hat{\beta}_k x_k \qquad \text{(6.2.2)}$$

$$SE(\hat{Y}(x_1, \ldots, x_k)) = \hat{\sigma}\sqrt{1 + x^T C x} \qquad \text{(6.2.3)}$$

where $x^T = [1, x_1, \ldots, x_k]$ and $C = (X^T X)^{-1}$. Note that the values $x_1, x_2, \ldots, x_k$ specify the subpopulation from which the Y value is to be obtained. The table-value for a two-sided $1 - \alpha$ prediction interval is $t_{1-\alpha/2:n-k-1}$. These are substituted into (6.2.1) to obtain a $1 - \alpha$ prediction interval for $Y(x_1, \ldots, x_k)$.

Prediction Interval for the Average of h Future Values

Suppose we want to estimate the *average* of h future Y values where the ith future Y value is to be chosen at random from the subpopulation determined by $X_1 = x_{i,1}, \ldots, X_k = x_{i,k}$ (the values $x_{i,1}, \ldots, x_{i,k}$ for $i = 1, \ldots, h$ are not necessarily the sample values) and is denoted by $Y_i(x_{i,1}, \ldots, x_{i,k})$. That is, we want a point and interval estimate of the average of $Y_i(x_{i,1}, \ldots, x_{i,k})$ for $i = 1, \ldots, h$. We denote this average by Y_A so

$$Y_A = \frac{1}{h} \sum_{i=1}^{h} Y_i(x_{i,1}, \ldots, x_{i,k})$$

The estimate and its standard error to use in (6.2.1) are in (6.2.4) and (6.2.5), respectively.

$$\hat{Y}_A = \frac{1}{h} \sum_{i=1}^{h} \hat{Y}_i(x_{i,1}, \ldots, x_{i,k}) = \hat{\beta}_0 + \hat{\beta}_1 \bar{x}_1 + \cdots + \hat{\beta}_k \bar{x}_k \qquad \text{(6.2.4)}$$

$$SE(\hat{Y}_A) = \hat{\sigma}\sqrt{\frac{1}{h} + \bar{x}^T C \bar{x}} \qquad \text{(6.2.5)}$$

where

$$x_i^T = [1, x_{i,1}, x_{i,2}, \ldots, x_{i,k}] \qquad \bar{x}^T = [1, \bar{x}_1, \ldots, \bar{x}_k] \qquad \text{(6.2.6)}$$

and

$$\bar{x}_j = \frac{1}{h} \sum_{i=1}^{h} x_{i,j} \qquad \text{for } j = 1, \ldots, k \qquad \text{(6.2.7)}$$

Also

$$\bar{x}^T = \frac{1}{h}[x_1^T + \ldots + x_h^T]$$

Note that x_i for $i = 1, 2, \ldots, h$ determines the h subpopulations for which we want to predict Y values. The values of $x_{i,1}, \ldots, x_{i,k}$ can be the same for all $i = 1, \ldots, h$, or some can be the same and some different. The formulas (6.2.4) and (6.2.5) are valid under either assumptions (A) or (B).

Prediction Interval for the Sum of h Future Values

Sometimes we want to estimate the *sum* of h future Y values rather than the *average*. That is, we want a point and interval estimate of the sum of $Y_i(x_{i,1}, \ldots, x_{i,k})$ for $i = 1, \ldots, h$. We denote this sum by Y_S so

$$Y_S = \sum_{i=1}^{h} Y_i(x_{i,1}, \ldots, x_{i,k}) \tag{6.2.8}$$

The estimate of Y_S and its standard error are obtained from those for Y_A in (6.2.4) and (6.2.5) by multiplying those results by h. Specifically,

$$\hat{Y}_S = \sum_{i=1}^{h} \hat{Y}_i(x_{i,1}, \ldots, x_{i,k}) \tag{6.2.9}$$

i.e.,

$$\hat{Y}_S = h[\hat{\beta}_0 + \hat{\beta}_1 \bar{x}_1 + \cdots + \hat{\beta}_k \bar{x}_k] \tag{6.2.10}$$

and

$$SE(\hat{Y}_S) = \hat{\sigma}\sqrt{h + h^2\, \bar{x}^T C \bar{x}} \tag{6.2.11}$$

where $\bar{x}^T$ is defined in (6.2.6). Note that $SE(\hat{Y}_S) = h\, SE(\hat{Y}_A)$.
   We illustrate the use of the preceding formulas in Task 6.2.1.

Task   6.2.1

Suppose that in Example 2.2.2 the agency that evaluates the performance of new cars also evaluates the performance of used cars (cars more than 1 year old). The objective is to predict Y, the maintenance costs of used cars the first year after they are purchased by a new owner. The predictor factors are

- X_1 = miles (in thousands) the car will be driven the first year after it is purchased
- X_2 = age (in months) of the car when it is purchased by the new owner
- X_3 = odometer reading of the car in thousands of miles at the time it was purchased

We suppose that assumptions (A) are valid where the population regression function of Y on X_1, X_2, X_3 is given by

$$\mu_Y(x_1, x_2, x_3) = \beta_0 + \beta_1 x_1 + \beta_2 x_2 + \beta_3 x_3$$

The records of used car sales over the past 2 years is the study population, and a simple random sample of size 42 was obtained from this population. The data appear in Table 6.2.1 and are also in the file **usedcars.dat** on the data disk. It is assumed that any cars purchased will be chosen at random from a population of cars that is similar to the population of cars from which the sample of size 42 is obtained; i.e., the target population is judged to be similar to the study population.

T A B L E **6.2.1**
Used Cars Data

Observation Number	Maintenance Cost (dollars) Y	Miles Driven (thousands) X_1	Age (months) X_2	Odometer Reading (thousands of miles) X_3
1	190	6	70	70.7
2	379	11	72	70.9
3	201	6	98	108.8
4	194	8	60	49.0
5	189	4	84	95.7
6	379	8	84	96.3
7	183	6	64	64.6
8	186	7	64	64.8
9	456	19	60	49.2
10	149	17	36	17.8
11	175	5	66	68.1
12	276	8	72	71.1
13	277	8	72	70.9
14	243	7	72	70.8
15	195	7	66	68.3
16	179	10	58	52.9
17	186	8	61	58.8
18	161	9	54	43.3
19	267	11	60	49.7
20	216	9	60	48.8
21	130	15	36	18.5
22	167	7	58	52.1
23	211	18	48	31.7
24	186	7	24	59.1
25	165	9	54	43.6
26	168	9	54	43.6
27	148	6	60	49.3
28	179	7	61	58.4
29	186	7	64	64.6
30	116	13	36	18.5

(Continued)

T A B L E 6.2.1
(Continued)

Observation Number	Maintenance Cost (dollars) Y	Miles Driven (thousands) X_1	Age (months) X_2	Odometer Reading (thousands of miles) X_3
31	312	9	72	71.1
32	168	6	61	58.1
33	190	7	28	49.7
34	235	5	84	95.6
35	195	8	29	48.3
36	140	9	48	32.2
37	209	9	64	68.4
38	607	14	72	95.8
39	181	5	66	68.6
40	176	7	61	58.1
41	201	10	61	58.8
42	279	8	72	70.3

A SAS output from a regression analysis of Y on X_1, X_2, and X_3 is shown in Exhibit 6.2.1 below.

E X H I B I T 6.2.1
SAS Output for Task 6.2.1

```
                        The SAS System

                              00:00 Saturday,  Jan 1, 1994


Model: MODEL1


        X'X Inverse, Parameter Estimates, and SSE


             INTERCEP         MILES          AGE       ODOMETER         MTCOST


INTERCEP   0.8985876124  -0.037785163  -0.005724835  -0.003282576  -198.2198953
MILES     -0.037785163    0.0028862847  -0.000127123   0.0003375752   22.355858102
AGE       -0.005724835   -0.000127123    0.0003194889  -0.000210015   -1.117485612
ODOMETER  -0.003282576    0.0003375752  -0.000210015    0.0002187719    4.8509091502
MTCOST    -198.2198953    22.355858102  -1.117485612    4.8509091502  94817.286025


Dependent Variable: MTCOST
```

E X H I B I T 6.2.1
(Continued)

Analysis of Variance

Source	DF	Sum of Squares	Mean Square	F Value	Prob>F
Model	3	253228.33302	84409.44434	33.829	0.0001
Error	38	94817.28603	2495.19174		
C Total	41	348045.61905			

Root MSE	49.95189	R-square	0.7276
Dep Mean	219.76190	Adj R-sq	0.7061
C.V.	22.73001		

Parameter Estimates

Variable	DF	Parameter Estimate	Standard Error	T for H0: Parameter=0	Prob > |T|
INTERCEP	1	-198.219895	47.35132929	-4.186	0.0002
MILES	1	22.355858	2.68362326	8.330	0.0001
AGE	1	-1.117486	0.89285271	-1.252	0.2184
ODOMETER	1	4.850909	0.73883552	6.566	0.0001

Below are two typical problems that the agency may want to solve.

1 Someone who plans to buy a used car that will be driven 9,000 miles next year asks the agency to predict the first-year maintenance cost of this car. The car is 36 months old and has 30,200 miles on the odometer.

We compute point and interval estimates of $Y(x_1, x_2, x_3)$, the first-year maintenance cost of a future randomly chosen car ($h = 1$) from the subpopulation with

$$X_1 = x_{1,1} = 9.0, \quad X_2 = x_{1,2} = 36, \quad X_3 = x_{1,3} = 30.2$$

For the point estimate we get $\hat{Y}(9.0, 36, 30.2) = 109.25$ by substituting $x_{1,1} = 9.0$, $x_{1,2} = 36$, and $x_{1,3} = 30.2$ into the estimated regression equation. Thus the best estimate of the first-year maintenance cost of this car is $109.25.
To compute a 95% prediction interval for $Y(9.0, 36, 30.2)$ we need

$$SE(\hat{Y}(9.0, 36, 30.2))$$

which is $\hat{\sigma}\sqrt{1 + x^T C x}$ in (6.2.3) with $h = 1$. The matrix C is given in Exhibit 6.2.1 (see the explanation preceding Problem 4.6.1.). Since $x^T = [1, 9.0, 36, 30.2]$, we get

$$x^T C x = 0.0998475$$

From Exhibit 6.2.1 we get $\hat{\sigma} = 49.95189$, and so

$$SE\left(\hat{Y}(9.0, 36, 30.2)\right) = \hat{\sigma}\sqrt{1 + x^T C x} = 52.3864$$

The table-value, obtained by linear interpolation in Table T-2 in Appendix T, is $t_{1-\alpha/2:n-k-1} = t_{0.975:38} = 2.025$. Substituting into (6.2.1) we get

$$C[3.17 \leq Y(9.0, 36, 30.2) \leq 215.33] = 0.95$$

If only an upper confidence bound is desired, then the purchaser has 97.5% confidence that the first-year maintenance cost of a car with $X_1 = 9$, $X_2 = 36$, $X_3 = 30.2$, randomly chosen from the study population, will not exceed \$215.33. To extrapolate this to the first-year maintenance cost of a car to be purchased next year is judgment-based inference using the preceding confidence interval.

2   A company plans to purchase two used cars. Car 1 will be driven 6,000 miles the first year and car 2 will be driven 15,000 miles. The two cars being considered for purchase have the following ages and odometer readings. Car 1 is 24 months old and has 48,900 miles on its odometer, whereas car 2 is 21 months old and has 32,100 miles on its odometer. The company wants to predict the *total* first-year maintenance cost of these two cars.

We denote the total first-year maintenance cost of these two cars by Y_S, and the estimate is

$$\hat{Y}_S = \hat{Y}_1(6.0, 24, 48.9) + \hat{Y}_2(15.0, 21, 32.1)$$

using (6.2.9) with $h = 2$. By substituting the values of the predictor variables into the estimated regression equation, we get

$$\hat{Y}_1(6.0, 24, 48.9) = 146.31 \quad and \quad \hat{Y}_2(15.0, 21, 32.1) = 269.36$$

so

$$\hat{Y}_S = 146.31 + 269.36 = 415.67$$

To obtain a 95% prediction interval for Y_S, we also need $SE(\hat{Y}_S)$, and to obtain this we use (6.2.11). Hence we need

$$\bar{x}^T C \bar{x}$$

for which we first need to calculate $\bar{x}$ in (6.2.6), where

$$\bar{x} = (1/2)[x_1 + x_2]$$

We note that

$$x_1^T = [1, 6.0, 24, 48.9] \quad and \quad x_2^T = [1, 15.0, 21, 32.1]$$

and hence

$$\bar{x}^T = (1/2)\{[1, 6.0, 24, 48.9] + [1, 15.0, 21, 32.1]\} = [1, 10.5, 22.5, 40.5]$$

The quantity $\bar{x}^T C \bar{x} = 0.2646767$, *and (note that* $h = 2$)

$$\sqrt{h + h^2 \, \bar{x}^T C \bar{x}} = \sqrt{2 + 4(0.2646767)} = 1.748916$$

From the computer output we have $\hat{\sigma} = 49.95189$. *So*

$$SE(\hat{Y}_S) = \hat{\sigma}\sqrt{h + h^2 \, \bar{x}^T C \bar{x}} = (49.95189)(1.748916) = 87.3617$$

Thus a 95% prediction interval for Y_S *is*

$$\hat{Y}_S - t_{0.975:38} SE(\hat{Y}_S) \le Y_S \le \hat{Y}_S + t_{0.975:38} SE(\hat{Y}_S)$$

This gives us

$$C[415.67 - (2.025)(87.3617) \le Y_S \le 415.67 + (2.025)(87.3617)] = 0.95$$

or

$$C[238.76 \le Y_S \le 592.58] = 0.95$$

*If only an upper confidence bound is needed, the company can be 97.5% confident that the **total** first-year maintenance cost of two cars to be chosen at random from this study population, with specified miles driven, age, and odometer readings, will not exceed $592.58. As usual, the extrapolation of these results to **next** year's cars is a subject matter inference.*

Problems   6.2

Problems 6.2.1–6.2.3 refer to Task 6.2.1, for which the data are in Table 6.2.1 and are also stored in the file **usedcars.dat** on the data disk. You may use the computer output in Exhibit 6.2.1.

6.2.1 **a** What is the estimated first-year maintenance cost of a used car that will be driven 12,000 miles during the first year after purchase if it is 22 months old and has 9,300 miles on its odometer at the time of purchase by a new owner?

 b What is the estimate of the *average* first-year maintenance cost of all used cars driven 12,000 miles the first year if they are 22 months old and have 9,300 miles on their odometers at the time of purchase by their respective new owners?

 c Obtain a 95% prediction interval for $Y(12.0, 22, 9.3)$ in part (a). In (6.2.3) we calculated the value of the quantity $x^T C x$, which is needed to compute this prediction interval, and it is equal to 0.1903.

 d Obtain a 95% confidence interval for $\mu_Y(12.0, 22, 9.3)$ in part (b).

 e Write a short report and explain how the results in parts (a)–(d) can be used to make decisions about the target population.

6.2.2 A company plans to buy two used cars for its sales staff who will drive the cars the following distances the first year after purchase: driver 1 will drive 9,000 miles, and driver 2 will drive 18,000 miles. The two cars being considered for purchase have the following ages and odometer readings:

Car 1:   age = 12 months, odometer reading = 13,700 miles

Car 2:   age = 20 months, odometer reading = 24,300 miles

The company wants to hold the estimated *total* first-year maintenance cost to a minimum. Which driver should be given which car? Explain your reasoning in detail.

6.2.3   In Problem 6.2.2 estimate the *total* first-year maintenance cost of the two cars if driver 1 is given car 1 and driver 2 is given car 2.

6.2.4   This problem refers to Task 3.4.1 where crystalline forms of certain chemical compounds are used in various electronic devices and it is more desirable to have large crystals than small ones.

Crystals of one particular compound are to be produced by a commercial process, and an investigator wants to examine the relationship between Y, the weight of a crystal in grams, and X, the time in hours for the crystal to grow to its final size. The following data are from a laboratory study in which 14 crystals of various sizes were obtained by allowing the crystals to grow for different preselected amounts of time. The data, along with the MINITAB output from a regression analysis of Y on X, are given in Exhibit 6.2.2. These are the same data that appear in Table 3.4.2 and are also stored in the file **crystal.dat** on the data disk. Assumptions (A) are presumed to be valid, and the data were obtained by sampling with preselected X values.

EXHIBIT 6.2.2

MINITAB Output for Problem 6.2.4

obsno	weight	time
1	0.08	2
2	1.12	4
3	4.43	6
4	4.98	8
5	4.92	10
6	7.18	12
7	5.57	14
8	8.40	16
9	8.81	18
10	10.81	20
11	11.16	22
12	10.12	24
13	13.12	26
14	15.04	28

E X H I B I T **6.2.2**
(Continued)

```
The regression equation is
weight = 0.001 + 0.503 time

Predictor          Coef         Stdev       t-ratio         p
Constant         0.0014        0.5994          0.00     0.998
time            0.50343       0.03520         14.30     0.000


s = 1.062          R-sq = 94.5%       R-sq(adj) = 94.0%


Analysis of Variance

SOURCE           DF            SS            MS          F         p
Regression        1        230.63        230.63     204.58     0.000
Error            12         13.53          1.13
Total            13        244.16


The C matrix is

  .318681318        -.016483516
 -.016483516         .001098901
```

a   If a crystal is allowed to grow for 15 hours, estimate $Y(15)$, its weight.

b   Obtain a 90% two-sided prediction interval for the weight of the crystal in part (a).

Problems (c), (d), (e), and (f) depend on the fact that the crystals are priced based on the time taken to grow them as well on their actual weight. Crystals that are grown for 8 hours or less are priced at $2 per gram; those that are grown between 8 hours and 16 hours are priced at $10 per gram; and those that are grown for more than 16 hours are priced at $16 per gram. These prices reflect the additional amount of operator intervention necessary to grow crystals for longer periods.

c   Estimate the *total* weight of three crystals where crystal 1 is to be grown for 3 hours, crystal 2 is to be grown for 5 hours, and crystal 3 is to be grown for 13 hours.

d   Obtain a lower bound L in part (c) such that you have 95% confidence that the total weight of all three crystals is greater than L.

e   A customer orders a crystal that is to grow for 20 hours. Find the point estimate of the dollar value of this crystal.

f   Obtain a lower bound L in part (e) such that you have 90% confidence that the dollar value of this crystal is greater than L.

g   Obtain a point estimate of the *average* weight of all crystals grown for 20 hours.

h   Obtain a 90% two-sided confidence interval for the *average* weight of the crystals in part (g).

6.2.5   A scientist is interested in Y, the weight in pounds of babies belonging to a certain ethnic group as a function of X, the age in days. A simple random sample was selected from the records of babies born in two hospitals in a certain large city for the past three years, and their weights and ages were recorded. The data are given in Table 6.2.2 and are also stored in the file **ethnic.dat** on the data disk.

   The target population that the scientist wants to investigate is the entire set of babies in the United States that belong to this ethnic group. The study population consists of the records in the hospitals previously referred to. A MINITAB output from a regression of Y on X, including a plot of Y versus X is given in Exhibit 6.2.3. Assumptions (A) are presumed valid.

T A B L E   **6.2.2**
Ethnic Babies Data

Observation Number	Weight Y	Age X	Observation Number	Weight Y	Age X
1	6.60	7	23	12.77	88
2	7.44	10	24	14.04	92
3	9.10	12	25	13.13	96
4	8.16	13	26	11.14	99
5	7.16	14	27	12.02	105
6	6.41	16	28	12.11	108
7	11.79	17	29	8.84	113
8	12.94	17	30	13.64	120
9	10.92	18	31	15.67	132
10	5.97	18	32	15.36	135
11	8.19	30	33	14.41	138
12	6.99	37	34	13.39	142
13	5.70	45	35	14.02	147
14	11.43	46	36	17.15	156
15	9.48	49	37	17.21	159
16	9.49	52	38	14.61	167
17	6.59	56	39	15.01	168
18	6.93	58	40	14.32	183
19	9.78	59	41	17.05	191
20	7.54	67	42	18.88	194
21	8.50	67	43	16.77	195
22	6.44	84	44	17.26	196

E X H I B I T 6.2.3
MINITAB Output for Problem 6.2.5.

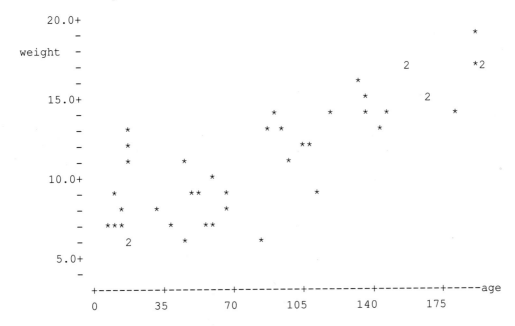

```
The regression equation is
weight = 6.81 + 0.0517 age


Predictor        Coef        Stdev     t-ratio         p
Constant       6.8124       0.5624       12.11     0.000
age          0.051738     0.005219        9.91     0.000


s = 2.104        R-sq = 70.1%      R-sq(adj) = 69.3%

Analysis of Variance

SOURCE          DF          SS            MS          F         p
Regression       1       434.90        434.90      98.29     0.000
Error           42       185.84          4.42
Total           43       620.74



The C matrix is
 0.07148029      -0.0005477867
-0.00054778       0.0000061549
```

a Exhibit the population parameter that represents the average weight of all babies in the study population who are 30 days old. Repeat this for babies who are 60 days old.

b What is the estimated average weight of babies who are 60 days old?

c Compute a 90% two-sided confidence interval for the quantity in part (b).

d Compute a 90% two-sided interval for the weight of a randomly chosen baby who is 60 days old.

6.3
Tolerance Intervals

Until now, in straight line regression, we have been concentrating on $\mu_Y(x)$, the mean of the Y values in the subpopulation where $X = x$, and on a randomly chosen value $Y(x)$ from this subpopulation. We now turn our attention to the actual distribution of the Y values in the subpopulation where $X = x$. In particular we are interested in finding a number $\lambda_p(x)$ such that a proportion p of the Y values in the subpopulation corresponding to $X = x$ will be less than this number (see Figure 6.3.1). Then a proportion $1 - p$ of the Y values in this subpopulation will be greater than $\lambda_p(x)$ and, for any p such that $0 \leq p \leq 1$, a proportion p will lie in the interval

$$[\lambda_{(1-p)/2}(x), \lambda_{(1+p)/2}(x)]$$

(see Figure 6.3.2). Note that in Figure 6.3.2 the area to the left of $\lambda_{(1-p)/2}(x)$ is $(1 - p)/2$ and the area to the left of $\lambda_{(1+p)/2}(x)$ is $(1 + p)/2$, so the area between $\lambda_{(1-p)/2}(x)$ and $\lambda_{(1+p)/2}(x)$ is equal to $(1 + p)/2 - (1 - p)/2 = p$. We illustrate with two examples.

F I G U R E   **6.3.1**

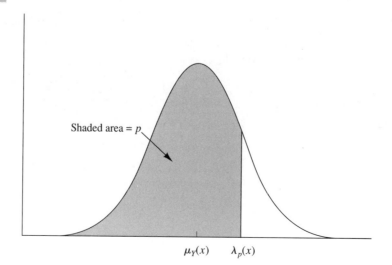

Shaded area = p

$\mu_Y(x)$ $\lambda_p(x)$

F I G U R E 6.3.2

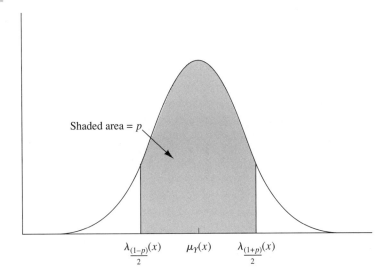

Shaded area = p.

$$\lambda_{\frac{(1-p)}{2}}(x) \qquad \mu_Y(x) \qquad \lambda_{\frac{(1+p)}{2}}(x)$$

E X A M P L E 6.3.1

Blood cholesterol level Y in adults who have taken a certain drug for a year tends to change with age X. Suppose that in a study population the regression function is given by $\mu_Y(x) = \beta_0 + \beta_1 x$ for $35 \le x \le 60$. Based on a simple random sample of size n, we can estimate $\mu_Y(x)$ for any specified age group in $35 \le x \le 60$. This gives us information about the average cholesterol levels for various age groups. For a given age group with $X = x$ and a given proportion p, we may want to determine a number $\lambda_p(x)$ such that *a proportion p of this subpopulation will have cholesterol values below this number*. For example, we may want to make statements of the form "99% (i.e., $p = 0.99$) of the individuals in this population who are 40 years old (i.e., $x = 40$) have cholesterol levels below $\lambda_{0.99}(40)$." The number $\lambda_{0.99}(40)$ gives us an idea of the upper extremes of cholesterol levels for individuals in the age group $x = 40$, and this information can be extremely useful in identifying individuals with abnormally high cholesterol levels. ∎

E X A M P L E 6.3.2

A company that manufactures steel rods is studying how X_1, the hardness of the steel rods, and X_2, the amount of carbon used in the alloy, affect Y, the strength of the rods. For a given value of $X_1 = x_1$ and $X_2 = x_2$, and for $p = 0.001$, the company wants to determine the number $\lambda_p(x_1, x_2) = \lambda_{0.001}(x_1, x_2)$ such that a proportion of only 0.001 (i.e., 0.1%) of the steel rods produced will have strength less than that number; i.e., 99.9% of the rods will have strength greater than $\lambda_{0.001}(x_1, x_2)$. ∎

Tolerance Points

Consider a $(k + 1)$-variable study population $\{(Y, X_1, \ldots, X_k)\}$. We are interested in the following quantities for the subpopulation with $X_1 = x_1, \ldots, X_k = x_k$.

1   The number

$$\lambda_p(x_1, \ldots, x_k) \tag{6.3.1}$$

such that a proportion p of the subpopulation Y values are below $\lambda_p(x_1, \ldots, x_k)$, and hence a proportion $1 - p$ of the subpopulation Y values are above $\lambda_p(x_1, \ldots, x_k)$. This is called the pth *tolerance point* or pth *percentile* for the subpopulation (i.e., the area between $-\infty$ and $\lambda_p(x_1, \ldots, x_k)$ is p) (see Figure 6.3.1.).

2   The two numbers $\lambda_{(1-p)/2}(x_1, \ldots, x_k)$ and $\lambda_{(1+p)/2}(x_1, \ldots, x_k)$ such that a proportion p of the subpopulation Y values are in the interval

$$[\lambda_{(1-p)/2}(x_1, \ldots, x_k), \lambda_{(1+p)/2}(x_1, \ldots, x_k)] \tag{6.3.2}$$

This is called a *two-sided* (symmetric) *population tolerance interval* (See Figure 6.3.2).

Under population assumptions (A) or (B), the pth subpopulation tolerance point corresponding to $X_1 = x_1, \ldots, X_k = x_k$ is

$$\lambda_p(x_1, \ldots, x_k) = \beta_0 + \beta_1 x_1 + \beta_2 x_2 + \ldots + \beta_k x_k + z_p \sigma \tag{6.3.3}$$

where z_p is the pth percentile of the standard Gaussian population. Because $\lambda_p(x_1, \ldots, x_k)$ in (6.3.3) contains unknown parameters β_i and σ, we must use sample values to compute point and interval estimates for it. So data are collected and $\hat{\beta}_i$ and $\hat{\sigma}$ are computed by formulas in Chapter 4. A point estimate of $\lambda_p(x_1, \ldots, x_k)$ is

$$\hat{\lambda}_p(x_1, \ldots, x_k) = \hat{\beta}_0 + \hat{\beta}_1 x_1 + \cdots + \hat{\beta}_k x_k + z_p \hat{\sigma} \tag{6.3.4}$$

We also compute a confidence interval for $\lambda_p(x_1, \ldots, x_k)$ of the form

$$C[L_p \leq \lambda_p(x_1, \ldots, x_k) \leq U_p] = 1 - \alpha \tag{6.3.5}$$

In (6.3.5) and elsewhere we use L_p for $L_p(x_1, \ldots, x_k)$ for ease of notation. Similarly we write U_p for $U_p(x_1, \ldots, x_k)$. We can attach one of the following three meanings to the confidence interval in (6.3.5).

1   We have confidence $1 - \alpha$ that the interval L_p to U_p includes $\lambda_p(x_1, \ldots, x_k)$.

2   If only an upper confidence bound is needed, then we have $1 - \alpha/2$ confidence that *at least* $100p\%$ of the Y values in the subpopulation corresponding to $X_1 = x_1, \ldots, X_k = x_k$ are below U_p because $100p\%$ of Y values in this subpopulation are below $\lambda_p(x_1, \ldots, x_k)$, which in turn is smaller than U_p with confidence $1 - \alpha/2$.

3   If only a lower confidence bound is needed, then we have $1 - \alpha/2$ confidence that *at least* $100(1 - p)\%$ of the Y values in the subpopulation corresponding to $X_1 = x_1, \ldots, X_k = x_k$ are above L_p because $100(1 - p)\%$ of the Y values are above $\lambda_p(x_1, \ldots, x_k)$, which in turn is greater than L_p with confidence $1 - \alpha/2$.

Also, using the Bonferroni method we can obtain numbers $L_{(1-p)/2}$ and $U_{(1+p)/2}$ such that

$$C[L_{(1-p)/2} \le \lambda_{(1-p)/2}(x_1, \ldots, x_k) \quad \text{and} \quad \lambda_{(1+p)/2}(x_1, \ldots, x_k) \le U_{(1+p)/2}] \ge 1 - \alpha$$

But, since a proportion p of the subpopulation values are between $\lambda_{(1-p)/2}(x_1, \ldots, x_k)$ and $\lambda_{(1+p)/2}(x_1, \ldots, x_k)$ we have confidence greater than or equal to $1 - \alpha$ that at least $100p\%$ of the Y values in the subpopulation with $X_1 = x_1, \ldots, X_k = x_k$ are between the values $L_{(1-p)/2}$ and $U_{(1+p)/2}$. For example, if $p = 0.80$, then $(1-p)/2 = 0.10$ and $(1+p)/2 = 0.90$, and we have confidence greater than $1 - \alpha$ that at least $100p\% = 80\%$ of the Y values in the subpopulation with $X_1 = x_1, \ldots, X_k = x_k$ will be between $L_{0.10}$ and $U_{0.90}$.

The instructions for computing L_p and U_p are in Box 6.3.1.

B O X **6.3.1** **Instructions for Computing L_p and U_p**

Assumptions (A) or (B) are presumed to be valid, and a sample of size n is selected from a $(k + 1)$-variable study population $\{(Y, X_1, \ldots, X_k)\}$. The sample values and the corresponding X matrix are

Sample			
Y	X_1	...	X_k
y_1	$x_{1,1}$	$\cdots$	$x_{1,k}$
y_2	$x_{2,1}$	$\cdots$	$x_{2,k}$
$\vdots$	$\vdots$	$\vdots$	$\vdots$
y_n	$x_{n,1}$	$\cdots$	$x_{n,k}$

$$X = \begin{bmatrix} 1 & x_{1,1} & \cdots & x_{1,k} \\ 1 & x_{2,1} & \cdots & x_{2,k} \\ \vdots & \vdots & \vdots & \vdots \\ 1 & x_{n,1} & \cdots & x_{n,k} \end{bmatrix}$$

1. The confidence coefficient $1 - \alpha$ is specified by the investigator.
2. The number p is specified by the investigator.
3. The subpopulation of interest has $X_1 = x_1, \ldots, X_k = x_k$ and, using these values, the vector x is defined by

$$x^T = [1, x_1, \ldots, x_k] \tag{6.3.6}$$

4. The statistics $\hat{\beta}_0, \hat{\beta}_1, \cdots, \hat{\beta}_k$ and $\hat{\sigma}$ are computed by the formulas in Chapter 4 using the sample data.
5. The matrix $C = (X^T X)^{-1}$ is computed where X is given above. The quantity A is computed where

$$A = \sqrt{x^T C x}$$

and where x is given in (6.3.6).

6 z_p is obtained from Table T-1 in Appendix T where z_p is the pth percentile point of a standard normal population.

7 δ_p is computed where $\delta_p = -z_p/A$.

8   Look up $t_{1-\alpha/2:n-k-1;(\delta_p)}$ and $t_{\alpha/2:n-k-1;(\delta_p)}$. The quantity $t_{\gamma:n-k-1;(\delta_p)}$ (for $\gamma = 1 - \alpha/2$ and for $\gamma = \alpha/2$) is obtained from Table T-8 in Appendix T and is the γ percentile point of the noncentral t distribution with $n-k-1$ degrees of freedom and noncentrality δ_p. The quantity δ_p is in (7). The quantity $t_{\gamma:n-k-1;(\delta_p)}$ can also be obtained from several statistical computing packages. It is useful to note that $t_{1-\alpha/2:n-k-1;(\delta)} = -t_{\alpha/2:n-k-1;(-\delta)}$.

9   Compute $g_{p,1-\alpha/2}$ and $g_{p,\alpha/2}$, where $g_{p,1-\alpha/2} = -At_{1-\alpha/2:n-k-1;(\delta_p)}$ and $g_{p,\alpha/2} = -At_{\alpha/2:n-k-1;(\delta_p)}$.

10 The confidence bounds L_p and U_p in (6.3.5) are given in (6.3.7) and (6.3.8), respectively.

$$L_p = \hat{\beta}_0 + \hat{\beta}_1 x_1 + \ldots + \hat{\beta}_k x_k + \hat{\sigma} g_{p,1-\alpha/2} \tag{6.3.7}$$

$$U_p = \hat{\beta}_0 + \hat{\beta}_1 x_1 + \ldots + \hat{\beta}_k x_k + \hat{\sigma} g_{p,\alpha/2} \tag{6.3.8}$$

To help understand these computations, we illustrate the procedures using an artificial example.

E X A M P L E 6.3.3

Assume that the data in Table 6.3.1 were obtained from a two-variable target population $\{(Y, X)\}$ by sampling with preselected X values. Suppose assumptions (A) are satisfied. The data are also stored in the file **table631.dat** on the data disk.

We want the following quantities.

a   A point estimate of $\lambda_{0.80}(3.0)$, the number such that 80% of the Y values in the subpopulation with $X = 3.0$ is less than that number.

b   A 95% two-sided confidence interval for $\lambda_{0.80}(3.0)$.

c   A point estimate of $\lambda_{0.20}(3.0)$, the number such that 20% of the subpopulation with $X = 3.0$ is less than that number.

d   A 95% two-sided confidence interval for $\lambda_{0.20}(3.0)$.

For this problem $k = 1$ and, as usual, the matrix X can be obtained from the column of x_i values given in Table 6.3.1 by attaching a column of 1's as the first column. A SAS output from a regression analysis of Y on X follows in Exhibit 6.3.1.

T A B L E **6.3.1**

Observation Number	Y	X
1	4.81	1.0
2	3.60	1.1
3	4.90	1.3
4	3.05	1.6
5	3.44	1.8
6	3.17	1.8
7	3.34	1.8
8	1.61	2.1
9	1.22	2.4
10	0.20	2.6
11	1.56	2.6
12	0.55	2.7
13	−2.56	2.9
14	−0.34	3.0
15	−2.56	3.5
16	−2.96	3.6
17	−1.04	4.1
18	−4.64	5.2

E X H I B I T **6.3.1**
SAS Output for Example 6.3.3

The SAS System

00:00 Saturday, Jan 1, 1994

Model: MODEL1

X'X Inverse, Parameter Estimates, and SSE

	INTERCEP	X	Y
INTERCEP	0.3598685805	−0.121455309	6.9909129346
X	−0.121455309	0.0484744028	−2.405464142
Y	6.9909129346	−2.405464142	17.808145181

Dependent Variable: Y

■ E X H I B I T   **6.3.1**
(Continued)

Analysis of Variance

Source	DF	Sum of Squares	Mean Square	F Value	Prob>F
Model	1	119.36728	119.36728	107.247	0.0001
Error	16	17.80815	1.11301		
C Total	17	137.17543			

Root MSE	1.05499	R-square	0.8702
Dep Mean	0.96389	Adj R-sq	0.8621
C.V.	109.45167		

Parameter Estimates

Variable	DF	Parameter Estimate	Standard Error	T for H0: Parameter=0	Prob > |T|
INTERCEP	1	6.990913	0.63287992	11.046	0.0001
X	1	-2.405464	0.23227667	-10.356	0.0001

The quantities needed for parts (a) and (b) as described in Box 6.3.1, are exhibited below.

1 $1 - \alpha = 0.95$, so $\alpha = 0.05$, $\alpha/2 = .025$, and $1 - \alpha/2 = 0.975$.

2 $p = 0.80$, so $1 - p = 0.20$.

3 The subpopulation for which the tolerance point is to be evaluated has $X = 3.0$, so $x^T = [1, \ 3.0]$.

4 From the computer output in Exhibit 6.3.1 we get

$$\hat{\beta}_0 = 6.9909 \qquad \hat{\beta}_1 = -2.4055 \qquad \hat{\sigma} = 1.055$$

5 From the matrix C and the vector $x^T = [1, \ 3.0]$ above, we compute

$$A = \sqrt{x^T C x} = \sqrt{0.0674} = 0.2596$$

6 $z_p = z_{0.80} = 0.8416$ (from SAS).

7 $\delta_p = \delta_{0.80} = -0.8416/0.2596 = -3.242$.

8 $t_{0.975:16;(-3.242)} = -1.254$ and $t_{0.025:16;(-3.242)} = -6.174$ from Table T-8 in Appendix T.

9 $g_{0.80,0.975} = -(0.2596)(-1.254) = 0.3255$ and
$g_{0.80,0.025} = -(0.2596)(-6.174) = 1.6028$.

Also $\hat{\beta}_0 + \hat{\beta}_1 x = 6.9909 + (-2.4055)(3.0) = -0.2256$. We are now in a position to calculate the needed quantities.

a   The point estimate of $\lambda_{0.80}(3.0)$ is

$$\hat{\lambda}_{0.80}(3.0) = -0.2256 + (0.8416)(1.055) = 0.6623$$

b   From (6.3.7) and (6.3.8) we get

$$L_{0.80} = -0.2256 + (0.3255)(1.055) = 0.1178$$

and

$$U_{0.80} = -0.2256 + (1.6028)(1.055) = 1.465$$

so a 95% confidence interval for $\lambda_{0.80}(3.0)$ is given by

$$C[0.1178 \le \lambda_{0.80}(3.0) \le 1.465] = 0.95$$

c   The point estimate of $\lambda_{0.20}(3.0)$, using (6.3.4) with $z_{0.20} = -0.8416$, is

$$\hat{\lambda}_{0.20}(3.0) = -0.2256 + (-0.8416)(1.055) = -1.113$$

d   To obtain $L_{0.20}$ and $U_{0.20}$ we get $\delta_{0.20} = -(-0.8416)/0.2596 = 3.242$. From Table T-8 in Appendix T we obtain $t_{0.975:16;(3.242)} = -t_{0.025:16;(-3.242)} = 6.174$, and $t_{0.025:16;(3.242)} = 1.254$. So $g_{0.20,0.975} = -(0.2596)(6.174) = -1.6028$, and $g_{0.20,0.025} = -(0.2596)(1.254) = -0.3255$. From (6.3.7) and (6.3.8) we get

$$L_{0.20} = -0.2256 - 1.6028(1.055) = -1.917$$

and

$$U_{0.20} = -0.2256 - 0.3255(1.055) = -0.5690$$

so a 95% confidence interval for $\lambda_{0.20}(3.0)$ is given by

$$C[-1.917 \le \lambda_{0.20}(3.0) \le -0.5690] = 0.95 \quad \blacksquare$$

Problems   6.3

6.3.1   **a**   In Example 6.3.3 compute a point estimate of $\lambda_{0.85}(3.0)$.

   b   In Example 6.3.3 compute a 90% two-sided confidence interval for $\lambda_{0.85}(3.0)$. Write a paragraph explaining what this means.

   c   In Example 6.3.3 compute a 90% two-sided confidence interval for $\lambda_{0.15}(3.0)$ and state in words what this means.

6.3.2   In Example 6.3.3 compute numbers L and U such that you have confidence greater than or equal to 90% that at least a proportion $p = 0.7$ of the subpopulation of Y values with $X = 3.0$ is between L and U.

6.3.3   This problem refers to Problem 6.2.5 where a scientist is interested in Y, the weight in pounds of babies belonging to a certain ethnic group, as a function of X, the age in days. A simple random sample was selected from the records of weights and ages of babies in two hospitals in a certain large city. The data are given in Table 6.2.2 and are also stored in the file **ethnic.dat** on the data disk. Assumptions (A) are presumed to hold, and the regression function is

$$\mu_Y(x) = \beta_0 + \beta_1 x \qquad \text{for} \qquad 5 \leq x \leq 200$$

A SAS output from a regression analysis of Y on X is given in Exhibit 6.3.2.

E X H I B I T **6.3.2**
SAS Output for Problem 6.3.3

The SAS System   00:00 Saturday, Jan 1, 1994

Model: MODEL1

X'X Inverse, Parameter Estimates, and SSE

	INTERCEP	AGE	WEIGHT
INTERCEP	0.0714802886	-0.000547787	6.8123997925
AGE	-0.000547787	6.1549067E-6	0.0517375917
WEIGHT	6.8123997925	0.0517375917	185.83817639

Dependent Variable: WEIGHT

Analysis of Variance

Source	DF	Sum of Squares	Mean Square	F Value	Prob>F
Model	1	434.90154	434.90154	98.289	0.0001
Error	42	185.83818	4.42472		
C Total	43	620.73972			

Root MSE	2.10350	R-square	0.7006	
Dep Mean	11.41705	Adj R-sq	0.6935	
C.V.	18.42422			

Parameter Estimates

Variable	DF	Parameter Estimate	Standard Error	Parameter=0	Prob > |T|
INTERCEP	1	6.812400	0.56238790	12.113	0.0001
AGE	1	0.051738	0.00521859	9.914	0.0001

a Parents of a 30-day-old baby who is in the study population bring their baby to the hospital and want to know if their baby is underweight. Do you consider this baby's weight to be a random observation from the study population? Explain.

b In part (a) would you compare this baby's weight with the average of all babies in the study population who are 30 days old to help answer their question? Explain.

c To help answer the parents' question in part (a), suppose the physician informs them that 90% of all babies in this study population who are 30 days old weigh more than c pounds. Which of the following subpopulation parameters is the quantity c?

   i $Y(30)$

   ii $\mu_Y(30)$

   iii $\lambda_{0.90}(30)$

   iv $\lambda_{0.10}(30)$

d Compute an appropriate interval for the correct quantity in part (c) such that we have 95% confidence that the interval contains this quantity. Write a short report for the physician to give to the parents.

e Suppose two other parents whose baby is in the study population bring their 60-day-old baby to the hospital and want to know if their baby is underweight. The physician weighs the baby and notices that it weighs 7.3 pounds. To determine how usual or unusual this weight is, which of the following parameters is of interest to the physician?

   i $Y(x)$

   ii $\mu_Y(x)$

   iii $\lambda_p(x)$

What is the correct value of x for the physician to use?

f Suppose that a baby's weight is considered to be unusually low if its weight is in the lower 5% of the subpopulation for its age. What value of p should be used in part (e) for the physician to consider the baby's weight unusual?

g Compute an appropriate 95% one-sided confidence bound for the proper quantity in part (e) and, in a written report, interpret this bound so the family can understand it.

6.4
Calibration and Regulation for Straight Line Regression

Most of the discussion in this book has been about the population regression function $\mu_Y(x)$ or about predicting Y as a function of X, and the statistical inferences are valid if assumptions (A) or (B) are satisfied. However, there are several very important applications where the primary interest is not predicting Y as a function of X, but rather to predict X as a function of Y. It may seem at first that we could use all the formulas developed in Chapter 3 and just interchange Y and X. But as we shall see, it is sometimes not that simple. Two such applications are discussed in this section where we want to predict X as a function of Y, but it is not appropriate to simply interchange the roles of Y and X. The applications are referred to as **calibration** and **regulation**, respectively.

We illustrate with two examples.

E X A M P L E 6.4.1

Determining small quantities of toxic substances in water samples is sometimes a very difficult problem for analytical chemists. Typically analytical methods do not recover all of a given toxic substance present in the water sample. Suppose for a given chemical procedure a chemist knows that the relationship between X, the actual amount of mercury present in water samples and Y, the amount of mercury recovered by the analytical method, is given by the regression function $\mu_Y(x) = \beta_0 + \beta_1 x$. A set of standard solutions containing *known* amounts $x_1, x_2, \ldots, x_n$ of mercury, is prepared and is subjected to chemical analysis. Then $y_1, y_2, \ldots, y_n$, the amounts of mercury recovered from these solutions, are recorded. Clearly data are obtained by sampling with preselected X values because the values $x_1, x_2, \ldots, x_n$ were preselected (solutions with *known* amounts of mercury were prepared and used). So it may be reasonable to suppose that assumptions (A) hold. The sample can be presumed to have been obtained from a target population of interest. Estimates of the parameters β_0, β_1, and σ can be calculated using the sample data $(y_1, x_1), \ldots, (y_n, x_n)$. Then a water sample containing an *unknown* amount x_0 of mercury is subjected to the same chemical analysis, and the amount of mercury recovered is measured to be y_0. The chemist wants to determine x_0, the actual amount of mercury present. Thus it is necessary to estimate x_0, the subpopulation from which a single sample value y_0 was obtained.   ∎

E X A M P L E 6.4.2

Suppose a company is investigating a new food supplement for increasing the weight of chickens. The research scientist decides that the regression function $\mu_Y(x) = \beta_0 + \beta_1 x$ relates the average weight Y gained in pounds and X, the time in weeks the chickens have been fed the new food supplement. An experiment is conducted with five groups of chickens, where each group of chickens is fed the new ration for a different amount of time, say $X = 2, 4, 6, 8$, and 10 weeks, respectively, and the weights Y recorded. The data are used to obtain estimates of β_0, β_1, and σ. Since the X values were preselected (the number of weeks each chicken was fed the new ration was preselected), it may be reasonable to suppose that assumptions (A) apply where data are obtained by preselected X values. The company wants to determine x_0 so that in their advertisements for the new product they can claim, "If you want your chickens to gain an *average* of 10 pounds, feed them the new supplement for x_0 weeks." In other words, the investigator wants to determine x_0 for which $\mu_Y(x_0) = 10$.   ∎

The two examples are slightly different. In Example 6.4.1 a chemist wants to determine x_0, the X value for the subpopulation from which a *single* sample value y_0 is observed. In Example 6.4.2 a scientist wants to determine x_0, the value of X for the subpopulation whose *mean* $\mu_Y(x_0)$ is equal to a specified value, say m_0. The first problem, determining x_0 for a given y_0, is referred to as a *calibration* problem, whereas the second problem, determining x_0 for a given value of $\mu_Y(x_0)$, is referred to as a *regulation* problem. Both are *inverse estimation* problems, i.e., estimating the value of X given some information about Y.

As stated earlier, the first thing that may cross your mind is to interchange Y and X, use the regression function of X on Y given by $\mu_X(y) = \beta_0^* + \beta_1^* y$, and use the formulas in Chapter 3 for point estimates and confidence intervals for $\mu_X(y)$ and $X(y)$. This is the correct way to proceed if the problem is such that assumptions (B) are met, but if the data are obtained by preselecting the X values, as they are in many cases (for instance in Examples 6.4.1 and 6.4.2), then assumptions (A) are not satisfied when Y and X are interchanged. It is easy to see why this procedure is not justified, because if X and Y were interchanged, assumptions (A) require the observed X values to be a simple random sample from a Gaussian population for each specified value of Y, but X values are preselected and thus cannot be a simple random sample from a Gaussian (or any other) population.

This section applies to problems when assumptions (A) are valid, data are obtained by preselected X values, and the unknown value x_0 corresponding to an observed value y_0 (or corresponding to a specified value m_0 of $\mu_Y(x)$) is to be determined.

We discuss calibration and regulation separately and exhibit formulas for point and confidence interval estimates for x_0 in each case.

Calibration

Suppose we are required to calibrate a new instrument, for example a thermometer. An experiment is conducted, and readings $y_1, y_2, \ldots, y_n$ on the new instrument are taken at *preselected, known* temperatures $x_1, x_2, \ldots, x_n$; thus the data are obtained by sampling with preselected X values. We suppose that assumptions (A) are valid where the population regression function is

$$\mu_Y(x) = \beta_0 + \beta_1 x$$

The data are $(y_1, x_1), \ldots, (y_n, x_n)$. From the data we compute $\hat{\beta}_0, \hat{\beta}_1, \hat{\mu}_Y(x)$, and $\hat{\sigma}$ by the formulas in Chapter 3. To use the new instrument, we need to observe a reading y_0 on it and determine the true temperature x_0. Actually we do this every-time we read any gauge, not just a thermometer. Another way to view this is as follows: let $y(x_0)$ denote the value of a random observation (i.e., a reading on a new thermometer to be calibrated) from a subpopulation determined by the unknown value x_0 (unknown *true* temperature). We assume that the subpopulation from which $y(x_0)$ was obtained is Gaussian with unknown mean $\mu_Y(x_0)$ and unknown standard deviation σ. The problem is to determine x_0, the X value for the subpopulation from which $y(x_0)$ was selected; i.e., to predict the true temperature x_0 when the reading on the new instrument is $y(x_0)$ (see Figure 6.4.1.).

F I G U R E **6.4.1**

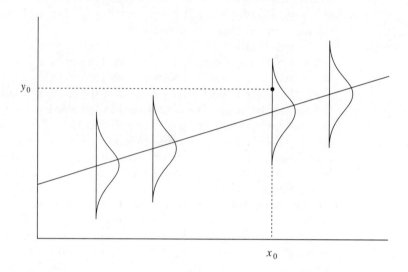

The estimated value for $Y(x_0)$ at the point x_0 (unknown) based on sample data is given by

$$\hat{Y}(x_0) = \hat{\beta}_0 + \hat{\beta}_1 x_0 \qquad \text{(6.4.1)}$$

Assuming that $\hat{\beta}_1 \neq 0$ and solving (6.4.1) for x_0, we get

$$x_0 = \frac{\hat{Y}(x_0) - \hat{\beta}_0}{\hat{\beta}_1} \qquad \text{(6.4.2)}$$

If the observed value y_0 is substituted for $\hat{Y}(x_0)$ in (6.4.2), we get the point estimate of x_0 as

$$\hat{x}_0 = \frac{y_0 - \hat{\beta}_0}{\hat{\beta}_1} \qquad \text{(6.4.3)}$$

To obtain a $1 - \alpha$ confidence region for x_0, we compute the quantities in Box 6.4.1.

B O X **6.4.1**   **Confidence Region Computations for Calibration Problems**

Carry out steps (1)–(5) below.

1   Compute

$$A = \hat{\beta}_1^2 - \frac{\hat{\sigma}^2 t_{1-\alpha/2:n-2}^2}{SSX}$$

where $SSX = \sum_{i=1}^{n}(x_i - \bar{x})^2$.

2   Compute

$$B = A(1 + \frac{1}{n}) + \frac{[y_0 - \bar{y}]^2}{SSX}$$

3   Compute

$$C = \hat{\beta}_1[y_0 - \bar{y}]$$

4   Compute

$$D = t_{1-\alpha/2:n-2}\hat{\sigma}$$

5   If $A \neq 0$ and $B > 0$, compute L and U where

$$L = \bar{x} + \frac{C - D\sqrt{B}}{A}$$

$$U = \bar{x} + \frac{C + D\sqrt{B}}{A}$$

A $1 - \alpha$ confidence region for x_0 is given by

a $-\infty < x_0 < \infty$		if $A < 0$ and $B \leq 0$
b $-\infty < x_0 \leq U$ and $L \leq x_0 < \infty$		if $A < 0$ and $B > 0$ **(6.4.4)**
c $L \leq x_0 \leq U$		if $A > 0$

If we obtain either (a) or (b) in (6.4.4), then the result will certainly be unsatisfactory because the confidence region is not a finite interval. However, even if we obtain the confidence interval (c), the result may still be unsatisfactory if the width $U - L$ is so large that we cannot make a useful decision about x_0. Note that the confidence interval in (c) results if and only if $A > 0$. This condition will hold when

$$\left[\frac{\hat{\beta}_1}{SE(\hat{\beta}_1)}\right]^2 > t_{1-\alpha/2:n-2}^2$$

which in turn implies that the $1 - \alpha$ confidence interval for β_1 does not include zero, and we have $1 - \alpha$ confidence that $\beta_1 \neq 0$. If $A \leq 0$, then this implies that a $1 - \alpha$ confidence interval for β_1 will include zero. This does not necessarily imply that $\beta_1 = 0$, but based on sample data we are unable to rule out this possibility at the $1 - \alpha$ confidence level. Clearly if β_1 is indeed 0, then the regression function of Y on X is $\mu_Y(x) = \beta_0$; i.e., x is not in the regression function, so y_0 does not contain any information about x_0. If the result is (a) or (b) in (6.4.4), then either the sample size is too small and a larger sample must be obtained to get a confidence interval as in (c), or β_1 is indeed zero in $\mu_Y(x) = \beta_0 + \beta_1 x$ and consequently y_0 does not contain any information about x_0. This latter situation does not occur in practice if the investigator knows, a priori, that $\beta_1 \neq 0$ in the regression function of Y on X.

The following example illustrates the computations described in Box 6.4.1.

E X A M P L E   6.4.3

A mercury-in-glass thermometer that is designed to measure temperatures between 95°F and 110°F is being calibrated to determine whether it is reliable enough for a physician to use. It is placed in a water bath at a known constant temperature X, and the corresponding thermometer reading Y is recorded. This is repeated at several known temperatures. The data from the calibration experiment are given in Table 6.4.1 and are also stored in the file **thermom.dat** on the data disk.

Clearly the data are obtained by sampling with preselected X values, and we suppose that assumptions (A) hold.

The thermometer is later used to measure the temperature of a patient in a hospital's emergency room. We want to estimate the true temperature x_0 of the patient if the thermometer reads $y_0 = 104$. We will also obtain a 95% two-sided confidence interval for the true temperature, x_0. MINITAB output from a regression analysis of Y on X is given in Exhibit 6.4.1.

From Table T-2 in Appendix T we get $t_{0.975;6} = 2.447$, and from the data we compute $\bar{x} = 103$, $\bar{y} = 102.975$, and $SSX = 168$. From the computer output we obtain

$$\hat{\beta}_0 = -3.140 \qquad \hat{\beta}_1 = 1.03024 \qquad \hat{\sigma} = 0.2344$$

Thus

$$\hat{Y}(x) = -3.140 + 1.03024x$$

T A B L E   6.4.1
Thermometer Calibration Data

Observation Number	Thermometer Reading Y	Known Temperature X
1	95.71	96
2	98.16	98
3	99.52	100
4	102.09	102
5	103.79	104
6	106.18	106
7	108.14	108
8	110.21	110

MINITAB Output for Example 6.4.3

```
The regression equation is
reading = - 3.14 + 1.03 knowntmp

Predictor          Coef        Stdev      t-ratio          p
Constant         -3.140        1.865        -1.68      0.143
knowntmp        1.03024      0.01809        56.96      0.000


s = 0.2344        R-sq = 99.8%       R-sq(adj) = 99.8%
```

and

$$\hat{x}_0 = \frac{104 - (-3.140)}{1.03024} = 103.995$$

To compute a 95% confidence region for x_0, we follow the instructions in Box 6.4.1. From steps (1)–(4) of Box 6.4.1 we get

$$A = 1.0594 \qquad B = 1.1981 \qquad C = 1.0560 \qquad D = 0.57358$$

Since $A \neq 0$ and $B > 0$, we compute L and U as described in step (5) of Box 6.4.1 and obtain

$$L = 103.4 \qquad \text{and} \qquad U = 104.6$$

Since $A > 0$, case (c) in Box 6.4.1 applies, and we get a 95% confidence interval for x_0 as

$$C[103.4 \leq x_0 \leq 104.6] = .95$$

Thus a physician can be 95% confident that the patient's temperature is between 103.4 and 104.6 degrees. ■

Regulation

This application is very similar to the calibration problem except that we want to determine x_0 corresponding to a specified *average* value of Y, i.e., corresponding to a specified value of $\mu_Y(x_0)$ denoted by m_0. We suppose that assumptions (A) hold with $\mu_Y(x) = \beta_0 + \beta_1 x$, and data are obtained by sampling with preselected X values.

Example 6.4.2 illustrates one such situation. Consider the following question: For how many weeks (i.e., what is the value of x_0?) should one feed the supplement to the chickens so that the *average* weight gain for the entire subpopulation of chickens is, say, 10 pounds? Here $m_0 = 10$ and x_0 is the unknown quantity to be estimated. Note that we are not trying to predict the value of x_0 corresponding to the observed weight y_0 of a single chicken; if we were, then we would apply the results from *calibration* just discussed. What we want here is to *regulate* the average weight gain of chickens by feeding them the supplement for an appropriate number of weeks, viz., x_0 weeks.

We estimate the value of x_0 corresponding to the specified average value m_0 of Y by setting $\mu_Y(x_0)$ equal to m_0 in the regression function $\mu_Y(x_0) = \beta_0 + \beta_1 x_0$ and solving for x_0. This gives

$$x_0 = \frac{m_0 - \beta_0}{\beta_1}$$

Then we substitute the estimates of β_0 and β_1, obtained by formulas in Chapter 3, and get $\hat{x}_0$, which is

$$\hat{x}_0 = \frac{m_0 - \hat{\beta}_0}{\hat{\beta}_1} \qquad \text{(6.4.5)}$$

Thus the estimate of x_0 is the same as in (6.4.3) except m_0 takes the place of y_0. A confidence region for x_0 is computed by following the instructions for calibration problems given in Box 6.4.1, except B and C are slightly modified. The B and C values to use in Box 6.4.1, for *regulation* problems are

$$B = \frac{A}{n} + \frac{(m_0 - \bar{y})^2}{SSX} \qquad \text{and} \qquad C = \hat{\beta}_1(m_0 - \bar{y}) \qquad \text{(6.4.6)}$$

In summary, the results in this section are used when:

1   The unknown value x_0 of X is to be predicted corresponding to an observed value y_0 of Y (calibration) or for a specified value m_0 of the average Y value for the subpopulation with $X = x_0$ (regulation).

2   The regression function is $\mu_Y(x) = \beta_0 + \beta_1 x$ and $\beta_1 \neq 0$.

3   Assumptions (A) are presumed to be valid and the data are obtained by sampling with preselected X values.

4   Remember, if assumptions (B) are satisfied, then the data are collected by simple random sampling. Hence, to estimate and to obtain confidence intervals for x_0 corresponding to an observed value y_0 of Y, you simply interchange Y and X and use the standard procedures in Chapter 3 for regressing X on Y. In this case x_0 is in fact $X(y_0)$, a randomly chosen x value from the subpopulation determined by $Y = y_0$.

Note   One-sided $1 - \alpha/2$ confidence bounds for x_0 cannot be obtained by using only the upper bound or only the lower bound from the $1 - \alpha$ confidence region for x_0 discussed in this section.

Problems   6.4

6.4.1   The temperature of a reaction chamber is regulated by adjusting a circular dial that has markings running from 0 to 100 engraved on it. The relationship between X, the dial readings, and Y, the temperature of the reaction chamber, is given by

$$\mu_Y(x) = \beta_0 + \beta_1 x$$

To examine this relationship an experiment is conducted in which the dial is set at various preselected levels and the actual temperature of the reaction chamber is measured. Assumptions (A) are presumed to be valid where the data are collected by sampling with preselected X values. The data are shown in Table 6.4.2 and are also stored in the file **chamber.dat** on the data disk.

The computer output from a regression of Y on X for these data is given in Exhibit 6.4.2. If it is desired to have the *average* temperature of the reaction chamber to be 400°F, compute the estimate of x_0, the dial setting that would achieve this objective.

T A B L E 6.4.2

Reaction Chamber Data.

Observation Number	Chamber Temperature $Y(°F)$	Dial Setting X
1	206.36	0
2	225.52	10
3	252.18	20
4	289.33	30
5	318.11	40
6	349.49	50
7	383.03	60
8	410.70	70
9	444.40	80
10	469.14	90
11	501.16	100

E X H I B I T 6.4.2

MINITAB Output for Problem 6.4.1

```
The regression equation is
chambtmp = 198 + 3.03 dialset

Predictor        Coef        Stdev      t-ratio         p
Constant      198.456        2.282        86.98     0.000
dialset       3.02982      0.03857        78.56     0.000

s = 4.045        R-sq = 99.9%        R-sq(adj) = 99.8%
```

6.4.2 In Problem 6.4.1 compute a 99% confidence region for x_0.

6.4.3 In Example 6.4.3 suppose that a patient's temperature is measured to be 100°. What is the estimate of her actual temperature?

6.4.4 In Problem 6.4.3 obtain a 90% confidence region for her actual temperature.

6.4.5 Consider Task 3.4.1 where crystalline forms of certain chemical compounds are used in various electronic devices and it is often more desirable to have large crystals than small ones. Crystals of one particular compound are to be produced by a commercial process, and an investigator wants to examine the relationship between Y, the weight in grams of a crystal, and X, the time in hours taken for the crystal to grow to its final size. The following data are from a laboratory study in which 14 crystals of various sizes were obtained by allowing the crystals to grow for different preselected amounts of time. Assumptions (A) are presumed to be valid where the X values are preselected, and the regression function of Y on X is

$$\mu_Y(x) = \beta_0 + \beta_1 x$$

The data are given in Table 3.4.2. For convenience, they are listed below and are also stored in the file **crystal.dat** on the data disk. A computer output for the regression of Y on X is given in Exhibit 6.4.3.

Crystal Data of Table 3.4.2

Crystal Number	Weight Y (grams)	Time X (hours)
1	0.08	2
2	1.12	4
3	4.43	6
4	4.98	8
5	4.92	10
6	7.18	12
7	5.57	14
8	8.40	16
9	8.81	18
10	10.81	20
11	11.16	22
12	10.12	24
13	13.12	26
14	15.04	28

A company orders one crystal that must weigh approximately 5 grams and wants to know what the cost will be. In order for the manufacturer to determine what to charge for the crystal, it wants to know the value of x_0 such that the average weight of crystals that are grown x_0 hours is 5 grams. Estimate the value of x_0.

6.4.6 In Problem 6.4.5 obtain a 90% confidence region for x_0.

6.4.7 In Problem 3.5.1 we considered a coal burning power plant located at a distance of 25 miles from a national park. The emissions from the power plant contain the gas

E X H I B I T   6.4.3

MINITAB Output for Problem 6.4.5

```
The regression equation is
weight = 0.001 + 0.503 time

Predictor          Coef        Stdev       t-ratio          p
Constant         0.0014       0.5994          0.00      0.998
time            0.50343      0.03520         14.30      0.000


s = 1.062        R-sq = 94.5%      R-sq(adj) = 94.0%


Analysis of Variance

SOURCE          DF           SS          MS          F          p
Regression       1       230.63      230.63     204.58      0.000
Error           12        13.53        1.13
Total           13       244.16
```

sulfur dioxide (SO_2), which is linked to acid rain. A certain fraction of the emitted SO_2 will be transported through the atmosphere to the national park. There is always a certain amount of background SO_2 that is present at the national park that is not emitted by the power plant. In order to assess the SO_2 contribution by the power plant to the national park, the SO_2 output X by the plant, in tons/hour, as well as the SO_2 concentrations Y at the national park, in micrograms/cubic meter, were recorded at various randomly selected times during a particular year. The data are given in Table 3.5.4 and are reproduced here for convenience. They are also stored in the file **SO2.dat** on the data disk.

Power Plant SO_2 Data

Observation Number	Y (μg/m^3)	X (tons/hour)
1	5.21	1.92
2	7.36	3.92
3	16.26	6.80
4	10.10	6.32
5	5.80	2.00
6	8.06	4.32
7	4.76	2.40
8	6.93	2.96
9	9.36	3.52
10	10.90	4.24
11	12.48	5.12
12	11.70	5.84
13	7.44	3.60
14	6.99	2.80

From these data we compute the following:

$$\sum_{i=1}^{14} x_i = 55.76 \qquad \sum_{i=1}^{14} y_i = 123.35$$

$$\sum_{i=1}^{14} x_i^2 = 253.9072 \qquad \sum_{i=1}^{14} y_i^2 = 1220.2711$$

$$\sum_{i=1}^{14} x_i y_i = 549.3552$$

In similar investigations assumptions (B) have been used so we presume they are valid for this problem. On a certain day, the SO_2 concentration measured at the park was $y_0 = 10.5$ micrograms/cubic meter. Predict the SO_2 emission rate, x_0, by the power plant on this day. Should you consider the model

$$\mu_Y(x) = \beta_0 + \beta_1 x$$

or the model

$$\mu_X(y) = \beta_0^* + \beta_1^* y$$

Note the assumptions!

6.4.8 In Problem 6.4.7 compute a 90% confidence region for x_0. Note the assumptions and use the proper model to compute the confidence interval.

6.5
Comparison of Several Straight Line Regressions—Identical, Parallel, and Intersecting Lines

In many applied problems we want to compare two or more population regression functions. Consider, for instance, the regression function relating annual salary (Y) and the number of years of experience (X) for computer programmers in California. We may want to examine the regression functions for female and male programmers separately. We may also want to compare the two regression functions and study the differences, if any, between them. In this section we discuss the comparison of several straight line regression functions. Questions of practical interest can often be answered by determining how much the lines differ in slope, intercept, or both. Then, based on these differences, practical decisions can be made. The procedures discussed in this section are sometimes referred to as *regression using dummy variables*. We illustrate with two examples.

E X A M P L E   **6.5.1**

A study is conducted to determine the relationship between age X and blood pressure Y of individuals between the ages of 30 and 50. It is assumed that the population regression function is a straight line, but both females and males were included in the study so it was decided that a separate regression function for each was appropriate. The models are given below.

Note   To simplify notation in this section, we use α and β instead of β_0 and β_1 for regression coefficients.

$$\text{Females: } \mu_Y^{(1)}(x) = \alpha_1 + \beta_1 x \qquad 30 \le x \le 50$$
$$\text{Males: } \mu_Y^{(2)}(x) = \alpha_2 + \beta_2 x \qquad 30 \le x \le 50$$

We want to study the two models separately and together. In examining the two models together, we want to determine whether the value of α_1 is close enough to the value of α_2, and the value of β_1 is close enough to the value of β_2 so that, for this problem, the relationship between age and blood pressure is considered to be the same for males as it is for females, in the range $30 \le x \le 50$.   ∎

E X A M P L E 6.5.2

Previous experience indicates that if a commercial fertilizer is applied to wheat fields, the yield Y in bushels per acre is linearly related to X, the amount of fertilizer applied per acre. An experiment is conducted in which three varieties of wheat are used. It is suspected that the fertilizer may affect the yield differently for each variety. So there are three regression functions as given below.

$$\text{Variety 1: } \mu_Y^{(1)} = \alpha_1 + \beta_1 x \qquad 0 \le x \le 4$$
$$\text{Variety 2: } \mu_Y^{(2)} = \alpha_2 + \beta_2 x \qquad 0 \le x \le 4$$
$$\text{Variety 3: } \mu_Y^{(3)} = \alpha_3 + \beta_3 x \qquad 0 \le x \le 4$$

An investigator may be interested in obtaining answers to the following questions.

1   Are the average yields of the three varieties of wheat the same when the same amount of fertilizer is applied to each variety? To help determine this it is sometimes recommended that one test whether the three regression functions are the same. The null hypothesis considered is

$$\text{NH: } \alpha_1 = \alpha_2 = \alpha_3 \quad \text{and} \quad \beta_1 = \beta_2 = \beta_3 \tag{6.5.1}$$

2   If the amount of fertilizer applied to each variety is increased (or decreased) by one unit, is the change in *average* yield the same for each variety? In terms of population parameters, this is true if and only if $\beta_1 = \beta_2 = \beta_3$, so to answer this question it is sometimes recommended that one should test whether or not the three regressions lines have the same slope (i.e., that the three lines are parallel). The null hypothesis considered is

$$\text{NH: } \beta_1 = \beta_2 = \beta_3 \tag{6.5.2}$$

3   Are the average yields for the three varieties the same when a specified amount x_0 of fertilizer is applied? To help answer this question it is sometimes recommended that one should test whether or not the three lines intersect in a common

specified point with $X = x_0$. The null hypothesis for this test is

$$\text{NH: } \mu_Y^{(1)}(x_0) = \mu_Y^{(2)}(x_0) = \mu_Y^{(3)}(x_0) \tag{6.5.3}$$

which is equivalent to

$$\text{NH: } \alpha_1 + \beta_1 x_0 = \alpha_2 + \beta_2 x_0 = \alpha_3 + \beta_3 x_0 \tag{6.5.4}$$

If the specified point x_0 is 0, then one would be testing whether or not the regression lines have the same intercept (i.e., whether or not the three varieties have the same average yield if no fertilizer is applied).

4   Most quantities of interest in applied problems such as these can be represented as linear combinations of the α_i and β_j given by

$$\boldsymbol{d}^T \boldsymbol{\beta} = \sum_{i=1}^{3} (a_i \alpha_i + b_i \beta_i) = a_1 \alpha_1 + b_1 \beta_1 + a_2 \alpha_2 + b_2 \beta_2 + a_3 \alpha_3 + b_3 \beta_3 \tag{6.5.5}$$

where $\boldsymbol{d}^T = [a_1, b_1, a_2, b_2, a_3, b_3]$, $\boldsymbol{\beta}^T = [\alpha_1, \beta_1, \alpha_2, \beta_2, \alpha_3, \beta_3]$, and the a_i and b_i are specified constants. For example, suppose the investigator is interested in the difference between the average yields of variety 1 and variety 2 when 1.5 units of fertilizer per acre are applied to fields growing each variety. This means that the investigator wants to examine

$$\mu_Y^{(1)}(1.5) - \mu_Y^{(2)}(1.5) = (\alpha_1 + 1.5\beta_1) - (\alpha_2 + 1.5\beta_2)$$
$$= \alpha_1 + 1.5\beta_1 - \alpha_2 - 1.5\beta_2 = \boldsymbol{d}^T \boldsymbol{\beta}$$

In this case $\boldsymbol{d}^T = [1, 1.5, -1, -1.5, 0, 0]$. Suppose, instead, the investigator is interested in the difference between the average yields of varieties 2 and 3 when 1.5 units of fertilizer per acre are applied to fields where variety 2 is grown, and 3.0 units of fertilizer per acre are applied to fields where variety 3 is grown. In this case the investigator is interested in

$$\mu_Y^{(2)}(1.5) - \mu_Y^{(3)}(3.0) = \alpha_2 + 1.5\beta_2 - \alpha_3 - 3.0\beta_3 = \boldsymbol{d}^T \boldsymbol{\beta}$$

which is the linear combination in (6.5.5) with $\boldsymbol{d}^T = [0, 0, 1, 1.5, -1, -3.0]$.   ∎

As stated above, statistical tests (of hypotheses) are often used to determine whether regression lines are identical, parallel, or intersect in a common (specified) point. We, however, recommend that tests not be used for these purposes because, as explained earlier in several places, tests alone do not give much information. Not only do tests give very little information, but it is inconceivable that the three varieties of wheat will have *exactly* the same average yield or have exactly the same values of α_i, β_i, etc. So rather than ask if the three lines are exactly identical, or have exactly the same slopes or intercepts, it seems more useful to consider the question: "What are the differences in the average yields, or what are the differences in the slopes or intercepts, of *each pair* of varieties?" With this information, an investigator can decide whether these differences are close enough to zero to be considered neglible for the problem under study. In fact, one would generally want to make a full examination of the regression lines by examining all *pairs* to see how different the lines are. This full examination requires several decisions, and one may want to have a specified confidence, say $1 - \alpha$, that *collectively* all of the decisions are correct. This can be done by using simultaneous confidence intervals, and will be explained next.

To illustrate procedures for answering the questions just discussed, we can consider an arbitrary number (say H) of regression lines, but to simplify notation we let $H = 3$ and consider only three lines. You should have no difficulty in extending the results to any value of H.

Suppose samples of sizes n_1, n_2, and n_3 are obtained from study populations 1, 2, and 3, respectively, and suppose that either assumptions (A) or (B) apply for each population. The sample values may be organized as shown in Table 6.5.1.

T A B L E **6.5.1**

Sample from Population 1		Sample from Population 2		Sample from Population 3	
$y_{1,1}$	$x_{1,1}$	$y_{1,2}$	$x_{1,2}$	$y_{1,3}$	$x_{1,3}$
$y_{2,1}$	$x_{2,1}$	$y_{2,2}$	$x_{2,2}$	$y_{2,3}$	$x_{2,3}$
$\vdots$	$\vdots$	$\vdots$	$\vdots$	$\vdots$	$\vdots$
$y_{n_1,1}$	$x_{n_1,1}$	$y_{n_2,2}$	$x_{n_2,2}$	$y_{n_3,3}$	$x_{n_3,3}$

We use the notation σ_h to represent the subpopulation standard deviation in population h for $h = 1, 2, 3$, and we assume that $\sigma_1 = \sigma_2 = \sigma_3$; this common value is denoted by σ.

The three straight line regression functions are

$$
\begin{aligned}
&\text{Model 1:} \quad \mu_Y^{(1)}(x) = \alpha_1 + \beta_1 x \qquad a \le x \le b \\
&\text{Model 2:} \quad \mu_Y^{(2)}(x) = \alpha_2 + \beta_2 x \qquad a \le x \le b \\
&\text{Model 3:} \quad \mu_Y^{(3)}(x) = \alpha_3 + \beta_3 x \qquad a \le x \le b
\end{aligned} \tag{6.5.6}
$$

To examine the three regression functions individually, we can write

$$ y_h = X_h \beta_h + e_h \qquad h = 1, 2, 3 \tag{6.5.7} $$

where

$$ y_h = \begin{bmatrix} y_{1,h} \\ y_{2,h} \\ \vdots \\ y_{n_h,h} \end{bmatrix} \qquad X_h = \begin{bmatrix} 1 & x_{1,h} \\ 1 & x_{2,h} \\ \vdots & \vdots \\ 1 & x_{n_h,h} \end{bmatrix} \qquad \beta_h = \begin{bmatrix} \alpha_h \\ \beta_h \end{bmatrix} $$

and

$$ C_h = (X_h^T X_h)^{-1} \tag{6.5.8} $$

All the procedures in Chapter 3 can be applied to each model separately to obtain point estimates and confidence intervals for α_h, β_h, and σ_h for $h = 1, 2, 3$ by regressing Y on X for each sample.

To examine the three regression functions *collectively* to determine how much they differ in slope, intercept, or both, we can compute confidence intervals for $\alpha_i - \alpha_j$, for $\beta_i - \beta_j$, and for $\mu_Y^{(i)}(x_0) - \mu_Y^{(j)}(x_0^*)$, for all i and j $(i \neq j)$, where x_0 and x_0^* are specified constants which may or may not be the same. These differences can all be written as special cases of the linear function $d^T \beta$ in (6.5.5).

A confidence interval for $d^T \beta$ is

$$d^T \hat{\beta} - \text{(table-value)} SE(d^T \hat{\beta}) \leq d^T \beta \leq d^T \hat{\beta} + \text{(table-value)} SE(d^T \hat{\beta}) \tag{6.5.9}$$

where

$$SE(d^T \hat{\beta}) = \hat{\sigma} \sqrt{d^T C d} \tag{6.5.10}$$

$$C = \begin{bmatrix} C_1 & 0 & 0 \\ 0 & C_2 & 0 \\ 0 & 0 & C_3 \end{bmatrix}$$

and C_h is given in (6.5.8). Also the estimate of σ is given by

$$\hat{\sigma} = \sqrt{\frac{(n_1 - 2)\hat{\sigma}_1^2 + (n_2 - 2)\hat{\sigma}_2^2 + (n_3 - 2)\hat{\sigma}_3^2}{(n_1 - 2) + (n_2 - 2) + (n_3 - 2)}}$$

$$= \sqrt{\frac{SSE(1) + SSE(2) + SSE(3)}{(n_1 - 2) + (n_2 - 2) + (n_3 - 2)}} \tag{6.5.11}$$

where $SSE(h)$ denotes the **sum of squared errors** for group h, for $h = 1, 2, 3$. The numerator of the expression under the square root sign in (6.5.11) is called the *pooled sum of squared errors* and is denoted by $SSE(pooled)$. More generally, when H straight line regressions are being compared, we have

$$SSE(pooled) = \sum_{h=1}^{H}(n_h - 2)\hat{\sigma}_h^2 = (n_1 - 2)\hat{\sigma}_1^2 + \cdots + (n_H - 2)\hat{\sigma}_H^2$$

$$= SSE(1) + \cdots + SSE(H) \tag{6.5.12}$$

The quantity

$$(n_1 - 2) + (n_2 - 2) + (n_3 - 2) = \sum_{h=1}^{3}(n_h - 2) = n - 2(3) = n - 6$$

(where $n = n_1 + n_2 + n_3$) in the denominator of the expression under the square root in (6.5.11) is called the *pooled degrees of freedom* (written $df(pooled)$) for estimating σ. More generally, when there are H straight regressions being compared, the pooled degrees of freedom for estimating σ is given by

$$df(pooled) = (n_1 - 2) + \cdots + (n_H - 2) = n - 2H \tag{6.5.13}$$

where $n = n_1 + \cdots + n_H$ is the total number of observations. The quantities $\hat{\alpha}_h$, $\hat{\beta}_h$, $\hat{\sigma}_h$, and C_h can be obtained by regressing Y on X for each sample.

For a confidence interval on a *single* linear function $d^T \beta$ in (6.5.5), the table-value is $t_{1-\alpha/2:df(pooled)}$ when H straight lines are being compared ($H = 3$ for the case we are considering here).

If confidence intervals are desired for m *distinct* linear combinations $d_i^T \beta$, for $i = 1, \ldots, m$ such that one has at least $1 - \alpha$ confidence that *all m intervals are simultaneously correct*, then one procedure uses (6.5.9) with

$$\text{table-value} = t_{1-\alpha/2m:df(pooled)} \qquad \text{(6.5.14)}$$

which is the $1 - \alpha/2m$ percentile of a student's t population with $df(pooled) = n - 2H$ degrees of freedom, which we denote by v. For other procedures see [23].

Note For obtaining the value of m in (6.5.14), two linear combinations of parameters, say θ_1 and θ_2, *are distinct* if there do not exist constants a and b such that $\theta_1 = a\theta_2 + b$. For instance, the linear combinations $\theta_1 = \beta_1 - \beta_2$ and $\theta_2 = 2\beta_1 - 2\beta_2$ are *not* distinct. Likewise, $\theta_1 = \beta_1 - \beta_2$ and $\theta_2 = 3(\beta_2 - \beta_1) + 6$ are not distinct. On the other hand, the linear combinations $\theta_1 = \alpha_1 - \beta_1$ and $\theta_2 = \alpha_1 - \beta_3$ are distinct. We leave these for you to verify.

The table-values $t_{1-\alpha/2m:v}$ are in Table T-4 in Appendix T for $m = 2, \ldots, 6$. The confidence statements for $d_i^T \beta$ for $i = 1, 2, \ldots, m$ are

$$C \left[\begin{array}{c} d_i^T \hat{\beta} - t_{1-\alpha/2m:v} SE(d_i^T \hat{\beta}) \leq d_i^T \beta \leq d_i^T \hat{\beta} + t_{1-\alpha/2m:v} SE(d_i^T \hat{\beta}) \\ \text{simultaneously for all } i = 1, \ldots, m \end{array} \right] \geq 1 - \alpha \quad \text{(6.5.15)}$$

If $m = 1$, then in (6.5.15) the $\geq$ sign is replaced with the $=$ sign. We can use (6.5.15) with $m = 6$ to obtain confidence intervals for

$$\alpha_1 - \alpha_2 \qquad \alpha_1 - \alpha_3 \qquad \alpha_2 - \alpha_3 \qquad \beta_1 - \beta_2 \qquad \beta_1 - \beta_3 \qquad \beta_2 - \beta_3 \quad \text{(6.5.16)}$$

for the three regression functions in (6.5.6), and we have confidence greater than or equal to $1 - \alpha$ that all *six* of the intervals are correct. By examining the confidence intervals for these six quantities, an investigator can determine how much any two of the regression lines differ in slope, intercept, or both.

To determine the difference between two straight line regression functions, say $\mu_Y^{(1)}(x)$ and $\mu_Y^{(2)}(x)$ for x values in the range $a \leq x \leq b$, it may be useful to compute simultaneous confidence intervals for the differences $\mu_Y^{(1)}(a) - \mu_Y^{(2)}(a)$ and $\mu_Y^{(1)}(b) - \mu_Y^{(2)}(b)$. If both these differences are acceptably small (at a given confidence level), then the investigator may conclude that the two lines can be considered to be equivalent for the problem under consideration. For instance, if for some constant d we have

$$|\mu_Y^{(1)}(a) - \mu_Y^{(2)}(a)| \leq d$$

and

$$|\mu_Y^{(1)}(b) - \mu_Y^{(2)}(b)| \leq d$$

where $a < b$, then the two regression functions differ by less than d units for all x in the interval $a \leq x \leq b$ (see Figures 6.5.1 and 6.5.2).

FIGURE **6.5.1**

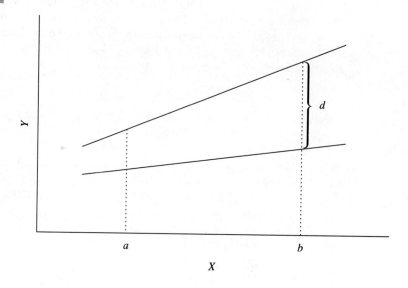

FIGURE **6.5.2**

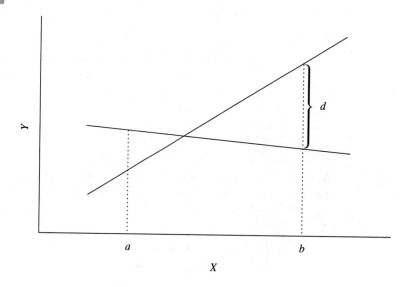

We illustrate the procedures discussed in this section with Example 6.5.3.

E X A M P L E 6.5.3

Various amounts X of a new food supplement were fed to three different breeds of chickens for 6 weeks to determine its effect on the hardness Y of egg shells. A straight line regression function is assumed to hold for each breed. The three regression functions are

Breed 1: $\mu_Y^{(1)}(x) = \alpha_1 + \beta_1 x$ $0 \le x \le 20$

Breed 2: $\mu_Y^{(2)}(x) = \alpha_2 + \beta_2 x$ $0 \le x \le 20$

Breed 3: $\mu_Y^{(3)}(x) = \alpha_3 + \beta_3 x$ $0 \le x \le 20$

The problem is to study these regression functions together and separately to determine how they differ in slope (change in average hardness of egg shells per unit of the new food supplement used), in intercept (average hardness of egg shells for each breed if the new food supplement is not used), or both. Assumptions (A) are presumed to hold, and the data are obtained by sampling with preselected X values. The investigator wants to be at least 95% confident that the decisions are simultaneously correct. The data are given in Table 6.5.2 and are also stored in the file **eggshell.dat** on the data disk.

A regression analysis is carried out for each breed and the results are summarized in the MINITAB output in Exhibit 6.5.1. From these data we get the following quantities:

$$\hat{\alpha}_1 = 5.9662 \quad \hat{\beta}_1 = 3.0494 \quad \hat{\sigma}_1 = 1.330 \quad SSE(1) = 17.7$$
$$\hat{\alpha}_2 = 6.4673 \quad \hat{\beta}_2 = 1.0948 \quad \hat{\sigma}_2 = 0.9015 \quad SSE(2) = 4.876$$
$$\hat{\alpha}_3 = 5.0225 \quad \hat{\beta}_3 = 0.26344 \quad \hat{\sigma}_3 = 1.284 \quad SSE(3) = 11.538$$

T A B L E 6.5.2
Eggshell Data

Observation Number	Breed 1 Y_1	X_1	Breed 2 Y_2	X_2	Breed 3 Y_3	X_3
1	8.42	1	9.86	3	6.52	2
2	14.68	3	9.54	3	5.11	5
3	21.42	5	11.96	4	7.75	7
4	25.45	6	12.46	5	6.84	8
5	27.14	7	11.38	6	7.65	10
6	30.53	8	14.69	8	9.49	15
7	34.51	9	16.48	9	7.03	16
8	34.52	9	20.11	12	9.41	18
9	33.24	10			12.01	20
10	39.63	11				
11	43.98	12				
12	47.77	14				

E X H I B I T 6.5.1
MINITAB Output for Example 6.5.3

REGRESSION ANALYSIS FOR BREED 1

The regression equation is
Y1 = 5.97 + 3.05 X1

Predictor	Coef	Stdev	t-ratio	p
Constant	5.9662	0.9289	6.42	0.000
X1	3.0494	0.1068	28.54	0.000

s = 1.330   R-sq = 98.8%   R-sq(adj) = 98.7%

Analysis of Variance

SOURCE	DF	SS	MS	F	p
Regression	1	1440.6	1440.6	814.61	0.000
Error	10	17.7	1.8		
Total	11	1458.3			

MATRIX C1

```
 0.487896719  -0.051102743
-0.051102743   0.006455083
```

REGRESSION ANALYSIS FOR BREED 2

The regression equation is
Y2 = 6.47 + 1.09 X2

Predictor	Coef	Stdev	t-ratio	p
Constant	6.4673	0.7386	8.76	0.000
X2	1.0948	0.1066	10.27	0.000

s = 0.9015   R-sq = 94.6%   R-sq(adj) = 93.7%

Analysis of Variance

SOURCE	DF	SS	MS	F	p
Regression	1	85.703	85.703	105.47	0.000
Error	6	4.876	0.813		
Total	7	90.579			

MATRIX C2

```
 0.67132867  -0.08741259
-0.08741259   0.01398601
```

E X H I B I T 6.5.1
(Continued)

REGRESSION ANALYSIS FOR BREED 3

```
The regression equation is
Y3 = 5.02 + 0.263 X3
```

Predictor	Coef	Stdev	t-ratio	p
Constant	5.0225	0.9193	5.46	0.000
X3	0.26344	0.07250	3.63	0.008

s = 1.284   R-sq = 65.4%   R-sq(adj) = 60.4%

Analysis of Variance

SOURCE	DF	SS	MS	F	p
Regression	1	21.761	21.761	13.20	0.008
Error	7	11.538	1.648		
Total	8	33.298			

```
 MATRIX C3

 0.512756910   -0.035790220
-0.035790220    0.003189227
```

To get the point estimate of σ we can use the formula in (6.5.11) to *pool* the estimates $\hat{\sigma}_h$. We get

$$\hat{\sigma} = \sqrt{\frac{17.7 + 4.876 + 11.538}{10 + 6 + 7}} = 1.218 \qquad \text{(6.5.17)}$$

Using (6.5.10), we get

$$SE(\hat{\alpha}_1 - \hat{\alpha}_2) = 1.218\sqrt{0.487897 + 0.671329} = 1.311$$
$$(\text{note } d^T = [1, 0, -1, 0, 0, 0])$$
$$SE(\hat{\alpha}_1 - \hat{\alpha}_3) = 1.218\sqrt{0.487897 + 0.512757} = 1.218$$
$$(\text{note } d^T = [1, 0, 0, 0, -1, 0])$$
$$SE(\hat{\alpha}_2 - \hat{\alpha}_3) = 1.218\sqrt{0.671329 + 0.512757} = 1.325$$
$$(\text{note } d^T = [0, 0, 1, 0, -1, 0])$$
$$SE(\hat{\beta}_1 - \hat{\beta}_2) = 1.218\sqrt{0.006455 + 0.013986} = 0.1741$$
$$(\text{note } d^T = [0, 1, 0, -1, 0, 0])$$
$$SE(\hat{\beta}_1 - \hat{\beta}_3) = 1.218\sqrt{0.006455 + 0.003189} = 0.1196$$
$$(\text{note } d^T = [0, 1, 0, 0, 0, -1])$$
$$SE(\hat{\beta}_2 - \hat{\beta}_3) = 1.218\sqrt{0.013986 + 0.003189} = 0.1596$$
$$(\text{note } d^T = [0, 0, 0, 1, 0, -1]).$$

To compute the appropriate table-value we observe that $1 - \alpha = 0.95$, $\alpha = 0.05$, $m = 6$, and $df\,(pooled) = v = n_1 + n_2 + n_3 - 6 = 23$. From Table T-4 in Appendix T the required table-value is $t_{1-\alpha/2m:v} = t_{0.995833:23} = 2.886$. So the confidence statement is

$$C \begin{bmatrix} -4.285 \leq \alpha_1 - \alpha_2 \leq 3.283 \\ -2.572 \leq \alpha_1 - \alpha_3 \leq 4.459 \\ -2.379 \leq \alpha_2 - \alpha_3 \leq 5.269 \\ 1.452 \leq \beta_1 - \beta_2 \leq 2.457 \\ 2.441 \leq \beta_1 - \beta_3 \leq 3.131 \\ 0.371 \leq \beta_2 - \beta_3 \leq 1.292 \end{bmatrix} \geq 0.95$$

We have at least 95% confidence that all six of these intervals are correct. On the basis of these, an investigator can decide, for the problem at hand, how the three lines, or any two of them, differ in slope, intercept, or both.

Suppose the investigator is willing to consider the regression lines to be equivalent for this problem provided that they do not differ by more than ten units in the interval $0 \leq x \leq 20$. We can compute simultaneous confidence intervals (say, with confidence coefficient equal to 0.95) for the differences

$$\mu_Y^{(1)}(0) - \mu_Y^{(2)}(0) \qquad \text{which is } \alpha_1 - \alpha_2$$
$$\mu_Y^{(1)}(0) - \mu_Y^{(3)}(0) \qquad \text{which is } \alpha_1 - \alpha_3$$
$$\mu_Y^{(2)}(0) - \mu_Y^{(3)}(0) \qquad \text{which is } \alpha_2 - \alpha_3$$
$$\mu_Y^{(1)}(20) - \mu_Y^{(2)}(20) \qquad \text{which is } \alpha_1 + 20\beta_1 - \alpha_2 - 20\beta_2$$
$$\mu_Y^{(1)}(20) - \mu_Y^{(3)}(20) \qquad \text{which is } \alpha_1 + 20\beta_1 - \alpha_3 - 20\beta_3$$
$$\mu_Y^{(2)}(20) - \mu_Y^{(3)}(20) \qquad \text{which is } \alpha_2 + 20\beta_2 - \alpha_3 - 20\beta_3$$

You should verify the following confidence statement.

$$C \begin{bmatrix} -4.285 \leq \mu_Y^{(1)}(0) - \mu_Y^{(2)}(0) \leq 3.283 \\ -2.572 \leq \mu_Y^{(1)}(0) - \mu_Y^{(3)}(0) \leq 4.459 \\ -2.379 \leq \mu_Y^{(2)}(0) - \mu_Y^{(3)}(0) \leq 5.269 \\ 31.743 \leq \mu_Y^{(1)}(20) - \mu_Y^{(2)}(20) \leq 45.439 \\ 52.5297 \leq \mu_Y^{(1)}(20) - \mu_Y^{(3)}(20) \leq 60.7961 \\ 11.8571 \leq \mu_Y^{(2)}(20) - \mu_Y^{(3)}(20) \leq 24.2869 \end{bmatrix} \geq 0.95$$

Using this information the investigator would perhaps conclude:

a   When no supplement is added, the average hardnesses differ by no more than 5.269 units, which is of no practical importance (less than 10 units) for this problem.

b   When 20 units of the supplement are used, the differences between the average hardnesses are greater than 10 units for each pair of breeds, which *is* of practical importance for this problem.

c The three regression lines are not equivalent for values of x in the range from 0 to 20 units. ∎

In Section 6.5 of the laboratory manuals we discuss computer programs we have written (supplied on the data disk) to carry out the computations discussed in this section.

Another problem of interest in this and similar examples is finding the point where two regression lines intersect. This is discussed in the next section.

Problems 6.5

6.5.1 This is a continuation of Task 3.4.1. Crystals of various sizes are used in electronic devices, and there is a linear relationship between Y, the weight in grams of a crystal, and X, the time in hours it takes the crystal to grow. Three procedures, denoted here as 1, 2, and 3, are used to grow the crystals, and we want to determine how the procedures differ. A straight line regression model is assumed to hold for each procedure with subpopulation standard deviation σ_h for $h = 1, 2, 3$.

$$\text{Procedure 1: } \mu_Y^{(1)}(x) = \alpha_1 + \beta_1 x \qquad 1 \le x \le 30$$
$$\text{Procedure 2: } \mu_Y^{(2)}(x) = \alpha_2 + \beta_2 x \qquad 1 \le x \le 30$$
$$\text{Procedure 3: } \mu_Y^{(3)}(x) = \alpha_3 + \beta_3 x \qquad 1 \le x \le 30$$

The data and a computer output containing relevant regression analyses are given in Exhibit 6.5.2. The data are also stored in the file **crystal3.dat** on the data disk. We suppose that assumptions (A) are valid and that the data are obtained by preselecting the X values.

a Exhibit the estimates of β_h and α_h for $h = 1, 2, 3$.

b Exhibit the estimates of σ_i for $i = 1, 2, 3$.

c Compute the estimate of σ. Assume $\sigma_1 = \sigma_2 = \sigma_3 = \sigma$.

d Exhibit the vector $\boldsymbol{d}^T$ for which $SE(\hat{\beta}_1 - \hat{\beta}_2) = \hat{\sigma}\sqrt{\boldsymbol{d}^T \boldsymbol{Cd}}$ in (6.5.10).

e Exhibit the vector $\boldsymbol{d}^T$ for which $SE(\hat{\alpha}_2 - \hat{\alpha}_3) = \hat{\sigma}\sqrt{\boldsymbol{d}^T \boldsymbol{Cd}}$ in (6.5.10).

f What value of m should be used to compute simultaneous confidence intervals for

$$\alpha_1 - \alpha_2 \text{ and } \beta_1 - \beta_2$$

such that one has confidence ≥ 0.90 that they are both correct?

g What value of m should be used to compute simultaneous confidence intervals for

$$\alpha_1 - \alpha_3, \ 2\alpha_1 - 2\alpha_3, \ \beta_1 - \beta_3, \text{ and } \beta_1 + \beta_2 - 2\beta_3$$

such that one has confidence ≥ 0.90 that all four intervals are correct?

E X H I B I T **6.5.2**
MINITAB Output for Problem 6.5.1

	Procedure 1		Procedure 2		Procedure 3	
obsno	Y1	X1	Y2	X2	Y3	X3
1	0.10	2	0.31	2	2.57	4
2	0.95	3	2.74	4	4.96	6
3	4.79	5	5.93	7	7.23	8
4	5.02	6	7.98	9	8.41	9
5	5.56	7	10.00	11	11.02	11
6	5.79	7	12.36	13	11.31	12
7	6.31	8	14.94	15	12.56	13
8	7.58	9	16.02	16	17.86	17
9	8.19	9	16.87	17	20.14	19
10	9.37	10	19.13	19	32.67	29

REGRESSION ANALYSIS FOR PROCEDURE 1

The regression equation is
Y1 = - 1.88 + 1.10 X1

Predictor	Coef	Stdev	t-ratio	p
Constant	-1.8817	0.4945	-3.80	0.005
X1	1.09814	0.07008	15.67	0.000

s = 0.5536   R-sq = 96.8%   R-sq(adj) = 96.5%

The matrix C1 is

```
 0.79807692   -0.10576923
-0.10576923    0.01602564
```

REGRESSION ANALYSIS FOR PROCEDURE 2

The regression equation is
Y2 = - 1.87 + 1.11 X2

Predictor	Coef	Stdev	t-ratio	p
Constant	-1.8710	0.1290	-14.50	0.000
X2	1.10611	0.01030	107.44	0.000

s = 0.1766   R-sq = 99.9%   R-sq(adj) = 99.9%

The matrix C2 is

```
 0.534172050   -0.038422305
-0.038422305    0.003400204
```

EXHIBIT 6.5.2
(Continued)

REGRESSION ANALYSIS FOR PROCEDURE 3

```
The regression equation is
Y3 = - 2.47 + 1.20 X3
```

Predictor	Coef	Stdev	t-ratio	p
Constant	-2.4721	0.2426	-10.19	0.000
X3	1.19883	0.01666	71.97	0.000

```
s = 0.3663      R-sq = 99.8%      R-sq(adj) = 99.8%
```

The matrix C3 is

```
 0.438792390   -0.026468156
-0.026468156    0.002067825
```

h What value of m should be used to compute simultaneous confidence intervals for

$$\mu_Y^{(1)}(2) - \mu_Y^{(2)}(2) \text{ and } \mu_Y^{(1)}(5) - \mu_Y^{(2)}(5)$$

such that one has confidence $\geq 1 - \alpha$ that they are both correct?

i Compute confidence intervals for the quantities in part (f) such that one has at least 90% confidence that the two intervals are simultaneously correct.

6.5.2 In Problem 6.5.1 suppose that an investigator is interested in comparing the three procedures. She decides that for the problem at hand, the slopes can be considered to be equivalent for her purposes if they differ by less than 0.30 unit in magnitude. She also decides that if the intercepts differ by less than 0.2 unit in magnitude they can be considered to be equivalent for this problem. Make a complete analysis and decide which, if any, of the regression lines are equivalent in intercept, slope, or both. Use an appropriate 90% simultaneous confidence statement to help make this decision. Write a short report explaining your results.

6.5.3 In Example 6.5.2 an investigator wants to determine the difference between the average yields of varieties 1 and 2 if the fields where variety 1 is grown receive no fertilizer and the fields where variety two is grown receive 1.5 units of fertilizer per acre. Exhibit the population parameters for which an investigator would want point estimates and confidence intervals to make this determination.

6.5.4 In Example 6.5.2 suppose there are six varieties to be compared, instead of three, and the investigator wants to examine the differences between each distinct pair of the α_i and each distinct pair of the β_j by obtaining confidence intervals for the differences of all distinct pairs with a simultaneous confidence coefficient of at least 0.95. What value of m should be used in (6.5.5)? Write out these m distinct quantities $\alpha_i - \alpha_j$ and $\beta_i - \beta_j$.

6.5.5   In Problem 6.5.1 suppose the investigator is willing to consider the population regression lines to be equivalent for practical purposes, provided that they do not differ by more than 2 units for $1 \le x \le 30$ hours. To help make a decision, exhibit the population quantities in which the investigator would be interested.

6.6

Intersection of Two Straight Line Regression Functions

As we indicated at the end of the previous section, there are situations when an investigator may want to find the point, say x_0, where two population regression lines intersect. We illustrate one such situation in Example 6.6.1.

E X A M P L E   6.6.1

Consider the problem in Example 2.2.2 where a company is studying the relationship between Y, the first-year maintenance cost of new cars, and X, the number of miles the car will be driven the first year. Suppose the company is interested in cars made by two manufacturers, say manufacturers 1 and 2. It is assumed that the population regression function of Y on X is a straight line for each make of car. The two regression functions are given by

$$\text{Manufacturer 1: } \mu_Y^{(1)}(x) = \alpha_1 + \beta_1 x \qquad a \le x \le b$$
$$\text{Manufacturer 2: } \mu_Y^{(2)}(x) = \alpha_2 + \beta_2 x \qquad a \le x \le b$$

The investigator wants to determine the value of X, say $X = x_0$, where the two lines intersect because to the right of x_0 one make of car will have lower average first-year maintenance costs, and to the left of x_0 the other make of car will have lower average first-year maintenance costs. If x_0 is outside the interval $[a, b]$, then cars made by one of the manufacturers have lower average first-year maintenance costs than the other everywhere in the interval. For example, in Figure 6.6.1 cars made by manufacturer 1 have lower average first-year maintenance costs if the miles driven is between x_0 and b, and cars made by manufacturer 2 have lower average first-year maintenance costs if the number of miles driven is between a and x_0. In Figure 6.6.2, cars made by manufacturer 1 have lower average first-year maintenance costs in the entire interval from a to b.   ∎

To find the point x_0 where the two regression lines intersect, we set $\mu_Y^{(1)}(x) = \mu_Y^{(2)}(x)$, solve for x, and denote the solution by x_0. We get

$$x_0 = \frac{\alpha_1 - \alpha_2}{\beta_2 - \beta_1}$$

Thus the two regression lines intersect at $X = x_0$ (we assume $\beta_1 \ne \beta_2$).

To compute point and confidence interval estimates of x_0 we obtain samples from populations 1 and 2. We suppose that assumptions (A) or (B) are valid for each

FIGURE 6.6.1

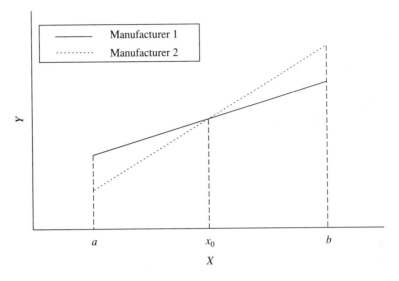

FIGURE 6.6.2

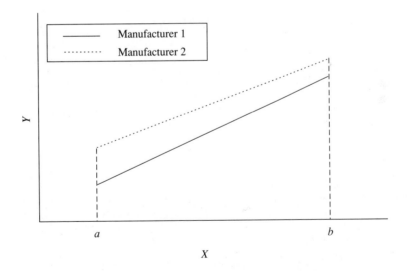

population, and we assume that the two subpopulation standard deviations are the same; i.e., $\sigma_1 = \sigma_2$, with their common value denoted by σ.

Let $(y_{i,1}, x_{i,1})$ for $i = 1, \ldots, n_1$ denote the n_1 sample values from population (1), and let $(y_{i,2}, x_{i,2})$ for $i = 1, \ldots, n_2$ denote the n_2 sample values from population (2). To get a point estimate of x_0, compute estimates of the unknown parameters for each population by formulas in Chapter 3. The point estimate of x_0 is

$$\hat{x}_0 = \frac{\hat{\alpha}_1 - \hat{\alpha}_2}{\hat{\beta}_2 - \hat{\beta}_1} \qquad \text{(6.6.1)}$$

In Box 6.6.1 we give the instructions for computing a $1 - \alpha$ confidence region for x_0.

B O X **6.6.1** **Instructions for Computing a Confidence Region for x_0, the Point of Intersection of Two Straight Line Regressions**

Carry out steps 1–8 below.

1 Compute $\bar{x}_1$, the mean of the X values from sample 1, and $\bar{x}_2$, the mean of the X values from sample 2. Also compute

$$SSX(1) = \sum_{i=1}^{n_1}(x_{i,1} - \bar{x}_1)^2 \qquad Q_1 = \sum_{i=1}^{n_1} x_{i,1}^2$$

and

$$SSX(2) = \sum_{i=1}^{n_2}(x_{i,2} - \bar{x}_2)^2 \qquad Q_2 = \sum_{i=1}^{n_2} x_{i,2}^2$$

2 Compute $SSE(1)$ and $SSE(2)$, the sum of squared errors for samples 1 and 2, respectively.

3 $\hat{\sigma}^2 = \dfrac{(n_1 - 2)\hat{\sigma}_1^2 + (n_2 - 2)\hat{\sigma}_2^2}{(n_1 - 2) + (n_2 - 2)} = \dfrac{SSE(1) + SSE(2)}{(n_1 - 2) + (n_2 - 2)}$

4 $A = [\hat{\beta}_1 - \hat{\beta}_2]^2 - \left[\dfrac{1}{SSX(1)} + \dfrac{1}{SSX(2)} \right] \hat{\sigma}^2 F_{1-\alpha:1,n_1+n_2-4}$

5 $B = [\hat{\alpha}_1 - \hat{\alpha}_2][\hat{\beta}_1 - \hat{\beta}_2] + \left[\dfrac{\bar{x}_1}{SSX(1)} + \dfrac{\bar{x}_2}{SSX(2)} \right] \hat{\sigma}^2 F_{1-\alpha:1,n_1+n_2-4}$

6 $C = (\hat{\alpha}_1 - \hat{\alpha}_2)^2 - \left[\dfrac{Q_1}{n_1 SSX(1)} + \dfrac{Q_2}{n_2 SSX(2)} \right] \hat{\sigma}^2 F_{1-\alpha:1,n_1+n_2-4}$

7 $D = B^2 - AC$

8 If $A \neq 0$ and $D > 0$, compute L and U where $L = -\dfrac{(B + \sqrt{D})}{A}$

and $U = -\dfrac{(B - \sqrt{D})}{A}$.

A $1 - \alpha$ confidence region for x_0 is

a	$-\infty < x_0 < \infty$		if $D \leq 0$
b	$-\infty < x_0 \leq U$	and $\quad L \leq x_0 < \infty$	if $D > 0$ and $A < 0$
c	$L \leq x_0 \leq U$		if $A > 0, D > 0$

$$\text{(6.6.2)}$$

The resulting confidence region can take any one of the forms (a), (b), or (c) in (6.6.2). If the result is given by (a), then the confidence region consists of the entire range between minus infinity and infinity. If the result is given by (b), then the confidence region consists of two disconnected intervals! Only case (c) results in a finite width confidence *interval*. If (a) or (b) is obtained, then the results are unsatisfactory and a larger sample is required to obtain a confidence interval as in (c). However, even in case (c), if the interval is too wide to draw useful conclusions, then the result is considered unsatisfactory and a larger sample is required to obtain a shorter confidence interval.

We illustrate the computations discussed above in Example 6.6.2.

E X A M P L E 6.6.2

In Example 6.5.3 suppose that we want to compare the hardness Y of egg shells for breeds 2 and 3 for values of X in the range from 2 to 20 units. To help make this comparison, we want to determine x_0, the point where the regression lines for breeds 2 and 3 intersect. We will find the point estimate of x_0 and a 95% confidence region for x_0. The data are given in Table 6.6.1 and are also stored in the file **eggshell.dat** on the data disk. Exhibit 6.6.1 gives a SAS output containing the results of regressing Y on X for breeds 2 and 3, respectively. Assumptions (A) are presumed to be valid, and the data are obtained by sampling with preselected X values.

We also compute

$$\bar{x}_2 = 6.25 \qquad \bar{x}_3 = 11.22 \qquad SSX(2) = 71.5 \qquad SSX(3) = 313.556 \qquad \textbf{(6.6.3)}$$

T A B L E 6.6.1

Observation Number	Breed 2		Breed 3	
	Y_2	X_2	Y_3	X_3
1	9.86	3	6.52	2
2	9.54	3	5.11	5
3	11.96	4	7.75	7
4	12.46	5	6.84	8
5	11.38	6	7.65	10
6	14.69	8	9.49	15
7	16.48	9	7.03	16
8	20.11	12	9.41	18
9			12.01	20

E X H I B I T **6.6.1**
SAS Output for Example 6.6.2

Regression of Y on X for Breed 2

The SAS System

00:00 Saturday, Jan 1, 1994

Model: MODEL1
Dependent Variable: Y2

Analysis of Variance

Source	DF	Sum of Squares	Mean Square	F Value	Prob>F
Model	1	85.70291	85.70291	105.466	0.0001
Error	6	4.87569	0.81261		
C Total	7	90.57860			

Root MSE	0.90145	R-square	0.9462	
Dep Mean	13.31000	Adj R-sq	0.9372	
C.V.	6.77274			

Parameter Estimates

Variable	DF	Parameter Estimate	Standard Error	T for H0: Parameter=0	Prob > |T|
INTERCEP	1	6.467343	0.73860086	8.756	0.0001
X2	1	1.094825	0.10660785	10.270	0.0001

Regression of Y on X for Breed 3

Model: MODEL1
Dependent Variable: Y3

Analysis of Variance

Source	DF	Sum of Squares	Mean Square	F Value	Prob>F
Model	1	21.76050	21.76050	13.202	0.0084
Error	7	11.53778	1.64825		
C Total	8	33.29829			

E X H I B I T **6.6.1**
(Continued)

Root MSE	1.28384	R-square	0.6535
Dep Mean	7.97889	Adj R-sq	0.6040
C.V.	16.09051		

Parameter Estimates

Variable	DF	Parameter Estimate	Standard Error	T for H0: Parameter=0	Prob > |T|
INTERCEP	1	5.022537	0.91932262	5.463	0.0009
X3	1	0.263437	0.07250283	3.633	0.0084

From the computer output for each breed we obtain

$$\hat{\alpha}_2 = 6.467343 \qquad \hat{\beta}_2 = 1.094825 \qquad \hat{\alpha}_3 = 5.022537 \qquad \hat{\beta}_3 = 0.263437 \text{ (6.6.4)}$$

$$\hat{\sigma}_2 = 0.90145 \qquad \hat{\sigma}_3 = 1.28384 \qquad SSE(2) = 4.87569 \qquad SSE(3) = 11.53778$$
(6.6.5)

Using (6.6.1) we get

$$\hat{x}_0 = \frac{(6.467343 - 5.022537)}{(0.263437 - 1.094825)} = -1.738 \qquad \text{(6.6.6)}$$

Hence the estimate of x_0 indicates that the two regression lines do not intersect in the range from 2 to 20 units of food supplement. We now compute a 95% confidence region for x_0 by following the instructions in Box 6.6.1 using subscripts 2 and 3 (breeds 2 and 3) in place of subscripts 1 and 2, respectively.

1 $\bar{x}_2 = 6.25 \qquad \bar{x}_3 = 11.22 \qquad SSX(2) = 71.5 \qquad SSX(3) = 313.556$
   $Q_2 = 384 \qquad Q_3 = 1447$

2 $SSE(2) = 4.87569 \qquad SSE(3) = 11.53778$

3 $\hat{\sigma}^2 = \dfrac{(n_2 - 2)\hat{\sigma}_2^2 + (n_3 - 2)\hat{\sigma}_3^2}{(n_2 - 2) + (n_3 - 2)} = \dfrac{SSE(2) + SSE(3)}{(n_2 - 2) + (n_3 - 2)} = 1.263$

The table-value is $F_{0.95:1,13} = 4.67$. Using this we get

4 $A = 0.5900$

5 $B = 1.9272$

6 $C = -4.8900$

7 $D = 6.6000$
   Since $A > 0$ and $D > 0$, the confidence region is of the form $L \le x_0 \le U$ where

8 $L = -7.62 \qquad U = 1.09$

The required confidence statement is

$$C[-7.62 \le x_0 \le 1.09] = 0.95$$

Thus we conclude that the average hardness of egg shells will be greater for breed 2 than for breed 3 when the amount of food supplement is in the range from 2 to 20 units. ∎

Caution   One-sided $1 - \alpha/2$ confidence bounds for x_0 cannot be obtained by using only the upper bound or only the lower bound from the $1 - \alpha$ confidence region for x_0 discussed in this section.

You will have undoubtedly noticed that the calculations required to compute a confidence region for x_0 are quite cumbersome. For this reason we provide a computer program on the data disk for carrying out these computations. The use of this program is explained in Section 6.6 of the laboratory manuals.

Problems   6.6

6.6.1   For the breed data in Example 6.5.3, estimate the point x_0 where the regression lines for breed 1 and breed 2 intersect.

6.6.2   For the breed data in Example 6.5.3, estimate the point x_0 where the regression lines for breed 1 and breed 3 intersect.

6.6.3   In Problem 6.6.1 compute a 95% confidence region for x_0. What conclusions can you draw about the average hardness of egg shells for breeds 1 and 2? Explain. Below are the values of A, B, and C required in the calculations of Box 6.6.1.

$$A = 3.69095 \qquad B = -0.101865 \qquad C = -7.09432$$

You should verify these values.

6.6.4   In Problem 6.5.1 estimate the point x_0 where the two lines representing procedures 1 and 2 intersect.

6.6.5   In Problem 6.6.4 find a 90% confidence region for x_0. What conclusions can you draw from this confidence region? Below are values of A, B, and C in Box 6.6.1. Be sure to check these values.

$$A = -0.009932 \qquad B = 0.07428 \qquad C = -0.685414$$

6.7
Maximum or Minimum of a Quadratic Regression Model

In many practical applications we want to find the value of the predictor variable X that would maximize or minimize the average response $\mu_Y(x)$. For instance, we may be interested in maximizing the average breaking strength of an alloy by controlling the amount of carbon in it. Or we may want to minimize the number of pests in agricultural plots by using suitable amounts of insecticides, etc. If the regression function of Y on X is a straight line over $a \leq x \leq b$, the interval of interest, then the maximum value or the minimum value of $\mu_Y(x)$ in the interval $a \leq x \leq b$ must occur either at a or at b. Thus the problem is easily solved in the case of straight line regression. However, in a number of applied problems, the regression function may not be a straight line, but it may be well approximated by a quadratic function

$$\mu_Y(x) = \beta_0 + \beta_1 x + \beta_2 x^2$$

In such situations, finding the value of x that would minimize or maximize the value of $\mu_Y(x)$ is not as simple as it is in the case of a straight line regression. This is the topic of our discussion in this section. We begin with an example.

E X A M P L E 6.7.1

Suppose an agricultural scientist is testing a new fertilizer to see how it affects the yield of corn. It is assumed that corn yield Y is related to the amount of fertilizer X through a quadratic regression function. This is reasonable because it is known that as increasing amounts of the fertilizer are applied the yield increases, but if too much fertilizer is used the yield will then decrease (see Figure 6.7.1). The scientist wants to determine x_0, the amount of fertilizer to apply, so that the average yield is a maximum.

F I G U R E 6.7.1

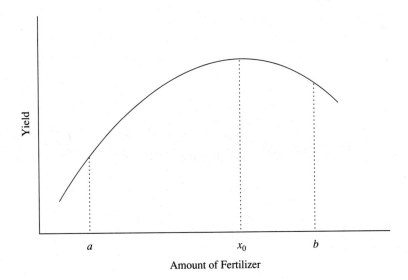

Yield

a x_0 b

Amount of Fertilizer

In some problems the quadratic function may be of the form shown in Figure 6.7.2, in which case the maximum will occur at one of the endpoints, a or b, of the interval $a \leq x \leq b$ of interest. In the figure the maximum occurs at a. ■

In certain problems it is the minimum point of a quadratic function that is of interest. This is illustrated in the following example.

F I G U R E   **6.7.2**

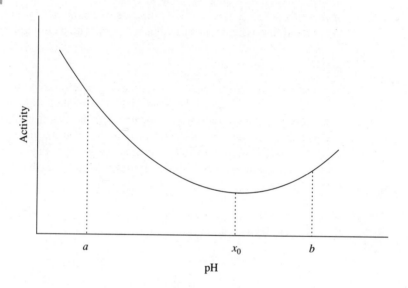

E X A M P L E   **6.7.2**

The activity level Y in a growing medium of a certain disease-causing bacteria is related to the pH level X of the growing medium according to a quadratic regression function similar to the one in Figure 6.7.2. A scientist wants to determine x_0 such that when the pH level of the growing medium equals x_0, the average activity level of the bacteria is a *minimum*. The results from such a study will be used to prepare drugs to slow down the progress of the disease in affected humans. ■

For this problem, assumptions (A) are presumed to hold with

$$\mu_Y(x) = \beta_0 + \beta_1 x + \beta_2 x^2$$

which we write as a multiple linear regression function

$$\mu_Y(x) = \beta_0 + \beta_1 x_1 + \beta_2 x_2$$

where we denote x by x_1 and x^2 by x_2. With a little algebra we can verify that

$$\mu_Y(x) = \beta_0 + \beta_1 x + \beta_2 x^2 = \beta_2 \left(x + \frac{\beta_1}{2\beta_2} \right)^2 + \left(\beta_0 - \frac{\beta_1^2}{4\beta_2} \right)$$

From this it is seen that $\mu_Y(x)$ attains a maximum at

$$x = \frac{-\beta_1}{2\beta_2}$$

when $\beta_2 < 0$, whereas $\mu_Y(x)$ attains a minimum at

$$x = \frac{-\beta_1}{2\beta_2}$$

when $\beta_2 > 0$ (see Figures 6.7.1 and 6.7.2). We denote the value where the quadratic function $\mu_Y(x)$ attains its maximum or minimum by x_0. Thus

$$x_0 = \frac{-\beta_1}{2\beta_2}$$

We want to obtain a point and confidence interval estimate of x_0. A sample of size n, denoted by $(y_1, x_1), \ldots, (y_n, x_n)$, is selected either by simple random sampling or by sampling with preselected X values, and the data are organized as in Table 6.7.1.

T A B L E **6.7.1**

Y	$X_1 = X$	$X_2 = X^2$
y_1	$x_{1,1} = x_1$	$x_{1,2} = x_1^2$
y_2	$x_{2,1} = x_2$	$x_{2,2} = x_2^2$
$\vdots$	$\vdots$	$\vdots$
y_n	$x_{n,1} = x_n$	$x_{n,2} = x_n^2$

From this we get the X matrix

$$X = \begin{bmatrix} 1 & x_{1,1} & x_{1,2} \\ 1 & x_{2,1} & x_{2,2} \\ \vdots & \vdots & \vdots \\ 1 & x_{n,1} & x_{n,2} \end{bmatrix} = \begin{bmatrix} 1 & x_1 & x_1^2 \\ 1 & x_2 & x_2^2 \\ \vdots & \vdots & \vdots \\ 1 & x_n & x_n^2 \end{bmatrix}$$

As usual we let C denote $(X^T X)^{-1}$ where C is a 3 by 3 matrix given by

$$C = \begin{bmatrix} c_{1,1} & c_{1,2} & c_{1,3} \\ c_{2,1} & c_{2,2} & c_{2,3} \\ c_{3,1} & c_{2,3} & c_{3,3} \end{bmatrix} \qquad \text{(6.7.1)}$$

Point estimates of β_0, β_1, β_2, $\mu_Y(x)$, and $\sigma = \sigma_{Y|X_1,X_2}$ are obtained by using the formulas in Section 4.4. The point estimate of x_0 is

$$\hat{x}_0 = -\frac{\hat{\beta}_1}{2\hat{\beta}_2} \qquad (6.7.2)$$

A $1 - \alpha$ confidence region for x_0 can be computed by following the instructions given in Box 6.7.1.

B O X **6.7.1**   **Instructions for Computing a Confidence Region for the Maximum or the Minimum Point of a Quadratic Regression Function**

Carry out steps 1-6:

1 Compute

$$T = \hat{\sigma}^2 F_{1-\alpha:1,n-3}$$

2 Compute

$$A = 4(\hat{\beta}_2^2 - c_{3,3}T)$$

3 Compute

$$B = 2(\hat{\beta}_1\hat{\beta}_2 - c_{2,3}T)$$

4 Compute

$$D = \hat{\beta}_1^2 - c_{2,2}T$$

In the preceding formulas, $c_{i,j}$ is the (i, j)th element of the matrix C in (6.7.1).

5 Compute

$$G = B^2 - AD$$

6 If $G > 0$ and $A \neq 0$, compute L and U, where

$$L = \frac{-B - \sqrt{G}}{A} \qquad \text{and} \qquad U = \frac{-B + \sqrt{G}}{A}$$

A $1 - \alpha$ confidence region for x_0 is given by

a	$-\infty < x_0 < \infty$	if $G \leq 0$
b	$-\infty < x_0 \leq U$ and $L \leq x_0 < \infty$	if $G > 0$ and $A < 0$ (6.7.3)
c	$L \leq x_0 \leq U$	if $G > 0$ and $A > 0$

If (a) or (b) is obtained in (6.7.3), the results will certainly be unsatisfactory and more observations will be required to obtain finite bounds such as in (c). However, (c) may also be unsatisfactory if the confidence interval is too wide for decision-making purposes.

The computations are illustrated in Example 6.7.3. We provide a computer program on the data disc for carrying out the preceding calculations, and the use of this program is explained in Section 6.7 of the laboratory manuals.

E X A M P L E 6.7.3

The output Y of an industrial process that manufactures sulfuric acid depends on X, the temperature at which the process is run. Past experience indicates that the regression function of output on temperature can be closely represented using a quadratic (i.e., a second-degree polynomial) function. To determine the temperature at which the maximum rate of production is achieved, a scientist carries out an experiment using several different process temperatures. The quantity of sulfuric acid produced in a day for each temperature is recorded. The data are given in Table 6.7.2 and are also stored in the file **sulfuric.dat** on the data disk. The (conceptual) target population can be defined as the collection of all possible pairs of numbers (Y, X), with Y equal to total daily output of sulfuric acid and X is any temperature that could be used. The sample can be considered to have been obtained from this target population using sampling with preselected X values.

T A B L E 6.7.2
Sulfuric Acid Data

Observation Number	Daily Acid Production Y (tons)	Temperature X (°C)
1	1.93	100
2	2.22	125
3	2.85	150
4	2.69	175
5	3.01	200
6	3.82	225
7	3.91	250
8	3.65	275
9	3.71	300
10	3.40	325
11	3.71	350
12	2.57	375
13	2.71	400

Assumptions (A) are presumed valid with

$$\mu_Y(x) = \beta_0 + \beta_1 x + \beta_2 x^2$$

We want a point estimate and a 95% confidence region for x_0, the temperature at which to run the process so the output will be maximized. A SAS output containing the results of the regression analysis is given in Exhibit 6.7.1. The predictor variables are $X_1 = X$ and $X_2 = X^2$. Exhibit 6.7.1 also contains a plot of the data. From the plot it seems reasonable to assume that a quadratic model is appropriate for this problem. From the computer output we get

$$\hat{x}_0 = \frac{(-0.0355190809)}{2(-0.000065223)} = 272.29$$

E X H I B I T 6.7.1
SAS Output for Example 6.7.3

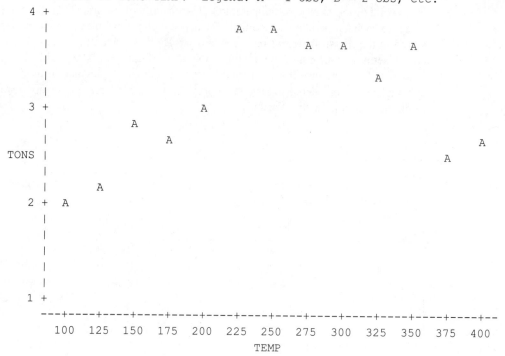

```
                              The SAS System
                                          00:00 Saturday, Jan 1, 1994
             Plot of TONS*TEMP.  Legend: A = 1 obs, B = 2 obs, etc.
    4 +
      |
      |                                A     A
      |                                   A     A        A
      |
      |                                      A
      |
    3 +                          A
      |                    A
      |                    A                                    A
 TONS |                                                   A
      |
      |        A
    2 + A
      |
      |
      |
      |
      |
    1 +
       ---+----+----+----+----+----+----+----+----+----+----+----+--
         100  125  150  175  200  225  250  275  300  325  350  375  400
                                   TEMP
```

```
                              The SAS System
                                          00:00 Saturday, Jan 1, 1994
Model: MODEL1
                 X'X Inverse, Parameter Estimates, and SSE
                     INTERCEP           TEMP            TEMP2           TONS
 INTERCEP      4.3206793207    -0.036563437    0.0000687313    -1.141878122
 TEMP         -0.036563437      0.0003284715   -6.393606E-7     0.0355190809
 TEMP2         0.0000687313    -6.393606E-7     1.2787213E-9   -0.000065223
 TONS         -1.141878122      0.0355190809   -0.000065223     0.8366037962
 Dependent Variable: TONS
```

EXHIBIT 6.7.1

(Continued)

Analysis of Variance

Source	DF	Sum of Squares	Mean Square	F Value	Prob>F
Model	2	4.28849	2.14424	25.630	0.0001
Error	10	0.83660	0.08366		
C Total	12	5.12509			

Root MSE	0.28924	R-square	0.8368	
Dep Mean	3.09077	Adj R-sq	0.8041	
C.V.	9.35822			

Parameter Estimates

Variable	DF	Parameter Estimate	Standard Error	T for H0: Parameter=0	Prob > |T|
INTERCEP	1	-1.141878	0.60122348	-1.899	0.0867
TEMP	1	0.035519	0.00524214	6.776	0.0001
TEMP2	1	-0.000065223	0.00001034	-6.306	0.0001

A glance at the plot confirms that the maximum occurs when the temperature is close to 272.

For confidence region calculations we need certain elements of the matrix C (the first three rows and the first three columns in the matrix labeled X'X Inverse) given in the computer output of Exhibit 6.7.1. The table-value that we need is $F_{0.95:1,10} = 4.96$. The quantities mentioned in Box 6.7.1 are

1	$T = 0.414955$	
2	$A = 0.0000000148936$	
3	$B = -0.00000410269$	**(6.7.4)**
4	$D = 0.00112530$	
5	$G = 0.0000000000000722759$	

Since $G > 0$ we get

$$L = 257.4 \quad \text{and} \quad U = 293.5$$

Thus the maximum yield is estimated to occur at the temperature 272°C. A 95% two-sided confidence interval for x_0, the temperature at which the maximum yield occurs, is given by

$$C[257.4 \leq x_0 \leq 293.5] = 0.95$$

Note These calculations are *very sensitive to rounding errors*, as is the case with most polynomial models, and it is advisable to keep as many significant figures as possible for all intermediate calculations. Most good computer programs for regression analysis use special numerical techniques to minimize problems due to rounding.   ■

Caution   One-sided $1 - \alpha/2$ confidence bounds for x_0 cannot be obtained by using only the upper bound or only the lower bound from the $1 - \alpha$ confidence region for x_0 discussed in this section.

Problems   6.7

6.7.1   A study was done to determine how Y, the crushing strength of concrete, is affected by X, the amount of sand (in cubic inches) used in the mixture for a fixed amount of cement. The data in Table 6.7.3 were obtained by crushing concrete cylinders made with various amounts of sand and measuring the strength, which is the number of tons of force that concrete cylinders withstood before crumbling. The data are also stored in the file **concrete.dat** on the data disk. Assumptions (A) are presumed to be valid with $\mu_Y(x)$ given by

$$\mu_Y(x) = \beta_0 + \beta_1 x + \beta_2 x^2 \qquad 1 \leq x \leq 60$$

The data may be considered to have been obtained by sampling with preselected X values from the (conceptual) target population consisting of all possible pairs of numbers (Y, X) (with Y equal to crushing strength and X equal to amount of sand added) that could be observed.

T A B L E   **6.7.3**
Concrete Crushing Strength Data

Observation Number	Strength Y (tons)	Amount of Sand X (cubic inches)
1	2.2	1
2	3.7	5
3	5.3	10
4	5.8	15
5	6.4	20
6	7.1	25
7	8.2	30
8	7.9	35
9	6.2	40
10	4.8	50
11	3.9	60

A MINITAB output from a regression analysis of Y on $X_1 = X$ and $X_2 = X^2$ is given in Exhibit 6.7.2.

E X H I B I T   **6.7.2**

MINITAB Output for Problem 6.7.1

```
The regression equation is
strength = 2.16 + 0.328  sand - 0.00515  sand^2

Predictor            Coef         Stdev      t-ratio          p
Constant           2.1628        0.4717         4.59      0.000
sand              0.32785       0.03714         8.83      0.000
sand^2         -0.0051511     0.0006031        -8.54      0.000

s = 0.6300       R-sq = 90.7%       R-sq(adj) = 88.4%

The C matrix is

 .5606449894334     -.0369011838509      .0004973796640
-.0369011838509      .0034763981595     -.0000540772111
 .0004973796640     -.0000540772111      .0000009164621
```

a   Plot Y against X and assess the appropriateness of the quadratic regression model.

b   Estimate $\beta_0, \beta_1, \beta_2, \sigma$.

c   Find a point estimate of x_0, the amount of sand that gives the maximum crushing strength.

d   Find a 90% confidence region for x_0. You are given the values of A, B, and D defined in Box 6.7.1, as follows:

$$A = 0.0001011 \qquad B = -0.00323 \qquad D = 0.102715$$

You must verify these calculations.

6.8
Linear Splines

In some applications the regression function cannot be adequately described by a straight line or a polynomial function over the entire range of interest. If $[a, b]$ denotes the interval of interest for X, it may be that for some value q (between a and b) the regression function is a straight line between a and q, and is a *different* straight line between q and b and that the two lines intersect at q (see Figure 6.8.1).

FIGURE **6.8.1**

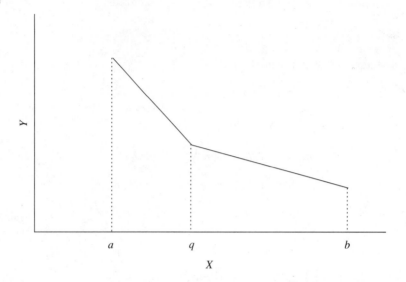

The point q (which is assumed to be known) where the two lines intersect is called a *knot-point*, and the graph consisting of the two lines is called a linear spline with a knot-point at q. We could consider linear splines with several knot-points, and we could consider polynomial splines with several knot points. However, in this book we discuss only linear splines with one knot-point. We illustrate with two examples.

EXAMPLE **6.8.1**

An investigator notices that for small companies the volume (Y) of sales in dollars tends to increase as a function of X, the dollars spent on advertising. The rate of increase in sales is rapid for the first several thousand dollars spent on advertising, but it slows down at some point. This suggests that a linear spline between a and b, similar to that shown in Figure 6.8.2, may be used as a prediction function to model the relationship between the average yearly sales and dollars spent on advertising.   ■

EXAMPLE **6.8.2**

A computer company leases a large number of personal computers to small businesses, and the company wants to predict next year's total maintenance cost of all computers leased. The computers are from 1 to 7 years old. It is known that the maintenance cost Y of a personal computer is a function of its age X. During the first 3 years the maintenance cost tends to increase less rapidly than during the next 4 years. This suggests that an appropriate model may be a linear spline regression function such as the one shown in Figure 6.8.3 with a knot-point at $q = 3$.   ■

F I G U R E **6.8.2**

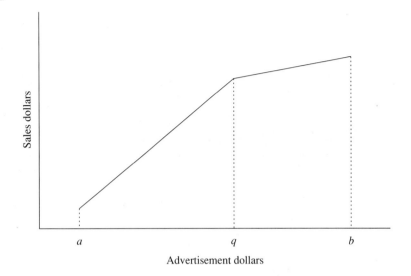

F I G U R E **6.8.3**

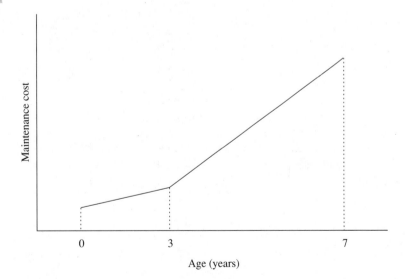

A linear spline regression function $\mu_Y(x)$ with a single knot-point at $x = q$ is defined by

$$\mu_Y(x) = \begin{cases} \mu_Y^{(1)}(x) = \alpha_1 + \beta_1 x & \text{for } a \le x \le q \\ \mu_Y^{(2)}(x) = \alpha_2 + \beta_2 x & \text{for } q \le x \le b \end{cases} \qquad (6.8.1)$$

Because the lines are connected at $x = q$, where q is a known number, it follows that $\mu_Y^{(1)}(q) = \mu_Y^{(2)}(q)$, and we get

$$\alpha_1 + q\beta_1 = \alpha_2 + q\beta_2$$

We solve for α_2 and get

$$\alpha_2 = \alpha_1 + q(\beta_1 - \beta_2) \qquad (6.8.2)$$

Thus, while it appears that there are four unknown parameters α_1, α_2, β_1, and β_2 in the model in (6.8.1), in fact only three parameters are needed because one of these parameters can be expressed in terms of the other three.

Suppose an investigator wants to estimate $\mu_Y(x)$ in (6.8.1) using sample data

$$(y_1, x_1), (y_2, x_2), \ldots, (y_n, x_n)$$

where $x_1, \ldots, x_m$ $(m < n)$ are in the interval $a \le x \le q$ and $x_{m+1}, \ldots, x_n$ are in the interval $q < x \le b$. The sample regression model is written out in detail below.

$$
\begin{aligned}
y_1 &= \alpha_1 + \beta_1 x_1 & + e_1 \\
&\ \ \vdots \\
y_m &= \alpha_1 + \beta_1 x_m & + e_m \\
y_{m+1} &= \alpha_2 + \beta_2 x_{m+1} & + e_{m+1} \\
&= \vdots \\
y_n &= \alpha_2 + \beta_2 x_n & + e_n
\end{aligned}
\qquad (6.8.3)
$$

Recall that $e_1, \ldots, e_n$ are the deviations of $y_1, \ldots, y_n$ from their respective subpopulation means. Assumptions (A) are presumed valid, and the data are obtained either by simple random sampling or by sampling with preselected X values. Substituting for α_2 using (6.8.2), the equations in (6.8.3) for $y_1, \ldots, y_n$ can be rewritten as follows:

$$
\begin{aligned}
y_1 &= \alpha_1 + \beta_1 x_1 & + e_1 \\
&\ \ \vdots \\
y_m &= \alpha_1 + \beta_1 x_m & + e_m \\
y_{m+1} &= \alpha_1 + q\beta_1 + \beta_2(x_{m+1} - q) & + e_{m+1} \\
&\ \ \vdots \\
y_n &= \alpha_1 + q\beta_1 + \beta_2(x_n - q) & + e_n
\end{aligned}
\qquad (6.8.4)
$$

To use the theory in Chapter 4 to compute point and interval estimates, the data are first organized as follows:

Y	X_1	X_2
y_1	x_1	0
y_2	x_2	0
$\vdots$	$\vdots$	$\vdots$
y_m	x_m	0
y_{m+1}	q	$x_{m+1} - q$
y_{m+2}	q	$x_{m+2} - q$
$\vdots$	$\vdots$	$\vdots$
y_n	q	$x_n - q$

(6.8.5)

Observe that the data for X_1 and X_2 in (6.8.5) correspond to the coefficients of β_1 and β_2 in (6.8.4). The X matrix is obtained by putting a column of 1's as the first column, values of X_1 in the second column, and the values of X_2 in the third column. Thus, in matrix notation, we can write (6.8.4) as

$$y = X\beta + e \qquad (6.8.6)$$

where

$$y = \begin{bmatrix} y_1 \\ y_2 \\ \vdots \\ y_n \end{bmatrix} \quad X = \begin{bmatrix} 1 & x_1 & 0 \\ 1 & x_2 & 0 \\ \vdots & \vdots & \vdots \\ 1 & x_m & 0 \\ 1 & q & x_{m+1} - q \\ \vdots & \vdots & \vdots \\ 1 & q & x_n - q \end{bmatrix} \quad \beta = \begin{bmatrix} \alpha_1 \\ \beta_1 \\ \beta_2 \end{bmatrix} \quad e = \begin{bmatrix} e_1 \\ e_2 \\ \vdots \\ e_n \end{bmatrix} \qquad (6.8.7)$$

Hence we get

$$\hat{\beta} = \begin{bmatrix} \hat{\alpha}_1 \\ \hat{\beta}_1 \\ \hat{\beta}_2 \end{bmatrix} = (X^T X)^{-1} X^T y \qquad (6.8.8)$$

and

$$\hat{\alpha}_2 = \hat{\alpha}_1 + q\hat{\beta}_1 - q\hat{\beta}_2 = d^T \hat{\beta} \qquad (6.8.9)$$

where $d^T = [1, q, -q]$. So from (6.8.1) we get

$$\hat{\mu}_Y(x) = \begin{cases} \hat{\mu}_Y^{(1)}(x) = \hat{\alpha}_1 + \hat{\beta}_1 x & \text{for } a \le x \le q \\ \hat{\mu}_Y^{(2)}(x) = \hat{\alpha}_2 + \hat{\beta}_2 x & \text{for } q \le x \le b \end{cases} \qquad (6.8.10)$$

Note that

$$\mu_Y^{(1)}(x) = \alpha_1 + \beta_1 x = d^T \beta \quad \text{where} \quad d^T = [1, x, 0] \text{ for } a \le x \le q$$

and

$$\mu_Y^{(2)}(x) = \alpha_2 + \beta_2 x = \alpha_1 + \beta_1 q + \beta_2(x - q) = d^T \beta$$
$$\text{where } d^T = [1, q, x - q] \text{ for } q \le x \le b$$

Since the spline model of this section is a special case of the multiple linear regression model, all formulas for point and confidence interval estimation discussed in Chapter 4 can be used for this model.

If we let C denote $(X^T X)^{-1}$ and let $\hat{\sigma}$ denote the estimate of the standard deviations of the subpopulations, then

$$\hat{\sigma} = \sqrt{\frac{SSE}{(n - 3)}} \tag{6.8.11}$$

where SSE is the error sum of squares. The standard errors to be used in confidence interval calculations for $\alpha_1, \alpha_2, \beta_1, \beta_2$, and $\mu_Y(x)$ are

$$SE(\hat{\alpha}_1) = \hat{\sigma}\sqrt{c_{1,1}} \quad SE(\hat{\beta}_1) = \hat{\sigma}\sqrt{c_{2,2}} \quad SE(\hat{\beta}_2) = \hat{\sigma}\sqrt{c_{3,3}} \tag{6.8.12}$$

$$SE(\hat{\alpha}_2) = \hat{\sigma}\sqrt{d^T C d} \quad (\text{where } d^T = [1, q, -q])$$
$$= \hat{\sigma}\sqrt{c_{1,1} + q^2 c_{2,2} + q^2 c_{3,3} + 2q c_{1,2} - 2q c_{1,3} - 2q^2 c_{2,3}} \tag{6.8.13}$$

and

$$SE(\hat{\mu}_Y(x)) = \hat{\sigma}\sqrt{d^T C d} \quad (\text{where } d^T = [1, x, 0])$$
$$= \hat{\sigma}[c_{1,1} + 2x c_{1,2} + x^2 c_{2,2}]^{1/2} \quad \text{for } a \le x \le q \tag{6.8.14}$$

$$SE(\hat{\mu}_Y(x)) = \hat{\sigma}\sqrt{d^T C d} \quad (\text{where } d^T = [1, q, x - q])$$
$$= \hat{\sigma}[c_{1,1} + q^2 c_{2,2} + (x - q)^2 c_{3,3} + 2q c_{1,2}$$
$$+ 2q(x - q)c_{2,3} + 2(x - q)c_{1,3}]^{1/2} \quad \text{for } q < x \le b \tag{6.8.15}$$

In the preceding equations, $c_{i,j}$ is the (i, j)th element of the matrix C. We illustrate the procedure in Example 6.8.3.

E X A M P L E 6.8.3

Consider the problem discussed in Example 6.8.1 where Y is annual sales and X is the number of dollars spent on advertising in one year. Data for this problem were obtained by simple random sampling of sales and advertising records of a study population consisting of several hundred small companies and are given in Table 6.8.1. These data are also stored in the file **sales.dat** on the data disk. Suppose that a spline model with a knot point at $q = 50$ is appropriate for this problem. We presume that assumptions (A) are valid. We exhibit the computations for a linear spline regression with a knot-point at $q = 50$ and find estimates of α_i and β_i and their standard errors. We will also exhibit the computations required to estimate $\mu_Y(65)$ and obtain the corresponding standard error.

T A B L E 6.8.1
Advertising and Sales Data

Observation Number	Sales Y (thousands of dollars)	Advertisement Budget X (thousands of dollars)
1	260	12
2	328	25
3	376	30
4	356	35
5	404	41
6	399	41
7	404	41
8	414	44
9	428	45
10	436	46
11	439	47
12	452	47
13	465	55
14	461	59
15	475	64
16	462	66
17	472	73
18	456	74
19	490	83
20	496	87

The population regression function is

$$\mu_Y(x) = \begin{cases} \mu_Y^{(1)}(x) = \alpha_1 + \beta_1 x & \text{for } 0 \le x \le 50 \\ \mu_Y^{(2)}(x) = \alpha_2 + \beta_2 x & \text{for } 50 \le x \le 100 \end{cases} \tag{6.8.16}$$

The **y** vector and the **X** matrix, described in (6.8.7), are

$$
y = \begin{bmatrix} 260 \\ 328 \\ 376 \\ 356 \\ 404 \\ 399 \\ 404 \\ 414 \\ 428 \\ 436 \\ 439 \\ 452 \\ 465 \\ 461 \\ 475 \\ 462 \\ 472 \\ 456 \\ 490 \\ 496 \end{bmatrix}
\qquad
X = \begin{bmatrix} 1 & 12 & 0 \\ 1 & 25 & 0 \\ 1 & 30 & 0 \\ 1 & 35 & 0 \\ 1 & 41 & 0 \\ 1 & 41 & 0 \\ 1 & 41 & 0 \\ 1 & 44 & 0 \\ 1 & 45 & 0 \\ 1 & 46 & 0 \\ 1 & 47 & 0 \\ 1 & 47 & 0 \\ 1 & 50 & 5 \\ 1 & 50 & 9 \\ 1 & 50 & 14 \\ 1 & 50 & 16 \\ 1 & 50 & 23 \\ 1 & 50 & 24 \\ 1 & 50 & 33 \\ 1 & 50 & 37 \end{bmatrix}
\qquad \textbf{(6.8.17)}
$$

The regression of Y on X_1 and X_2 (where X_1 and X_2 are columns 2 and 3, respectively, of the preceding matrix X) can be obtained by the formulas in Chapter 4. Exhibit 6.8.1 gives a MINITAB output containing the results of a regression analysis of Y on X_1 and X_2. A plot of Y against X for the data in Table 6.8.1 is also given.

The estimates of the regression coefficients are

$$\hat{\alpha}_1 = 201.45 \qquad \hat{\beta}_1 = 5.0218 \qquad \hat{\beta}_2 = 0.9658$$

From these, and using (6.8.9), we get

$$\hat{\alpha}_2 = \hat{\alpha}_1 + q\hat{\beta}_1 - q\hat{\beta}_2$$

or

$$\hat{\alpha}_2 = 201.45 + 50(5.0218) - 50(0.9658) = 404.25$$

Furthermore, the estimate of σ is 11.05.

The standard errors of the estimated regression coefficients are obtained from (6.8.12) and (6.8.13) and are as follows:

$$SE(\hat{\alpha}_1) = 11.70 \qquad SE(\hat{\beta}_1) = 0.2875 \qquad SE(\hat{\beta}_2) = 0.2398 \qquad SE(\hat{\alpha}_2) = 15.28$$

E X H I B I T **6.8.1**
MINITAB Output for Example 6.8.3

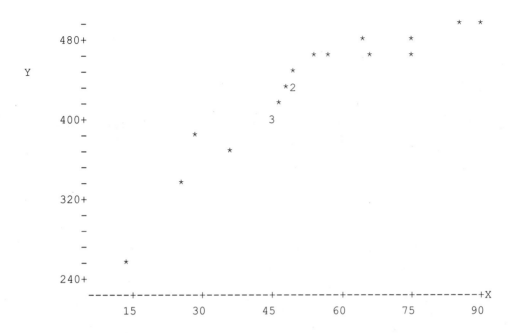

```
The regression equation is
  Y = 201 + 5.02 X1 + 0.966 X2

Predictor       Coef        Stdev      t-ratio        p
Constant       201.45       11.70        17.22     0.000
X1             5.0218       0.2875       17.47     0.000
X2             0.9658       0.2398        4.03     0.001

s = 11.05      R-sq = 96.8%      R-sq(adj) = 96.5%

The C matrix is

 1.1211817651        -.0266383640        .0082330906
 -.0266383640         .0006769450       -.0002816381
  .0082330906        -.0002816381        .0004711620
```

The standard errors for $\hat{\alpha}_1$, $\hat{\beta}_1$, and $\hat{\beta}_2$ can also be obtained directly from Exhibit 6.8.1. From these we can obtain confidence intervals for α_1, β_1, α_2, and β_2 in the usual manner.

To obtain the estimate of $\mu_Y(65)$ we first note that 65 is greater than 50 and so $\mu_Y(65) = \mu_Y^{(2)}(65) = \alpha_2 + 65\beta_2$. Hence $\hat{\mu}_Y(65) = \hat{\alpha}_2 + 65\hat{\beta}_2 = 467.03$. To obtain $SE(\hat{\mu}_Y(65))$ we apply the formula in (6.8.15). Hence $SE(\hat{\mu}_Y(65)) = \hat{\sigma}\sqrt{d^T C d}$ where $d^T = [1, q, x - q] = [1, 50, 65 - 50] = [1, 50, 15]$. You should verify that $d^T C d = 0.0803$. Thus we get $SE(\hat{\mu}_Y(65)) = 11.05\sqrt{0.0803} = 3.13$.

The data disk contains a program written by us that can be used to do the calculations required in this section. The use of this program is explained in Section 6.8 of the laboratory manuals.   ■

Problems   6.8

6.8.1   Consider the sulfuric acid data in Example 6.7.3, reproduced for convenience in Table 6.8.2. Assumptions (A) are presumed to hold, and the data are obtained by sampling with preselected X values. Suppose $\mu_Y(x)$ is a linear spline regression function given in (6.8.1) with a knot-point at $q = 270$. Exhibit the X matrix in (6.8.7) for these data.

6.8.2   A MINITAB output for the regression of Y on X_1 and X_2 (see (6.8.5)–(6.8.7)) in Problem 6.8.1 is given in Exhibit 6.8.2.

   a   Exhibit the point estimate and a 95% confidence interval for σ.

T A B L E   **6.8.2**
Sulfuric Acid Data

Observation Number	Daily Acid Production Y (tons)	Temperature X (°C)
1	1.93	100
2	2.22	125
3	2.85	150
4	2.69	175
5	3.01	200
6	3.82	225
7	3.91	250
8	3.65	275
9	3.71	300
10	3.40	325
11	3.71	350
12	2.57	375
13	2.71	400

E X H I B I T **6.8.2**

MINITAB Output for Problem 6.8.2

```
The regression equation is
Y = 0.780 + 0.0120 X1 - 0.0103 X2


Predictor          Coef        Stdev     t-ratio          p
Constant         0.7798       0.3269        2.39      0.038
X1              0.012033     0.001616        7.45      0.000
X2             -0.010350     0.002179       -4.75      0.000


s = 0.2795       R-sq = 84.8%       R-sq(adj) = 81.7%


The C matrix is


     1.36823595149        -.00647157429        .00401126291
     -.00647157429         .00003341185       -.00002697841
      .00401126291        -.00002697841        .00006075838
```

b Exhibit point estimates and 90% confidence intervals for α_1, β_1, α_2, and β_2.

c Exhibit the point estimate and 95% confidence interval for $\mu_Y(175)$.

d Exhibit the point estimate and 95% confidence interval for $\mu_Y(375)$.

6.8.3 A quadratic regression function $\mu_Y(x) = \beta_0 + \beta_1 x + \beta_2 x^2$ was fitted to the data in Problem 6.8.1. The corresponding MINITAB output with $X1 = X$ and $X2 = X^2$ is given in Exhibit 6.8.3. Which fits better, a spline function or a quadratic regression function? Explain.

E X H I B I T **6.8.3**

MINITAB Output for Problem 6.8.3

```
The regression equation is
Y = - 1.14 + 0.0355 X1 -0.000065 X2

Predictor          Coef        Stdev     t-ratio          p
Constant        -1.1419       0.6012       -1.90      0.087
X1              0.035519     0.005242        6.78      0.000
X2           -0.00006522   0.00001034       -6.31      0.000


s = 0.2892       R-sq = 83.7%       R-sq(adj) = 80.4%


The C matrix is


   4.320679320679321        -.036563436563437        .000068731268731
   -.036563436563437         .000328471528472       -.000000639360639
    .000068731268731        -.000000639360639        .000000001278721
```

6.8.4 **a** Obtain the residual plots corresponding to the regression functions in Problems 6.8.2 and 6.8.3.

b Study these plots carefully and decide which of the two models results in residuals that are consistent with assumptions (A) for regression.

6.9
Exercises

6.9.1 Past research indicates that in track-and-field events such as 100-m, 200-m, and 400-m races, the performances of athletes during practice a day before a race and their performances during the race are closely related. The data given in Table 6.9.1 were obtained from a simple random sample of female athletes chosen from Division I schools. X is time, in seconds, to run 400 m in practice, and Y is time, in seconds to run 400 m during a race. The data are also stored in the file **track.dat** on the floppy disk. Assumptions (B) are presumed to hold.

A MINITAB output containing the regression of Y on X is given in Exhibit 6.9.1.

a Which of the following quantities can be validly estimated from the preceding data: $\mu_Y(x)$, β_0, β_1, σ, σ_Y, σ_X, μ_Y, and μ_X ?

b Find the point estimate of each parameter in (a) for which a valid point estimate can be computed.

T A B L E **6.9.1**
Track Data

Observation Number	Race Times (seconds) Y	Practice Times (seconds) X
1	50.97	51.85
2	49.47	51.05
3	52.07	52.47
4	51.17	51.82
5	51.38	52.94
6	50.26	51.33
7	51.11	51.90
8	49.15	50.17
9	49.97	51.24
10	49.51	50.43
11	50.63	52.00
12	49.63	50.88
13	50.81	51.31

EXHIBIT 6.9.1

MINITAB Output for Exercise 6.9.1

```
The regression equation is
racetime  = - 2.23 + 1.02 practime


Predictor          Coef        Stdev     t-ratio          p
Constant         -2.226        7.537       -0.30      0.773
practime         1.0234       0.1464        6.99      0.000


s = 0.3949       R-sq = 81.6%       R-sq(adj) = 80.0%


Analysis of Variance


SOURCE           DF           SS           MS          F          p
Regression        1       7.6268       7.6268      48.90      0.000
Error            11       1.7158       0.1560
Total            12       9.3426


The C matrix is

   364.1887817      -7.0712953
    -7.0712953       0.1373293
```

c   If an athlete runs the 400-m course in 52.08 seconds during practice a day before the race, predict her time for running 400 m during the race. Also compute a 95% lower prediction bound for her time.

d   Four athletes are chosen to form a team to run a 4 by 400-m relay where each of the four athletes runs 400 m. Their times for 400 m during practice were 51.32, 52.07, 52.58, 51.96 seconds, respectively. Predict this team's total time for the relay during the race and compute a 99% lower bound for it.

e   In (d) compute a 95% two-sided prediction interval for total team time.

6.9.2 The relationship between systolic blood pressure Y and weight X is known to be approximately linear for a population of people who have taken a certain medication for 1 year. The weights are in the range from 120 pounds to 250 pounds. A simple random sample of individuals was chosen from this population, and their weights and blood pressures were recorded. The data are shown in Table 6.9.2 and are also stored in the file **bpweight.dat** on the data disk. Assumptions (B) are presumed to hold.

A SAS output containing the results of a regression analysis of Y on X is given in Exhibit 6.9.2.

a   Estimate $\lambda_{0.99}(210)$, the number such that 99% of the people in the population who weigh 210 pounds have blood pressure lower than this. Find a 95% two-sided confidence interval for $\lambda_{0.99}(210)$.

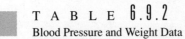

T A B L E 6.9.2
Blood Pressure and Weight Data

Observation Number	Blood Pressure Y	Weight X
1	127	175
2	120	189
3	149	245
4	140	233
5	107	126
6	128	194
7	163	247
8	146	234
9	146	232
10	124	160
11	101	142
12	129	178
13	120	176
14	127	205
15	98	132
16	120	188
17	151	245
18	105	126
19	110	160
20	120	176

E X H I B I T 6.9.2
SAS Output for Exercise 6.9.2

```
                              The SAS System

                                        00:00 Saturday, Jan  1, 1994

Model: MODEL1


                X'X Inverse, Parameter Estimates, and SSE


                     INTERCEP             WEIGHT                   BP


      INTERCEP     1.175733109       -0.005983168       47.731630338
      WEIGHT      -0.005983168        0.0000318          0.4189124085
      BP          47.731630338        0.4189124085     624.47022328


Dependent Variable: BP
```

EXHIBIT **6.9.2**
(Continued)

Analysis of Variance

Source	DF	Sum of Squares	Mean Square	F Value	Prob>F
Model	1	5518.47978	5518.47978	159.067	0.0001
Error	18	624.47022	34.69279		
C Total	19	6142.95000			

Root MSE	5.89006	R-square	0.8983
Dep Mean	126.55000	Adj R-sq	0.8927
C.V.	4.65433		

Parameter Estimates

Variable	DF	Parameter Estimate	Standard Error	T for H0: Parameter=0	Prob > |T|
INTERCEP	1	47.731630	6.38666283	7.474	0.0001
WEIGHT	1	0.418912	0.03321491	12.612	0.0001

b A person who weighs 240 pounds has a blood pressure of 210. Compute an appropriate 80% confidence interval to help decide whether his blood pressure is in the upper 5% of the blood pressures of all people in this population who weigh 240 pounds.

c Calculate a 95% confidence interval for $\lambda_{0.99}(160)$, which is the number such that 99% of the people in the population who weigh 160 pounds will have blood pressure below this number.

6.9.3 In certain chemical assays the unknown concentration of a given compound in a chemical solution is measured indirectly by making use of the fact that when a beam of light of a given intensity passes through this chemical solution, the amount Y of light transmitted decreases with increasing concentrations X of the compound in the solution. The measurement system is calibrated by using solutions of known concentration and recording the intensity of transmitted light. The calibration data are given in Table 6.9.3 and are also stored in the file **assay.dat** on the data disk.

Here X is concentration in parts per million (ppm) and Y is intensity of transmitted light in appropriate units. Suppose that assumptions (A) apply and the data are obtained by sampling with preselected X values (i.e., preparing solutions with specified concentrations). The population regression function is of the form

$$\mu_Y(x) = \beta_0 + \beta_1 x$$

T A B L E   **6.9.3**

Chemical Assay Data

Observation Number	Intensity of Light Y	Concentration (parts per million) X
1	102.930	0
2	99.971	1
3	100.601	2
4	81.673	3
5	88.179	4
6	90.340	5
7	82.643	6
8	80.984	7
9	70.672	8
10	76.386	9
11	68.437	10
12	67.204	11
13	63.110	12
14	53.971	13
15	62.395	14
16	54.063	15
17	48.696	16
18	53.478	17
19	55.605	18
20	28.731	19
21	49.159	20

The (conceptual) study population is the collection of pairs of numbers (Y, X) (where Y is intensity of transmitted light and X is concentration) obtained from all possible solutions that could have been prepared. A MINITAB output containing the results of a regression analysis of Y on X is given in Exhibit 6.9.3.

E X H I B I T   **6.9.3**

MINTAB Output for Exercise 6.9.3

```
The regression equation is
intensty = 101 - 3.04 concentr

Predictor        Coef       Stdev      t-ratio          p
Constant      100.840       2.555        39.46      0.000
concentr      -3.0400      0.2186       -13.91      0.000

s = 6.066      R-sq = 91.1%      R-sq(adj) = 90.6%
```

EXHIBIT **6.9.3**
(Continued)

```
Analysis of Variance

SOURCE        DF           SS           MS          F          p
Regression    1        7116.2       7116.2     193.42     0.000
Error         19        699.0         36.8
Total         20       7815.2


The C Matrix is

  .177489177      -.012987013
 -.012987013       .001298701
```

a Estimate $\mu_Y(x)$, the average amount of light transmitted when the concentration is x ppm.

b A solution containing an unknown concentration x_0 of the compound gives a Y reading of 55 units. Estimate the unknown concentration x_0 and also compute a 95% confidence region for x_0.

6.9.4 The hardness Y of steel ball bearings is related to the rate X at which they were cooled after they were made. Data were collected to determine this relationship by making ball bearings using various known rates of cooling and then measuring the hardness. Data obtained from a simple random sample of a week's output of ball bearings (which is the study population) at each rate of cooling are given in Table 6.9.4 and are also stored in the file **ballbear.dat** on the data disk.

Suppose assumptions (A) hold and the data were obtained by sampling with preselected X values. The population regression function is

$$\mu_Y(x) = \beta_0 + \beta_1 x \qquad \text{(6.9.1)}$$

A MINITAB output containing the results of a regression analysis of Y on X is given in Exhibit 6.9.4.

a Plot Y versus X.

b Estimate $\mu_Y(x)$.

c If the manufacturer wants to produce ball bearings with an average hardness of 35.00 units, estimate the required cooling rate x_0 and also compute a 95% confidence region for x_0.

d Calculate a 95% confidence interval for $\lambda_{0.01}(25)$, the number such that 1% of all ball bearings made using a cooling rate of 25°C will have a hardness less than that number (i.e., 99% of all ball bearings made using a cooling rate of 25°C will have a hardness greater than that number).

e Write a short report summarizing the results of parts (a), (b), (c), and (d). The language of the report should be such that people who are not well acquainted with the meaning of confidence intervals, tolerance intervals, etc., can understand the results.

T A B L E 6.9.4
Ball Bearing Data

Observation Number	Hardness Index Y	Rate of cooling (deg C per minute) X
1	48.60	10
2	47.80	10
3	47.60	15
4	46.70	15
5	46.20	20
6	45.70	20
7	46.55	25
8	46.57	25
9	46.49	30
10	41.82	30
11	41.40	35
12	42.10	35
13	42.01	40
14	41.67	40
15	38.96	45
16	40.97	45
17	38.71	50
18	37.00	50
19	35.88	55
20	36.25	55
21	39.23	60
22	34.18	60
23	34.59	65
24	37.56	65
25	33.49	70
26	33.93	70
27	31.02	75
28	31.57	75
29	26.99	80
30	28.38	80

E X H I B I T 6.9.4
MINITAB Output for Exercise 6.9.4

```
The regression equation is
hardness = 51.9 - 0.271 coolrate


Predictor      Coef        Stdev      t-ratio        p
Constant     51.8648      0.6631       78.22     0.000
coolrate     -0.27113     0.01328     -20.41     0.000
```

E X H I B I T 6.9.4
(Continued)

```
s = 1.572          R-sq = 93.7%      R-sq(adj) = 93.5%
```

Analysis of Variance

SOURCE	DF	SS	MS	F	p
Regression	1	1029.1	1029.1	416.61	0.000
Error	28	69.2	2.5		
Total	29	1098.3			

The C matrix is

```
 .17797619048      -.00321428571
-.00321428571       .00007142857
```

6.9.5 We want to determine whether or not Y, the number of days of prison sentence for thefts whose total value is under $1,000 dollars for first time offenders, is related to X, the amount of money stolen. Sample data from thefts falling in this category in three different states in the United States, denoted as states 1, 2, and 3, are obtained from last year's police records in each state. These data are given in Table 6.9.5 and are also stored in the file **prison.dat** on the data disk.

T A B L E 6.9.5
Prison Data

Observation Number	Days in Prison Y	Dollars Stolen X	State
1	44	367	1
2	81	855	1
3	43	284	1
4	40	305	1
5	38	215	1
6	44	308	1
7	49	433	1
8	51	455	1
9	49	454	1
10	57	429	1
11	47	345	1
12	37	167	1
13	67	689	1
14	55	499	1
15	43	538	2

(Continued)

T A B L E   **6.9.5**

(Continued)

Observation Number	Days in Prison Y	Dollars Stolen X	State
16	32	290	2
17	53	759	2
18	55	734	2
19	40	499	2
20	44	541	2
21	42	474	2
22	51	940	2
23	35	314	2
24	39	351	2
25	51	703	2
26	37	459	2
27	50	732	3
28	52	556	3
29	53	960	3
30	39	134	3
31	55	826	3
32	53	738	3
33	37	403	3
34	46	511	3
35	45	699	3
36	49	778	3
37	37	530	3
38	35	140	3
39	38	429	3
40	48	554	3
41	50	672	3
42	30	125	3
43	29	124	3

There are three populations, one for each of the three states. The population regression function for state (i) is

$$\mu_Y^{(i)}(x) = \alpha_i + \beta_i x \qquad \text{for } i = 1, 2, 3$$

The subpopulation standard deviation for state (i) is σ_i for $i = 1, 2, 3$. We suppose that assumptions (A) are valid for each state and that data are obtained for each state by simple random sampling. The study and target populations are the same for this problem because the investigator wants to study the police records for these three states for the past 2 years. A MINITAB output that contains the results of regression analysis for each state is given in Exhibit 6.9.5. In this Exhibit Y_1, X_1 denote the response variable and the predictor variable for state (1). Likewise Y_2, X_2 and Y_3, X_3 refer to states (2) and (3), respectively.

E X H I B I T **6.9.5**
MINITAB Output for Exercise 6.9.5

REGRESSION ANALYSIS FOR STATE (1)

The regression equation is
$Y1 = 23.5 + 0.0643\ X1$

Predictor	Coef	Stdev	t-ratio	p
Constant	23.472	1.941	12.10	0.000
X1	0.064322	0.004312	14.92	0.000

$s = 2.823$ R-sq = 94.9%   R-sq(adj) = 94.5%

The C matrix is:

```
    .472391779577      -.000967008598
   -.000967008598       .000002332148
```

ymean = 50.1429
xmean = 414.643
SSY   = 1869.71
SSX   = 428789
SXY   = 27580.7

REGRESSION ANALYSIS FOR STATE (2)

The regression equation is
$Y2 = 24.5 + 0.0345\ X2$

Predictor	Coef	Stdev	t-ratio	p
Constant	24.496	2.805	8.73	0.000
X2	0.034543	0.004820	7.17	0.000

$s = 3.171$ R-sq = 83.7%   R-sq(adj) = 82.1%

The C matrix is:

```
    .782652676944      -.001271104532
   -.001271104532       .000002310399
```

y mean = 43.5000
x mean = 550.167
SSY   = 617.000
SSX   = 432826
SXY   = 14951.0

E X H I B I T **6.9.5**
(Continued)

REGRESSION ANALYSIS FOR STATE (3)

The regression equation is
Y3 = 29.3 + 0.0278 X3

Predictor	Coef	Stdev	t-ratio	p
Constant	29.309	2.246	13.05	0.000
X3	0.027802	0.003844	7.23	0.000

s = 4.089   R-sq = 77.7%   R-sq(adj) = 76.2%

The C matrix is:

      .3015545702755   -.0004630712260
     -.0004630712260   .0000008834262

y mean = 43.8824
x mean = 524.176
SSY   = 1125.76
SSX   = 1131957
SXY   = 31470.4

a Estimate $\mu_Y^{(1)}(x) = \alpha_1 + \beta_1 x$; $\mu_Y^{(2)}(x) = \alpha_2 + \beta_2 x$; $\mu_Y^{(3)}(x) = \alpha_3 + \beta_3 x$; and $\sigma_1, \sigma_2, \sigma_3$.

b If we assume that $\sigma_1 = \sigma_2 = \sigma_3$ and denote their common value by σ, estimate this common standard deviation.

c If $|\mu_Y^{(i)}(x) - \mu_Y^{(j)}(x)| \le 20$ days for i and j and for all x between 200 and 1,000 dollars, then the population regression function of Y on X for state (i) would be considered to be equivalent to that for state (j) for the purposes of this problem. Compute appropriate 95% simultaneous confidence intervals that an investigator can use to help decide whether the three states (or any two states) have equivalent population regression functions.

d What population parameters must be examined to help determine how much the average sentences differ for each pair of states when $1,000 is stolen?

6.9.6 An investigator wants to study how the salaries of high school teachers who teach in the public school system of a large city are related to experience (in years employed) and determine what the differences are, if any, between the salaries of male teachers and female teachers. A simple random sample of male teachers and a simple random sample of female teachers are chosen, and their monthly salaries Y and years of experience X are recorded. The data for both males and females are given in Table 6.9.6. The data are also stored in the file **salaries.dat** on the data disk. The study populations of items in this problem are all male teachers in this school system and all female teachers in the school system, respectively, the year that the sample was collected. These are also the target populations of items for this problem.

Suppose that assumptions (A) are valid for each population (male and female). The population regression functions for males and females are

$$\text{females:} \quad \mu_Y^{(1)}(x) = \alpha_1 + \beta_1 x$$

$$\text{males:} \quad \mu_Y^{(2)}(x) = \alpha_2 + \beta_2 x$$

The results of a regression analysis, done separately for males and females, is given in Exhibit 6.9.6.

T A B L E **6.9.6**

Salary Data

Observation Number	Salary (thousands of dollars) Y	Experience (years) X	Sex (1 = females 2 = males)
1	25.1	0	1
2	41.3	17	1
3	29.6	5	1
4	40.7	15	1
5	36.1	9	1
6	40.2	15	1
7	34.5	8	1
8	28.9	5	1
9	37.1	13	1
10	42.0	20	1
11	36.7	11	1
12	24.8	1	1
13	33.0	6	2
14	33.4	7	2
15	54.9	23	2
16	53.2	20	2
17	46.8	18	2
18	61.2	27	2
19	40.9	11	2
20	38.9	10	2
21	61.7	29	2
22	53.5	23	2
23	30.7	4	2
24	53.2	22	2
25	58.6	25	2
26	37.8	9	2
27	58.3	25	2

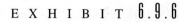

E X H I B I T 6.9.6

MINITAB Output for Exercise 6.9.6

```
                    REGRESSION ANALYSIS FOR FEMALES


The regression equation is
salary = 25.3 + 0.954 yearsexp


Predictor        Coef        Stdev       t-ratio          p
Constant       25.2873       0.7711        32.79      0.000
yearsexp       0.95422       0.06626       14.40      0.000


s = 1.398       R-sq = 95.4%      R-sq(adj) = 94.9%



The C matrix is:


        0.3043641       -0.0222888
       -0.0222888        0.0022476


mean y = 34.7500
mean x = 9.91667
SSY    = 424.650
SSX    = 444.917
SXY    = 424.550

                    REGRESSION ANALYSIS FOR MALES


The regression equation is
 salary = 25.6 + 1.28 yearsexp


Predictor        Coef        Stdev       t-ratio          p
Constant       25.6121       0.6921        37.01      0.000
yearsexp       1.28154       0.03618       35.42      0.000


s = 1.154       R-sq = 99.0%      R-sq(adj) = 98.9%


 The C matrix is:


        0.35984004      -0.01697915
       -0.01697915       0.00098335


y mean = 47.7400
x mean = 17.2667
SSY    = 1687.46
SSX    = 1016.93
SXY    = 1303.24
```

a   Plot the estimated regression lines for both males and females on the same graph.

b   What is the difference between the average salaries for males and females with the same number of years of experience? Express your answer in terms of population parameters.

c   What is the difference between the average starting salaries of males and females? Express your answer in terms of population parameters. Find a point estimate and a 95% confidence interval for this quantity.

d   What population parameters would you examine to determine whether the disparity between male and female salaries remains constant at every experience level (years) or whether it changes with years of experience for the years from 0 to 30 (i.e., $0 \leq x \leq 30$)? Find a 95% confidence interval for this quantity. Explain.

e   What population parameters would you examine to determine whether the salaries ever become equal in the range $0 \leq x \leq 30$? Explain.

f   The investigator will conclude that there is evidence of a *systematic* salary differential if the difference between the male and female average annual salaries exceeds $500 anywhere in the range $0 \leq x \leq 30$.

   i   Write the population parameters needed to determine whether there is a systematic salary differential between males and females.

   ii   Do the data provide evidence of a systematic salary differential between males and females (i.e., estimate the quantities in (i))? If so, in which direction?

   iii   Compute appropriate 95% simultaneous confidence intervals to help arrive at a conclusion in (i).

6.9.7   The gas mileage Y in miles per gallon (mpg) for cars depends on X, the speed in miles per hour (mph) at which they are driven. A study was conducted to evaluate the gas mileage, at various speeds, of cars (with four cylinders) made by two leading manufacturers, 1 and 2. The data are given in Table 6.9.7 and are also stored in the file **mpg.dat** on the data disk.

   We suppose that assumptions (A) are valid for both populations under consideration. The data were obtained by sampling with preselected X values. The regression functions are

$$\mu_Y^{(1)}(x) = \alpha_1 + \beta_1 x$$
$$\mu_Y^{(2)}(x) = \alpha_2 + \beta_2 x$$

where x is in the interval $25 \leq x \leq 65$. A MINITAB output is given in Exhibit 6.9.7 to help you answer questions of interest.

   List the population parameters that must be estimated to answer each of questions (a), (b), and (c).

a   What is the difference between the average gas mileages for cars made by manufacturer 1 and manufacturer 2 when driven at the speed of 25 miles per hour?

b   What is the difference between the average gas mileages for cars made by manufacturer 1 and manufacturer 2 when driven at the speed of 60 miles per hour?

T A B L E **6.9.7**

Miles per Gallon Data

Observation Number	Mpg Y	Mph X	Manufacturer
1	30.4	30	1
2	28.9	35	1
3	28.6	40	1
4	29.2	45	1
5	27.1	50	1
6	26.8	55	1
7	26.6	60	1
8	25.1	65	1
9	28.2	30	2
10	28.0	35	2
11	27.2	40	2
12	26.9	45	2
13	27.6	50	2
14	27.0	55	2
15	26.1	60	2
16	26.5	65	2

E X H I B I T **6.9.7**

MINITAB Output for Exercise 6.9.7

REGRESSION ANALYSIS FOR CARS OF MANUFACTURER (1)

```
The regression equation is
mile/gal = 34.2 - 0.134 mile/hr


Predictor         Coef        Stdev      t-ratio       p
Constant       34.1821       0.8804       38.83     0.000
mile/hr       -0.13357       0.01802       -7.41     0.000


s = 0.5839      R-sq = 90.2%      R-sq(adj) = 88.5%


 The C matrix is


        2.27380943         -0.04523810
       -0.04523810          0.00095238


y mean = 27.8375
x mean = 47.5000
SSY    = 20.7787
SSX    = 1050.00
SXY    = -140.250
```

EXHIBIT 6.9.7

(Continued)

REGRESSION ANALYSIS FOR CARS OF MANUFACTURER (1)

```
The regression equation is
mile/gal = 29.6 - 0.507 mile/hr
```

Predictor	Coef	Stdev	t-ratio	p
Constant	29.5964	0.5931	49.90	0.000
mile/hr	-0.05071	0.01214	-4.18	0.006

s = 0.3933   R-sq = 74.4%   R-sq(adj) = 70.2%

The C matrix is:

```
    2.27380943        -0.04523810
   -0.04523810         0.00095238
```

```
y mean = 27.1875
x mean = 47.5000
SSY    = 3.62875
SSX    = 1050.00
SXY    =-53.2500
```

c   What is the difference between the slopes of the two population regression lines?

d   Obtain point estimates for the quantities of interest in parts (a), (b), and (c).

e   Obtain 90% two-sided confidence intervals for the quantities of interest in parts (a), (b), and (c). You must decide whether one-at-a-time intervals or simultaneous intervals are appropriate.

f   Estimate where the two lines intersect; i.e., compute the point estimate of x_0 such that $\mu_Y^{(1)}(x_0) = \mu_Y^{(2)}(x_0)$. Also compute a 90% two-sided confidence region for x_0.

g   Write a short report discussing which population of cars, those made by manufacturer 1 or 2, gives better gas mileage on the average, and by how much, for highway driving (i.e., for driving at speeds between 50 and 65 miles per hour).

6.9.8   A study was conducted to understand the relationship between the nickel-to-iron ratio Y in oat plants and X, their age in days after emergence. From past experience it is known that a polynomial of degree less than or equal to 3 will give an adequate fit. Assumptions (A) are presumed to be valid, and the data are obtained by sampling with preselected X values. The population regression function is

$$\mu_Y(x) = \beta_0 + \beta_1 x + \beta_2 x^2 + \beta_3 x^3$$

where some of the β_i's may be negligible. The data are given in Table 6.9.8 and are also stored in the file **nickel.dat** on the data disk.

T A B L E   **6.9.8**

Nickel-to-Iron Ratio Data

Observation Number	Nickel-to-Iron Ratio Y	Age (days) X
1	0.08	0
2	0.71	5
3	0.69	10
4	0.96	15
5	1.02	20
6	1.13	25
7	1.16	30
8	1.16	35
9	1.13	40
10	1.19	45
11	1.25	50
12	1.17	55
13	1.24	60
14	1.08	65
15	1.07	70
16	1.02	75
17	0.73	80

The SAS output in Exhibit 6.9.8 contains the regression analyses of Y on X corresponding to polynomial models of degrees 3, 2, and 1, respectively. In the computer output, $X1$ denotes X, $X2$ denotes X^2, and $X3$ denotes X^3.

a   Fit polynomial models of degrees 3, 2, 1, and 0, in this order, and evaluate how good each model is by computing the mean squared error (*MSE*) for each model. Plot each fitted regression function along with the observed data. Which model would you choose after examining the plots and each *MSE*?

b   Estimate all the parameters for the model chosen in (a).

c   Use the model chosen in (a) and exhibit the population parameter that is the average nickel-to-iron ratio of oat plants at the end of 12 days after emergence.

d   Estimate the quantity of interest in (c). Also compute a two-sided 80% confidence interval for this quantity.

e   Compute a two-sided 80% confidence interval for the nickel-to-iron ratio of a single oat plant 12 days after emergence if this plant is to be chosen at random from all plants 12 days after emergence. Use the model chosen in (a).

E X H I B I T 6.9.8
SAS Output for Exercise 6.9.8

The SAS System

REGRESSION OF Y ON X1, X2, X3

00:00 Saturday, Jan 1, 1994

Model: MODEL1
Dependent Variable: RATIO

Analysis of Variance

Source	DF	Sum of Squares	Mean Square	F Value	Prob>F
Model	3	1.23880	0.41293	37.829	0.0001
Error	13	0.14191	0.01092		
C Total	16	1.38071			

Root MSE	0.10448	R-square	0.8972
Dep Mean	0.98765	Adj R-sq	0.8735
C.V.	10.57853		

Parameter Estimates

Variable	DF	Parameter Estimate	Standard Error	T for H0: Parameter=0	Prob > |T|
INTERCEP	1	0.262497	0.08255493	3.180	0.0072
X	1	0.052611	0.00921737	5.708	0.0001
X2	1	-0.000837	0.00027267	-3.071	0.0089
X3	1	0.000003402	0.00000224	1.520	0.1523

REGRESSION OF Y ON X1, X2

Model: MODEL2
Dependent Variable: RATIO

Analysis of Variance

Source	DF	Sum of Squares	Mean Square	F Value	Prob>F
Model	2	1.21357	0.60678	50.825	0.0001
Error	14	0.16714	0.01194		
C Total	16	1.38071			

E X H I B I T 6.9.8
(Continued)

Root MSE	0.10926	R-square	0.8789
Dep Mean	0.98765	Adj R-sq	0.8617
C.V.	11.06305		

Parameter Estimates

Variable	DF	Parameter Estimate	Standard Error	T for H0: Parameter=0	Prob > |T|
INTERCEP	1	0.333942	0.07098660	4.704	0.0003
X	1	0.039938	0.00411591	9.703	0.0001
X2	1	-0.000429	0.00004964	-8.642	0.0001

REGRESSION OF Y ON X1

Model: MODEL3
Dependent Variable: RATIO

Analysis of Variance

Source	DF	Sum of Squares	Mean Square	F Value	Prob>F
Model	1	0.32189	0.32189	4.560	0.0496
Error	15	1.05881	0.07059		
C Total	16	1.38071			

Root MSE	0.26568	R-square	0.2331
Dep Mean	0.98765	Adj R-sq	0.1820
C.V.	26.90064		

Parameter Estimates

Variable	DF	Parameter Estimate	Standard Error	T for H0: Parameter=0	Prob > |T|
INTERCEP	1	0.762941	0.12338876	6.183	0.0001
AGE	1	0.005618	0.00263066	2.135	0.0496

6.9.9 The growth rate of newly hatched turkeys is known to depend on the amount of vitamin A in the diet. An experiment is conducted to evaluate this relationship. Twelve newly hatched turkeys were grouped into six pairs, and each pair was put on a diet consisting of a specified amount of vitamin A. The average weight gain Y in pounds per week (for the first three weeks) and the vitamin A dosage Z for each turkey are given in Table 6.9.9 and are also stored in the file **turkey.dat** on the data disk.

T A B L E 6.9.9
Turkey Growth Data

Observation Number	Weight Gain Y (lb/week)	Vitamin A Z (units/g of diet)
1	0.169	1.5
2	0.137	1.5
3	0.219	3.0
4	0.221	3.0
5	0.278	6.0
6	0.289	6.0
7	0.328	12.0
8	0.317	12.0
9	0.287	24.0
10	0.336	24.0
11	0.274	48.0
12	0.286	48.0

From previous investigations it is known that the regression function

$$\beta_0 + \beta_1 \log_{10} z + \beta_2 (\log_{10} z)^2$$

is appropriate for modeling the relation between weight gain and vitamin dosage. The population under study is $\{(Y, Z)\}$, but if we let $X = \log_{10}(Z)$, then the regression function becomes

$$\mu_Y(x) = \beta_0 + \beta_1 x + \beta_2 x^2 \qquad (6.9.2)$$

Thus it is convenient to work with the population $\{(Y, X)\}$ as the study population. We suppose that assumptions (A) hold for the study population $\{(Y, X)\}$ (at least approximately). A MINITAB output obtained by regressing Y on X and X^2 is shown in Exhibit 6.9.9. In the computer output $X1$ denotes X and $X2$ denotes X^2. The sum of squares for pure error = 0.001908.

a Were the data for this study obtained by simple random sampling or by sampling with preselected X values? Explain.

b Estimate $\mu_Y(x)$.

E X H I B I T 6.9.9

MINITAB Output for Exercise 6.9.9

```
The regression equation is
Y = 0.0850 + 0.379 X1 - 0.156 X2


Predictor          Coef        Stdev      t-ratio          p
Constant        0.08498      0.01585        5.36      0.000
X1              0.37935      0.03987        9.52      0.000
X2             -0.15577      0.02089       -7.46      0.000


s = 0.01636      R-sq = 94.4%      R-sq(adj) = 93.2%


Analysis of Variance


SOURCE          DF          SS          MS          F          p
Regression       2     0.040583     0.020291      75.85      0.000
Error            9     0.002408     0.000268
Total           11.    0.042990


The C matrix is


   0.93869811       -2.10459471       0.97548175
  -2.10459471        5.94146967      -3.02917099
   0.97548175       -3.02917099       1.63092554
```

c What parameter would you test equal to zero to test whether or not a straight line regression model is as good as the quadratic model in (6.9.2) for predicting Y?

d Carry out the test in (c). Find the P-value and reject NH if it is less than 0.05. What is your conclusion based on this test?

e Estimate the number x_0 for which $\mu_Y(x_0)$, the average growth rate, is a maximum.

f Calculate a 95% confidence region for x_0 in (e).

g Calculate a 95% confidence region for z_0, the dosage of vitamin A at which the average growth rate $\mu_Y(z)$ is a maximum.

6.9.10 The weight of a newborn baby increases quite rapidly during the first 100 days after birth, and after that it slows down somewhat. Data for seventeen babies were obtained from the past 3 years' records for babies born at a certain hospital. Suppose that assumptions (A) are valid where the data were obtained by simple random sampling. The regression function $\mu_Y(x)$ is given by the following spline model with a knot-point at $x = 100$.

$$\mu_Y(x) = \begin{cases} \alpha_1 + \beta_1 x & 0 < x \le 100 \\ \alpha_2 + \beta_2 x & 100 < x \le 200 \end{cases} \qquad \text{(6.9.3)}$$

The data are given in Table 6.9.10 and are also stored in the file **babywt.dat** on the data disk. A MINITAB output is given in Exhibit 6.9.10 to help answer various questions of interest.

a   Estimate $\alpha_1, \beta_1, \alpha_2, \beta_2$, and σ. Plot y_i against x_i. Exhibit the X matrix in (6.8.7).

b   In (a) exhibit the population quantities needed to determine the difference between the *average growth rate* of newborn babies during the first 100 days and the *average growth rate* during the next 100 days.

c   In (b) compute a 90% two-sided confidence interval for the quantity of interest.

d   Instead of using the preceding spline model, assume that a quadratic model given by

$$\mu_Y(x) = \beta_0 + \beta_1 x + \beta_2 x^2$$

holds for $0 \le x \le 200$. Estimate this regression function. Exhibit 6.9.11 contains a MINITAB output for a quadratic regression model along with the fits, residuals, and standardized residuals. Plot the residuals. Compare this residual plot with the residual plot in (a). Which model (a quadratic or the spline) is better for this problem? Give your reasons in a short report.

T A B L E **6.9.10**
Weights of Newborn Babies.

Observation Number	Weight Y (pounds)	Age X (days)
1	7.5	7
2	7.9	12
3	8.4	18
4	10.1	45
5	11.5	67
6	12.8	88
7	13.4	92
8	13.9	99
9	13.5	105
10	14.5	108
11	14.7	120
12	15.8	147
13	16.1	156
14	16.5	159
15	16.9	167
16	17.3	183
17	17.6	195

E X H I B I T **6.9.10**
MINITAB Output for Exercise 6.9.10

```
The regression equation is
Y = 7.08 + 0.0673 X1 + 0.0423 X2

Predictor        Coef        Stdev      t-ratio         p
Constant       7.0810       0.1356       52.21      0.000
X1            0.067321     0.001825      36.89      0.000
X2            0.042319     0.001873      22.60      0.000

s = 0.2178      R-sq = 99.6%      R-sq(adj) = 99.6%

Analysis of Variance

SOURCE        DF          SS          MS         F         p
Regression     2       175.155      87.577    1845.97   0.000
Error         14         0.664       0.047
Total         16       175.819

The C matrix is

  .38767760595    -.00453033498    .00096764899
 -.00453033498     .00007021359   -.00003688170
  .00096764899    -.00003688170    .00007392925
```

E X H I B I T **6.9.11**
MINITAB Output for (d) in Exercise 6.9.10

```
The regression equation is
Y = 6.92 + 0.0801 X1 -0.000128 X2

Predictor        Coef        Stdev      t-ratio         p
Constant       6.9199       0.1528       45.30      0.000
X1            0.080080     0.003443      23.26      0.000
X2         -0.00012802   0.00001694      -7.56      0.000

s = 0.2229      R-sq = 99.6%      R-sq(adj) = 99.5%
```

E X H I B I T 6.9.11

(Continued)

Analysis of Variance

SOURCE	DF	SS	MS	F	p
Regression	2	175.124	87.562	1763.12	0.000
Error	14	0.695	0.050		
Total	16	175.819			

ROW	Y	X1	X2	fits	residual	stdresid
1	7.5	7	49	7.4742	0.025764	0.14462
2	7.9	12	144	7.8625	0.037526	0.20115
3	8.4	18	324	8.3199	0.080087	0.41234
4	10.1	45	2025	10.2643	-0.164321	-0.78694
5	11.5	67	4489	11.7107	-0.210653	-1.00600
6	12.8	88	7744	12.9756	-0.175645	-0.84295
7	13.4	92	8464	13.2038	0.196205	0.94185
8	13.9	99	9801	13.5932	0.306801	1.47206
9	13.5	105	11025	13.9170	-0.416988	-1.99805
10	14.5	108	11664	14.0754	0.424573	2.03240
11	14.7	120	14400	14.6861	0.013862	0.06599
12	15.8	147	21609	15.9254	-0.125438	-0.59224
13	16.1	156	24336	16.2971	-0.197060	-0.93540
14	16.5	159	25281	16.4163	0.083673	0.39861
15	16.9	167	27889	16.7231	0.176897	0.85592
16	17.3	183	33489	17.2875	0.012501	0.06499
17	17.6	195	38025	17.6678	-0.067781	-0.39812

Applications of Regression II

7.1
Overview

In Chapter 6 we discussed several practical applications of inference procedures, developed in Chapters 3 and 4, for straight line regression and multiple linear regression. In this chapter we consider two further important applications of regression: *subset analysis* and *growth curves*. Section 7.2 gives a brief introduction to subset analysis, which is often called *selection of variables*. In Sections 7.3 and 7.4 we discuss various methods of performing a subset analysis so that you can become acquainted with the terminology and some of the commonly used procedures. We do not discuss some of the complex theoretical and conceptual issues that underly model building. Section 7.5 introduces growth curve models and presents appropriate inference procedures associated with them.

7.2
Subset Analysis and Variable Selection

To motivate the material in this section, we consider Example 4.4.2. In that example we were interested in predicting the GPA (Y) of a student at the end of the freshman year based on the values of SATmath (X_1), SATverbal (X_2), HSmath(X_3), and HSenglish (X_4). Suppose that the regression function of Y on X_1, X_2, X_3, and X_4 is given by

$$\mu_Y(x_1, x_2, x_3, x_4) = \beta_0 + \beta_1 x_1 + \beta_2 x_2 + \beta_3 x_3 + \beta_4 x_4 \qquad (7.2.1)$$

and that the predictions based on (7.2.1) are sufficiently accurate for the needs of the admissions director. It may happen that other prediction functions, based on all or only some of the variables X_1, X_2, X_3, and X_4, may also be sufficiently accurate for the problem at hand. If a prediction function such as

$$P_Y(x_3, x_4) = \beta_0 + \beta_3 x_3 + \beta_4 x_4 \qquad (7.2.2)$$

based only on HSmath and HSenglish, is as good (or nearly as good) as the regression function in (7.2.1), which is based on all four predictors, then the director of admissions can base decisions solely on the high school grades of the applicants, who can then be exempted from having to take the SAT. Keep in mind the possibility that there may be other predictors, such as the ACT scores, which, together with the predictors in (7.2.1), might improve the predictions substantially. Here we are concerned only with the four predictors SATmath, SATverbal, HSmath, and HSenglish. The prediction function given in (7.2.2), which may not be the *best* function for predicting Y using X_3 and X_4, (i.e., it may not be the *regression* function of Y on X_3 and X_4), is called a **subset model** because it is obtained by using a *subset* of the predictors in (7.2.1). Thus the director of admissions may be interested in examining the performance of various prediction functions based on subsets of $\{X_1, X_2, X_3, X_4\}$ and eventually selecting one or more prediction functions that may be satisfactory for the problem. This process of examining subset models and selecting one or more suitable prediction functions is often called **subset selection**, **subset analysis**, or **selection of variables**.

The general problem of subset selection can be described as follows. We know from Chapter 4 that when the $(k + 1)$-variable population $\{(Y, X_1, \ldots, X_k)\}$ satisfies population assumptions (A) or (B), the best function for predicting Y using $X_1, \ldots, X_k$ as predictors, is the regression function of Y on $X_1, \ldots, X_k$; i.e.,

$$\mu_Y(x_1, \ldots, x_k) = \beta_0 + \beta_1 x_1 + \cdots + \beta_k x_k \tag{7.2.3}$$

If the investigator decides that σ is sufficiently small for the problem under study (where σ denotes $\sigma_{Y|X_1, \ldots, X_k}$), then the regression function in (7.2.3) will in fact be an adequate prediction function for this problem. However, investigators are often interested in examining prediction functions based on various *subsets* of the full set $\{X_1, \ldots, X_k\}$ for a variety of reasons. Two of the reasons are:

■   To get some insight into how various subsets of predictors contribute to the prediction of Y.

■   To find adequate prediction functions based on subsets of the predictor variables that are easier or less expensive to measure than the full set $\{X_1, \ldots, X_k\}$.

Suppose $\{(Y, X_1, \ldots, X_k)\}$ satisfies population assumptions (B). Then the regression function of Y on $X_1, \ldots, X_k$ is the one given in (7.2.3); i.e., it is a multiple *linear* regression function. In this case the regression function of Y on any subset $\{X_1, \ldots, X_m\}$ of m predictors $(0 < m < k)$ is also a multiple *linear* regression function of the form

$$\mu_Y^{(A)}(x_1, \ldots, x_m) = \beta_0^A + \beta_1^A x_1 + \cdots + \beta_m^A x_m \tag{7.2.4}$$

However, if $\{(Y, X_1, \ldots, X_k)\}$ satisfies population assumptions (A) but fails to satisfy population assumptions (B), then although the regression function of Y on the k predictors $X_1, \ldots, X_k$ is still of the form given in (7.2.3), the regression function of Y on the m predictors $X_1, \ldots, X_m$ $(m < k)$ is *not* necessarily of the form given in (7.2.4). Nevertheless, it is useful to consider **linear prediction functions** of the form

$$\beta_0^A + \beta_1^A x_1 + \cdots + \beta_m^A x_m \tag{7.2.5}$$

based on various subsets of predictors. We can relabel the predictor variables suitably so that the subset under consideration can always be written as the first m predictor variables $\{X_1, \ldots, X_m\}$.

In Sections 7.3 and 7.4 we discuss exploratory statistical procedures for finding linear prediction functions based on subsets of $\{X_1, \ldots, X_k\}$ that compare favorably with the full model regression function given in (7.2.3). The investigator may eventually choose a subset model (which may be the full model), or the investigator may choose several subset models, based on considerations such as cost of obtaining the values of the predictors, simplicity and interpretability of the prediction function, adequacy of the predictions, etc.

Commonly Used Procedures for Subset Analysis

Subset analysis generally requires a considerable amount of computing, and even with the availability of high-speed computers, the amount of computing must sometimes be taken into consideration in deciding which of the several available methods to use. A large amount of material has been written on this subject during the past three decades, but there are very few rigorous results of an inferential nature; most of the procedures are merely descriptive or exploratory. Four general procedures that are commonly used in subset analysis are:

1 All-subsets regression

2 Forward selection procedure

3 Backward elimination procedure

4 Stepwise regression procedure

The *all-subsets regression* procedure is discussed in Section 7.3, and the remaining three procedures are discussed in Section 7.4.

Problems 7.2

7.2.1 Give two examples from your field of specialization where the investigator might be interested in a subset analysis. In each case, explain clearly the purpose that would lead to the consideration of the subset selection problem.

7.2.2 Suppose, in a research problem, the form of the regression function $\mu_Y(x_1, x_2)$ of Y on X_1, X_2 is not known. However, the investigator thinks that the linear prediction function $P_Y(x_1, x_2) = \beta_0^A + \beta_1^A x_1 + \beta_2^A x_2$ is an adequate approximation to the regression function $\mu_Y(x_1, x_2)$, so $P_Y(x_1, x_2)$ is used in place of $\mu_Y(x_1, x_2)$. Explain in detail what this means.

7.3
All-Subsets Regression

The notation for subset analysis can get extremely cumbersome since there are many models to examine and each model may contain several β coefficients. For example, consider the two models

$$(1) \quad \beta_0 + \beta_1 x_1 + \beta_2 x_2$$

and

$$(2) \quad \beta_0 + \beta_1 x_1 + \beta_3 x_3 \tag{7.3.1}$$

When studied together these should be written as

$$(3) \quad \beta_0^{(1)} + \beta_1^{(1)} x_1 + \beta_2^{(1)} x_2$$

and

$$(4) \quad \beta_0^{(2)} + \beta_1^{(2)} x_1 + \beta_3^{(2)} x_3 \tag{7.3.2}$$

to indicate that, in general, the β_i in (1) are not the same as the β_i in (2), etc. However, because there are many models to consider in subset analysis, and to avoid complicated notation, we simply write models (3) and (4) as in (7.3.1). *You should remember that the β_i in different models such as (1) and (2) are generally different.*

We begin with the premise that the investigator is interested in examining subsets of the predictors $X_1, \ldots, X_k$, to determine how good each subset is for predicting the response variable Y. Some of the predictor variables may be *derived quantities* based on other predictor variables. For example, X_3 may stand for X_1^2 and X_4 may stand for $X_1 X_2$, etc. We assume that the best prediction function (i.e., the regression function) using all of $X_1, \ldots, X_k$ is of the form

$$\beta_0 + \beta_1 x_1 + \cdots + \beta_k x_k \tag{7.3.3}$$

Our objective is to determine how well Y may be predicted by each of the subset prediction functions of the form

$$\beta_0 + \beta_1 x_1 + \cdots + \beta_m x_m \tag{7.3.4}$$

where $\{X_1, \ldots, X_m\}$ is a subset of $\{X_1, \ldots, X_k\}$. Although we have used the first m predictors $X_1, X_2, \ldots, X_m$ for notational simplicity, we could use any m predictors, not just the first m. Based on suitable criteria, the investigator will select one or more of these subset prediction functions for further consideration. The subset prediction functions considered are all linear in the predictor variables X_i as well as the parameters. However, since a predictor variable X_i may itself be a quantity computed from other predictor variables—for instance, X_3 might be X_1^2, etc.—this is not a serious limitation. As we pointed out earlier, you should bear in mind that the prediction function in (7.3.4) may not be the regression function of Y on $X_1, \ldots, X_m$; i.e., it may not be the *best* function for predicting Y using $X_1, \ldots, X_m$, but it may be an *adequate* prediction function for the problem.

When there are k predictor variables in all, the number of possible subset prediction functions of the form (7.3.4) is 2^k, which can be quite large. For instance, if $k = 10$, then the number of possible subset models is $2^{10} = 1,024$ (including the constant model that uses none of the k predictors). When $k = 20$, the number of subset models to be considered is $2^{20} = 1,048,576$. Clearly this presents a non-trivial computational problem. Fortunately, efficient computing methods exist, and many statistical software packages have the capability of performing the *all-subsets* regression analysis if k is not too large. The particular computer being used to perform the calculations will determine the feasibility of the computations for different values of k.

In principle the all-subsets regression analysis proceeds as follows:

1   Each of the 2^k subset models is fitted to the sample data by the method of least squares using the formulas in Chapter 4.

2   For each subset model, a *criterion measure* is calculated which summarizes how good that model is for predicting Y.

3   A table or a graph that exhibits the predictors in each model, along with the *criterion measure* calculated for that model, is constructed.

4   This table or graph is examined by the investigator, and one or more models are selected based on many considerations such as the adequacy of the predictions based on the model, costs of collecting data, subject matter knowledge regarding the reasonableness of the various models, etc. These models comprise the *short list*.

5   The models on this short list may be subjected to further scrutiny, which may include various residual analyses, the diagnostic procedures of Chapter 5, or validation based on additional sample data, etc. The investigator may finally select one or more of these models as a *tentative* model that is satisfactory for the problem under study. In some cases the investigator may not be interested in actually using any of these models for the purposes of prediction. Instead he or she may formulate some hypothesis or theory, or modify an existing hypothesis or theory, as a result of the examination of the subset models.

Steps (4) and (5) typically require the participation of both the investigator and the statistician (in some cases the same individual will play both roles) and involve subjective judgments. *No rigorous measure of confidence can be attached to the final conclusions at this stage, but a new study can be designed to validate or invalidate the conclusions arrived at as a result of the subset analysis.* Subset analysis is necessarily an iterative procedure, and only after considerable effort has been expended can an investigator reach the stage of being able to make valid probability or confidence statements regarding the conclusions. Thus, in many applications, subset analysis may be viewed as "data in search of a model."

Many different candidates have been proposed for the *criterion measure* mentioned in step (2) above. We discuss the most popular among these, which are: Root mean square of residuals, R-square, adjusted R-square, and Mallow's C_p criterion. Each one of these criteria may be given theoretical justification depending on the objective of the subset analysis and population and sample assumptions, but we do not discuss the justifications here. If you are interested you should consult [6], and [13]. Because subset analysis is merely an empirical or descriptive analysis of all the subsets and involves no strict statistical inferences, any one of these criteria generally leads to essentially the same collection of subset models for further scrutiny.

Root Mean Squared Error (s)

To use this criterion we must calculate the square root of the mean squared error for each subset model. As before, consider a subset of *any* m predictors chosen from $X_1, \ldots, X_k$. For concreteness these m predictors will be denoted by $X_1, \ldots, X_m$ (after relabeling the predictor factors if necessary). For the model based on the subset of m predictors $X_1, \ldots, X_m$, the linear prediction function

$$\beta_0 + \beta_1 x_1 + \cdots + \beta_m x_m \tag{7.3.5}$$

is fitted to the sample data using the least squares method of Chapter 4. The fitted model may be written as

$$\hat{\beta}_0 + \hat{\beta}_1 x_1 + \cdots + \hat{\beta}_m x_m \tag{7.3.6}$$

The mean squared error $MSE(X_1, \ldots, X_m)$ corresponding to this model, which we simply write as MSE when there is no possibility of confusion, can then be calculated using the formula in (4.4.15), which gives

$$MSE(X_1, \ldots, X_m) = MSE = \frac{\sum_{i=1}^{n}[y_i - (\hat{\beta}_0 + \hat{\beta}_1 x_{i,1} + \cdots + \hat{\beta}_m x_{i,m})]^2}{(n - p)} \tag{7.3.7}$$

where p is the number of β parameters in the subset model (which is $m + 1$ for the model in (7.3.5)) and n is the number of observations in the sample. The root mean square of the residuals for the subset model in (7.3.5), denoted by $s(X_1, \ldots, X_m)$ or simply as s when the subset model under consideration is unambiguous, is then

$$s(X_1, \ldots, X_m) = s = \sqrt{MSE(X_1, \ldots, X_m)} \tag{7.3.8}$$

Models that have small values of s are selected for inclusion in the *short list*.

In the special case when the $(m + 1)$-variable population $\{(Y, X_1, \ldots, X_m)\}$ satisfies assumptions (A) or assumptions (B), the quantity s is an estimate of the subpopulation standard deviation $\sigma_{Y|X_1, \ldots, X_m}$; otherwise this may not be true. Throughout this book we have used subpopulation standard deviations as a measure of how good a regression function is for predicting Y, and in general, regression functions with smaller standard deviations are better than those with larger standard deviations. This is why s (which can be viewed as an estimate of σ when assumptions (A) or (B) are satisfied) is sometimes used to distinguish between prediction functions.

R-Square (R^2)

To use this criterion, we calculate the quantity called R-square (written R^2) for each subset model. For the model based on the subset of m predictors $X_1, \ldots, X_m$, the linear prediction function in (7.3.5) is fitted to the sample data using the method of least squares. The fitted model may be written as in (7.3.6). The sum of squared errors $SSE(X_1, \ldots, X_m)$ for this subset model and the (corrected) total sum of squares of Y, denoted as usual by SSY, are calculated using the formulas

$$SSE(X_1, \ldots, X_m) = SSE = \sum_{i=1}^{n} [y_i - (\hat{\beta}_0 + \hat{\beta}_1 x_{i,1} + \cdots + \hat{\beta}_m x_{i,m})]^2 \qquad \textbf{(7.3.9)}$$

and

$$SSY = \sum_{i=1}^{n} (y_i - \bar{y})^2 \qquad \textbf{(7.3.10)}$$

respectively. Then the value of R^2 corresponding to the subset of predictors $X_1, \ldots, X_m$ is calculated using

$$R^2(X_1, \ldots, X_m) = R^2 = \frac{SSY - SSE(X_1, \ldots, X_m)}{SSY} \qquad \textbf{(7.3.11)}$$

Models with large values of R^2 (equivalently, models with small values of SSE) are included in the short list. No subset model will have an R^2 larger than that of the full model, which includes all k predictor variables, but there may exist subset models with R^2 values that are nearly equal to that of the full model.

Note that when assumptions (B) hold (data must be obtained by simple random sampling), R^2 is in fact used to estimate the multiple coefficient of determination $\rho^2_{Y(X_1,\ldots,X_m)}$ of Y with the predictors $X_1, \ldots, X_m$. Otherwise R^2 may not be a valid estimate of $\rho^2_{Y(X_1,\ldots,X_m)}$. Throughout this book we have stated that $\rho^2_{Y(X_1,\ldots,X_m)}$ should not be used as a measure of how good $X_1, \ldots, X_m$ are for predicting Y. However, to compare one model (say, model A) with another (say, model B), we can compare ρ^2_A with ρ^2_B or we can compare σ_A with σ_B since ρ^2_A is less than ρ^2_B if and only if σ_A is greater than σ_B. This fact follows from the definition of $\rho^2_{Y(X_1,\ldots,X_k)}$ given in (4.9.8).

Adjusted R-Square (Adj-R^2)

To use this criterion we must calculate a quantity called adjusted R-square (written as adj-R^2) for each subset model. For the model based on the subset of predictors $X_1, \ldots, X_m$, the linear model in (7.3.5) is fitted to the sample data using least squares. The fitted model is in (7.3.6). The mean squared error MSE for this subset model is computed as in (7.3.7). Then the adjusted R-square for this model, denoted by adj-$R^2(X_1, \ldots, X_m)$, or simply by adj-R^2 when the subset model under consideration is unambiguous, is calculated using

$$adj\text{-}R^2(X_1, \ldots, X_m) = adj\text{-}R^2 = \frac{MSY - MSE(X_1, \ldots, X_m)}{MSY} \qquad \textbf{(7.3.12)}$$

where $MSY = SSY/(n-1)$. Subset models with large values of adj-R^2 (equivalently, models with small values of MSE) are included in the short list.

Note that when assumptions (B) are satisfied, *adj-R^2* is in fact used as an alternative to R^2 to estimate the multiple coefficient of determination $\rho_{Y(X_1,\ldots,X_m)}^2$ of Y with the predictors $X_1,\ldots,X_m$; otherwise this may not be a valid estimate of $\rho_{Y(X_1,\ldots,X_m)}^2$.

Note The quantities *SSE*, *MSE*, and *SSY* can be obtained from an ANOVA table by regressing Y on $X_1,\ldots,X_m$. Also, you should observe that whenever one model is better than a second model according to the adjusted R-square criterion, then it is also better than the second model according to the root mean squared error (s) criterion, and vice versa. Thus the adjusted R-square criterion is equivalent to the root mean squared error (s) criterion in selecting models, but the R-square criterion is not equivalent to the s criterion or the adjusted R-square criterion. If n is very large relative to m, then all three are essentially equivalent.

Mallow's C_p Criterion

A measure that is quite widely used in subset selection analysis is the C_p criterion measure, originally proposed by C. Mallows [21]. Loosely speaking, this quantity is an estimate of the average squared prediction error relative to the estimate of $\sigma_{Y|X_1,\ldots,X_k}$ if the fitted model

$$\hat{\beta}_0 + \hat{\beta}_1 x_1 + \cdots + \hat{\beta}_m x_m \qquad \text{(7.3.13)}$$

is used to estimate $\mu_Y(X_1,\ldots,X_k)$ for values of $X_1,\ldots,X_k$ in the sample. The subscript p in C_p refers to the number of β parameters in the subset model, which is $m+1$ in (7.3.13). The measure C_p is calculated for each of the 2^k possible subset models. For the model based on the predictors $X_1,\ldots,X_m$, the linear model in (7.3.5) is fitted to the sample data using least squares. The fitted model is given in (7.3.6). The sum of squared errors $SSE(X_1,\ldots,X_m)$ for this subset model, denoted simply by *SSE*, is calculated as in (7.3.9). Then the measure C_p corresponding to this subset is denoted by $C_p(X_1,\ldots,X_m)$, or simply by C_p when the set of predictors $X_1,\ldots,X_m$ does not need to be identified explicitly, and is calculated using the formula

$$C_p = C_p(X_1,\ldots,X_m) = \frac{SSE(X_1,\ldots,X_m)}{\hat{\sigma}_{Y|X_1,\ldots,X_k}^2} + 2p - n \qquad \text{(7.3.14)}$$

where $\hat{\sigma}_{Y|X_1,\ldots,X_k}^2$ is the estimated variance of the subpopulation of Y values determined by using all of the predictors $X_1,\ldots,X_k$ and, as before, p is the number of β parameters in the subset model under consideration, and n is the sample size. For the model in (7.3.5) the value of p is $m+1$. Models with small values of C_p are included in the short list.

When the population $\{(Y,X_1,\ldots,X_k)\}$ satisfies assumptions (A), it can be shown that the quantity $C_p - p$ is a measure of the tendency of the fitted model

$$\hat{\beta}_0 + \hat{\beta}_1 x_1 + \cdots + \hat{\beta}_m x_m \qquad \text{(7.3.15)}$$

corresponding to the subset of predictors $X_1, \ldots, X_m$, to underestimate or overestimate the subpopulation mean $\mu_Y(x_1, \ldots, x_k)$, which is used to predict Y when all k predictor factors are used. Negative values of $C_p - p$ should be regarded as zero because it can be shown that $C_p - p$ is an estimate of a population quantity that can never be negative. This tendency for underestimation or overestimation is referred to as bias. Thus $C_p - p$ is a measure of bias. Some authors recommend selection of models for which both C_p and $C_p - p$ are small.

C_p-Plot

As discussed above, in selecting subset models for further consideration we may want to examine the values of C_p as well as $C_p - p$. This is conveniently carried out by plotting the values of C_p for the different models and superimposing the straight line $C_p = p$ on this plot. The values of $C_p - p$ can be judged visually by examining how close the various points are to the line $C_p = p$. Good subset models should have small values for C_p and also should have $C_p - p$ close to zero (as stated earlier, negative values of $C_p - p$ are interpreted as zero).

Formulas for s, adj-R^2, and C_p in Terms of R^2

The quantities s, R^2, adj-R^2, and C_p, for a subset model with p parameters β are interrelated, and the relationship of R^2 to the various measures is

$$adj\text{-}R^2(X_1, \ldots, X_m) = 1 - \frac{n-1}{n-p}[1 - R^2(X_1, \ldots, X_m)] \qquad \textbf{(7.3.16)}$$

$$s(X_1, \ldots, X_m) = \sqrt{\frac{SSY[1 - R^2(X_1, \ldots, X_m)]}{n-p}} \qquad \textbf{(7.3.17)}$$

$$C_p(X_1, \ldots, X_m) = \frac{(n-k-1)[1 - R^2(X_1, \ldots, X_m)]}{[1 - R^2(X_1, \ldots, X_k)]} + 2p - n \quad \textbf{(7.3.18)}$$

Recall that p is the number of β parameters in the subset model under consideration, and it equals $m + 1$ for the model with m predictor variables and an intercept term (see (7.3.5)). Also, in (7.3.18), $R^2(X_1, \ldots, X_k)$ is the value of R^2 corresponding to the model consisting of all k predictors $X_1, \ldots, X_k$; i.e.,

$$R^2(X_1, \ldots, X_k) = \frac{SSY - SSE(X_1, \ldots, X_k)}{SSY}$$

When comparing two or more subset models with the same number of predictor variables (i.e., the same value for p, the number of beta coefficients in the model), the four criteria—s, R^2, adj-R^2, C_p—will lead to the same ordering of the subset models. However, when comparing two or more subset models with *different* numbers of predictor variables (different values of p), these four criteria, in general, will not lead to the same ordering of the subset models (except, of course, for the s and the *adj-R²* criteria, which will always lead to the same ordering of all subset models).

We illustrate the use of each of the preceding measures using the GPA data of Example 4.4.2.

E X A M P L E   7.3.1

To illustrate the results of this section, we carry out an all-subsets regression analysis for the GPA data of Example 4.4.2. Recall that Y = GPA, X_1 = SATmath, X_2 = SATverbal, X_3 = HSmath, and X_4 = HSenglish.

Table 7.3.1 displays the values of s, R^2, *adj-R²*, C_p, and $C_p - p$ for each subset of the predictors $\{X_1, X_2, X_3, X_4\}$, and Figure 7.3.1 shows the C_p plot.

From Table 7.3.1 we see that if we use R^2 as the criterion measure, the best one-variable model uses X_1, which is SATmath; the second-best one-variable model uses X_3, which is HSmath; the best two-variable model uses X_1 and X_2—i.e., SATmath and SATverbal; the second-best two-variable model uses X_1 and X_3—i.e., SATmath and HSmath, etc.

The model with the smallest value of C_p is model (12) in Table 7.3.1, which uses X_1, X_2, and X_3. By the C_p criterion this model would be chosen as the best model. The value of $C_p - p$ computed for this model is -0.8, but recall that $C_p - p$ is estimating a population quantity that can never be negative, so we set $C_p - p = 0$

T A B L E   7.3.1
s, R^2, *adj-R²*, C_p, and $C_p - p$ for all 16 Subset Models for the GPA Data

Model	Predictors in the Model	p	s	R^2	*adj-R²*	C_p	$C_p - p$
(1)	None	1	0.6218	not defined	not defined	83.9	82.9
(2)	X_1	2	0.3370	72.2	70.6	12.4	10.4
(3)	X_2	2	0.4837	42.7	39.5	42.4	40.4
(4)	X_3	2	0.4595	48.3	45.4	36.7	34.7
(5)	X_4	2	0.5079	36.8	33.3	48.4	46.4
(6)	X_1, X_2	3	0.2858	81.1	78.9	5.3	2.3
(7)	X_1, X_3	3	0.2998	79.2	76.7	7.2	4.2
(8)	X_1, X_4	3	0.3447	72.5	69.3	14.0	11.0
(9)	X_2, X_3	3	0.3971	63.5	59.2	23.2	20.2
(10)	X_2, X_4	3	0.4351	56.2	51.0	30.6	27.6
(11)	X_3, X_4	3	0.3771	67.1	63.2	19.5	16.5
(12)	X_1, X_2, X_3	4	0.2621	85.0	82.2	3.2	-0.8
(13)	X_1, X_2, X_4	4	0.2945	81.1	77.6	7.3	3.3
(14)	X_1, X_3, X_4	4	0.3014	80.2	76.5	8.2	4.2
(15)	X_2, X_3, X_4	4	0.3477	73.7	68.7	14.8	10.8
(16)	X_1, X_2, X_3, X_4	5	0.2685	85.3	81.4	5.0	0.0

FIGURE 7.3.1

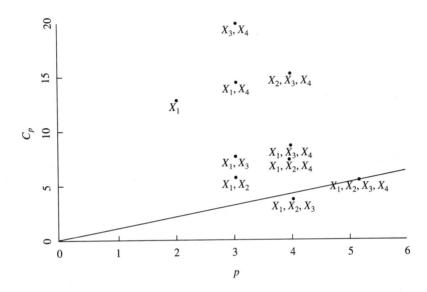

for this model. The observed negative value of $C_p - p$ actually suggests that the bias for model (12) is close to zero if not equal to zero. This model also happens to be the model with the smallest s as well as the largest adj-R^2. Model (16) has the largest value of R^2, but as we mentioned earlier, no subset model can have a larger R^2. Based on Table 7.3.1 and Figure 7.3.1, the director of admissions may decide to restrict attention to models (6), (7), (12), (13), (14), and (16). ∎

The entries in Table 7.3.1 can be calculated using formulas (7.3.8), (7.3.11), (7.3.12), and (7.3.14). However, we typically use a suitable statistical package—e.g., SAS, SPSS, BMDP, SPlus, MINITAB, etc.—for obtaining these quantities. In Section 7.3 of the laboratory manuals we demonstrate how to use the computer to obtain the quantities referred to above.

As stated earlier, computations for all-subsets regressions quickly become infeasible as the number of predictor variables increases. When it becomes infeasible or expensive to carry out an all-subsets regression analysis, we can resort to less expensive (but less desirable) methods. We discuss some such methods in the next section.

Problems 7.3

7.3.1 An investigator is interested in understanding the relationship, if any, between the analytical skills of young gifted children and the following variables:

X_1 = father's IQ

X_2 = mother's IQ

X_3 = age in months when the child first said "mommy" or "daddy"

X_4 = age in months when the child first counted to 10 successfully

X_5 = average number of hours per week the child's mother or father read to the child

X_6 = average number of hours per week the child watched an educational program on TV during the past 3 months

X_7 = average number of hours per week the child watched cartoons on TV during the past 3 months

The analytical skills are evaluated using a standard testing procedure, and the score Y on this test is used as the response variable. The model to be examined using all seven predictor variables is

$$\mu_Y(x_1, \ldots, x_7) = \beta_0 + \beta_1 x_1 + \cdots + \beta_7 x_7$$

Data were collected from schools in a large city on a set of thirty-six children who were identified as gifted children soon after they reached the age of four. These data are given in Table 7.3.2, and they are also stored in the file **gifted.dat** on the data disk.

T A B L E 7.3.2
Gifted Children Data

Child	Y	X_1	X_2	X_3	X_4	X_5	X_6	X_7
1	159	115	117	18	26	1.9	3.00	2.00
2	164	117	113	20	37	2.5	1.75	3.25
3	154	115	118	20	32	2.2	2.75	2.50
4	157	113	131	12	24	1.7	2.75	2.25
5	156	110	109	17	34	2.2	2.25	2.50
6	150	113	109	13	28	1.9	1.25	3.75
7	155	118	119	19	24	1.8	2.00	3.00
8	161	117	120	18	32	2.3	2.25	2.50
9	163	111	128	22	28	2.1	1.00	4.00
10	162	122	120	18	27	2.1	2.25	2.75
11	154	111	117	19	32	2.2	1.75	3.75
12	159	112	120	20	33	2.3	2.00	2.75
13	167	119	126	20	35	2.2	0.75	4.00
14	155	120	114	22	21	1.7	2.50	2.50
15	159	114	129	17	27	1.8	1.50	3.75
16	159	111	118	18	29	2.0	1.75	3.25
17	160	111	115	21	32	2.3	1.75	3.25
18	154	115	111	18	32	2.2	2.00	3.00

(Continued)

T A B L E 7.3.2
(Continued)

Child	Y	X_1	X_2	X_3	X_4	X_5	X_6	X_7
19	160	126	111	12	35	2.2	2.75	1.75
20	151	115	109	21	29	2.0	2.00	2.75
21	166	114	124	15	39	2.4	2.00	2.75
22	161	115	122	20	30	2.1	2.00	3.50
23	162	115	118	15	33	2.3	2.25	2.75
24	169	112	121	23	36	2.3	2.00	2.75
25	160	115	124	18	35	2.3	2.00	3.50
26	161	117	118	20	31	2.3	2.00	3.00
27	166	116	128	17	36	2.4	1.25	3.75
28	163	114	119	22	36	2.4	2.25	2.75
29	159	116	123	15	24	1.8	1.50	3.50
30	155	111	117	13	29	2.1	1.75	3.50
31	155	112	117	17	25	2.0	2.75	2.25
32	157	115	111	10	31	2.2	1.75	3.50
33	151	111	101	17	29	2.1	3.00	2.25
34	162	119	113	23	28	2.1	1.25	3.50
35	164	111	121	18	36	2.3	1.00	4.50
36	159	114	123	20	30	2.2	1.75	3.25

These data are considered to be a simple random sample from a study population of gifted children. In Exhibit 7.3.1 we give a MINITAB computer output for this problem containing the results from a subset analysis. Only the values of s, R^2, adj-R^2, and C_p are given for the best five subset models for each value of m, where m is the number of predictors in the subset model. This computer output is interpreted as follows.

When considering subset models with a *single* predictor, the best single variable is X_2 (using any of the four criterion functions s, R^2, adj-R^2, and C_p). Look across the line that contains the first 1 in the column labeled Vars (variables), where the 1 stands for the model with a single variable, until you come to the symbol X. Notice that X is below X2, so X_2 is the best single variable for predicting Y. The values of R^2, adj-R^2, C_p, and s for the subset model containing X_2 only are

$$R^2 = 32.6 \qquad adj\text{-}R^2 = 30.7 \qquad C_p = 43.3 \qquad s = 3.8557$$

respectively. The second best variable for predicting Y when there is only one predictor is X_4. Look across the line that contains the second 1 under the column labeled Vars until you come to the symbol X. This X is under X4, so X_4 is the second best variable, among all single variable subset models, for predicting Y. The third best single variable is X_5, etc.

For the best subset models that contain two predictors, look under the column labeled Vars until you come to the first 2 (the 2 stands for models with two predictor variables). Look across that line until you come to two X's. These X's are under X2 and X5, so X_2 and X_5 are the best two predictors in subset models that contain

E X H I B I T 7.3.1

MINITAB Output for All-Subsets Regression for Gifted Children Data

```
Best Subsets Regression of Y
```

Vars	R-sq	Adj. R-sq	C-p	s	1	2	3	4	5	6	7
1	32.6	30.7	43.3	3.8557	X						
1	29.6	27.5	46.7	3.9411			X				
1	27.6	25.5	49.0	3.9976				X			
1	13.7	11.2	64.5	4.3638						X	
1	7.2	4.4	71.8	4.5259		X					
2	62.9	60.7	11.5	2.9038	X			X			
2	60.8	58.4	13.9	2.9864	X	X					
2	37.8	34.0	39.5	3.7607	X	X					
2	36.7	32.9	40.7	3.7930	X	X					
2	36.3	32.5	41.2	3.8045	X					X	
3	68.7	65.8	7.0	2.7080	X	X		X			
3	66.7	63.5	9.3	2.7961	X	X	X				
3	64.6	61.3	11.5	2.8790		X	X	X			
3	64.5	61.2	11.7	2.8855	X	X		X			
3	63.7	60.3	12.6	2.9174	X			X	X		
4	70.8	67.0	6.7	2.6598	X	X	X	X			
4	70.7	66.9	6.8	2.6636	X				X	X	X
4	70.4	66.6	7.1	2.6765	X	X	X		X		
4	70.0	66.2	7.5	2.6933	X	X		X	X		
4	69.3	65.3	8.3	2.7257	X	X		X	X		
5	73.4	68.9	5.8	2.5805	X	X		X	X	X	
5	71.8	67.1	7.5	2.6562	X	X	X	X	X		
5	71.6	66.9	7.7	2.6649		X	X		X	X	X
5	71.4	66.7	7.9	2.6733	X	X	X		X	X	
5	71.4	66.6	8.0	2.6759	X	X	X	X		X	
6	74.4	69.1	6.6	2.5726	X	X	X		X	X	X
6	73.5	68.0	7.6	2.6178	X	X		X	X	X	X
6	73.3	67.8	7.8	2.6270	X	X	X	X		X	X
6	72.5	66.8	8.7	2.6670	X	X	X	X	X	X	
6	71.9	66.1	9.4	2.6943		X	X	X	X	X	X
7	75.0	68.7	8.0	2.5905	X	X	X	X	X	X	X

exactly two predictors. You can look at the corresponding values of R^2, $adj\text{-}R^2$, C_p, and s to see how good that model is for predicting Y.

In Exhibit 7.3.2 we give a SAS computer output (edited for easier reading) for this problem containing the results from a subset analysis. To save space we have given the values of s, R^2, $adj\text{-}R^2$, and C_p for only the best five subset models for each value of m, where m is the number of predictors in the subset model.

E X H I B I T **7.3.2**

SAS Output for All-Subsets Regression for Gifted Children Data

SAS   Jan 1, Saturday, 1994

N = 36   Regression Models for Dependent Variable: Y

Number in Model	R-square	Adjusted R-square	C(p)	Root MSE	Variables in Model
1	0.32631738	0.30650318	43.32170	3.8557360	X2
1	0.29616080	0.27545965	46.69338	3.9410900	X4
1	0.27583227	0.25453322	48.96623	3.9975988	X5
1	0.13709230	0.11171266	64.47818	4.3637694	X6
1	0.07176563	0.04446462	71.78209	4.5259364	X3
2	0.62913424	0.60665753	11.46498	2.9038253	X2 X5
2	0.60775185	0.58397923	13.85566	2.9863628	X2 X4
2	0.37795937	0.34025994	39.54782	3.7607242	X2 X3
2	0.36724948	0.32890096	40.74525	3.7929609	X1 X2
2	0.36338752	0.32480494	41.17704	3.8045184	X2 X6
3	0.68725000	0.65792969	6.96730	2.7079632	X1 X2 X5
3	0.66655505	0.63529459	9.28112	2.7961223	X1 X2 X4
3	0.64649418	0.61335301	11.52404	2.8790047	X2 X3 X4
3	0.64490698	0.61161701	11.70150	2.8854607	X2 X3 X5
3	0.63700347	0.60297255	12.58516	2.9173956	X2 X5 X6
4	0.70770862	0.66999361	6.67990	2.6597832	X1 X2 X3 X4
4	0.70687611	0.66905367	6.77298	2.6635683	X2 X5 X6 X7
4	0.70401433	0.66582264	7.09295	2.6765390	X1 X2 X3 X5
4	0.70029934	0.66162829	7.50830	2.6932836	X1 X2 X5 X6
4	0.69303401	0.65342549	8.32061	2.7257333	X1 X2 X4 X5
5	0.73375503	0.68938087	5.76776	2.5804720	X1 X2 X5 X6 X7
5	0.71790384	0.67088781	7.54002	2.6561772	X1 X2 X3 X4 X5
5	0.71604225	0.66871596	7.74815	2.6649270	X2 X3 X5 X6 X7
5	0.71424574	0.66662002	7.94902	2.6733438	X1 X2 X3 X5 X6
5	0.71370085	0.66598432	8.00994	2.6758914	X1 X2 X3 X4 X6
6	0.74418888	0.69126244	6.60119	2.5726446	X1 X2 X3 X5 X6 X7
6	0.73512383	0.68032186	7.61472	2.6178305	X1 X2 X4 X5 X6 X7
6	0.73325500	0.67806638	7.82367	2.6270493	X1 X2 X3 X4 X6 X7
6	0.72508170	0.66820205	8.73749	2.6669932	X1 X2 X3 X4 X5 X6
6	0.71943293	0.66138457	9.36906	2.6942533	X2 X3 X4 X5 X6 X7
7	0.74956601	0.68695751	8.00000	2.5905185	X1 X2 X3 X4 X5 X6 X7

To read this table, suppose that you want to examine a model that contains m predictors. Select the number m and look under the column heading `Number in Model` for that value of m. Then the best five models (in order from first best, second best, ... , fifth best) are given under the column heading `Variables in Model`. For example, to find the variables in the third best model when the model contains $m = 2$ predictors, go down the column with heading `Number in Model` until you come to the number 2; then go down to the third 2 (because we want the third best model using two predictors); the value of R^2 is $0.378 = 37.8\%$. Go across this line to the column with heading `Variables in Model` and you see that the variables are X_2 and X_3. Hence the third best model that contains two predictor variables is $\beta_0 + \beta_2 x_2 + \beta_3 x_3$.

Answer (a)–(i) using both the MINITAB and the SAS exhibits and notice that both give the same answer.

a Find the predictor variable that results in the best prediction function (for the sample data at hand) among all subset models containing only *one* predictor.

b Find the two predictor variables that result in the best prediction function (for the sample data at hand) among all subset models containing exactly *two* predictors.

c Find the three predictor variables that result in the best prediction function (for the sample data at hand) among all subset models containing exactly *three* predictors.

d Find a short list of the three best models using the criterion R^2.

e Find a short list of the three best models using the criterion *adj-R^2*.

f Find a short list of the three best models using the criterion s.

g Find a short list of the three best models using the criterion C_p.

h Prepare a short list consisting of four models using C_p and $C_p - p$ together.

i Write a short report summarizing the results of parts (a) through (h).

7.3.2 Consider the GPA data of Example 4.4.2. Suppose that the model that we want to examine for this study is

$$\mu_Y(x_1, \ldots, x_6) = \beta_0 + \beta_1 x_1 + \cdots + \beta_6 x_6$$

where $X_1 =$ SATmath, $X_2 =$ SATverbal, $X_3 =$ HSmath, $X_4 =$ HSenglish, $X_5 = X_1 X_2$, $X_6 = X_3 X_4$. The raw data, which include the values of Y, X_1, X_2, X_3, and X_4, are in Table 7.3.3 along with the values of the derived variables $X_5 = X_1 X_2$ and $X_6 = X_3 X_4$. These data are also stored in the file **table733.dat** on the data disk.

A MINITAB output is given in Exhibit 7.3.3. It contains the results of an all-subsets regression analysis based on the six predictor variables $X_1, \ldots, X_6$. Only the results for the best five subset models for each subset size are given. A SAS output is given in Exhibit 7.3.4. It contains the results of an all-subsets regression analysis based on the six predictor variables $X_1, \ldots, X_6$. Only the results for the best five subset models for each subset size are given. We give a MINITAB output as well as a SAS output so that you can compare them against each other and become familiar with both. Answer (a)–(i) using either the MINITAB output or the SAS output. Of course your answers will be the same either way.

T A B L E 7.3.3

Subject	Y	X_1	X_2	X_3	X_4	$X_5 = X_1X_2$	$X_6 = X_3X_4$
1	1.97	321	247	2.30	2.63	79287	6.0490
2	2.74	718	436	3.80	3.57	313048	13.5660
3	2.19	358	578	2.98	2.57	206924	7.6586
4	2.60	403	447	3.58	2.21	180141	7.9118
5	2.98	640	563	3.38	3.48	360320	11.7624
6	1.65	237	342	1.48	2.14	81054	3.1672
7	1.89	270	472	1.67	2.64	127440	4.4088
8	2.38	418	356	3.73	2.52	148808	9.3996
9	2.66	443	327	3.09	3.20	144861	9.8880
10	1.96	359	385	1.54	3.46	138215	5.3284
11	3.14	669	664	3.21	3.37	444216	10.8177
12	1.96	409	518	2.77	2.60	211862	7.2020
13	2.20	582	364	1.47	2.90	211848	4.2630
14	3.90	750	632	3.14	3.49	474000	10.9586
15	2.02	451	435	1.54	3.20	196185	4.9280
16	3.61	645	704	3.50	3.74	454080	13.0900
17	3.07	791	341	3.20	2.93	269731	9.3760
18	2.63	521	483	3.59	3.32	251643	11.9188
19	3.11	594	665	3.42	2.70	395010	9.2340
20	3.20	653	606	3.69	3.52	395718	12.9888

a   Find the predictor variable that results in the best one-variable subset model.

b   Find the two predictor variables that result in the best two-variable subset model.

c   Find the three predictor variables that result in the best three-variable subset model.

d   Find a short list of the three best models using the R^2 criterion.

e   Find a short list of the three best models using the adj-R^2 criterion.

f   Find a short list of the three best models using the s criterion.

g   Find a short list of the three best models using the C_p criterion.

h   Find a short list of three models using C_p and $C_p - p$ together.

i   Write a short report summarizing the results of parts (a) through (h).

E X H I B I T 7.3.3
MINITAB Output for Problem 7.3.2

```
Best Subsets Regression of Y
```

Vars	R-sq	Adj. R-sq	C-p	s	X1	X2	X3	X4	X5	X6
1	81.4	80.4	10.7	0.27515					X	
1	72.2	70.6	24.0	0.33700	X					
1	65.9	64.0	33.0	0.37286						X
1	48.3	45.4	58.4	0.45949			X			
1	42.7	39.5	66.4	0.48367		X				
2	86.3	84.7	5.6	0.24301				X	X	
2	85.9	84.3	6.3	0.24675					X	X
2	85.8	84.1	6.4	0.24757		X			X	
2	84.3	82.4	8.6	0.26075	X				X	
2	81.5	79.3	12.6	0.28271				X	X	
3	89.8	87.9	2.7	0.21638		X	X		X	
3	88.4	86.2	4.7	0.23106		X			X	X
3	88.0	85.8	5.2	0.23458	X		X		X	
3	87.1	84.7	6.5	0.24308	X				X	X
3	86.7	84.2	7.1	0.24728				X	X	X
4	90.4	87.9	3.8	0.21657	X	X	X		X	
4	89.9	87.2	4.5	0.22264		X	X		X	X
4	89.8	87.1	4.6	0.22339	X	X	X	X		
4	89.3	86.4	5.4	0.22903		X		X	X	X
4	88.7	85.7	6.2	0.23518	X	X			X	X
5	90.6	87.2	5.6	0.22264	X	X	X		X	X
5	90.4	87.0	5.8	0.22407	X	X	X	X	X	
5	90.1	86.5	6.3	0.22837		X	X	X	X	X
5	89.7	86.1	6.8	0.23223	X	X		X	X	X
5	88.1	83.9	9.0	0.24944	X		X	X	X	X
6	91.0	86.8	7.0	0.22607	X	X	X	X	X	X

E X H I B I T 7.3.4
SAS Output for Problem 7.3.2

N = 20   Regression Models for Dependent Variable: Y

Number in Model	R-square	Adjusted R-square	C(p)	Root MSE	Variables in Model
1	0.81449296	0.80418702	10.66302	0.27514656	X5
1	0.72170925	0.70624865	23.99888	0.33700262	X1
1	0.65934389	0.64041855	32.96268	0.37285672	X6
1	0.48264669	0.45390483	58.35946	0.45949153	X3
1	0.42677523	0.39492941	66.38990	0.48366690	X2
2	0.86333578	0.84725764	5.64282	0.24300941	X3 X5
2	0.85909950	0.84252297	6.25170	0.24674704	X5 X6
2	0.85815613	0.84146861	6.38729	0.24757169	X2 X5
2	0.84265718	0.82414626	8.61497	0.26074689	X1 X5
2	0.81503604	0.79327558	12.58497	0.28270874	X4 X5
3	0.89802358	0.87890300	2.65712	0.21637647	X2 X3 X5
3	0.88371683	0.86191373	4.71344	0.23105671	X2 X5 X6
3	0.88014407	0.85767108	5.22696	0.23457943	X1 X3 X5
3	0.87129761	0.84716591	6.49846	0.24308237	X1 X5 X6
3	0.86681811	0.84184651	7.14230	0.24727644	X3 X4 X5
4	0.90422221	0.87868146	3.76619	0.21657430	X1 X2 X3 X5
4	0.89878613	0.87179576	4.54752	0.22263556	X2 X3 X5 X6
4	0.89809996	0.87092662	4.64615	0.22338895	X2 X3 X4 X5
4	0.89288831	0.86432520	5.39522	0.22903030	X2 X4 X5 X6
4	0.88706066	0.85694350	6.23283	0.23517824	X1 X2 X5 X6
5	0.90553163	0.87179293	5.57799	0.22263802	X1 X2 X3 X5 X6
5	0.90430939	0.87013417	5.75366	0.22407365	X1 X2 X3 X4 X5
5	0.90060360	0.86510489	6.28630	0.22837126	X2 X3 X4 X5 X6
5	0.89721211	0.86050215	6.77376	0.23223470	X1 X2 X4 X5 X6
5	0.88141776	0.83906695	9.04389	0.24943992	X1 X3 X4 X5 X6
6	0.90955296	0.86780818	7.00000	0.22607141	X1 X2 X3 X4 X5 X6

7.4
Alternative Methods for Subset Selection

In this section we discuss three methods of variable selection that require fewer computations than the method of all-subsets regression. These methods are

1   Forward selection

2   Backward elimination

3   Stepwise regression

Unlike the all-subsets regression procedure, none of these methods generally examines all the possible subset models, and they should be used only when computing facilities are not adequate for handling the all-subsets regression procedure. But like the all-subsets regression procedure, these methods are to be regarded as only exploratory or descriptive in nature. No rigorous confidence statements regarding the conclusions are generally possible. Thus we regard these methods as mainly suitable for exploratory analysis whether or not assumptions (A) or (B) are satisfied. Nevertheless these three methods, particularly stepwise regression, have been found to be quite useful in subset analysis. The drawback of not being able to obtain confidence statements regarding the results may be overcome by conducting additional studies and by obtaining new data for validation (or invalidation) of the conclusions arrived at with the help of these procedures. Thus a tentative model can be obtained using the procedures discussed here and in the previous section, and this tentative model can then be evaluated by selecting and using a new sample from the same population.

The forward selection and the backward elimination procedures are special cases of the stepwise regression procedure. They are computationally somewhat less taxing than stepwise regression. However, for purely exploratory purposes, stepwise regression is the recommended procedure among the three being considered here. Nevertheless, we first discuss the forward selection and the backward elimination methods, mainly because they facilitate the understanding of the stepwise regression procedure. An additional reason for discussing them in this book is that some researchers occasionally use them and consequently journal articles contain references to these methods.

Forward Selection

This procedure begins with the simplest function, viz.,

$$\beta_0 \tag{7.4.1}$$

and successively adds one variable at a time to the model in such a way that at each step the variable added is the best variable that can be added. To make ideas concrete, we describe the forward selection algorithm using $k = 4$; i.e., the total number of predictors under consideration is four. We label them X_1, X_2, X_3, and X_4 as usual. We now describe the forward selection algorithm.

At each step of the algorithm we will have a current model, and we will choose a predictor variable not already included in the current model as the best candidate variable for adding to that model. Any criterion measure—s, R^2, adj-R^2, or C_p— described in Section 7.3 can be used, and each measure will select the same best candidate variable. Whether or not this candidate variable is actually added to the current model depends on whether a computed quantity, which we denote by F_C, exceeds a criterion value, which we denote by F-in. This criterion value is chosen by the investigator to somewhat correspond to a tabled F-value with 1 degree of freedom in the numerator and $n - (m + 1)$ degrees of freedom in the denominator ($m + 1$ is the number of βs in the model under consideration). You will note in Table T-5 in Appendix T, that if $n - m - 1$ is larger than 10, then $F_{0.95:1,n-m-1}$ is close to 4.0 so 4.0 is a commonly used value for F-in. More will be said about this later.

The criterion value F-in is denoted differently in different statistical computing packages. MINITAB uses the name `FENTER` to refer to the criterion value F-in. Other statistical packages use other names. Some statistical packages such as SAS use criteria related to P-values in place of F-table values.

The algorithm proceeds as follows. We start with the model

$$\beta_0 \tag{7.4.2}$$

as the current model. We calculate SSY, the sum of squared errors when using $\bar{y}$ to predict Y. Actually $\bar{y}$ is $\hat{\beta}_0$, the least squares estimate of β_0 for the model in (7.4.2).

We define a calculation (or set of calculations) as a step if a predictor variable is added to the current model.

Step 1 For each predictor variable X_j, $j = 1, \ldots, 4$, fit the model

$$\beta_0 + \beta_j x_j \tag{7.4.3}$$

by least squares and obtain $SSE(X_j)$, the sum of squared errors when using $\hat{\beta}_0 + \hat{\beta}_j x_j$ to predict Y. Choose the variable X_j that results in the smallest value for $SSE(X_j)$ as the best candidate variable to be added to the current model. Note that the variable X_j with the smallest value for $SSE(X_j)$ at this step is the same variable X_j with the largest value of $R^2(X_j)$, or adj-$R^2(X_j)$, or the smallest value of $s(X_j)$, or $C_p(X_j)$. Thus any of these criteria can be used in place of SSE to identify the best candidate variable at each step. They all give the same result. However, to simplify discussion, we use the smallest SSE. Suppose, for concreteness, this variable is X_1. Calculate

$$F_C = \frac{SSY - SSE(X_1)}{MSE(X_1)} \tag{7.4.4}$$

where $MSE(X_1) = SSE(X_1)/(n - 2)$. If F_C is less than or equal to F-in, then the algorithm stops and the original model in (7.4.2) is the final model.

If F_C is greater than F-in, then add X_1 to the current model. In this case in step 1 the variable X_1 is added and the resulting model, which contains variable X_1, is

$$\beta_0 + \beta_1 x_1 \tag{7.4.5}$$

Proceed to step 2.

Step 2   The revised current model is

$$\beta_0 + \beta_1 x_1$$

The predictor variables not in the current model at this step are X_2, X_3, and X_4. For $j = 2, 3, 4$, fit the model

$$\beta_0 + \beta_1 x_1 + \beta_j x_j \tag{7.4.6}$$

and obtain $SSE(X_1, X_j)$. Choose the variable X_j that results in the smallest value for $SSE(X_1, X_j)$ as the best candidate variable to be added to the current model. Suppose this variable is X_2. Calculate

$$F_C = \frac{SSE(X_1) - SSE(X_1, X_2)}{MSE(X_1, X_2)} \tag{7.4.7}$$

where $MSE(X_1, X_2) = SSE(X_1, X_2)/(n - 3)$.

If F_C is less than or equal to F-*in*, the algorithm stops and chooses the model in (7.4.5) as the final model (i.e., the algorithm chooses the model in step 1 as the final model). If F_C is greater than F-*in*, then add X_2 to the current model. In this case in step 2 the variable X_2 is added to the current model and the resulting model, which contains variables X_1, X_2, is

$$\beta_0 + \beta_1 x_1 + \beta_2 x_2 \tag{7.4.8}$$

Proceed to step 3.

Step 3   The revised current model is

$$\beta_0 + \beta_1 x_1 + \beta_2 x_2$$

The predictor variables not in the current model at this stage are X_3 and X_4. For $j = 3, 4$, fit the model

$$\beta_0 + \beta_1 x_1 + \beta_2 x_2 + \beta_j x_j \tag{7.4.9}$$

and obtain $SSE(X_1, X_2, X_j)$. Choose the variable X_j that results in the smallest value for $SSE(X_1, X_2, X_j)$ as the best candidate variable to be added to the current model. Suppose this variable is X_3. Calculate

$$F_C = \frac{SSE(X_1, X_2) - SSE(X_1, X_2, X_3)}{MSE(X_1, X_2, X_3)} \tag{7.4.10}$$

where $MSE(X_1, X_2, X_3) = SSE(X_1, X_2, X_3)/(n - 4)$. If F_C is less than or equal to F-*in*, then the algorithm stops and chooses the model in (7.4.8) as the final model (i.e., the algorithm chooses the model in step 2 as the final model).

If F_C is greater than F-*in*, then add X_3 to the current model. In this case in step 3 the variable X_3 is added and the resulting model, which contains variables X_1, X_2, X_3, is

$$\beta_0 + \beta_1 x_1 + \beta_2 x_2 + \beta_3 x_3 \tag{7.4.11}$$

Proceed to step 4.

Step 4 At step 4 the revised current model is

$$\beta_0 + \beta_1 x_1 + \beta_2 x_2 + \beta_3 x_3$$

The only predictor variable not in the current model at this stage is X_4. Fit the model

$$\beta_0 + \beta_1 x_1 + \beta_2 x_2 + \beta_3 x_3 + \beta_4 x_4 \qquad\text{(7.4.12)}$$

and obtain $SSE(X_1, X_2, X_3, X_4)$. Calculate

$$F_C = \frac{SSE(X_1, X_2, X_3) - SSE(X_1, X_2, X_3, X_4)}{MSE(X_1, X_2, X_3, X_4)} \qquad\text{(7.4.13)}$$

where $MSE(X_1, X_2, X_3, X_4) = SSE(X_1, X_2, X_3, X_4)/(n - 5)$.

If F_C is less than or equal to $F\text{-}in$, then the algorithm stops and chooses the model in (7.4.11) as the final model (i.e., the algorithm chooses the model in step 3 as the final model). If F_C is greater than $F\text{-}in$, then add X_4 to the current model and, since all the predictor variables are already included in the current model, there is no need to proceed further and the algorithm stops. In this case at step 4 the model is the one given in (7.4.14) and is the final model.

$$\beta_0 + \beta_1 x_1 + \beta_2 x_2 + \beta_3 x_3 + \beta_4 x_4 \qquad\text{(7.4.14)}$$

Although we used $k = 4$ for simplicity of explanation, the procedure can be extended in an obvious manner to handle any number of predictor variables. We now illustrate this procedure in Example 7.4.1.

E X A M P L E 7.4.1

Consider the GPA data of Example 4.4.2. We apply the forward selection procedure to choose a subset (which may turn out to be the full set in some problems) of the four predictor variables. The sums of squares and mean squares required during various steps of the forward selection algorithm may be obtained using a statistical package such as SAS, SPlus, BMDP, SPSS, MINITAB, etc. In Table 7.4.1 we give the error sums of squares and mean squares for all subsets, although in a real problem these would not all be computed (that is the advantage of the forward selection procedure over the all-subsets regression procedure). We select $F\text{-}in = 4.0$ and start with the model

$$\beta_0 \qquad\text{(7.4.15)}$$

with no predictor variables as the current model.

Step 1 The variables not in the model are X_1, X_2, X_3, X_4. We regress Y on each of these variables and obtain

$$SSE(X_1) = 2.0443$$
$$SSE(X_2) = 4.2108$$
$$SSE(X_3) = 3.8004$$
$$SSE(X_4) = 4.6439$$

T A B L E **7.4.1**
Sums of Squares and Mean Squares for All 16 Subset Models for the GPA Data

Model	Predictors in the Model	Sum of Squares	Mean Square
(1)	None	7.3458	0.3866
(2)	X_1	2.0443	0.1136
(3)	X_2	4.2108	0.2339
(4)	X_3	3.8004	0.2111
(5)	X_4	4.6439	0.2580
(6)	X_1, X_2	1.3884	0.0817
(7)	X_1, X_3	1.5282	0.0899
(8)	X_1, X_4	2.0199	0.1188
(9)	X_2, X_3	2.6812	0.1577
(10)	X_2, X_4	3.2179	0.1893
(11)	X_3, X_4	2.4180	0.1422
(12)	X_1, X_2, X_3	1.0992	0.0687
(13)	X_1, X_2, X_4	1.3881	0.0868
(14)	X_1, X_3, X_4	1.4532	0.0908
(15)	X_2, X_3, X_4	1.9344	0.1209
(16)	X_1, X_2, X_3, X_4	1.0815	0.0721

The predictor variable that leads to the smallest value for SSE is X_1. We calculate

$$F_C = \frac{SSY - SSE(X_1)}{MSE(X_1)} = \frac{7.3458 - 2.0443}{0.1136} = 46.67 \qquad \textbf{(7.4.16)}$$

Since $F_C > F\text{-}in = 4.0$, we add X_1 to the current model. So in step 1 the variable X_1 is added to the current model and the resulting model, which contains variable X_1, is

$$\beta_0 + \beta_1 x_1 \qquad \textbf{(7.4.17)}$$

Proceed to step 2.

Step 2   The revised current model is

$$\beta_0 + \beta_1 x_1$$

The variables not in the current model are X_2, X_3, and X_4. We add each of these variables in turn to the current model in (7.4.17) and calculate the corresponding sums of squared errors.

$$SSE(X_1, X_2) = 1.3884$$
$$SSE(X_1, X_3) = 1.5282$$
$$SSE(X_1, X_4) = 2.0199$$

Hence the best predictor variable for adding to the current model is X_2. We calculate

$$F_C = \frac{SSE(X_1) - SSE(X_1, X_2)}{MSE(X_1, X_2)} = \frac{2.0443 - 1.3884}{0.0817} = 8.03 \qquad \textbf{(7.4.18)}$$

Because $F_C > F\text{-}in = 4.0$, we add X_2 to the current model. So in step 2 the variable X_2 is added and the resulting model, which contains the variables X_1, X_2, is

$$\beta_0 + \beta_1 x_1 + \beta_2 x_2 \qquad (7.4.19)$$

Proceed to step 3.

Step 3 The revised current model is

$$\beta_0 + \beta_1 x_1 + \beta_2 x_2$$

The variables not in the current model are X_3 and X_4. We add each of these variables in turn to the current model in (7.4.19) and calculate the corresponding *SSE*.

$$SSE(X_1, X_2, X_3) = 1.0992$$
$$SSE(X_1, X_2, X_4) = 1.3881$$

Hence the best predictor variable for adding to the current model is X_3. We calculate

$$F_C = \frac{SSE(X_1, X_2) - SSE(X_1, X_2, X_3)}{MSE(X_1, X_2, X_3)} = \frac{1.3884 - 1.0992}{0.0687} = 4.21 \qquad (7.4.20)$$

Because $F_C > F\text{-}in = 4.0$, we add X_3 to the current model. So in step 3 the variable X_3 is added and the resulting model, which contains the variables X_1, X_2, X_3, is

$$\beta_0 + \beta_1 x_1 + \beta_2 x_2 + \beta_3 x_3 \qquad (7.4.21)$$

Proceed to step 4.

Step 4 The revised current model is

$$\beta_0 + \beta_1 x_1 + \beta_2 x_2 + \beta_3 x_3$$

The only variable not in the current model is X_4. We add X_4 to the current model in (7.4.21) and calculate the corresponding *SSE*.

$$SSE(X_1, X_2, X_3, X_4) = 1.0815 \qquad (7.4.22)$$

Also,

$$F_C = \frac{SSE(X_1, X_2, X_3) - SSE(X_1, X_2, X_3, X_4)}{MSE(X_1, X_2, X_3, X_4)} = \frac{1.0992 - 1.0815}{0.0721}$$

$$= 0.25 \qquad (7.4.23)$$

Because $F_C < F\text{-}in = 4.0$, we do not add X_4 to the current model. Thus the model in (7.4.21) is the final model selected by the forward selection algorithm using $F\text{-}in = 4.0$.

Next we summarize the variable added and the variables in the model at each step.

Step	Variables Added	Variables in the Model
Start	—	None
1	X_1	X_1
2	X_2	X_1, X_2
3	X_3	X_1, X_2, X_3 ∎

Some comments are in order.

a We regard the forward selection procedure as a purely exploratory analysis technique, whether or not assumptions (A) or (B) hold, because, even when these assumptions are satisfied, no measure of confidence is available that can be used to judge whether the final model is an adequate model.

b The decision to add or not add a candidate variable to the current model at any particular step is sometimes based on an F-test, discussed in Section 4.9, which compares two competing models where one model is nested in the other.

  To see this in more detail, suppose that the current model in step 1 is

$$\beta_0 + \beta_1 x_1$$

Suppose that in step 2 the best candidate variable to add to this model is X_2. To determine if X_2 should be added to the current model, we decide whether or not the model

$$\beta_0 + \beta_1 x_1 + \beta_2 x_2$$

is better than the model

$$\beta_0 + \beta_1 x_1$$

If we decide $\beta_0 + \beta_1 x_1 + \beta_2 x_2$ is better than $\beta_0 + \beta_1 x_1$, then the variable X_2 is added. If we decide $\beta_0 + \beta_1 x_1 + \beta_2 x_2$ is no better than $\beta_0 + \beta_1 x_1$, then the variable X_2 is not added. Many variable selection procedures make this decision based on a test of

$$\text{NH: } \beta_2 = 0 \qquad \text{against} \qquad \text{AH: } \beta_2 \neq 0$$

In Section 4.9 (under appropriate assumptions) we discussed an F-test for testing the preceding null hypothesis. According to this F-test, NH is rejected if a computed F, denoted by F_C, is larger than a tabled F value; viz., $F_{1-\alpha:1,dfd}$ where the degrees of freedom for the numerator is 1 and dfd is the degrees of freedom for the denominator. If NH is rejected, then based on the test the procedure will conclude that $\beta_2 \neq 0$ and that the model $\beta_0 + \beta_1 x_1 + \beta_2 x_2$ is better than the model $\beta_0 + \beta_1 x_1$. Thus the variable X_2 is added to the model and the variables in the model at this step are X_1 and X_2; the resulting model, which is the new current model, is $\beta_0 + \beta_1 x_1 + \beta_2 x_2$. If F_C is less than or equal to the tabled F-value, then NH is not rejected and, based on the test, the procedure will conclude that the model $\beta_0 + \beta_1 x_1 + \beta_2 x_2$ is no better than the model $\beta_0 + \beta_1 x_1$. Hence X_2 is not added to the current model. Thus the current model is still $\beta_0 + \beta_1 x_1$.

Whereas the forward selection procedure is *guided* by a statistical test, it is not a *valid F*-test, and the actual Type-I error is not known even though we use an α value of our choice. It is quite common (but arbitrary) to carry out this test using $\alpha = 0.05$. It may be verified that the tabled *F*-values for carrying out this test are roughly in the neighborhood of 4.0 when there are enough degrees of freedom for the denominator. For this reason *F-in* is often taken to be equal to 4.0. Some authors recommend using *F-in* = 2.0. In principle, any nonnegative value may be chosen as the value for *F-in*. However, the final model chosen by this procedure is generally dependent on the value of *F-in* that is used. It is clear that larger values of *F-in* tend to result in the selection of a final model with a smaller number of variables. You should convince yourself of this.

One reason why the preceding test is not valid, even if population assumptions (A) or (B) are satisfied for the population $\{(Y, X_1, \ldots, X_k)\}$, is that we look at all models $\beta_0 + \beta_1 x_1 + \beta_j x_j$ for $j = 2, 3, 4$ to find the best candidate model. Say it is $\beta_0 + \beta_1 x_1 + \beta_2 x_2$; then the forward selection procedure tests

$$\text{NH: } \beta_2 = 0 \qquad \text{against} \qquad \text{AH: } \beta_2 \neq 0$$

to decide whether the model $\beta_0 + \beta_1 x_1 + \beta_2 x_2$ is better than the model $\beta_0 + \beta_1 x_1$. *The test is not a valid F-test because the data are used to determine which model to test.* In Section 4.9 the test is a valid *F*-test if the data are not used to determine which test to make, i.e., if we decide to test $\beta_2 = 0$ before looking at the data.

c   Notice that the forward selection procedure does not examine each and every subset model, and it is quite possible that there exist other subsets of variables that predict Y as well as, or better than, the chosen subsets. This is the price we pay for the savings in computing that results from not examining every possible model.

d   Some users of this procedure interpret the order in which the variables are entered into the model as the order of importance of the variables. This is an incorrect interpretation, particularly in view of the fact that the concept of importance is usually not precisely defined. For instance, the following situation occurs not infrequently. $\sigma_{Y|X_1}$ is the smallest and $\sigma_{Y|X_2}$ the second smallest among the quantities $\sigma_{Y|X_i}$, $i = 1, \ldots, k$, whereas $\sigma_{Y|X_2,X_3}$ is substantially smaller than $\sigma_{Y|X_1,X_3}$. Thus X_1 alone is a better predictor of Y than X_2 alone, but X_2 in the presence of X_3 is more useful as a predictor of Y than X_1 is in the presence of X_3. Hence it is usually not possible to precisely define a meaningful order of importance based on statistical considerations alone.

e   Instead of selecting the candidate variable to add to the current model by using the sum of squared errors, we can use any of the criterion measures R^2, *adj-R^2*, s, or C_p discussed in Section 7.3. This will not alter the candidate variable to add at any step.

f   Notice that some of the predictor variables can be derived variables, i.e., known functions of the *basic* predictor variables. For instance, we could have started the forward selection procedure in Example 7.4.1 with the set $\{X_1, X_2, X_3, X_4, X_5, X_6\}$ as the full set of predictors, where $X_5 = X_1^2 =$ (SATmath)^2, $X_6 = X_3 X_4 =$ (HSmath) $\times$ (HSenglish), etc.

Backward Elimination

Whereas the forward selection procedure begins with the constant model (i.e., the model β_0 that includes no predictor variables), the backward elimination procedure begins with the model that includes *all* of the available predictor variables, viz.,

$$\beta_0 + \beta_1 x_1 + \cdots + \beta_k x_k \tag{7.4.24}$$

and proceeds by successively removing from the model one variable at a time in such a way that, at each step, the variable removed is the variable contributing the least to the prediction of Y at that step. The details of the algorithm are described next using $k = 4$ for simplicity. Thus the predictor variables are X_1, X_2, X_3, and X_4.

At each step of the algorithm (we define a step as a calculation or set of calculations in which a variable is deleted) we will have a current model, and we will label a predictor variable included in the current model as the best candidate variable for deletion from the model. Whether or not this candidate variable is actually deleted from the current model depends on whether a computed quantity, denoted by F_C, is smaller than a criterion value which we call *F-out*. This criterion value is chosen by the investigator to somewhat correspond to a tabled F-value with 1 degree of freedom in the numerator and $n - (m + 1)$ degrees of freedom in the denominator ($m + 1$ is the number of βs in the current model).

The criterion value *F-out* is denoted differently in different statistical computing packages. MINITAB uses the name FREMOVE to refer to the criterion value *F-out*. Other statistical packages use other names. Some statistical packages such as SAS use criteria related to P-values in place of F-table values.

We start with the model

$$\beta_0 + \beta_1 x_1 + \beta_2 x_2 + \beta_3 x_3 + \beta_4 x_4 \tag{7.4.25}$$

as the current model. Fit this model by the method of least squares and calculate the quantity $SSE(X_1, X_2, X_3, X_4)$, the sum of squared errors when using $\hat{\beta}_0 + \hat{\beta}_1 x_1 + \hat{\beta}_2 x_2 + \hat{\beta}_3 x_3 + \hat{\beta}_4 x_4$ to predict Y.

Step 1   In step 1 the variables in the model are X_1, X_2, X_3, X_4. For each predictor variable $X_j, j = 1, \ldots, 4$, fit the model obtained by deleting this predictor variable from the current model and calculate the corresponding SSE. In the case $k = 4$, this leads us to consider the following four models.

$$\beta_0 + \beta_1 x_1 + \beta_2 x_2 + \beta_3 x_3 \qquad \text{i.e., } X_4 \text{ is omitted} \tag{7.4.26}$$
$$\beta_0 + \beta_1 x_1 + \beta_2 x_2 + \beta_4 x_4 \qquad \text{i.e., } X_3 \text{ is omitted} \tag{7.4.27}$$
$$\beta_0 + \beta_1 x_1 + \beta_3 x_3 + \beta_4 x_4 \qquad \text{i.e., } X_2 \text{ is omitted} \tag{7.4.28}$$
$$\beta_0 + \beta_2 x_2 + \beta_3 x_3 + \beta_4 x_4 \qquad \text{i.e., } X_1 \text{ is omitted} \tag{7.4.29}$$

and the corresponding *SSE* are

$$SSE(X_1, X_2, X_3) \qquad SSE(X_1, X_2, X_4) \qquad SSE(X_1, X_3, X_4) \qquad SSE(X_2, X_3, X_4)$$

respectively. Suppose the smallest among these *SSE* is $SSE(X_1, X_2, X_3)$. This means that if we want to delete one of the predictors in the current model, the best candidate to delete will be X_4 because the remaining three predictors, X_1, X_2, and X_3, are the best among all three predictor subset models of the current model. Calculate

$$F_C = \frac{SSE(X_1, X_2, X_3) - SSE(X_1, X_2, X_3, X_4)}{MSE(X_1, X_2, X_3, X_4)} \tag{7.4.30}$$

where $MSE(X_1, X_2, X_3, X_4) = SSE(X_1, X_2, X_3, X_4)/(n - 5)$.

If F_C is greater than *F-out*, then the algorithm stops and chooses the model in (7.4.25) as the final model. In this case no variables are deleted in step 1, and the variables in the model are X_1, X_2, X_3, and X_4.

If F_C is less than or equal to *F-out*, then delete X_4 from the current model. In this case the variable X_4 is deleted in step 1 and the model, which contains the variables X_1, X_2, X_3, is

$$\beta_0 + \beta_1 x_1 + \beta_2 x_2 + \beta_3 x_3 \tag{7.4.31}$$

Proceed to step 2.

Step 2 In step 2 the variables in the model are X_1, X_2, and X_3, so the revised current model is

$$\beta_0 + \beta_1 x_1 + \beta_2 x_2 + \beta_3 x_3$$

For each predictor variable $X_j, j = 1, 2, 3$, fit the model obtained by deleting this predictor variable from the current model and calculate the corresponding *SSE*. This step leads us to consider the following three models

$$\beta_0 + \beta_1 x_1 + \beta_2 x_2 \qquad \text{i.e., } X_3 \text{ is omitted from (7.4.31)} \tag{7.4.32}$$
$$\beta_0 + \beta_1 x_1 + \beta_3 x_3 \qquad \text{i.e., } X_2 \text{ is omitted from (7.4.31)} \tag{7.4.33}$$
$$\beta_0 + \beta_2 x_2 + \beta_3 x_3 \qquad \text{i.e., } X_1 \text{ is omitted from (7.4.31)} \tag{7.4.34}$$

and the corresponding *SSE* are $SSE(X_1, X_2), SSE(X_1, X_3)$, and $SSE(X_2, X_3)$, respectively. Suppose the smallest among these *SSE* is $SSE(X_1, X_2)$. This means that if we want to delete one of the predictors in the current model, which is given in (7.4.31), the best candidate to delete will be X_3. Calculate

$$F_C = \frac{SSE(X_1, X_2) - SSE(X_1, X_2, X_3)}{MSE(X_1, X_2, X_3)} \tag{7.4.35}$$

where $MSE(X_1, X_2, X_3) = SSE(X_1, X_2, X_3)/(n - 4)$.

If F_C is greater than *F-out*, then the algorithm stops and chooses the model in (7.4.31) as the final model.

If F_C is less than or equal to *F-out*, then delete X_3 from the current model. In this case the variable X_3 is deleted in step 2 and the model, which contains the variables

X_1, X_2, is

$$\beta_0 + \beta_1 x_1 + \beta_2 x_2 \tag{7.4.36}$$

Proceed to step 3.

Step 3 In step 3 the variables in the model are X_1, X_2, so the revised current model is

$$\beta_0 + \beta_1 x_1 + \beta_2 x_2$$

For each predictor variable $X_j, j = 1, 2$, fit the model obtained by deleting this predictor variable from the current model and calculate the corresponding *SSE*. At this step, we are led to consider the following two models.

$$\beta_0 + \beta_1 x_1 \qquad \text{i.e., } X_2 \text{ is omitted from (7.4.36)} \tag{7.4.37}$$
$$\beta_0 + \beta_2 x_2 \qquad \text{i.e., } X_1 \text{ is omitted from (7.4.36)} \tag{7.4.38}$$

and the corresponding *SSE* are $SSE(X_1)$ and $SSE(X_2)$, respectively. Suppose the smaller of these two *SSE* is $SSE(X_1)$. This means that if we want to delete one of the predictors in the current model, which is in (7.4.36), the best candidate to delete will be X_2. Calculate

$$F_C = \frac{SSE(X_1) - SSE(X_1, X_2)}{MSE(X_1, X_2)} \tag{7.4.39}$$

where $MSE(X_1, X_2) = SSE(X_1, X_2)/(n - 3)$.

If F_C is greater than *F-out*, then the algorithm stops and chooses the model in (7.4.36) as the final model.

If F_C is less than or equal to *F-out*, then delete X_2 from the current model. In this case the variable X_2 is deleted in step 3 and the model, which contains variable X_1, is

$$\beta_0 + \beta_1 x_1 \tag{7.4.40}$$

Proceed to step 4.

Step 4 In step 4 the variable in the model is X_1, so the revised current model is

$$\beta_0 + \beta_1 x_1$$

The only predictor variable in the current model at this step is X_1. If this variable is deleted, then the resulting model is β_0. The *SSE* when Y is predicted using $\hat{\beta}_0$ (which is equal to $\bar{y}$) is *SSY*. Calculate

$$F_C = \frac{SSY - SSE(X_1)}{MSE(X_1)} \tag{7.4.41}$$

where $MSE(X_1) = SSE(X_1)/(n - 2)$.

If F_C is greater than *F-out*, then the algorithm stops and chooses the model in (7.4.40) as the final model.

If F_C is less than or equal to *F-out*, then delete X_1 from the current model. In this case the variable X_1 is deleted in step 4 and there are no variables left in the model.

Because there are no predictor variables in the current model, there is no need to proceed further, and so the algorithm stops and declares the model

$$\beta_0 \tag{7.4.42}$$

to be the final model.

Although we used $k = 4$ for simplicity of explanation, the procedure can be extended in an obvious manner to handle any number of predictor variables.

We now illustrate this procedure with an example.

E X A M P L E 7.4.2

Consider the GPA data of Example 4.4.2. We apply the backward elimination procedure to choose a subset (which may turn out to be the full set in some problems) of the four predictor variables.

The required sums of squares and mean squares can be obtained using a statistical package. However, all of the sums of squares needed are given in Table 7.4.1.

We select $F\text{-}out = 4.0$ and start with the model

$$\beta_0 + \beta_1 x_1 + \beta_2 x_2 + \beta_3 x_3 + \beta_4 x_4 \tag{7.4.43}$$

as the current model. The value of $SSE(X_1, X_2, X_3, X_4)$ is 1.0815.

Step 1 The variables in the current model are X_1, X_2, X_3, X_4. The models that would result when one of these four predictors is deleted from the current model are

$$\beta_0 + \beta_1 x_1 + \beta_2 x_2 + \beta_3 x_3 \qquad \text{i.e., } X_4 \text{ is omitted} \tag{7.4.44}$$
$$\beta_0 + \beta_1 x_1 + \beta_2 x_2 + \beta_4 x_4 \qquad \text{i.e., } X_3 \text{ is omitted} \tag{7.4.45}$$
$$\beta_0 + \beta_1 x_1 + \beta_3 x_3 + \beta_4 x_4 \qquad \text{i.e., } X_2 \text{ is omitted} \tag{7.4.46}$$
$$\beta_0 + \beta_2 x_2 + \beta_3 x_3 + \beta_4 x_4 \qquad \text{i.e., } X_1 \text{ is omitted} \tag{7.4.47}$$

respectively. The SSE corresponding to these four models are

$$SSE(X_1, X_2, X_3) = 1.0992$$
$$SSE(X_1, X_2, X_4) = 1.3881$$
$$SSE(X_1, X_3, X_4) = 1.4532$$
$$SSE(X_2, X_3, X_4) = 1.9344$$

Thus we see that the model using X_1, X_2, X_3 to predict Y, which is obtained by deleting X_4 from the current model in (7.4.43), leads to the smallest SSE among the four models given in (7.4.44)–(7.4.47). Hence X_4 is the best candidate for deletion at this step. We calculate

$$F_C = \frac{SSE(X_1, X_2, X_3) - SSE(X_1, X_2, X_3, X_4)}{MSE(X_1, X_2, X_3, X_4)} = \frac{1.0992 - 1.0815}{0.0721}$$
$$= 0.25 \tag{7.4.48}$$

Since $F_C < F\text{-}out = 4.0$, we delete X_4 from the current model. Thus the variable X_4 is deleted in step 1 and the model, which contains variables X_1, X_2, X_3, is

$$\beta_0 + \beta_1 x_1 + \beta_2 x_2 + \beta_3 x_3 \qquad \text{(7.4.49)}$$

Proceed to step 2.

Step 2 The revised current model is

$$\beta_0 + \beta_1 x_1 + \beta_2 x_2 + \beta_3 x_3$$

The variables in the model are X_1, X_2, and X_3. We delete each of these variables in turn from the current model in (7.4.49) and calculate the corresponding SSE for the resulting models.

$$
\begin{aligned}
SSE(X_1, X_2) &= 1.3884 &\quad \text{i.e., } X_3 \text{ is omitted} \\
SSE(X_1, X_3) &= 1.5282 &\quad \text{i.e., } X_2 \text{ is omitted} \\
SSE(X_2, X_3) &= 2.6812 &\quad \text{i.e., } X_1 \text{ is omitted}
\end{aligned}
$$

Hence the best predictor variable to delete from the current model is X_3. We calculate

$$F_C = \frac{SSE(X_1, X_2) - SSE(X_1, X_2, X_3)}{MSE(X_1, X_2, X_3)} = \frac{1.3884 - 1.0992}{0.0687} = 4.21 \qquad \text{(7.4.50)}$$

Because $F_C > F\text{-}out = 4$, we do not delete X_3 from the current model.

Thus in step 2 no variable is deleted, so the algorithm stops and chooses the model in (7.4.49) as the final model. A summary follows:

Step	Variable Deleted	Variables in the Model
Start	—	X_1, X_2, X_3, X_4
1	X_4	X_1, X_2, X_3

Remarks

1 We regard the backward elimination procedure as a purely exploratory analysis technique whether or not assumptions (A) or (B) hold because, even if these assumptions are satisfied, no statistical measure of confidence is available that can be used to judge whether the final model is an adequate one.

2 Although in the example presented the forward and the backward selection procedures lead to the same final model, this is generally not the case. We illustrate this later with an example.

3 The decision to delete or not delete a candidate variable from the current model at any particular step is sometimes based on the computed F-value of a test that compares the two competing models (where one model is nested in the other) as discussed in Section 4.9. See the discussion in part (b) on page 526. While the procedure is *guided* by a statistical test, it is not a *valid* test (it could perhaps be called a *pseudo test*), and the actual Type-I error is not known even though we use an α value of our choice. It is quite common (but arbitrary) to carry out this pseudo test using $\alpha = 0.05$. It can be verified that the corresponding table F-values for carrying out this test are roughly in the neighborhood of 4.0 when there are enough degrees of freedom for the denominator. For this reason F-*out* is often taken to be equal to 4.0. Some authors recommend using F-*out* = 2.0. In principle, any nonnegative value can be chosen as the value for F-*out*. However, the final model chosen by this procedure may depend on the value of F-*out* that is chosen. It is clear that larger values of F-*out* will tend to result in the selection of a final model with a smaller number of variables.

4 Notice that this procedure does not examine each and every subset model, and so it is quite possible that there exist other subset models that predict Y as well as, or better than, the chosen subset or subsets. This is the price we pay for the savings in computing that results from not examining every possible model.

5 Some users of this procedure interpret the order in which the variables are deleted as an indication of the order of importance of the variables, with the first variable deleted as the least important, etc. This interpretation is incorrect and could lead to erroneous conclusions. For example, it may be the case that $\sigma_{Y|X_1}$ is the smallest among $\sigma_{Y|X_i}$, $i = 1, 2, 3, 4$ and may actually be very close to $\sigma_{Y|X_1, X_2, X_3, X_4}$, and yet X_1 could be deleted first by the backward elimination procedure.

6 Instead of selecting the candidate variable to delete from the current model by using the sum of squared errors, we could use any of the criterion measures R^2, *adj-R^2*, s, or C_p discussed in Section 7.3. This will not alter the candidate variable to be deleted.

Stepwise Regression

The stepwise regression procedure is a combination of the forward selection procedure and the backward elimination procedure. In fact, there are several variations of this procedure, all of which are referred to by the name *stepwise regression*. Here we discuss only one of the versions in detail. Again we consider the case $k = 4$ for concreteness, but the generalization to any number of predictors is obvious.

Suppose the predictor variables are X_1, X_2, X_3, and X_4. We start with an initial or current model that may be the model with no predictors.

$$\beta_0 \tag{7.4.51}$$

or the full model

$$\beta_0 + \beta_1 x_1 + \beta_2 x_2 + \beta_3 x_3 + \beta_4 x_4 \tag{7.4.52}$$

or *any other subset model*. The algorithm will proceed in two stages.

1 In stage 1 the backward elimination procedure is used to delete *as many* of the variables as possible.

2 In stage 2 the forward selection procedure is used *once* to add one variable if possible.

Then the stepwise procedure continues deleting variables and adding a variable (repeating stage 1, then stage 2, etc.) until a final model is obtained.

A step is defined as a calculation (or a set of calculations) when a predictor variable is added to, or deleted from, the current model.

The algorithm proceeds as follows. Select any model to start with and call it the current model. Investigators may be able to use their subject matter knowledge to decide on an initial model. Select two constants, one called *F-in* and the other called *F-out*, to be used in making decisions during various stages of the algorithm, regarding whether or not a predictor variable should be added to the current model or deleted from the current model. A word of caution applies here. *F-out must always be less than or equal to F-in; otherwise the procedure could end up in an infinite loop of adding and deleting the same predictor variable.* You should convince yourself of this possibility.

Stage 1   Start with the current model and perform the backward elimination procedure as many times as is necessary until no more variables can be deleted. If the current model is β_0, omit this stage and go to stage 2.

Stage 2   Start with the final model of stage 1 and perform the forward selection procedure *once*. If a predictor variable is added to the current model, then go back to stage 1 with this revised current model. If no predictor is added to the current model at this stage, then the procedure terminates because no variable can be added to the current model and no variable can be removed from the current model. In this case the current model is selected as the final model.

Comments

1   It is customary to choose as the initial model the model in (7.4.51) or the full model in (7.4.52), but any model can be the initial model. The investigator may feel that certain predictor variables should be together in the starting model, and so these variables can be included in the initial model. Both *F-in* and *F-out* must be specified by the analyst. They are both set equal to 4.0 by default in many statistical packages. We say more about this in Section 7.4 of the laboratory manuals.

2   If *F-out* = 0, then no variable can be deleted at any stage. You should verify this. So setting *F-out* = 0 and *F-in* equal to any other suitable value, say 4.0, and taking β_0 as the initial model is equivalent to performing the forward selection procedure. This is the way in which you would perform the forward selection procedure in MINITAB.

3   If *F-in* = ∞ (in practice, use a very large number, say 100,000), then no variable can enter the model at any stage. You should convince yourself of this. Hence setting *F-in* equal to a very large number and *F-out* equal to any suitable value, say 4.0, and taking as the initial model the one consisting of all the predictor variables is equivalent to performing the backward elimination procedure. In fact, this is the way in which you would perform the backward elimination procedure in MINITAB.

4 Because a stepwise regression procedure is a combination of the forward selection and backward elimination procedures, it almost always leads to a better model (or at least as good a model) than either of the other two procedures. There is usually some extra computation involved, but the potential benefits outweigh the additional cost involved.

5 As a final remark, we should point out that usually none of the three methods discussed (forward selection, backward elimination, or stepwise regression) examines all possible subsets of the predictors. This is in fact the most attractive feature of these methods—they are computationally less expensive than the all-subsets regression procedure that requires an examination of all possible subsets. The price we pay for using the computationally cheaper methods, however, is that there is no guarantee that they will find the best model from all the possible subset models according to any of the criteria discussed in Section 7.3.

All of the procedures, including the all-subsets regression procedure, should be used with caution and only in an exploratory sense, i.e., to gain insight into the way different subsets of predictor variables contribute to predicting Y for the data at hand.

Authors' Recommendation

For selection of variables, we recommend that the all-subsets regressions procedure be used whenever the computing facilities are adequate rather than the forward selection, the backward elimination, or the stepwise regression procedure.

E X A M P L E **7.4.3**

We illustrate the stepwise regression procedure using the GPA data of Example 4.4.2. We shall use *F-in* $= 4.0$ and *F-out* $= 3.0$ (note that we have made sure *F-out* is not bigger than *F-in*). The sums of squares needed for this example are given in Table 7.4.1. Of course, in a real problem these would not all be computed. (This is the reason we use stepwise regression rather than the all-subsets regression procedure.) In a problem with only $k = 4$ predictors, you should always use the all-subsets regression procedure, but we use stepwise regression for illustration.

We start with the initial model

$$\beta_0 \tag{7.4.53}$$

Stage 1 There are no variables to delete and so we proceed to stage 2.

Stage 2 We perform the forward selection procedure *once*. The best candidate to add to the current model is X_1 because $SSE(X_1)$ is smaller than $SSE(X_2)$, $SSE(X_3)$, and $SSE(X_4)$. To determine whether or not X_1 should be added to the current model we compute

$$F_C = \frac{SSY - SSE(X_1)}{MSE(X_1)} = \frac{7.3458 - 2.0443}{0.1136} = 46.67 \qquad \text{(7.4.54)}$$

Because $F_C > F\text{-}in = 4.0$, we add X_1 to the current model. So in step 1 X_1 is added and the model, which contains variable X_1, is

$$\beta_0 + \beta_1 x_1 \qquad \text{(7.4.55)}$$

We now go back to stage 1.

Stage 1 (Repeated) The revised current model is

$$\beta_0 + \beta_1 x_1$$

We examine the possibility of removing one or more variables from the current model by applying the backward elimination procedure. The only candidate for removal is X_1 because it is the only predictor variable in the current model. To decide whether or not X_1 should be removed from the current model, we need to compare the quantity

$$F_C = \frac{SSY - SSE(X_1)}{MSE(X_1)} = \frac{7.3458 - 2.0443}{0.1136} = 46.67 \qquad \text{(7.4.56)}$$

with $F\text{-}out$. Because $F_C > F\text{-}out = 3$, we conclude that X_1 cannot be deleted from the current model and the backward elimination algorithm terminates. Thus we go to stage 2 again.

Stage 2 (Repeated) We now examine the possibility of adding a variable to the current model by applying the forward selection algorithm once. The best candidate for addition is X_2 because $SSE(X_1, X_2) = 1.3884$ is smaller than $SSE(X_1, X_3)$ and $SSE(X_1, X_4)$. To decide whether or not X_2 should actually be added to the current model we compute

$$F_C = \frac{SSE(X_1) - SSE(X_1, X_2)}{MSE(X_1, X_2)} = \frac{2.0443 - 1.3884}{0.0817} = 8.03 \qquad \text{(7.4.57)}$$

Because F_C is greater than $F\text{-}in = 4.0$, we add X_2 to the current model. Thus in step 2 the variable X_2 is added and the model, which contains variables X_1, X_2, is

$$\beta_0 + \beta_1 x_1 + \beta_2 x_2 \qquad \text{(7.4.58)}$$

Now we return to stage 1.

Stage 1 (Repeated) The revised current model is

$$\beta_0 + \beta_1 x_1 + \beta_2 x_2$$

We examine the possibility of removing one or more variables from the current model. The candidate variable for removal is X_2 because $SSE(X_1)$ is smaller than $SSE(X_2)$. To decide whether or not X_2 should actually be removed from the current model we compute

$$F_C = \frac{SSE(X_1) - SSE(X_1, X_2)}{MSE(X_1, X_2)} = 8.03 \qquad \text{(7.4.59)}$$

as in (7.4.57). Because $F_C > F\text{-}out = 3.0$, we cannot remove X_2 from the current model and the backward elimination algorithm terminates. This sends us back to stage 2.

Stage 2 (Repeated)   We now apply the forward selection algorithm *once*. The predictor variables not in the current model are X_3 and X_4. Because $SSE(X_1, X_2, X_3) = 1.0992$ is smaller than $SSE(X_1, X_2, X_4) = 1.3881$, the candidate variable for addition is X_3. To decide whether or not X_3 should actually be added to the current model we compute

$$F_C = \frac{SSE(X_1, X_2) - SSE(X_1, X_2, X_3)}{MSE(X_1, X_2, X_3)} = \frac{1.3884 - 1.0992}{0.0687} = 4.21 \qquad \text{(7.4.60)}$$

Because $F_C > F\text{-}in = 4.0$, we add X_3 to the current model. Thus in step 3 the variable X_3 is added and the model, which contains variables X_1, X_2, X_3, is

$$\beta_0 + \beta_1 x_1 + \beta_2 x_2 + \beta_3 x_3 \qquad \text{(7.4.61)}$$

We now go back to stage 1.

Stage 1 (Repeated)   The revised current model is

$$\beta_0 + \beta_1 x_1 + \beta_2 x_2 + \beta_3 x_3$$

Again we examine the possibility of deleting one or more variables from the current model. The candidate variable for deletion is X_3 because $SSE(X_1, X_2)$ is smaller than either of $SSE(X_1, X_3)$ or $SSE(X_2, X_3)$. To decide whether or not X_3 should actually be deleted from the current model we compute

$$F_C = \frac{SSE(X_1, X_2) - SSE(X_1, X_2, X_3)}{MSE(X_1, X_2, X_3)} = 4.21 \qquad \text{(7.4.62)}$$

as in (7.4.60). Because $F_C > F\text{-}out = 3.0$, we do not delete X_3 from the current model and the backward elimination algorithm terminates. This sends us back to stage 2.

Stage 2 (Repeated)   Now apply the forward selection procedure *once*. The only variable not in the current model is X_4. To decide whether or not X_4 should be added to the current model we compute

$$F_C = \frac{SSE(X_1, X_2, X_3) - SSE(X_1, X_2, X_3, X_4)}{MSE(X_1, X_2, X_3, X_4)} = \frac{1.0992 - 1.0815}{0.0721} \qquad \text{(7.4.63)}$$

$$= 0.25$$

Because $F_C < F\text{-}in = 4.0$, we cannot add X_4 to the current model and the forward selection algorithm terminates. Because both the backward elimination and the forward selection algorithms have terminated, the current model given in (7.4.61) becomes the final model chosen by the stepwise regression procedure, using $F\text{-}in = 4.0$ and $F\text{-}out = 3.0$ and starting with β_0 given in (7.4.53) as the initial model.

A summary of the results of each step follows:

Step	Variable Added or Deleted	Variables in the Model
Start	—	None
1	X_1 added	X_1
2	X_2 added	X_1, X_2
3	X_3 added	X_1, X_2, X_3

∎

E X A M P L E **7.4.4**

We now again apply the stepwise regression procedure to the GPA data, this time using as our initial model

$$\beta_0 + \beta_3 x_3 + \beta_4 x_4 \tag{7.4.64}$$

This initial model may have been chosen based on the investigator's experience and knowledge. Let us again use $F\text{-}in = 4$ and $F\text{-}out = 3.0$.

Stage 1   The variables X_3, X_4 are in the model. The candidate for deletion is X_4 because $SSE(X_3)$ is smaller than $SSE(X_4)$. To decide whether or not X_4 should actually be deleted from the current model we compute

$$F_C = \frac{SSE(X_3) - SSE(X_3, X_4)}{MSE(X_3, X_4)} = \frac{3.8004 - 2.4180}{0.1422} = 9.72 \tag{7.4.65}$$

Because $F_C > F\text{-}out = 3.0$, we do not delete X_4 from the current model. Instead we proceed to stage 2.

Stage 2   We now apply the forward selection algorithm *once*. The variables not in the current model are X_1 and X_2. The candidate variable for adding to the current model is X_1 because $SSE(X_1, X_3, X_4)$ is smaller than $SSE(X_2, X_3, X_4)$. We compute

$$F_C = \frac{SSE(X_3, X_4) - SSE(X_1, X_3, X_4)}{MSE(X_1, X_3, X_4)} = \frac{2.4180 - 1.4532}{0.0908} = 10.63 \tag{7.4.66}$$

Because F_C is greater than $F\text{-}in = 4.0$, we add X_1 to the current model. Thus in step 1 the variable X_1 is added and the model, which contains variables X_1, X_3, X_4, is

$$\beta_0 + \beta_1 x_1 + \beta_3 x_3 + \beta_4 x_4 \tag{7.4.67}$$

We go to stage 1 and apply the backward elimination procedure.

Stage 1 (Repeated)   The revised current model is

$$\beta_0 + \beta_1 x_1 + \beta_3 x_3 + \beta_4 x_4$$

In this case the candidate for deletion is X_4 because $SSE(X_1, X_3)$ is smaller than $SSE(X_1, X_4)$ and $SSE(X_3, X_4)$. We compute

$$F_C = \frac{SSE(X_1, X_3) - SSE(X_1, X_3, X_4)}{MSE(X_1, X_3, X_4)} = \frac{1.5282 - 1.4532}{0.0908} = 0.83 \qquad \textbf{(7.4.68)}$$

Because F_C is less than $F\text{-}out = 3.0$, we delete X_4 from the model. Thus in step 2 the variable X_4 is deleted and the model, which contains X_1, X_3, is

$$\beta_0 + \beta_1 x_1 + \beta_3 x_3 \qquad \textbf{(7.4.69)}$$

We continue to apply the backward elimination algorithm to see if we can delete any more variables.

Stage 1 (Continued)   The revised current model is

$$\beta_0 + \beta_1 x_1 + \beta_3 x_3$$

The candidate variable for deletion is X_3 because $SSE(X_1)$ is smaller than $SSE(X_3)$. We compute

$$F_C = \frac{SSE(X_1) - SSE(X_1, X_3)}{MSE(X_1, X_3)} = \frac{2.0443 - 1.5282}{0.0899} = 5.74 \qquad \textbf{(7.4.70)}$$

Because F_C is greater than $F\text{-}out = 3.0$, X_3 cannot be deleted from the current model. We now apply the forward selection algorithm *once*.

Stage 2 (Repeated)   Variables not in the current model are X_2 and X_4. The candidate variable for addition is X_2 because $SSE(X_1, X_2, X_3)$ is smaller than $SSE(X_1, X_3, X_4)$. We compute

$$F_C = \frac{SSE(X_1, X_3) - SSE(X_1, X_2, X_3)}{MSE(X_1, X_2, X_3)} = \frac{1.5282 - 1.0992}{0.0687} = 6.24 \qquad \textbf{(7.4.71)}$$

Because F_C is greater than $F\text{-}in = 4.0$, X_2 is added to the model. Thus in step 3 the variable X_2 is added and the model, which contains variables X_1, X_2, X_3, is

$$\beta_0 + \beta_1 x_1 + \beta_2 x_2 + \beta_3 x_3 \qquad \textbf{(7.4.72)}$$

We now see whether variables can be deleted by the backward elimination procedure.

Stage 1 (Repeated)   The revised current model is

$$\beta_0 + \beta_1 x_1 + \beta_2 x_2 + \beta_3 x_3$$

The candidate variable for deletion is X_3 because $SSE(X_1, X_2)$ is smaller than $SSE(X_1, X_3)$ and $SSE(X_2, X_3)$. We compute

$$F_C = \frac{SSE(X_1, X_2) - SSE(X_1, X_2, X_3)}{MSE(X_1, X_2, X_3)} = \frac{1.3884 - 1.0992}{0.0687} = 4.21 \qquad \textbf{(7.4.73)}$$

Because F_C is greater than $F\text{-}out = 3.0$, X_3 cannot be deleted from the current model. We proceed to apply the forward selection algorithm *once*.

Stage 2 (Repeated)   The only candidate for addition is X_4. We compute

$$F_C = \frac{SSE(X_1, X_2, X_3) - SSE(X_1, X_2, X_3, X_4)}{MSE(X_1, X_2, X_3, X_4)} = \frac{1.0992 - 1.0815}{0.0721} \quad \text{(7.4.74)}$$

$$= 0.25$$

Because F_C is smaller than $F\text{-}in = 4.0$, we cannot add X_4 to the current model. Since no more variables can be added or deleted from the current model given in (7.4.72), the stepwise regression algorithm terminates and chooses the model

$$\beta_0 + \beta_1 x_1 + \beta_2 x_2 + \beta_3 x_3 \quad \text{(7.4.75)}$$

as the final model.

A summary of the steps of the stepwise procedure for this example follows:

Step	Variable Added or Deleted	Variables in the Model
Start	—	X_3, X_4
1	X_1 added	X_1, X_3, X_4
2	X_4 deleted	X_1, X_3
3	X_2 added	X_1, X_2, X_3

Note   In this particular problem the stepwise regression procedure results in the same final model whether we use β_0 or $\beta_0 + \beta_3 x_3 + \beta_4 x_4$ as the initial model. In general this will not be the case because the final model may depend on which initial model is used. Also, for the GPA data, we find that all three variable selection procedures—forward selection, backward elimination, and stepwise regression—lead to the same final model in the examples discussed if the same values of $F\text{-}in$ and $F\text{-}out$ are used. This is not always the case as you can see from the following example. ■

E X A M P L E **7.4.5**

To illustrate the computations we use the small data set given in Table 7.4.2. Here Y is the response variable, and X_1, X_2, and X_3 are the predictor variables. The total number of observations is 10. (In real problems we do not recommend selection procedures with such a small data set.)

We will carry out a stepwise regression analysis for this data using $F\text{-}in = 3.0$ and $F\text{-}out = 3.0$. For convenience, we list all the pertinent sums of squares and mean squares in Table 7.4.3.

We start with the initial model

$$\beta_0 \quad \text{(7.4.76)}$$

Stage 1   There are no variables to delete and so we proceed to stage 2.

Stage 2 We perform the forward selection procedure *once*. The best candidate to add to the current model is X_1 because $SSE(X_1)$ is smaller than $SSE(X_2)$ and $SSE(X_3)$. To determine whether or not X_1 should be added to the current model we compute

$$F_C = \frac{SSY - SSE(X_1)}{MSE(X_1)} = \frac{8.5090 - 5.3710}{0.6714} = 4.67 \qquad \textbf{(7.4.77)}$$

Because $F_C > F\text{-}in = 3.0$, we add X_1 to the current model. Thus in step 1 the variable X_1 is added and the model, which contains variable X_1, is

$$\beta_0 + \beta_1 x_1 \qquad \textbf{(7.4.78)}$$

We now go back to stage 1.

T A B L E **7.4.2**

Observation Number	Y	X_1	X_2	X_3
1	12.5	7.0	1.7	5.7
2	11.4	6.8	2.0	5.0
3	9.7	1.7	2.1	3.8
4	11.4	3.8	2.1	4.7
5	10.7	3.8	3.3	2.7
6	12.9	3.3	4.1	3.0
7	10.6	3.3	2.6	4.3
8	10.7	3.2	2.5	3.5
9	10.5	2.2	4.0	2.4
10	11.7	5.2	2.9	4.1

T A B L E **7.4.3**
Sums of Squares and Mean Squares for All 8 Subset Models for the Data in Table 7.4.2

Model	Predictors in the Model	Sum of Squares	Mean Square
(1)	None	8.5090	0.9454
(2)	X_1	5.3710	0.6714
(3)	X_2	8.4160	1.0520
(4)	X_3	7.5759	0.9470
(5)	X_1, X_2	3.6315	0.5188
(6)	X_1, X_3	5.1229	0.7318
(7)	X_2, X_3	2.4325	0.3475
(8)	X_1, X_2, X_3	2.0764	0.3461

Stage 1 (Repeated)   The revised current model is

$$\beta_0 + \beta_1 x_1$$

In this stage we examine the possibility of removing one or more variables from the current model by applying the backward elimination procedure. The only candidate for removal is X_1 because it is the only predictor variable in the current model. To decide whether or not X_1 should be removed from the current model we need to compare the quantity

$$F_C = \frac{SSY - SSE(X_1)}{MSE(X_1)} = \frac{8.5090 - 5.3710}{0.6714} = 4.67 \qquad \text{(7.4.79)}$$

with *F-out*. Since $F_C >$ *F-out*, we conclude that X_1 cannot be deleted from the current model, and the backward elimination algorithm terminates. Thus we go to stage 2 again.

Stage 2 (Repeated)   We examine the possibility of adding a variable to the current model by applying the forward selection algorithm *once*. The best candidate for addition is X_2 because $SSE(X_1, X_2) = 3.6315$ is smaller than $SSE(X_1, X_3) = 5.1229$. To decide whether or not X_2 should actually be added to the current model we compute

$$F_C = \frac{SSE(X_1) - SSE(X_1, X_2)}{MSE(X_1, X_2)} = \frac{5.3710 - 3.6315}{0.5188} = 3.35 \qquad \text{(7.4.80)}$$

Because F_C is greater than *F-in* = 3.0, we add X_2 to the current model.

Thus in step 2 the variable X_2 is added and the model, which contains variables X_1, X_2, is

$$\beta_0 + \beta_1 x_1 + \beta_2 x_2 \qquad \text{(7.4.81)}$$

Now we return to stage 1.

Stage 1 (Repeated)   The revised current model is

$$\beta_0 + \beta_1 x_1 + \beta_2 x_2$$

We now attempt to remove one or more variables from the current model. The candidate variable for removal is X_2 because $SSE(X_1)$ is smaller than $SSE(X_2)$. To decide whether or not X_2 should actually be removed from the current model we compute

$$F_C = \frac{SSE(X_1) - SSE(X_1, X_2)}{MSE(X_1, X_2)} = \frac{5.3710 - 3.6315}{0.5188} = 3.35 \qquad \text{(7.4.82)}$$

as in (7.4.80). Because $F_C >$ *F-out* = 3.0, we cannot remove X_2 from the current model, and the backward elimination algorithm terminates. This sends us back to stage 2.

Stage 2 (Repeated)   We now apply the forward selection algorithm *once*. The only predictor variable not in the current model is X_3. To decide whether or not X_3 should actually be added to the current model we compute

$$F_C = \frac{SSE(X_1, X_2) - SSE(X_1, X_2, X_3)}{MSE(X_1, X_2, X_3)} = \frac{3.6315 - 2.0764}{0.3461} = 4.49 \qquad \text{(7.4.83)}$$

Since $F_C > F\text{-}in = 3.0$, we add X_3 to the current model. Thus in step 3 the variable X_3 is added and the model, which contains variables X_1, X_2, X_3, is

$$\beta_0 + \beta_1 x_1 + \beta_2 x_2 + \beta_3 x_3 \qquad \text{(7.4.84)}$$

We now go back to stage 1.

Stage 1 (Repeated)   The revised current model is

$$\beta_0 + \beta_1 x_1 + \beta_2 x_2 + \beta_3 x_3$$

Again we examine the possibility of deleting one or more variables from the current model. The candidate variable for deletion is X_1 because $SSE(X_2, X_3)$ is smaller than either $SSE(X_1, X_3)$ or $SSE(X_1, X_2)$. To decide whether or not X_1 should actually be deleted from the current model we compute

$$F_C = \frac{SSE(X_2, X_3) - SSE(X_1, X_2, X_3)}{MSE(X_1, X_2, X_3)} = \frac{2.4325 - 2.0764}{0.3461} = 1.03 \qquad \text{(7.4.85)}$$

Because $F_C < F\text{-}out = 3.0$, we delete X_1 from the current model. Thus in step 4 the variable X_1 is deleted and the model, which contains the variables X_2, X_3, is

$$\beta_0 + \beta_2 x_2 + \beta_3 x_3 \qquad \text{(7.4.86)}$$

We continue with stage 1 to see whether any more variables can be removed from the model.

Stage 1 (Continued)   The revised current model is

$$\beta_0 + \beta_2 x_2 + \beta_3 x_3$$

The candidate variable for deletion is X_2 since $SSE(X_3)$ is smaller than $SSE(X_2)$. To decide whether or not X_2 should actually be deleted from the current model we compute

$$F_C = \frac{SSE(X_3) - SSE(X_2, X_3)}{MSE(X_2, X_3)} = \frac{7.5759 - 2.4325}{0.3475} = 14.80 \qquad \text{(7.4.87)}$$

Because $F_C > F\text{-}out = 3.0$, we cannot delete X_2 from the current model. This sends us back to stage 2.

Stage 2 (Repeated)   The current model is

$$\beta_0 + \beta_2 x_2 + \beta_3 x_3$$

Apply the forward selection procedure *once*. The only variable not in the current model is X_1. To decide whether or not X_1 should be added to the current model we compute

$$F_C = \frac{SSE(X_2, X_3) - SSE(X_1, X_2, X_3)}{MSE(X_1, X_2, X_3)} = \frac{2.4325 - 2.0764}{0.3461} = 1.03 \qquad \textbf{(7.4.88)}$$

Since $F_C < F\text{-}in = 3.0$, we cannot add X_1 to the current model, and the forward selection algorithm terminates. Because both the backward elimination and the forward selection algorithms have terminated, the current model given in (7.4.86) becomes the final model chosen by the stepwise regression procedure, using $F\text{-}in = 3.0$ and $F\text{-}out = 3.0$ and starting with (7.4.76) as the initial model.

A summary of the steps of the stepwise procedure for this example follows:

Step	Variable Added or Deleted	Variables in the Model
Start	—	None
1	X_1 added	X_1
2	X_2 added	X_1, X_2
3	X_3 added	X_1, X_2, X_3
4	X_1 deleted	X_2, X_3

You can easily verify that the forward selection method with $F\text{-}in = 3.0$ will lead to the full model as the final model, whereas the backward elimination method with $F\text{-}out = 3.0$ will pick the model in (7.4.86) as the final model. Hence the backward elimination and stepwise regression procedures agree for this problem, but the forward selection algorithm leads to a different model.   ∎

Caution

Note that even though the variable selection procedures discussed in this section automatically pick a best final subset model, this final model is not chosen after examining *all* the possible subset models. Typically it is chosen after examining only a fraction of all the possible subset models. Moreover, the model chosen by the algorithm very much depends on the choice of values for $F\text{-}in$ and $F\text{-}out$. Smaller values for $F\text{-}in$ (larger values for $F\text{-}out$) generally lead to final models with a larger number of predictor variables, whereas larger values of $F\text{-}in$ (smaller values of $F\text{-}out$) generally lead to final models with a smaller number of predictor variables. In practice, it is advisable to try different values for $F\text{-}in$ and $F\text{-}out$. Moreover, you should examine not only the final models from these procedures but also the models chosen at each stage of the algorithm. The bottom line is this: When conducting a variable selection analysis, you should examine as many subset models as possible—when it is not practical to examine every subset model—and construct a short list of subset models for further consideration.

Realistically, additional data need to be collected to validate the results of subset analyses. The additional data may or may not support the results of such analyses, and it is quite possible that the chosen models will need to be revised or discarded in favor of other models. This iterative nature of scientific inquiry should come as no surprise.

When we have a large amount of data, the validity of a tentative model can be examined by applying **data splitting** methods whereby the data are divided (randomly) into two sets, with one of the sets used to carry out variable selection procedures and the other set used to assess the validity of the models in the chosen short list.

Problems 7.4

7.4.1 Consider the GPA problem in Example 4.4.2. The sums of squares and mean squares are in Table 7.4.1.

a Apply the forward selection procedure with F-$in = 2.0$. What variables are in the final model?

b Repeat (a) with F-$in = 5.0$.

c Apply the backward elimination procedure with F-$out = 2.0$. What variables are in the final model?

d Repeat (c) with F-$out = 5.0$.

e Apply the stepwise regression procedure with F-$in = F$-$out = 2.0$. List the variables in the model at each step if the initial model is β_0.

f In (e) what variables are in the final model if the initial model is $\beta_0 + \beta_1 x_1$?

g In (e) what variables are in the final model if the initial model is $\beta_0 + \beta_4 x_4$?

h Repeat (e) with F-$in = F$-$out = 4.0$.

i Repeat (e) with F-$in = 2.0$ and F-$out = 4.0$. What's wrong?

j Repeat (e) with F-$in = 4.0$ and F-$out = 2.0$.

7.4.2 Consider Example 7.4.5. The data are given in Table 7.4.2 and the sums of squares and mean squares are given in Table 7.4.3.

a Apply the forward selection procedure with F-$in = 5.0$. What variables are in the final model?

b Repeat (a) with F-$in = 2.0$. What variables are in the final model?

c Apply the backward elimination procedure with F-$out = 4.0$. What variables are in the final model?

d Repeat (c) with F-$out = 2.0$. What variables are in the final model?

e Apply the stepwise regression procedure with F-$in = F$-$out = 2.0$. List the variables in the model at each step if the initial model is $\beta_0 + \beta_1 x_1$.

f In (e) list the variables in the model at each step if the initial model is $\beta_0 + \beta_3 x_3$.

g Repeat (e) with F-$in = F$-$out = 4.0$.

h Repeat (e) with F-$in = 2.0$ and F-$out = 4.0$. What's wrong?

i Repeat (f) with F-$in = 4.0$ and F-$out = 2.0$.

7.4.3   In Exercise 4.12.4 an investigator is interested in studying how Y, the height at age 18 years of males belonging to a group of people who have lived in mountain isolation for several generations, is related to the following variables.

$$X_1 = \text{length at birth}$$
$$X_2 = \text{mother's height at age 18}$$
$$X_3 = \text{father's height at age 18}$$
$$X_4 = \text{maternal grandmother's height at age 18}$$
$$X_5 = \text{maternal grandfather's height at age 18}$$
$$X_6 = \text{paternal grandmother's height at age 18}$$
$$X_7 = \text{paternal grandfather's height at age 18}$$

All heights and lengths are in inches. A simple random sample of 20 males of age 18 or more was drawn, and all the preceding information was recorded. The data are given in Table 7.4.4 and are also stored in the file **age18.dat** on the data disk. In Exhibit 7.4.1 is a SAS computer output (edited for ease of presentation) for all-subsets regressions and includes R^2, $adj\text{-}R^2$, C_p, s, and SSE.

T A B L E 7.4.4

Heights at Age 18 of a Sample of 20 males

Observation Number	Y	X_1	X_2	X_3	X_4	X_5	X_6	X_7
1	67.2	19.7	60.5	70.3	65.7	69.3	65.7	67.3
2	69.1	19.6	64.9	70.4	62.6	69.6	64.6	66.4
3	67.0	19.4	65.4	65.8	66.2	68.8	64.0	69.4
4	72.4	19.4	63.4	71.9	60.7	68.0	64.9	67.1
5	63.6	19.7	65.1	65.1	65.5	65.5	61.8	70.9
6	72.7	19.6	65.2	71.1	63.5	66.2	67.3	68.6
7	68.5	19.8	64.3	67.9	62.4	71.4	63.4	69.4
8	69.7	19.7	65.3	68.8	61.5	66.0	62.4	67.7
9	68.4	19.7	64.5	68.7	63.9	68.8	62.3	68.8
10	70.4	19.9	63.4	70.3	65.9	69.0	63.7	65.1
11	67.5	18.9	63.3	70.4	63.7	68.2	66.2	68.5
12	73.3	20.8	66.2	70.2	65.4	66.6	61.7	64.0
13	70.0	20.3	64.9	68.8	65.2	70.2	62.4	67.0
14	69.8	19.7	63.5	70.3	63.1	64.4	65.1	67.0
15	63.6	19.9	62.0	65.5	64.1	67.7	62.1	66.5
16	64.3	19.6	63.5	65.2	63.9	70.0	64.2	64.5
17	68.5	21.3	66.1	65.4	64.8	68.4	66.4	70.8
18	70.5	20.1	64.8	70.2	65.3	65.5	63.7	66.9
19	68.1	20.2	62.6	68.6	63.7	69.8	66.7	68.0
20	66.1	19.2	62.2	67.3	63.6	70.9	63.6	66.7

E X H I B I T 7.4.1
SAS Output for Problem 7.4.3

SAS   0:00 Saturday, Jan 1, 1994

N = 20   Regression Models for Dependent Variable: Y

In	R-square	Adj Rsq	C(p)	Root MSE	SSE	Variables in Model
1	0.626215	0.605449	38.1	1.73969	54.477	X3
1	0.159555	0.112864	105.6	2.60865	122.491	X2
1	0.067723	0.015930	118.9	2.74748	135.875	X1
1	0.063297	0.011258	119.5	2.75399	136.520	X5
1	0.053640	0.001064	120.9	2.76815	137.928	X7
1	0.045767	-.007246	122.1	2.77964	139.075	X4
1	0.037081	-.016414	123.3	2.79226	140.341	X6
2	0.850862	0.833316	7.5778	1.13075	21.736	X2 X3
2	0.793939	0.769697	15.8136	1.32914	30.032	X1 X3
2	0.635011	0.592072	38.8077	1.76894	53.195	X3 X5
2	0.628645	0.584956	39.7289	1.78430	54.123	X3 X6
2	0.627145	0.583280	39.9459	1.78790	54.342	X3 X7
2	0.626973	0.583087	39.9708	1.78831	54.367	X3 X4
2	0.274930	0.189628	90.9055	2.49323	105.676	X2 X7
2	0.242256	0.153109	95.6	2.54879	110.438	X2 X6
2	0.216180	0.123965	99.4	2.59228	114.238	X2 X4
2	0.176363	0.079465	105.2	2.65730	120.041	X2 X5
2	0.169445	0.071733	106.2	2.66844	121.050	X1 X2
2	0.169122	0.071372	106.2	2.66896	121.097	X1 X4
2	0.125925	0.023092	112.5	2.73746	127.393	X5 X7
2	0.121524	0.018174	113.1	2.74434	128.034	X1 X7
2	0.120306	0.016812	113.3	2.74624	128.212	X1 X5
2	0.116103	0.012116	113.9	2.75279	128.824	X6 X7
2	0.113808	0.009550	114.2	2.75637	129.158	X1 X6
2	0.110820	0.006211	114.6	2.76101	129.594	X4 X5
2	0.107410	0.002400	115.1	2.76630	130.091	X5 X6
2	0.099637	-.006288	116.3	2.77832	131.224	X4 X7
2	0.071884	-.037306	120.3	2.82081	135.269	X4 X6

E X H I B I T 7.4.1
(Continued)

3	0.904959	0.887139	1.7509	0.93045	13.852	X1 X2 X3
3	0.857221	0.830450	8.6578	1.14043	20.809	X2 X3 X7
3	0.856767	0.829911	8.7234	1.14224	20.876	X2 X3 X5
3	0.853147	0.825612	9.2471	1.15659	21.403	X2 X3 X6
3	0.851003	0.823067	9.5573	1.16500	21.716	X2 X3 X4
3	0.803560	0.766727	16.4216	1.33768	28.630	X1 X3 X4
3	0.797093	0.759048	17.3573	1.35952	29.573	X1 X3 X7
3	0.795603	0.757278	17.5729	1.36450	29.790	X1 X3 X5
3	0.795326	0.756950	17.6129	1.36543	29.830	X1 X3 X6
3	0.636360	0.568177	40.6126	1.82001	52.999	X3 X5 X6
3	0.635405	0.567043	40.7509	1.82240	53.138	X3 X4 X5
3	0.635290	0.566907	40.7675	1.82268	53.155	X3 X5 X7
3	0.631290	0.562156	41.3462	1.83265	53.738	X3 X6 X7
3	0.629232	0.559713	41.6439	1.83776	54.038	X3 X4 X6
3	0.628118	0.558390	41.8051	1.84052	54.200	X3 X4 X7
3	0.437918	0.332528	69.3239	2.26275	81.921	X2 X6 X7
3	0.334408	0.209610	84.3000	2.46230	97.007	X2 X4 X7
3	0.291041	0.158111	90.5746	2.54125	103.328	X2 X5 X7
3	0.281658	0.146968	91.9322	2.55802	104.695	X2 X4 X6
3	0.278528	0.143252	92.3849	2.56358	105.151	X1 X2 X7
3	0.258781	0.119803	95.2	2.59843	108.029	X2 X5 X6
3	0.252943	0.112870	96.1	2.60864	108.880	X1 X2 X4
3	0.251404	0.111042	96.3	2.61133	109.105	X1 X2 X6
3	0.232792	0.088940	99.0	2.64359	111.817	X2 X4 X5
3	0.223356	0.077735	100.4	2.65980	113.192	X1 X4 X7
3	0.219631	0.073312	100.9	2.66617	113.735	X1 X4 X5
3	0.200984	0.051168	103.6	2.69783	116.453	X5 X6 X7
3	0.199202	0.049053	103.9	2.70084	116.713	X1 X4 X6
3	0.196018	0.045271	104.3	2.70621	117.177	X1 X6 X7
3	0.187802	0.035515	105.5	2.72000	118.374	X1 X2 X5
3	0.182330	0.029017	106.3	2.72915	119.172	X1 X5 X7
3	0.173828	0.018921	107.5	2.74330	120.411	X4 X5 X7
3	0.172748	0.017638	107.7	2.74509	120.568	X1 X5 X6
3	0.147581	-.012248	111.3	2.78653	124.236	X4 X6 X7
3	0.142786	-.017941	112.0	2.79436	124.935	X4 X5 X6

EXHIBIT 7.4.1

(Continued)

4	0.910211	0.886267	2.9910	0.93404	13.086	X1	X2	X3	X5
4	0.908997	0.884729	3.1667	0.94033	13.263	X1	X2	X3	X4
4	0.906468	0.881526	3.5325	0.95330	13.632	X1	X2	X3	X7
4	0.906074	0.881027	3.5895	0.95531	13.689	X1	X2	X3	X6
4	0.865298	0.829377	9.4892	1.14404	19.632	X2	X3	X6	X7
4	0.862259	0.825529	9.9288	1.15687	20.075	X2	X3	X5	X7
4	0.858558	0.820840	10.4643	1.17231	20.615	X2	X3	X5	X6
4	0.857225	0.819151	10.6572	1.17782	20.809	X2	X3	X4	X7
4	0.857085	0.818974	10.6775	1.17840	20.829	X2	X3	X4	X5
4	0.853370	0.814269	11.2149	1.19361	21.371	X2	X3	X4	X6
4	0.805732	0.753927	18.1073	1.37389	28.314	X1	X3	X4	X7
4	0.805623	0.753789	18.1231	1.37428	28.330	X1	X3	X4	X5
4	0.805446	0.753564	18.1488	1.37490	28.355	X1	X3	X4	X6
4	0.800836	0.747726	18.8157	1.39110	29.027	X1	X3	X6	X7
4	0.798134	0.744303	19.2066	1.40050	29.421	X1	X3	X5	X7
4	0.796628	0.742396	19.4245	1.40571	29.640	X1	X3	X5	X6
4	0.637474	0.540801	42.4514	1.87681	52.837	X3	X5	X6	X7
4	0.636675	0.539788	42.5671	1.87888	52.953	X3	X4	X5	X6
4	0.635780	0.538655	42.6965	1.88119	53.083	X3	X4	X5	X7
4	0.632148	0.534054	43.2221	1.89055	53.613	X3	X4	X6	X7
4	0.472904	0.332345	66.2620	2.26306	76.822	X2	X4	X6	X7
4	0.453379	0.307613	69.0870	2.30460	79.668	X2	X5	X6	X7
4	0.439511	0.290047	71.0935	2.33365	81.689	X1	X2	X6	X7
4	0.357523	0.186196	82.9557	2.49851	93.638	X1	X2	X4	X7
4	0.350314	0.177064	83.9988	2.51249	94.689	X2	X4	X5	X7
4	0.312278	0.128886	89.5019	2.58499	100.232	X1	X2	X4	X6
4	0.298047	0.110860	91.5608	2.61160	102.306	X2	X4	X5	X6
4	0.295590	0.107747	91.9164	2.61616	102.665	X1	X2	X5	X7
4	0.281952	0.090473	93.8895	2.64137	104.652	X1	X4	X5	X7
4	0.276871	0.084037	94.6	2.65069	105.393	X1	X4	X6	X7
4	0.272666	0.078711	95.2	2.65839	106.006	X1	X2	X4	X5
4	0.269408	0.074584	95.7	2.66434	106.480	X1	X2	X5	X6
4	0.268330	0.073218	95.9	2.66630	106.638	X1	X5	X6	X7
4	0.254977	0.056305	97.8	2.69052	108.584	X1	X4	X5	X6
4	0.232757	0.028159	101.0	2.73035	111.822	X4	X5	X6	X7

EXHIBIT 7.4.1
(Continued)

5	0.913517	0.882630	4.5126	0.94885	12.604	X1 X2 X3 X4 X5
5	0.911339	0.879674	4.8278	0.96073	12.922	X1 X2 X3 X5 X7
5	0.911009	0.879227	4.8755	0.96251	12.970	X1 X2 X3 X5 X6
5	0.910902	0.879082	4.8909	0.96309	12.986	X1 X2 X3 X4 X7
5	0.909781	0.877559	5.0532	0.96913	13.149	X1 X2 X3 X4 X6
5	0.909639	0.877368	5.0737	0.96989	13.170	X1 X2 X3 X6 X7
5	0.869013	0.822232	10.9516	1.16774	19.091	X2 X3 X5 X6 X7
5	0.865307	0.817202	11.4879	1.18415	19.631	X2 X3 X4 X6 X7
5	0.862326	0.813157	11.9191	1.19718	20.065	X2 X3 X4 X5 X7
5	0.858969	0.808601	12.4048	1.21169	20.555	X2 X3 X4 X5 X6
5	0.809813	0.741889	19.5169	1.40710	27.719	X1 X3 X4 X6 X7
5	0.807197	0.738339	19.8954	1.41674	28.100	X1 X3 X4 X5 X7
5	0.807047	0.738135	19.9171	1.41729	28.122	X1 X3 X4 X5 X6
5	0.801236	0.730249	20.7578	1.43847	28.969	X1 X3 X5 X6 X7
5	0.637946	0.508641	44.3832	1.94142	52.768	X3 X4 X5 X6 X7
5	0.488252	0.305485	66.0414	2.30814	74.585	X2 X4 X5 X6 X7
5	0.485651	0.301955	66.4177	2.31400	74.964	X1 X2 X4 X6 X7
5	0.455610	0.261185	70.7642	2.38061	79.342	X1 X2 X5 X6 X7
5	0.375894	0.152999	82.2978	2.54896	90.961	X1 X2 X4 X5 X7
5	0.345401	0.111615	86.7096	2.61049	95.405	X1 X4 X5 X6 X7
5	0.331501	0.092752	88.7206	2.63805	97.431	X1 X2 X4 X5 X6
6	0.914988	0.875751	6.2999	0.97626	12.390	X1 X2 X3 X4 X5 X7
6	0.914078	0.874422	6.4314	0.98147	12.523	X1 X2 X3 X4 X5 X6
6	0.913690	0.873855	6.4876	0.98368	12.579	X1 X2 X3 X5 X6 X7
6	0.913647	0.873792	6.4938	0.98393	12.586	X1 X2 X3 X4 X6 X7
6	0.869082	0.808658	12.9417	1.21151	19.081	X2 X3 X4 X5 X6 X7
6	0.810470	0.722995	21.4218	1.45769	27.623	X1 X3 X4 X5 X6 X7
6	0.502828	0.273364	65.9325	2.36091	72.461	X1 X2 X4 X5 X6 X7
7	0.917060	0.868679	8.0000	1.00366	12.088	X1 X2 X3 X4 X5 X6 X7

a Find the predictor variable that results in the best prediction function (for the sample data at hand) among all subset models containing only one predictor.

b Find the two predictor variables that result in the best prediction function (for the sample data at hand) among all subset models containing exactly two variables.

c Find the three predictor variables that result in the best prediction function (for the sample data at hand) among all subset models containing exactly three variables.

d Find a short list of the three best models using the criterion R^2 in Section 7.3.

e   Find a short list of the three best models using the criterion *adj-R²* in Section 7.3.

f   Find a short list of the three best models using the criterion *s* in Section 7.3.

g   Find a short list of the three best models using the criterion C_p in Section 7.3.

h   Find a short list of four best models using C_p and $C_p - p$ together.

i   Write a short report giving your findings using parts (a) through (h).

7.4.4   **a**   In Problem 7.4.3 apply the forward selection procedure with *F-in* = 5.0. What variables are in the final model?

b   Repeat (a) with *F-in* = 4.0. What variables are in the final model?

c   Apply the backward elimination procedure with *F-out* = 5.0. What variables are in the final model?

d   Repeat (c) with *F-out* = 4.0. What variables are in the final model?

e   Apply the stepwise regression procedure with *F-in* = *F-out* = 4.0. List the variables that are in the model in each step if the initial model is $\beta_0 + \beta_1 x_1$.

f   In (e) list the variables that are in the model at each step if the initial model is β_0.

g   Repeat (e) with *F-in* = *F-out* = 3.0.

h   Repeat (e) with *F-in* = 4.0 and *F-out* = 2.0.

7.5
Growth Curves

In this section we discuss an important class of statistical models known as **growth curve models** or **longitudinal models**. These models may be viewed as modifications of the multiple regression model. We introduce this topic with some examples.

E X A M P L E 7.5.1

Consider the population of all premature babies (let *M* represent the number of such babies) born in the state of New York over the past 5 years. An investigator is interested in studying how premature babies grow (how their weight *Y* changes) as a function of time *t* from day 5 to day 50 after birth. Let us consider one baby in this population, whom we denote as baby *I*. The observed weight of this baby at time *t* will be denoted by $Y_I(t)$, and its 'true' weight at time *t* will be denoted by $\mu_{Y_I}(t)$, or simply $\mu_I(t)$ for ease of notation. Suppose that the 'true' weight of baby *I* as a function of time *t* is given by

$$\mu_I(t) = \alpha_I + \beta_I t \tag{7.5.1}$$

and that $Y_I(t)$, the observed weight at time *t*, is related to the 'true' weight according to the model

$$Y_I(t) = \mu_I(t) + E_{I,t} = \alpha_I + \beta_I t + E_{I,t} \tag{7.5.2}$$

The quantity $E_{I,t}$ represents the difference between the observed weight and the 'true' weight at time t for baby I.

In many applications it is reasonable to regard $E_{I,t}$ as a randomly chosen number from a Gaussian population with mean zero and standard deviation σ_E. In such a case, the function $\mu_I(t)$ is the regression function of the weight of baby I on the predictor variable t, and it is called the *growth curve* of baby I. The reason that $Y_I(t)$, the observed weight of baby I at time t, does not equal $\mu_I(t)$, the 'true' weight of baby I at time t, is that the 'true' weight may not be observed exactly but is observed with error; this error includes measurement error as well as other uncontrollable random fluctuations. In practice the 'true' weight of baby I as a function of time t may not be of the form $\mu_I(t) = \alpha_I + \beta_I t$ *exactly*, but such a function may be an adequate approximation to the growth curve for baby I.

Clearly, there is a growth curve for each of the M babies in the population, so there are M growth curves

$$\mu_I(t) = \alpha_I + \beta_I t \qquad \text{for } I = 1, \ldots, M \tag{7.5.3}$$

The investigator may be interested in the growth curve of some of the individual babies, or in the average of the growth curves of all M babies, or both. The average of the growth curves of all M babies is denoted by $\mu_Y(t)$ and is called the **population growth curve**. It is given by

$$\mu_Y(t) = \frac{1}{M} \sum_{I=1}^{M} \mu_I(t) = \alpha + \beta t \tag{7.5.4}$$

where

$$\alpha = \frac{1}{M} \sum_{i=1}^{M} \alpha_I \qquad \text{and} \qquad \beta = \frac{1}{M} \sum_{i=1}^{M} \beta_I \tag{7.5.5}$$

Note   To be consistent with our notation for regression functions thus far, we should write Equation (7.5.1) using β_0, β_1, etc. However, we need one set of β coefficients for each item (baby) in the population, thus making it necessary to use double subscripts. Because this would lead to unnecessarily complicated notation, we have used α_I and β_I.

To estimate the population intercept and slope, α and β in (7.5.5), we select a simple random sample of m babies from the population of M babies and measure the weight of each baby at k preselected times $t_1, \ldots, t_k$. From these data we calculate point estimates and confidence intervals for α_i and β_i, the coefficients in the growth curve for the ith baby in the sample, for $i = 1, \ldots, m$, and from these we obtain point estimates and confidence intervals for the population regression coefficients α and β.

To summarize, there is a population of M growth curves (corresponding to the M babies), and a simple random sample of m of them is selected (they correspond to the m babies in the sample). Values of each of these m regression functions are measured at the same k preselected times $t_1, \ldots, t_k$. From these data we compute point estimates and confidence intervals for parameters in each of the m growth curves (i.e., for each of the m babies in the sample) and for the parameters in the population growth curve in (7.5.4).

To graphically illustrate the situation, we have plotted, in Figure 7.5.1, the growth curves, $\mu_i(t)$, $i = 1, 2, 3$, for three babies in the population, and also the population growth curve $\mu_Y(t)$, i.e., the average of all M growth curves. ∎

F I G U R E 7.5.1

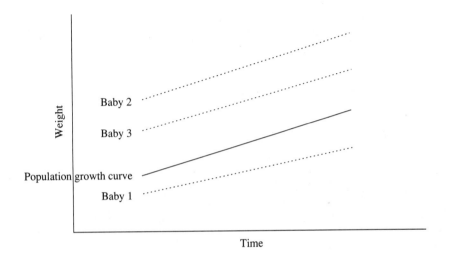

E X A M P L E 7.5.2
Suppose an investigator is interested in the change in blood pressure (Y) of individuals, in a population of M individuals, as a function of the number of minutes (t) after they are placed in a certain stressful situation, such as being subjected to very loud noise. In this example the growth curves are the regression functions relating blood pressures of the individuals in the population to elapsed time since the introduction of stress. Suppose that the ith individual's true blood pressure as a function of time t is given by

$$\mu_I(t) = \alpha_I + \beta_I t + \gamma_I t^2 \qquad 0 \le t \le 30$$

Each individual in the population has a growth curve that is a quadratic function of t but perhaps with different coefficients α_I, β_I, and γ_I. The average of all M growth curves is called the population growth curve and it is given by

$$\mu_Y(t) = \alpha + \beta t + \gamma t^2$$

where

$$\alpha = \frac{1}{M}\sum_{I=1}^{M}\alpha_I, \quad \beta = \frac{1}{M}\sum_{I=1}^{M}\beta_I, \text{ and } \gamma = \frac{1}{M}\sum_{I=1}^{M}\gamma_I \quad \blacksquare$$

In Examples 7.5.1 and 7.5.2 the predictor variable is time. The response variable Y is measured on each subject in the sample at several preselected times. However, the distinguishing feature of growth curve models is not that the predictor variable is time but that the Y values are measured or observed on each of the m sample subjects more than once, usually under different conditions.

This type of model is sometimes referred to as a *repeated measurements* model. The different conditions are different times in Examples 7.5.1 and 7.5.2. This is not the case in the following example, which involves a growth curve model but time is not the predictor variable.

E X A M P L E 7.5.3

A company evaluates Y, the number of miles per gallon (mpg) that different makes and models of new cars get when X milliliters/gallon (ml/g for short) of a gasoline additive is used. The company wants to make statements such as, "On the average this make of automobile gets $\mu_Y(x)$ miles per gallon when x ml/g of the additive is used." Suppose that the mpg for the ith car in the population, when x ml/g of the additive is used, is given by the regression function $\mu_{Y_I}(x)$, or $\mu_I(x)$ for short, where

$$\mu_I(x) = \alpha_I + \beta_I x \qquad 50 \leq x \leq 90$$

The population growth curve is the average of M regression functions of all M cars in the population, and it is given by

$$\mu_Y(x) = \alpha + \beta x$$

where

$$\alpha = \frac{1}{M} \sum_{I=1}^{M} \alpha_I \qquad \text{and} \qquad \beta = \frac{1}{M} \sum_{I=1}^{M} \beta_I$$

To obtain point estimates and confidence intervals for α and β, the company selects a simple random sample of m automobiles of this make and model. The relationship of Y and X for the ith car sampled is given by the regression function

$$\mu_i(x) = \alpha_i + \beta_i x$$

The number of miles per gallon of the ith car in the sample is measured by driving it a specified distance on k different occasions with k preselected amounts $x_1, \ldots, x_k$ of the additive. From these data, estimates of α_i and β_i are computed for each i (each car in the sample). Using these we then compute point and confidence interval estimates for the population parameters α, β, and $\mu_Y(x)$.

Observe that the number of miles per gallon is measured on each car in the sample under k different conditions—viz., with k different amounts of the additive—and so a growth curve model is appropriate here. However, note that the predictor variable here is the amount of the additive used, not time. ∎

E X A M P L E   7.5.4

An agronomist is interested in studying the average height of a new variety of wheat as a function of time t after planting. It is assumed that the relationship between height Y and time t in days after planting, where $20 \le t \le 60$ for plant I in the population of M plants, is given by

$$\mu_I(t) = \alpha_I + \beta_I t + \gamma_I t^2$$

The corresponding population growth curve is given by

$$\mu_Y(t) = \alpha + \beta t + \gamma t^2$$

where

$$\alpha = \frac{1}{M} \sum_{I=1}^{M} \alpha_I \qquad \beta = \frac{1}{M} \sum_{I=1}^{M} \beta_I \qquad \gamma = \frac{1}{M} \sum_{I=1}^{M} \gamma_I$$

To obtain point estimates and confidence intervals for the quantities α, β, γ, and $\mu_Y(t)$, the agronomist selects a simple random sample of m wheat plants and measures the height of the ith plant in the sample, $i = 1, \ldots, m$, at k specified times $t_1, \ldots, t_k$. From these data, estimates of the growth curves are obtained for each of the m plants in the sample. These in turn lead to point and confidence interval estimates of the population growth curve.   ■

Models of the type discussed in Examples 7.5.1–7.5.4 are called **random coefficient growth curve models**. In this section we give formulas for estimating $\mu_i(t)$, the **individual growth curves** for each of the m individuals in the sample, and also for estimating the population growth curve $\mu_Y(t)$. Keep in mind that we will use the symbol t for the predictor variable and refer to it as *time* for simplicity of explanations, and also because it is in this context that growth curve models are commonly used. However, as pointed out in Example 7.5.3, predictor variables other than time may also make sense in certain situations. We now state the assumptions underlying random coefficient growth curve models.

Assumptions for the Growth Curve Model

Notation   There is a population consisting of M items (in the preceding examples the items are babies, cars, wheat plants, or individuals). For each item in the population there is a regression function of the response variable Y on the predictor variable t, and this is the growth curve for that item. The regression function for the ith item in the population is denoted by $\mu_{Y_I}(t)$, or $\mu_I(t)$ for short. The average of the M individual regression functions in the population is denoted by $\mu_Y(t)$, and it is given by

$$\mu_Y(t) = \frac{1}{M} \sum_{I=1}^{M} \mu_I(t)$$

The function $\mu_Y(t)$ is called the *population growth curve* or the *population growth function*. In the mathematical theory of growth curves, M is infinite.

Any linear regression function is a candidate for modeling growth curves of population items. Some examples with the number of parameters $p = 2$, 3, and 4 are

$$\mu_I(t) = \alpha_I + \beta_I t \quad (p = 2)$$
$$\mu_I(t) = \alpha_I + \beta_I t + \gamma_I t^2 \quad (p = 3)$$
$$\mu_I(t) = \alpha_I + \beta_I t + \gamma_I t^2 + \delta_I t^3 \quad (p = 4)$$
$$\mu_I(t) = \alpha_I + \beta_I t + \gamma_I \log(t) \quad (p = 3)$$
$$\mu_I(t) = \alpha_I + \beta_I e^t \quad (p = 2)$$

B O X **7.5.1**   **Assumptions for Growth Curve Models**

(Population) Assumption 1   The regression function for the ith population item is a *linear* regression function with p unknown parameters. We choose a second-degree polynomial model (quadratic model) for illustration; i.e., for population item I the growth curve is

$$\mu_I(t) = \alpha_I + \beta_I t + \gamma_I t^2 \tag{7.5.6}$$

Note that $\mu_I(t)$ has the same form for each item in the population, but with perhaps different α_I, β_I, and γ_I values, because each item grows at its own rate. The population growth curve at time t is

$$\mu_Y(t) = \alpha + \beta t + \gamma t^2 \tag{7.5.7}$$

where $\alpha = \frac{1}{M} \sum_{I=1}^{M} \alpha_I$, $\beta = \frac{1}{M} \sum_{I=1}^{M} \beta_I$, and $\gamma = \frac{1}{M} \sum_{I=1}^{M} \gamma_I$.

(Population) Assumption 2   The population of regression coefficients $\{(\alpha_I, \beta_I, \gamma_I), I = 1, \ldots, M\}$ is (approximately) a Gaussian population.

(Sample) Assumption 3   A simple random sample of m items is selected from the population of M items.

(Sample) Assumption 4   For each of the m chosen items, the value of Y is measured at k preselected times $t_1, \ldots, t_k$. The recorded Y value for sample item i at time t_j is denoted by $y_{i,j}$. A schematic representation of the sample data is given in Table 7.5.1.

The observed responses $y_{i,j}$ $(j = 1, \ldots, k)$, for sample item i, are given by

$$y_{i,j} = \mu_i(t_j) + e_{i,j}$$

i.e.,

$$y_{i,j} = \alpha_i + \beta_i t_j + \gamma_i t_j^2 + e_{i,j} \tag{7.5.8}$$

where $\mu_i(t_j) = \alpha_i + \beta_i t_j + \gamma_i t_j^2$ is the 'true' response of sample item i at time t_j, and $e_{i,j}$ is an error that includes measurement error and other random errors due to unknown and uncontrolled disturbances.

(Sample) Assumption 5 $\{e_{i,j}, i = 1, \ldots, m; j = 1, \ldots, k\}$ is a simple random sample from a Gaussian population with zero mean and standard deviation σ_E. As usual the $\{e_{i,j}\}$ are not observable.

(Sample) Assumption 6   The quantities $t_1, \ldots, t_k$ are measured without error.

Even though we are using a quadratic model for explaining the methods for analyzing growth curves, the methods presented in this section are easily adapted for application to any linear regression function described in Chapters 3 and 4.

Inferences for Growth Curve Models

One of the main objectives in a growth curve study is to make inferences about the population growth curve. This includes inferences about the unknown parameters α, β, γ, etc. in the population growth curve, as well as certain functions of them. We begin our discussion by examining the data for each item in the sample (see Table 7.5.1). As stated, we work with a quadratic growth curve model although the procedures given can be adapted very easily for any linear growth curve model (linear in the unknown parameters).

T A B L E 7.5.1

A Schematic Representation of the Sample Data for a Growth Curve Study

Item	Response at Time t_1	$\cdots$	Response at Time t_j	$\cdots$	Response at Time t_k
1	$y_{1,1}$	$\cdots$	$y_{1,j}$	$\cdots$	$y_{1,k}$
2	$y_{2,1}$	$\cdots$	$y_{2,j}$	$\cdots$	$y_{2,k}$
$\vdots$	$\vdots$	$\vdots$	$\vdots$	$\vdots$	$\vdots$
i	$y_{i,1}$	$\cdots$	$y_{i,j}$	$\cdots$	$y_{i,k}$
$\vdots$	$\vdots$	$\vdots$	$\vdots$	$\vdots$	$\vdots$
m	$y_{m,1}$	$\cdots$	$y_{m,j}$	$\cdots$	$y_{m,k}$

The observed value $y_{1,j}$ for sample item 1 (say baby 1 in Example 7.5.1) at time t_j is given by

$$y_{1,j} = \alpha_1 + \beta_1 t_j + \gamma_1 t_j^2 + e_{1,j} \qquad j = 1, \ldots, k \qquad \text{(7.5.9)}$$

where $e_{1,j}$ are random errors from a Gaussian population with zero mean and standard deviation σ_E. This is a multiple linear regression model with $X_1 = t$ and $X_2 = t^2$, so using the methods in Chapter 4 we can estimate α_1, β_1, and γ_1. The data for sample item 1 are

$$
\begin{array}{ccc}
y_1 & t & t^2 \\
\hline
y_{1,1} & t_1 & t_1^2 \\
y_{1,2} & t_2 & t_2^2 \\
\vdots & \vdots & \vdots \\
y_{1,k} & t_k & t_k^2
\end{array}
\qquad \text{(7.5.10)}
$$

and the model equations in (7.5.9) for the first sample item can be written in matrix notation as

$$
y_1 = X_1 \beta_1 + e_1 \qquad \text{(7.5.11)}
$$

where

$$
y_1 = \begin{bmatrix} y_{1,1} \\ y_{1,2} \\ \vdots \\ y_{1,k} \end{bmatrix}
\quad
X_1 = \begin{bmatrix} 1 & t_1 & t_1^2 \\ 1 & t_2 & t_2^2 \\ \vdots & \vdots & \vdots \\ 1 & t_k & t_k^2 \end{bmatrix}
\quad
\beta_1 = \begin{bmatrix} \alpha_1 \\ \beta_1 \\ \gamma_1 \end{bmatrix}
\quad
e_1 = \begin{bmatrix} e_{1,1} \\ e_{1,2} \\ \vdots \\ e_{1,k} \end{bmatrix}
\qquad \text{(7.5.12)}
$$

The point estimate of β_1, using (4.4.8), is given by

$$
\hat{\beta}_1 = \begin{bmatrix} \hat{\alpha}_1 \\ \hat{\beta}_1 \\ \hat{\gamma}_1 \end{bmatrix} = (X_1^T X_1)^{-1} X_1^T y_1 \qquad \text{(7.5.13)}
$$

Hence the estimated growth curve for sample item 1 is given by

$$
\hat{\mu}_1(t) = \hat{\alpha}_1 + \hat{\beta}_1 t + \hat{\gamma}_1 t^2 \qquad \text{(7.5.14)}
$$

The observed value $y_{2,j}$ for sample item 2 at time t_j is given by

$$
y_{2,j} = \alpha_2 + \beta_2 t_j + \gamma_2 t_j^2 + e_{2,j} \qquad j = 1, \ldots, k \qquad \text{(7.5.15)}
$$

where $e_{2,j}$ are random errors from a Gaussian population with zero mean and standard deviation σ_E. This is a multiple linear regression model, so using the methods of Chapter 4 we can estimate α_2, β_2, and γ_2. The data for sample item 2 are

$$
\begin{array}{ccc}
y_2 & t & t^2 \\
\hline
y_{2,1} & t_1 & t_1^2 \\
y_{2,2} & t_2 & t_2^2 \\
\vdots & \vdots & \vdots \\
y_{2,k} & t_k & t_k^2
\end{array}
\qquad \text{(7.5.16)}
$$

and the model equations in (7.5.15) for the 2nd sample item can be written in matrix notation as

$$
y_2 = X_2 \beta_2 + e_2 \qquad \text{(7.5.17)}
$$

where

$$y_2 = \begin{bmatrix} y_{2,1} \\ y_{2,2} \\ \vdots \\ y_{2,k} \end{bmatrix} \quad X_2 = \begin{bmatrix} 1 & t_1 & t_1^2 \\ 1 & t_2 & t_2^2 \\ \vdots & \vdots & \vdots \\ 1 & t_k & t_k^2 \end{bmatrix} \quad \beta_2 = \begin{bmatrix} \alpha_2 \\ \beta_2 \\ \gamma_2 \end{bmatrix} \quad e_2 = \begin{bmatrix} e_{2,1} \\ e_{2,2} \\ \vdots \\ e_{2,k} \end{bmatrix} \qquad \text{(7.5.18)}$$

The point estimate of β_2, using (4.4.8), is given by

$$\hat{\beta}_2 = \begin{bmatrix} \hat{\alpha}_2 \\ \hat{\beta}_2 \\ \hat{\gamma}_2 \end{bmatrix} = (X_2^T X_2)^{-1} X_2^T y_2 \qquad \text{(7.5.19)}$$

Hence the estimated growth curve for sample item 2 is given by

$$\hat{\mu}_2(t) = \hat{\alpha}_2 + \hat{\beta}_2 t + \hat{\gamma}_2 t^2 \qquad \text{(7.5.20)}$$

We proceed along similar lines for each of the m sample items. The observed value $y_{m,j}$ for sample item m at time t_j is given by

$$y_{m,j} = \alpha_m + \beta_m t_j + \gamma_m t_j^2 + e_{m,j} \qquad j = 1, \ldots, k \qquad \text{(7.5.21)}$$

The data for sample item m are

y_m	t	t^2
$y_{m,1}$	t_1	t_1^2
$y_{m,2}$	t_2	t_2^2
$\vdots$	$\vdots$	$\vdots$
$y_{m,k}$	t_k	t_k^2

$$\text{(7.5.22)}$$

and the model equations in (7.5.21) for sample item m can be written in matrix notation as

$$y_m = X_m \beta_m + e_m \qquad \text{(7.5.23)}$$

where

$$y_m = \begin{bmatrix} y_{m,1} \\ y_{m,2} \\ \vdots \\ y_{m,k} \end{bmatrix} \quad X_m = \begin{bmatrix} 1 & t_1 & t_1^2 \\ 1 & t_2 & t_2^2 \\ \vdots & \vdots & \vdots \\ 1 & t_k & t_k^2 \end{bmatrix} \quad \beta_m = \begin{bmatrix} \alpha_m \\ \beta_m \\ \gamma_m \end{bmatrix} \quad e_m = \begin{bmatrix} e_{m,1} \\ e_{m,2} \\ \vdots \\ e_{m,k} \end{bmatrix} \qquad \text{(7.5.24)}$$

The point estimate of β_m, using (4.4.8), is given by

$$\hat{\beta}_m = \begin{bmatrix} \hat{\alpha}_m \\ \hat{\beta}_m \\ \hat{\gamma}_m \end{bmatrix} = (X_m^T X_m)^{-1} X_m^T y_m \qquad \text{(7.5.25)}$$

Hence the estimated growth curve for sample item m is given by

$$\hat{\mu}_m(t) = \hat{\alpha}_m + \hat{\beta}_m t + \hat{\gamma}_m t^2 \qquad \text{(7.5.26)}$$

Note   Because of the way the sample observations are obtained, the responses for each item in the sample are observed at the same k times $t_1, \ldots, t_k$. Hence $X_1 = X_2 = \cdots = X_m$. We denote this common matrix by X. Thus

$$X = \begin{bmatrix} 1 & t_1 & t_1^2 \\ 1 & t_2 & t_2^2 \\ \vdots & \vdots & \vdots \\ 1 & t_k & t_k^2 \end{bmatrix} \qquad \text{a } k \times 3 \text{ matrix} \tag{7.5.27}$$

Point Estimates for the Population Growth Curve

The parameters determining the population growth curve are in the vector β where $\beta = [\alpha, \beta, \gamma]^T$ and α, β, and γ are as in (7.5.7). Their estimates are

$$\hat{\beta}^T = [\hat{\alpha}, \hat{\beta}, \hat{\gamma}], \hat{\alpha} = \frac{1}{m} \sum_{i=1}^{m} \hat{\alpha}_i, \hat{\beta} = \frac{1}{m} \sum_{i=1}^{m} \hat{\beta}_i, \text{ and } \hat{\gamma} = \frac{1}{m} \sum_{i=1}^{m} \hat{\gamma}_i \tag{7.5.28}$$

with $\hat{\alpha}_i$, $\hat{\beta}_i$, and $\hat{\gamma}_i$ being the estimated values of the coefficients in the growth curve for sample item i. Hence the estimated population growth curve is

$$\hat{\mu}_Y(t) = \hat{\alpha} + \hat{\beta}t + \hat{\gamma}t^2 \tag{7.5.29}$$

Confidence intervals

Many questions about the population growth curve can be answered using point estimates and confidence intervals for θ where

$$\theta = a^T \beta = a_0 \alpha + a_1 \beta + a_2 \gamma \tag{7.5.30}$$

by appropriately specifying the elements of the vector

$$a^T = [a_0, a_1, a_2] \tag{7.5.31}$$

A $1 - \alpha$ confidence interval for $\theta = a^T \beta$ for a specified vector a is given by

$$a^T \hat{\beta} - t_{1-\alpha/2:m-1} SE(a^T \hat{\beta}) \leq a^T \beta \leq a^T \hat{\beta} + t_{1-\alpha/2:m-1} SE(a^T \hat{\beta}) \tag{7.5.32}$$

with

$$SE(a^T \hat{\beta}) = \sqrt{\frac{a^T G a}{m}} \tag{7.5.33}$$

and

$$G = \frac{1}{m-1} \sum_{i=1}^{m} (\hat{\beta}_i - \hat{\beta})(\hat{\beta}_i - \hat{\beta})^T \tag{7.5.34}$$

where

$$\hat{\beta}_i = [\hat{\alpha}_i, \hat{\beta}_i, \hat{\gamma}_i]^T \qquad (7.5.35)$$

for $i = 1, 2, \ldots, m$ and

$$\hat{\beta} = [\hat{\alpha}, \hat{\beta}, \hat{\gamma}]^T \qquad (7.5.36)$$

In particular, the formula in (7.5.32) can be used for obtaining confidence intervals for the following quantities:

a   for α, by choosing $a^T = [1, 0, 0]$

b   for β, by choosing $a^T = [0, 1, 0]$

c   for γ, by choosing $a^T = [0, 0, 1]$

d   for $\mu_Y(t)$, by choosing $a^T = [1, t, t^2]$

e   for $\mu_Y(t_1) - \mu_Y(t_2)$, by choosing $a^T = [0, t_1 - t_2, t_1^2 - t_2^2]$

We illustrate the use of the preceding formulas in Example 7.5.5.

E X A M P L E   7.5.5

A scientist working for a pharmaceutical company is interested in the concentrations of a certain drug at various times after it is injected into the bloodstream. In particular, she wants to determine the amount of time it takes for the body to eliminate the drug so that the drug concentrations reach near-zero levels in the circulatory system. To understand this, she selects a simple random sample of 24 human subjects from the study population. A fixed dose of the drug is injected into the bloodstream of each subject, and blood samples are collected at the end of 1 hour, 2 hours, 3 hours, and 4 hours after injection. The blood samples are analyzed in a laboratory, and the drug concentrations are determined in milligrams per liter (mg/l for short). The data are presented in Table 7.5.2 and they are also stored in the file **drugconc.dat** on the data disk.

Assume that in the range from 1 to 4 hours after injection, the growth curve for each subject in the population is a quadratic function, so for subject i in the sample the growth curve is given by

$$\mu_i(t) = \alpha_i + \beta_i t + \gamma_i t^2$$

We regress Y (concentration) on t (time) and t^2 and obtain estimates $\hat{\alpha}_i$, $\hat{\beta}_i$, and $\hat{\gamma}_i$ for $i = 1, \ldots, 24$ (i.e., for each subject). One of the objectives in this study is to estimate the population growth curve, $\mu_Y(t) = \alpha + \beta t + \gamma t^2$, and obtain a 95% confidence interval for $\mu_Y(3) = \alpha + 3\beta + 9\gamma$, the (population) average concentration of the drug at the end of 3 hours after injection.

We assume that the 24 subjects ($m = 24$) constitute a simple random sample from a well-defined population of subjects of interest (e.g., population of persons for whom the symptoms might suggest the use of this particular drug). We further suppose that assumptions in Box 7.5.1 for the random coefficient growth model are satisfied. We have $k = 4$, $p = 3$, $m = 24$, and the population growth curve is

$$\mu_Y(t) = \alpha + \beta t + \gamma t^2 \qquad 1 \leq t \leq 4 \qquad (7.5.37)$$

T A B L E **7.5.2**

Drug Concentration Data (in milligrams/liter)

Subject	1 Hour	2 Hours	3 Hours	4 Hours
1	10.55	4.11	2.00	1.02
2	10.47	4.30	2.15	1.11
3	9.46	3.81	1.78	0.94
4	9.27	3.72	1.92	0.95
5	9.37	3.75	1.95	0.97
6	9.67	4.28	1.96	1.04
7	10.58	3.95	2.30	1.08
8	9.96	3.73	1.86	1.01
9	9.84	3.92	2.00	1.05
10	10.20	4.20	1.96	1.03
11	9.45	4.18	2.18	1.02
12	9.64	4.04	2.08	0.96
13	10.03	4.01	2.08	1.04
14	9.81	3.65	1.97	0.97
15	10.74	4.41	2.07	1.03
16	10.08	3.80	1.86	0.99
17	10.00	3.84	2.07	0.95
18	9.73	3.94	1.93	0.96
19	9.64	4.24	2.11	1.06
20	10.40	4.11	2.07	1.01
21	10.34	4.20	2.21	1.14
22	10.09	4.35	1.91	1.07
23	9.51	3.74	1.87	0.99
24	9.63	3.77	1.96	1.01

Note that the X matrix is given by

$$X = \begin{bmatrix} 1 & 1 & 1 \\ 1 & 2 & 4 \\ 1 & 3 & 9 \\ 1 & 4 & 16 \end{bmatrix} \tag{7.5.38}$$

Also

$$y_1^T = [10.55, 4.11, 2.00, 1.02]$$

which is the first row of Table 7.5.2. The data for subject 1 are in this first row, and

$$y_2^T = [10.47, 4.30, 2.15, 1.11]$$

which is the second row of Table 7.5.2. The data for subject 2 are in this second row, etc. Thus we perform the regressions of y on t and t^2 for each sample subject and get $\hat{\alpha}_i$, $\hat{\beta}_i$, and $\hat{\gamma}_i$ for $i = 1, \ldots, 24$. From these, we compute the quantities $\hat{\beta}$ and G using (7.5.28) and (7.5.34), respectively. The results are

$$\hat{\beta} = \begin{bmatrix} 17.6035 \\ -9.0499 \\ 1.2350 \end{bmatrix} \qquad G = \begin{bmatrix} 0.695830 & -0.436192 & 0.067454 \\ -0.436192 & 0.298748 & -0.048488 \\ 0.067454 & -0.048488 & 0.008067 \end{bmatrix} \qquad \textbf{(7.5.39)}$$

Thus

$$\hat{\mu}_Y(t) = 17.6035 - 9.0499t + 1.2350t^2 \qquad 1 \le t \le 4$$

To compute a 95% confidence interval for the (population) average drug concentration 3 hours after injection, we let $a = [1, 3, 9]^T$ and use (7.5.32). We get $t_{0.975:23} = 2.069$, $\hat{\mu}_Y(3) = 1.569$, and $SE(a^T \hat{\beta}) = 0.02642$. This leads to the confidence statement

$$C[1.569 - 2.069(0.02642) \le \mu_Y(3) \le 1.569 + 2.069(0.02642)] = 0.95$$

which, when simplified, is

$$C[1.514 \le \mu_Y(3) \le 1.623] = 0.95$$

From this we have 95% confidence that the *average* drug concentration in the blood for the study population of subjects, 3 hours after injection, is between 1.514 and 1.623 mg/l. ∎

Remarks There are many other questions that are often of interest in growth curve studies. For instance, in Example 7.5.5 the investigator may be interested in the *distribution* of the drug concentrations 3 hours after injection. This involves not only the estimation of the *mean* drug concentration 3 hours after injection (which is computed using (7.5.29) with $t = 3$) but also the estimation of the standard deviation of the drug concentration 3 hours after injection. In some applications the investigator may be interested in a *tolerance interval* for this distribution. In other applications the investigator may want to predict a future value of the response of an item in the sample, to predict the value of the response of an item not in the sample, and so on. The procedures for answering these and other interesting questions may be found in more advanced textbooks.

In Section 7.5 of the laboratory manuals we discuss a macro we have supplied on the data disk that can be used to perform the calculations needed in this section to obtain point estimates and confidence intervals for $a^T \beta$ for specified vectors a. In particular, this program can be used to compute point estimates and confidence intervals for α, β, γ, $\mu_Y(t)$, $\mu_Y(t_1) - \mu_Y(t_2)$, etc.

We pointed out earlier that even though the procedures for growth curves are explained using a quadratic growth curve model, they can be easily adapted for other situations involving multiple linear regression models. In the next example we illustrate the procedure when the growth curve for each subject in the population is a straight line as a function of time.

E X A M P L E **7.5.6**

To establish a growth curve for the ramus bone (a bone in the jaw) in young boys, a random sample of 20 boys was selected and the ramus height measured in millimeters at the ages of $8, 8\frac{1}{2}, 9$, and $9\frac{1}{2}$ years. See [10]. The data are given in Table 7.5.3 and are also stored in the file **ramus.dat** on the data disk. Assume that in the range from 8 to 10 years of age, the growth curve for the ramus bone of each boy in the population is a straight line, so for boy i in the sample the growth curve is given by $\mu_i(t) = \alpha_i + \beta_i t$. We regress Y (ramus height) on t (age) and obtain estimates $\hat{\alpha}_i$ and $\hat{\beta}_i$ for $i = 1, \ldots, 20$ (i.e., for each boy). The principle objective in this study is to estimate the population growth curve and to obtain a 95% confidence interval for β, the average growth rate of the ramus bone. We assume the 20 boys ($m = 20$) constitute a simple random sample from a well-defined population of boys in this age group. We further suppose that assumptions in Box 7.5.1 for the random coefficient growth curve model are satisfied. We have $k = 4, p = 2, m = 20$, and the population growth curve is

$$\mu_Y(t) = \alpha + \beta t \qquad 8 \le t \le 10 \tag{7.5.40}$$

T A B L E **7.5.3**

Height of 20 Boys' Ramus Bones

Boy	Age 8	Age $8\frac{1}{2}$	Age 9	Age $9\frac{1}{2}$
1	47.8	48.8	49.0	49.7
2	46.4	47.3	47.7	48.4
3	46.3	46.8	47.8	48.5
4	45.1	45.3	46.1	47.2
5	47.6	48.5	48.9	49.3
6	52.5	53.2	53.3	53.7
7	51.2	53.0	54.3	54.5
8	49.8	50.0	50.3	52.7
9	48.1	50.8	52.3	54.4
10	45.0	47.0	47.3	48.3
11	51.2	51.4	51.6	51.9
12	48.5	49.2	53.0	55.5
13	52.1	52.8	53.7	55.0
14	48.2	48.9	49.3	49.8
15	49.6	50.4	51.2	51.8
16	50.7	51.7	52.7	53.3
17	47.2	47.7	48.4	49.5
18	53.3	54.6	55.1	55.3
19	46.2	47.5	48.1	48.4
20	46.3	47.6	51.3	51.8

Also note that the X matrix is given by

$$X = \begin{bmatrix} 1 & 8.0 \\ 1 & 8.5 \\ 1 & 9.0 \\ 1 & 9.5 \end{bmatrix} \qquad \text{(7.5.41)}$$

Furthermore

$$y_1^T = [47.8, 48.8, 49.0, 49.7]$$

which is the first row of Table 7.5.3; i.e., the data for boy 1, and

$$y_2^T = [46.4, 47.3, 47.7, 48.4]$$

which is the second row of Table 7.5.3; i.e., the data for boy 2, etc. So we regress Y on t for each sample subject and get $\hat{\alpha}_i$, $\hat{\beta}_i$ for $i = 1, \ldots, 20$. From these we compute the quantities $\hat{\beta}$ and G using (7.5.28) and (7.5.34), respectively. The results are

$$\hat{\beta} = \begin{bmatrix} 33.7475 \\ 1.8660 \end{bmatrix} \qquad G = \begin{bmatrix} 103.959 & -11.525 \\ -11.525 & 1.358 \end{bmatrix}$$

Thus

$$\hat{\mu}_Y(t) = 33.7475 + 1.8660t \qquad 8 \le t \le 10$$

To compute a 95% confidence interval for β, we let $a = [0, 1]^T$ and use (7.5.32). We get $a^T \hat{\beta} = \hat{\beta} = 1.866$, and $SE(\hat{\beta}) = 0.2606$. This leads to the confidence statement (rounding to two decimals)

$$C[1.32 \le \beta \le 2.41] = 0.95$$

Thus we have 95% confidence that the average growth rate of the ramus bone for this population of boys is between 1.32 and 2.41 millimeters per year for $8 \le t \le 10$. ∎

Problems 7.5

7.5.1 An agricultural experiment station investigator is interested in studying the growth pattern of pumpkins as a function of time. He monitors the weights (in pounds) of 24 pumpkins (selected by simple random sampling from a large pumpkin patch) each week, from the time they are 4 weeks old until they are 12 weeks old. The study population is all pumpkins in this patch, and the target population is the set of all pumpkins in similar patches. The data are given in Table 7.5.4 and are also stored in the file **pumpkin.dat** on the data disk.

Suppose that a quadratic growth curve model holds for each pumpkin growing on the given patch; i.e., (7.5.8) holds. In this case the population growth curve has the form

$$\mu_Y(t) = \alpha + \beta t + \gamma t^2$$

where α, β, and γ are unknown parameters. We assume that all the conditions in Box 7.5.1 are satisfied.

a   What is the value of k? What is the value of m? What is the value of p?

b   What are the values of t_i for $i = 1, 2, \ldots, k$?

c   Display the X matrix.

d   Display the vectors $\mathbf{y}_3$ and $\mathbf{y}_{10}$.

e   Write out the regression model for pumpkin 15.

f   We have computed $\hat{\boldsymbol{\beta}} = [\hat{\alpha}, \hat{\beta}, \hat{\gamma}]^T$ and $SE(\hat{\gamma})$. They are

$$\hat{\boldsymbol{\beta}} = [-18.9433, 6.1739, -0.2459]^T$$

and

$$SE(\hat{\gamma}) = 0.006821$$

Use these values to obtain a 90% confidence interval for γ.

g   Find $\hat{\mu}_Y(t)$.

T A B L E **7.5.4**
Pumpkin Growth Data

Pumpkin	4 Weeks	6 Weeks	8 Weeks	10 Weeks	12 Weeks
1	2.1	9.6	15.7	20.3	21.6
2	3.1	11.3	18.5	23.6	26.6
3	3.5	11.3	18.4	22.5	25.3
4	0.2	5.9	9.5	12.0	12.1
5	2.8	11.8	17.8	22.8	24.0
6	2.1	8.7	13.3	16.8	17.4
7	2.9	11.3	15.7	19.6	20.4
8	0.3	6.3	11.6	14.6	16.7
9	0.8	7.5	13.2	16.6	16.8
10	3.6	11.4	18.3	21.6	23.9
11	1.6	8.8	13.5	16.0	16.3
12	2.2	9.8	15.6	19.3	20.3
13	1.5	8.5	13.8	16.9	17.9
14	3.0	10.0	16.3	20.2	21.1
15	0.2	6.6	11.5	15.5	17.4
16	1.3	8.9	13.0	17.5	18.4
17	2.9	11.0	16.6	22.6	24.7
18	1.8	9.6	15.6	19.2	21.3
19	0.5	7.5	13.0	16.2	17.8
20	2.0	8.7	12.8	14.6	14.7
21	2.3	10.6	15.9	20.4	20.6
22	0.2	6.8	11.7	14.6	16.2
23	2.1	9.7	15.1	18.2	20.6
24	1.6	8.8	15.4	18.5	20.1

   h Find $\hat{\mu}_Y(8)$.

   i Display a if $a^T\beta = \beta$.

   j Display a if $a^T\beta = \mu_Y(t)$.

   k An investigator wants to determine what the change in the average weight (in pounds) of pumpkins is from week 4 to week 12. What population parameter(s) must be estimated to estimate this change in average weight?

7.6
Exercises

7.6.1 An investigator wants to study the relationship between height and mineral composition of foliage in Japanese larch (trees). She has obtained a simple random sample of 26 trees from a plantation of Japanese larch and recorded the following information on each tree. (See Leyton, *Plant and Soil* **7**: 167–177, 1956.)

$$Y = \text{height of the tree in centimeters}$$
$$X_1 = \text{percentage content of nitrogen}$$
$$X_2 = \text{percentage content of phosporus}$$
$$X_3 = \text{percentage content of potassium}$$
$$X_4 = \text{percentage content of residual ash}$$

The mineral composition data were obtained from dried, ground, new needles collected immediately below the terminal shoot. The data are given in Table 7.6.1 and are also stored in the file **larch.dat** on the data disk.

    The sums of squared errors, *SSE*, and mean squared errors *MSE* for each subset model follow:

Variables in model	Sum of Squares	Mean Squares
none	227954	9118
X1	75363	3140
X2	91404	3809
X3	85123	3547
X4	93614	3901
X1,X2	47090	2047
X1,X3	36771	1599
X1,X4	53503	2326
X2,X3	63875	2777
X2,X4	65810	2861

RexLox =1

X3,X4	61929	2693
X1,X2,X3	31858	1448
X1,X2,X4	40782	1854
X1,X3,X4	33403	1518
X2,X3,X4	52828	2401
X1,X2,X3,X4	30122	1434

a Carry out a variable selection analysis by the method of all-subsets regression using each of the following criteria: s, R^2, and adj-R^2. Find a short list of three best subset models in each case.

b Carry out a variable selection analysis by the method of all-subsets regression using the C_p criterion.

   i Display in a table the value of C_p that corresponds to each subset model.

T A B L E **7.6.1**
Japanese Larch Tree Data

Observation Number	Y	X_1	X_2	X_3	X_4
1	351	2.20	0.417	1.35	1.79
2	249	2.10	0.354	0.90	1.08
3	171	1.52	0.208	0.71	0.47
4	373	2.88	0.335	0.90	1.48
5	321	2.18	0.314	1.26	1.09
6	191	1.87	0.271	1.15	0.99
7	225	1.52	0.164	0.83	0.85
8	291	2.37	0.302	0.89	0.94
9	284	2.06	0.373	0.79	0.80
10	213	1.84	0.265	0.72	0.77
11	138	1.89	0.192	0.46	0.46
12	213	2.45	0.221	0.76	0.95
13	151	1.88	0.186	0.52	0.95
14	130	1.93	0.207	0.60	0.92
15	93	1.80	0.157	0.67	0.60
16	95	1.81	0.195	0.47	0.57
17	147	1.49	0.165	0.66	0.80
18	88	1.53	0.226	0.68	0.66
19	65	1.43	0.224	0.44	0.45
20	120	1.54	0.271	0.51	0.95
21	72	1.13	0.187	0.38	0.63
22	160	1.63	0.200	0.62	1.10
23	72	1.36	0.211	0.71	0.47
24	252	1.76	0.283	0.96	0.96
25	310	2.53	0.284	0.85	1.39
26	336	2.59	0.303	1.02	0.95

ii Exhibit a plot of the C_p values. On this plot show the line through the origin with unit slope for comparing the values of $C_p - p$ for the various subset models.

iii Exhibit the best subset model according to the C_p criterion.

iv Exhibit any other models that are almost as good as the best model.

c Carry out a variable selection analysis for the preceding data using the forward selection procedure. Use $F\text{-}in = 3.0$. List the variables in the model at each step.

d Carry out a variable selection analysis for the preceding data using the backward elimination procedure with $F\text{-}out = 3.0$. List the variables in the model at each step.

e Carry out a variable selection analysis for the preceding data using stepwise regression with $F\text{-}in = F\text{-}out = 3.0$. Use β_0 as the initial model. List the variables in the model at each step.

f Summarize the results of parts (a)–(e) in a short written report.

7.6.2 Consider the data for the heights of the ramus bone in Table 7.5.3. Assume that in the range from 8 to 10 years of age, the growth curve for each boy (and for the population) is a quadratic function of time; i.e., $\mu_Y(t) = \alpha + \beta t + \gamma t^2, 8 \le t \le 10$. Suppose assumptions in Box 7.5.1 for a random coefficients growth curve model are satisfied. The results of some computations that you need follow:

$$\hat{\beta}^T = [\hat{\alpha}, \hat{\beta}, \hat{\gamma}] = [27.8742, 3.3863, -0.0916]$$
$$SE(\hat{\beta}) = 3.668$$
$$SE(\hat{\gamma}) = 0.2114$$

a What are the values of m, k, and p?

b Display the X matrix.

c Display the vector a such that $a^T\beta = \alpha$.

d Estimate the population growth curve.

e Construct 95% two sided confidence intervals for β and γ.

f Express $\mu_Y(t)$ in terms of α, β, γ, and t.

g Express $\mu_Y(8.5)$ in terms of α, β, and γ.

h Based on the confidence interval for γ obtained in (e), is it reasonable to conclude that the population growth curve is, for all practical purposes, a straight line for the ages of interest $8 \le t \le 10$? You may suppose that if $|\gamma| < 0.002$, it may be considered negligible for this problem.

8

Alternate Assumptions
for Regression

8.1
Overview

The inference procedures for simple and multiple linear regression we have discussed so far are based on assumptions (A) or (B) given in Chapters 3 and 4. In those chapters we also discussed various diagnostic tools for examining the validity of these assumptions. In situations where we know or suspect that one or more of the required assumptions are not satisfied, it is useful to have alternative valid approaches. In this chapter we discuss alternate sets of assumptions under which valid inferences are possible even if assumptions (A) and/or assumptions (B) do not hold. Section 8.2 introduces procedures that can be used when the assumption of homogeneity of subpopulation standard deviations (variances) is not satisfied. When the Gaussian assumption does not hold for subpopulations, the assumptions discussed in Section 8.3 for the case of straight line regression and the corresponding inference procedures may apply. Our presentation of the topics in this chapter is necessarily limited in scope and hence is only introductory. Sections 8.2 and 8.3 in the laboratory manuals discuss the use of the computer to perform the calculations needed in this chapter.

8.2
Straight Line Regression with Unequal Subpopulation Standard Deviations

The procedures for regression analysis discussed in Chapters 3 and 4 were based on the assumption of *homogeneity of standard deviations*, sometimes referred to as *homogeneity of variances*; i.e., it was assumed that the standard deviations, or equivalently, the variances, of all the subpopulations are the same. This assumption may never be satisfied *exactly* in practical applications, but it is often a reasonable approximation. However, there are situations when the assumption of equal standard

deviations is not appropriate. For such situations, alternate assumptions and proce-
~~d~~ ~~and in this section we discuss~~ one of the alternate procedures for

~~values corresponding to X = 50, ...~~
equal, we have used the symbol $\sigma_{Y|X}$ (σ for short) for their common value. In this
section we consider the situation where the $\sigma_Y(x)$ are not all equal, but where *the
relative values of the standard deviations of the different subpopulations are known.*
This amounts to the assumption that

$$\sigma_Y(x) = \sigma_0 g(x) \tag{8.2.1}$$

where σ_0 is an *unknown* constant and $g(x)$ is a *known* function of x. To illustrate, let
us suppose that $\sigma_Y(x) = \sigma_0\sqrt{x}$. Then $\sigma_Y(4) = \sigma_0\sqrt{4} = 2\sigma_0$, and $\sigma_Y(9) = \sigma_0\sqrt{9} =
3\sigma_0$ so that $\sigma_Y(9)/\sigma_Y(4) = 3\sigma_0/2\sigma_0 = 1.5$, which is a known constant. Thus the
standard deviation of the subpopulation corresponding to $X = 9$ is 1.5 times as big
as the standard deviation of the subpopulation corresponding to $X = 4$.

If the function $g(x)$ is equal to 1 for all allowable values of x, then the subpopulation standard deviations are all the same and are equal to σ_0. In that case σ_0 will
be equal to $\sigma_{Y|X}$, the common standard deviation of all the subpopulations.

In this section we discuss inference procedures for straight line regression when
the assumptions in Box 8.2.1 hold.

B O X **8.2.1**   **Weighted Regression Assumptions for Straight Line Regression**

Notation   A two-variable population $\{(Y, X)\}$ is the study population under investigation.

(Population) Assumption 1   The mean $\mu_Y(x)$ of the subpopulation of Y values
for specified x is

$$\mu_Y(x) = \beta_0 + \beta_1 x \tag{8.2.2}$$

where β_0 and β_1 are unknown parameters.

(Population) Assumption 2   The standard deviation of the Y values in the subpopulation determined by $X = x$ is $\sigma_Y(x)$ where

$$\sigma_Y(x) = \sigma_0 g(x) \tag{8.2.3}$$

and σ_0 is an *unknown* positive constant; $g(x)$ is a *known* function of x such
that $g(x) > 0$ for all allowable values of x.

(Population) Assumption 3   Each subpopulation of Y values, determined by
specified values of X, is Gaussian.

(Sample) Assumption 4   A sample (of size n) is selected either by simple random sampling or by preselecting X values.

(Sample) Assumption 5   All sample values y_i, x_i for $i = 1, \ldots, n$ are observed
without error.

Note Observe that *weighted regression assumptions* in Box 8.2.1 are identical to assumptions (A) for regression *with the exception of population assumption 2.* Under weighted regression assumptions the standard deviations of subpopulations are allowed to be different, but the ratio of the standard deviation of any one subpopulation, say with $X = x_1$, relative to any other subpopulation, say with $X = x_2$, is assumed to be known and equal to $g(x_1)/g(x_2)$; this is actually population assumption 2 in Box 8.2.1. Under assumptions (A) and (B) of Chapters 3 and 4, all the subpopulation standard deviations are the same, so weighted regression reduces to ordinary regression.

The Method of Weighted Least Squares

In Chapter 3, we estimated the parameters β_0, β_1 by the method of least squares; viz., we found the quantities $\hat{\beta}_0, \hat{\beta}_1$ that minimize the sum of squares of prediction errors given by

$$\sum_{i=1}^{n}[y_i - (\beta_0 + \beta_1 x_i)]^2 \tag{8.2.4}$$

The estimates $\hat{\beta}_0, \hat{\beta}_1$ we obtained were called *least squares estimates.* The quantities y_i, x_i are the data values corresponding to sample item i.

When subpopulation standard deviations are unequal but weighted regression assumptions are satisfied, the estimates of β_0, β_1 given in Chapter 3 are not the best estimates, and the method of least squares needs to be modified. The prediction errors are first *weighted* by dividing each prediction error by a factor proportional to the corresponding subpopulation standard deviation. This ensures that the method of estimation will give more weight to observations from subpopulations with smaller standard deviations because these observations are more reliable, and less weight will be given to observations from subpopulations with larger standard deviations because these observations are less reliable. The weighted prediction error corresponding to sample item i, when $\beta_0 + \beta_1 x_i$ is used to predict y_i, is denoted by $e_i^{(w)}$, and it is given by

$$e_i^{(w)} = \frac{y_i - (\beta_0 + \beta_1 x_i)}{g(x_i)} \tag{8.2.5}$$

Note that the denominator of the right-hand side of (8.2.5) is a quantity that is proportional to the standard deviation of the subpopulation of Y values with $X = x_i$. Thus the weighted prediction error corresponding to the ith sample observation, when $\beta_0 + \beta_1 x_i$ is used to predict y_i, is obtained by weighting the prediction error $y_i - (\beta_0 + \beta_1 x_i)$ by the quantity $1/g(x_i)$, i.e., by a quantity that is inversely proportional to the corresponding subpopulation standard deviation.

The best estimates of β_0, β_1 under weighted regression assumptions are obtained by minimizing the *sum of squares*

$$\sum_{i=1}^{n}\{e_i^{(w)}\}^2$$

of weighted prediction errors. The resulting estimates of β_0, β_1 are called **weighted**

$\hat{e}_i^{(w)}$ by the equation

$$\hat{e}_i^{(w)} = \frac{y_i - (\hat{\beta}_0^{(w)} + \hat{\beta}_1^{(w)} x_i)}{g(x_i)} \qquad (8.2.6)$$

and call this quantity the *weighted residual* for sample item i. We then have

$$WSSE(X) = \sum_{i=1}^{n} \{\hat{e}_i^{(w)}\}^2 \qquad (8.2.7)$$

$$= \sum_{i=1}^{n} \left[\frac{y_i - (\hat{\beta}_0^{(w)} + \hat{\beta}_1^{(w)} x_i)}{g(x_i)} \right]^2 \qquad (8.2.8)$$

$$= \sum_{i=1}^{n} w_i \left[y_i - (\hat{\beta}_0^{(w)} + \hat{\beta}_1^{(w)} x_i) \right]^2 \qquad (8.2.9)$$

where

$$w_i = \left[\frac{1}{g(x_i)} \right]^2 \qquad (8.2.10)$$

The quantities w_i are called *weights*.

To distinguish between weighted least squares estimates of β_0, β_1 discussed in this section and the estimates of β_0, β_1 discussed in Chapter 3, we refer to the estimates discussed in Chapter 3 as *unweighted least squares estimates* or *ordinary least squares (OLS) estimates* of β_0, β_1. Note that if the subpopulation standard deviations are all equal (i.e., if $g(x) = 1$), then the WLS estimates of β_0, β_1, given in (8.2.11), are the same as the OLS estimates given in (4.4.8).

Point Estimation and Confidence Intervals

We now discuss point and confidence interval estimation for straight line regression when weighted regression assumptions hold. As usual we let $\beta = [\beta_0, \beta_1]^T$. The weighted least squares estimate of β is denoted by $\hat{\beta}^{(w)} = [\hat{\beta}_0^{(w)}, \hat{\beta}_1^{(w)}]^T$. It can be proved that

$$\hat{\beta}^{(w)} = (X^T W X)^{-1} X^T W y \qquad (8.2.11)$$

where

$$X = \begin{bmatrix} 1 & x_1 \\ \vdots & \vdots \\ 1 & x_i \\ \vdots & \vdots \\ 1 & x_n \end{bmatrix} \qquad y = \begin{bmatrix} y_1 \\ \vdots \\ y_i \\ \vdots \\ y_n \end{bmatrix} \qquad (8.2.12)$$

and

$$W = \begin{bmatrix} w_1 & 0 & \cdots & 0 \\ 0 & w_2 & \cdots & 0 \\ \vdots & \vdots & \ddots & \vdots \\ 0 & 0 & \cdots & w_n \end{bmatrix} \qquad (8.2.13)$$

Note that the diagonal elements of W are the weights $w_1, w_2, \ldots, w_n$, where $w_i = [1/g(x_i)]^2$ and the off-diagonal elements of W are all zero.

An estimate of the standard deviation $\sigma_Y(x)$ of the subpopulation corresponding to $X = x$ is

$$\hat{\sigma}_Y(x) = g(x)\hat{\sigma}_0 \qquad (8.2.14)$$

where

$$\hat{\sigma}_0 = \sqrt{WMSE(X)} \qquad (8.2.15)$$

is an estimate of σ_0 and

$$WMSE(X) = \frac{WSSE(X)}{(n-2)} \qquad (8.2.16)$$

The quantity $WSSE(X)$ given in (8.2.7)–(8.2.9) is called the *weighted sum of squared errors*, and it has $n - 2$ degrees of freedom associated with it. The corresponding quantity $WMSE(X)$ in (8.2.16) is called the **weighted mean squared error**.

To compute confidence intervals for

$$\beta_0, \beta_1 \qquad (8.2.17)$$

$$Y(x), \mu_Y(x) \qquad (8.2.18)$$

and

$$a^T\beta = a_0\beta_0 + a_1\beta_1 \qquad (8.2.19)$$

we need their point estimates and corresponding standard errors. Point estimates of β_0, β_1 are obtained from (8.2.11). The corresponding standard errors are given by

$$SE(\hat{\beta}_{i-1}^{(w)}) = \sqrt{WMSE(X)c_{ii}^{(w)}} \qquad (8.2.20)$$

$$= \hat{\sigma}_0\sqrt{c_{ii}^{(w)}} \qquad \text{for } i = 1, 2 \qquad (8.2.21)$$

where $c_{ii}^{(w)}$ is the ith diagonal element of the 2 by 2 matrix $\boldsymbol{C}^{(w)}$, given by

$$= (X^T W X)$$

Point estimates and standard errors for the quantities in (8.2.18) and (8.2.19) are given in (8.2.24)–(8.2.29).

$$\hat{Y}^{(w)}(x) = \hat{\beta}_0^{(w)} + \hat{\beta}_1^{(w)} x \qquad (8.2.24)$$

$$SE(\hat{Y}^{(w)}(x)) = \hat{\sigma}_0 \sqrt{[g(x)]^2 + \boldsymbol{x}^T \boldsymbol{C}^{(w)} \boldsymbol{x}} \qquad (8.2.25)$$

$$\hat{\mu}_Y^{(w)}(x) = \hat{\beta}_0^{(w)} + \hat{\beta}_1^{(w)} x \qquad (8.2.26)$$

$$SE(\hat{\mu}_Y^{(w)}(x)) = \hat{\sigma}_0 \sqrt{\boldsymbol{x}^T \boldsymbol{C}^{(w)} \boldsymbol{x}} \qquad (8.2.27)$$

$$\boldsymbol{a}^T \hat{\boldsymbol{\beta}}^{(w)} = a_0 \hat{\beta}_0^{(w)} + a_1 \hat{\beta}_1^{(w)} \qquad (8.2.28)$$

$$SE(\boldsymbol{a}^T \hat{\boldsymbol{\beta}}^{(w)}) = \hat{\sigma}_0 \sqrt{\boldsymbol{a}^T \boldsymbol{C}^{(w)} \boldsymbol{a}} \qquad (8.2.29)$$

where $\boldsymbol{x} = [1, x]^T$ and $\boldsymbol{a} = [a_0, a_1]^T$ are specified. As usual, we use $SE(\hat{Y}^{(w)}(x))$ to denote $SE(\hat{Y}^{(w)}(x) - Y(x))$.

Confidence intervals for the quantities in (8.2.17)–(8.2.19) can be computed using (4.6.1) with the point estimates and standard errors given in (8.2.11), (8.2.21), and (8.2.24)–(8.2.29). Confidence intervals for σ_0 have the same form as those in (4.6.13) and (4.6.14), with $\sigma_{Y|X}$ replaced by σ_0, $\hat{\sigma}_{Y|X}$ by $\hat{\sigma}_0$, and $SSE(X)$ by $WSSE(X)$. Confidence intervals and tests for the subpopulation standard deviation $\sigma_Y(x) = \sigma_0 g(x)$ can be obtained from those for σ_0 because $g(x)$ is a known multiplier. Example 8.2.1 illustrates the computations.

E X A M P L E **8.2.1**

A study was conducted to understand the relationship, if any, between Y, the levels of carbon monoxide (CO) in the air (measured in parts per million) and X, the number (in thousands) of automobiles in various U.S. cities that do not have an ongoing clean air program. Thirteen cities were chosen using simple random sampling from the study population (which is also the target population in this problem), which consists of all cities in the United States that have a population of more than 50,000 and do not have an ongoing clean air program. Data for these thirteen cities are given in Table 8.2.1 and are also stored in the file **carbmon.dat** on the data disk. It is known that the subpopulation standard deviations $\sigma_Y(X)$ are not all the same, but the investigator expects the weighted regression assumptions in Box 8.2.1 to hold with $\mu_Y(x) = \beta_0 + \beta_1 x$ and $\sigma_Y(x) = \sigma_0 g(x)$ for $100 \le x \le 1200$, where σ_0 is an unknown constant and $g(x) = \sqrt{x}$. To illustrate the formulas of this section, we compute point estimates and confidence intervals for β_0 and β_1 using the method of weighted least squares.

T A B L E **8.2.1**
Carbon Monoxide Data

City	CO Y (in ppm)	Number of Automobiles X (in thousands)
1	5817	873
2	1063	109
3	2616	398
4	2018	353
5	3147	506
6	7210	1026
7	4339	862
8	5153	742
9	4450	786
10	5591	896
11	2747	377
12	3712	720
13	2354	655

The matrices y, X, and W are (note $w_i = [1/g(x_i)]^2 = 1/x_i$):

$$
y = \begin{bmatrix} 5817 \\ 1063 \\ 2616 \\ 2018 \\ 3147 \\ 7210 \\ 4339 \\ 5153 \\ 4450 \\ 5591 \\ 2747 \\ 3712 \\ 2354 \end{bmatrix}
\qquad
X = \begin{bmatrix} 1 & 873 \\ 1 & 109 \\ 1 & 398 \\ 1 & 353 \\ 1 & 506 \\ 1 & 1026 \\ 1 & 862 \\ 1 & 742 \\ 1 & 786 \\ 1 & 896 \\ 1 & 377 \\ 1 & 720 \\ 1 & 655 \end{bmatrix}
\qquad (8.2.30)
$$

$$\begin{bmatrix} \frac{1}{873}, 0, 0, 0, 0, 0, 0, 0, 0, 0, 0, 0, 0, 0 \\ 0 & \frac{1}{} & 0\ 0\ 0\ 0\ 0.\ 0.\ 0,\ 0,\ 0,\ 0,\ 0,\ 0 \end{bmatrix}$$

$$\boldsymbol{W} = \begin{bmatrix} 0, 0, 0, 0, \frac{1}{506}, 0, 0, 0, 0, 0, 0, 0, 0, 0 \\ 0, 0, 0, 0, 0, \frac{1}{1026}, 0, 0, 0, 0, 0, 0, 0, 0 \\ 0, 0, 0, 0, 0, 0, \frac{1}{862}, 0, 0, 0, 0, 0, 0, 0 \\ 0, 0, 0, 0, 0, 0, 0, \frac{1}{742}, 0, 0, 0, 0, 0, 0 \\ 0, 0, 0, 0, 0, 0, 0, 0, \frac{1}{786}, 0, 0, 0, 0, 0 \\ 0, 0, 0, 0, 0, 0, 0, 0, 0, \frac{1}{896}, 0, 0, 0 \\ 0, 0, 0, 0, 0, 0, 0, 0, 0, 0, \frac{1}{377}, 0, 0 \\ 0, 0, 0, 0, 0, 0, 0, 0, 0, 0, 0, \frac{1}{720}, 0 \\ 0, 0, 0, 0, 0, 0, 0, 0, 0, 0, 0, 0, \frac{1}{655} \end{bmatrix} \tag{8.2.31}$$

We get

$$\hat{\boldsymbol{\beta}}^{(w)} = (\boldsymbol{X}^T \boldsymbol{W} \boldsymbol{X})^{-1} \boldsymbol{X}^T \boldsymbol{W} \boldsymbol{y} = \begin{bmatrix} 371.620 \\ 5.466 \end{bmatrix} \tag{8.2.32}$$

Thus $\hat{\beta}_0^{(w)} = 371.620$ and $\hat{\beta}_1^{(w)} = 5.466$.

To calculate the standard errors of $\hat{\beta}_0^{(w)}$ and $\hat{\beta}_1^{(w)}$ we need $\boldsymbol{C}^{(w)} = (\boldsymbol{X}^T \boldsymbol{W} \boldsymbol{X})^{-1}$ and $\hat{\sigma}_0$. First we obtain

$$\boldsymbol{C}^{(w)} = (\boldsymbol{X}^T \boldsymbol{W} \boldsymbol{X})^{-1} = \begin{bmatrix} 114.596 & -0.179 \\ -0.179 & 0.000401 \end{bmatrix} \tag{8.2.33}$$

Next we calculate $WSSE(X)$, the weighted sum of squared errors, using the formula in (8.2.9). We get

$$\begin{aligned} WSSE(X) &= \frac{1}{873}[5817 - (371.620 + 5.466 \times 873)]^2 \\ &\quad + \frac{1}{109}[1063 - (371.620 + 5.466 \times 109)]^2 \\ &\quad + \cdots \\ &\quad + \cdots \\ &\quad + \frac{1}{655}[2354 - (371.620 + 5.466 \times 655)]^2 \\ &= 8498.69 \end{aligned}$$

From this we obtain

$$WMSE(X) = \frac{WSSE(X)}{(n-2)} = \frac{8498.69}{11} = 772.61 \text{ (to two decimal places)} \tag{8.2.34}$$

and consequently

$$\hat{\sigma}_0 = \sqrt{WMSE(X)} = \sqrt{772.61} = 27.8 \text{ (rounded to one decimal place)} \tag{8.2.35}$$

Using (8.2.20) we get $SE(\hat{\beta}_0^{(w)}) = 297.6$ and $SE(\hat{\beta}_1^{(w)}) = 0.5569$. Hence a two-sided 90% confidence interval for β_0 is given by the confidence statement

$$C[\hat{\beta}_0^{(w)} - t_{0.95:11}SE(\hat{\beta}_0^{(w)}) \le \beta_0 \le \hat{\beta}_0^{(w)} + t_{0.95:11}SE(\hat{\beta}_0^{(w)})]$$
$$= C[371.62 - 1.796 \times 297.6 \le \beta_0 \le 371.62 + 1.796 \times 297.6]$$
$$= C[-162.9 \le \beta_0 \le 906.1] = 0.90$$

Likewise, a two-sided 90% confidence interval for β_1 is given by the confidence statement

$$C[\hat{\beta}_1^{(w)} - t_{0.95:11}SE(\hat{\beta}_1^{(w)}) \le \beta_1 \le \hat{\beta}_1^{(w)} + t_{0.95:11}SE(\hat{\beta}_1^{(w)})]$$
$$= C[5.466 - 1.796 \times 0.5569 \le \beta_1 \le 5.466 + 1.796 \times 0.5569]$$
$$= C[4.466 \le \beta_1 \le 6.466] = 0.90$$

For the purpose of illustration we calculate $\hat{Y}(x)$, $\hat{\sigma}_Y(x)$, and $SE(\hat{Y}(x))$ for $x = 300$.

$$\hat{Y}(300) = \hat{\beta}_0^{(w)} + \hat{\beta}_1^{(w)}300 = 2011.48 \tag{8.2.36}$$
$$\hat{\sigma}_Y(300) = \hat{\sigma}_0 g(300) = (27.8)(\sqrt{300}) = 481.51 \tag{8.2.37}$$
$$SE(\hat{Y}(300)) = \hat{\sigma}_0\sqrt{[g(300)]^2 + x^T C^{(w)} x} \tag{8.2.38}$$

$$= 27.8\sqrt{300 + [1 \quad 300]C^{(w)}\begin{bmatrix} 1 \\ 300 \end{bmatrix}} = 514.9$$

With this information we can use (4.6.1) and calculate confidence intervals for $Y(300)$.

Finally we illustrate how to obtain a two-sided 80% confidence interval for $\sigma_Y(300)$. This is done by first computing a two-sided 80% confidence interval for σ_0 using (4.6.13) with $WSSE(X)$ in place of $SSE(X)$. We get

$$C\left[\sqrt{\frac{WSSE(X)}{\chi^2_{1-\alpha/2:n-2}}} \le \sigma_0 \le \sqrt{\frac{WSSE(X)}{\chi^2_{\alpha/2:n-2}}}\right]$$

$$= C\left[\sqrt{\frac{8498.69}{\chi^2_{0.9:11}}} \le \sigma_0 \le \sqrt{\frac{8498.69}{\chi^2_{0.1:11}}}\right]$$

$$= C\left[\sqrt{\frac{8498.69}{17.275}} \le \sigma_0 \le \sqrt{\frac{8498.69}{5.578}}\right]$$

$$= C\left[22.18 \le \sigma_0 \le 39.03\right] = 0.80$$

Hence, by multiplying each term in the preceding confidence statement by $g(300) = \sqrt{300}$, we get

$$C[22.18 \ g(300) \le \sigma_0 g(300) \le 39.03 \ g(300)] = 0.80$$

i.e.,

$$C[384.17 \le \sigma_Y(300) \le 676.02] = 0.80 \quad \blacksquare$$

Most computer packages will perform a weighted least squares regression analy-
~~sis with the weights. In Section 8.2~~ of the laboratory manuals we show

~~Exhibit 8.2.1, which is obtained using~~
weighted regression program. The output is very similar to the output from an ordinary (unweighted) regression analysis. The data used are from Example 8.2.1.

Note the weights w_i for performing a weighted least squares regression are $w_i = [1/g(x_i)]^2 = 1/x_i$ for this problem.

The values of $\hat{\beta}_0^{(w)}$ and $\hat{\beta}_1^{(w)}$ are given in (8.2.39) and (8.2.40), respectively.
Compare these with the values in (8.2.32). $WSSE(X) = 8499$ and $WMSE(X) = 773$
are given in (8.2.41) under the headings SS and MS, respectively. Thus, $\hat{\sigma}_0$ may be
obtained as $\hat{\sigma}_0 = \sqrt{WMSE(X)} = \sqrt{773} = 27.8$, the same as in (8.2.35), to within
rounding error. The matrix $C^{(w)}$ is given in (8.2.42).

E X H I B I T   8.2.1
MINITAB Output for Example 8.2.1

Row	CO	cars	weights
1	5817	873	0.0011455
2	1063	109	0.0091743
3	2616	398	0.0025126
4	2018	353	0.0028329
5	3147	506	0.0019763
6	7210	1026	0.0009747
7	4339	862	0.0011601
8	5153	742	0.0013477
9	4450	786	0.0012723
10	5591	896	0.0011161
11	2747	377	0.0026525
12	3712	720	0.0013889
13	2354	655	0.0015267

```
The regression equation is
CO = 372 + 5.47 cars
```

Predictor	Coef	Stdev	t-ratio	p	
Constant	371.6	297.6	1.25	0.238	**(8.2.39)**
cars	5.4662	0.5569	9.82	0.000	**(8.2.40)**

EXHIBIT 8.2.1
(Continued)

Analysis of Variance

SOURCE	DF	SS	MS	F	p
Regression	1	74445	74445	96.36	0.000
Error	11	8499	773		
Total	12	82944			

(8.2.41)

The weighted C matrix is

$$
\begin{array}{rr}
114.5957 & -0.1794 \\
-0.1794 & 0.000401
\end{array}
$$

(8.4.42)

Problems 8.2

8.2.1 Consider Problem 3.5.1 where an investigator is studying the association between sulfur dioxide (SO_2) concentrations in a national park and the rate of emission of SO_2 by a coal burning power plant 25 miles away. A certain fraction of the emitted SO_2 will be transported by winds to the national park. At the national park, there is always a certain amount of background SO_2 that is not emitted by the power plant. The SO_2 emissions (X, in tons/hour) by the power plant and the SO_2 concentrations at the national park (Y, in micrograms/cubic meter, or mg/m^3) were recorded at various randomly selected times during a particular year. The data are given in Table 8.2.2 and are also stored in the file named **so2.dat** on the data disk. Suppose that weighted regression assumptions in Box 8.2.1 are valid with

$$\mu_Y(x) = \beta_0 + \beta_1 x \qquad (8.2.43)$$

and $\sigma_Y(x) = \sigma_0 g(x) = \sigma_0 x$ where β_0, β_1, and σ_0 are unknown constants. So $g(x) = x$ and the weights are $w_i = [1/g(x_i)]^2 = 1/x_i^2$. We use weighted regression to obtain point estimates and confidence intervals for the unknown parameters.

The computer output in Exhibit 8.2.2 lists the data along with the weights and the results from a weighted regression analysis. Two weights, those for sample items 2 and 5, have not been computed.

a Compute the weights for items 2 and 5.

b Verify that the weights are all correct.

c What are the weighted least squares estimates for β_0 and β_1?

d Compare the estimates in (c) with the unweighted estimates obtained in Problem 3.5.2.

e State what an appropriate target population might be for this problem. What is the study population?

T A B L E 8.2.2

	(micrograms...)	
1	5.21	1.92
2	7.36	3.92
3	16.26	6.80
4	10.10	6.32
5	5.80	2.00
6	8.06	4.32
7	4.76	2.40
8	6.93	2.96
9	9.36	3.52
10	10.90	4.24
11	12.48	5.12
12	11.70	5.84
13	7.44	3.60
14	6.99	2.80

E X H I B I T 8.2.2

MINITAB Output for Problem 8.2.1

Row	Y	X	weights
1	5.21	1.92	0.271267
2	7.36	3.92	********
3	16.26	6.80	0.021626
4	10.10	6.32	0.025036
5	5.80	2.00	********
6	8.06	4.32	0.053584
7	4.76	2.40	0.173611
8	6.93	2.96	0.114134
9	9.36	3.52	0.080708
10	10.90	4.24	0.055625
11	12.48	5.12	0.038147
12	11.70	5.84	0.029321
13	7.44	3.60	0.077160
14	6.99	2.80	0.127551

```
The regression equation is
Y = 1.72 + 1.78 X
```

Predictor	Coef	Stdev	t-ratio	p
Constant	1.7214	0.7683	2.24	0.045
X	1.7762	0.2415	7.36	0.000

Analysis of Variance

SOURCE	DF	SS	MS	F	p
Regression	1	6.0298	6.0298	54.10	0.000
Error	12	1.3375	0.1115		
Total	13	7.3672			

```
The weighted C matrix is

   5.29681    -1.54690
  -1.54690     0.52319
```

8.2.2   In Problem 8.2.1 estimate the mean and the standard deviation of the SO_2 concentrations at the park associated with an emission rate of 3.0 tons/hour at the power plant.

8.2.3   In Problem 8.2.1 what is the population parameter that represents the difference between the average SO_2 concentration at the park associated with a power plant emission rate of 5.0 tons/hour, and that associated with a power plant emission rate of 2.5 tons/hour?

8.2.4   Estimate the difference in Problem 8.2.3 and compute a two-sided 95% confidence interval for the difference.

8.2.5   Discuss whether or not claims can be made to the effect that the SO_2 emissions at the power plant cause the SO_2 concentrations at the national park to increase. In particular, can we conclude, on the basis of these data, that the SO_2 concentrations at the national park will decrease if the power plant is shut down?

8.2.6   Compute a 90% two-sided confidence interval for σ_0.

8.2.7   Compute a 90% two-sided confidence interval for $\sigma_Y(4.00)$.

subpopulation of Y values is Gaussian. In some problems ... ...
that the subpopulations are not Guassian (not even approximately), or a residual
analysis of the sample data may cast doubt on this assumption. In such cases it
is useful to have alternative, valid inference procedures available. In this section
we discuss one such alternative for straight line regression, called *Theil's method*
because it was first proposed by H. Theil [34].

We call the assumptions underlying Theil's method for straight line regression
non-Gaussian assumptions, and they are given in Box 8.3.1.

B O X **8.3.1**   **Non-Gaussian Assumptions for Straight Line Regression**

Notation   Let $\{(Y, X)\}$ be a two-variable study population.

(Population) Assumption 1   For each distinct value x of X in the population,
the mean $\mu_Y(x)$ of the corresponding subpopulation of Y values is given by

$$\mu_Y(x) = \beta_0 + \beta_1 x \qquad a \le x \le b \qquad \textbf{(8.3.1)}$$

where β_0 and β_1 are unknown parameters.

(Population) Assumption 2   Each subpopulation of Y values, determined by
the distinct values of X, is symmetric and continuous.

(Sample) Assumption 3   The sample of size n is selected either by simple
random sampling or by preselecting the X values.

(Sample) Assumption 4   The values of y_i and x_i for $i = 1, 2, \ldots, n$ are observed without error.

We make a few comments about these assumptions.

1   The term *continuous subpopulation* in population assumption 2 means that the
response variable Y is a continuous variable such as weight, height, time, etc.
and not a discrete variable such as counts of numbers of people, homes, days,
etc. In particular, when Y is a continuous variable, no two of its values will be
the same if they are measured sufficiently precisely.

2   In population assumption 2, the requirement that the subpopulation of Y values
be symmetric for each X value means that for every value of Y in the subpopulation that is d units below the mean $\mu_Y(x) = \beta_0 + \beta_1 x$, there is a corresponding
Y value d units above the mean. Subpopulations of Y values for different values
of X may be different with respect to their mean values, or standard deviations,
or other characteristics, but they must all be symmetric. For example they can all
be Gaussian (which is symmetric) with different means and different standard
deviations.

3   The subpopulation of Y values determined by the X values need not be Gaussian.

4 If the subpopulations of Y values happen to be Gaussian for each X, their standard deviations need not all be the same, nor do their relative magnitudes need to be known as in weighted regression assumptions in Box 8.2.1.

Point Estimation

We now explain the procedure for estimating θ, a linear combination of β_0 and β_1, given by

$$\theta = a_0\beta_0 + a_1\beta_1 \tag{8.3.2}$$

where a_0 and a_1 are specified constants. Note that β_0 is obtained from θ by setting $a_0 = 1$ and $a_1 = 0$ in (8.3.2), whereas β_1 is obtained from θ by setting $a_0 = 0$ and $a_1 = 1$. Also $\mu_Y(x)$ is obtained from θ by letting $a_0 = 1$ and $a_1 = x$, and $\mu_Y(x_1) - \mu_Y(x_2)$ is obtained from θ by setting $a_0 = 0$ and $a_1 = x_1 - x_2$.

Let $(y_1, x_1), (y_2, x_2), \ldots, (y_n, x_n)$ be a sample of size n arranged according to increasing values of x; i.e., $x_1 < x_2 < \cdots < x_n$. If several y values are available for a given x value, we let y_i be their mean so we can assume that the x_i's are distinct. If n is an odd number, say $n = 2m + 1$, then discard the middle observation so there are now $2m$ observations (y_i, x_i) for $i = 1, \ldots, 2m$. Of course if n is even, no observation is discarded and $n = 2m$. The observations are arranged as in Table 8.3.1. Remember that $x_1 < x_2 < \cdots < x_{2m}$ and that no two x values are the same. From Table 8.3.1 we compute the quantities z, w, u, v, and t, which are exhibited in Table 8.3.2.

T A B L E 8.3.1

Column 1	Column 2	Column 3	Column 4
y_1	x_1	y_{m+1}	x_{m+1}
y_2	x_2	y_{m+2}	x_{m+2}
$\vdots$	$\vdots$	$\vdots$	$\vdots$
y_m	x_m	y_{2m}	x_{2m}

T A B L E 8.3.2

z	w	u	v	t
$z_1 = y_{m+1} - y_1$	$w_1 = x_{m+1} - x_1$	$u_1 = y_1 x_{m+1}$	$v_1 = y_{m+1} x_1$	$t_1 = u_1 - v_1$
$z_2 = y_{m+2} - y_2$	$w_2 = x_{m+2} - x_2$	$u_2 = y_2 x_{m+2}$	$v_2 = y_{m+2} x_2$	$t_2 = u_2 - v_2$
$\vdots$	$\vdots$	$\vdots$	$\vdots$	$\vdots$
$z_m = y_{2m} - y_m$	$w_m = x_{2m} - x_m$	$u_m = y_m x_{2m}$	$v_m = y_{2m} x_m$	$t_m = u_m - v_m$

Compute q_i^* where

$$q_1, q_2, \ldots, q_m \text{ ...... } q_1 \quad {}_{12} \qquad {}_{*m}$$

then the middle number q_{k+1} is the estimate of $a_0\beta_0 + a_1\beta_1$. If m is an even number (i.e., $m = 2k$), then $(q_k + q_{k+1})/2$, the average of the two middle numbers, is the estimate of $a_0\beta_0 + a_1\beta_1$.

Confidence Intervals

A confidence interval for $a_0\beta_0 + a_1\beta_1$ may not be available with confidence coefficient exactly equal to a specified value $1 - \alpha$, so we find confidence intervals with confidence coefficients as close to $1 - \alpha$ as the procedure allows. To do this we follow the instructions in Box 8.3.2.

BOX **8.3.2**

1 Let $m = \dfrac{n}{2}$ if n is even and $m = \dfrac{n-1}{2}$ if n is odd.

2 For this value of m, examine the numbers given in row m of Table T-6 in Appendix T. These are the confidence coefficients (which are ≥ 0.50) for which a two-sided confidence interval is available. Choose one of these confidence coefficients, say $1 - \alpha$, and proceed to step 3.

3 Go across row m in Table T-6 and select the value of r corresponding to the confidence coefficient $1 - \alpha$ chosen in step 2.

4 Using this value of r, compute $m - r + 1$.

5 For the values of r and $m - r + 1$, select q_r and q_{m-r+1} where $q_1 < q_2 < \ldots < q_m$ are obtained by ordering the q_i^* in (8.3.3).

6 A $1 - \alpha$ two-sided confidence interval for $a_0\beta_0 + a_1\beta_1$ is given by the confidence statement

$$C[q_r \leq a_0\beta_0 + a_1\beta_1 \leq q_{m-r+1}] = 1 - \alpha \qquad \text{(8.3.4)}$$

Example 8.3.1 illustrates the relevant computations.

E X A M P L E **8.3.1**

A random sample of 20 college professors was selected, and their annual salaries (Y) in thousands of dollars and number of years (X) of experience were recorded. The data are given in Table 8.3.3 and are stored in the file **profsal.dat** on the data disk. The investigator believes that neither assumptions (A) nor (B) hold, but the assumptions in Box 8.3.1 are appropriate. We compute a confidence interval for $\mu_Y(10)$, the average annual salary of all college professors with 10 years'

TABLE **8.3.3**

Professors' Salary Data

Observation Number	Annual Salary Y (in thousands of dollars)	Experience X (in years)
1	63	19
2	48	14
3	50	14
4	47	9
5	41	7
6	44	10
7	43	7
8	66	20
9	78	28
10	59	16
11	49	12
12	65	21
13	67	21
14	58	13
15	40	8
16	69	22
17	58	15
18	71	20
19	51	12
20	49	13

experience, with confidence coefficient as close to 90% as possible. We exhibit the computations to obtain the q_i.

The observations are rearranged so that X values occur in increasing order. The ordered data along with the original data are given in Table 8.3.4. Note that some of the X values are repeated. For instance, the value $X = 7$ occurs twice with the corresponding Y values being 41 and 43. In each such case we compute the mean of the Y values corresponding to the same X value, which results in the condensed data set shown in Table 8.3.5.

T A B L E **8.3.4**

Observation Number	Salary Y	Years X	Observation Number	Salary Y	Years X
1	63	19	5	41	7
2	48	14	7	43	7
3	50	14	15	40	8
4	47	9	4	47	9
5	41	7	6	44	10
6	44	10	11	49	12
7	43	7	19	51	12
8	66	20	14	58	13
9	78	28	20	49	13
10	59	16	2	48	14
11	49	12	3	50	14
12	65	21	17	58	15
13	67	21	10	59	16
14	58	13	1	63	19
15	40	8	8	66	20
16	69	22	18	71	20
17	58	15	12	65	21
18	71	20	13	67	21
19	51	12	16	69	22
20	49	13	9	78	28

T A B L E **8.3.5**
Condensed Data from Table 8.3.4

Y mean	X
42	7
40	8
47	9
44	10
50	12
53.5	13
49	14
58	15
59	16
63	19
68.5	20
66	21
69	22
78	28

Since $n = 14$, which is an even number, $m = n/2 = 7$. Thus we divide the data in Table 8.3.5 into four columns as in Table 8.3.1. We get

Column 1	Column 2	Column 3	Column 4
42	7	58	15
40	8	59	16
47	9	63	19
44	10	68.5	20
50	12	66	21
53.5	13	69	22
49	14	78	28

First we compute z and w in Table 8.3.2 and get

z	w
16	8
19	8
16	10
24.5	10
16	9
15.5	9
29	14

Next we compute u, v, and t in Table 8.3.2 and get

u	v	t
630	406	224
640	472	168
893	567	326
880	685	195
1050	792	258
1177	897	280
1372	1092	280

Next we compute the q_i^* in (8.3.3). Note that we have taken $a_0 = 1$ and $a_1 = 10$

q^*
48.0000
44.7500
48.6000
44.0000
46.4444
48.3333
40.7143

Next we order the q_i^* from smallest to largest and get the q_i as follows:

40.7143   44.0000   44.7500   46.4444   48.0000   48.3333   48.6000

Because $m = 7$ is an odd number, the middle number, namely q_4, is 46.4444. Hence the estimated value of $\mu_Y(10)$ is 46 (rounded to the nearest thousand).

Look in Table T-6 in Appendix T across row $m = 7$ and find that the confidence coefficient that is nearest to 0.90 is 0.88 and the corresponding value of r is 2. Hence $m - r + 1 = 6$. Thus the 88% two-sided confidence bounds for $\mu_Y(10)$ are $q_2 = 44.0000$ and $q_6 = 48.3333$. We obtain the confidence statement (rounding to the nearest thousand)

$$C[44 \leq \mu_Y(10) \leq 48] = 0.88 \quad \blacksquare \qquad (8.3.5)$$

Problems   8.3

8.3.1   Consider Problem 8.2.1 where an investigator is studying the relationship of sulfur dioxide concentrations, Y, in a national park, and the sulfur dioxide emission rate, X, by a coal burning power plant 25 miles away. Suppose that the regression function of Y on X is of the form

$$\mu_Y(x) = \beta_0 + \beta_1 x$$

and that assumptions in Box 8.3.1 are satisfied. Some of the computations required for this problem are in Tables 8.3.6–8.3.8. The asterisks (****) indicate that some values have not been computed, and you will be asked to supply them.

a   What is the value of n?

b   What is the value of m?

c   Compute the missing values for w, v, and t in Table 8.3.7.

d   Compute the missing values for q_i^* and q_i in Table 8.3.8.

T A B L E 8.3.6

Computations for Table 8.3.1

1	5.21	1.92	7.36	3.92
2	5.80	2.00	10.90	4.24
3	4.76	2.40	8.06	4.32
4	6.99	2.80	12.48	5.12
5	6.93	2.96	11.70	5.84
6	9.36	3.52	10.10	6.32
7	7.44	3.60	16.26	6.80

T A B L E 8.3.7

Computations for Table 8.3.2

Row	z	w	u	v	t
1	2.15	2.00	20.4232	14.1312	6.2920
2	5.10	****	24.5920	****	****
3	3.30	1.92	20.5632	19.3440	1.2192
4	5.49	2.32	35.7888	34.9440	0.8448
5	4.77	2.88	40.4712	34.6320	5.8392
6	0.74	2.80	59.1552	****	23.6032
7	8.82	3.20	50.5920	58.5360	−7.9440

T A B L E 8.3.8

Computations for Point Estimate and Confidence Interval for β_0

Row	q_i^*	q_i
1	3.14600	−2.48250
2	****	****
3	0.63500	0.63500
4	****	****
5	2.02750	2.02750
6	8.42971	3.14600
7	−2.48250	8.42971

e Use Theil's method and estimate β_0.

f Use Theil's method and estimate β_1.

g Use Theil's method and obtain a confidence interval for β_0 with a confidence coefficient as close to 90% as possible.

h Use Theil's method and obtain a confidence interval for β_1 with a confidence coefficient as close to 90% as possible.

    i   Compare the estimates in parts (e) and (f) with the estimates obtained using weighted regression in Problem 8.2.1

    j   Use Theil's method and estimate the mean SO_2 concentration at the park associated with an emission rate of 5.0 tons/hour at the power plant.

    k   Write out the population parameters that represent the difference between the average SO_2 concentration at the park corresponding to a power plant SO_2 emission rate of 5.0 tons/hour and that corresponding to a power plant SO_2 emission rate of 2.5 tons/hour.

    l   Use Theil's method and estimate the difference in part (k) and compute a two-sided confidence interval for it with a confidence coefficient as close to 90% as possible.

8.3.2   Can the assumptions in Box 8.3.1 be satisfied for straight line regression if assumptions (A) are satisfied? Discuss.

8.3.3   Can the assumptions in Box 8.3.1 be satisfied for straight line regression if assumptions (B) are satisfied? Discuss.

8.3.4   Can assumptions (B) be satisfied for straight line regression if assumptions in Box 8.3.1 are satisfied? Discuss.

8.3.5   Can assumptions (A) be satisfied for straight line regression if assumptions in Box 8.3.1 are satisfied? Discuss.

8.3.6   Can the assumptions in Box 8.2.1 be satisfied for straight line regression if assumptions (B) are satisfied? Discuss.

8.3.7   Can the assumptions in Box 8.2.1 be satisfied for straight line regression if assumptions (A) are satisfied? Discuss.

8.3.8   Can the assumptions in Box 8.3.1 be satisfied for straight line regression if assumptions in Box 8.2.1 are satisfied? Discuss.

8.3.9   Can the assumptions in Box 8.2.1 be satisfied for straight line regression if assumptions in Box 8.3.1 are satisfied? Discuss.

8.4
Exercises

8.4.1   The texture score Y of a soybean product (soyburger, a meat substitute) depends to some extent on the percent X of a filler material used. Typically the texture score for a batch of this product is obtained by asking a trained panel of food experts to assign scores (from 0 to 10) and then taking the average of these individual rating scores. Texture scores greater than 7 indicate an acceptable product. The ultimate objective is to find the smallest amount of the filler material that will result in an acceptable texture score for the final product. Consequently a food engineer is interested in studying the relationship between the texture score and the percent of filler used. To do this, he makes several batches of soyburger with different amounts of filler material and obtains the texture scores for each batch. The data are given in Table 8.4.1 and are also stored in the file **soyburgr.dat** on the data disk.

T A B L E   **8.4.1**
Soyburger Data

Batch Number	Texture Score Y	Filler Material X (percent)
1	2.5	0.5
2	2.9	1.0
3	3.4	1.5
4	3.7	2.0
5	4.3	2.5
6	4.5	3.0
7	4.9	3.5
8	5.8	4.0
9	6.4	4.5
10	6.8	5.0
11	6.5	5.5
12	8.0	6.0
13	8.4	6.5
14	8.5	7.0
15	7.4	7.5
16	9.9	8.0

Suppose the regression function of Y on X is of the form

$$\mu_Y(x) = \beta_0 + \beta_1 x \qquad (8.4.1)$$

and that the weighted regression assumptions in Box 8.2.1 hold with $g(x) = x^2$ so that the weights are given by $w_i = 1/x_i^4$. A computer output for a weighted regression analysis of Y on X obtained using MINITAB is given in Exhibit 8.4.1. Additionally we also give the quantities SSX, SSY, SXY, $\bar{x}$, and $\bar{y}$ required to compute the ordinary least squares estimates of β_0 and β_1.

a   Perform an ordinary regression of Y on X and calculate the residuals and standardized residuals. Examine appropriate plots. Based on these plots, do you think any of assumptions (A) appear to be violated? In particular, does it appear that the homogeneity of variance assumption holds?

Answer parts (b)–(f) using weighted regression.

b   Estimate the regression function of Y on X given in (8.4.1).

c   Obtain a two-sided 99% confidence interval for the mean texture score of all batches of soyburger made with 6% filler material.

d   Obtain a two-sided 90% confidence interval for the texture score of a single batch of soyburger to be made with 6% filler material.

e   Obtain a two-sided 80% confidence interval for the standard deviation of the texture scores of all batches of soyburger made with 6% filler material.

f   Estimate the proportion of batches of soyburger made with 6% filler material that will have texture scores in the acceptable range (a score of 7 or greater).

E X H I B I T   8.4.1
MINITAB Output for Exercise 8.4.1

```
ROW       texture     filler        weights

  1         2.5        0.5      16.0000000000
  2         2.9        1.0       1.0000000000
  3         3.4        1.5       0.1975308657
  4         3.7        2.0       0.0625000000
  5         4.3        2.5       0.0255999994
  6         4.5        3.0       0.0123456791
  7         4.9        3.5       0.0066638901
  8         5.8        4.0       0.0039062500
  9         6.4        4.5       0.0024386526
 10         6.8        5.0       0.0016000000
 11         6.5        5.5       0.0010928215
 12         8.0        6.0       0.0007716049
 13         8.4        6.5       0.0005602045
 14         8.5        7.0       0.0004164931
 15         7.4        7.5       0.0003160494
 16         9.9        8.0       0.0002441406
```

```
The regression equation is
texture = 2.07 + 0.858 filler

Predictor       Coef       Stdev     t-ratio          p
Constant     2.06979     0.01139      181.71      0.000
filler       0.85776     0.01883       45.56      0.000

Analysis of Variance

SOURCE         DF           SS          MS          F          p
Regression      1      0.74544     0.74544    2075.50      0.000
Error          14      0.00503     0.00036
Total          15      0.75047

The weighted C matrix is

   0.361228   -0.547297
  -0.547297    0.987004


     SSY = 74.2544;    SSX = 85.00;    SXY = 77.525;
     mean of y = 5.86875;              mean of x = 4.25;
```

g What might be an appropriate target population (of items and of numbers) for this problem? What is the study population?

8.4.2 The final exam scores (Y) in a particular statistics course are related to the midterm test scores (X) according to a straight line regression model. A random sample of 24 students during a particular semester yielded the data in Table 8.4.2, which are also in the file **exam.dat** on the data disk. Suppose that an investigator believes assumptions (A) for straight line regression are satisfied. A computer output containing the results of an ordinary regression analysis of Y on X is given in Exhibit 8.4.2.

T A B L E **8.4.2**
Exam Scores Data

Student	Final Exam Score (Y)	Midterm Exam Score (X)
1	40	44
2	47	48
3	41	49
4	41	50
5	43	52
6	42	53
7	50	54
8	87	58
9	61	61
10	74	66
11	75	75
12	89	76
13	72	77
14	69	78
15	78	80
16	78	83
17	92	84
18	84	85
19	85	86
20	99	87
21	89	90
22	83	91
23	96	95
24	100	99

E X H I B I T 8.4.2
MINITAB Output for Exercise 8.4.2

```
The regression equation is
final = -6.13 + 1.08 midterm
```

Predictor	Coef	Stdev	t-ratio	p
Constant	-6.132	8.147	-0.75	0.460
midterm	1.0820	0.1106	9.78	0.000

```
s = 9.093      R-sq = 81.3%    R-sq(adj) = 80.5%
```

Analysis of Variance

SOURCE	DF	SS	MS	F	p
Regression	1	7910.9	7910.9	95.68	0.000
Error	22	1819.0	82.7		
Total	23	9730.0			

ROW	final	midterm	fits	stdresid	nscores
1	40	44	41.477	-0.17675	0.15498
2	47	48	45.805	0.14045	0.73025
3	41	49	46.887	-0.68940	-0.73025
4	41	50	47.969	-0.81308	-0.87320
5	43	52	50.133	-0.82653	-1.03703
6	42	53	51.215	-1.06442	-1.49944
7	50	54	52.297	-0.26458	-0.26024
8	87	58	56.626	3.46287	1.95007
9	61	61	59.872	0.12790	0.60110
10	74	66	65.282	0.98187	1.03703
11	75	75	75.020	-0.00225	0.48148
12	89	76	76.102	1.45102	1.49944
13	72	77	77.184	-0.58364	-0.48148
14	69	78	78.266	-1.04414	-1.23521
15	78	80	80.430	-0.27446	-0.36854
16	78	83	83.676	-0.64404	-0.60110
17	92	84	84.758	0.82320	0.87320
18	84	85	85.840	-0.20961	0.05147
19	85	86	86.922	-0.21944	-0.05147
20	99	87	88.004	1.25817	1.23521
21	89	90	91.250	-0.25961	-0.15498
22	83	91	92.332	-1.07989	-1.95007
23	96	95	96.661	-0.07752	0.36854
24	100	99	100.989	-0.11806	0.26024

a Estimate the regression function $\mu_Y(x) = \beta_0 + \beta_1 x$.

b Plot the standardized residuals r_i against the fitted values $\hat{\mu}_Y(x_i)$. Does this plot suggest that any of the assumptions are violated?

c Obtain a Gaussian rankit-plot of r_i. Do the standardized residuals appear to be a simple random sample from a Gaussian population with zero mean and unit standard deviation?

d Suppose that from the plot in part (b) and other considerations, an investigator believes that assumptions (A) are not valid, and she decides to use Theil's method to estimate β_0, β_1, and $\mu_Y(x)$. Some results to help you to do the computations for Theil's method for regression are given in Exhibit 8.4.3.

E X H I B I T **8.4.3**
Some Calculations for Theil's Method of Regression

Results for Table 8.3.1

ROW	column1	column2	column3	column4
1	40	44	72	77
2	47	48	69	78
3	41	49	78	80
4	41	50	78	83
5	43	52	92	84
6	42	53	84	85
7	50	54	85	86
8	87	58	99	87
9	61	61	89	90
10	74	66	83	91
11	75	75	96	95
12	89	76	100	99

EXHIBIT **8.4.3**
(Continued)

Results for Table 8.3.2

ROW	z	w	u	v	t
1	32	33	3080	3168	-88
2	22	30	3666	3312	354
3	37	31	3280	3822	-542
4	37	33	3403	3900	-497
5	49	32	3612	4784	-1172
6	42	32	3570	4452	-882
7	35	32	4300	4590	-290
8	12	29	7569	5742	1827
9	28	29	5490	5429	61
10	9	25	6734	5478	1256
11	21	20	7125	7200	-75
12	11	23	8811	7600	1211

i Estimate the regression function $\mu_Y(x) = \beta_0 + \beta_1 x$ using Theil's method.

ii Using Theil's method obtain two-sided confidence intervals for β_0 and β_1 with confidence coefficient as close to 85% as possible.

iii Using Theil's method predict the final exam score of a student who obtained 75 points on the midterm exam.

iv Using Theil's method obtain a two-sided confidence interval for $\mu_Y(75)$, the average final exam score of all students who obtain 75 point on the midterm examination. Use a confidence coefficient as close to 85% as possible.

e Explain what might be an appropriate target population for this problem. What is the study population?

9

Nonlinear Regression

9.1
Overview

In Chapter 2 we made a distinction between linear and nonlinear regression functions. Chapters 3 through 8 were concerned with inferences for, and applications of, simple and multiple *linear* regression models. Recall that in a linear regression model the regression function $\mu_Y(x_1, \ldots, x_k)$ is a *linear function of the unknown parameters,* whereas in a nonlinear regression model the regression function is *not a linear function of the unknown parameters*. In this chapter we present some commonly used nonlinear regression models, discuss point and interval estimation for unknown parameters, and provide some examples. In Section 9.2 we list several commonly used regression models. Section 9.3 gives the statistical assumptions underlying the inference procedures for nonlinear regression models and discusses parameter estimation, confidence intervals, and tests for them. In Section 9.4 we illustrate how nonlinear regression functions can sometimes be reformulated as linear regression functions by applying suitable transformations. Section 9.5 contains chapter exercises. Our discussion of nonlinear regression is limited, and we give only a brief introduction to the subject.

The computations required for nonlinear regression analyses are not feasible without the use of a computer, and most major statistical packages have routines for nonlinear regression. In the laboratory manuals we explain the use of the computer for nonlinear regression analysis.

9.2
Some Commonly Used Families
of Nonlinear Regression Functions

While simple and multiple linear regression functions are adequate for modeling a wide variety of relationships between response variables and predictor variables, many situations require nonlinear functions. Certain types of nonlinear regression functions have served, and will continue to serve, as useful models for describing various physical and biological systems. We list a few of these situations for the case of a single predictor variable.

1   The following functions have been considered in modeling the relationship between crop yield Y and the spacing between rows of plants, concentration Y of a drug in the bloodstream and time X after the drug is injected when this concentration is measured, the rate Y of a chemical reaction and the amount X of catalyst used, and many other relationships.

$$\mu_Y(x) = \frac{1}{(\beta_1 + \beta_2 x)^{\beta_3}} \tag{9.2.1}$$

$$\mu_Y(x) = \frac{1}{\beta_1 + \beta_2 x + \beta_3 x^2} \tag{9.2.2}$$

$$\mu_Y(x) = \frac{1}{\beta_1 + \beta_2 x^{\beta_3}} \tag{9.2.3}$$

Typical members of the families of curves (9.2.1)–(9.2.3) are displayed in Figures 9.2.1–9.2.3, respectively.

F I G U R E   **9.2.1**
Three members of the family of curves $\mu_Y(x) = \frac{1}{(\beta_1 + \beta_2 x)^{\beta_3}}$

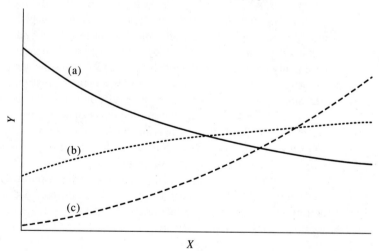

(a) $\beta_1 = 0.6, \beta_2 = 1, \beta_3 = -1$;   (b) $\beta_1 = 0.1, \beta_2 = 1, \beta_3 = 0.3$;
(c) $\beta_1 = 0.2, \beta_2 = 1, \beta_3 = 2$.

F I G U R E   9.2.2

Three members of the family of curves $\mu_Y(x) = \dfrac{1}{\beta_1 + \beta_2 x + \beta_3 x^2}$

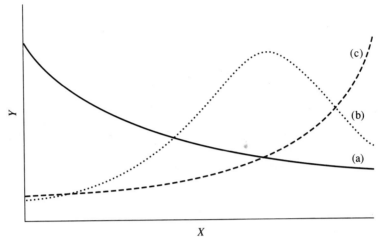

(a) $\beta_1 = 1, \beta_2 = 3, \beta_3 = -0.2$;   (b) $\beta_1 = 8.94, \beta_2 = -22.4, \beta_3 = 16$;
(c) $\beta_1 = 8, \beta_2 = -8, \beta_3 = 1$.

F I G U R E   9.2.3

Three members of the family of curves $\mu_Y(x) = \dfrac{1}{\beta_1 + \beta_2 x^{\beta_3}}$

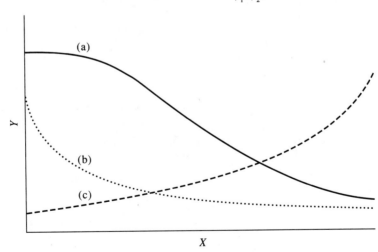

(a) $\beta_1 = 1, \beta_2 = 6, \beta_3 = 3$;   (b) $\beta_1 = 1.2, \beta_2 = 9, \beta_3 = 0.9$;
(c) $\beta_1 = 10, \beta_2 = -8.8, \beta_3 = 0.5$.

2   S-shaped curves, often referred to as sigmoidal curves, arise in various applications, including bioassay, signal detection theory, engineering, and economics. Various types of growth data often conform to sigmoidal curves. Some of the nonlinear regression functions that have been used in such situations include

$$\mu_Y(x) = \beta_1 e^{-e^{-(\beta_2 + \beta_3 x)}} \tag{9.2.4}$$

$$\mu_Y(x) = \frac{\beta_1}{1 + e^{-(\beta_2 + \beta_3 x)}} \tag{9.2.5}$$

and

$$\mu_Y(x) = \frac{\beta_1}{[1 + e^{-(\beta_2 + \beta_3 x)}]^{\beta_4}} \tag{9.2.6}$$

The model in (9.2.4) is often called the *Gompertz model*, the model in (9.2.5) is usually referred to as a *logistic regression model*, and the model in (9.2.6) is called *Richard's model*. Typical curves belonging to the families (9.2.4)–(9.2.6) are shown in Figures 9.2.4–9.2.6, respectively.

F I G U R E 9.2.4

Three members of the family of curves $\mu_Y(x) = \beta_1 e^{-e^{-(\beta_2 + \beta_3 x)}}$

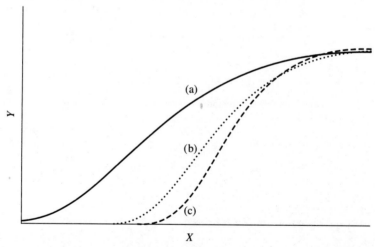

(a) $\beta_1 = 1, \beta_2 = -1.5, \beta_3 = 5$;   (b) $\beta_1 = 1, \beta_2 = -3.5, \beta_3 = 7$;
(c) $\beta_1 = 1, \beta_2 = -5, \beta_3 = 9$.

F I G U R E **9.2.5**

Three members of the family of curves $\mu_Y(x) = \dfrac{\beta_1}{1+e^{-(\beta_2+\beta_3 x)}}$

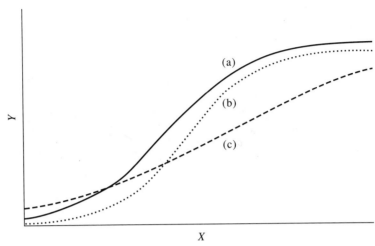

(a) $\beta_1 = 1, \beta_2 = -3.22, \beta_3 = 8$;
(b) $\beta_1 = 0.95, \beta_2 = -4.61, \beta_3 = 10$;
(c) $\beta_1 = 1, \beta_2 = -2.3, \beta_3 = 4$.

F I G U R E **9.2.6**

Three members of the family of curves $\mu_Y(x) = \dfrac{\beta_1}{[1+e^{-(\beta_2+\beta_3 x)}]^{\beta_4}}$

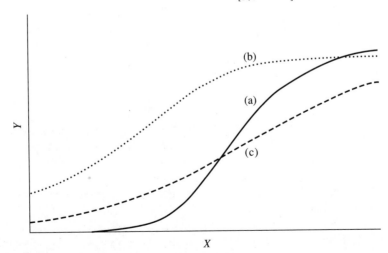

(a) $\beta_1 = 1, \beta_2 = -3.22, \beta_3 = 8, \beta_4 = 3.33$;
(b) $\beta_1 = 0.95, \beta_2 = -4.61, \beta_3 = 10, \beta_4 = 0.33$;
(c) $\beta_1 = 1, \beta_2 = -2.3, \beta_3 = 4, \beta_4 = 1.25$.

3   When the response variable Y steadily increases (decreases) with the independent variable X but the magnitude of the *rate of increase (decrease)* becomes smaller and smaller, with the response variable ultimately approaching a constant value called the *asymptote*, the family of curves defined by

$$\mu_Y(x) = \beta_1 + \beta_2 e^{-\beta_3 x} \tag{9.2.7}$$

has been found to provide useful nonlinear regression models. Three members of the family in (9.2.7) are displayed in Figure 9.2.7. Typical applications where such models are useful include the study of yield as a function of rate of application of fertilizer, mortality rate as a function of time, amount of chemical converted in a reaction as a function of time, etc.

F I G U R E   **9.2.7**

Three members of the family of curves $\mu_Y(x) = \beta_1 + \beta_2 e^{-\beta_3 x}$

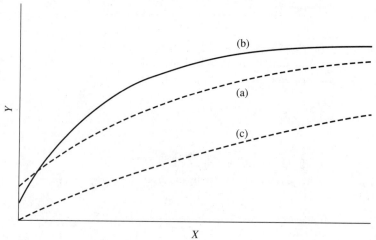

(a) $\beta_1 = 1, \beta_2 = -0.9, \beta_3 = 4$;   (b) $\beta_1 = 1, \beta_2 = -0.8, \beta_3 = 2$;
(c) $\beta_1 = 1, \beta_2 = -1, \beta_3 = 0, 9$.

Numerous other useful families of nonlinear regression functions exist, but we have limited ourselves to presenting some of the simplest and the most commonly used functions. Note that although all the applications just discussed involve only a single predictor variable, the models can be extended to the case of multiple predictor variables in a variety of ways. For more information on useful nonlinear regression models, you should refer to the book by Ratkowsky [29].

In the next section we turn our attention to the statistical assumptions for nonlinear regression and procedures for point estimation, confidence intervals, and tests.

Problems   9.2

9.2.1   Find two examples from your field of specialization where the regression function of a response variable with one or more predictor variables is a *nonlinear* function of unknown parameters.

9.2.2   In Problem 9.2.1 investigate the possibility that one of the several nonlinear regression functions given in (9.2.1)–(9.2.7) might provide an appropriate model.

9.2.3   Which of the following regression functions are linear and which are nonlinear? Here $\beta_0, \beta_1, \beta_2,$ and β_3 represent unknown parameters.

a $\mu_Y(x) = \beta_0 + \beta_1 e^{2x}$

b $\mu_Y(x) = \beta_0 + \beta_1 x + \beta_2 \log(x) \quad (x > 0)$

c $\mu_Y(x) = 6\beta_0 + x^{\beta_1} \quad (x > 0)$

d $\mu_Y(x) = \beta_0 + \beta_1 e^x + \beta_2 e^{-x}$

e $\mu_Y(x) = \beta_0 + \beta_1 e^{\beta_2 x} + \sin(\beta_3 x)$

9.3
Statistical Assumptions and Inferences for Nonlinear Regression

The most commonly used set of assumptions for nonlinear regression is the same as assumptions (A) for linear regression, the only exception being that the regression function $\mu_Y(x_1, \ldots, x_k)$ is a *nonlinear function of the unknown parameters* instead of a linear function of the parameters. For the record the complete set of assumptions is given in Box 9.3.1.

B O X   **9.3.1**   **Assumptions (A) for Nonlinear Regression**

The $(k+1)$-variable population $\{(Y, X_1, \ldots, X_k)\}$ is the study population under investigation.

(Population) Assumption 1   The mean of the subpopulation of Y values determined by $X_1 = x_1, \ldots, X_k = x_k$ is denoted by $\mu_Y(x_1, \ldots, x_k)$, and is a *nonlinear function of unknown parameters*. At times we find it useful to write $\mu_Y(x_1, \ldots, x_k; \beta_1, \ldots, \beta_p)$ for the regression function to emphasize the fact that it depends on the parameters $\beta_1, \ldots, \beta_p$.

(Population) Assumption 2   The standard deviations of the Y values are the same for each subpopulation determined by specified values of the predictor variables $X_1, \ldots, X_k$. This common standard deviation of all the subpopulations is denoted by $\sigma_{Y|X_1, \ldots, X_k}$, but for simplicity of notation we simply write σ when there is no possibility of confusion.

(Population) Assumption 3   Each subpopulation of Y values, determined by specified values of the predictor variables $X_1, \ldots, X_k$ is Gaussian.

(Sample) Assumption 4   The sample (of size n) is selected either by simple random sampling or by sampling with preselected values of $X_1, \ldots, X_k$.

(Sample) Assumption 5   All sample values $y_i, x_{i,1}, \ldots, x_{i,k}$ for $i = 1, \ldots, n$ are observed without error.

Remarks   Assumptions (A) for nonlinear regression are the same as assumptions (A) for linear regression except (population) assumption 1. Sample assumption 5 states that all sample values are observed without error. However, if only the response variable Y is measured with error, the procedures discussed in this chapter are still applicable for making inferences about the unknown parameters β_i. However, the standard error of the parameter estimates will tend to be larger and the confidence intervals will tend to be wider. *But if the predictor variables are measured with errors (that are not negligible), then the inference procedures discussed in this chapter are generally not applicable.*

Parameter Estimation

In practice, depending on the application, we assume that the form of the nonlinear regression function $\mu_Y(x_1, \ldots, x_k)$ is known but it contains unknown parameters $\beta_1, \ldots, \beta_p$. For instance, based on subject matter knowledge we may know that $\mu_Y(x)$ is of the form $\beta_1 + \beta_2 e^{\beta_3 x}$, but we may not know the values of β_1, β_2, and β_3. Observe that in the case of a linear regression function, i.e., when $\mu_Y(x_1, \ldots, x_k)$ has the form

$$\mu_Y(x_1, \ldots, x_k) = \beta_0 + \beta_1 x_1 + \cdots + \beta_k x_k$$

the value of p is equal to $k + 1$ (equal to k if there is no intercept in the model). However, in nonlinear regression models, it is not always the case that p is equal to $k + 1$ or k; for the nonlinear function $\mu_Y(x) = \beta_1 + \beta_2 e^{\beta_3 x}$, we have $k = 1$ but $p = 3$.

We now discuss point estimation for the unknown parameters in a nonlinear regression function using the method of least squares.

Least Squares Estimates of $\beta_1, \ldots, \beta_p$

A popular method for estimating the unknown parameters in a nonlinear regression function is the *method of least squares*. This is described in Section 4.4 for multiple linear regression. According to this method, the estimates of $\beta_1, \ldots, \beta_p$ are obtained by minimizing the quantity $\sum_{i=1}^{n} e_i^2$, the sum of squares of errors of prediction, where e_i is given by

$$e_i = y_i - \mu_Y(x_{i,1}, \ldots, x_{i,k})$$

As usual, the least squares estimates of $\beta_1, \ldots, \beta_p$ are denoted by $\hat{\beta}_1, \ldots, \hat{\beta}_p$. The estimated value of the subpopulation mean $\mu_Y(x_1, \ldots, x_k)$ is denoted by

$\hat{\mu}_Y(x_1, \ldots, x_k)$. It is referred to as the **fitted value** corresponding to $x_1, \ldots, x_k$, and is obtained by substituting the least squares estimates of the parameters into the regression function. This is algebraically expressed by the equation

$$\hat{\mu}_Y(x_1, \ldots, x_k) = \mu_Y(x_1, \ldots, x_k; \hat{\beta}_1, \ldots, \hat{\beta}_p) \tag{9.3.1}$$

The quantity $\hat{e}_i$ defined by

$$\hat{e}_i = y_i - \hat{\mu}_Y(x_{i,1}, \ldots, x_{i,k}) \tag{9.3.2}$$

is called the **residual** corresponding to sample item i.

The minimum value for the sum of squares of errors of prediction corresponding to the least squares estimates $\hat{\beta}_1, \ldots, \hat{\beta}_p$ is denoted by SSE, an abbreviation for the more complete notation $SSE(X_1, \ldots, X_k)$. Thus

$$SSE = \sum_{i=1}^{n} \hat{e}_i^2 = \sum_{i=1}^{n} [y_i - \hat{\mu}_Y(x_{i,1}, \ldots, x_{i,k})]^2 \tag{9.3.3}$$

and, as in linear regression, we refer to SSE as the *sum of squared errors*. The quantity MSE, which is an abbreviation for the more complete notation $MSE(X_1 \ldots, X_k)$, is given by

$$MSE = \frac{SSE}{(n - p)} \tag{9.3.4}$$

and is called the *mean squared error*, and it is an unbiased estimate of σ^2. The corresponding estimate of σ is given by

$$\hat{\sigma} = \sqrt{\frac{SSE}{n - p}} = \sqrt{MSE} \tag{9.3.5}$$

Computation of Least Squares Estimates

In the case of multiple linear regression, the least squares estimates of the parameters $\beta_1, \ldots, \beta_p$ can be computed quite easily using formula (4.4.8). However, the estimation of parameters in nonlinear regression models usually requires the use of *iterative methods* on digital computers, and explicit formulas for the estimates are generally not available. Most commonly available statistical software packages provide routines for calculating $\hat{\beta}_1, \ldots, \hat{\beta}_p$. The use of the computer for nonlinear regression analysis is discussed in the laboratory manual that accompanies the book.

To use any of these nonlinear regression programs, you must supply, in addition to the data, a set of starting values or initial guesses for $\beta_1, \ldots, \beta_p$. It is often helpful if the starting values are close to the actual least squares estimates $\hat{\beta}_i$. However, you may not have such initial guesses. Sometimes you can obtain good initial estimates of $\hat{\beta}_i$, or at least the signs of $\hat{\beta}_i$, based on theoretical considerations or by plotting the sample data.

In Example 9.3.1 we use the statistical package SAS for estimating the parameters for a specific nonlinear regression problem.

E X A M P L E 9.3.1

It is well known that when a beam of light is passed through a chemical solution, a certain fraction of the incident light will be absorbed or reflected and the remainder will be transmitted. The intensity of the transmitted light decreases as the concentration of the chemical solution increases. This fact is often used to determine concentrations of various chemicals in solutions.

In Table 9.3.1 the data are results from an experiment in which several solutions of known concentrations of pure chemical were used to measure the amount of transmitted light to determine the relationship between the optical readings Y and the concentrations X. The data are also stored in the file **light.dat** on the data disk.

T A B L E 9.3.1
Light Data

Observation Number	Optical Reading Y (in arbitrary units)	Concentration X (in milligrams/liter)
1	2.86	0.0
2	2.64	0.0
3	1.57	1.0
4	1.24	1.0
5	0.45	2.0
6	1.02	2.0
7	0.65	3.0
8	0.18	3.0
9	0.15	4.0
10	0.01	4.0
11	0.04	5.0
12	0.36	5.0

Suppose assumptions (A) for nonlinear regression are satisfied with $\mu_Y(x)$ given by

$$\mu_Y(x) = \beta_1 + \beta_2 e^{-\beta_3 x} \tag{9.3.6}$$

where β_1, β_2, and β_3 are unknown parameters. On the basis of the data above, we want to estimate β_1, β_2, and β_3. Our initial guesses supplied to the computer program are $\beta_1 = 0.0$, $\beta_2 = 2.0$, and $\beta_3 = 0.5$. These initial guesses were obtained by plotting the data and interpreting geometrically the meanings of the parameters β_1, β_2, and β_3. Exhibit 9.3.1 shows the output from the statistical package SAS using a routine called NLIN (short for *Nonlin*ear regression). A plot of the data points along with the fitted curve is shown in Figure 9.3.1.

E X H I B I T 9.3.1

SAS Output for Example 9.3.1

The SAS System   0:00 Saturday,   Jan 1, 1994

Non-Linear Least Squares Summary Statistics Dependent Variable LIGHT

Source	DF	Sum of Squares	Mean Square	
Regression	3	20.542872863	6.847624288	
Residual	9	0.460427137	0.051158571	**(9.3.7)**
Uncorrected Total	12	21.003300000		
(Corrected Total)	11	10.605891667		

Parameter	Estimate	Asymptotic Std. Error	Asymptotic 95% Confidence Interval Lower	Upper	
BETA1	0.028763192	0.17163881268	-0.3595140815	0.4170404648	**(9.3.8)**
BETA2	2.723273503	0.21054950823	2.2469733725	3.1995736340	**(9.3.9)**
BETA3	0.682773200	0.14160078051	0.3624472546	1.0030991454	**(9.3.10)**

F I G U R E 9.3.1

Light Data

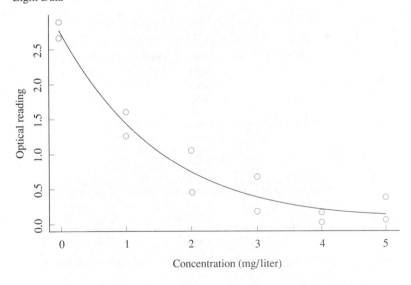

The final parameter estimates $\hat{\beta}_1$, $\hat{\beta}_2$, and $\hat{\beta}_3$, are given in (9.3.8)–(9.3.10). They are (after rounding to four decimals)

$$\hat{\beta}_1 = 0.0288 \qquad \hat{\beta}_2 = 2.7233 \qquad \hat{\beta}_3 = 0.6828 \qquad \textbf{(9.3.11)}$$

The estimate of σ^2 is given in (9.3.7) as part of an analysis of variance table along the row labeled `Residual` in the column labeled `Mean Square` and is equal to 0.0512. Thus $\hat{\sigma} = \sqrt{0.0512} = 0.2263$. ∎

We now turn our attention to confidence intervals and test for nonlinear regression.

(Approximate) Confidence Intervals and Tests of Hypotheses

Exact confidence interval procedures or exact hypothesis tests are generally not available for parameters in nonlinear regression models. However, approximate inference procedures are available. In practice, the computations required for carrying out approximate hypothesis tests or obtaining approximate confidence intervals are best performed using a suitable computer program. Any computer program for calculating the estimates of parameters in a nonlinear regression function usually outputs an *approximate standard error* (ASE), sometimes also referred to as an *asymptotic standard error,* for each parameter estimate. The approximation is usually quite good if the number of observations in the sample is large. Confidence intervals for $\beta_1, \ldots, \beta_p$ may be computed using (4.6.1), and tests may be carried out according to Box 4.7.1, with approximate standard errors in place of exact standard errors. Confidence intervals and tests for $\sigma = \sigma_{Y|X_1,\ldots,X_k}$ can be computed using (4.6.13) and Box 4.7.2, respectively, without any modifications. The degrees of freedom to use for table-values are, as usual, equal to

$$n - p = n - \text{(number of } \beta \text{ parameters in the model)}$$

The details are as follows. Suppose $\beta_1, \ldots, \beta_p$ are unknown parameters in a nonlinear regression function. An *approximate* $100(1 - \alpha)\%$ confidence interval for β_j is given by the confidence statement

$$C[\hat{\beta}_j - t_{1-\alpha/2:n-p}ASE(\hat{\beta}_j) \le \beta_j \le \hat{\beta}_j + t_{1-\alpha/2:n-p}ASE(\hat{\beta}_j)] \approx 1 - \alpha \quad \textbf{(9.3.12)}$$

where $ASE(\hat{\beta}_j)$ is the approximate standard error for $\hat{\beta}_j$. An approximate level α test of the hypothesis

$$\text{NH: } \beta_j = q \qquad \text{versus} \qquad \text{AH: } \beta_j \ne q \quad \textbf{(9.3.13)}$$

(where q is a specified number) is conducted as follows:

$$\text{Compute } t_C = \frac{\hat{\beta}_j - q}{ASE(\hat{\beta}_j)}. \qquad \text{Reject NH if } |t_C| > t_{1-\alpha/2:n-p} \quad \textbf{(9.3.14)}$$

One-sided tests of hypotheses are handled in the usual manner. We illustrate these computations in Example 9.3.2.

E X A M P L E 9.3.2

Here we continue with Example 9.3.1. From the computer output in Exhibit 9.3.1 (see (9.3.8)–(9.3.10)) under the column labeled `Asymptotic Std. Error`) we get

$$ASE(\hat{\beta}_1) = 0.1716 \qquad ASE(\hat{\beta}_2) = 0.2105 \qquad ASE(\hat{\beta}_3) = 0.1416 \qquad \text{(9.3.15)}$$

a Suppose we want to test NH: $\beta_3 = 0$ against AH: $\beta_3 \neq 0$ using $\alpha = 0.05$. Note that if β_3 is indeed zero, then $\mu_Y(x)$ does not depend on the predictor variable X. To carry out the test we calculate

$$t_C = \frac{0.6828 - 0}{0.1416} = 4.82 \qquad \text{(9.3.16)}$$

Because $t_{0.975:9} = 2.262$, we reject the null hypothesis at the 5% level. The (approximate) P-value for the test is between 0 and 0.001 (using Table T-2 in Appendix T with 9 degrees of freedom).

b If we want to test NH: $\beta_2 \geq 3$ against AH: $\beta_2 < 3$ using $\alpha = 0.10$, we calculate

$$t_C = \frac{2.7233 - 3}{0.2105} = -1.314 \qquad \text{(9.3.17)}$$

Because $t_{0.90:9} = 1.383$, we cannot reject the null hypothesis at $\alpha = 0.10$. The approximate P-value for this test is between 0.1 and 0.2.

c A two-sided 90% confidence interval for β_3 is $\hat{\beta}_3 \pm t_{0.95:9}$ ASE $(\hat{\beta}_3)$, leading to the confidence statement

$$C[0.4232 \leq \beta_3 \leq 0.9424] \approx 0.90 \qquad \text{(9.3.18)}$$

d To compute an approximate 90% upper confidence bound for β_2 we first compute an approximate 80% two-sided confidence interval for β_2, and then use only the upper limit. We get

$$C[2.4322 \leq \beta_2 \leq 3.0144] \approx 0.80$$

from which it follows that

$$C[\beta_2 \leq 3.0144] \approx 0.90$$

e Suppose we wish to compute an approximate 95% two-sided confidence interval for σ. From (4.6.13) we have

$$C\left[\sqrt{\frac{SSE}{\chi^2_{0.975:9}}} \leq \sigma \leq \sqrt{\frac{SSE}{\chi^2_{0.025:9}}} \right] \approx 0.95$$

From (9.3.7) (under the column labeled `Sum of Squares`) we get $SSE = 0.46043$. From Table T-3 in Appendix T we obtain $\chi^2_{0.025:9} = 2.700$ and $\chi^2_{0.975:9} = 19.023$. Thus the required confidence interval is given by

$$C[0.156 \leq \sigma \leq 0.413] \approx 0.95 \quad \blacksquare$$

Note that the SAS output gives asymptotic 95% two-sided confidence intervals for each of the three parameters (see (9.3.8)–(9.3.10)) under the column labeled `Asymptotic 95% Confidence Interval`).

In general, due to the approximate nature of the inference procedures for nonlinear regression problems, the actual confidence coefficients associated with the confidence intervals and P-values for tests discussed in this section may be quite different from the stated values. If critical decisions have to be made, the investigator should consult a professional statistician.

Problems   9.3

9.3.1   Consider the experiment discussed in Example 9.3.1. If we know that β_2 is nonnegative, show that no matter what the concentration of the chemical is in the solution, the average optical reading will never be less than β_1. Obtain an approximate 95% two-sided confidence interval for this parameter. Also obtain one-at-a-time 95% two-sided confidence intervals for β_2 and β_3.

9.3.2   In nondestructive testing of aluminum blocks, an electromagnetic probe is used to detect flaws below the surface. The sensitivity Y of the probe is known to be related to the thickness X of the wire used to construct the coil in the probe. An investigator interested in understanding this relationship has collected the data given in Table 9.3.2, which are also stored in the file **coil.dat** on the data disk.

T A B L E 9.3.2
Coil Data

Observation Number	Sensitivity Y (a unitless quantity)	Wire Thickness X (in millimeters)
1	1.51	0.05
2	1.49	0.06
3	1.47	0.07
4	1.43	0.08
5	1.35	0.09
6	1.19	0.10
7	0.96	0.11
8	0.85	0.12
9	0.65	0.13
10	0.64	0.14
11	0.58	0.15
12	0.56	0.16
13	0.52	0.17
14	0.53	0.18
15	0.49	0.19
16	0.50	0.20

For the thickness values considered in the experiment, suppose assumptions (A) hold with the regression function of Y on X given by

$$\mu_Y(x) = \beta_1(1 - e^{-e^{-(\beta_2 + \beta_3 x)}}) \qquad (9.3.19)$$

Using the SAS output given in Exhibit 9.3.2, answer the following questions.

a   What are the least squares estimates of the parameters β_1, β_2, β_3, and σ?

b   Plot the estimated regression function $\hat{\mu}_Y(x)$ and on the same graph show the observed data points. You may do the plot manually or by using a computer program of your choice. Verify that, on the average, the sensitivity decreases as the thickness increases.

c   If β_1 and β_3 are known to be positive, then show that no matter how thin the wire used, the average sensitivity can never exceed $\beta_1(1 - e^{-e^{-\beta_2}})$. Denote this upper bound for the sensitivity by θ. Estimate θ.

d   Obtain one-at-a-time 95% two-sided approximate confidence intervals for β_1 and β_2. Using these estimates, obtain an approximate confidence interval for θ with confidence coefficient greater than or equal to 0.90 (use the Bonferroni method).

E X H I B I T 9.3.2
SAS Output for Problem 9.3.2

```
                    The SAS System              0:00 Saturday,  Jan 1, 1994


Non-Linear Least Squares Summary Statistics     Dependent Variable Sensitvy


Source                     DF  Sum of Squares       Mean Square
Regression                  3    15.975543503       5.325181168
Residual                   13     0.136656497       0.010512038
Uncorrected Total          16    16.112200000
(Corrected Total)          15     2.569800000
```

Parameter	Estimate	Asymptotic Std. Error	Asymptotic 95% Confidence Interval	
			Lower	Upper
BETA1	1.94834511	0.4724494031	0.9276808412	2.969009384
BETA2	-1.26991572	0.6808804835	-2.740867911	0.201036258
BETA3	14.36306754	2.7037907546	8.5218862331	20.204248838

9.3.3   An experiment was performed to evaluate the rate of absorption and the rate of removal of a certain drug administered intravenously on human subjects. The investigator collects blood samples at half hour intervals for six hours. The concentrations Y of the drug in the blood are obtained by a laboratory analysis of the blood samples that were taken at preselected times X after the drug was injected. The data for one particular subject are given in Table 9.3.3 and are also stored in the file **absorpt.dat** on the data disk.

T A B L E   9.3.3
Drug Absorption Data

Observation Number	Concentration Y (in micrograms/deciliter)	Time X (in hours)
1	1.8	0.5
2	2.9	1.0
3	6.0	1.5
4	8.8	2.0
5	6.6	2.5
6	3.8	3.0
7	2.9	3.5
8	1.5	4.0
9	1.1	4.5
10	0.5	5.0
11	1.1	5.5
12	0.2	6.0

Suppose assumptions (A) in Box 9.3.1 for nonlinear regression are satisfied with $\mu_Y(x)$ given by

$$\mu_Y(x) = \frac{1}{\beta_1 + \beta_2 x + \beta_3 x^2} \qquad (9.3.20)$$

The coefficients β_1, β_2, and β_3 may be different for different subjects, but here we are considering the data for a single subject. Use the computer output given in Exhibit 9.3.3 to do the following.

a Obtain the least squares estimates of the parameters $\beta_1, \beta_2, \beta_3$, and σ.

b Plot Y against X and on the same graph show a plot of the estimated regression function.

c Using the sample regression function, estimate the number of hours it takes for the drug to reach peak concentration in the bloodstream. Use the graph in part (b) to verify your answer.

d Estimate the number of hours required for the body to eliminate enough of the drug so that the average drug concentration in the bloodstream falls below 1 microgram/deciliter. Use the graph in part (b) to verify your answer.

E X H I B I T **9.3.3**
SAS Output for Problem 9.3.3

```
                    The SAS System              0:00 Saturday, Jan 1, 1994

Non-Linear Least Squares Summary Statistics     Dependent Variable concentr


     Source                  DF Sum of Squares    Mean Square
     Regression               3   195.23040871    65.07680290
     Residual                 9     1.22959129     0.13662125
     Uncorrected Total       12   196.46000000
     (Corrected Total)       11    81.14000000


Parameter      Estimate       Asymptotic            Asymptotic 95%
                              Std Error          Confidence Interval
                                               Lower          Upper
  BETA1    0.8183215325  0.06534472402  0.67050023924  0.96614282570
  BETA2   -.6776058218   0.06162515696 -.81701279819  -.53819884543
  BETA3    0.1632703464  0.01450101285  0.13046649717  0.19607419557
```

9.4
Linearizable Models

In some situations it may be possible to transform a nonlinear regression function $\mu_Y(x)$ using appropriate transformations of the response variable, the predictor variables, the parameters, or any combination of these, such that the transformed function is linear in the unknown parameters. If the transformed variables satisfy assumptions (A) or (B) for multiple linear regression, then the results of Chapters 3 and 4 can be applied to the transformed problem. Using the results for the transformed problem, we can often obtain results for the original problem.

As an example, consider the model

$$\mu_Y(x) = \beta_1^* e^{-\beta_2^* x} \tag{9.4.1}$$

where we know that β_1^* is positive. This model is a special case of the model given in (9.2.7) with the β_1 term set to zero. By taking the logarithm to the base e of both sides, we get the transformed function $\ln[\mu_Y(x)] = \ln(\beta_1^*) - \beta_2^* x$. Now let $\ln(\beta_1^*) = \beta_0$ and $-\beta_2^* = \beta_1$. We thus have

$$\ln[\mu_Y(x)] = \beta_0 + \beta_1 x$$

which is linear in the unknown parameters. This suggests that if we set $Z = \ln(Y)$, the regression function of Z on X will be approximately linear and will be given by

$$\mu_Z(x) \approx \beta_0 + \beta_1 x \tag{9.4.2}$$

Thus we can use the theory of Chapter 3 to study this approximate regression function of Z on X and make approximate inferences about the parameters $\beta_0 = \ln(\beta_1^*)$ and $\beta_1 = -\beta_2^*$. This in turn will lead to inferences about β_1^* and β_2^*, the parameters of interest in the original problem.

More specifically, if the data are $(y_1, x_1), \ldots, (y_n, x_n)$, we let $z_i = \ln(y_i)$ and get at the transformed data $(z_1, x_1), \ldots, (z_n, x_n)$. If the investigator is confident that the transformed data satisfy assumptions (A) or (B) for straight line regression (at least approximately), then the theory of Chapter 3 can be used to draw inferences about $\mu_Z(x)$ in (9.4.2). Thus the estimates of β_0 and β_1 in (9.4.2) are those given in Chapter 3 (see (3.4.8) and (3.4.9)).

$$\hat{\beta}_1 = \frac{\sum (z_i - \bar{z})(x_i - \bar{x})}{\sum (x_i - \bar{x})^2} \tag{9.4.3}$$

and

$$\hat{\beta}_0 = \bar{z} - \hat{\beta}_1 \bar{x} \tag{9.4.4}$$

where $\bar{z} = \dfrac{1}{n} \sum_{i=1}^{n} z_1$. So we get

$$\hat{\beta}_2^* = -\hat{\beta}_1 \tag{9.4.5}$$

and

$$\hat{\beta}_1^* = exp(\bar{z} - \hat{\beta}_1 \bar{x}) \tag{9.4.6}$$

Some examples of linearizable models and their linear representations are given in Table 9.4.1. You are encouraged to think of other examples.

In some cases an investigator is not confident that assumptions (A) or (B) hold (even approximately) for the transformed variables. In these cases, the parameter

T A B L E 9.4.1

Original Regression Function	Linearizing Transformation	Suggested Transformation of Y
$\mu_Y^*(x) = \beta_1^* e^{\beta_2^* x}$	$\ln[\mu_Y^*(x)] = \ln(\beta_1^*) + \beta_2^* x = \beta_0 + \beta_1 x$ where $\beta_0 = \ln(\beta_1^*)$ and $\beta_1 = \beta_2^*$	$Z = \ln(Y),\ \text{for } Y > 0$
$\mu_Y^*(x) = (\beta_1^*)^x$	$\ln[\mu_Y^*(x)] = x \ln(\beta_1^*) = \beta_1 x$ where $\beta_1 = \ln(\beta_1^*)$	$Z = \ln(Y),\ \text{for } Y > 0$
$\mu_Y^*(x) = \dfrac{1}{\beta_1^* - \beta_2^* x}$	$\dfrac{1}{\mu_Y^*(x)} = \beta_1^* - \beta_2^* x = \beta_0 + \beta_1 x$ where $\beta_0 = \beta_1^*$ and $\beta_1 = -\beta_2^*$	$Z = \dfrac{1}{Y},\ \text{for } Y \neq 0$
$\mu_Y^*(x) = \dfrac{1}{1 + e^{-(\beta_1^* + \beta_2^* x)}}$	$\ln\left(\dfrac{\mu_Y^*(x)}{1 - \mu_Y^*(x)}\right) = \beta_1^* + \beta_2^* x = \beta_0 + \beta_1 x$ where $\beta_0 = \beta_1^*$ and $\beta_1 = \beta_2^*$	$Z = \ln\left(\dfrac{Y}{1 - Y}\right)$ $for\ 0 < Y < 1$

estimates obtained by performing a linear regression analysis on the transformed data may be useful as starting values for nonlinear regression programs. This is a commonly used strategy. We illustrate this in Example 9.4.1.

E X A M P L E 9.4.1

A study was conducted using several subjects to determine the relationship between optical signal contrasts and visual responses in humans. Each subject was asked to view an image of a photographic slide projected on a screen and determine whether or not a specified object was present in the image. Ten different slides were used, each containing the specified object, but the optical contrast X between the object and the background was different on each slide. The ten slides were presented in random order, and each slide was seen by the subject 100 times. The response recorded was the proportion Y of times the subject reported having seen the object. Data for one particular subject are given in Table 9.4.2 and are also stored in the file **contrast.dat** on the data disk. Suppose that assumptions (A) in Box 9.3.1 are satisfied with the regression function $\mu_Y(x)$ given by

$$\mu_Y(x) = \frac{1}{1 + e^{-(\beta_1 + \beta_2 x)}} \tag{9.4.7}$$

This function is a special case of the function in (9.2.5). Although this is a nonlinear regression function, we observe that $\ln[\mu_Y(x)/(1 - \mu_Y(x))]$ is equal to $\beta_1 + \beta_2 x$, a linear function of the unknown parameters. Hence we apply the transformation $z_i = \ln[y_i/(1 - y_i)]$ to the original data, which leads to the transformed data set $(z_1, x_1), \ldots, (z_{10}, x_{10})$ given in Table 9.4.3. The transformed Y variable is denoted by Z. The regression function of Z on X should be, at least approximately, a straight line function; i.e.,

$$\mu_Z(x) \approx \beta_1 + \beta_2 x \tag{9.4.8}$$

T A B L E 9.4.2
Contrast Data

Observation Numbers	Proportion of Time the Subject Saw the Object Y	Constant Between the Object and the Background X
1	0.02	0.000
2	0.06	0.005
3	0.10	0.010
4	0.18	0.015
5	0.35	0.020
6	0.56	0.025
7	0.78	0.030
8	0.86	0.035
9	0.94	0.040
10	0.99	0.045

TABLE **9.4.3**

Transformed Contrast Data

Observation	Z	X
1	−3.89182	0.000
2	−2.75154	0.005
3	−2.19722	0.010
4	−1.51635	0.015
5	−0.61904	0.020
6	0.24116	0.025
7	1.26567	0.030
8	1.81529	0.035
9	2.75154	0.040
10	4.59512	0.045

We fit a straight line to the transformed data and obtain

$$\hat{\beta}_1 = -3.9627 \qquad \hat{\beta}_2 = 174.755 \qquad \text{(9.4.9)}$$

The fitted straight line and the observations $\{(z_i, x_i)\}$ are displayed in Figure 9.4.1. The original data points $\{(y_i, x_i)\}$ and the estimated regression function

$$\hat{\mu}_Y(x) = \frac{1}{1 + e^{-(\hat{\beta}_1 + \hat{\beta}_2 x)}} = \frac{1}{1 + e^{-(-3.9627 + 174.755x)}} \qquad \text{(9.4.10)}$$

are shown in Figure 9.4.2.

FIGURE **9.4.1**

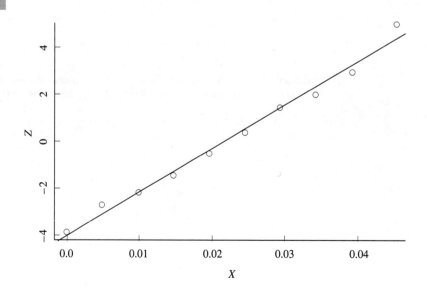

F I G U R E   **9.4.2**

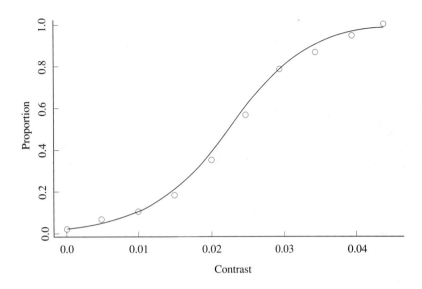

We can also estimate β_1 and β_2 directly, without using a linearizing transformation, by making use of a suitable nonlinear regression program as discussed in Section 9.3. In Exhibit 9.4.1 we present a SAS output for fitting the nonlinear regression function in (9.4.7).

E X H I B I T   **9.4.1**
SAS Output for Example 9.4.1

```
                        The SAS System  0:00 Saturday,  Jan 1, 1994
Non-Linear Least Squares Summary Statistics    Dependent Variable Y


    Source                DF Sum of Squares    Mean Square
    Regression             2   3.6923266027    1.8461633013
    Residual               8   0.0018733973    0.0002341747       (9.4.11)

    Uncorrected Total     10   3.6942000000
    (Corrected Total)      9   1.3516400000


Parameter     Estimate      Asymptotic            Asymptotic 95%
                            Std. Error        Confidence Interval
                                              Lower        Upper


    B1       -4.0261321    0.1290043165        -4.3236     -3.7286       (9.4.12)
    B2      171.6643713    5.2990073124       159.4447    183.8840       (9.4.13)
```

From the SAS output we see that the estimates $\hat{\beta}_1 = -4.0261$ and $\hat{\beta}_2 = 171.6644$, given in (9.4.12) and (9.4.13), respectively, differ only slightly from the estimates obtained from the linearization approach. A plot of the original data points $\{(y_i, x_i)\}$ and the two estimated regression curves (one obtained by linearization and the other by nonlinear regression analysis) are displayed in Figure 9.4.3.

You may experiment with different starting values for β_1 and β_2, but you will see that the estimates obtained from the linearization approach lead to good starting values. ■

F I G U R E **9.4.3**

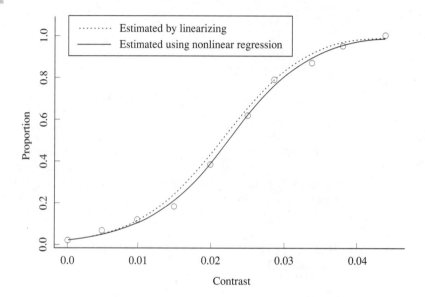

Proportion vs. Contrast. Legend: .... Estimated by linearizing; ——— Estimated using nonlinear regression

Problems 9.4

9.4.1 Consider the experiment discussed in Problem 9.3.3.

a Verify that $1/(\mu_Y(x))$ is a linear function of the unknown parameters. Use this fact to arrive at a suitable transformation of the response variable Y so that linear regression techniques may be used with the transformed data.

b Analyze the data for Problem 9.3.3 after applying the transformation in part (a) to obtain estimates of the unknown parameters β_1, β_2, and β_3. You can use the MINITAB output containing the results from a regression of $Z = 1/Y$ on X and X^2 given in Exhibit 9.4.2.

c Compare the estimates in part (b) with those obtained in Problem 9.3.3.

EXHIBIT 9.4.2
MINITAB Output for Problem 9.4.1

```
Note:   Z = 1/Y.            Y = concentr.    X = time


The regression equation is
Z = 1.40 - 1.29 X + 0.276 X^2


Predictor          Coef      Stdev    t-ratio          p
Constant         1.4026     0.7977       1.76      0.113
X               -1.2903     0.5642      -2.29      0.048
X^2             0.27640    0.08450       3.27      0.010


s = 0.7718         R-sq + 74.3%    R-sq(adj) = 68.6%


Analysis of Variance

Source       DF         SS        MS          F        p
Regression    2    15.5380    7.7690      13.04    0.002
Error         9     5.3611    0.5957
Total        11    20.8991
```

9.4.2 Consider the data of Problem 9.3.2. For illustrative purposes, suppose we know that the actual value of β_1 is 2.0, so the regression function of Y on X is

$$\mu_Y(x) = 2(1 - e^{-e^{-(\beta_2 + \beta_3 x)}})$$

a   Verify that $-\ln\left[-\ln\left(1 - \frac{\mu_Y(x)}{2}\right)\right]$ is linear in the unknown parameters. Use this fact to arrive at a suitable transformation of the response variable Y so that linear regression techniques may be applied to the transformed data.

b   Analyze the data for Problem 9.3.2 after applying the transformation in part (a) to obtain estimates of the unknown parameters β_2 and β_3. You may use the MINITAB output containing the results from a regression of $Z = -\ln\left[-\ln\left(1 - \frac{Y}{2}\right)\right]$ on X given in Exhibit 9.4.3. Note that the regression function of Z on X is *approximately* equal to

$$\mu_Z(x) = \beta_2 + \beta_3 x$$

c   Compare your answers in part (b) with those obtained in Problem 9.3.2.

■ E X H I B I T **9.4.3**

■ MINITAB Output for Problem 9.4.2

```
The regression equation is
Z = -1.08 + 13.0 X


Predictor          Coef         Stdev         t-ratio          p
Constant        -1.0801        0.1274           -8.48      0.000
X               13.0413        0.9566           13.63      0.000


s = 0.1764      R-sq = 93.0%        R-sq(adj) = 92.5%


Analysis of Variance

SOURCE       DF          SS           MS           F          P
Regression    1      5.7826       5.7826      185.88      0.000
Error        14      0.4355       0.0311
Total        15      6.2181
```

9.5
Exercises

9.5.1 The concentration Y of a drug in human serum, after injection into the bloodstream, will reach a peak value in a certain amount of time X, after which it will begin to diminish with time. To understand how long after injecting a particular drug the peak concentration will be attained, and how much longer after that the concentration will return to normal levels, an experiment was conducted using human subjects. The concentration of the drug in the serum was measured immediately following injection and then after time lapses of 30 minutes, 1 hour, 2 hours, 4 hours, 8 hours, 16 hours, 24 hours, and 48 hours on each of the subjects. Table 9.5.1 gives the data for one of the subjects. These data are also stored in the file **serum.dat** on the data disk. Suppose that assumptions (A) for nonlinear regression are satisfied with the regression function of Y on X equal to

$$\mu_Y(x) = \frac{1}{\beta_1 + \beta_2 x + \beta_3 x^2} + \beta_4$$

This function is a slight generalization of the function in (9.2.2). Do the following (you can use the computer output given in Exhibit 9.5.1 whenever necessary).

a Plot the concentration against time.

b What are the estimated values of β_1, β_2, β_3, and β_4?

c Display the fitted curve in the graph of part (a).

d Using the fitted curve, estimate the time it takes for the drug concentration to reach a maximum value for this particular subject.

e Using the fitted curve, estimate the time it takes for the drug concentration to drop below 50 milligrams/liter for this particular subject.

T A B L E 9.5.1
Serum Data

Observation Number	Drug Concentration in Serum (in mg/liter) Y	Time after Injection (in hours) X
1	67	0.0
2	70	0.5
3	82	1.0
4	85	2.0
5	104	4.0
6	129	8.0
7	171	16.0
8	162	24.0
9	18	48.0

E X H I B I T 9.5.1
SAS Output for Problem 9.5.1

```
                         The SAS System   0:00 Saturday, Jan 1, 1994
Non-Linear Least Squares Summary Statistics   Dependent Variable Y

   Source                DF  Sum of Squares    Mean Square

   Regression             4   106518.91654     26629.72914
   Residual               5       85.08346        17.01669
   Uncorrected Total      9   106604.00000
   (Corrected Total)      8    18988.00000


Parameter     Estimate      Asymptotic            Asymptotic 95%
                            Std Error          Confidence Interval
                                             Lower          Upper
   BETA1    0.00656892    0.001475260     0.00277670   0.0103611407
   BETA2   -0.00028246    0.000112385    -0.00057135   0.0000064315
   BETA3    0.00000749    0.000003193    -0.00000071   0.0000157026
   BETA4  -80.65213231   34.151101655  -168.43906150   7.1347968761
```

9.5.2 An investigator conducted a laboratory study to evaluate the ability of human observers to detect the presence of layered haze in the atmosphere. She generated several photographic slides of a scenic vista (Hopi Point, Grand Canyon) with various levels of layered haze artificially superimposed using computer imagery, causing different levels of degradation in the visibility of the scene. Subjects were shown the projected images on a screen and were asked to respond whether or not they saw any layered haze. There were eight levels of haze quantified by optical contrast values of 0, 0.005, 0.008, 0.012, 0.015, 0.020, 0.030, and 0.040, respectively. The contrast value of 0 corresponds to a slide of the scene with no superimposed haze. The slides were shown in a randomized order. Each subject saw each slide 100 times during the experiment. The data pertaining to one of the subjects are listed in Table 9.5.2 and are also stored in the file **haze.dat** on the data disk.

T A B L E   **9.5.2**

Haze Data

Observation Number	Proportion of Times Subject Reported Presence of Layered Haze Y	Contrast Value of Slide X
1	0.34	0.000
2	0.44	0.005
3	0.50	0.008
4	0.59	0.012
5	0.72	0.015
6	0.93	0.020
7	0.99	0.030
8	1.00	0.040

Suppose that the regression function of Y on X is given by

$$\mu_Y(x) = \frac{\beta_1}{[1 + e^{-(\beta_2 + \beta_3 x)}]^{\beta_4}}$$

and that assumptions (A) for nonlinear regression are satisfied. Do the following (you can use the computer output given in Exhibit 9.5.2 wherever necessary).

a Plot Y versus X.

b What are the estimated values of β_1, β_2, β_3, and β_4? What is the estimated value of the subpopulation standard deviation σ?

c Display the fitted curve along with the sample data in a graph.

d Using the fitted curve, estimate the contrast value x_0 such that for values of X greater than or equal to x_0, this subject will report seeing the haze at least 90% of the time; i.e., $\mu_Y(x_0) \geq 0.90$. *Hint:* Locate the value 0.90 along the Y axis and find the corresponding value x_0 along the X axis using the fitted curve. You can also do this algebraically by solving the following equation for x_0:

$$0.90 = \frac{\hat{\beta}_1}{[1 + e^{-(\hat{\beta}_2 + \hat{\beta}_3 x_0)}]^{\hat{\beta}_4}}$$

e Using the fitted curve, estimate the proportion of affirmative responses by the subject even when shown the slide with no superimposed haze. (This is referred to as false alarm rate.) Also construct an approximate 85% confidence interval for this subject's false alarm rate. (*Hint:* First compute an approximate 85% simultaneous confidence interval for β_1, β_2, and β_4 using the Bonferroni method.)

E X H I B I T **9.5.2**

SAS Output for Problem 9.5.2

```
                    The SAS System    0:00 Saturday, Jan 1, 1994

Non-Linear Least Squares Summary Statistics    Dependent Variable Y

     Source                    DF Sum of Squares    Mean Square

     Regression                4   4.2698528454     1.0674632113
     Residual                  4   0.0008471546     0.0002117887
     Uncorrected Total         8   4.2707000000
     (Corrected Total)         7   0.4756875000
```

Parameter	Estimate	Asymptotic Std Error	Asymptotic 95% Confidence Interval Lower	Upper
BETA1	0.995000	0.01029050	0.9664292	1.0235704
BETA2	-115.929282	10.75382750	-145.7862897	-86.0722743
BETA3	5402.347473	571.24393936	38 414938	6988.3534528
BETA4	0.009437	0.00077196	0.0072935	0.0115800

A

Answers to Selected Problems and Exercises

1.6.1 **a** $\hat{\mu}_Y = 6.989.$ $6.140 \le \mu_Y \le 7.838$; the width is 1.698.
  b $C[6.140 \le \mu_Y \le 7.838] = 0.80.$
  c A 95% lower confidence bound for μ_Y is 5.889.
  d $C[5.889 \le \mu_Y] = 0.95.$
  g NH is not rejected at the $\alpha = 0.05$ level.
  h $C[2.93 \le \sigma y \le 4.54] = 0.90.$ Accept NH in (g).
  i A 99% two-sided confidence interval for μ_Y is [5.205, 8.773]. The width is 3.568.

1.7.1 **a** Not defined.   **b** $f(16)/f(36) = 0.188386.$ **c** Not defined.
  d $f(34) + f(13) = 7563.3096.$ **e** Not defined.

1.7.2 **a** $\mu_Y(6, 1) = \beta_0 + 6\beta_1 + \beta_2.$ **b** $\mu_Y(15, -4) = \beta_0 + 15\beta_1 - 64\beta_2.$
  c Yes, no, no.

1.8.1 **a** $A^T = \begin{bmatrix} 9 & 4 & 3 \\ 4 & 16 & 8 \\ 3 & 8 & 12 \end{bmatrix}$

  b $C + B^T$ is defined.
    $B + C$ is not defined.
    $B + C^T$ is defined.
    AC is defined.
    CA is not defined.
    $B - C^T$ is defined.

  c $C + B^T = \begin{bmatrix} 24 & 27 & 22 & 28 \\ 38 & 59 & 32 & 22 \\ 34 & 35 & 62 & 58 \end{bmatrix}$

  $B + C^T = \begin{bmatrix} 24 & 38 & 34 \\ 27 & 59 & 35 \\ 22 & 32 & 62 \\ 28 & 22 & 58 \end{bmatrix}$

  $AC = \begin{bmatrix} 297 & 409 & 352 & 359 \\ 680 & 780 & 700 & 624 \\ 600 & 653 & 695 & 550 \end{bmatrix}$

$$B - C^T = \begin{bmatrix} 0 & -10 & -28 \\ -19 & 3 & -25 \\ -12 & -6 & -20 \\ -16 & -18 & 4 \end{bmatrix}$$

1.8.2 $\hat{\beta} = \begin{bmatrix} 2 \\ -1 \end{bmatrix}$

1.8.3 **a** $X^T X = \begin{bmatrix} 4 & 10 \\ 10 & 30 \end{bmatrix}$

  b $X^T y = \begin{bmatrix} 12 \\ 31 \end{bmatrix}$

  c $(X^T X)^{-1} = \begin{bmatrix} 1.5 & -0.5 \\ -0.5 & 0.2 \end{bmatrix}$

  d $\hat{\beta} = \begin{bmatrix} 2.5 \\ 0.2 \end{bmatrix}$

1.10.1 **Target population of items**: The collection of all state employees last year.
Target population of numbers: The collection of numbers of days of sick leave each state employee took last year.
Study population of items and of numbers can be defined to be the same as the target population of items and of numbers for this problem. The **parameter** to be studied is the average number of days of sick leave taken by state employees last year.

1.10.3 **Target population of items**: The collection of all bottles of aspirin that will be manufactured next month by the company.
Target population of numbers: The response variable in this problem is an attribute variable, viz., damaged or undamaged. We assign a score of 1 for bottles that arrive damaged at the retail store and a score of 0 for undamaged bottles. In this way we can convert attribute data to numerical data. The target population of numbers is the set of scores (0 or 1) assigned to each bottle of aspirin to be manufactured next month.
Study population of items: Set of all aspirin bottles manufactured by the company last month.
Study population of numbers: Set of scores (0 or 1) of all bottles of aspirin manufactured by the company last month.
Parameter of interest: The proportion of bottles to be manufactured next month that will arrive damaged at the retail store. Note that this proportion is in fact the average of the 0, 1 scores assigned to the aspirin bottles!

1.10.6 The proportion of the population that is less than 17.5 is 0.8554. The proportion of the population between 12 and 16 is equal to 0.4977. The proportion of the population that is greater than 15 is equal to 0.7611.

1.10.7 **Target population of items**: All hospital patients who will stay at that hospital next year.
Target population of numbers: The number of days each patient will stay at the hospital next year.
If $\mu = 8$ and $\sigma = 2$, then the proportion of patients who will stay less than 7 days is equal to 0.3085.

1.10.9 NH: $\mu_Y \leq 540$ *vs* AH: $\mu_Y > 540$

1.10.10 $t_C = 19.1663$

  Reject NH.

1.10.11 The *P*-value is < 0.0005.

1.10.12 $C[594.6 \leq \mu_Y] = 0.95$

1.10.13 **a** Not defined. **b** Defined. **c** Defined. **d** Defined.

1.10.14 **a** Cannot be done.

b $C - B^T = \begin{bmatrix} 10 & 0 & 0 \\ -1 & -7 & 4 \end{bmatrix}$

c $AB = \begin{bmatrix} 16 & 0 \\ 16 & 9 \\ 17 & 8 \end{bmatrix}$

d $CA^T = \begin{bmatrix} 76 & 46 & 37 \\ 1 & 19 & 35 \end{bmatrix}$

1.10.15 $K = \begin{bmatrix} 11237 & 212.08 \\ 212.08 & 4.34 \end{bmatrix}$

1.10.16 $K^{-1} = \begin{bmatrix} 0.00114 & -0.05572 \\ -0.05572 & 2.95214 \end{bmatrix}$

1.10.18 $\mu_Y(3, -1) = 39$

1.10.19 $\beta_0 + \beta_1 + 3\beta_2 - 2\beta_3$

1.10.20 **a** linear in β_0 and β_1.

b linear in β_1 and β_2 but not linear in β_0.

c linear in β_0 and β_2 but not linear in β_1.

d linear in all the β_i, $i = 0, 1, 2, 3$, simultaneously.

2.2.4 $P_Y(490) = 2.6956$, $P_Y(625) = 3.349$

2.2.5 0.242

2.2.6 $x \geq 2.676/.00484 = 552.892562$, say 553 rounded to the nearest integer.

2.4.4 Eleven cars in the population were driven 14,300 miles the first year. Eleven cars in the population were driven 7,100 miles the first year.

2.4.5 Yes, there are 17 cars that were driven 9,200 miles the first year.

2.4.6 There are 22 cars that were driven 8,700 miles during the first year. The mean maintenance cost for these cars is $433.318, and the standard deviation is $29.2177.

2.4.8 There are 18 cars in the subpopulation with $X_2 = 10,000$. The mean maintenance cost for this subpopulation is $464.222, and the standard deviation is $30.5024.

2.4.11 **a** Simple random sampling. **b** Sampling with preselected X values.

3.2.6 The sample in Problem 3.2.1 is preferred because the sample size is larger.

3.4.1 $\hat{\sigma} = 0.916696$, $\hat{\beta}_1 = 0.530253$, $\hat{\beta}_0 = -0.227365$

3.4.2 (1) On the average, the crystals grow 0.530 gram per hour.
(2) If a crystal is allowed to grow for 15 hours, its predicted size (weight) is 7.72643 grams.
(3) The additional dollars that the 24-hour crystal is expected to fetch is $138.62.

3.4.3 $\hat{\mu}_Y(19) = 9.847$ grams. $\hat{\mu}_Y(25) = 13.029$ grams. For 40 hours it is not safe to use the estimated regression function based on these data because the data do not include 40 hours or any number close to 40 hours. Thus an excessive amount of *extrapolation* would be needed and the results would not be reliable in general.

3.4.4 **a** $\hat{\beta}_1 = 0.03291$, $\hat{\beta}_0 = 158.49$, $\hat{\mu}_Y(x) = 158.49 + 0.03291x$

b The buyer would be interested in $Y(13000)$ and not $\mu_Y(13000)$. $\hat{Y}(13000) = \$586.28$. This is also the estimate for $\mu_Y(13000)$.

c Here the quantity of interest is $\mu_Y(16000)$.

3.4.5 $\bar{x} = 6.00$, $\bar{y} = 68.7308$, $SXY = 1494.0$, $SSX = 364$, $SSY = 6295.12$, $\hat{\beta}_0 = 44.10$, $\hat{\beta}_1 = 4.10$, $SSE = 163.1$, σ is unknown but $\hat{\sigma} = 2.607$

3.4.6   **a**

Row	y*	x*
1	-6.90	-12
2	-5.99	-10
3	-3.11	-8
4	-1.86	-6
5	-1.81	-4
6	-0.11	-2
7	-1.71	0
8	1.70	2
9	2.42	4
10	4.38	6
11	4.38	8
12	4.73	10
13	5.95	12
14	8.10	14

b $SSY = 265.947$, $SSX = 910$, $SXY = 482.53$, $\hat{\beta}_1^* = 0.5302$, $\hat{\sigma}^* = 0.9167$

3.5.1   Yes. A straight line regression seems reasonable.

3.5.2 $\hat{\beta}_0 = 1.54$, $\hat{\beta}_1 = 1.82$, $\hat{\sigma} = 1.514$

3.5.5   The plot is not inconsistent with Gaussian assumptions.

3.6.2   **a** $\hat{\beta}_1 = 0.03131$. Required difference in Problem 3.6.1 is estimated to be \$156.55.

    b $C[126 \le 5000\beta_1 \le 187] = 0.90$

    c   \$1,470.23

    d $C[548.22 \le \mu_Y(12{,}500) \le 588.48] = 0.90$

    e $C[38.59 \le \sigma \le 59.69] = 0.80$. The interval is too wide to make a decision.

3.6.3   You have 90% confidence that the two confidence intervals

$$-0.019 \le \beta_0 \le 0.228 \text{ and } .958 \le \beta_1 \le 1.02$$

are simultaneously correct. An investigator must decide if these are close enough to 0 and 1, respectively, in the problem being studied.

3.6.4   **a**   Use the following confidence statement to help make a decision.
$C[0.4407 \le \beta_1 \le 0.5662] = 0.90$.

    b $C[5.16 \le Y(15) \le 9.95] = 0.95$

    c $C[2.61 \le Y(10) \le 7.46 \quad \text{and} \quad 10.07 \le Y(25) \le 15.10] \ge 0.90$

3.7.1   **e**   NH: $\beta_1 = 0$ and AH: $\beta_1 \ne 0$.
The P-value is between 0.001 and 0; i.e., $0 < P < 0.001$. Therefore NH is rejected and we conclude (at $\alpha = 0.05$) that shelf life is related to storage temperature.

    f   Zero degrees Celsius is very much outside the data range of temperatures, and so you should not extrapolate the results to zero.

    g $C[705.7982 \le \mu_Y(15) \le 733.1072] = 0.95$

    h $C[-15.3951 \le \beta_1 \le -12.1117] = 0.95$

    Hence conclude that shelf life is related to storage temperature.

    i   NH: $\mu_Y(13) \le 650$ versus AH: $\mu_Y(13) > 650$. We get $t_C = \frac{746.9595 - 650}{7.7122} = 12.5723$. The P-value is between 0 and 0.0005. Thus the NH will be rejected at $\alpha = 0.05$.

    j $C[730.6 \le \mu_Y(13) \le 763.3] = 0.95$. We can conclude that $\mu_Y(13)$ is greater than 650.

3.8.1   **a**   Analysis of Variance

SOURCE	DF	SS	MS	F	P
Regression	1	91645.2	91645.2	315.37	< 0.01
Error	16	4649.7	290.6		
Total	17	96294.9			

b   The *P*-value is smaller than 0.01.

c $t_C = -17.758$

The *P*-value is between 0 and 0.001.

e   From (b) we can conclude that β_1 is not equal to zero (if we use $\alpha = 0.05$ or $\alpha = 0.01$ for instance).

3.8.2   **a**   Analysis of Variance

SOURCE	DF	SS	MS	F	P
Regression	1	9337.7	9337.7	1161.31	0.000
Error	22	176.9	8.0		
Total	23	9514.6			

b $F_C = 1162.81$. The *P*-value is less than 0.001. The NH: $\beta_1 = 0$ would be rejected at any commonly used α level.

c $t_C = 34.1$. The *P*-value is less than 0.001 so reject NH for $\alpha > .001$.

d $t_C^2 = (34.1)^2 = 1161.31 = F_C$ except for rounding errors.

e   For any usual α value, NH: $\beta_1 = 0$ would be rejected.

f $C[1.48 \le \beta_1 \le 1.74] = 0.99$

3.9.2   **a**   Based on this plot the regression of *Y* on *X* seems to be linear.

b $\hat{\rho}_{Y,X} = 0.99$

   i   No

   ii   No

   iii   Yes

c $\hat{\sigma}_Y = 20.34$, and $\hat{\sigma} = 2.836$. $\hat{\sigma}$ can be used to decide whether *X* is an adequate predictor of *Y*.

3.10.1   **a** $\hat{\beta}_0 = 15.7$, $\hat{\beta}_1 = 0.414$, and $C[43.94 \le \mu_Y(72) \le 47.07] = 0.85$.

c   The investigator concludes that soil temperature is not an important factor relative to the objectives.

d   NH: $5\beta_1 \le 5$ and AH: $5\beta_1 > 5$, or NH: $\beta_1 \le 1$ and AH: $\beta_1 > 1$. $t_C = -6.01$, $n = 14$, so $df = 12$. *P*-value is greater than 0.9995. NH cannot be rejected using $\alpha = 0.10$.

e $\hat{Y}(75) = 46.75$. No valid prediction interval is available for $Y(75)$.

f   None available

g $\hat{\sigma}_{V|U} = 3.649$

3.10.2   **a** $\hat{\mu}_Y(u) = \hat{\beta}_0^* + \hat{\beta}^*u = 276 + 0.0259u$

b   This is $\mu_Y(x) = \beta_0 + \beta_1 x$ with $x = 12,500$. No valid point or interval estimate of $\mu_Y(x)$ is available since no valid point estimate of β_0 or β_1 is available.

c $\mu_Y(u) = \beta_0^* + \beta_1^* u$ with $u = 12,500$. $\hat{\mu}_Y(12,500) = \hat{\beta}_0^* + \hat{\beta}_1^*(12500) = \$599.60$

d   No valid 90% confidence interval is available.

e   A valid confidence statement is $C[\$567.60 \le \mu_Y(12,500) \le \$631.50] = 0.95$.

3.11.1   **b** $\hat{\beta}_1 = 0.504$ and $\hat{\sigma} = 1.02$

c β_1

d $C[0.482 \le \beta_1 \le 0.525] = 0.80$

e $C[0.78 \le \sigma \le 1.52] = 0.90$

f   The investigator would conclude that $\mu_Y(x)$ is an adequate prediction function for this problem.

3.11.2   The investigator would conclude that the chemical analysis provides unbiased estimates for As.

3.12.2　**a**　If assumptions (B) are valid, then all of the parameters listed have valid estimates.

$$\hat{\beta}_0 = 1.20260$$

$$\hat{\beta}_1 = 0.00257942$$

$$\hat{\sigma} = 0.298319$$

$$\hat{\mu}_Y(x) = 1.20260 + 0.00257942x$$

$$\hat{\mu}_Y = \bar{y} = 2.523$$

$$\hat{\mu}_X = \bar{x} = 511.90$$

$$\hat{\sigma}_X = 169.2$$

$$\hat{\sigma}_Y = 0.519$$

$$\hat{\rho}_{Y,X} = 0.841$$

c

$$\sigma: \quad \sqrt{\frac{SSE}{15.507}}, \quad \sqrt{\frac{SSE}{2.733}}, \quad i.e., \ (0.214, 0.510)$$

$$\sigma_Y: \quad \sqrt{\frac{SSY}{16.919}}, \quad \sqrt{\frac{SSY}{3.325}}, \quad i.e., \ (0.3787, 0.8543)$$

$$\sigma_X: \quad \sqrt{\frac{SSX}{16.919}}, \quad \sqrt{\frac{SSX}{3.325}}, \quad i.e., \ (123.4, 278.4)$$

f　$50\beta_1$

g　$C[0.06122 \le 50\beta_1 \le 0.19672] = 0.95$

h　$C(0.12244 \le 100\beta_1 \le 0.39344] = 0.95$

3.12.3　**b**　The regression equation is

$$\hat{\mu}_Y(x) = 90.71 + 1.4272x$$

d　$SE(\hat{\beta}_0) = 17.42$ and $SE(\beta_1) = 0.1777$

e　$\hat{\sigma} = 29.56$

f　$C[\beta_0 \le 120.916] = 0.95$

g　$C[0.9158 \le \beta_1 \le 1.9386] = 0.99$

No. It appears that β_1 is not zero because the confidence interval above excludes zero.

h　$C[22.34 \le \sigma \le 43.72] = 0.95$

k　An upper 97.5% confidence bound for $\mu_Y(25)$ is 154.58.

l　$\hat{\rho}_{Y,X} = 0.884$. $C[0.8 \le \rho_{Y,X} \le 0.95] = 0.95$. We conclude that the regression function $\mu_Y(x) = \beta_0 + \beta_1 x$, using fat intake as the predictor variable, is better than using μ_Y for predicting cholesterol.

m　$t_C = -3.2234$. The P-value is greater than 0.995. The null hypothesis cannot be rejected using an α value of 0.01.

n　**Accept** the hypothesis that β_1 is less than 2!

3.12.4　**a**　$\hat{\sigma}$ is 29.56 and $\hat{\sigma}_Y$ is 61.6.

b The analysis of variance table is

```
Analysis of Variance
SOURCE       DF        SS        MS        F        P
Regression    1      56368     56368     64.50    0.000
Error        18      15731       874
Total        19      72099
```

c $F_C = 64.50$

d The *P*-value is less than 0.01.

e $C[1.67 \leq \sigma_Y/\sigma \leq 3.20] = 0.95$.

3.12.5 **a** NH: $\beta_1 \leq 0.8$ AH: $\beta_1 > 0.8$.

b $t_C = 3.53$, *P*-value < 0.005.

c The manufacturer's claim seems to be incorrect.

d $C[45.8 \leq 50\beta_1 \leq 96.9] = 0.99$

4.2.1 **b** $E_I = 0.300$ for population item $I = 962$. $E_I = 0.900$ for population item $I = 1376$.

d $\sigma_Y(240,18) = 1.7076$

e $\sigma_E(240,18) = 1.7076$ including items 962 and 1376 in the calculations.

f Yes.

4.2.2 41.0

4.2.4 $\mu_Y(280,18) = 33.0$

4.4.2 $X^T X = \begin{bmatrix} 10 & 2410 & 138 \\ 2410 & 587900 & 32860 \\ 138 & 32860 & 2020 \end{bmatrix}$ $X^T Y = \begin{bmatrix} 297.7 \\ 73497.0 \\ 3910.6 \end{bmatrix}$

4.4.3 $(X^T X)^{-1} = \begin{bmatrix} 16.3005 & -0.0504 & -0.2930 \\ -0.0504 & 0.0002 & 0.0006 \\ -0.2930 & 0.0006 & 0.0107 \end{bmatrix}$ $\hat{\beta} = \begin{bmatrix} -0.64546 \\ 0.18721 \\ -1.06533 \end{bmatrix}$

4.4.5 $\hat{\mu}_Y = 29.77$, $\hat{\mu}_{X_1} = 241$, $\hat{\mu}_{X_2} = 13.80$
$\hat{\sigma}_Y = 7.80$, $\hat{\sigma}_{X_1} = 28.07$, $\hat{\sigma}_{X_2} = 3.58$

4.4.6 $\hat{\sigma} = 1.117$

4.4.8 $\mu_Y(280,19)$

4.4.10 $\hat{\mu}_Y(280,18) = 32.595$

4.4.12 About 4.55% of the plastic containers have a strength greater than 31 if they are manufactured with a temperature of 250° and a pressure of 16.

4.5.2 Based on the plots we conclude the data are *not inconsistent* with the assumption that $\{(Y, X_1, X_2, X_3, X_4)\}$ is Gaussian.

4.5.3 There is no reason to seriously doubt that the set of standardized residuals is a simple random sample from a standard Gaussian population.

4.6.2 $\hat{\mu}_{X_1} = 4033.2$, $\hat{\mu}_{X_2} = 2.5588$, $\hat{\mu}_{X_3} = 1889.1$, $\hat{\mu}_Y = 619.41$.
$\hat{\sigma}_{X_1} = 1082.2$, $\hat{\sigma}_{X_2} = 1.5013$, $\hat{\sigma}_{X_3} = 692.68$, $\hat{\sigma}_Y = 334.9\hat{\ }$.

4.6.3 $\hat{\beta}_0 = -358.4$, $\hat{\beta}_1 = 0.0751$, $\hat{\beta}_2 = 55.09$, $\hat{\beta}_3 = 0.2811$.
$SE(\hat{\beta}_0) = 198.7$, $SE(\hat{\beta}_1) = 0.1361$, $SE(\hat{\beta}_2) = 29.05$, $SE(\hat{\beta}_3) = 0.2261$.
$\hat{\mu}_Y(x_1, x_2, x_3) = -358.4 + 0.0751x_1 + 55.09x_2 + 0.2811x_3$

4.6.5 $\hat{\sigma}_Y = 334.92$

4.6.7 $C[\mu_Y(4000, x_2, x_3) - \mu_Y(3000, x_2, x_3) \leq \$306.12] = 0.95$

4.6.8 $C[\mu_Y(x_1, x_2, 2000) - \mu_Y(x_1, x_2, 1500) \leq \$332.43] = 0.95$

4.6.9 $\hat{Y}(3200, 6, 2800) = \$999.60$

4.6.10 $C[Y(3200, 6, 2800) \leq 1361.46] = 0.90$

4.6.11 $C[\mu_Y(3200, 6, 2800) \leq 1314.89] = 0.90$

4.6.13 $\hat{\sigma} = 135.42$ and $C[108.21 \leq \sigma \leq 181.01] = 0.95$

4.7.1   **b**   NH: $\beta_1 = 0$ AH: $\beta_1 \neq 0$. The P-value is 0.585.

4.7.3 $X_C^2 = 550163/(50)^2 = 220.07$. The P-value is greater than 0.99.

4.8.1 P-value < 0.01.

4.8.2 P-value < 0.01.

4.8.3 P-value < 0.01.

4.9.1   **c** $\hat{\rho}_{Y(x_1,x_2,x_3)}^2 = 0.905$
$\hat{\rho}_{Y(x_4,x_5,x_6,x_7)|x_1,x_2,x_3}^2 = 0.127$
$\hat{\rho}_{Y(x_1,x_2,\ldots,x_7)}^2 = 0.917$

    d $\dfrac{\hat{\sigma}_B}{\hat{\sigma}_A} = 0.927$

4.10.1   **a**   **i** $C[0.255 \leq \sigma_{Y|x_1} \leq 0.498] = 0.95$, $C[0.347 \leq \sigma_{Y|x_3} \leq 0.679] = 0.95$
           ii $C[0.376 \leq \sigma_{Y|x_1}/\sigma_{Y|x_3} \leq 1.435] = 0.90$
    b   **i** $C[0.365 \leq \sigma_{Y|x_2} \leq 0.715] = 0.95$, $C[0.384 \leq \sigma_{Y|x4} \leq 0.751] = 0.95$
           ii $C[0.486 \leq \sigma_{Y|x_2}/\sigma_{Y|x_4} \leq 1.862] = 0.90$

4.11.1   **a** $m = 4$
    b $n = 24$, $n_1 = 6$, $n_2 = 6$, $n_3 = 6$, and $n_4 = 6$
    c $\hat{\mu}_1 = 11.567$, $\hat{\mu}_2 = 10.033$, $\hat{\mu}_3 = 8.217$, $\hat{\mu}_4 = 6.450$
        $\hat{\sigma}_1 = 2.342$, $\hat{\sigma}_2 = 2.655$, $\hat{\sigma}_3 = 2.070$, $\hat{\sigma}_4 = 2.145$
    e $\hat{\theta}_1 = -0.075$, $\hat{\theta}_2 = 0.1083$, $\hat{\theta}_3 = 0.0083$, and $\hat{\theta}_4 = -0.0417$
    h   1.91 units
    i   Yes, for the X values studied.
    j $F_C = 0.011$. The P-value is greater than 0.20.

4.12.1   **c** $\hat{\mu}_{log_{10}(C)}^{(t)}(30) = 0.0404$

    d   For this problem a quadratic term cannot be ignored.

    e   NH: $\beta_2 = 0$ against NH: $\beta_2 \neq 0$. The P-value is less than 0.001.

4.12.2   **b**   **i** $\hat{\beta}_0^{(A)} = 200.68$, $\hat{\beta}_1^{(A)} = 0.014623$, $\hat{\sigma}_{Y|x} = 22.29$.
            $\mu_Y^{(A)}(x) = \hat{Y}(x) = 200.68 + 0.014623x$.
        ii SS (pure error) $= 1804.436$, df (pure error) $= 8$, MS (pure error) $= 225.55$.
        v   Yes
        vi   NH: The regression function is $\mu_Y(x) = \beta_0 + \beta_1 x$ against AH: The regression
            function is not $\mu_Y(x) = \beta_0 + \beta_1 x$. The P-value is between 0.025 and 0.05.
    c $\hat{\beta}_0 = 258.72$, $\hat{\beta}_1 = 0.003309$, $\hat{\beta}_2 = 0.00000042$, $\hat{\sigma}_{Y|x_1,x_2} = 12.85$

4.12.3   **c**   By preselecting X_1 and X_2 values.

    e $\hat{\beta}_0 = 1.697$, $\hat{\beta}_1 = -0.02030$, $\hat{\beta}_2 = -0.5477$, $\hat{\beta}_3 = 0.0151$. $\hat{\sigma}_{Y|x_1,x_2,x_3} = 0.3724$
        $\hat{\mu}_Y(x_1, x_2, x_3) = \hat{Y}(x_1, x_2, x_3) = 1.697 - 0.0203x_1 - 0.5477x_2 + 0.0151x_3$

  .f   NH: $\beta_2 = \beta_3 = 0$. AH: at least one of β_2, β_3 is not zero. The *P*-value is less than 0.01.

  g $\hat{\mu}_Y(65, 10, 650) = 4.72$ and $C[4.44 \le \mu_Y(65, 10, 650) \le 4.99] = 0.95$

4.12.4 **b** **ii** $\hat{\sigma}_A = 1.004$

      iii

β_i	Lower	Upper
β_0	−137.014	−19.5212
β_1	0.237369	2.50626
β_2	0.348319	1.21653
β_3	0.7555	1.34732
β_4	−0.494088	0.254260
β_5	−0.192064	0.374936
β_6	−0.263157	0.439846
β_7	−0.439239	0.23575

      v $C[2.106 \le \sigma_Y \le 4.045] = 0.95$, $C[0.720 \le \sigma_A \le 1.657] = 0.95$, and $C[1.271 \le \sigma_Y/\sigma_A \le 5.618] \ge 0.90$

      vi 0.917

      vii 69.347 inches

      ix $C[65.98 \le \mu_Y(20, 60, 72, 61, 71, 62, 70)] = 0.95$

      x $C[66.5 \le \mu_Y(20, 60, 72, 61, 71, 62, 70)] = 0.95$

      xi Standard errors are not the same.

  c **i** $\hat{\beta}_0^C = -61.20$, $\hat{\beta}_2^C = 0.8947$, $\hat{\beta}_3^C = 1.0556$, $\hat{\sigma}_{Y|X_2, X_3} = 1.131$.

      ii $\hat{\beta}_0^D = -78.23$, $\hat{\beta}_1^D = 1.3503$, $\hat{\beta}_2^D = 0.6925$, $\hat{\beta}_3^D = 1.1025$. $\hat{\sigma}_{Y|X_1, X_2, X_3} = 0.9305$.

      iv $\mu_Y^{(C)}(x_2 + 1, x_3) - \mu_Y^{(C)}(x_2, x_3) = \beta_2^C$. A 95% lower confidence bound for β_2^C is 0.587.

      vii $\mu_Y^{(C)}(x_2, x_3 + 1) - \mu_Y^{(C)}(x_2, x_3) = \beta_3$. A 95% upper confidence bound for β_3^C is 1.26.

      viii NH: $\beta_3^C > 1$. AH: $\beta_3^C \le 1$. The *P*-value is greater than 0.20.

      x $\hat{\rho}_{Y(x_3)|x_2}^2 = 0.84$

      xi $C[0.888 \le \sigma_{Y|x_2, x_3} \le 1.583] = 0.90$

      xii $\hat{Y}(60, 72) = 68.5$, $C[65.5 \le Y(60,72) \le 71.4] = 0.95$

      xiii $\hat{\mu}_Y^{(C)}(60, 72) = 68.5$, $C[67.04 \le \mu_Y^{(C)}(60, 72) \le 69.93] = 0.90$

  d **ii** $C[67.1 \le Y^{(C)}(20, 60, 72) \le 72.3] = 0.95$

      iii $C[68.0 \le \mu_Y^{(C)}(20, 60, 72) \le 71.4] = 0.95$

      iv $\hat{\rho}_{Y(x_1)|x_2,x_3}^2 = 0.3627$

5.2.1 **a** A plot of Y against X reveals no outliers.

  b Observation 12 is a candidate for a possible outlier and should be examined.

5.2.2 For observation 23, the studentized deleted residual is $T_{23} = 21.6335$ and the standardized residual is $r_i = 5.5577$.

5.3.1 **a** $n = 36$. $p = 3$. $2p/n = 0.166667$.   **b** Observation 13 has a hat value of 0.191368. Observation 21 has a hat value of 0.200206.

5.3.3 No change. No change.

5.4.1 **a** $h_{3,3,} = 0.14463$

 b $DFFITS_4 = 0.94814$

 c $c_4 = 0.26849$

 d $\hat{Y}(3.5) = 17.4884$

 e $\hat{r}_3 = -0.64454$

 f $\hat{e}_3 = -1.08058$

 g $T_6 = -0.60789$

5.4.2 **a** 2.52849. Yes

 b 0.81405. Yes

 c 0.26849. No

 d 0.94814. Yes

5.4.3 **a** $\hat{Y}(13) = 21.562$

 b $\hat{Y}_{(-6)}(13) = 21.76$

 c $\hat{\sigma} = 3.476$

 d $\hat{\sigma}_{(-6)} = 1.967$

5.4.4 $T_6 = -3.08081$ in Exhibit 5.4.5.

5.4.5 $h_{6,6} = 0.97319$ in (5.3.3) and in Exhibit 5.4.5.

5.6.1 **a** No.

 b Sample items 12 and 16 should be examined further as possible outliers.

 c Items 5, 12, 16, and 17 have DFFITS larger than $2\sqrt{p/n} = 1.26$ and should be examined as influential observations. Cook's distance for item 12 is greater than $F_{.5:8,12} = 0.97$. Hence items 5, 12, 16, and 17 should be examined.

 e No

6.2.1 **a** $\hat{Y}(12.0, 22, 9.3) = \$90.58$

 b $\hat{\mu}_Y(12.0, 22, 9.3) = \$90.58$

 c $C[\$0.00 \le Y(12.0, 22, 9.3) \le \$200.93] = 0.95$

 d $C[\$46.46 \le \mu_Y(12.0, 22, 9.3) \le \$134.70] = 0.95$

6.2.3 $\$355.74$

6.2.4 **a** $\hat{Y}(15) = 7.553$ grams

 b $C[5.59 \le Y(15) \le 9.51] = 0.90$

 c $\hat{Y}(3) = 1.512$, $\hat{Y}(5) = 2.519$, and $\hat{Y}(13) = 6.546$, so $\hat{Y}_s = 10.577$.

 d $C[6.66 \le Y_s] = 0.95$

 g $\hat{\mu}_Y(20) = 10.07$ grams

 h $C[9.47 \le \mu_Y(20) \le 10.67] = 0.90$

6.2.5 **a** $\mu_Y(30), \mu_Y(60)$

 b $\hat{\mu}_Y(60) = 9.92$

 c $C[9.33 \le \mu_Y(60) \le 10.51] = 0.90$

6.3.1 **a** $\hat{\lambda}_{0.85}(3.0) = 0.868$

 b We have 90% confidence that at least a proportion $p = 0.85$ of these subpopulation Y values is between 0.392 and 1.574.

 c We have 90% confidence that $\lambda_{0.15}(3.0)$ is between -2.024 and -0.843.

6.3.2 We have confidence of at least 90% that a proportion 0.7 of the subpopulation values at $X = 3.0$ are between L and U where $L = -2.024$ and $U = 1.574$.

6.3.3 **b** No. It is better to compare this baby's weight with $\lambda_p(30)$, for some suitable percentile p.

c **iv** since $k = \lambda_{0.10}(30)$.

d $\hat{\lambda}_{0.10}(30) = 5.67$ pounds and $C[4.454 \leq \lambda_{0.10}(30) \leq 6.604] = 0.95$.

e $\lambda_p(x)$, $x = 60$

f $p = 0.05$

g We have confidence of 95% that at most 5% of babies who are 60 days old weigh less than 5.45 pounds.

6.4.1 $\hat{x}_0 = 66.5$ degrees

6.4.2 $C[65.07 \leq x_0 \leq 68.03] = 0.99$

6.4.3 The estimate of her actual temperature is 100.1 degrees.

6.4.4 $C[99.63 \leq x_0 \leq 100.59] = 0.90$

6.4.5 $\hat{x}_0 = 9.93$

6.4.6 $C[8.65 \leq x_0 \leq 11.05] = 0.90$

6.4.7 $\hat{X}(y_0) = \hat{X}(10.5) = 4.718$. You should consider the model $\mu_X(y) = \beta_0^* + \beta_1^* y$ since assumptions (B) apply.

6.4.8 $C[3.340 \leq X(10.5) \leq 6.096] = 0.90$

6.5.1 **a** $\hat{\alpha}_1 = -1.882$, $\hat{\beta}_1 = 1.098$
$\hat{\alpha}_2 = -1.871$, $\hat{\beta}_2 = 1.106$
$\hat{\alpha}_3 = -2.472$, $\hat{\beta}_3 = 1.199$

b $\hat{\sigma}_1 = 0.5536$, $\hat{\sigma}_2 = 0.1766$, $\hat{\sigma}_3 = 0.3663$

c $\hat{\sigma} = 0.3966$

d $d^T = \begin{bmatrix} 0 & 1 & 0 & -1 & 0 & 0 \end{bmatrix}$

e $d^T = \begin{bmatrix} 0 & 0 & 1 & 0 & -1 & 0 \end{bmatrix}$

f $m = 2$

g $m = 3$

h $m = 2$

i One has at least 90% confidence that the following are correct.

$$-0.9554 \leq \alpha_1 - \alpha_2 \leq 0.9340$$
$$-0.1220 \leq \beta_1 - \beta_2 \leq 0.1061$$

6.5.2 The confidence intervals are

$$-1.18876 \leq \alpha_1 - \alpha_2 \leq 1.16731$$
$$-0.54474 \leq \alpha_1 - \alpha_3 \leq 1.72542$$
$$-0.40567 \leq \alpha_2 - \alpha_3 \leq 1.60779$$
$$-0.15022 \leq \beta_1 - \beta_2 \leq 0.13428$$
$$-0.23798 \leq \beta_1 - \beta_3 \leq 0.03659$$
$$-0.16820 \leq \beta_2 - \beta_3 \leq -0.01726$$

The rates of growth are "equivalent" for this problem. Results for intercepts are not conclusive and a larger sample size is need.

6.5.3 $\mu_Y^{(1)}(0) - \mu_Y^{(2)}(1.5) = \alpha_1 - \alpha_2 - 1.5\beta_2$

6.5.4 $m = 30$

6.5.5 The investigator would like to know whether the following inequalities are true.

$$|\mu_Y^{(1)}(1) - \mu_Y^{(2)}(1)| \le 2, \quad |\mu_Y^{(1)}(30) - \mu_Y^{(2)}(30)| \le 2$$
$$|\mu_Y^{(1)}(1) - \mu_Y^{(3)}(1)| \le 2, \quad |\mu_Y^{(1)}(30) - \mu_Y^{(3)}(30)| \le 2$$
$$|\mu_Y^{(2)}(1) - \mu_Y^{(3)}(1)| \le 2, \quad |\mu_Y^{(2)}(30) - \mu_Y^{(3)}(30)| \le 2$$

6.6.1 $\hat{x}_0 = 0.256$

6.6.2 $\hat{x}_0 = -0.3387$

6.6.3 $C[-1.359 \le x_0 \le 1.414] = 0.95$

6.6.4 $\hat{x}_0 = -1.346$

6.6.5 The confidence region is $-\infty$ to $+\infty$.

6.7.1 **b** $\hat{\beta}_0 = 2.1628, \hat{\beta}_1 = 0.32785, \hat{\beta}_2 = -0.005151, \hat{\sigma} = 0.6300$

c $\hat{x}_0 = 31.8$

d $C[29.9 \le x_0 \le 34.0] = 0.90$

6.8.1

$$X = \begin{bmatrix} 1 & 100 & 0 \\ 1 & 125 & 0 \\ 1 & 150 & 0 \\ 1 & 175 & 0 \\ 1 & 200 & 0 \\ 1 & 225 & 0 \\ 1 & 250 & 0 \\ 1 & 270 & 5 \\ 1 & 270 & 30 \\ 1 & 270 & 55 \\ 1 & 270 & 80 \\ 1 & 270 & 105 \\ 1 & 270 & 130 \end{bmatrix}$$

6.8.2 **a** $\hat{\sigma} = 0.2795$. Also $C[0.1953 \le \sigma \le 0.4905] = 0.95$.

b $\hat{\alpha}_1 = 0.780, \quad C[0.187 \le \alpha_1 \le 1.37] = 0.90$
$\hat{\beta}_1 = 0.0120, \quad C[0.0091 \le \beta_1 \le 0.0150] = 0.90$
$\hat{\alpha}_2 = 6.823, \quad C[5.531 \le \alpha_2 \le 8.115] = 0.90$
$\hat{\beta}_2 = -0.0103, \quad C[-0.0143 \le \beta_2 \le -0.00640] = 0.90$

c $\hat{\mu}_Y(175) = 2.89, \quad C[2.6641 \le \mu_Y(175) \le 3.1071] = 0.95$
$\hat{\mu}_Y(375) = 2.94, \quad C[2.606 \le \mu_Y(375) \le 3.279] = 0.95$

6.8.4 Neither model appears to be inconsistent with these data.

6.9.1 **a**   All of these parameters can be validly estimated.

b $\hat{\mu}_Y = 50.472, \hat{\mu}_X = 51.492. \hat{\sigma}_Y = 0.882, \hat{\sigma}_X = 0.779, \hat{\sigma} = 0.3949. \hat{\beta}_0 = -2.226,$
$\hat{\beta}_1 = 1.0234$

c $\hat{Y}(52.08) = 51.074$. We have 95% confidence that $Y(52.08) \ge 50.322$.

d $\hat{Y}_S = 203.896$ seconds. We have 99% confidence that $Y_S \ge 201.320$.

e $C[201.810 \le Y_S \le 205.982] = 0.95$

6.9.2 **a** $\hat{\lambda}_{0.99}(210) = 149.4. C[145.2 \le \lambda_{0.99}(210) \le 156.9] = 0.95$

c $C[124.14 \le \lambda_{0.99}(160) \le 136.12] = 0.95$

6.9.3 **a** $\hat{\mu}_Y(x) = 101 - 3.04x$

b $\hat{x}_0 = 15.08, C[10.8 \le x_0 \le 19.6] = 0.95$

6.9.4 **b** $\hat{\mu}_Y(x) = 51.9 - 0.27x$

c $\hat{x}_0 = 62.20. C[59.59 \le x_0 \le 65.17] = 0.95$

d $C[39.85 \le \lambda_{0.01}(25) \le 42.45] = 0.95$

6.9.5   **a** $\hat{\mu}_Y^{(1)}(x) = 23.5 + 0.0643x$, $\hat{\mu}_Y^{(2)}(x) = 24.5 + 0.0345x$, $\hat{\mu}_Y^{(3)}(x) = 29.3 + 0.0278x$.
$\hat{\sigma}_1 = 2.823$, $\hat{\sigma}_2 = 3.171$, $\hat{\sigma}_3 = 4.089$

     b $\hat{\sigma} = 3.476$

     d $\mu_Y^{(1)}(1000) - \mu_Y^{(2)}(1000)$
$\mu_Y^{(1)}(1000) - \mu_Y^{(3)}(1000)$
$\mu_Y^{(2)}(1000) - \mu_Y^{(3)}(1000)$

6.9.6   **b** $\mu_Y^{(1)}(x) - \mu_Y^{(2)}(x) = \alpha_1 - \alpha_2 + \beta_1 x - \beta_2 x$

     c $\mu_Y^{(1)}(0) - \mu_Y^{(2)}(0) = \alpha_1 - \alpha_2$. $C[-2.46 \le \alpha_1 - \alpha_2 \le 1.81] = 0.95$

     d $\beta_1 - \beta_2$. $C[-0.476 \le \beta_1 - \beta_2 \le -0.178] = 0.95$

6.9.7   **a** $\mu_Y^{(1)}(25) - \mu_Y^{(2)}(25)$

     b $\mu_Y^{(1)}(60) - \mu_Y^{(2)}(60)$

     c $\beta_1 - \beta_2$

     d $\hat{\mu}_Y^{(1)}(25) - \hat{\mu}_Y^{(2)}(25) = 2.515$, $\hat{\mu}_Y^{(1)}(60) - \hat{\mu}_Y^{(2)}(60) = -0.386$, $\hat{\beta}_1 - \hat{\beta}_2 = -0.083$

     f $\hat{x}_0 = 55.3448$ and $C[49.88 \le x_0 \le 65.20] = 0.90$

6.9.8   **a**   *MSE* (cubic) = 0.01092
MSE (quadratic) = 0.01194
MSE (linear) = 0.07059
With these mean squared errors, and after examining the plots, we choose the quadratic model.

     b $\hat{\beta}_0 = 0.334$, $\hat{\beta}_1 = 0.0399$, $\hat{\beta}_2 = -0.000429$, $\hat{\sigma}_{quadratic} = \hat{\sigma} = 0.1093$

     c $\mu_Y(12) = \beta_0 + 12\beta_1 + 144\beta_2$

     d $\hat{\mu}_Y(12) = 0.7514$ and $C[0.696 \le \mu_Y(12) \le 0.807] = 0.80$

     e $C[0.594 \le Y(12) \le 0.909] = 0.80$

6.9.9   **a**   By preselecting X values.

     b $\hat{\mu}_Y(x) = 0.085 + 0.379x - 0.156x^2$

     c   NH: $\beta_2 = 0$. AH: $\beta_2 \ne 0$

     d   The *P*-value is less than 0.001, so we reject NH at the 0.05 level.

     e $x_0 = 1.218$

     f $C[1.128 \le x_0 \le 1.366] = 0.95$

     g $C[13.42 \le z_0 \le 23.24] = 0.95$

6.9.10   **a** $\hat{\alpha}_1 = 7.08$, $\hat{\beta}_1 = 0.06732$, $\hat{\alpha}_2 = 9.581$, $\hat{\beta}_2 = 0.0423$

     b $\beta_1 - \beta_2$

     c $\hat{\beta}_1 - \hat{\beta}_2 = 0.025$, $C[0.0193 \le \beta_1 - \beta_2 \le 0.0307] = 0.90$

     d $\hat{\beta}_0 = 6.92$, $\hat{\beta}_1 = 0.08$, $\hat{\beta}_2 = -0.00013$. *MSE* (spline) = 0.047, *MSE*(quadratic) = 0.050. The two models fit equally well according to the data. The residual plots indicate that neither model is inconsistent with assumptions (A).

7.3.1   **a** X_2

     b X_2, X_5

     c X_1, X_2, X_5

     d $X_1, X_2, X_3, X_4, X_5, X_6, X_7$
$X_1, X_2, X_3, X_5, X_6, X_7$
$X_1, X_2, X_4, X_5, X_6, X_7$

     e $X_1, X_2, X_3, X_5, X_6, X_7$
X_1, X_2, X_5, X_6, X_7
$X_1, X_2, X_3, X_4, X_5, X_6, X_7$

f   Same as (**e**).

g X_1, X_2, X_5, X_6, X_7
$X_1, X_2, X_3, X_5, X_6, X_7$
X_1, X_2, X_3, X_4

h X_1, X_2, X_5, X_6, X_7
$X_1, X_2, X_3, X_5, X_6, X_7$
$X_1, X_2, X_4, X_5, X_6, X_7$
$X_1, X_2, X_3, X_4, X_5, X_6, X_7$

i   Variables X_1, X_2, X_5, X_6, X_7 occur in most of the better fitting models. The model with X_1, X_2, X_5, X_6, X_7 appears to be a good model based on the criteria R^2, adj-R^2, s, C_p, and $C_p - p$.

7.4.1   **a** X_1, X_2, X_3

   b X_1, X_2

   c X_1, X_2, X_3

   d X_1, X_2

   f X_1, X_2, X_3

   g X_1, X_2, X_3

7.4.2   **a**   None

   b X_1, X_2, X_3

   c X_2, X_3

   d X_2, X_3

   h   *F-in* must be greater than or equal to *F-out*.

   i

Step	Variables in Model
1	X_3
2	None
3	X_1

7.4.4   **a** X_1, X_2, X_3

   b X_1, X_2, X_3

   c X_1, X_2, X_3

   d X_1, X_2, X_3

   e

Step	Variables in Model
1	X_1
2	None
3	X_3
4	X_2, X_3
5	X_1, X_2, X_3

   g

Step	Variables in Model
1	X_1
2	None
3	X_3
4	X_2, X_3
5	X_1, X_2, X_3

7.5.1   **a** $k = 5, m = 24, p = 3$

b $t_1 = 4, t_2 = 6, t_3 = 8, t_4 = 10, t_5 = 12$

c
$$X = \begin{bmatrix} 1 & 4 & 16 \\ 1 & 6 & 36 \\ 1 & 8 & 64 \\ 1 & 10 & 100 \\ 1 & 12 & 144 \end{bmatrix}$$

d $y_3 = [3.5 \quad 11.3 \quad 18.4 \quad 22.5 \quad 25.3]^T$
$y_{10} = [3.6 \quad 11.4 \quad 18.3 \quad 21.6 \quad 23.9]^T$

e $y_{15,j} = \alpha_{15} + \beta_{15}t_j + \gamma_{15}t_j^2 + e_{15,j}$ for $j = 1, 2, 3, 4, 5$.

f $C[-0.258 \le \gamma \le -0.234] = 0.90$

g $\hat{\mu}_Y(t) = -18.94 + 6.17t - 0.246t^2$

h $\hat{\mu}(8) = 14.710$

i $a^T = [0, 1, 0]$

j $a^T = [1, t, t^2]$

k $8\beta + 128\gamma$

7.6.1   **a**   If s is used as the criterion, then the short list of the three best subset models is (X_1, X_2, X_3, X_4); (X_1, X_2, X_3); (X_1, X_3, X_4).
If R^2 is used as the criterion, then the short list of three best subset models is (X_1, X_2, X_3, X_4); (X_1, X_2, X_3); (X_1, X_3, X_4).
If adj-R^2 is used as the criterion, then the short list of three best subset models is (X_1, X_2, X_3, X_4); (X_1, X_2, X_3); (X_1, X_3, X_4).

b   **i**

C_p	Variables in Model
30.5	X_1
37.3	X_3
41.7	X_2
43.3	X_4
5.6	X_1, X_3
12.8	X_1, X_2
17.3	X_1, X_4
23.2	X_3, X_4
24.5	X_2, X_3
25.9	X_2, X_4
4.2	X_1, X_2, X_3
5.3	X_1, X_3, X_4
10.4	X_1, X_2, X_4
18.8	X_2, X_3, X_4
5.0	X_1, X_2, X_3, X_4

iii X_1, X_2, X_3

iv (X_1, X_2, X_3, X_4); (X_1, X_3, X_4); (X_1, X_3)

e

Step	Variables in Model
1	1
2	1, 3
3	1, 2, 3

      f   Variables X_1 and X_3 appear in all well-fitting models. The model with X_1, X_2, X_3 is a good final candidate.

7.6.2   **a** $m = 20$, $k = 4$, $p = 3$

      b

$$X = \begin{bmatrix} 1 & 8.0 & 64.00 \\ 1 & 8.5 & 72.25 \\ 1 & 9.0 & 81.00 \\ 1 & 9.5 & 90.25 \end{bmatrix}$$

      c $a^T = [1, 0, 0]$

      d $\mu_Y(t) = 27.87 + 3.39t - 0.092t^2$

      e $C[-3.864 \le \beta \le 12.751] = 0.95$; $C[-0.534 \le \gamma \le 0.351] = 0.95$

      f $\mu_Y(t) = \alpha + \beta t + \gamma t^2$

      g $\mu_Y(8.5) = \alpha + 8.5\beta + 72.25\gamma$

      h   No. Need more data.

8.2.1   **a**   0.0651 and 0.2500, respectively.

      c $\hat{\beta}_0^{(w)} = 1.72$; $\hat{\beta}_1^{(w)} = 1.78$

      d $\hat{\beta}_0 = 1.54$; $\hat{\beta}_1 = 1.82$

8.2.2 $\hat{\mu}_Y^{(w)}(3.0) = 7.0499$; $\hat{\sigma}_0 = 0.334$; $\hat{\sigma}_Y(3.0) = 1.002$

8.2.3 $2.5\hat{\beta}_1$

8.2.4 $C[3.125 \le 2.5\beta_1 \le 5.756] = 0.95$

8.2.6 $C[0.252 \le \sigma_0 \le 0.506] = 0.90$

8.2.7 $C[1.008 \le \sigma_Y(4.00) \le 2.024] = 0.90$

8.3.1   **a** $n = 14$

      b $m = 7$

      c $w_2 = 2.24$; $v_2 = 21.8000$; $v_6 = 35.552$. $t_2 = 2.792$

      d $q_2^* = 1.24643$; $q_4^* = 0.36414$; $q_2 = 0.36414$; $q_4 = 1.24643$

      e $\hat{\beta}_0 = 1.24643$

      f $\hat{\beta}_1 = 1.71875$

      g $C[0.364 \le \beta_0 \le 3.146] = 0.88$

      h $C[1.075 \le \beta_1 \le 2.366] = 0.88$

      i   By Theil's method, $\hat{\beta}_0 = 1.24643$ and $\hat{\beta}_1 = 1.71875$. By weighted least squares, $\hat{\beta}_0 = 1.72$ and $\hat{\beta}_1 = 1.78$.

      j $\hat{\mu}_Y(5.0) = 10.3087$

      k $\mu_Y(5.0) - \mu_Y(2.5) = 2.5\beta_1$

      l $2.5\hat{\beta}_1 = 4.297$; $C[2.69 \le 2.5\beta_1 \le 5.92] = 0.88$

8.3.2   Yes. A Gaussian population is symmetric and continuous.

8.3.3   Yes

8.3.4   Yes, but not necessarily

8.3.5   Yes

8.3.6   Yes, if $g(x) = 1$.

8.3.7   Yes, if $g(x) = 1$.

8.3.8 Yes

8.3.9 Yes

8.4.1 **a** Ordinary least squares (OLS) estimate is $\hat{\mu}_Y(x) = 1.99 + 0.912x$.

b $\hat{\mu}_Y^{(w)}(x) = 2.07 + 0.858x$

c $C[6.911 \leq \mu_Y(6) \leq 7.522] = 0.99$

d $C[6.0016 \leq Y(6) \leq 8.432] = 0.90$

e $C[0.0155 \leq \sigma_0 \leq 0.0254] = 0.80$
$C[0.556 \leq \sigma_Y(6.0) \leq 0.915] = 0.80$

f An estimate of $P[Y \geq 7]$ is 0.6245.

8.4.2 **a** $\hat{\mu}_Y(x) = -6.132 + 1.082x$

b No, except student 8 has a large standardized residual $= 3.46$.

d **i** $-3.208 + 1.0099x$

ii $C[-15.06 \leq \beta_0 \leq 11.8] = 0.85$
$C[0.733 \leq \beta_1 \leq 1.121] = 0.85$

iii $\hat{Y}(75) = \hat{\mu}_Y(75) = 73.74$

iv $C[70.875 \leq \mu_Y(75) \leq 77.24] = 0.85$

9.2.3 **a** This is a linear regression function.

b This is a linear regression function.

c This is a nonlinear regression function. It is linear in β_0 but is nonlinear in β_1.

d This is a linear regression function.

e This is a nonlinear regression function. It is simultaneously linear in β_0 and β_1, but it is nonlinear in β_2 and in β_3.

9.3.1 $C[-0.3595 \leq \beta_1 \leq 0.4170] \approx 0.95$
$C[2.2470 \leq \beta_2 \leq 3.1996] \approx 0.95$

and

$C[0.3624 \leq \beta_3 \leq 1.0031] \approx 0.95$

9.3.2 **a** $\hat{\beta}_1 = 1.9483$
$\hat{\beta}_2 = -1.2699$
$\hat{\beta}_3 = 14.3631$
$\hat{\sigma} = 0.1025$

c 1.8929671

d $C[0.5182 \leq \theta \leq 2.9690] \geq 0.90$ (approximately).

9.3.3 **a** $\hat{\beta}_1 = 0.8183215325$, $\hat{\beta}_2 = -0.6776058218$, $\hat{\beta}_3 = 0.1632703464$, and $\hat{\sigma} = 0.3696$.

c We seek the value x_0 such that the regression curve attains its maximum at $X = x_0$. This point is given by $\hat{x}_0 = 2.0751038$.

d Let x_c represent the elapsed time at which the drug concentration in the bloodstream falls below 1 microgram/deciliter (after attaining the peak concentration). We obtain $\hat{x}_c = 4.402935787$.

9.4.1 **a** The linearizing transformation is $Z = 1/Y$.

b $\hat{\beta}_1 = 1.4026$, $\hat{\beta}_2 = -1.2903$, $\hat{\beta}_3 = 0.27640$

9.4.2 **a** The linearizing transformation is $Z = -\ln\left[-\ln\left(1 - \frac{Y}{2}\right)\right]$.

b The estimates of β_2 and β_3 obtained after linearization are $\hat{\beta}_2 = -1.081$ and $\hat{\beta}_3 = 13.0413$.

9.5.1 **b** $\hat{\beta}_1 = 0.00656892$, $\hat{\beta}_2 = -0.00028246$, $\hat{\beta}_3 = 0.00000749$, $\hat{\beta}_4 = -80.65213231$

9.5.2 **b** $\hat{\beta}_1 = 0.995$, $\hat{\beta}_2 = -115.929282$, $\hat{\beta}_3 = 5402.347473$, $\hat{\beta}_4 = 0.009437$

Also $\hat{\sigma} = 0.0145530$.

d The required value of x_0 is obtained to be .01949075756.

e The required quantity is equal to θ where $\theta = \dfrac{\beta_1}{[1 + e^{-\beta_2}]^{\beta_4}}$.

$C[0.1786 \leq \theta \leq 0.5464] \geq 0.85$ (approximately).

Bibliography

[1] Belsley, D. A., Kuh, E., and Welsch, R. E. (1980). *Regression Diagnostics: Identifying Influential Data and Sources of Collinearity.* New York: Wiley.

[2] Cook, R. D. and Weisberg, S. (1982). *Residuals and Influence in Regression.* New York: Chapman and Hall.

[3] Daniel, C. and Wood, F. S. (1980). *Fitting Equations to Data: Computer Analysis of Multifactor Data,* 2nd edition. New York: Wiley.

[4] David, F. N. (1954). Tables of the Correlation Coefficient. London: Cambridge University Press, issued by *Biometrika* office.

[5] Devore, J., and Peck, R. (1986). *Statistics: The Exploration and Analysis of Data.* St. Paul, MN: West.

[6] Draper, N. R. and Smith, H. (1981). *Applied Regression Analysis.* 2nd edition. New York: Wiley.

[7] Fuller, W. A. (1987). *Measurement Error Models.* New York: Wiley.

[8] Furnival, G. M. (1971). All possible regressions with less computation. *Technometrics. 13,* 403–408.

[9] Gallant, A. R. (1987). *Nonlinear Statistical Models.* New York: Wiley.

[10] Graybill, F. A. (1976). *Theory and Application of the Linear Model.* North Scituate, MA: Duxbury Press.

[11] Graybill, F. A. (1961). *An Introduction to Linear Statistical Models: Volume 1.* New York: McGraw-Hill.

[12] Gunst, R. F. and Mason, R. L. (1980). *Regression Analysis and Its Applications: A Data-Oriented Approach.* New York: Marcel Dekker.

[12] Hoaglin, D. C. and Welsch, R. E. (1978). The hat matrix in regression and ANOVA. *The American Statistician.* 32, 17–22.

[13] Hocking, R. R. (1976). The analysis and selection of variables in linear regression. *Biometrics. 32,* 1–49.

[14] Hocking, R. R. (1985). *The Analysis of Linear Models.* Monterey, CA: Brooks/Cole.

[15] Hogg, R. A. and Craig, A. T. (1970). *Introduction to Mathematical Statistics.* 3rd edition. New York: McMillan.

[16] Khazanie, R. (1990). *Elementary Statistics in a World of Applications.* 3rd edition. Glenview, IL: Scott, Foresman and Company.

[17] Kleinbaum, D. G., Kupper, L. L., and Muller, K. E. (1988). *Applied Regression Analysis and Other Multivariable Methods.* 2nd edition. Boston: PWS-Kent.

[18] Kohler, H. (1985). *Statistics for Business and Economics.* Glenview, IL: Scott, Foresman and Company.

[19] Larsen, R. J. and Marx, M. L. (1986). *An Introduction to Mathematical Statistics and Its Applications.* Englewood Cliffs, NJ: Prentice Hall.

[20] Madansky, A. (1959). The fitting of straight lines when both variables are subject to error. *Journal of the American Statistical Association. 54,* 173–205.

[21] Mallows, C. L. (1973). Some comments on C_p. *Technometrics. 15,* 661–675.

[22] Mendenhall, W. (1993). *Beginning Statistics—A to Z.* Belmont, CA: Duxbury Press.

[23] Miller, R. G. Jr. (1981). *Simultaneous Statistical Inference.* 2nd edition. New York: Springer-Verlag.

[24] Montgomery, D. C. and Peck, E. A. (1992). *Introduction to Linear Regression Analysis.* 2nd edition. New York: Wiley.

[25] Mood, A. M., Graybill, F. A., and Boes, D. C. (1974). *Introduction to the Theory of Statistics.* 3rd ed. New York: McGraw-Hill.

[26] Mosteller, F. and Tukey, J. W. (1977). *Data Analysis and Regression: A Second Course in Statistics.* Reading, MA: Addison-Wesley.

[27] Myers, R. H. (1990). *Classical and Modern Regression with Applications.* 2nd edition. Boston: PWS Kent.

[28] Neter, J., Wasserman, W., and Kutner, M. H. (1989). *Applied Linear Regression Models.* 2nd edition. Homewood, IL: Richard D. Irwin.

[29] Ratkowsky, D. A. (1983). *Nonlinear Regression Modeling: a unified practical approach.* New York: Marcel Dekker.

[30] Rawlings. J. O. (1988). *Applied Regression Analysis: A Research Tool.* Pacific Grove, CA: Wadsworth & Brooks/Cole.

[31] Rice, J. A. (1988). *Mathematical Statistics and Data Analysis.* Pacific Grove, CA: Wadsworth and Brooks/Cole.

[32] Scheaffer, R. L., Mendenhall, W., and Ott, L. (1986). *Elementary Survey Sampling.* 3rd edition. Boston: Prindle, Weber, and Schmidt.

[33] Snedecor, G. W. and Cochran, W. G. (1980). *Statistical Methods.* 7th edition. Ames, IA: Iowa State University Press.

[34] Theil, H. (1971). *Principles of Econometrics.* New York: Wiley.

[35] Tukey, J. W. (1977). *Exploratory Data Analysis.* Reading, MA: Addison-Wesley.

[36] Weisberg, S. (1985). *Applied Linear Regression.* 2nd edition. New York: Wiley.

[37] Wonnacott, T. H. and Wonnacott, R. J. (1981). *Regression: A Second Course in Statistics.* New York: Wiley.

[38] Younger, M. S. (1985). *A First Course in Linear Regression.* 2nd edition. Boston: Prindle, Weber, and Schmidt.

D

Data Sets

Table D-1   Car Data
Table D-2   Car2 Data
Table D-3   Grades Data
Table D-4   Plastic Data

Data Sets

T A B L E **D·1**
Car Data

carno I	mtcost Y	price X_1	miles X_2	carno I	mtcost Y	price X_1	miles X_2	carno I	mtcost Y	price X_1	miles X_2
1	551	36400	12400	33	444	18000	9500	65	482	18500	11900
2	661	15200	15400	34	772	22000	16700	66	454	18900	6400
3	679	14100	16000	35	485	19000	11100	67	592	16000	13000
4	561	22500	12100	36	592	26400	12400	68	599	26900	12500
5	497	20600	11200	37	598	28700	13700	69	378	15200	8100
6	409	22900	6500	38	480	10300	10600	70	411	21500	6200
7	700	23300	15700	39	561	16200	12200	71	427	10300	3500
8	466	30100	8800	40	669	28100	15900	72	418	16600	8100
9	493	9500	11900	41	426	16300	10000	73	420	23600	7100
10	499	17100	12000	42	539	18100	11000	74	417	17600	6400
11	568	18100	13200	43	858	19500	17700	75	387	28700	7600
12	658	27000	15000	44	490	18100	9300	76	461	29200	8700
13	543	19100	13100	45	377	13200	7900	77	416	18500	9600
14	390	16100	7300	46	358	16500	5100	78	656	16100	14000
15	719	24600	15900	47	468	17000	11600	79	521	15700	11000
16	663	24200	14100	48	472	33200	8900	80	709	22800	16200
17	558	21000	11600	49	554	22900	13200	81	462	30700	7600
18	626	30700	14600	50	399	19400	4500	82	697	20600	15900
19	560	26700	12900	51	424	28100	7800	83	668	12600	14100
20	579	25300	13500	52	400	27700	4700	84	509	25000	12400
21	718	25600	16700	53	685	24800	14900	85	516	16500	11400
22	680	13700	15100	54	449	19000	9300	86	439	23500	9900
23	434	17300	8800	55	436	29600	10300	87	561	12900	13200
24	412	25300	8500	56	621	13000	13100	88	534	15400	12800
25	569	17600	11900	57	547	16400	12700	89	483	20700	10200
26	505	16100	10500	58	602	19300	13700	90	515	11500	11700
27	489	20400	9900	59	396	23100	7800	91	560	24900	13600
28	561	18000	11600	60	479	23800	8800	92	415	24600	6200
29	661	27200	15500	61	696	16700	15400	93	639	30700	14400
30	479	20700	10200	62	622	15200	13800	94	494	29800	11000
31	638	15500	13500	63	657	31600	14900	95	484	22000	10700
32	546	18000	11500	64	529	19200	11500	96	393	12600	6200

T A B L E D - 1

Car Data (Continued)

carno I	mtcost Y	price X_1	miles X_2	carno I	mtcost Y	price X_1	miles X_2	carno I	mtcost Y	price X_1	miles X_2
97	445	15200	10400	147	462	17600	7800	197	440	10700	7100
98	553	29100	13700	148	578	19200	12300	198	579	14600	13600
99	417	35800	3900	149	426	24400	8700	199	440	20700	9100
100	689	23800	15300	150	425	26100	6000	200	574	16200	13600
101	662	28100	15300	151	523	11000	10500	201	487	16100	11100
102	628	25800	13300	152	647	17000	13900	202	801	20800	17200
103	453	29500	7500	153	510	14900	10900	203	608	22700	13600
104	562	25500	13800	154	549	26800	11500	204	415	18400	8700
105	537	9400	13300	155	547	30700	13500	205	483	15200	9500
106	425	10300	7200	156	603	14300	14500	206	437	20200	8700
107	483	21600	11600	157	397	24000	5500	207	513	26900	11300
108	418	17000	8500	158	617	17900	14300	208	434	16500	9400
109	608	21700	14200	159	383	16600	7500	209	633	18400	14000
110	500	16400	12400	160	477	29900	11200	210	613	22400	14900
111	810	21700	17300	161	547	16100	11400	211	639	16400	15100
112	468	13000	7800	162	666	20100	14300	212	809	25800	16600
113	740	10100	16400	163	442	26500	10600	213	841	32500	17300
114	424	15600	9900	164	370	20400	5800	214	656	12900	14700
115	389	19000	6500	165	711	16400	15000	215	392	15200	8100
116	525	15800	13000	166	371	21800	6300	216	728	27900	16700
117	560	17000	11900	167	383	11300	8300	217	399	18400	6400
118	484	11700	9900	168	476	20100	8200	218	554	33000	12800
119	428	20600	9800	169	480	22100	11900	219	472	20800	10500
120	595	22900	13000	170	531	20700	11700	220	489	12400	11700
121	527	7600	10700	171	412	20600	9300	221	629	21100	14100
122	598	15600	12800	172	562	20100	13000	222	536	13700	11500
123	572	14600	12700	173	587	16700	13100	223	454	13600	8300
124	537	13100	12500	174	539	15300	10900	224	478	15700	9700
125	525	15000	12300	175	554	15500	13100	225	434	13800	9200
126	495	25800	10400	176	570	14900	12100	226	541	21100	12800
127	492	13500	9200	177	723	23500	15300	227	438	20500	8000
128	489	21500	9400	178	636	12600	13700	228	467	24200	10700
129	698	29400	15000	179	551	15900	11200	229	598	19200	12800
130	456	16000	10200	180	561	13900	11800	230	624	12800	14900
131	522	20900	12300	181	744	11900	16900	231	410	18700	3900
132	533	26000	12300	182	498	15100	11900	232	459	11900	10100
133	415	32600	6400	183	677	25700	14500	233	490	20700	12000
134	446	23200	5600	184	523	26100	12100	234	529	32800	10700
135	456	9800	10500	185	514	18900	12400	235	673	17100	16000
136	441	14400	9900	186	456	15500	9000	236	481	32300	9800
137	456	23300	10900	187	588	20600	13100	237	402	20800	7800
138	451	17300	10000	188	393	13200	8700	238	481	16300	9000
139	539	22500	12400	189	405	22800	6400	239	560	12700	12100
140	494	22200	10500	190	447	18000	10600	240	629	16300	13600
141	450	9600	9600	191	668	21000	15200	241	683	20500	16000
142	643	26800	14100	192	768	17900	16600	242	560	20600	13200
143	367	10800	5400	193	488	15600	11000	243	536	9500	11500
144	426	23900	4100	194	817	27800	17200	244	454	26100	10000
145	395	25700	8200	195	523	9500	12600	245	523	24900	10400
146	608	17400	13600	196	478	28600	11000	246	519	9100	11400

T A B L E D - 1

Car Data (Continued)

carno I	mtcost Y	price X_1	miles X_2	carno I	mtcost Y	price X_1	miles X_2	carno I	mtcost Y	price X_1	miles X_2
247	713	21000	15000	297	662	18000	14300	347	486	29900	11500
248	444	12900	6100	298	468	19700	11200	348	502	13500	11600
249	525	23000	12200	299	443	14000	10300	349	591	22300	12700
250	443	20100	6600	300	679	21100	15000	350	612	23500	14100
251	610	13500	12900	301	515	28600	11300	351	476	16500	11500
252	465	17000	8700	302	423	19200	8800	352	548	13900	13400
253	734	27400	15600	303	614	31900	13400	353	410	15800	7300
254	458	19400	10700	304	402	19900	8800	354	483	17700	9600
255	428	25500	8400	305	641	27400	14100	355	464	24200	11200
256	540	10400	11800	306	439	15200	10500	356	418	14400	6600
257	499	12300	9500	307	469	21200	10100	357	458	25600	9800
258	470	18600	9800	308	607	22500	13400	358	537	14500	11000
259	616	13400	14600	309	421	25200	4500	359	457	7700	9000
260	599	14700	13600	310	493	13600	11100	360	387	12900	8600
261	511	25700	11100	311	402	14000	9400	361	676	26600	14400
262	576	20600	12200	312	485	17900	9000	362	527	21700	12900
263	676	30300	14400	313	416	19100	5000	363	445	23300	6800
264	513	26200	10800	314	473	18700	11500	364	474	12200	9300
265	359	19600	6500	315	558	14000	13600	365	700	20900	15100
266	620	27400	13600	316	528	23200	12600	366	457	18800	8300
267	389	13100	4200	317	458	28800	11100	367	433	17600	9100
268	455	19000	8800	318	500	15200	11100	368	563	14300	11600
269	595	26000	12700	319	455	12800	10200	369	416	18500	4600
270	597	26800	12600	320	491	19800	11000	370	461	14400	8000
271	527	23800	10900	321	539	18200	12900	371	617	25700	13700
272	537	19600	12700	322	458	17500	7300	372	489	25800	10000
273	471	21800	10000	323	433	28800	7200	373	624	18400	14100
274	475	26800	9400	324	478	11800	9800	374	558	28700	11600
275	505	14500	12500	325	555	11700	12700	375	587	18100	12700
276	408	9900	8800	326	389	19700	8800	376	591	18200	14100
277	477	12400	10300	327	622	16600	13300	377	492	12600	9300
278	540	22000	12800	328	493	19500	9600	378	400	22700	6200
279	374	7200	6900	329	669	12000	15400	379	450	21200	10800
280	439	14700	9900	330	536	23800	11800	380	503	19600	10400
281	518	11800	10400	331	583	21600	12200	381	569	21300	12600
282	503	17700	11800	332	649	21700	15100	382	637	12900	14000
283	525	19400	11900	333	444	21800	7100	383	644	20400	15300
284	429	19900	10100	334	514	22500	10500	384	613	18500	13200
285	473	27500	11600	335	406	19600	8600	385	370	26500	4000
286	522	15800	11200	336	481	24700	10600	386	671	32200	14900
287	560	22000	12100	337	391	31800	8200	387	445	15800	9800
288	411	20700	5700	338	454	21100	8700	388	735	21700	15900
289	809	15800	18100	339	494	26700	11000	389	571	23800	12400
290	650	19200	13800	340	517	26300	12800	390	458	21300	10900
291	499	19500	12000	341	559	29200	12500	391	725	17400	15800
292	484	12500	10800	342	486	23800	9100	392	394	15600	6900
293	460	14500	8100	343	439	20600	10200	393	501	15000	10300
294	729	19600	16000	344	444	24300	9200	394	538	14900	11200
295	452	21200	8400	345	436	10600	8000	395	486	20600	9400
296	436	18000	9200	346	809	12900	17900	396	478	22700	8500

T A B L E D-1

Car Data (Continued)

carno I	mtcost Y	price X_1	miles X_2	carno I	mtcost Y	price X_1	miles X_2	carno I	mtcost Y	price X_1	miles X_2
397	671	19400	15700	447	896	13700	18000	497	480	15800	10400
398	634	26200	14100	448	527	25300	10900	498	583	19000	14200
399	854	24800	17600	449	507	16400	11300	499	446	24600	6800
400	456	12900	11200	450	433	19600	9300	500	736	13300	16200
401	434	15000	10500	451	389	13100	8700	501	635	23400	13900
402	612	22000	14000	452	437	13200	9800	502	426	28100	9700
403	457	28200	10500	453	523	24100	10700	503	536	15500	10800
404	398	16600	8700	454	412	14900	8300	504	369	17600	4200
405	451	21600	7800	455	469	12600	9600	505	524	14200	10900
406	618	12800	14300	456	745	17800	16200	506	443	21700	7300
407	394	23700	3800	457	405	17400	8900	507	703	11300	15900
408	825	16300	17100	458	584	28300	12900	508	560	25800	12000
409	518	16400	10300	459	657	18300	15000	509	597	18700	13900
410	452	24800	9800	460	475	20700	11500	510	512	14900	10600
411	608	15700	13700	461	359	19400	5200	511	523	29800	12100
412	681	19500	15000	462	382	24500	8100	512	727	23000	16200
413	773	19500	17500	463	460	15100	9900	513	740	24900	16700
414	697	29700	15000	464	683	21600	14800	514	591	12200	13400
415	397	12500	7700	465	646	20300	13700	515	557	22600	13400
416	583	13600	12500	466	562	24600	11500	516	487	14900	11900
417	525	18800	10600	467	756	22900	17000	517	567	13400	13500
418	451	18000	9800	468	410	22200	2900	518	500	21600	10000
419	464	21800	11500	469	667	32600	14900	519	550	18000	12300
420	488	11500	10100	470	430	17200	10400	520	569	22200	12900
421	445	22500	9700	471	452	11300	8700	521	387	9400	8400
422	444	8800	9000	472	459	12300	8400	522	469	13800	10800
423	414	21900	8700	473	577	22700	12600	523	469	21900	10800
424	472	12300	10500	474	694	19600	15600	524	617	10800	13500
425	599	25100	12900	475	412	26900	3300	525	662	15800	14600
426	469	13400	8400	476	489	22600	11200	526	463	11400	8800
427	432	16300	8300	477	368	23800	6600	527	458	21100	8400
428	395	27400	6500	478	534	14800	11900	528	480	17500	11000
429	513	18000	11100	479	551	25400	12200	529	673	13700	14800
430	454	20600	10400	480	557	22400	11900	530	436	15700	6300
431	434	13000	8200	481	468	9500	10600	531	640	28100	14600
432	489	17700	9300	482	457	25300	7100	532	484	14200	9600
433	493	17400	10500	483	363	13300	5900	533	496	19400	11000
434	456	21600	10200	484	376	15400	6000	534	433	17200	9600
435	524	11500	11900	485	456	11200	9300	535	474	13000	11400
436	395	15400	7800	486	601	14900	13100	536	623	12300	14200
437	432	19100	8100	487	422	23000	9200	537	524	17600	11800
438	545	22300	12500	488	602	11500	12600	538	437	28100	10500
439	430	22200	5800	489	418	28400	3400	539	617	11500	13100
440	373	14100	7300	490	485	14300	11600	540	469	16300	8700
441	549	20900	11800	491	404	25200	7800	541	676	16200	14800
442	434	23700	7400	492	714	21900	15800	542	365	18100	5500
443	546	23000	11400	493	462	24800	9300	543	386	27500	8500
444	522	18000	11900	494	571	14500	13100	544	644	20400	15000
445	501	22700	11500	495	452	25900	10000	545	436	15700	7200
446	372	11200	3800	496	394	25200	8300	546	595	21200	13800

T A B L E D - 1

Car Data (Continued)

carno *I*	mtcost *Y*	price X_1	miles X_2	carno *I*	mtcost *Y*	price X_1	miles X_2	carno *I*	mtcost *Y*	price X_1	miles X_2
547	589	18300	13300	597	464	22200	11100	647	446	32100	10700
548	506	16000	11000	598	651	30100	15600	648	466	8800	9700
549	706	13400	14900	599	460	13100	8200	649	382	12600	6100
550	560	15900	11900	600	771	21900	16100	650	446	21200	9500
551	493	11100	10100	601	526	22800	10500	651	479	11100	10300
552	533	30200	11900	602	575	17700	11900	652	645	27700	14100
553	606	28000	13900	603	511	32500	10500	653	380	16000	7600
554	604	11000	13500	604	535	22100	11600	654	541	28100	12300
555	604	24600	12800	605	433	12300	9000	655	424	16100	5200
556	680	32100	15500	606	676	21600	15600	656	631	11600	14300
557	526	23700	11500	607	555	13300	12600	657	555	23400	11600
558	549	11000	11900	608	390	12900	6100	658	458	35200	8500
559	520	22600	10400	609	531	27000	12400	659	650	24500	14900
560	421	19800	8500	610	414	19800	4800	660	757	23700	16300
561	635	22600	13700	611	656	21700	14400	661	470	21700	11200
562	360	16700	5300	612	430	8300	8500	662	401	23500	6400
563	476	20300	9500	613	409	16200	8600	663	490	30200	11200
564	465	10700	8400	614	828	21800	17300	664	352	23800	3200
565	471	22900	11100	615	359	22600	4700	665	452	23900	11000
566	546	18500	12300	616	581	21700	12900	666	520	24500	12400
567	450	18800	9900	617	475	17900	9400	667	485	8600	9700
568	629	25000	14600	618	433	31200	10000	668	445	18000	7900
569	726	9900	15400	619	401	25000	7900	669	503	11900	12200
570	574	12300	12100	620	445	28200	9300	670	628	17800	14600
571	426	21500	9700	621	437	14200	10000	671	727	22900	16300
572	541	9700	13200	622	531	15300	12600	672	496	16800	10800
573	806	21500	17600	623	374	10900	7400	673	547	32700	12300
574	690	28200	15400	624	636	12000	13700	674	474	10900	8200
575	397	26400	8300	625	425	17500	7100	675	483	17800	10800
576	557	25500	12200	626	624	21900	14000	676	538	17000	13100
577	473	21800	8400	627	390	11500	6700	677	508	21600	9800
578	551	21700	11300	628	495	24100	11700	678	586	21000	13400
579	516	13700	11300	629	415	27100	5100	679	444	25400	8000
580	493	10400	10700	630	355	29200	4800	680	489	20900	10400
581	496	20900	11800	631	724	26000	15600	681	448	26100	10000
582	447	21400	7300	632	540	19800	12400	682	426	26000	9400
583	474	24400	8800	633	430	21000	9400	683	455	13700	9800
584	796	13300	16500	634	640	27400	14200	684	467	20000	8700
585	442	18800	9700	635	499	23800	10900	685	640	9200	14800
586	547	19900	13500	636	419	21900	9100	686	478	27300	11200
587	564	18800	13300	637	502	21700	10100	687	490	20200	12200
588	591	16300	12800	638	807	13400	17800	688	461	13200	8200
589	475	30200	11000	639	387	8400	6700	689	515	27500	10200
590	422	20500	8000	640	526	23000	12500	690	627	25800	13500
591	519	17600	10300	641	620	18100	14000	691	552	19900	11600
592	716	18500	15100	642	663	22600	14700	692	514	16300	11400
593	482	17200	11000	643	457	15300	10200	693	370	16000	7300
594	521	22400	11600	644	482	12400	10000	694	428	29200	3500
595	421	17900	9000	645	620	11300	13000	695	380	33600	3400
596	506	26800	11500	646	367	17600	6500	696	533	28500	12000

T A B L E D-1

Car Data (Continued)

carno I	mtcost Y	price X_1	miles X_2	carno I	mtcost Y	price X_1	miles X_2	carno I	mtcost Y	price X_1	miles X_2
697	579	13800	12500	747	500	12900	10900	797	551	21100	12300
698	369	22900	5400	748	444	9300	10200	798	605	16400	14200
699	546	13400	11500	749	711	19000	15600	799	768	25000	17100
700	510	20300	12000	750	695	12300	15300	800	418	25900	7100
701	533	18400	10700	751	487	27600	11700	801	522	25000	11600
702	601	33700	13400	752	818	18400	16800	802	441	20600	10300
703	396	16000	7700	753	362	30900	6200	803	432	13200	9500
704	477	14700	10300	754	566	16800	12600	804	430	20300	7700
705	439	21600	8700	755	398	21500	7700	805	657	19000	14300
706	464	24600	9200	756	427	24300	3500	806	462	17600	10300
707	639	22000	13600	757	776	15900	17400	807	394	19300	7800
708	435	15200	9900	758	531	23400	11100	808	614	20000	13400
709	548	18200	13000	759	532	14700	12400	809	440	19600	7500
710	475	17800	10300	760	446	10800	10200	810	582	22000	12700
711	493	23600	9300	761	762	22900	16700	811	448	23200	8600
712	505	14000	10100	762	517	11000	11800	812	525	25900	11000
713	728	18000	15500	763	459	19300	9900	813	534	20800	12000
714	627	16200	14400	764	654	20900	14500	814	519	16300	11700
715	624	31500	13900	765	402	11100	9200	815	565	15400	13500
716	777	33600	17100	766	456	19400	10300	816	426	24900	7600
717	716	15700	15700	767	439	26400	9100	817	662	22300	14300
718	386	32300	5900	768	504	22000	12200	818	665	20600	14800
719	523	21900	12700	769	471	29900	11100	819	417	33400	7800
720	553	15000	11400	770	483	24800	11000	820	560	30000	12900
721	430	14600	7900	771	620	13600	15000	821	418	15500	3400
722	515	21700	10500	772	451	14800	10100	822	441	16200	10500
723	574	9200	13600	773	610	16100	12900	823	465	22200	7500
724	524	26100	11900	774	417	14600	8700	824	544	16000	13000
725	541	18200	12700	775	562	26400	12500	825	531	26000	10700
726	657	18800	14900	776	386	14600	8500	826	439	22800	9600
727	399	19500	7800	777	605	17000	14000	827	401	11000	8700
728	453	13500	7400	778	458	28700	9400	828	499	15400	10900
729	410	13100	9100	779	444	22500	9700	829	495	11900	11900
730	364	17600	1600	780	465	14700	10100	830	648	13100	13800
731	461	16300	8200	781	742	13500	15800	831	616	24000	12900
732	502	16100	12500	782	401	26000	8000	832	513	32300	10900
733	518	27000	11700	783	606	16600	13800	833	568	10800	11800
734	437	20500	5900	784	618	26000	13100	834	494	24800	10800
735	400	22000	5800	785	564	14900	11800	835	734	18300	16800
736	483	18500	11300	786	736	18400	16200	836	695	35100	14900
737	430	12600	7500	787	516	21000	10500	837	373	27400	6000
738	426	14000	8500	788	587	22100	12700	838	642	10700	14200
739	488	25900	11800	789	620	18300	13900	839	481	16700	9500
740	474	10100	8900	790	428	16400	9000	840	565	26600	12300
741	586	13000	12900	791	575	14100	13800	841	437	17900	7900
742	490	20200	11800	792	528	13800	11800	842	599	21300	13900
743	546	23800	13000	793	438	23300	5300	843	574	31900	12800
744	484	14600	11000	794	485	20700	11100	844	417	16400	10000
745	612	15000	14200	795	731	17000	15400	845	440	15900	9500
746	404	14500	8900	796	456	21900	10000	846	438	14800	4800

T A B L E **D - 1**

Car Data (Continued)

carno I	mtcost Y	price X_1	miles X_2	carno I	mtcost Y	price X_1	miles X_2	carno I	mtcost Y	price X_1	miles X_2
847	417	15000	8700	897	496	20200	11600	947	419	18500	5500
848	462	19300	7500	898	766	12200	16200	948	431	28800	7800
849	568	13300	12700	899	672	14500	14700	949	416	22100	7900
850	577	13100	12100	900	621	9600	13200	950	607	25800	13900
851	438	22300	9600	901	596	16800	14100	951	431	25800	6800
852	374	10800	7700	902	542	25100	13200	952	434	17900	6600
853	435	13600	6700	903	369	16200	3900	953	453	16400	9500
854	445	10600	10300	904	449	20800	10200	954	523	18800	11700
855	719	23000	16200	905	639	13900	14500	955	492	20000	11000
856	693	21100	15300	906	803	28200	17500	956	463	23400	9200
857	573	13200	14100	907	504	14000	10600	957	444	17000	7700
858	617	17100	14800	908	664	19700	14500	958	472	26200	9300
859	442	23700	10100	909	765	19500	16100	959	587	12700	12400
860	646	14600	14600	910	582	17600	14300	960	457	37100	10600
861	450	19500	10700	911	551	19700	11300	961	527	27900	12200
862	425	13200	7900	912	644	22100	14400	962	397	13100	7000
863	439	23900	6900	913	646	34900	14800	963	481	21500	8700
864	565	23200	12200	914	440	25800	9200	964	680	16200	15100
865	396	20500	5100	915	508	19300	12500	965	432	23200	9600
866	672	13000	15300	916	554	22000	13000	966	462	16300	11100
867	682	11300	15400	917	368	30700	3700	967	426	26500	6200
868	484	11200	9300	918	803	20700	17000	968	476	10600	11600
869	493	19500	10400	919	489	18100	11000	969	439	15500	9500
870	472	32100	10300	920	394	15400	3400	970	497	18200	9500
871	622	36600	14600	921	432	14000	9200	971	565	19500	11800
872	670	16800	15500	922	506	18300	10600	972	607	17400	13400
873	440	31500	10700	923	594	18800	13400	973	530	11000	12600
874	601	22100	13300	924	497	16800	11900	974	487	10500	11500
875	404	15900	4800	925	598	16300	13200	975	745	18000	16200
876	533	22800	11600	926	393	21500	7400	976	646	21100	15100
877	470	26500	8600	927	577	21100	13200	977	774	20800	16400
878	626	16100	13800	928	620	17300	14000	978	607	16200	13200
879	555	29700	12500	929	402	11800	6300	979	456	18100	10900
880	445	22100	9200	930	455	12200	10700	980	743	15900	15900
881	518	13300	12000	931	412	10400	7000	981	566	21500	12100
882	516	13100	10300	932	773	9600	17400	982	416	12700	9900
883	378	13500	4000	933	417	23400	7500	983	439	23500	3900
884	649	23400	14100	934	484	18300	11000	984	503	16000	12500
885	639	14000	14100	935	638	25900	14300	985	609	20200	13200
886	631	29800	13600	936	617	14100	14900	986	443	17500	5400
887	424	25300	7400	937	648	15600	15300	987	423	21800	7400
888	607	13300	14000	938	504	29600	12200	988	554	7900	12900
889	425	19000	7400	939	588	15300	13700	989	418	18300	7300
890	503	34600	10700	940	479	27200	11700	990	560	24300	12700
891	654	25600	14000	941	738	17400	15600	991	551	17000	12800
892	569	12800	12800	942	395	19100	5000	992	519	21600	10500
893	538	14100	12300	943	716	14700	15800	993	552	25300	12100
894	457	26400	10400	944	393	29300	6700	994	458	23200	6900
895	439	7500	7700	945	553	8900	12300	995	442	25600	8400
896	573	14500	12000	946	414	18000	7900	996	485	20500	9300

T A B L E D-1
Car Data (Continued)

carno	mtcost	price	miles	carno	mtcost	price	miles	carno	mtcost	price	miles
I	Y	X_1	X_2	I	Y	X_1	X_2	I	Y	X_1	X_2
997	458	15600	7400	1047	489	23100	11500	1097	512	22700	12600
998	451	9500	9200	1048	462	16700	11400	1098	434	16000	6200
999	646	33500	14900	1049	569	22900	12500	1099	692	25300	14800
1000	806	24300	18000	1050	627	14600	15000	1100	535	29300	11600
1001	475	17400	8400	1051	518	16400	11900	1101	521	20200	10600
1002	460	17200	10900	1052	491	26500	9300	1102	435	22600	8500
1003	608	22200	14100	1053	476	22900	10000	1103	730	10400	15400
1004	460	29100	8700	1054	858	9100	18500	1104	559	26200	12900
1005	380	18300	8200	1055	425	26000	8800	1105	516	12100	12300
1006	597	14300	13500	1056	491	21000	10200	1106	570	27900	12300
1007	428	20200	6900	1057	549	18300	12700	1107	495	24300	10700
1008	433	12100	9200	1058	568	16400	12100	1108	551	20000	12200
1009	398	8200	8900	1059	738	23300	15500	1109	542	22400	13000
1010	488	14300	11700	1060	700	17000	14700	1110	507	22300	10300
1011	525	18400	12000	1061	490	15900	11300	1111	555	15500	11300
1012	670	21800	15800	1062	481	19100	9900	1112	415	26100	3900
1013	837	22300	17400	1063	521	25300	10800	1113	433	23400	6400
1014	430	22300	9200	1064	722	13500	15500	1114	614	24200	13200
1015	861	14800	18200	1065	427	18200	10300	1115	743	20200	16400
1016	448	26700	8300	1066	658	11300	15100	1116	468	25500	11000
1017	377	15700	7500	1067	518	28700	10100	1117	385	30500	5400
1018	497	20000	12300	1068	433	16500	6500	1118	460	18500	7900
1019	465	10800	9200	1069	620	28200	14300	1119	428	24600	10000
1020	460	27900	10900	1070	493	11400	11000	1120	453	23700	9800
1021	624	22100	13400	1071	448	15000	8000	1121	681	23300	15100
1022	440	16700	9000	1072	531	23700	10900	1122	514	10900	10600
1023	555	25300	11900	1073	473	16400	9700	1123	516	28800	11200
1024	551	14100	12000	1074	427	19400	6800	1124	508	17900	9800
1025	502	12200	10300	1075	412	26500	7600	1125	438	12500	8500
1026	364	12000	5900	1076	487	16200	9700	1126	573	38300	12600
1027	717	20700	15400	1077	540	12700	13100	1127	432	12500	10300
1028	759	11800	16900	1078	453	20200	10400	1128	447	15200	9200
1029	622	16500	14000	1079	447	17600	8200	1129	363	13400	4900
1030	645	9500	14000	1080	693	17000	15100	1130	665	27900	14700
1031	540	26500	11500	1081	528	32300	12100	1131	662	17800	14900
1032	441	33300	6300	1082	624	27600	14100	1132	652	21800	14200
1033	463	28200	11400	1083	449	30700	8500	1133	713	13900	15200
1034	506	21800	10900	1084	498	17300	10400	1134	461	14900	9700
1035	498	9800	9800	1085	675	19400	14300	1135	650	28700	14400
1036	497	26900	12200	1086	473	25700	8900	1136	495	27400	12000
1037	667	25700	14200	1087	577	19400	12100	1137	496	15600	10400
1038	475	9300	11300	1088	614	11700	13100	1138	425	29500	8800
1039	416	26400	7100	1089	525	23700	12800	1139	772	15500	17200
1040	567	15300	14000	1090	511	15100	10300	1140	554	17800	12100
1041	376	12800	7700	1091	381	14600	4500	1141	431	16500	8900
1042	546	10500	11800	1092	424	16300	3700	1142	568	18100	12100
1043	455	20900	8900	1093	596	23700	14000	1143	487	14600	9700
1044	485	20400	9400	1094	406	20600	7200	1144	672	24400	14700
1045	490	9600	11000	1095	478	23000	11600	1145	384	16000	5700
1046	445	20800	6800	1096	879	22600	17900	1146	539	24800	13000

T A B L E **D-1**

Car Data (Continued)

carno *I*	mtcost *Y*	price X_1	miles X_2	carno *I*	mtcost *Y*	price X_1	miles X_2	carno *I*	mtcost *Y*	price X_1	miles X_2
1147	546	15600	11800	1179	466	11000	10700	1211	753	23100	16800
1148	572	16000	13800	1180	520	15000	12500	1212	440	24200	9600
1149	454	27300	9800	1181	727	20500	15600	1213	756	23200	16900
1150	475	24900	11300	1182	554	18200	12800	1214	516	14900	10000
1151	793	20300	16700	1183	427	8700	5400	1215	564	17500	13500
1152	645	17200	14500	1184	520	20500	12900	1216	477	24700	11800
1153	399	16400	7500	1185	824	25600	17000	1217	454	23700	10500
1154	482	20100	8900	1186	674	25800	15500	1218	425	23700	8300
1155	407	14600	5500	1187	457	21000	8000	1219	661	18000	15400
1156	859	30600	17400	1188	525	8200	12900	1220	443	18300	10200
1157	425	23000	7700	1189	433	17400	8600	1221	479	14200	11700
1158	583	15200	13800	1190	437	13800	9500	1222	482	17000	10700
1159	397	17700	7100	1191	458	25200	9000	1223	666	25500	14100
1160	480	18000	9900	1192	514	28500	10600	1224	477	22300	9800
1161	372	24500	7100	1193	749	25500	15900	1225	445	20600	7100
1162	397	27400	7100	1194	452	20900	8600	1226	463	15600	10100
1163	619	38300	14800	1195	589	22700	14200	1227	623	20800	13200
1164	548	23900	13600	1196	398	22800	8000	1228	578	11100	14200
1165	509	18100	10000	1197	410	14300	8100	1229	663	35800	14500
1166	460	16800	10700	1198	597	20900	13900	1230	466	18100	9200
1167	401	11400	8700	1199	639	13600	14000	1231	449	15500	8200
1168	711	16600	15600	1200	527	25500	10600	1232	535	16600	12200
1169	587	21500	12500	1201	453	32000	9900	1233	524	11900	13000
1170	772	20500	16000	1202	511	23400	10000	1234	461	16500	7900
1171	544	19600	13500	1203	383	11000	6800	1235	476	28100	8800
1172	366	14800	5700	1204	474	20700	8700	1236	925	20300	18500
1173	407	25500	9400	1205	597	20000	13000	1237	390	10600	4000
1174	629	14400	13500	1206	650	9600	15300	1238	381	23600	8000
1175	756	17200	15900	1207	548	10500	11800	1239	464	30700	7300
1176	403	15500	8700	1208	453	24500	6800	1240	563	21100	12000
1177	491	11200	9500	1209	452	27400	9700	1241	602	22300	14000
1178	541	16300	12900	1210	460	14700	9700	1242	582	14600	13100

T A B L E **D-2**

Car2 Data

subpop	miles	ycount	ymean	ystdevn	subpop	miles	ycount	ymean	ystdevn
1	1600	1	364.000	0.0000	10	4000	3	379.333	8.2192
2	2900	1	410.000	0.0000	11	4100	1	426.000	0.0000
3	3200	1	352.000	0.0000	12	4200	2	379.000	10.0000
4	3300	1	412.000	0.0000	13	4500	3	400.333	16.3571
5	3400	4	402.500	16.2711	14	4600	1	416.000	0.0000
6	3500	3	427.333	0.4714	15	4700	2	379.500	20.5000
7	3700	2	396.000	28.0000	16	4800	4	402.750	30.2107
8	3800	2	383.000	11.0000	17	4900	1	363.000	0.0000
9	3900	5	410.000	22.7860	18	5000	2	405.500	10.5000

T A B L E D · 2
Car2 Data (Continued)

subpop	miles	ycount	ymean	ystdevn	subpop	miles	ycount	ymean	ystdevn
19	5100	3	389.667	23.6972	70	10200	13	462.538	21.7702
20	5200	2	391.500	32.5000	71	10300	20	474.800	30.4197
21	5300	2	399.000	39.0000	72	10400	15	483.600	28.2154
22	5400	5	398.200	31.1024	73	10500	19	483.053	32.4548
23	5500	4	397.000	20.0499	74	10600	14	492.714	28.5943
24	5600	1	446.000	0.0000	75	10700	18	485.667	31.4148
25	5700	3	387.000	18.4932	76	10800	10	491.500	24.9048
26	5800	3	400.000	24.4949	77	10900	16	497.813	29.1359
27	5900	4	387.500	30.0208	78	11000	23	493.087	20.3254
28	6000	3	391.333	23.8374	79	11100	13	487.000	21.3181
29	6100	3	405.333	27.5358	80	11200	13	493.538	28.6104
30	6200	7	405.857	22.1387	81	11300	11	511.909	28.6117
31	6300	4	412.500	28.2710	82	11400	9	510.444	34.0820
32	6400	7	417.714	18.3203	83	11500	18	512.611	30.7971
33	6500	6	392.000	24.8395	84	11600	19	518.632	33.5311
34	6600	4	415.750	28.9860	85	11700	11	502.091	18.3226
35	6700	4	401.250	19.6007	86	11800	18	530.333	28.2686
36	6800	7	432.857	22.0287	87	11900	21	525.571	29.5515
37	6900	5	418.600	30.4998	88	12000	13	526.923	27.0284
38	7000	2	404.500	7.5000	89	12100	15	557.400	17.8878
39	7100	11	421.000	24.0568	90	12200	15	535.267	29.5916
40	7200	4	425.000	11.6833	91	12300	15	542.133	20.6619
41	7300	9	419.222	33.9044	92	12400	12	540.500	28.5409
42	7400	8	423.000	26.3439	93	12500	16	546.187	31.6726
43	7500	9	425.111	31.2888	94	12600	12	555.250	28.7232
44	7600	5	413.400	29.4455	95	12700	14	563.857	21.8791
45	7700	9	408.778	24.8765	96	12800	14	560.714	27.6701
46	7800	12	420.250	25.9362	97	12900	17	567.059	30.0362
47	7900	10	426.600	24.7919	98	13000	13	560.615	29.2668
48	8000	10	428.600	25.7301	99	13100	13	582.615	29.8832
49	8100	7	410.286	27.2014	100	13200	14	584.857	29.1471
50	8200	11	438.909	32.9088	101	13300	6	590.167	31.8194
51	8300	9	422.444	25.9534	102	13400	11	594.818	22.6387
52	8400	10	450.800	25.2974	103	13500	13	586.538	32.5544
53	8500	12	428.083	25.7082	104	13600	13	594.385	29.4867
54	8600	7	429.286	27.4628	105	13700	10	611.900	26.8903
55	8700	22	433.318	29.2177	106	13800	10	603.900	30.0248
56	8800	13	439.923	29.4134	107	13900	9	614.667	16.9247
57	8900	9	443.778	32.4817	108	14000	16	621.187	22.4450
58	9000	10	450.300	20.2092	109	14100	17	629.706	26.1304
59	9100	6	437.833	24.0792	110	14200	11	618.091	27.9820
60	9200	17	445.059	20.1742	111	14300	11	638.909	26.9796
61	9300	15	468.467	24.0551	112	14400	7	652.571	17.0198
62	9400	11	451.545	31.2100	113	14500	7	649.286	22.2371
63	9500	13	462.923	24.2121	114	14600	8	633.625	13.9458
64	9600	11	452.455	24.3511	115	14700	6	671.333	13.9483
65	9700	14	459.429	20.3355	116	14800	9	656.778	25.8963
66	9800	16	465.812	23.3217	117	14900	13	657.692	27.3703
67	9900	14	453.571	22.2477	118	15000	11	671.364	30.9876
68	10000	18	464.222	30.5024	119	15100	10	674.200	24.1073
69	10100	12	473.667	26.2467	120	15200	2	690.500	22.5000

T A B L E D - 2
Car2 Data (Continued)

subpop	miles	ycount	ymean	ystdevn		subpop	miles	ycount	ymean	ystdevn
121	15300	9	675.111	25.2034		137	16900	3	753.000	6.4807
122	15400	10	696.300	26.7060		138	17000	3	794.333	28.4292
123	15500	7	696.143	29.5317		139	17100	3	790.000	25.0200
124	15600	9	707.333	27.2356		140	17200	3	796.667	18.6250
125	15700	3	695.667	18.6250		141	17300	3	826.333	12.7105
126	15800	5	713.400	23.8462		142	17400	4	811.250	37.5791
127	15900	8	721.375	28.0532		143	17500	2	788.000	15.0000
128	16000	5	707.200	38.0126		144	17600	2	830.000	24.0000
129	16100	2	768.000	3.0000		145	17700	1	858.000	0.0000
130	16200	8	735.375	16.4236		146	17800	1	807.000	0.0000
131	16300	2	742.000	15.0000		147	17900	2	844.000	35.0000
132	16400	3	752.333	15.3695		148	18000	2	851.000	45.0000
133	16500	1	796.000	0.0000		149	18100	1	809.000	0.0000
134	16600	2	788.500	20.5000		150	18200	1	861.000	0.0000
135	16700	6	752.167	26.0027		151	18500	2	891.500	33.5000
136	16800	3	768.333	35.9660						

T A B L E D - 3
Grades Data

student	hours	score	student	hours	score	student	hours	score	student	hours	score	student	hours	score
1	0	46	27	0	47	53	0	42	79	0	42	105	0	44
2	0	48	28	0	47	54	0	52	80	0	44	106	0	46
3	0	42	29	0	42	55	0	47	81	0	42	107	0	41
4	0	44	30	0	49	56	0	45	82	0	48	108	0	48
5	0	42	31	0	46	57	0	47	83	0	43	109	0	45
6	0	39	32	0	41	58	0	46	84	0	47	110	0	46
7	0	49	33	0	45	59	0	46	85	0	48	111	0	46
8	0	48	34	0	43	60	0	49	86	0	48	112	0	46
9	0	48	35	0	48	61	0	43	87	0	44	113	0	45
10	0	46	36	0	51	62	0	42	88	0	44	114	0	43
11	0	44	37	0	47	63	0	45	89	0	43	115	0	41
12	0	42	38	0	42	64	0	42	90	0	47	116	0	50
13	0	50	39	0	44	65	0	40	91	0	45	117	0	47
14	0	46	40	0	41	66	0	46	92	0	46	118	0	44
15	0	49	41	0	44	67	0	48	93	0	38	119	0	45
16	0	47	42	0	51	68	0	50	94	0	46	120	0	41
17	0	43	43	0	46	69	0	39	95	0	41	121	0	44
18	0	45	44	0	45	70	0	46	96	0	47	122	0	42
19	0	40	45	0	42	71	0	49	97	0	43	123	0	47
20	0	45	46	0	47	72	0	48	98	0	42	124	0	45
21	0	43	47	0	43	73	0	45	99	0	44	125	0	51
22	0	48	48	0	49	74	0	49	100	0	48	126	0	49
23	0	47	49	0	45	75	0	46	101	0	44	127	0	44
24	0	45	50	0	42	76	0	42	102	0	41	128	0	42
25	0	44	51	0	44	77	0	43	103	0	48	129	0	43
26	0	41	52	0	47	78	0	47	104	0	48	130	0	39

T A B L E D - 3
Grades Data (Continued)

student	hours	score	student	hours	score	student	hours	score	student	hours	score	student	hours	score
131	0	43	182	0	41	233	1	47	284	1	48	335	1	42
132	0	44	183	0	45	234	1	45	285	1	46	336	1	53
133	0	45	184	0	43	235	1	53	286	1	54	337	1	49
134	0	42	185	0	41	236	1	45	287	1	53	338	1	48
135	0	43	186	0	47	237	1	48	288	1	51	339	1	49
136	0	42	187	0	44	238	1	52	289	1	51	340	1	47
137	0	47	188	0	45	239	1	51	290	1	51	341	1	51
138	0	46	189	0	45	240	1	54	291	1	41	342	1	49
139	0	47	190	0	47	241	1	50	292	1	52	343	1	44
140	0	50	191	0	45	242	1	48	293	1	43	344	1	52
141	0	37	192	0	45	243	1	50	294	1	48	345	1	49
142	0	41	193	0	44	244	1	47	295	1	47	346	1	50
143	0	45	194	0	39	245	1	49	296	1	49	347	1	48
144	0	42	195	0	47	246	1	51	297	1	49	348	1	44
145	0	45	196	0	40	247	1	47	298	1	52	349	1	50
146	0	44	197	0	45	248	1	55	299	1	48	350	1	46
147	0	44	198	0	46	249	1	50	300	1	51	351	1	54
148	0	42	199	0	48	250	1	45	301	1	51	352	1	52
149	0	45	200	0	45	251	1	48	302	1	47	353	1	51
150	0	46	201	1	52	252	1	50	303	1	50	354	1	47
151	0	46	202	1	46	253	1	52	304	1	46	355	1	46
152	0	42	203	1	46	254	1	50	305	1	53	356	1	48
153	0	51	204	1	53	255	1	52	306	1	48	357	1	53
154	0	50	205	1	53	256	1	49	307	1	47	358	1	50
155	0	41	206	1	45	257	1	50	308	1	46	359	1	49
156	0	41	207	1	49	258	1	51	309	1	46	360	1	44
157	0	49	208	1	50	259	1	46	310	1	51	361	1	53
158	0	43	209	1	48	260	1	45	311	1	51	362	1	48
159	0	46	210	1	49	261	1	51	312	1	47	363	1	43
160	0	47	211	1	47	262	1	50	313	1	49	364	1	43
161	0	46	212	1	52	263	1	49	314	1	46	365	1	47
162	0	44	213	1	47	264	1	46	315	1	49	366	1	50
163	0	47	214	1	53	265	1	49	316	1	51	367	1	48
164	0	47	215	1	51	266	1	49	317	1	53	368	1	43
165	0	44	216	1	50	267	1	54	318	1	48	369	1	54
166	0	49	217	1	52	268	1	45	319	1	49	370	1	55
167	0	43	218	1	48	269	1	56	320	1	49	371	1	55
168	0	43	219	1	51	270	1	45	321	1	50	372	1	49
169	0	46	220	1	46	271	1	50	322	1	52	373	1	45
170	0	46	221	1	47	272	1	52	323	1	50	374	1	46
171	0	48	222	1	50	273	1	51	324	1	52	375	1	51
172	0	49	223	1	49	274	1	46	325	1	45	376	1	45
173	0	49	224	1	55	275	1	46	326	1	50	377	1	49
174	0	46	225	1	46	276	1	48	327	1	49	378	1	50
175	0	43	226	1	47	277	1	45	328	1	49	379	1	51
176	0	52	227	1	49	278	1	48	329	1	49	380	1	51
177	0	47	228	1	47	279	1	48	330	1	52	381	1	50
179	0	45	229	1	46	280	1	53	331	1	46	382	1	50
179	0	45	230	1	46	281	1	45	332	1	52	383	1	52
180	0	42	231	1	46	282	1	48	333	1	51	384	1	56
181	0	43	232	1	49	283	1	47	334	1	49	385	1	47

T A B L E   **D - 3**

Grades Data (Continued)

student	hours	score	student	hours	score	student	hours	score	student	hours	score	student	hours	score
386	1	50	437	2	54	488	2	53	539	2	52	590	2	53
387	1	48	438	2	56	489	2	53	540	2	52	591	2	52
388	1	46	439	2	53	490	2	53	541	2	52	592	2	51
389	1	48	440	2	49	491	2	54	542	2	49	593	2	57
390	1	50	441	2	54	492	2	51	543	2	56	594	2	54
391	1	48	442	2	56	493	2	46	544	2	56	595	2	49
392	1	51	443	2	58	494	2	55	545	2	54	596	2	50
393	1	50	444	2	51	495	2	53	546	2	53	597	2	56
394	1	53	445	2	52	496	2	57	547	2	58	598	2	54
395	1	45	446	2	54	497	2	53	548	2	54	599	2	52
396	1	47	447	2	50	498	2	47	549	2	49	600	2	51
397	1	51	448	2	48	499	2	50	550	2	48	601	3	59
398	1	46	449	2	53	500	2	53	551	2	55	602	3	63
399	1	47	450	2	55	501	2	55	552	2	53	603	3	64
400	1	51	451	2	55	502	2	50	553	2	57	604	3	59
401	2	49	452	2	53	503	2	56	554	2	55	605	3	54
402	2	53	453	2	54	504	2	56	555	2	57	606	3	60
403	2	50	454	2	53	505	2	53	556	2	54	607	3	57
404	2	55	455	2	57	506	2	50	557	2	51	608	3	51
405	2	49	456	2	52	507	2	52	558	2	51	609	3	58
406	2	51	457	2	50	508	2	52	559	2	52	610	3	56
407	2	57	458	2	51	509	2	51	560	2	60	611	3	58
408	2	54	459	2	53	510	2	60	561	2	55	612	3	54
409	2	49	460	2	59	511	2	56	562	2	53	613	3	58
410	2	54	461	2	56	512	2	57	563	2	50	614	3	57
411	2	50	462	2	54	513	2	50	564	2	53	615	3	62
412	2	52	463	2	51	514	2	54	565	2	53	616	3	58
413	2	50	464	2	58	515	2	54	566	2	55	617	3	60
414	2	50	465	2	52	516	2	53	567	2	54	618	3	53
415	2	53	466	2	50	517	2	48	568	2	51	619	3	54
416	2	49	467	2	51	518	2	55	569	2	49	620	3	59
417	2	53	468	2	52	519	2	50	570	2	55	621	3	57
418	2	56	469	2	55	520	2	55	571	2	50	622	3	53
419	2	52	470	2	55	521	2	47	572	2	50	623	3	61
420	2	54	471	2	59	522	2	55	573	2	54	624	3	60
421	2	53	472	2	58	523	2	52	574	2	49	625	3	58
422	2	50	473	2	51	524	2	50	575	2	54	626	3	58
423	2	54	474	2	57	525	2	55	576	2	52	627	3	60
424	2	50	475	2	51	526	2	59	577	2	57	628	3	57
425	2	51	476	2	50	527	2	56	578	2	51	629	3	55
426	2	57	477	2	57	528	2	56	579	2	52	630	3	53
427	2	54	478	2	54	529	2	57	580	2	52	631	3	54
428	2	52	479	2	52	530	2	55	581	2	54	632	3	57
429	2	49	480	2	53	531	2	47	582	2	52	633	3	52
430	2	56	481	2	59	532	2	53	583	2	54	634	3	58
431	2	55	482	2	51	533	2	55	584	2	54	635	3	57
432	2	55	483	2	53	534	2	50	585	2	55	636	3	60
433	2	45	484	2	55	535	2	51	586	2	56	637	3	55
434	2	47	485	2	53	536	2	58	587	2	49	638	3	54
435	2	51	486	2	50	537	2	56	588	2	52	639	3	54
436	2	49	487	2	55	538	2	56	589	2	55	640	3	55

T A B L E D - 3
Grades Data (Continued)

student	hours	score	student	hours	score	student	hours	score	student	hours	score	student	hours	score
641	3	56	692	3	54	743	3	56	794	3	57	845	4	61
642	3	53	693	3	63	744	3	57	795	3	54	846	4	61
643	3	55	694	3	53	745	3	60	796	3	54	847	4	64
644	3	58	695	3	53	746	3	63	797	3	55	848	4	63
645	3	55	696	3	56	747	3	55	798	3	60	849	4	60
646	3	56	697	3	59	748	3	59	799	3	57	850	4	58
647	3	58	698	3	61	749	3	58	800	3	60	851	4	59
648	3	57	699	3	55	750	3	62	801	4	63	852	4	62
649	3	62	700	3	55	751	3	62	802	4	60	853	4	58
650	3	56	701	3	56	752	3	55	803	4	63	854	4	61
651	3	61	702	3	60	753	3	58	804	4	62	855	4	59
652	3	59	703	3	57	754	3	54	805	4	59	856	4	62
653	3	55	704	3	57	755	3	59	806	4	58	857	4	58
654	3	63	705	3	58	756	3	61	807	4	61	858	4	61
655	3	56	706	3	58	757	3	55	808	4	61	859	4	61
656	3	60	707	3	58	758	3	61	809	4	62	860	4	63
657	3	56	708	3	57	759	3	53	810	4	61	861	4	62
658	3	61	709	3	58	760	3	55	811	4	62	862	4	66
659	3	53	710	3	61	761	3	51	812	4	63	863	4	67
660	3	56	711	3	61	762	3	56	813	4	59	864	4	63
661	3	54	712	3	55	763	3	59	814	4	58	865	4	58
662	3	61	713	3	52	764	3	57	815	4	60	866	4	65
663	3	57	714	3	56	765	3	60	816	4	59	867	4	63
664	3	53	715	3	59	766	3	51	817	4	67	868	4	56
665	3	64	716	3	54	767	3	60	818	4	58	869	4	57
666	3	57	717	3	57	768	3	61	819	4	62	870	4	60
667	3	58	718	3	54	769	3	60	820	4	63	871	4	65
668	3	54	719	3	56	770	3	55	821	4	62	872	4	55
669	3	59	720	3	55	771	3	61	822	4	65	873	4	63
670	3	58	721	3	55	772	3	52	823	4	62	874	4	57
671	3	54	722	3	57	773	3	56	824	4	64	875	4	64
672	3	57	723	3	58	774	3	58	825	4	62	876	4	60
673	3	56	724	3	56	775	3	59	826	4	57	877	4	63
674	3	61	725	3	59	776	3	54	827	4	63	878	4	63
675	3	53	726	3	57	777	3	56	828	4	60	879	4	65
676	3	58	727	3	54	778	3	59	829	4	64	880	4	61
677	3	56	728	3	58	779	3	60	830	4	67	881	4	58
678	3	59	729	3	56	780	3	57	831	4	59	882	4	63
679	3	59	730	3	62	781	3	58	832	4	61	883	4	58
680	3	56	731	3	57	782	3	59	833	4	63	884	4	61
681	3	51	732	3	59	783	3	57	834	4	64	885	4	59
682	3	58	733	3	59	784	3	60	835	4	65	886	4	61
683	3	58	734	3	59	785	3	59	836	4	61	887	4	58
684	3	54	735	3	56	786	3	50	837	4	61	888	4	60
685	3	56	736	3	55	787	3	55	838	4	58	889	4	59
686	3	54	737	3	60	788	3	53	839	4	64	890	4	60
687	3	57	738	3	59	789	3	53	840	4	64	891	4	65
688	3	59	739	3	54	790	3	53	841	4	61	892	4	64
689	3	54	740	3	54	791	3	57	842	4	60	893	4	58
690	3	49	741	3	58	792	3	57	843	4	64	894	4	60
691	3	59	742	3	57	793	3	59	844	4	58	895	4	59

T A B L E D - 3
Grades Data (Continued)

student	hours	score	student	hours	score	student	hours	score	student	hours	score	student	hours	score
896	4	63	947	4	62	998	4	61	1049	5	62	1100	5	65
897	4	62	948	4	59	999	4	57	1050	5	61	1101	5	68
898	4	63	949	4	60	1000	4	60	1051	5	66	1102	5	65
899	4	63	950	4	65	1001	5	59	1052	5	68	1103	5	68
900	4	64	951	4	64	1002	5	66	1053	5	62	1104	5	68
901	4	56	952	4	62	1003	5	67	1054	5	68	1105	5	62
902	4	60	953	4	64	1004	5	69	1055	5	68	1106	5	62
903	4	65	954	4	62	1005	5	67	1056	5	67	1107	5	66
904	4	60	955	4	61	1006	5	70	1057	5	63	1108	5	65
905	4	63	956	4	57	1007	5	66	1058	5	69	1109	5	63
906	4	54	957	4	65	1008	5	65	1059	5	71	1110	5	65
907	4	59	958	4	64	1009	5	69	1060	5	65	1111	5	62
908	4	61	959	4	60	1010	5	62	1061	5	69	1112	5	65
909	4	63	960	4	61	1011	5	63	1062	5	70	1113	5	69
910	4	58	961	4	62	1012	5	65	1063	5	64	1114	5	65
911	4	60	962	4	62	1013	5	65	1064	5	65	1115	5	69
912	4	57	963	4	57	1014	5	67	1065	5	64	1116	5	64
913	4	65	964	4	58	1015	5	68	1066	5	60	1117	5	63
914	4	61	965	4	59	1016	5	66	1067	5	64	1118	5	67
915	4	59	966	4	58	1017	5	65	1068	5	64	1119	5	63
916	4	62	967	4	61	1018	5	61	1069	5	62	1120	5	66
917	4	64	968	4	55	1019	5	64	1070	5	62	1121	5	67
918	4	64	969	4	65	1020	5	61	1071	5	66	1122	5	64
919	4	57	970	4	62	1021	5	63	1072	5	60	1123	5	65
920	4	57	971	4	68	1022	5	63	1073	5	64	1124	5	63
921	4	59	972	4	61	1023	5	68	1074	5	62	1125	5	71
922	4	62	973	4	63	1024	5	69	1075	5	67	1126	5	66
923	4	63	974	4	60	1025	5	57	1076	5	67	1127	5	67
924	4	62	975	4	58	1026	5	71	1077	5	64	1128	5	64
925	4	57	976	4	61	1027	5	65	1078	5	67	1129	5	64
926	4	63	977	4	59	1028	5	70	1079	5	65	1130	5	64
927	4	62	978	4	60	1029	5	62	1080	5	66	1131	5	64
928	4	53	979	4	55	1030	5	68	1081	5	66	1132	5	66
929	4	61	980	4	59	1031	5	64	1082	5	65	1133	5	69
930	4	57	981	4	57	1032	5	63	1083	5	62	1134	5	68
931	4	63	982	4	58	1033	5	67	1084	5	63	1135	5	64
932	4	62	983	4	59	1034	5	63	1085	5	72	1136	5	61
933	4	61	984	4	67	1035	5	66	1086	5	66	1137	5	67
934	4	66	985	4	61	1036	5	62	1087	5	66	1138	5	61
935	4	57	986	4	60	1037	5	68	1088	5	65	1139	5	61
936	4	62	987	4	63	1038	5	61	1089	5	66	1140	5	58
937	4	65	988	4	56	1039	5	66	1090	5	64	1141	5	66
938	4	61	989	4	62	1040	5	66	1091	5	63	1142	5	65
939	4	64	990	4	59	1041	5	69	1092	5	68	1143	5	61
940	4	55	991	4	68	1042	5	66	1093	5	65	1144	5	65
941	4	58	992	4	60	1043	5	59	1094	5	66	1145	5	65
942	4	60	993	4	66	1044	5	63	1095	5	66	1146	5	64
943	4	58	994	4	66	1045	5	67	1096	5	62	1147	5	67
944	4	58	995	4	62	1046	5	67	1097	5	63	1148	5	68
945	4	59	996	4	60	1047	5	65	1098	5	67	1149	5	59
946	4	58	997	4	66	1048	5	70	1099	5	64	1150	5	66

T A B L E D - 3
Grades Data (Continued)

student	hours	score	student	hours	score	student	hours	score	student	hours	score	student	hours	score
1151	5	60	1202	6	66	1253	6	70	1304	6	69	1355	6	70
1152	5	67	1203	6	65	1254	6	67	1305	6	65	1356	6	75
1153	5	67	1204	6	69	1255	6	71	1306	6	62	1357	6	72
1154	5	62	1205	6	71	1256	6	64	1307	6	73	1358	6	61
1155	5	62	1206	6	64	1257	6	70	1308	6	66	1359	6	73
1156	5	65	1207	6	74	1258	6	65	1309	6	73	1360	6	69
1157	5	62	1208	6	69	1259	6	63	1310	6	70	1361	6	66
1158	5	70	1209	6	68	1260	6	69	1311	6	74	1362	6	68
1159	5	65	1210	6	73	1261	6	73	1312	6	69	1363	6	66
1160	5	64	1211	6	71	1262	6	66	1313	6	65	1364	6	70
1161	5	62	1212	6	67	1263	6	67	1314	6	70	1365	6	66
1162	5	67	1213	6	70	1264	6	67	1315	6	71	1366	6	74
1163	5	72	1214	6	71	1265	6	69	1316	6	69	1367	6	66
1164	5	61	1215	6	75	1266	6	69	1317	6	73	1368	6	63
1165	5	63	1216	6	73	1267	6	73	1318	6	70	1369	6	70
1166	5	66	1217	6	67	1268	6	72	1319	6	68	1370	6	75
1167	5	66	1218	6	66	1269	6	65	1320	6	68	1371	6	68
1168	5	63	1219	6	68	1270	6	68	1321	6	67	1372	6	68
1169	5	63	1220	6	66	1271	6	71	1322	6	65	1373	6	70
1170	5	63	1221	6	69	1272	6	72	1323	6	67	1374	6	76
1171	5	62	1222	6	69	1273	6	66	1324	6	65	1375	6	71
1172	5	69	1223	6	74	1274	6	71	1325	6	73	1376	6	66
1173	5	62	1224	6	71	1275	6	72	1326	6	71	1377	6	69
1174	5	67	1225	6	63	1276	6	71	1327	6	70	1378	6	72
1175	5	67	1226	6	71	1277	6	70	1328	6	74	1379	6	72
1176	5	68	1227	6	72	1278	6	64	1329	6	65	1380	6	66
1177	5	67	1228	6	69	1279	6	65	1330	6	71	1381	6	65
1178	5	64	1229	6	70	1280	6	70	1331	6	71	1382	6	67
1179	5	61	1230	6	73	1281	6	69	1332	6	68	1383	6	70
1180	5	67	1231	6	71	1282	6	72	1333	6	67	1384	6	72
1181	5	62	1232	6	67	1283	6	69	1334	6	72	1385	6	66
1182	5	65	1233	6	66	1284	6	68	1335	6	67	1386	6	71
1183	5	61	1234	6	67	1285	6	72	1336	6	70	1387	6	72
1184	5	61	1235	6	71	1286	6	70	1337	6	66	1388	6	69
1185	5	68	1236	6	71	1287	6	72	1338	6	68	1389	6	65
1186	5	71	1237	6	66	1288	6	69	1339	6	65	1390	6	69
1187	5	64	1238	6	67	1289	6	63	1340	6	70	1391	6	68
1188	5	62	1239	6	69	1290	6	70	1341	6	70	1392	6	69
1189	5	69	1240	6	68	1291	6	68	1342	6	72	1393	6	73
1190	5	59	1241	6	67	1292	6	66	1343	6	67	1394	6	73
1191	5	65	1242	6	69	1293	6	68	1344	6	68	1395	6	68
1192	5	62	1243	6	69	1294	6	69	1345	6	66	1396	6	71
1193	5	63	1244	6	68	1295	6	76	1346	6	69	1397	6	71
1194	5	64	1245	6	70	1296	6	67	1347	6	68	1398	6	68
1195	5	65	1246	6	72	1297	6	66	1348	6	70	1399	6	69
1196	5	68	1247	6	66	1298	6	65	1349	6	69	1400	6	75
1197	5	61	1248	6	66	1299	6	72	1350	6	71	1401	7	76
1198	5	67	1249	6	71	1300	6	68	1351	6	70	1402	7	80
1199	5	69	1250	6	67	1301	6	71	1352	6	71	1403	7	72
1200	5	66	1251	6	67	1302	6	66	1353	6	69	1404	7	77
1201	6	68	1252	6	67	1303	6	70	1354	6	70	1405	7	69

T A B L E D - 3
Grades Data (Continued)

student	hours	score	student	hours	score	student	hours	score	student	hours	score	student	hours	score
1406	7	66	1457	7	74	1508	7	77	1559	7	71	1610	8	79
1407	7	70	1458	7	74	1509	7	67	1560	7	69	1611	8	83
1408	7	72	1459	7	73	1510	7	75	1561	7	70	1612	8	83
1409	7	74	1460	7	75	1511	7	76	1562	7	74	1613	8	73
1410	7	76	1461	7	72	1512	7	77	1563	7	71	1614	8	78
1411	7	75	1462	7	72	1513	7	70	1564	7	73	1615	8	77
1412	7	75	1463	7	75	1514	7	74	1565	7	73	1616	8	78
1413	7	75	1464	7	74	1515	7	67	1566	7	71	1617	8	77
1414	7	75	1465	7	69	1516	7	75	1567	7	69	1618	8	79
1415	7	67	1466	7	69	1517	7	70	1568	7	76	1619	8	78
1416	7	75	1467	7	76	1518	7	75	1569	7	71	1620	8	80
1417	7	73	1468	7	73	1519	7	70	1570	7	73	1621	8	77
1418	7	73	1469	7	77	1520	7	74	1571	7	72	1622	8	76·
1419	7	76	1470	7	71	1521	7	74	1572	7	75	1623	8	76
1420	7	74	1471	7	78	1522	7	76	1573	7	76	1624	8	79
1421	7	74	1472	7	71	1523	7	79	1574	7	71	1625	8	71
1422	7	72	1473	7	69	1524	7	76	1575	7	75	1626	8	81
1423	7	70	1474	7	70	1525	7	71	1576	7	69	1627	8	78
1424	7	73	1475	7	75	1526	7	74	1577	7	74	1628	8	77
1425	7	73	1476	7	72	1527	7	73	1578	7	75	1629	8	77
1426	7	71	1477	7	71	1528	7	71	1579	7	72	1630	8	73
1427	7	74	1478	7	70	1529	7	73	1580	7	74	1631	8	74
1428	7	78	1479	7	77	1530	7	74	1581	7	70	1632	8	77
1429	7	71	1480	7	75	1531	7	79	1582	7	74	1633	8	73
1430	7	74	1481	7	76	1532	7	78	1583	7	71	1634	8	76
1431	7	70	1482	7	73	1533	7	72	1584	7	71	1635	8	81
1432	7	69	1483	7	76	1534	7	79	1585	7	73	1636	8	74
1433	7	65	1484	7	74	1535	7	72	1586	7	77	1637	8	77
1434	7	69	1485	7	72	1536	7	73	1587	7	73	1638	8	77
1435	7	69	1486	7	73	1537	7	74	1588	7	78	1639	8	76
1436	7	75	1487	7	72	1538	7	75	1589	7	76	1640	8	77
1437	7	73	1488	7	75	1539	7	73	1590	7	75	1641	8	77
1438	7	71	1489	7	71	1540	7	75	1591	7	75	1642	8	75
1439	7	73	1490	7	77	1541	7	72	1592	7	70	1643	8	82
1440	7	70	1491	7	74	1542	7	69	1593	7	72	1644	8	79
1441	7	68	1492	7	70	1543	7	74	1594	7	72	1645	8	77
1442	7	74	1493	7	78	1544	7	72	1595	7	73	1646	8	75
1443	7	72	1494	7	73	1545	7	70	1596	7	70	1647	8	76
1444	7	76	1495	7	68	1546	7	73	1597	7	77	1648	8	76
1445	7	73	1496	7	76	1547	7	72	1598	7	67	1649	8	79
1446	7	74	1497	7	73	1548	7	71	1599	7	68	1650	8	80
1447	7	73	1498	7	70	1549	7	77	1600	7	70	1651	8	81
1448	7	70	1499	7	70	1550	7	72	1601	8	75	1652	8	75
1449	7	76	1500	7	73	1551	7	71	1602	8	76	1653	8	79
1450	7	72	1501	7	70	1552	7	70	1603	8	78	1654	8	75
1451	7	75	1502	7	72	1553	7	72	1604	8	80	1655	8	77
1452	7	74	1503	7	77	1554	7	74	1605	8	79	1656	8	79
1453	7	79	1504	7	75	1555	7	71	1606	8	70	1657	8	84
1454	7	69	1505	7	76	1556	7	77	1607	8	77	1658	8	78
1455	7	77	1506	7	73	1557	7	80	1608	8	77	1659	8	79
1456	7	69	1507	7	75	1558	7	70	1609	8	80	1660	8	76

T A B L E D-3
Grades Data (Continued)

student	hours	score	student	hours	score	student	hours	score	student	hours	score	student	hours	score
1661	8	74	1712	8	72	1763	8	71	1814	9	84	1865	9	80
1662	8	78	1713	8	80	1764	8	78	1815	9	86	1866	9	82
1663	8	74	1714	8	79	1765	8	81	1816	9	83	1867	9	81
1664	8	79	1715	8	80	1766	8	81	1817	9	81	1868	9	84
1665	8	74	1716	8	79	1767	8	77	1818	9	79	1869	9	85
1666	8	74	1717	8	75	1768	8	78	1819	9	81	1870	9	77
1667	8	81	1718	8	76	1769	8	78	1820	9	77	1871	9	83
1668	8	75	1719	8	80	1770	8	71	1821	9	87	1872	9	81
1669	8	80	1720	8	78	1771	8	72	1822	9	79	1873	9	82
1670	8	71	1721	8	78	1772	8	77	1823	9	78	1874	9	88
1671	8	80	1722	8	78	1773	8	77	1824	9	78	1875	9	82
1672	8	74	1723	8	75	1774	8	75	1825	9	81	1876	9	75
1673	8	78	1724	8	80	1775	8	76	1826	9	82	1877	9	84
1674	8	74	1725	8	77	1776	8	79	1827	9	81	1878	9	83
1675	8	74	1726	8	84	1777	8	78	1828	9	80	1879	9	82
1676	8	81	1727	8	78	1778	8	76	1829	9	83	1880	9	77
1677	8	73	1728	8	75	1779	8	74	1830	9	77	1881	9	84
1678	8	79	1729	8	77	1780	8	73	1831	9	83	1882	9	78
1679	8	79	1730	8	76	1781	8	81	1832	9	80	1883	9	78
1680	8	78	1731	8	73	1782	8	73	1833	9	78	1884	9	81
1681	8	74	1732	8	77	1783	8	77	1834	9	83	1885	9	82
1682	8	73	1733	8	76	1784	8	80	1835	9	80	1886	9	84
1683	8	79	1734	8	79	1785	8	73	1836	9	87	1887	9	82
1684	8	79	1735	8	78	1786	8	83	1837	9	84	1888	9	78
1685	8	77	1736	8	73	1787	8	81	1838	9	87	1889	9	85
1686	8	82	1737	8	75	1788	8	74	1839	9	84	1890	9	82
1687	8	82	1738	8	76	1789	8	74	1840	9	79	1891	9	78
1688	8	79	1739	8	80	1790	8	76	1841	9	78	1892	9	77
1689	8	69	1740	8	76	1791	8	75	1842	9	81	1893	9	79
1690	8	74	1741	8	74	1792	8	77	1843	9	80	1894	9	77
1691	8	75	1742	8	74	1793	8	79	1844	9	78	1895	9	85
1692	8	75	1743	8	81	1794	8	75	1845	9	74	1896	9	80
1693	8	77	1744	8	75	1795	8	78	1846	9	85	1897	9	81
1694	8	81	1745	8	72	1796	8	80	1847	9	79	1898	9	82
1695	8	74	1746	8	74	1797	8	82	1848	9	79	1899	9	78
1696	8	78	1747	8	83	1798	8	74	1849	9	80	1900	9	81
1697	8	79	1748	8	77	1799	8	81	1850	9	82	1901	9	80
1698	8	76	1749	8	75	1800	8	79	1851	9	81	1902	9	83
1699	8	76	1750	8	76	1801	9	79	1852	9	85	1903	9	82
1700	8	78	1751	8	74	1802	9	86	1853	9	82	1904	9	81
1701	8	75	1752	8	76	1803	9	83	1854	9	79	1905	9	82
1702	8	78	1753	8	80	1804	9	83	1855	9	79	1906	9	76
1703	8	77	1754	8	73	1805	9	78	1856	9	78	1907	9	81
1704	8	78	1755	8	77	1806	9	84	1857	9	82	1908	9	78
1705	8	75	1756	8	73	1807	9	79	1858	9	86	1909	9	75
1706	8	80	1757	8	79	1808	9	83	1859	9	75	1910	9	77
1707	8	80	1758	8	82	1809	9	82	1860	9	87	1911	9	83
1708	8	78	1759	8	76	1810	9	79	1861	9	84	1912	9	85
1709	8	73	1760	8	74	1811	9	82	1862	9	85	1913	9	82
1710	8	79	1761	8	78	1812	9	84	1863	9	80	1914	9	80
1711	8	76	1762	8	74	1813	9	83	1864	9	82	1915	9	79

T A B L E D · 3
Grades Data (Continued)

student	hours	score	student	hours	score	student	hours	score	student	hours	score	student	hours	score
1916	9	82	1967	9	78	2018	10	84	2069	10	86	2120	10	88
1917	9	82	1968	9	78	2019	10	89	2070	10	79	2121	10	86
1918	9	83	1969	9	80	2020	10	86	2071	10	86	2122	10	85
1919	9	85	1970	9	80	2021	10	83	2072	10	88	2123	10	80
1920	9	83	1971	9	79	2022	10	86	2073	10	81	2124	10	90
1921	9	80	1972	9	76	2023	10	91	2074	10	84	2125	10	84
1922	9	79	1973	9	81	2024	10	87	2075	10	87	2126	10	87
1923	9	81	1974	9	80	2025	10	89	2076	10	82	2127	10	85
1924	9	80	1975	9	80	2026	10	83	2077	10	81	2128	10	87
1925	9	83	1976	9	86	2027	10	82	2078	10	85	2129	10	83
1926	9	78	1977	9	85	2028	10	83	2079	10	87	2130	10	85
1927	9	80	1978	9	83	2029	10	84	2080	10	86	2131	10	86
1928	9	78	1979	9	78	2030	10	88	2081	10	85	2132	10	90
1929	9	78	1980	9	84	2031	10	84	2082	10	82	2133	10	89
1930	9	75	1981	9	79	2032	10	87	2083	10	83	2134	10	85
1931	9	81	1982	9	81	2033	10	84	2084	10	82	2135	10	82
1932	9	88	1983	9	82	2034	10	87	2085	10	87	2136	10	84
1933	9	83	1984	9	82	2035	10	87	2086	10	86	2137	10	91
1934	9	77	1985	9	82	2036	10	84	2087	10	85	2138	10	86
1935	9	81	1986	9	77	2037	10	81	2088	10	89	2139	10	82
1936	9	79	1987	9	77	2038	10	88	2089	10	84	2140	10	86
1937	9	82	1988	9	79	2039	10	84	2090	10	79	2141	10	84
1938	9	85	1989	9	83	2040	10	83	2091	10	81	2142	10	88
1939	9	80	1990	9	81	2041	10	89	2092	10	85	2143	10	87
1940	9	83	1991	9	81	2042	10	85	2093	10	88	2144	10	92
1941	9	83	1992	9	81	2043	10	88	2094	10	82	2145	10	89
1942	9	77	1993	9	78	2044	10	88	2095	10	85	2146	10	81
1943	9	80	1994	9	84	2045	10	90	2096	10	86	2147	10	91
1944	9	81	1995	9	85	2046	10	85	2097	10	87	2148	10	85
1945	9	83	1996	9	81	2047	10	84	2098	10	83	2149	10	85
1946	9	82	1997	9	79	2048	10	81	2099	10	82	2150	10	89
1947	9	80	1998	9	81	2049	10	86	2100	10	83	2151	10	85
1948	9	84	1999	9	84	2050	10	82	2101	10	84	2152	10	84
1949	9	83	2000	9	81	2051	10	80	2102	10	86	2153	10	84
1950	9	77	2001	10	85	2052	10	84	2103	10	77	2154	10	87
1951	9	86	2002	10	79	2053	10	86	2104	10	85	2155	10	81
1952	9	85	2003	10	81	2054	10	85	2105	10	87	2156	10	82
1953	9	81	2004	10	88	2055	10	82	2106	10	86	2157	10	82
1954	9	80	2005	10	81	2056	10	85	2107	10	88	2158	10	84
1955	9	81	2006	10	81	2057	10	84	2108	10	82	2159	10	88
1956	9	78	2007	10	87	2058	10	87	2109	10	81	2160	10	84
1957	9	73	2008	10	83	2059	10	84	2110	10	81	2161	10	82
1958	9	83	2009	10	83	2060	10	83	2111	10	85	2162	10	88
1959	9	84	2010	10	87	2061	10	83	2112	10	89	2163	10	86
1960	9	78	2011	10	87	2062	10	90	2113	10	79	2164	10	86
1961	9	77	2012	10	83	2063	10	83	2114	10	84	2165	10	82
1962	9	84	2013	10	82	2064	10	86	2115	10	82	2166	10	87
1963	9	80	2014	10	92	2065	10	88	2116	10	87	2167	10	87
1964	9	83	2015	10	82	2066	10	83	2117	10	85	2168	10	86
1965	9	79	2016	10	87	2067	10	89	2118	10	82	2169	10	85
1966	9	76	2017	10	87	2068	10	83	2119	10	88	2170	10	84

T A B L E D - 3
Grades Data (Continued)

student	hours	score	student	hours	score	student	hours	score	student	hours	score	student	hours	score
2171	10	85	2222	11	86	2273	11	87	2324	11	86	2375	11	87
2172	10	85	2223	11	90	2274	11	90	2325	11	87	2376	11	88
2173	10	87	2224	11	91	2275	11	89	2326	11	87	2377	11	94
2174	10	88	2225	11	88	2276	11	91	2327	11	86	2378	11	87
2175	10	83	2226	11	90	2277	11	92	2328	11	86	2379	11	89
2176	10	86	2227	11	91	2278	11	87	2329	11	87	2380	11	89
2177	10	89	2228	11	86	2279	11	95	2330	11	88	2381	11	87
2178	10	82	2229	11	92	2280	11	92	2331	11	91	2382	11	90
2179	10	87	2230	11	87	2281	11	85	2332	11	89	2383	11	88
2180	10	78	2231	11	87	2282	11	93	2333	11	88	2384	11	84
2181	10	89	2232	11	93	2283	11	89	2334	11	93	2385	11	94
2182	10	89	2233	11	88	2284	11	85	2335	11	87	2386	11	87
2183	10	80	2234	11	90	2285	11	95	2336	11	87	2387	11	86
2184	10	86	2235	11	91	2286	11	90	2337	11	81	2388	11	86
2185	10	91	2236	11	83	2287	11	90	2338	11	87	2389	11	88
2186	10	86	2237	11	93	2288	11	89	2339	11	92	2390	11	90
2187	10	88	2238	11	86	2289	11	89	2340	11	91	2391	11	88
2188	10	82	2239	11	91	2290	11	90	2341	11	84	2392	11	89
2189	10	85	2240	11	86	2291	11	92	2342	11	85	2393	11	91
2190	10	86	2241	11	91	2292	11	88	2343	11	93	2394	11	85
2191	10	82	2242	11	91	2293	11	94	2344	11	87	2395	11	86
2192	10	90	2243	11	89	2294	11	91	2345	11	83	2396	11	90
2193	10	86	2244	11	89	2295	11	85	2346	11	86	2397	11	93
2194	10	83	2245	11	96	2296	11	89	2347	11	94	2398	11	83
2195	10	85	2246	11	88	2297	11	90	2348	11	87	2399	11	92
2196	10	85	2247	11	93	2298	11	86	2349	11	89	2400	11	91
2197	10	86	2248	11	86	2299	11	92	2350	11	93	2401	12	93
2198	10	83	2249	11	89	2300	11	94	2351	11	95	2402	12	95
2199	10	81	2250	11	92	2301	11	90	2352	11	90	2403	12	87
2200	10	85	2251	11	86	2302	11	86	2353	11	92	2404	12	96
2201	11	85	2252	11	84	2303	11	91	2354	11	92	2405	12	97
2202	11	85	2253	11	87	2304	11	93	2355	11	90	2406	12	95
2203	11	91	2254	11	86	2305	11	86	2356	11	89	2407	12	90
2204	11	93	2255	11	89	2306	11	92	2357	11	86	2408	12	93
2205	11	90	2256	11	83	2307	11	85	2358	11	88	2409	12	90
2206	11	91	2257	11	91	2308	11	96	2359	11	90	2410	12	94
2207	11	89	2258	11	93	2309	11	90	2360	11	90	2411	12	88
2208	11	92	2259	11	88	2310	11	87	2361	11	90	2412	12	96
2209	11	90	2260	11	86	2311	11	89	2362	11	91	2413	12	89
2210	11	91	2261	11	88	2312	11	89	2363	11	89	2414	12	97
2211	11	93	2262	11	90	2313	11	87	2364	11	88	2415	12	94
2212	11	95	2263	11	91	2314	11	89	2365	11	90	2416	12	92
2213	11	85	2264	11	89	2315	11	88	2366	11	88	2417	12	94
2214	11	86	2265	11	91	2316	11	91	2367	11	91	2418	12	98
2215	11	82	2266	11	92	2317	11	90	2368	11	88	2419	12	96
2216	11	85	2267	11	92	2318	11	85	2369	11	90	2420	12	94
2217	11	91	2268	11	92	2319	11	90	2370	11	88	2421	12	95
2218	11	88	2269	11	89	2320	11	89	2371	11	86	2422	12	92
2219	11	91	2270	11	89	2321	11	89	2372	11	88	2423	12	95
2220	11	85	2271	11	92	2322	11	86	2373	11	89	2424	12	89
2221	11	88	2272	11	88	2323	11	89	2374	11	85	2425	12	89

T A B L E D - 3
Grades Data (Continued)

student	hours	score	student	hours	score	student	hours	score	student	hours	score	student	hours	score
2426	12	96	2461	12	97	2496	12	99	2531	12	90	2566	12	96
2427	12	87	2462	12	91	2497	12	89	2532	12	92	2567	12	92
2428	12	96	2463	12	93	2498	12	98	2533	12	95	2568	12	93
2429	12	92	2464	12	90	2499	12	91	2534	12	93	2569	12	95
2430	12	93	2465	12	90	2500	12	91	2535	12	90	2570	12	94
2431	12	97	2466	12	91	2501	12	97	2536	12	92	2571	12	91
2432	12	97	2467	12	92	2502	12	93	2537	12	95	2572	12	93
2433	12	93	2468	12	97	2503	12	95	2538	12	89	2573	12	94
2434	12	91	2469	12	97	2504	12	97	2539	12	91	2574	12	96
2435	12	93	2470	12	91	2505	12	87	2540	12	91	2575	12	99
2436	12	93	2471	12	90	2506	12	94	2541	12	91	2576	12	98
2437	12	95	2472	12	96	2507	12	93	2542	12	93	2577	12	93
2438	12	92	2473	12	90	2508	12	95	2543	12	91	2578	12	91
2439	12	94	2474	12	94	2509	12	94	2544	12	89	2579	12	95
2440	12	92	2475	12	89	2510	12	93	2545	12	96	2580	12	97
2441	12	93	2476	12	98	2511	12	99	2546	12	97	2581	12	89
2442	12	91	2477	12	94	2512	12	96	2547	12	95	2582	12	96
2443	12	93	2478	12	88	2513	12	89	2548	12	90	2583	12	94
2444	12	89	2479	12	93	2514	12	90	2549	12	92	2584	12	93
2445	12	86	2480	12	95	2515	12	95	2550	12	96	2585	12	95
2446	12	95	2481	12	100	2516	12	96	2551	12	91	2586	12	85
2447	12	95	2482	12	92	2517	12	95	2552	12	94	2587	12	90
2448	12	96	2483	12	93	2518	12	91	2553	12	90	2588	12	93
2449	12	100	2484	12	90	2519	12	91	2554	12	97	2589	12	90
2450	12	94	2485	12	94	2520	12	95	2555	12	88	2590	12	94
2451	12	92	2486	12	94	2521	12	90	2556	12	93	2591	12	96
2452	12	90	2487	12	93	2522	12	92	2557	12	89	2592	12	94
2453	12	95	2488	12	91	2523	12	90	2558	12	96	2593	12	92
2454	12	92	2489	12	94	2524	12	95	2559	12	93	2594	12	92
2455	12	92	2490	12	94	2525	12	90	2560	12	94	2595	12	89
2456	12	99	2491	12	92	2526	12	93	2561	12	92	2596	12	92
2457	12	95	2492	12	89	2527	12	94	2562	12	92	2597	12	91
2458	12	93	2493	12	90	2528	12	92	2563	12	93	2598	12	98
2459	12	95	2494	12	90	2529	12	95	2564	12	94	2599	12	87
2460	12	94	2495	12	90	2530	12	94	2565	12	90	2600	12	91

T A B L E D - 4
Plastic Data

I	Y	X_1	X_2	I	Y	X_1	X_2	I	Y	X_1	X_2	I	Y	X_1	X_2
1	30.7	240	16	8	36.4	280	12	15	24.2	220	14	22	30.6	260	16
2	24.7	250	18	9	22.2	200	12	16	18.8	210	18	23	41.0	270	10
3	30.6	260	16	10	20.7	210	14	17	36.2	260	10	24	29.5	220	10
4	32.8	240	10	11	30.7	250	14	18	32.6	270	16	25	28.2	240	14
5	20.7	240	20	12	40.6	280	10	19	24.6	240	18	26	20.2	220	18
6	34.5	260	16	13	20.5	210	18	20	43.0	280	10	27	20.5	210	20
7	41.9	290	12	14	34.2	290	18	21	41.0	300	16	28	24.6	200	10

T A B L E D - 4

Plastic Data (Continued)

I	Y	X_1	X_2	I	Y	X_1	X_2	I	Y	X_1	X_2	I	Y	X_1	X_2
29	29.2	260	16	80	25.7	260	20	131	36.6	300	18	182	20.6	200	14
30	34.8	260	12	81	41.0	290	14	132	32.7	240	14	183	22.2	220	16
31	33.9	240	10	82	26.2	200	10	133	30.6	280	20	184	43.8	300	14
32	41.8	270	10	83	26.6	220	12	134	23.7	230	16	185	21.9	210	16
33	18.6	210	18	84	32.6	240	10	135	27.0	200	10	186	26.7	230	12
34	33.2	260	12	85	34.2	250	10	136	36.2	300	18	187	32.2	290	20
35	31.9	270	18	86	27.3	240	16	137	17.2	210	18	188	29.5	250	14
36	24.2	250	20	87	16.2	210	20	138	28.5	250	18	189	28.5	220	14
37	22.2	220	18	88	33.8	260	16	139	22.5	230	20	190	42.5	280	12
38	27.5	240	16	89	30.7	210	10	140	34.8	290	18	191	27.7	250	16
39	32.2	260	14	90	16.6	210	20	141	30.2	260	16	192	20.1	220	20
40	32.6	250	12	91	39.8	270	12	142	33.7	280	16	193	34.1	280	18
41	34.6	280	16	92	28.2	220	10	143	36.5	240	10	194	26.6	250	18
42	27.3	230	14	93	36.2	280	14	144	41.5	300	14	195	40.2	270	10
43	30.2	230	12	94	36.2	280	14	145	34.7	230	10	196	28.4	280	20
44	24.2	240	18	95	29.7	260	16	146	25.5	210	10	197	26.7	220	10
45	24.4	250	18	96	26.2	210	10	147	26.4	220	10	198	18.1	200	18
46	22.8	200	12	97	44.1	300	12	148	24.1	200	12	199	24.7	230	16
47	28.7	250	14	98	33.5	260	14	149	43.5	300	12	200	36.5	290	18
48	32.2	280	18	99	32.8	260	14	150	28.7	260	16	201	33.5	270	14
49	28.7	260	16	100	30.8	270	18	151	30.2	260	16	202	38.2	300	16
50	44.7	300	14	101	20.6	200	14	152	32.2	240	10	203	33.5	270	16
51	35.3	290	18	102	32.7	280	18	153	33.7	260	12	204	36.7	270	12
52	32.2	250	14	103	33.7	280	16	154	46.7	300	12	205	26.4	270	20
53	37.3	280	14	104	31.7	250	12	155	38.1	260	10	206	24.2	240	20
54	26.5	220	14	105	33.0	270	18	156	37.8	270	14	207	18.4	210	16
55	36.5	280	16	106	26.7	260	18	157	32.6	250	12	208	40.7	300	12
56	41.5	290	12	107	26.8	220	12	158	34.1	270	16	209	40.7	300	18
57	42.6	300	12	108	34.4	290	16	159	28.2	270	20	210	30.7	240	10
58	38.5	280	14	109	34.7	290	16	160	42.7	280	12	211	17.5	220	20
59	36.2	300	18	110	30.7	260	14	161	38.7	300	20	212	24.2	240	18
60	21.3	200	14	111	28.8	270	20	162	40.6	300	14	213	19.2	220	18
61	30.1	220	10	112	30.5	230	12	163	38.7	290	14	214	37.0	260	12
62	26.7	200	12	113	20.7	220	18	164	26.7	250	16	215	35.8	280	18
63	22.8	230	18	114	20.3	220	18	165	25.7	230	14	216	23.7	240	18
64	30.2	260	18	115	24.7	220	14	166	35.5	250	10	217	35.7	270	12
65	23.8	230	20	116	22.2	210	14	167	29.7	240	12	218	24.6	240	18
66	31.7	250	12	117	27.9	210	10	168	41.8	300	16	219	22.3	200	12
67	26.2	240	16	118	29.0	210	10	169	32.7	250	12	220	22.6	230	18
68	32.7	280	16	119	35.5	280	14	170	29.9	260	18	221	28.2	210	10
69	18.6	220	20	120	26.1	220	14	171	41.3	290	12	222	30.6	240	12
70	39.5	270	10	121	34.4	280	14	172	25.8	200	12	223	34.7	270	12
71	23.7	240	18	122	26.5	250	20	173	25.3	220	14	224	21.5	240	20
72	40.4	300	12	123	32.2	260	14	174	26.5	240	18	225	20.4	200	12
73	34.7	270	18	124	27.9	260	20	175	24.3	240	18	226	30.6	270	18
74	31.7	280	18	125	22.6	240	20	176	20.6	230	20	227	26.6	240	16
75	27.2	250	16	126	22.7	210	12	177	28.6	230	12	228	36.5	260	14
76	22.8	220	16	127	26.6	220	12	178	15.7	200	18	229	35.5	260	12
77	37.5	290	16	128	26.2	210	10	179	40.1	270	10	230	17.3	210	20
78	29.3	270	20	129	28.1	210	10	180	28.7	280	20	231	24.2	200	10
79	30.3	250	14	130	25.3	250	20	181	37.3	270	12	232	34.2	300	20

T A B L E   D - 4

Plastic Data (Continued)

I	Y	X_1	X_2	I	Y	X_1	X_2	I	Y	X_1	X_2	I	Y	X_1	X_2
233	34.3	250	10	284	30.8	240	12	335	23.5	230	18	386	34.5	290	20
234	19.7	220	18	285	25.2	250	18	336	36.6	290	16	387	40.7	290	10
235	22.7	240	18	286	30.6	230	10	337	24.5	210	14	388	30.2	240	14
236	37.7	300	16	287	26.6	240	16	338	28.3	250	16	389	36.7	280	12
237	39.9	300	16	288	28.5	230	16	339	24.2	230	18	390	30.2	240	12
238	32.5	260	18	289	32.3	270	16	340	29.3	220	10	391	22.2	240	20
239	38.2	270	10	290	37.5	300	16	341	22.7	210	18	392	16.8	200	18
240	22.2	200	12	291	25.8	240	20	342	30.6	250	14	393	20.1	200	16
241	32.2	240	10	292	30.7	260	14	343	22.2	210	14	394	30.4	290	20
242	23.2	210	12	293	39.3	270	10	344	40.7	260	10	395	34.7	300	18
243	32.3	260	14	294	31.5	240	12	345	16.6	200	18	396	18.7	220	18
244	18.7	210	16	295	38.2	300	16	346	42.3	290	10	397	28.5	230	14
245	34.2	280	16	296	30.7	270	18	347	31.0	260	18	398	30.3	260	16
246	22.7	230	18	297	20.2	220	20	348	27.8	220	14	399	25.7	220	12
247	32.4	280	16	298	17.8	200	20	349	28.7	220	10	400	38.7	270	10
248	28.7	270	18	299	42.2	280	10	350	34.2	270	16	401	31.2	260	14
249	25.3	230	16	300	34.5	230	10	351	24.7	210	12	402	34.7	240	12
250	28.1	220	12	301	42.8	300	12	352	41.8	280	12	403	25.2	220	12
251	24.7	200	10	302	30.5	240	14	353	35.5	270	14	404	12.4	200	20
252	18.7	200	20	303	33.2	280	16	354	31.7	280	18	405	21.9	230	20
253	30.3	280	20	304	24.1	240	20	355	20.2	220	18	406	32.5	240	12
254	39.0	260	10	305	29.5	240	14	356	14.7	200	18	407	22.4	240	18
255	21.3	210	16	306	30.6	230	10	357	32.7	290	20	408	27.3	260	20
256	32.2	280	18	307	31.0	250	16	358	41.8	290	14	409	28.5	210	12
257	24.4	230	14	308	22.7	220	20	359	17.7	210	18	410	31.7	270	16
258	23.5	220	14	309	24.6	200	10	360	32.7	280	16	411	34.5	270	16
259	32.1	260	16	310	26.2	260	20	361	31.0	230	12	412	34.8	280	16
260	38.2	270	12	311	26.2	250	20	362	29.0	260	20	413	34.2	270	14
261	29.5	230	10	312	24.2	200	12	363	30.6	240	12	414	35.2	290	16
262	29.7	270	18	313	34.4	300	18	364	24.2	230	16	415	23.7	210	12
263	22.2	230	18	314	20.7	230	18	365	26.5	230	16	416	35.7	300	18
264	20.4	230	18	315	27.8	230	16	366	35.3	300	20	417	25.9	250	20
265	31.5	270	18	316	37.2	290	14	367	41.2	290	10	418	33.7	250	10
266	45.5	300	10	317	13.7	200	20	368	18.5	200	18	419	29.2	280	20
267	39.5	300	16	318	32.2	270	16	369	17.7	200	16	420	34.5	260	14
268	42.2	300	14	319	39.3	280	12	370	31.7	260	14	421	32.8	250	12
269	24.7	220	18	320	36.6	290	16	371	34.7	270	12	422	36.7	290	14
270	14.4	210	20	321	28.7	250	14	372	34.6	250	10	423	32.7	220	10
271	25.5	260	20	322	34.8	250	10	373	40.2	300	16	424	26.2	210	10
272	28.7	230	10	323	30.2	250	14	374	30.1	240	14	425	26.2	230	14
273	42.2	300	12	324	39.2	290	12	375	40.7	290	12	426	22.7	210	14
274	26.8	210	10	325	37.7	270	10	376	26.8	230	14	427	33.5	300	20
275	32.1	270	18	326	36.2	270	12	377	30.7	230	14	428	24.4	240	16
276	42.5	300	16	327	21.7	240	20	378	28.7	240	18	429	43.9	290	10
277	24.7	210	10	328	34.1	240	10	379	32.7	260	18	430	34.5	280	18
278	31.9	240	12	329	31.9	230	10	380	34.7	260	10	431	32.6	260	14
279	30.8	280	20	330	34.7	300	20	381	48.5	300	10	432	40.7	280	10
280	15.2	210	20	331	30.7	220	12	382	30.1	230	12	433	29.3	250	16
281	20.7	210	16	332	24.2	220	16	393	35.5	290	16	434	38.2	290	14
282	28.2	250	18	333	26.4	240	14	384	31.5	250	14	435	39.9	290	14
283	24.5	230	20	334	38.4	300	14	385	19.8	200	18	436	24.2	210	14

T A B L E D - 4
Plastic Data (Continued)

I	Y	X_1	X_2	I	Y	X_1	X_2	I	Y	X_1	X_2	I	Y	X_1	X_2
437	28.2	230	12	488	33.9	250	12	539	40.8	300	14	590	27.7	220	10
438	27.3	220	12	489	23.8	200	14	540	30.3	270	18	591	34.5	250	12
439	42.1	280	10	490	28.7	270	18	541	24.2	250	20	592	22.2	220	16
440	29.5	240	12	491	28.5	250	20	542	20.6	210	16	593	40.2	300	14
441	27.7	230	12	492	44.8	300	10	543	28.7	230	12	594	22.4	210	12
442	32.7	230	12	493	33.8	230	10	544	18.7	230	20	595	33.8	270	18
443	30.4	280	18	494	29.5	260	18	545	38.2	280	12	596	32.7	250	10
444	16.4	210	18	495	25.7	220	12	546	37.5	280	12	597	24.2	210	12
445	35.5	300	18	496	27.7	260	18	547	15.7	210	20	598	43.3	290	10
446	33.9	290	20	497	21.7	230	18	548	34.2	280	16	599	21.2	230	18
447	34.2	250	10	498	25.5	220	14	549	36.2	260	12	600	22.3	240	20
448	17.0	200	20	499	33.5	280	18	550	33.7	270	14	601	35.7	270	12
449	28.7	230	16	500	24.8	210	12	551	24.5	220	16	602	19.7	210	16
450	33.5	290	20	501	42.4	300	10	552	24.2	220	14	603	36.6	260	10
451	30.7	290	20	502	43.7	300	10	553	17.7	220	20	604	36.6	270	12
452	21.5	210	16	503	33.0	230	10	554	36.8	270	12	605	18.7	210	18
453	33.7	300	20	504	44.2	300	12	555	32.6	290	20	606	46.5	300	10
454	28.6	220	10	505	30.6	280	20	556	41.0	280	12	607	40.7	290	16
455	34.7	290	16	506	27.9	230	14	557	17.5	210	18	608	25.8	230	18
456	31.3	250	14	507	36.2	290	16	558	30.7	270	16	609	25.7	230	14
457	33.3	290	20	508	34.4	260	10	559	24.2	200	10	610	36.2	290	18
458	23.0	200	14	509	39.9	280	12	560	27.0	210	12	611	45.0	290	10
459	27.7	250	16	510	21.9	200	14	561	19.7	230	20	612	42.6	290	10
460	26.7	230	18	511	36.6	280	14	562	36.7	280	18	613	21.8	210	18
461	30.5	250	16	512	39.7	300	14	563	37.2	270	10	614	23.9	200	12
462	20.2	200	14	513	25.9	210	12	564	32.5	270	18	615	27.5	250	18
463	36.7	280	14	514	20.5	200	16	565	22.3	210	14	616	45.8	300	12
464	18.2	210	18	515	37.8	250	10	566	23.9	210	14	617	38.4	290	12
465	24.4	210	10	516	38.8	300	16	567	42.6	300	12	618	30.2	270	18
466	30.8	230	10	517	29.7	250	14	568	24.8	220	14	619	29.8	210	10
467	24.7	210	16	518	37.3	260	10	569	26.2	260	20	620	17.5	200	18
468	20.8	200	14	519	13.5	200	20	570	29.2	270	18	621	26.7	240	14
469	45.8	290	10	520	37.5	270	10	571	18.4	200	14	622	25.9	200	10
470	34.2	280	16	521	41.5	300	12	572	36.5	250	10	623	23.3	220	16
471	40.2	280	12	522	38.5	270	12	573	28.5	210	10	624	27.2	220	10
472	37.0	270	14	523	33.7	290	18	574	25.7	240	16	625	26.7	230	12
473	25.8	220	16	524	24.7	240	16	575	34.2	290	18	626	32.2	260	14
474	28.7	240	14	525	29.2	230	10	576	36.5	280	18	627	20.8	230	20
475	31.5	240	10	526	23.9	240	20	577	43.7	300	10	628	21.2	210	14
476	18.2	200	16	527	30.3	240	12	578	43.5	290	10	629	20.7	240	20
477	37.9	300	18	528	36.2	250	10	579	41.5	280	10	630	35.2	280	14
478	43.3	300	12	529	20.2	200	16	580	21.7	210	14	631	23.7	210	12
479	32.7	240	10	530	40.1	280	12	581	23.5	250	20	632	39.0	270	12
480	34.2	300	20	531	38.1	270	12	582	24.3	200	10	633	25.7	210	10
481	30.4	270	16	532	15.2	200	18	583	21.8	220	20	634	25.0	210	14
482	39.3	290	14	533	28.6	270	20	584	26.1	240	18	635	36.5	300	20
483	38.5	290	18	534	32.2	280	20	585	14.7	210	20	636	18.7	230	20
484	30.5	220	12	535	31.8	250	16	586	28.2	260	18	637	38.5	300	18
485	35.7	290	16	536	28.2	250	16	587	37.8	260	12	638	28.3	230	12
486	20.2	230	20	537	42.2	290	12	588	27.8	240	18	639	26.5	210	14
487	34.3	260	12	538	29.2	250	14	589	25.7	250	18	640	43.8	290	12

T A B L E D - 4
Plastic Data (Continued)

I	Y	X_1	X_2	I	Y	X_1	X_2	I	Y	X_1	X_2	I	Y	X_1	X_2
641	27.5	250	16	692	28.3	240	14	743	25.5	200	10	794	30.4	250	12
642	36.3	290	16	693	27.5	260	20	744	33.9	260	14	795	36.7	280	12
643	16.7	200	16	694	35.0	260	14	745	27.3	210	10	796	31.0	220	10
644	36.7	300	18	695	32.7	250	10	746	37.8	290	18	797	37.7	280	12
645	42.5	300	14	696	42.6	290	10	747	32.7	270	14	798	27.7	270	20
646	18.3	210	18	697	39.9	270	10	748	22.7	220	14	799	12.7	200	20
647	34.7	290	18	698	35.2	300	18	749	26.7	260	20	800	34.6	270	14
648	17.2	220	20	699	22.2	200	14	750	31.8	270	20	801	32.5	250	14
649	26.6	260	20	700	29.7	280	20	751	29.8	250	18	802	24.4	220	12
650	26.7	220	10	701	25.0	240	20	752	35.9	300	20	803	37.8	300	20
651	32.5	250	16	702	29.7	260	16	753	24.4	260	20	804	31.8	220	10
652	15.5	200	20	703	38.2	280	14	754	30.5	270	20	805	35.9	290	18
653	40.5	270	10	704	32.4	260	12	755	39.8	290	16	806	23.8	220	18
654	25.5	230	14	705	30.1	270	20	756	38.3	280	12	807	31.3	260	16
655	39.7	280	10	706	27.5	220	10	757	38.1	280	14	808	34.6	280	16
656	41.5	290	10	707	20.2	200	14	758	34.7	260	16	809	22.7	230	16
657	33.8	250	14	708	28.6	230	12	759	38.3	300	16	810	40.1	290	14
658	20.2	210	16	709	30.5	210	10	760	27.8	250	20	811	24.1	210	14
659	30.7	260	20	710	20.3	230	20	761	32.7	260	12	812	38.7	300	14
660	28.2	240	14	711	42.7	300	10	762	29.8	220	12	813	26.2	220	12
661	20.2	210	16	712	17.5	210	20	763	32.2	240	10	814	43.0	290	12
662	25.9	220	14	713	18.6	210	18	764	15.3	200	20	815	33.0	250	14
663	45.9	300	10	714	15.9	200	20	765	31.8	230	12	816	31.5	290	20
664	33.9	270	16	715	26.2	220	12	766	30.5	250	18	817	26.8	240	16
665	29.3	240	14	716	20.2	230	20	767	36.7	290	14	818	40.6	290	12
666	32.4	270	14	717	28.7	270	20	768	42.7	290	10	819	32.6	270	16
667	32.7	300	20	718	17.7	200	16	769	32.2	250	12	820	44.6	300	10
668	31.2	270	16	719	30.7	250	18	770	38.7	270	14	821	38.6	270	10
669	18.6	220	20	720	21.5	230	20	771	37.5	270	12	822	39.5	280	10
670	35.9	280	16	721	26.6	210	10	772	28.8	250	16	823	24.7	230	20
671	32.5	220	10	722	30.2	250	14	773	22.7	200	12	824	24.7	220	12
672	30.5	260	18	723	25.0	230	18	774	36.7	260	10	825	28.2	220	10
673	40.5	270	12	724	39.8	280	14	755	43.0	300	14	826	35.7	260	10
674	26.7	250	16	725	18.7	220	20	776	42.7	270	10	827	16.5	200	20
675	33.0	240	12	726	39.0	290	16	777	27.7	240	14	828	20.8	210	16
676	36.7	270	16	727	38.7	280	10	778	26.6	230	14	829	46.2	300	10
677	29.5	230	12	728	20.3	200	14	779	38.8	280	12	830	41.9	280	10
678	41.9	300	14	729	34.2	250	10	780	32.1	240	12	831	32.7	260	14
679	19.7	230	20	730	23.3	240	20	781	35.5	270	12	832	32.7	260	12
680	18.2	200	18	731	30.2	260	16	782	25.7	240	16	833	35.0	280	18
681	26.3	250	18	732	26.7	270	20	783	23.2	200	10	834	33.7	250	10
682	29.2	240	12	733	28.2	250	16	784	23.5	240	20	835	18.7	220	18
683	26.5	220	16	734	26.7	240	20	785	22.7	200	10	836	30.6	250	14
684	28.5	260	20	735	29.9	250	16	786	27.8	200	10	837	28.1	230	14
685	23.3	230	18	736	22.6	230	18	787	26.5	200	10	838	42.2	290	10
686	21.5	200	14	737	28.2	250	16	788	24.3	210	12	839	42.1	300	14
687	29.8	260	20	738	22.5	210	18	789	34.7	300	18	840	23.7	200	10
688	35.8	270	16	739	26.2	250	18	790	16.7	210	18	841	26.3	240	16
689	20.7	200	18	740	34.6	300	20	791	19.7	200	14	842	33.3	240	10
690	25.5	240	16	741	21.7	230	18	792	40.2	280	10	843	42.5	290	12
691	32.3	280	18	742	37.7	280	12	793	23.5	200	12	844	23.7	250	20

T A B L E D - 4
Plastic Data (Continued)

I	Y	X_1	X_2	I	Y	X_1	X_2	I	Y	X_1	X_2	I	Y	X_1	X_2
845	22.4	200	10	896	35.7	290	16	947	31.5	280	20	998	38.2	270	10
846	38.5	250	10	897	20.4	240	20	948	26.2	240	18	999	38.5	290	16
847	22.7	200	16	898	45.0	300	12	949	36.1	250	10	1000	27.7	220	10
848	40.2	300	14	899	35.9	250	10	950	36.2	300	18	1001	40.2	290	12
849	28.2	220	10	900	20.7	220	16	951	26.7	220	16	1002	22.4	220	14
850	40.2	290	14	901	14.8	200	20	952	22.6	220	16	1003	30.7	230	10
851	22.5	220	20	902	22.4	230	16	953	27.0	240	18	1004	28.4	250	14
852	36.8	300	18	903	30.7	280	18	954	21.9	220	18	1005	40.7	300	14
853	42.5	290	14	904	27.7	230	12	955	44.5	290	12	1006	39.8	300	18
854	27.8	210	12	905	32.6	240	10	956	26.1	200	10	1007	24.6	210	12
855	26.1	250	20	906	30.7	250	12	957	38.7	290	12	1008	19.5	220	20
856	24.2	210	12	907	26.2	250	18	958	33.3	250	12	1009	40.3	300	14
857	34.5	240	12	908	21.2	200	12	959	37.9	260	10	1010	34.3	290	18
858	25.0	220	16	909	30.2	270	18	960	34.2	280	18	1011	42.7	290	14
859	18.7	200	14	910	18.2	220	20	961	18.8	220	20	1012	37.0	290	18
860	26.8	250	18	911	31.5	270	16	962	25.3	240	18	1013	34.2	270	14
861	18.2	200	16	912	28.1	240	16	963	36.4	290	14	1014	31.7	240	10
862	24.6	210	12	913	26.3	210	10	964	18.4	230	20	1015	32.8	280	18
863	30.7	290	20	914	38.7	300	14	965	39.5	280	12	1016	29.0	250	18
864	32.5	280	20	915	27.0	230	16	966	40.5	290	14	1017	34.1	290	20
865	31.3	240	12	916	34.5	280	20	967	20.7	200	12	1018	42.7	300	16
866	33.5	240	10	917	19.7	220	18	968	38.6	300	16	1019	32.7	270	14
867	18.1	210	20	918	28.8	240	14	969	40.5	300	16	1020	35.2	270	12
868	26.7	240	14	919	35.7	260	10	970	34.4	270	12	1021	29.0	240	16
869	31.2	290	20	920	34.6	250	10	971	36.1	280	16	1022	36.2	260	10
870	32.2	230	10	921	16.7	200	16	972	40.4	290	10	1023	22.5	210	16
871	14.2	200	20	922	24.7	250	20	973	38.5	270	14	1024	37.0	280	16
872	36.1	300	20	923	27.5	270	20	974	33.5	290	18	1025	33.3	280	18
873	22.6	200	12	924	30.7	270	16	975	31.2	280	18	1026	39.3	300	16
874	30.2	230	10	925	27.0	250	20	976	18.2	210	18	1027	29.9	240	14
875	16.3	210	20	926	33.8	280	20	977	24.3	220	14	1028	34.2	260	12
876	38.2	300	16	927	29.7	270	18	978	22.6	220	16	1029	39.5	290	14
877	18.7	210	16	928	24.6	220	14	979	12.7	200	20	1030	24.5	240	20
878	34.1	250	12	929	36.7	270	10	980	34.7	280	16	1031	44.1	290	10
879	21.5	210	14	930	15.7	210	20	981	22.2	230	18	1032	27.5	230	12
880	36.3	280	14	931	22.3	220	16	982	28.4	270	18	1033	44.6	300	10
881	24.6	250	20	932	38.7	290	12	983	26.4	230	12	1034	36.6	270	12
882	36.4	270	10	933	26.3	220	12	984	26.2	260	20	1035	36.7	240	10
883	20.4	220	16	934	34.2	260	12	985	30.2	230	10	1036	27.2	240	14
884	24.5	230	18	935	44.2	300	10	986	34.2	270	14	1037	33.5	260	12
885	36.7	290	20	936	40.2	300	14	987	32.2	270	18	1038	21.7	210	14
886	26.6	210	10	937	24.8	200	10	988	18.6	200	16	1039	26.7	270	20
887	26.4	250	16	938	24.8	230	16	989	21.5	220	16	1040	15.5	200	18
888	21.7	220	16	939	42.2	300	12	990	20.6	230	20	1041	30.5	240	16
889	20.7	210	20	940	32.5	230	12	991	33.7	300	20	1042	28.7	210	12
890	30.7	260	16	941	16.1	200	20	992	47.0	300	10	1043	22.2	230	20
891	31.0	270	20	942	28.6	220	10	993	22.7	220	16	1044	28.2	270	20
892	30.6	270	18	943	30.2	240	12	994	24.6	220	14	1045	39.7	290	12
893	20.5	220	20	944	30.2	250	16	995	21.5	200	12	1046	30.4	260	14
894	22.2	210	14	945	32.7	290	18	996	36.2	270	14	1047	37.3	300	18
895	14.7	210	20	946	34.1	260	14	997	27.5	220	12	1048	22.2	200	12

T A B L E D - 4
Plastic Data (Continued)

I	Y	X_1	X_2	I	Y	X_1	X_2	I	Y	X_1	X_2	I	Y	X_1	X_2
1049	38.6	270	10	1100	33.7	290	18	1151	26.2	220	14	1202	34.7	260	12
1050	24.5	220	18	1101	23.5	200	10	1152	30.2	230	10	1203	31.7	290	20
1051	16.2	200	20	1102	42.7	300	12	1153	25.2	240	16	1204	36.8	260	10
1052	33.2	270	14	1103	26.2	210	12	1154	29.5	270	20	1205	23.9	230	18
1053	22.7	240	20	1104	24.8	250	20	1155	19.5	220	18	1206	24.5	200	14
1054	17.7	210	18	1105	20.8	220	18	1156	23.2	230	16	1207	20.6	210	16
1055	40.3	290	12	1106	36.6	300	18	1157	40.6	280	10	1208	36.4	300	16
1056	31.5	260	14	1107	43.8	280	10	1158	31.3	280	20	1209	25.5	240	18
1057	31.2	250	12	1108	28.2	220	12	1159	21.7	200	12	1210	34.7	250	10
1058	34.3	300	20	1109	21.0	200	16	1160	24.3	250	20	1211	30.4	240	10
1059	30.5	260	20	1110	25.3	210	12	1161	40.8	280	10	1212	32.5	230	10
1060	31.2	240	10	1111	31.7	290	20	1162	28.7	260	18	1213	34.2	260	12
1061	16.7	200	18	1112	25.5	230	16	1163	38.7	280	10	1214	25.7	250	18
1062	42.3	300	12	1113	26.7	230	14	1164	40.5	280	12	1215	40.5	290	16
1063	35.0	240	10	1114	24.7	250	18	1165	21.7	220	16	1216	37.3	290	16
1064	20.2	230	20	1115	37.7	290	14	1166	26.8	260	20	1217	14.3	200	20
1065	27.5	240	14	1116	14.7	200	18	1167	34.8	300	20	1218	32.7	270	16
1066	26.6	230	14	1117	26.7	210	10	1168	20.7	210	14	1219	46.5	290	10
1067	31.7	260	14	1118	22.7	240	18	1169	21.0	220	20	1220	26.7	250	18
1068	27.9	240	16	1119	16.2	200	18	1170	36.1	260	12	1221	35.8	250	12
1069	14.6	200	20	1120	23.5	230	16	1171	31.8	240	14	1222	18.7	200	14
1070	30.2	280	20	1121	26.3	260	20	1172	19.2	230	20	1223	24.7	230	14
1071	21.5	220	18	1122	30.7	240	10	1173	35.3	260	12	1224	28.7	280	20
1072	35.0	250	12	1123	32.2	290	20	1174	28.2	260	20	1225	17.5	200	16
1073	34.7	280	14	1124	28.2	240	14	1175	22.1	230	20	1226	25.5	210	12
1074	26.2	220	12	1125	28.3	270	20	1176	32.2	250	12	1227	23.7	250	20
1075	24.2	230	16	1126	34.7	270	14	1177	42.1	290	12	1228	24.1	220	16
1076	37.8	280	16	1127	26.1	210	12	1178	20.7	200	12	1229	18.2	210	20
1077	44.2	300	10	1128	38.8	290	14	1179	38.3	270	10	1230	24.2	220	14
1078	20.7	230	20	1129	16.8	210	20	1180	30.2	270	18	1231	34.7	280	20
1079	23.7	220	14	1130	22.7	220	14	1181	26.7	·240	16	1232	26.6	260	20
1080	19.5	200	14	1131	38.7	290	18	1182	30.7	240	12	1233	18.7	200	16
1081	35.5	300	20	1132	44.3	300	10	1183	32.2	270	16	1234	16.7	220	20
1082	38.6	280	12	1133	40.5	300	18	1184	35.0	290	20	1235	24.7	230	14
1083	19.9	210	18	1134	37.5	290	14	1185	25.2	230	14	1236	30.2	280	20
1084	32.6	280	18	1135	24.8	240	18	1186	20.7	230	18	1237	16.7	220	20
1085	30.2	240	12	1136	20.4	210	14	1187	34.6	300	20	1238	19.5	230	20
1086	44.7	280	10	1137	19.3	200	16	1188	37.7	270	10	1239	32.2	250	12
1087	29.3	230	12	1138	36.5	270	16	1189	34.5	240	10	1240	32.3	250	12
1088	30.2	250	14	1139	18.6	200	16	1190	26.7	260	18	1241	32.8	290	20
1089	26.2	240	16	1140	30.2	220	10	1191	24.7	200	14	1242	20.6	220	18
1090	27.3	250	18	1141	19.9	220	20	1192	20.2	200	14	1243	29.7	230	10
1091	22.1	200	14	1142	40.7	270	12	1193	34.6	270	14	1244	16.6	200	18
1092	30.3	230	10	1143	34.2	260	14	1194	24.7	210	10	1245	27.9	250	18
1093	33.3	270	16	1144	33.8	240	12	1195	19.2	200	14	1246	20.6	220	18
1094	34.2	250	12	1145	32.1	280	20	1196	22.6	210	14	1247	36.5	290	20
1095	33.9	280	18	1146	28.7	240	12	1197	30.7	280	20	1248	19.5	210	18
1096	25.3	200	10	1147	44.5	300	14	1198	37.2	300	16	1249	22.2	210	16
1097	24.2	230	16	1148	29.5	270	18	1199	40.2	280	10	1250	39.2	300	14
1098	16.2	210	20	1149	23.3	210	14	1200	29.0	220	12	1251	40.2	290	12
1099	27.2	260	18	1150	36.6	260	10	1201	35.8	240	10	1252	42.2	290	10

T A B L E D - 4
Plastic Data (Continued)

I	Y	X_1	X_2	I	Y	X_1	X_2	I	Y	X_1	X_2	I	Y	X_1	X_2
1253	36.6	280	14	1304	20.7	200	14	1355	37.5	280	14	1406	24.7	240	18
1254	34.3	270	14	1305	16.4	200	16	1356	35.5	290	18	1407	18.8	200	16
1255	44.7	290	12	1306	28.5	240	16	1357	41.3	300	14	1408	44.5	280	10
1256	22.5	220	18	1307	28.4	260	16	1358	36.7	260	14	1409	21.7	200	12
1257	18.2	220	20	1308	15.7	200	18	1359	28.2	240	16	1410	34.5	250	14
1258	24.1	230	18	1309	28.6	250	16	1360	16.2	210	20	1411	31.5	260	16
1259	42.5	280	10	1310	24.7	220	12	1361	15.5	210	20	1412	34.2	240	10
1260	29.7	280	20	1311	32.5	260	16	1362	22.1	220	18	1413	32.6	260	14
1261	28.8	230	12	1312	28.1	250	18	1363	38.7	280	12	1414	16.6	210	20
1262	41.2	300	12	1313	28.2	230	12	1364	32.7	270	20	1415	34.2	290	20
1263	45.3	300	10	1314	32.1	230	10	1365	17.9	210	20	1416	26.2	230	14
1264	31.5	230	10	1315	23.0	230	20	1366	40.3	280	10	1417	31.5	250	12
1265	22.5	200	16	1316	23.9	220	16	1367	16.2	200	18	1418	23.5	210	12
1266	19.3	210	18	1317	27.7	270	20	1368	38.2	280	12	1419	22.1	210	16
1267	18.5	210	20	1318	36.7	250	12	1369	38.5	280	16	1420	27.9	220	12
1268	38.2	290	16	1319	19.7	200	14	1370	26.7	210	14	1421	35.2	260	10
1269	23.2	250	20	1320	37.5	300	18	1371	22.6	210	14	1422	35.8	290	20
1270	19.7	210	16	1321	27.5	230	14	1372	37.2	280	12	1423	31.7	240	10
1271	28.3	220	10	1322	23.3	200	12	1373	22.7	250	20	1424	29.3	260	18
1272	29.8	240	16	1323	34.6	260	12	1374	27.5	210	10	1425	29.9	230	12
1273	39.5	290	12	1324	26.2	230	14	1375	14.2	200	20	1426	38.7	280	16
1274	36.3	270	12	1325	23.8	210	16	1376	25.9	240	18	1427	26.1	230	16
1275	16.4	220	20	1326	33.7	260	12	1377	36.8	290	16	1428	35.8	260	14
1276	38.2	280	12	1327	14.2	200	20	1378	30.8	260	16	1429	22.7	200	10
1277	26.5	240	20	1328	37.0	250	10	1379	25.2	210	10	1430	24.5	210	16
1278	28.6	260	18	1329	22.8	240	20	1380	31.8	260	18	1431	22.8	210	14
1279	36.3	300	18	1330	33.0	260	16	1381	16.7	210	20	1432	23.7	200	10
1280	24.7	260	20	1331	34.2	300	20	1382	33.0	280	20	1433	32.2	270	16
1281	24.2	200	10	1332	28.8	220	10	1383	30.8	250	14	1434	39.7	300	14
1282	37.9	270	12	1333	22.7	250	20	1384	42.2	300	12	1435	23.7	230	16
1283	17.3	200	18	1334	42.8	290	10	1385	36.8	280	14	1436	35.9	260	12
1284	16.7	210	18	1335	28.2	260	18	1386	35.9	270	14	1437	28.7	220	14
1285	24.2	240	18	1336	23.0	210	16	1387	14.7	200	20	1438	28.5	200	10
1286	39.2	280	10	1337	29.9	270	20	1388	33.2	250	10	1439	13.2	200	20
1287	40.7	290	10	1338	36.7	270	10	1389	18.5	200	20	1440	40.2	290	12
1288	21.7	240	20	1339	34.6	260	12	1390	38.8	270	10	1441	32.5	240	14
1289	39.8	260	10	1340	29.0	230	14	1391	40.2	280	10	1442	44.2	290	10
1290	39.0	300	18	1341	38.7	260	12	1392	28.7	250	20	1443	32.1	250	14
1291	26.7	220	12	1342	35.3	250	10	1393	28.2	260	18	1444	28.6	270	20
1292	43.2	300	10	1343	28.6	250	16	1394	26.2	250	18	1445	39.5	300	14
1293	40.7	300	12	1344	32.8	270	16	1395	29.7	240	12	1446	36.2	300	20
1294	32.4	250	10	1345	19.5	200	16	1396	25.5	250	20	1447	23.5	240	18
1295	36.7	290	16	1346	33.5	250	12	1397	25.7	210	10	1448	35.3	280	16
1296	16.2	200	18	1347	33.5	250	10	1398	38.2	260	10	1449	20.2	220	18
1297	34.5	270	18	1348	36.7	300	16	1399	22.6	240	20	1450	26.4	260	18
1298	19.5	210	16	1349	25.7	260	20	1400	17.2	200	16	1451	31.9	280	20
1299	36.5	270	14	1350	20.3	210	16	1401	28.7	200	10	1452	36.5	250	12
1300	18.2	210	18	1351	34.7	280	14	1402	36.1	290	18	1453	32.2	260	16
1301	26.5	200	12	1352	23.0	220	18	1403	38.5	300	20	1454	27.5	260	18
1302	28.5	240	18	1353	31.3	270	18	1404	38.4	280	10	1455	41.3	280	10
1303	33.3	260	14	1354	32.6	280	18	1405	27.2	230	12	1456	32.6	290	20

T A B L E D - 4
Plastic Data (Continued)

I	Y	X_1	X_2	I	Y	X_1	X_2	I	Y	X_1	X_2	I	Y	X_1	X_2
1457	36.2	290	16	1506	18.2	220	20	1555	18.4	220	18	1604	29.5	250	16
1458	34.2	290	18	1507	32.7	300	20	1556	32.2	240	12	1605	25.5	220	12
1459	23.2	240	18	1508	38.5	260	10	1557	22.2	220	16	1606	38.1	290	16
1460	32.4	300	20	1509	44.5	300	12	1558	36.2	280	16	1607	31.5	280	18
1461	36.5	260	12	1510	17.7	220	20	1559	42.7	300	10	1608	18.2	200	16
1462	28.8	260	18	1511	22.6	200	12	1560	40.5	260	10	1609	37.7	290	14
1463	20.7	220	16	1512	44.2	300	10	1561	28.7	230	10	1610	33.2	290	18
1464	34.3	280	16	1513	41.7	290	10	1562	38.2	290	14	1611	24.7	260	20
1465	30.7	280	18	1514	28.4	240	12	1563	21.5	230	18	1612	38.7	300	16
1466	38.2	300	18	1515	32.2	290	20	1564	30.7	250	12	1613	32.7	250	16
1467	25.5	250	18	1516	26.5	230	18	1565	18.3	200	16	1614	36.2	270	12
1468	21.0	210	18	1517	36.2	260	10	1566	24.7	240	16	1615	27.0	220	14
1469	21.3	230	20	1518	35.3	270	14	1567	30.5	230	14	1616	19.0	200	18
1470	34.7	260	10	1519	38.6	290	14	1568	20.5	200	18	1617	31.0	240	14
1471	29.7	250	14	1520	28.7	240	12	1569	28.4	230	10	1618	38.2	270	10
1472	43.9	300	12	1521	41.7	300	12	1570	24.2	250	20	1619	36.1	270	14
1473	33.2	300	20	1522	40.6	300	14	1571	40.8	290	12	1620	47.8	300	10
1474	28.2	230	14	1523	22.7	210	12	1572	23.5	220	16	1621	38.1	300	18
1475	39.0	280	14	1524	42.2	290	10	1573	36.2	290	16	1622	44.7	300	10
1476	31.9	250	14	1525	37.9	290	16	1574	25.8	210	14	1623	25.9	230	16
1477	34.7	250	14	1526	21.2	220	16	1575	25.0	200	12	1624	29.5	260	16
1478	19.9	200	16	1527	39.7	280	10	1576	21.2	240	20	1625	38.2	290	14
1479	27.7	240	14	1528	36.2	270	12	1577	41.7	300	12	1626	31.7	270	16
1480	35.0	270	16	1529	37.9	280	14	1578	24.6	230	16	1627	32.2	280	18
1481	30.1	260	18	1530	22.5	200	14	1579	37.7	300	16	1628	28.5	220	12
1482	32.3	290	20	1531	40.5	280	14	1580	26.2	240	16	1629	44.5	290	10
1483	36.7	300	16	1532	40.1	300	16	1581	16.3	200	18	1630	22.2	240	20
1484	37.5	260	10	1533	33.7	270	14	1582	36.3	260	10	1631	28.3	260	18
1485	31.9	260	16	1534	19.0	210	20	1583	30.2	280	20	1632	22.7	230	16
1486	35.7	280	14	1535	23.5	210	14	1584	41.7	290	10	1633	28.2	230	12
1487	40.6	290	12	1536	38.5	260	12	1585	24.6	250	20	1634	30.2	270	20
1488	33.5	280	16	1537	19.3	220	20	1586	26.5	210	12	1635	32.7	290	18
1489	29.7	230	10	1538	14.4	200	18	1587	25.2	260	20	1636	24.6	230	16
1490	35.5	260	10	1539	22.4	250	20	1588	28.2	270	20	1637	36.2	280	14
1491	35.7	280	14	1540	22.2	240	20	1589	18.3	220	20	1638	42.5	270	10
1492	46.7	290	10	1541	20.2	210	18	1590	34.8	270	14	1639	35.5	280	16
1493	24.5	200	12	1542	20.1	210	18	1591	22.2	230	18	1640	28.6	260	18
1494	40.7	280	14	1543	32.3	240	10	1592	32.4	290	18	1641	30.1	250	16
1495	29.9	220	10	1544	29.5	280	20	1593	27.7	260	18	1642	46.5	300	12
1496	34.6	290	18	1545	28.6	240	14	1594	22.3	230	18	1643	34.6	290	18
1497	13.7	200	20	1546	48.7	300	10	1595	38.6	300	16	1644	38.6	280	12
1498	27.2	270	20	1547	21.8	200	16	1596	38.3	290	14	1645	28.6	240	14
1499	29.8	230	14	1548	37.0	300	20	1597	24.3	230	16	1646	32.5	270	20
1500	38.7	250	10	1549	28.1	260	20	1598	30.5	220	10	1647	38.6	290	14
1501	46.1	300	10	1550	21.3	220	18	1599	26.6	250	18	1648	39.7	290	12
1502	23.7	220	14	1551	26.2	230	16	1600	20.2	210	16	1649	14.6	200	20
1503	19.2	210	16	1552	23.2	220	14	1601	43.5	300	10	1650	19.8	210	20
1504	28.7	250	16	1553	31.3	230	10	1602	24.2	210	12				
1505	17.9	200	18	1554	26.3	230	14	1603	35.7	300	18				

T

Tables

Table T-1   Percentiles of a Standard Gaussian Population

Table T-2   Percentiles of a Student's t Population

Table T-3   Percentiles of a Chi-Square Population

Table T-4   Student's t for m Simultaneous Confidence Intervals for $m = 2, 3, 4, 5, 6$

Table T-5   Percentiles of Snedecor's F Population

Table T-6   Table for Obtaining Confidence Bounds for $a_0 \beta_0 + a_1 \beta_1$ Using Theil's Method

Table T-7   Charts for Confidence Bounds for the Simple Correlation Coefficient

Table T-8   Selected Percentiles of the Noncentral t

T A B L E I·1

Percentiles of a Standard Gaussian Population

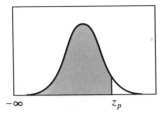

$-\infty$ z_p

The entries in this table are z_p, where p is the proportion of values of a standard Gaussian population that are between $-\infty$ and z_p.

z_p	.00	.01	.02	.03	.04	.05	.06	.07	.08	.09
.00	.5000	.5040	.5080	.5120	.5160	.5199	.5239	.5279	.5319	.5359
.10	.5398	.5438	.5478	.5517	.5557	.5596	.5636	.5675	.5714	.5753
.20	.5793	.5832	.5871	.5910	.5948	.5987	.6026	.6064	.6103	.6141
.30	.6179	.6217	.6255	.6293	.6331	.6368	.6406	.6443	.6480	.6517
.40	.6554	.6591	.6628	.6664	.6700	.6736	.6772	.6808	.6844	.6879
.50	.6915	.6950	.6985	.7019	.7054	7088	.7123	.7157	.7190	.7224
.60	.7257	.7291	.7324	.7357	.7389	.7422	.7454	.7486	.7517	.7549
.70	.7580	.7611	.7642	.7673	.7704	.7734	.7764	.7794	.7823	.7852
.80	.7881	.7910	.7939	.7967	.7995	.8023	.8051	.8078	.8106	.8133
.90	.8159	.8186	.8212	.8238	.8264	.8289	.8315	.8340	.8365	.8389
1.00	.8413	.8438	.8461	.8485	.8508	.8531	.8554	.8577	.8599	.8621
1.10	.8643	.8665	.8686	.8708	.8729	.8749	.8770	.8790	.8810	.8830
1.20	.8849	.8869	.8888	.8907	.8925	.8944	.8962	.8980	.8997	.9015
1.30	.9032	.9049	.9066	.9082	.9099	.9115	.9131	.9147	.9162	.9177
1.40	.9192	.9207	.9222	.9236	.9251	.9265	.9279	.9292	.9306	.9319
1.50	.9332	.9345	.9357	.9370	.9382	.9394	.9406	.9418	.9429	.9441
1.60	.9452	.9463	.9474	.9484	.9495	.9505	.9515	.9525	.9535	.9545
1.70	.9554	.9564	.9573	.9582	.9591	.9599	.9608	.9616	.9625	.9633
1.80	.9641	.9649	.9656	.9664	.9671	.9678	.9686	.9693	.9699	.9706
1.90	.9713	.9719	.9726	.9732	.9738	.9744	.9750	.9756	.9761	.9767
2.00	.9772	.9778	.9783	.9788	.9793	.9798	.9803	.9808	.9812	.9817
2.10	.9821	.9826	.9830	.9834	.9838	.9842	.9846	.9850	.9854	.9857
2.20	.9861	.9864	.9868	.9871	.9875	.9878	.9881	.9884	.9887	.9890
2.30	.9893	.9896	.9898	.9901	.9904	.9906	.9909	.9911	.9913	.9916
2.40	.9918	.9920	.9922	.9925	.9927	.9929	.9931	.9932	.9934	.9936
2.50	.9938	.9940	.9941	.9943	.9945	.9946	.9948	.9949	.9951	.9952
2.60	.9953	.9955	.9956	.9957	.9959	.9960	.9961	.9962	.9963	.9964
2.70	.9965	.9966	.9967	.9968	.9969	.9970	.9971	.9972	.9973	.9974
2.80	.9974	.9975	.9976	.9977	.9977	.9978	.9979	.9979	.9980	.9981
2.90	.9981	.9982	.9982	.9983	.9984	.9984	.9985	.9985	.9986	.9986
3.00	.9987	.9987	.9987	.9988	.9988	.9989	.9989	.9989	.9990	.9990
3.10	.9990	.9991	.9991	.9991	.9992	.9992	.9992	.9992	.9993	.9993
3.20	.9993	.9993	.9994	.9994	.9994	.9994	.9994	.9995	.9995	.9995
3.30	.9995	.9995	.9995	.9996	.9996	.9996	.9996	.9996	.9996	.9997
3.40	.9997	.9997	.9997	.9997	.9997	.9997	.9997	.9997	.9997	.9998
3.50	.9998	.9998	.9998	.9998	.9998	.9998	.9998	.9998	.9998	.9998

T A B L E T-2
Percentiles of a Student's *t* Population

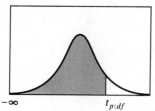

$$t_{p:df}$$

The entries in this table are $t_{p:df}$, where p is the proportion of the population values that are between $-\infty$ and $t_{p:df}$, for $p = .80, .90, .95, .975, .99, .995, .9995$.

df	.80	.90	.95	.975	.99	.995	.9995
1	1.376	3.078	6.314	12.706	31.821	63.657	636.619
2	1.061	1.886	2.920	4.303	6.965	9.925	31.599
3	.978	1.638	2.353	3.182	4.541	5.841	12.924
4	.941	1.533	2.132	2.776	3.747	4.604	8.610
5	.920	1.476	2.015	2.571	3.365	4.032	6.869
6	.906	1.440	1.943	2.447	3.143	3.707	5.959
7	.896	1.415	1.895	2.365	2.998	3.499	5.408
8	.889	1.397	1.860	2.306	2.896	3.355	5.041
9	.883	1.383	1.833	2.262	2.821	3.250	4.781
10	.879	1.372	1.812	2.228	2.764	3.169	4.587
11	.876	1.363	1.796	2.201	2.718	3.106	4.437
12	.873	1.356	1.782	2.179	2.681	3.055	4.318
13	.870	1.350	1.771	2.160	2.650	3.012	4.221
14	.868	1.345	1.761	2.145	2.624	2.977	4.140
15	.866	1.341	1.753	2.131	2.602	2.947	4.073
16	.865	1.337	1.746	2.120	2.583	2.921	4.015
17	.863	1.333	1.740	2.110	2.567	2.898	3.965
18	.862	1.330	1.734	2.101	2.552	2.878	3.922
19	.861	1.328	1.729	2.093	2.539	2.861	3.883
20	.860	1.325	1.725	2.086	2.528	2.845	3.850
21	.859	1.323	1.721	2.080	2.518	2.831	3.819
22	.858	1.321	1.717	2.074	2.508	2.819	3.792
23	.858	1.319	1.714	2.069	2.500	2.807	3.768
24	.857	1.318	1.711	2.064	2.492	2.797	3.745
25	.856	1.316	1.708	2.060	2.485	2.787	3.725
26	.856	1.315	1.706	2.056	2.479	2.779	3.707
27	.855	1.314	1.703	2.052	2.473	2.771	3.690
28	.855	1.313	1.701	2.048	2.467	2.763	3.674
29	.854	1.311	1.699	2.045	2.462	2.756	3.659
30	.854	1.310	1.697	2.042	2.457	2.750	3.646
40	.851	1.303	1.684	2.021	2.423	2.704	3.551
50	.849	1.299	1.676	2.009	2.403	2.678	3.496
60	.848	1.296	1.671	2.000	2.390	2.660	3.460
100	.845	1.290	1.660	1.984	2.364	2.626	3.390
∞	.842	1.282	1.645	1.960	2.327	2.576	3.292

T A B L E I · 3

Percentiles of a Chi-Square Population

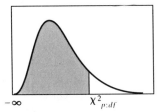

$\chi^2_{p:df}$

$-\infty$

The entries in this table are $x^2_{p:df}$, where p is the proportion of the population values that are between 0 and $\chi^2_{p:df}$, for $p = .01, .025, .05, .10, .50, .90, .95, .975, .99$.

df	.01	.025	.05	.10	.50	.90	.95	.975	.99
4	0.297	0.484	0.711	1.064	3.357	7.779	9.488	11.143	13.277
5	0.554	0.831	1.145	1.610	4.351	9.236	11.070	12.833	15.086
6	0.872	1.237	1.635	2.204	5.348	10.645	12.592	14.449	16.812
7	1.239	1.690	2.167	2.833	6.346	12.017	14.067	16.013	18.475
8	1.646	2.180	2.733	3.490	7.344	13.362	15.507	17.535	20.090
9	2.088	2.700	3.325	4.168	8.343	14.684	16.919	19.023	21.666
10	2.558	3.247	3.940	4.865	9.342	15.987	18.307	20.483	23.209
11	3.053	3.816	4.575	5.578	10.341	17.275	19.675	21.920	24.725
12	3.571	4.404	5.226	6.304	11.340	18.549	21.026	23.337	26.217
13	4.107	5.009	5.892	7.042	12.340	19.812	22.362	24.736	27.688
14	4.660	5.629	6.571	7.790	13.339	21.064	23.685	26.119	29.141
15	5.229	6.262	7.261	8.547	14.339	22.307	24.996	27.488	30.578
16	5.812	6.908	7.962	9.312	15.338	23.542	26.296	28.845	32.000
17	6.408	7.564	8.672	10.085	16.338	24.769	27.587	30.191	33.409
18	7.015	8.231	9.390	10.865	17.338	25.989	28.869	31.526	34.805
19	7.633	8.907	10.117	11.651	18.338	27.204	30.144	32.852	36.191
20	8.260	9.591	10.851	12.443	19.337	28.412	31.410	34.170	37.566
21	8.897	10.283	11.591	13.240	20.337	29.615	32.671	35.479	38.932
22	9.542	10.982	12.338	14.041	21.337	30.813	33.924	36.781	40.289
23	10.196	11.689	13.091	14.848	22.337	32.007	35.172	38.076	41.638
24	10.856	12.401	13.848	15.659	23.337	33.196	36.415	39.364	42.980
25	11.524	13.120	14.611	16.473	24.337	34.382	37.652	40.646	44.314
26	12.198	13.844	15.379	17.292	25.336	35.563	38.885	41.923	45.642
27	12.879	14.573	16.151	18.114	26.336	36.741	40.113	43.195	46.963
28	13.565	15.308	16.928	18.939	27.336	37.916	41.337	44.461	48.278
29	14.256	16.047	17.708	19.768	28.336	39.087	42.557	45.722	49.588
30	14.953	16.791	18.493	20.599	29.336	40.256	43.773	46.979	50.892
40	22.164	24.433	26.509	29.051	39.335	51.805	55.758	59.342	63.691
50	29.707	32.357	34.764	37.689	49.335	63.167	67.505	71.420	76.154
60	37.485	40.482	43.188	46.459	59.335	74.397	79.082	83.298	88.379

TABLE T-4

Student's t for $m = 2$ Simultaneous Confidence Intervals

The entries in this table are $t_{1-\alpha/2m:df}$ for obtaining $m = 2$ two-sided simultaneous confidence intervals with confidence coefficient $1 - \alpha$.

df	Confidence Coefficient = $1 - \alpha$						
	.60	.80	.90	.95	.98	.99	.999
1	3.078	6.314	12.706	25.452	63.657	127.321	1273.239
2	1.886	2.920	4.303	6.205	9.925	14.089	44.705
3	1.638	2.353	3.182	4.177	5.841	7.453	16.326
4	1.533	2.132	2.776	3.495	4.604	5.598	10.306
5	1.476	2.015	2.571	3.163	4.032	4.773	7.976
6	1.440	1.943	2.447	2.969	3.707	4.317	6.788
7	1.415	1.895	2.365	2.841	3.499	4.029	6.082
8	1.397	1.860	2.306	2.752	3.355	3.833	5.617
9	1.383	1.833	2.262	2.685	3.250	3.690	5.291
10	1.372	1.812	2.228	2.634	3.169	3.581	5.049
11	1.363	1.796	2.201	2.593	3.106	3.497	4.863
12	1.356	1.782	2.179	2.560	3.055	3.428	4.716
13	1.350	1.771	2.160	2.533	3.012	3.372	4.597
14	1.345	1.761	2.145	2.510	2.977	3.326	4.499
15	1.341	1.753	2.131	2.490	2.947	3.286	4.417
16	1.337	1.746	2.120	2.473	2.921	3.252	4.346
17	1.333	1.740	2.110	2.458	2.898	3.222	4.286
18	1.330	1.734	2.101	2.445	2.878	3.197	4.233
19	1.328	1.729	2.093	2.433	2.861	3.174	4.187
20	1.325	1.725	2.086	2.423	2.845	3.153	4.146
21	1.323	1.721	2.080	2.414	2.831	3.135	4.110
22	1.321	1.717	2.074	2.405	2.819	3.119	4.077
23	1.319	1.714	2.069	2.398	2.807	3.104	4.047
24	1.318	1.711	2.064	2.391	2.797	3.091	4.021
25	1.316	1.708	2.060	2.385	2.787	3.078	3.996
26	1.315	1.706	2.056	2.379	2.779	3.067	3.974
27	1.314	1.703	2.052	2.373	2.771	3.057	3.954
28	1.313	1.701	2.048	2.368	2.763	3.047	3.935
29	1.311	1.699	2.045	2.364	2.756	3.038	3.918
30	1.310	1.697	2.042	2.360	2.750	3.030	3.902
40	1.303	1.684	2.021	2.329	2.704	2.971	3.788
50	1.299	1.676	2.009	2.311	2.678	2.937	3.723
60	1.296	1.671	2.000	2.299	2.660	2.915	3.681
100	1.290	1.660	1.984	2.276	2.626	2.871	3.598
∞	1.282	1.645	1.960	2.242	2.576	2.808	3.482

T A B L E T·4 (Continued)

Student's *t* for *m* = 3 Simultaneous Confidence Intervals

The entries in this table are $t_{1-\alpha/2m:df}$ for obtaining *m* = 3 two-sided simultaneous confidence intervals with confidence coefficient $1-\alpha$.

df	\.60	\.80	\.90	\.95	\.98	\.99	\.999
			Confidence Coefficient = $1-\alpha$				
1	4.705	9.514	19.081	38.188	95.490	190.984	1909.859
2	2.457	3.677	5.339	7.649	12.186	17.277	54.759
3	2.046	2.821	3.740	4.857	6.741	8.575	18.709
4	1.880	2.501	3.186	3.961	5.167	6.254	11.438
5	1.791	2.337	2.912	3.534	4.456	5.247	8.693
6	1.736	2.237	2.749	3.287	4.058	4.698	7.314
7	1.698	2.169	2.642	3.128	3.806	4.355	6.503
8	1.671	2.122	2.566	3.016	3.632	4.122	5.973
9	1.650	2.086	2.510	2.933	3.505	3.954	5.602
10	1.634	2.057	2.466	2.870	3.409	3.827	5.329
11	1.621	2.035	2.431	2.820	3.334	3.728	5.120
12	1.610	2.017	2.403	2.779	3.273	3.649	4.955
13	1.601	2.001	2.380	2.746	3.223	3.584	4.822
14	1.594	1.989	2.360	2.718	3.181	3.530	4.712
15	1.587	1.977	2.343	2.694	3.146	3.484	4.620
16	1.581	1.968	2.328	2.673	3.115	3.444	4.542
17	1.577	1.959	2.316	2.655	3.089	3.410	4.475
18	1.572	1.952	2.304	2.639	3.065	3.380	4.416
19	1.568	1.945	2.294	2.625	3.045	3.354	4.365
20	1.565	1.940	2.285	2.613	3.026	3.331	4.320
21	1.562	1.934	2.278	2.601	3.010	3.310	4.279
22	1.559	1.929	2.270	2.591	2.995	3.291	4.243
23	1.556	1.925	2.264	2.582	2.982	3.274	4.210
24	1.554	1.921	2.258	2.574	2.970	3.258	4.181
25	1.552	1.918	2.252	2.566	2.959	3.244	4.154
26	1.550	1.914	2.247	2.559	2.949	3.231	4.129
27	1.548	1.911	2.243	2.552	2.939	3.219	4.107
28	1.546	1.908	2.238	2.546	2.930	3.208	4.086
29	1.544	1.906	2.234	2.541	2.922	3.198	4.067
30	1.543	1.903	2.231	2.536	2.915	3.189	4.049
40	1.532	1.885	2.204	2.499	2.862	3.122	3.925
50	1.526	1.875	2.188	2.477	2.831	3.083	3.853
60	1.522	1.868	2.178	2.463	2.811	3.057	3.807
100	1.513	1.854	2.158	2.435	2.771	3.007	3.716
∞	1.501	1.834	2.128	2.394	2.714	2.936	3.589

T A B L E T·4   (Continued)

Student's *t* for *m* = 4 Simultaneous Confidence Intervals

The entries in this table are $t_{1-\alpha/2m:df}$ for obtaining *m* = 4 two-sided simultaneous confidence intervals with confidence coefficient $1 - \alpha$.

df	\.60	\.80	\.90	\.95	\.98	\.99	\.999
			Confidence Coefficient = $1 - \alpha$				
1	6.314	12.706	25.452	50.923	127.321	254.647	2546.479
2	2.920	4.303	6.205	8.860	14.089	19.962	63.234
3	2.353	3.182	4.177	5.392	7.453	9.465	20.604
4	2.132	2.776	3.495	4.315	5.598	6.758	12.312
5	2.015	2.571	3.163	3.810	4.773	5.604	9.235
6	1.943	2.447	2.969	3.521	4.317	4.981	7.708
7	1.895	2.365	2.841	3.335	4.029	4.595	6.814
8	1.860	2.306	2.752	3.206	3.833	4.334	6.234
9	1.833	2.262	2.685	3.111	3.690	4.146	5.830
10	1.812	2.228	2.634	3.038	3.581	4.005	5.533
11	1.796	2.201	2.593	2.981	3.497	3.895	5.306
12	1.782	2.179	2.560	2.934	3.428	3.807	5.128
13	1.771	2.160	2.533	2.896	3.372	3.735	4.984
14	1.761	2.145	2.510	2.864	3.326	3.675	4.865
15	1.753	2.131	2.490	2.837	3.286	3.624	4.766
16	1.746	2.120	2.473	2.813	3.252	3.581	4.682
17	1.740	2.110	2.458	2.793	3.222	3.543	4.609
18	1.734	2.101	2.445	2.775	3.197	3.510	4.547
19	1.729	2.093	2.433	2.759	3.174	3.481	4.491
20	1.725	2.086	2.423	2.744	3.153	3.455	4.443
21	1.721	2.080	2.414	2.732	3.135	3.432	4.399
22	1.717	2.074	2.405	2.720	3.119	3.412	4.361
23	1.714	2.069	2.398	2.710	3.104	3.393	4.326
24	1.711	2.064	2.391	2.700	3.091	3.376	4.294
25	1.708	2.060	2.385	2.692	3.078	3.361	4.265
26	1.706	2.056	2.379	2.684	3.067	3.346	4.239
27	1.703	2.052	2.373	2.676	3.057	3.333	4.215
28	1.701	2.048	2.368	2.669	3.047	3.321	4.193
29	1.699	2.045	2.364	2.663	3.038	3.310	4.172
30	1.697	2.042	2.360	2.657	3.030	3.300	4.154
40	1.684	2.021	2.329	2.616	2.971	3.227	4.020
50	1.676	2.009	2.311	2.591	2.937	3.184	3.944
60	1.671	2.000	2.299	2.575	2.915	3.156	3.895
100	1.660	1.984	2.276	2.544	2.871	3.102	3.799
∞	1.645	1.960	2.242	2.498	2.808	3.024	3.664

T A B L E   T · 4   (Continued)

Student's *t* for *m* = 5 Simultaneous Confidence Intervals

The entries in this table are $t_{1-\alpha/2m:df}$ for obtaining *m* = 5 two-sided simultaneous confidence intervals with confidence coefficient $1 - \alpha$.

df	\multicolumn{7}{c}{Confidence Coefficient = $1 - \alpha$}
	.60	.80	.90	.95	.98	.99	.999
1	7.916	15.895	31.821	63.657	159.153	318.309	3183.099
2	3.320	4.849	6.965	9.925	15.764	22.327	70.700
3	2.605	3.482	4.541	5.841	8.053	10.215	22.204
4	2.333	2.999	3.747	4.604	5.951	7.173	13.034
5	2.191	2.757	3.365	4.032	5.030	5.893	9.678
6	2.104	2.612	3.143	3.707	4.524	5.208	8.025
7	2.046	2.517	2.998	3.499	4.207	4.785	7.063
8	2.004	2.449	2.896	3.355	3.991	4.501	6.442
9	1.973	2.398	2.821	3.250	3.835	4.297	6.010
10	1.948	2.359	2.764	3.169	3.716	4.144	5.694
11	1.928	2.328	2.718	3.106	3.624	4.025	5.453
12	1.912	2.303	2.681	3.055	3.550	3.930	5.263
13	1.899	2.282	2.650	3.012	3.489	3.852	5.111
14	1.887	2.264	2.624	2.977	3.438	3.787	4.985
15	1.878	2.249	2.602	2.947	3.395	3.733	4.880
16	1.869	2.235	2.583	2.921	3.358	3.686	4.791
17	1.862	2.224	2.567	2.898	3.326	3.646	4.714
18	1.855	2.214	2.552	2.878	3.298	3.610	4.648
19	1.850	2.205	2.539	2.861	3.273	3.579	4.590
20	1.844	2.197	2.528	2.845	3.251	3.552	4.539
21	1.840	2.189	2.518	2.831	3.231	3.527	4.493
22	1.835	2.183	2.508	2.819	3.214	3.505	4.452
23	1.832	2.177	2.500	2.807	3.198	3.485	4.415
24	1.828	2.172	2.492	2.797	3.183	3.467	4.382
25	1.825	2.167	2.485	2.787	3.170	3.450	4.352
26	1.822	2.162	2.479	2.779	3.158	3.435	4.324
27	1.819	2.158	2.473	2.771	3.147	3.421	4.299
28	1.817	2.154	2.467	2.763	3.136	3.408	4.275
29	1.814	2.150	2.462	2.756	3.127	3.396	4.254
30	1.812	2.147	2.457	2.750	3.118	3.385	4.234
40	1.796	2.123	2.423	2.704	3.055	3.307	4.094
50	1.787	2.109	2.403	2.678	3.018	3.261	4.014
60	1.781	2.099	2.390	2.660	2.994	3.232	3.962
100	1.769	2.081	2.364	2.626	2.946	3.174	3.862
∞	1.751	2.054	2.327	2.576	2.879	3.091	3.720

T A B L E   T-4   (Continued)

Student's *t* for *m* = 6 Simultaneous Confidence Intervals

The entries in this table are $t_{1-\alpha/2m:df}$ for obtaining *m* = 6 two-sided simultaneous confidence intervals with confidence coefficient $1 - \alpha$.

df	Confidence Coefficient = $1 - \alpha$						
	.60	**.80**	**.90**	**.95**	**.98**	**.99**	**.999**
1	9.514	19.081	38.188	76.390	190.984	381.971	3819.719
2	3.677	5.339	7.649	10.886	17.277	24.464	77.450
3	2.821	3.740	4.857	6.232	8.575	10.869	23.602
4	2.501	3.186	3.961	4.851	6.254	7.529	13.653
5	2.337	2.912	3.534	4.219	5.247	6.138	10.053
6	2.237	2.749	3.287	3.863	4.698	5.398	8.292
7	2.169	2.642	3.128	3.636	4.355	4.944	7.272
8	2.122	2.566	3.016	3.479	4.122	4.640	6.616
9	2.086	2.510	2.933	3.364	3.954	4.422	6.160
10	2.057	2.466	2.870	3.277	3.827	4.259	5.827
11	2.035	2.431	2.820	3.208	3.728	4.132	5.574
12	2.017	2.403	2.779	3.153	3.649	4.031	5.375
13	2.001	2.380	2.746	3.107	3.584	3.948	5.215
14	1.989	2.360	2.718	3.069	3.530	3.880	5.084
15	1.977	2.343	2.694	3.036	3.484	3.822	4.974
16	1.968	2.328	2.673	3.008	3.444	3.773	4.881
17	1.959	2.316	2.655	2.984	3.410	3.730	4.801
18	1.952	2.304	2.639	2.963	3.380	3.692	4.731
19	1.945	2.294	2.625	2.944	3.354	3.660	4.671
20	1.940	2.285	2.613	2.927	3.331	3.630	4.617
21	1.934	2.278	2.601	2.912	3.310	3.604	4.569
22	1.929	2.270	2.591	2.899	3.291	3.581	4.527
23	1.925	2.264	2.582	2.886	3.274	3.560	4.488
24	1.921	2.258	2.574	2.875	3.258	3.540	4.454
25	1.918	2.252	2.566	2.865	3.244	3.523	4.422
26	1.914	2.247	2.559	2.856	3.231	3.507	4.393
27	1.911	2.243	2.552	2.847	3.219	3.492	4.367
28	1.908	2.238	2.546	2.839	3.208	3.479	4.343
29	1.906	2.234	2.541	2.832	3.198	3.466	4.320
30	1.903	2.231	2.536	2.825	3.189	3.454	4.300
40	1.885	2.204	2.499	2.776	3.122	3.372	4.154
50	1.875	2.188	2.477	2.747	3.083	3.324	4.071
60	1.868	2.178	2.463	2.729	3.057	3.293	4.017
100	1.854	2.158	2.435	2.692	3.007	3.232	3.913
∞	1.834	2.128	2.394	2.639	2.936	3.145	3.766

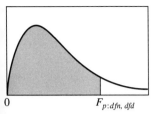

$$0 \qquad F_{p:dfn,\,dfd}$$

T A B L E T - 5
Percentiles of Snedecor's *F* Population

The entries in this table are $F_{p:dfn,dfd}$, where p is the proportion of the population values that are between 0 and $F_{p:dfn,dfd}$, for $p = .50, .90, .95, .975, .99$.

dfn = degrees of freedom for the numerator
dfd = degrees of freedom for the denominator

dfd	p	1	2	3	4	5	6	7	8	9	10	12	14	16	18	20
2	.500	.67	1.00	1.13	1.21	1.25	1.28	1.30	1.32	1.33	1.35	1.36	1.37	1.38	1.39	1.39
2	.900	8.53	9.00	9.16	9.24	9.29	9.33	9.35	9.37	9.38	9.39	9.41	9.42	9.43	9.44	9.44
2	.950	18.51	19.00	19.16	19.25	19.30	19.33	19.35	19.37	19.38	19.40	19.41	19.42	19.43	19.44	19.45
2	.975	38.51	39.00	39.17	39.25	39.30	39.33	39.36	39.37	39.39	39.40	39.41	39.43	39.44	39.44	39.45
2	.990	98.50	99.00	99.17	99.25	99.30	99.33	99.36	99.37	99.39	99.40	99.42	99.43	99.44	99.44	99.45
3	.500	.59	.88	1.00	1.06	1.10	1.13	1.15	1.16	1.17	1.18	1.20	1.21	1.21	1.22	1.23
3	.900	5.54	5.46	5.39	5.34	5.31	5.28	5.27	5.25	5.24	5.23	5.22	5.20	5.20	5.19	5.18
3	.950	10.13	9.55	9.28	9.12	9.01	8.94	8.89	8.85	8.81	8.79	8.74	8.71	8.69	8.67	8.66
3	.975	17.44	16.04	15.44	15.10	14.88	14.73	14.62	14.54	14.47	14.42	14.34	14.28	14.23	14.20	14.17
3	.990	34.12	30.82	29.46	28.71	28.24	27.91	27.67	27.49	27.35	27.23	27.05	26.92	26.83	26.75	26.69
4	.500	.55	.83	.94	1.00	1.04	1.06	1.08	1.09	1.10	1.11	1.13	1.13	1.14	1.15	1.15
4	.900	4.54	4.32	4.19	4.11	4.05	4.01	3.98	3.95	3.94	3.92	3.90	3.88	3.86	3.85	3.84
4	.950	7.71	6.94	6.59	6.39	6.26	6.16	6.09	6.04	6.00	5.96	5.91	5.87	5.84	5.82	5.80
4	.975	12.22	10.65	9.98	9.61	9.36	9.20	9.07	8.98	8.90	8.84	8.75	8.68	8.63	8.59	8.56
4	.990	21.20	18.00	16.69	15.98	15.52	15.21	14.98	14.80	14.66	14.55	14.37	14.25	14.15	14.08	14.02
5	.500	.53	.80	.91	.96	1.00	1.02	1.04	1.05	1.06	1.07	1.09	1.09	1.10	1.11	1.11
5	.900	4.06	3.78	3.62	3.52	3.45	3.40	3.37	3.34	3.32	3.30	3.27	3.25	3.23	3.22	3.21
5	.950	6.61	5.79	5.41	5.19	5.05	4.95	4.88	4.82	4.77	4.74	4.68	4.64	4.60	4.58	4.56
5	.975	10.01	8.43	7.76	7.39	7.15	6.98	6.85	6.76	6.68	6.62	6.52	6.46	6.40	6.36	6.33
5	.990	16.26	13.27	12.06	11.39	10.97	10.67	10.46	10.29	10.16	10.05	9.89	9.77	9.68	9.61	9.55
6	.500	.51	.78	.89	.94	.98	1.00	1.02	1.03	1.04	1.05	1.06	1.07	1.08	1.08	1.08
6	.900	3.78	3.46	3.29	3.18	3.11	3.05	3.01	2.98	2.96	2.94	2.90	2.88	2.86	2.85	2.84
6	.950	5.99	5.14	4.76	4.53	4.39	4.28	4.21	4.15	4.10	4.06	4.00	3.96	3.92	3.90	3.87
6	.975	8.81	7.26	6.60	6.23	5.99	5.82	5.70	5.60	5.52	5.46	5.37	5.30	5.24	5.20	5.17
6	.990	13.75	10.92	9.78	9.15	8.75	8.47	8.26	8.10	7.98	7.87	7.72	7.60	7.52	7.45	7.40
7	.500	.51	.77	.87	.93	.96	.98	1.00	1.01	1.02	1.03	1.04	1.05	1.06	1.06	1.07
7	.900	3.59	3.26	3.07	2.96	2.88	2.83	2.78	2.75	2.72	2.70	2.67	2.64	2.62	2.61	2.59
7	.950	5.59	4.74	4.35	4.12	3.97	3.87	3.79	3.73	3.68	3.64	3.57	3.53	3.49	3.47	3.44
7	.975	8.07	6.54	5.89	5.52	5.29	5.12	4.99	4.90	4.82	4.76	4.67	4.60	4.54	4.50	4.47
7	.990	12.25	9.55	8.45	7.85	7.46	7.19	6.99	6.84	6.72	6.62	6.47	6.36	6.27	6.21	6.16
8	.500	.50	.76	.86	.91	.95	.97	.99	1.00	1.01	1.02	1.03	1.04	1.05	1.05	1.05
8	.900	3.46	3.11	2.92	2.81	2.73	2.67	2.62	2.59	2.56	2.54	2.50	2.48	2.45	2.44	2.42
8	.950	5.32	4.46	4.07	3.84	3.69	3.58	3.50	3.44	3.39	3.35	3.28	3.24	3.20	3.17	3.15
8	.975	7.57	6.06	5.42	5.05	4.82	4.65	4.53	4.43	4.36	4.30	4.20	4.13	4.08	4.03	4.00
8	.990	11.26	8.65	7.59	7.01	6.63	6.37	6.18	6.03	5.91	5.81	5.67	5.56	5.48	5.41	5.36
9	.500	.49	.75	.85	.91	.94	.96	.98	.99	1.00	1.01	1.02	1.03	1.03	1.04	1.04
9	.900	3.36	3.01	2.81	2.69	2.61	2.55	2.51	2.47	2.44	2.42	2.38	2.35	2.33	2.31	2.30
9	.950	5.12	4.26	3.86	3.63	3.48	3.37	3.29	3.23	3.18	3.14	3.07	3.03	2.99	2.96	2.94
9	.975	7.21	5.71	5.08	4.72	4.48	4.32	4.20	4.10	4.03	3.96	3.87	3.80	3.74	3.70	3.67
9	.990	10.56	8.02	6.99	6.42	6.06	5.80	5.61	5.47	5.35	5.26	5.11	5 01	4.92	4.86	4.81
10	.500	.49	.74	.85	.90	.93	.95	.97	.98	.99	1.00	1.01	1.02	1.03	1.03	1.03
10	.900	3.29	2.92	2.73	2.61	2.52	2.46	2.41	2.38	2.35	2.32	2.28	2.26	2.23	2.22	2.20
10	.950	4.96	4.10	3.71	3.48	3.33	3.22	3.14	3.07	3.02	2.98	2.91	2.86	2.83	2.80	2.77
10	.975	6.94	5.46	4.83	4.47	4.24	4.07	3.95	3.85	3.78	3.72	3.62	3.55	3.50	3.45	3.42
10	.990	10.04	7.56	6.55	5.99	5.64	5.39	5.20	5.06	4.94	4.85	4.71	4.60	4.52	4.46	4.41

T A B L E T-5 (Continued)

Percentiles of Snedecor's *F* Population

The entries in this table are $F_{p:dfn,dfd}$, where *p* is the proportion of the population values that are between 0 and $F_{p:dfn,dfd}$, for *p* = .50, .90, .95, .975, .99.

dfn = degrees of freedom for the numerator
dfd = degrees of freedom for the denominator

dfp	p	1	2	3	4	5	6	7	8	9	10	12	14	16	18	20
11	.500	.49	.74	.84	.89	.93	.95	.96	.98	.99	.99	1.01	1.01	1.02	1.02	1.03
11	.900	3.23	2.86	2.66	2.54	2.45	2.39	2.34	2.30	2.27	2.25	2.21	2.18	2.16	2.14	2.12
11	.950	4.84	3.98	3.59	3.36	3.20	3.09	3.01	2.95	2.90	2.85	2.79	2.74	2.70	2.67	2.65
11	.975	6.72	5.26	4.63	4.28	4.04	3.88	3.76	3.66	3.59	3.53	3.43	3.36	3.30	3.26	3.23
11	.990	9.65	7.21	6.22	5.67	5.32	5.07	4.89	4.74	4.63	4.54	4.40	4.29	4.21	4.15	4.10
12	.500	.48	.73	.84	.89	.92	.94	.96	.97	.98	.99	1.00	1.01	1.01	1.02	1.02
12	.900	3.18	2.81	2.61	2.48	2.39	2.33	2.28	2.24	2.21	2.19	2.15	2.12	2.09	2.08	2.06
12	.950	4.75	3.89	3.49	3.26	3.11	3.00	2.91	2.85	2.80	2.75	2.69	2.64	2.60	2.57	2.54
12	.975	6.55	5.10	4.47	4.12	3.89	3.73	3.61	3.51	3.44	3.37	3.28	3.21	3.15	3.11	3.07
12	.990	9.33	6.93	5.95	5.41	5.06	4.82	4.64	4.50	4.39	4.30	4.16	4.05	3.97	3.91	3.86
13	.500	.48	.73	.83	.88	.92	.94	.96	.97	.98	.98	1.00	1.00	1.01	1.01	1.02
13	.900	3.14	2.76	2.56	2.43	2.35	2.28	2.23	2.20	2.16	2.14	2.10	2.07	2.04	2.02	2.01
13	.950	4.67	3.81	3.41	3.18	3.03	2.92	2.83	2.77	2.71	2.67	2.60	2.55	2.51	2.48	2.46
13	.975	6.41	4.97	4.35	4.00	3.77	3.60	3.48	3.39	3.31	3.25	3.15	3.08	3.03	2.98	2.95
13	.990	9.07	6.70	5.74	5.21	4.86	4.62	4.44	4.30	4.19	4.10	3.96	3.86	3.78	3.72	3.66
14	.500	.48	.73	.83	.88	.91	.94	.95	.96	.97	.98	.99	1.00	1.01	1.01	1.01
14	.900	3.10	2.73	2.52	2.39	2.31	2.24	2.19	2.15	2.12	2.10	2.05	2.02	2.00	1.98	1.96
14	.950	4.60	3.74	3.34	3.11	2.96	2.85	2.76	2.70	2.65	2.60	2.53	2.48	2 44	2.41	2.39
14	.975	6.30	4.86	4.24	3.89	3.66	3.50	3.38	3.29	3.21	3.15	3.05	2.98	2.92	2.88	2.84
14	.990	8.86	6.51	5.56	5.04	4.69	4.46	4.28	4.14	4.03	3.94	3.80	3.70	3.62	3.56	3.51
15	.500	.48	.73	.83	.88	.91	.93	.95	.96	.97	.98	.99	1.00	1.00	1.01	1.01
15	.900	3.07	2.70	2.49	2.36	2.27	2.21	2.16	2.12	2.09	2.06	2.02	1.99	1.96	1.94	1.92
15	.950	4.54	3.68	3.29	3.06	2.90	2.79	2.71	2.64	2.59	2.54	2.48	2.42	2.38	2.35	2.33
15	.975	6.20	4.77	4.15	3.80	3.58	3.41	3.29	3.20	3.12	3.06	2.96	2.89	2.84	2.79	2.76
15	.990	8.68	6.36	5.42	4.89	4.56	4.32	4.14	4.00	3.89	3.80	3.67	3.56	3.49	3.42	3.37
16	.500	.48	.72	.82	.88	.91	.93	.95	.96	.97	.97	.99	.99	1.00	1.00	1.01
16	.900	3.05	2.67	2.46	2.33	2.24	2.18	2.13	2.09	2.06	2.03	1.99	1.95	1.93	1.91	1.89
16	.950	4.49	3.63	3.24	3.01	2.85	2.74	2.66	2.59	2.54	2.49	2.42	2.37	2.33	2.30	2.28
16	.975	6.12	4.69	4.08	3.73	3.50	3.34	3.22	3.12	3.05	2.99	2.89	2.82	2.76	2.72	2.68
16	.990	8.53	6.23	5.29	4.77	4.44	4.20	4.03	3.89	3.78	3.69	3.55	3.45	3.37	3.31	3.26
17	.500	.47	.72	.82	.87	.91	.93	.94	.96	.96	.97	.98	.99	1.00	1.00	1.01
17	.900	3.03	2.64	2.44	2.31	2.22	2.15	2.10	2.06	2.03	2.00	1.96	1.93	1.90	1.88	1.86
17	.950	4.45	3.59	3.20	2.96	2.81	2.70	2.61	2.55	2.49	2.45	2.38	2.33	2.29	2.26	2.23
17	.975	6.04	4.62	4.01	3.66	3.44	3.28	3.16	3.06	2.98	2.92	2.82	2.75	2.70	2.65	2.62
17	.990	8.40	6.11	5.18	4.67	4.34	4.10	3.93	3.79	3.68	3.59	3.46	3.35	3.27	3.21	3.16
18	.500	.47	.72	.82	.87	.90	.93	.94	.95	.96	.97	.98	.99	1.00	1.00	1.00
18	.900	3.01	2.62	2.42	2.29	2.20	2.13	2.08	2.04	2.00	1.98	1.93	1.90	1.87	1.85	1.84
18	.950	4.41	3.55	3.16	2.93	2.77	2.66	2.58	2.51	2.46	2.41	2.34	2.29	2.25	2.22	2.19
18	.975	5.98	4.56	3.95	3.61	3.38	3.22	3.10	3.01	2.93	2.87	2.77	2.70	2.64	2.60	2.56
18	.990	8.29	6.01	5.09	4.58	4.25	4.01	3.84	3.71	3.60	3.51	3.37	3.27	3.19	3.13	3.08
19	.500	.47	.72	.82	.87	.90	.92	.94	.95	.96	.97	.98	.99	.99	1.00	1.00
19	.900	2.99	2.61	2.40	2.27	2.18	2.11	2.06	2.02	1.98	1.96	1.91	1.88	1.85	1.83	1.81
19	.950	4.38	3.52	3.13	2.90	2.74	2.63	2.54	2.48	2.42	2.38	2.31	2.26	2.21	2.18	2.16
19	.975	5.92	4.51	3.90	3.56	3.33	3.17	3.05	2.96	2.88	2.82	2.72	2.65	2.59	2.55	2.51
19	.990	8.18	5.93	5.01	4.50	4.17	3.94	3.77	3.63	3.52	3.43	3.30	3.19	3.12	3.05	3.00

T A B L E T · 5 (Continued)
Percentiles of Snedecor's *F* Population

The entries in this table are $F_{p:dfn,dfd}$, where p is the proportion of the population values that are between 0 and $F_{p:dfn,dfd}$, for $p = .50, .90, .95, .975, .99$.

dfn = degrees of freedom for the numerator
dfd = degrees of freedom for the denominator

dfd	p	1	2	3	4	5	6	7	8	9	10	12	14	16	18	20
20	.500	.47	.72	.82	.87	.90	.92	.94	.95	.96	.97	.98	.99	.99	1.00	1.00
20	.900	2.97	2.59	2.38	2.25	2.16	2.09	2.04	2.00	1.96	1.94	1.89	1.86	1.83	1.81	1.79
20	.950	4.35	3.49	3.10	2.87	2.71	2.60	2.51	2.45	2.39	2.35	2.28	2.22	2.18	2.15	2.12
20	.975	5.87	4.46	3.86	3.51	3.29	3.13	3.01	2.91	2.84	2.77	2.68	2.60	2.55	2.50	2.46
20	.990	8.10	5.85	4.94	4.43	4.10	3.87	3.70	3.56	3.46	3.37	3.23	3.13	3.05	2.99	2.94
30	.500	.47	.71	.81	.86	.89	.91	.93	.94	.95	.96	.97	.97	.98	.99	.99
30	.900	2.88	2.49	2.28	2.14	2.05	1.98	1.93	1.88	1.85	1.82	1.77	1.74	1.71	1.69	1.67
30	.950	4.17	3.32	2.92	2.69	2.53	2.42	2.33	2.27	2.21	2.16	2.09	2.04	1.99	1.96	1.93
30	.975	5.57	4.18	3.59	3.25	3.03	2.87	2.75	2.65	2.57	2.51	2.41	2.34	2.28	2.23	2.20
30	.990	7.56	5.39	4.51	4.02	3.70	3.47	3.30	3.17	3.07	2.98	2.84	2.74	2.66	2.60	2.55
40	.500	.46	.71	.80	.85	.89	.91	.92	.93	.94	.95	.96	.97	.97	.98	.98
40	.900	2.84	2.44	2.23	2.09	2.00	1.93	1.87	1.83	1.79	1.76	1.71	1.68	1.65	1.62	1.61
40	.950	4.08	3.23	2.84	2.61	2.45	2.34	2.25	2.18	2.12	2.08	2.00	1.95	1.90	1.87	1.84
40	.975	5.42	4.05	3.46	3.13	2.90	2.74	2.62	2.53	2.45	2.39	2.29	2.21	2.15	2.11	2.07
40	.990	7.31	5.18	4.31	3.83	3.51	3.29	3.12	2.99	2.89	2.80	2.66	2.56	2.48	2.42	2.37
50	.500	.46	.70	.80	.85	.88	.90	.92	.93	.94	.95	.96	.97	.97	.98	.98
50	.900	2.81	2.41	2.20	2.06	1.97	1.90	1.84	1.80	1.76	1.73	1.68	1.64	1.61	1.59	1.57
50	.950	4.03	3.18	2.79	2.56	2.40	2.29	2.20	2.13	2.07	2.03	1.95	1.89	1.85	1.81	1.78
50	.975	5.34	3.97	3.39	3.05	2.83	2.67	2.55	2.46	2.38	2.32	2.22	2.14	2.08	2.03	1.99
50	.990	7.17	5.06	4.20	3.72	3.41	3.19	3.02	2.89	2.78	2.70	2.56	2.46	2.38	2.32	2.27
60	.500	.46	.70	.80	.85	.88	.90	.92	.93	.94	.94	.96	.96	.97	.97	.98
60	.900	2.79	2.39	2.18	2.04	1.95	1.87	1.82	1.77	1.74	1.71	1.66	1.62	1.59	1.56	1.54
60	.950	4.00	3.15	2.76	2.53	2.37	2.25	2.17	2.10	2.04	1.99	1.92	1.86	1.82	1.78	1.75
60	.975	5.29	3.93	3.34	3.01	2.79	2.63	2.51	2.41	2.33	2.27	2.17	2.09	2.03	1.98	1.94
60	.990	7.08	4.98	4.13	3.65	3.34	3.12	2.95	2.82	2.72	2.63	2.50	2.39	2.31	2.25	2.20
70	.500	.46	.70	.80	.85	.88	.90	.92	.93	.94	.94	.95	.96	.97	.97	.98
70	.900	2.78	2.38	2.16	2.03	1.93	1.86	1.80	1.76	1.72	1.69	1.64	1.60	1.57	1.55	1.53
70	.950	3.98	3.13	2.74	2.50	2.35	2.23	2.14	2.07	2.02	1.97	1.89	1.84	1.79	1.75	1.72
70	.975	5.25	3.89	3.31	2.97	2.75	2.59	2.47	2.38	2.30	2.24	2.14	2.06	2.00	1.95	1.91
70	.990	7.01	4.92	4.07	3.60	3.29	3.07	2.91	2.78	2.67	2.59	2.45	2.35	2.27	2.20	2.15
80	.500	.46	.70	.80	.85	.88	.90	.91	.93	.93	.94	.95	.96	.97	.97	.97
80	.900	2.77	2.37	2.15	2.02	1.92	1.85	1.79	1.75	1.71	1.68	1.63	1.59	1.56	1.53	1.51
80	.950	3.96	3.11	2.72	2.49	2.33	2.21	2.13	2.06	2.00	1.95	1.88	1.82	1.77	1.73	1.70
80	.975	5.22	3.86	3.28	2.95	2.73	2.57	2.45	2.35	2.28	2.21	2.11	2.03	1.97	1.92	1.88
80	.990	6.96	4.88	4.04	3.56	3.26	3.04	2.87	2.74	2.64	2.55	2.42	2.31	2.23	2.17	2.12
90	.500	.46	.70	.79	.85	.88	.90	.91	.92	.93	.94	.95	.96	.97	.97	.97
90	.900	2.76	2.36	2.15	2.01	1.91	1.84	1.78	1.74	1.70	1.67	1.62	1.58	1.55	1.52	1.50
90	.950	3.95	3.10	2.71	2.47	2.32	2.20	2.11	2.04	1.99	1.94	1.86	1.80	1.76	1.72	1.69
90	.975	5.20	3.84	3.26	2.93	2.71	2.55	2.43	2.34	2.26	2.19	2.09	2.02	1.95	1.91	1.86
90	.990	6.93	4.85	4.01	3.53	3.23	3.01	2.84	2.72	2.61	2.52	2.39	2.29	2.21	2.14	2.09
100	.500	.46	.70	.79	.84	.88	.90	.91	.92	.93	.94	.95	.96	.97	.97	.97
100	.900	2.76	2.36	2.14	2.00	1.91	1.83	1.78	1.73	1.69	1.66	1.61	1.57	1.54	1.52	1.49
100	.950	3.94	3.09	2.70	2.46	2.31	2.19	2.10	2.03	1.97	1.93	1.85	1.79	1.75	1.71	1.68
100	.975	5.18	3.83	3.25	2.92	2.70	2.54	2.42	2.32	2.24	2.18	2.08	2.00	1.94	1.89	1.85
100	.990	6.90	4.82	3.98	3.51	3.21	2.99	2.82	2.69	2.59	2.50	2.37	2.27	2.19	2.12	2.07

T A B L E T-6

Table for Obtaining Confidence Bounds for $a_0\beta_0 + a_1\beta_1$ using Theil's Method.

This table has values of r and m that can be used to compute confidence intervals for $a_0\beta_0 + a_1\beta_1$ using Theil's method. The table entries are the allowable confidence coefficients which range from .50 to .99 for each value of m.

m	r 1	2	3	4	5	6	7	8	9	10
4	0.88									
5	0.94	0.63								
6	0.97	0.78								
7	0.98	0.88	0.55							
8	0.99	0.93	0.71							
9		0.96	0.82							
10		0.98	0.89	0.66						
11		0.99	0.93	0.77						
12		0.99	0.96	0.85	0.61					
13			0.98	0.91	0.73					
14			0.99	0.94	0.82	0.58				
15			0.99	0.96	0.88	0.70				
16				0.98	0.92	0.79	0.55			
17				0.99	0.95	0.86	0.67			
18				.099	0.97	0.90	0.76	0.52		
19					0.98	0.94	0.83	0.64		
20					0.99	0.96	0.88	0.74		
21					0.99	0.97	0.92	0.81	0.62	
22						0.98	0.95	0.87	0.71	
23						0.99	0.97	0.91	0.79	0.60

T A B L E I·7

Chart Giving Confidence Bounds for the Population Correlation Coefficient, $\rho_{Y.X}, = \rho$, Given the Sample Correlation Coefficient $\hat{\rho}_{Y.X} = r$.

Confidence coefficient for the two-sided interval is 0.95.

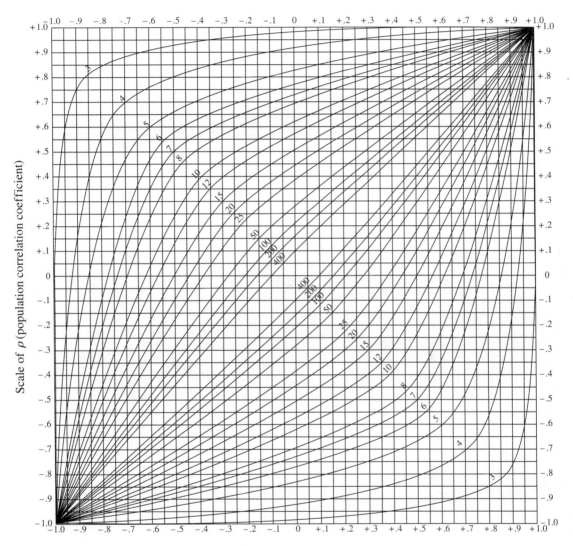

Scale of ρ (population correlation coefficient)

Scale of r (sample correlation coefficient)

Reprinted with permission from *Biometrika Tables for Statisticians,* Volume I, edited by E. S. Pearson and H. O. Hartley.

T A B L E T·7 (Continued)

Chart Giving Confidence Bounds for the Population Correlation Coefficient, $\rho_{Y,X} = \rho$, Given the Sample Correlation Coefficient $\hat{\rho}_{Y,X} = r$.

Confidence coefficient for the two-sided interval is 0.99.

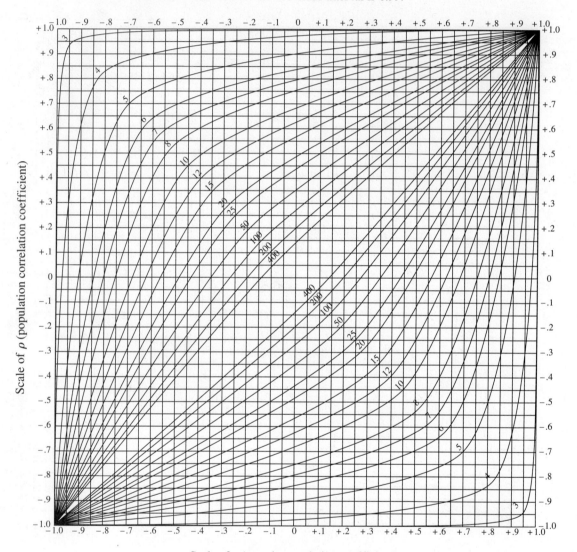

Scale of ρ (population correlation coefficient)

Scale of r (sample correlation coefficient)

Reprinted with permission from *Biometrika Tables for Statisticians,* Volume I, edited by E. S. Pearson and H. O. Hartley.

T A B L E T-8

Selected Percentiles of the Noncentral *t*

The entries in this table are $t_{\gamma:df;(\delta)}$, where γ is the proportion of the population values that are between $-\infty$ and $t_{\gamma:df;(\delta)}$, for selected degrees of freedom (df) and selected values of δ.

$t_{0.975:16;(-3.242)} = -1.254$	$t_{0.025:16;(-3.242)} = -6.174$
$t_{0.95:16;(-3.992)} = -2.255$	$t_{0.05:16;(-3.992)} = -6.568$
$t_{.975:42;(-6.099)} = -3.983$	$t_{.025:42;(-6.099)} = -8.847$
$t_{.90:42;(-6.099)} = -4.687$	$t_{.10:42;(-6.099)} = -7.824$
$t_{.95:42;(9.847)} = 12.699$	$t_{.05:42;(9.847)} = 7.733$
$t_{.975:18;(-9.112)} = -6.319$	$t_{.025:18;(-9.112)} = -14.119$
$t_{.90:18;(-4.469)} = -3.072$	$t_{.10:18;(-4.469)} = -6.416$
$t_{.975:28;(-9.350)} = -6.750$	$t_{.025:28;(-9.350)} = -13.380$
$t_{.975:18;(-4.673)} = -2.573$	$t_{.025:18;(-4.673)} = -7.952$
$t_{.975:18;(-8.483)} = -5.808$	$t_{.025:18;(-8.483)} = -13.227$

Note: $t_{1-\alpha:n-k-1;(\delta)} = -t_{\alpha:n-k-1;(-\delta)}$.

Index

A

Adequate predictor 182, 183, 295
Adj$-\rho^2$ 302
Adj$-R^2$ 507, 509
All-subsets regression 504
  Adj$-R^2$ 507, 509
  C_p 508, 509
  Root mean square (s) 506, 509
  R-square (R^2) 506, 509
Analysis of variance 178, 283
Assumptions A and B 109, 110,
  111, 232, 233, 293, 310

B

Basic observable variables 221
Best estimates of β_0, β_1 114
Best estimates of β_0, β_1, . . . , β_k
  237
Best prediction function 114
Bonferroni 40, 419

C

Calibration 425, 427
Cause and effect 1
Change of units 125
Chebyshev's theorem 16, 88
Checking Assumptions 132
Coefficient of correlation 17, 181,
  185
  confidence interval 187
  point estimate 186
  properties of 186
Coefficient of determination 181,
  185, 291, 296
  point estimate 301, 303
  point estimate (alternate) 302,
    303

properties of 298
tests 306
Comparison of several straight
  line regressions 436
Comparison of two regression
  functions (nested case) 291
Comparison of two regression
  functions (nonnested case)
  309
Confidence coefficient (level) 24,
Confidence intervals 22, 24, 31,
  161, 262
  equal-tailed 25, 265
  for $a_0\beta_0 + a_1\beta_1$ 161
  for $a_0\beta_0 + a_1\beta_1 + \ldots + a_k\beta_k$
    262
  for β_i 161, 262,
  for lack-of-fit 322
  for linear combination of β_i
    161, 262
  for maximum (minimum) of a
    quadratic regression model
    460
  for mean 28, 31
  for μ_Y, 28, 31
  for $\mu_Y(x)$ 161
  for $\mu_Y(x_1, \ldots, x_k)$ 262
  for nonlinear regression 610
  for point of intersection of two
    straight line regressions
    452
  for ratio of sigmas 187, 188,
    303, 311
  for ratio of standard deviations
    187, 188, 303, 311
  for σ 31, 163, 264
  for σ_B/σ_A 303, 311
  for σ_Y/σ 187

  for σ_Y 31
  for spline regression 470
  for $Y(x)$ 161
  for $Y(x_1, \ldots, x_k)$ 262
  lower bound 26, 263
  one-at-a-time 37
  one-sided 25, 163
  simultaneous 36, 38, 40, 170
  symmetric 26
  upper bound 26, 263
Controllable and noncontrollable
  factors 301
Conversation 41, 87, 128, 190,
  205, 269, 273, 334, 389
Cook's distance 372, 384
Correlation 17, 181, 185

D

Data splitting 545
Degrees of freedom 241
Derived variables 221
DFFITS 372, 373, 384
Diagnostics 156, 351, 383

E

Error of prediction 74
Error sum of squares 113
Estimate of regression coefficients
  in multiple linear regression
    238
  in nonlinear regression 606,
    607
  in straight line regression 114
  in straight line regression
    through the origin 210
  in spline regression 469
Estimate of sigma 115, 240
Explanatory variable 74

F

Fitted values 139, 148, 157, 252
Functional notation 47
Functions 47
  of many variables 48
  linear 48
  multivariate 48

G

Gaussian assumptions 110, 111,
  232, 233
  checking 143, 145, 252, 253
Gaussian (normal) population 2,
  10, 62, 88, 103, 110, 111,
  233, 258
Gaussian rankit-plot 143, 145,
  155, 252, 253
Gaussian scores (nscores) 143,
  145, 253, 258
Growth curves 551
  assumptions 555
  confidence intervals 560
  point estimates 560

H

Hand-held calculator 2
Hat values 136, 148, 365, 384
Histograms 13, 64
Homogeneity of standard devia-
  tions (variances) 110, 139

I

Identical regression lines 436
Ill-conditioning 392
Independent variables 48
Inference 1, 3, 20, 22
Influential observations 371
  Cook's distance 372
  DFFITS 372, 373
Intersection of two regression
  lines 436, 450

L

Laboratory manual 2, 3, 116, 146
Lack-of-fit 143, 318
  confidence intervals 322
  point estimates 322
  pure error 325
  sum of squares 330
  tests 329
Least absolute deviations 107
Least squares 107, 112, 113, 235
Leverages 365
Linear combination of Gaussian
  variables 63, 147, 156, 160
Linear functions 48, 239
Linear splines 465

M

Matrices 50
  addition 54
  equality 52
  inversion 60
  multiplication 54
  special 58
    diagonal matrix 59
    identity 59
    symmetric 60
    zero 59
  subtraction 54
  transposition 52
Maximum (minimum) of
  quadratic regression function
  456
Mean 4, 12, 15
Mean squared error 115
  for lack-of-fit 330
  for nonlinear regression 607
Measurement errors 110, 194,
  234
  assumptions 195
  Berkson model 195, 197
  classical errors in variables
    model 195, 197
  definition 194, 195
  in predictor variables 194
  in response variable 194, 197
Measures of goodness of prediction 294
Minimum of quadratic regression
  function 456
Minitab 2, 116, 146, 157, 242,
  258
Model 3, 10
Multicollinearity 392, 396, 397
Multiple linear regression 219
Multivariate Gaussian 62, (see
  Gaussian)
Multivariate populations 16

N

Noncentral t 420
Non-Gaussian assumptions 584
Nonlinear regression 96, 599
  assumptions 605
  confidence intervals 610
  families of functions 600
  least squares 607
  linearizing transformation 616
  point estimation 606
  tests 610
Normal equations 237
Notation 8, 47, 115, 220, 222,
  292
Nscores 143, 145, 156, 160, 258

O

Outliers 133, 352

P

Parallel regression lines 436
Parameters 3, 14, 16, 103, 223
Point estimation 22, 112, 235
  for lack-of-fit 322
  of β_i 114, 117, 236
  of correlation 23
  for intersection of two regres-
    sion lines 452
  of linear combination of β_i
    114, 118, 239
  for maximum (minimum) of a
    quadratic regression model
    460
  of mean 22, 31
  of multiple coefficient of deter-
    mination 301
  for β_i in nonlinear regression
    606
  of standard deviation 23, 31,
    118, 240
  unbiased 23
Polynomial regression 222
Population 1, 3, 4, 77, See Gaussian
  bivariate 7
  conceptual 4
  imagined 4
  items 4
  k-variate 7
  k-variable 7
  multivariate 7, 16
  parameters 4, 14, 103, 223
  real 4
  regression function 117
  regression model 111, 117, 224
  study 9, 82, 110
  target 9, 74, 82
  trivariate 7
  univariate 7
Power of test 33
Predicted values 139
Prediction 1, 73, 80, 99, 219
Prediction error 74, 112, 224
Prediction function 74, 81, 84,
  113, 117
  adequacy 295
Prediction interval 29, 403
  for average of h future values
    405
  for sum of h future values 406
Predictor variable (factor) 74, 81
P-value 32

R

Rankit plot 154, 252, 253
Regression 1, 73
  assumptions 109, 132, 232
  diagnostics 351
  function 83, 84, 86
  intercept 99
  linear 96
  multiple 86, 219
  nonlinear 96, 599
  polynomial 222
  sample regression model 112,
    117
  slope 99
  straight line 86, 99, 109
  through the origin 210
  using matrices 237
Regulation 425, 431
Residuals 113, 148, 154, 157,
    240, 251, 351, 384
Residual sum of squares 240
Response variable 74

S

Sample regression function 117
Sample regression model 112, 117
Samples 1, 3, 20, 21, 104, 224
  simple random 21, 94, 110,
    111, 233, 234
  with preselected X values 94,
    110, 233
SAS 2, 116, 146, 157, 242, 258
Selection of variables 502
Significance test 32
Size of test 32
Spline regression 465
Standard deviation 12, 15, 294
Standard errors 27
  for $\hat{\beta}_i$, $\hat{\mu}_Y(x)$, $\hat{Y}(x)$, and
    $a_0\hat{\beta}_0 + a_1\hat{\beta}_1$ 161

for $\hat{\beta}_i$, $\hat{\mu}_Y(x_1, \ldots, x_k)$,
    $\hat{Y}(x_1, \ldots, x_k)$, and
    $a_0\hat{\beta}_0 + \cdots + a_k\hat{\beta}_k$ 263
  for lack-of-fit 324
  in multiple linear regression
    263
  in nonlinear regression 610
  in straight line regression 161
  in spline regression 470
Standardized residuals 133, 136,
    137, 146, 148, 151, 154,
    157, 252, 351, 352, 384
Statistical computing package 2
Statistical inference 3
Straight line regression 99
Studentized deleted residuals 353,
    354, 384
Subpopulation 82, 85, 225
Subset analysis 501
Subset selection 501
  backward elimination 520, 528,
    533
  forward selection 520, 533
  stepwise regression 520, 533
Sum of crossproducts 115
Sum of squared errors 113, 115,
    178, 240, 440
Sum of squares
  corrected total 178, 283
  due to error 179, 284, 440
  due to regression 179, 284
  for pure error 325
  uncorrected total 178

T

Task 89, 118, 122, 163, 167, 174,
    175, 184, 200, 202, 225,
    246, 265, 406
Tests 22, 30, 33, 171, 278
  for β_i 172, 278
  for linear combination of β_i
    173, 278

for multiple correlation
    coefficients 306
  for nonlinear regression 610
  for σ 173, 278
  relationship with confidence in-
    tervals 34
  simultaneous 36
Theil's method for regression 584
  assumptions 584
  confidence intervals 586
  point estimation 585
Tolerance intervals 416
Tolerance points 418

U

Unequal subpopulation standard
    deviations 571

V

Variables (basic and derived) 221
Variable selection 501
  all-subsets regression 503, 504
  backward elimination 503, 520,
    528
  forward selection 503, 520
  stepwise regression 503, 520,
    533
Variance 13
Variance inflation factor 398
Vectors 50

W

Weighted least squares 573
Weighted regression 571
  assumptions 572
  confidence intervals 574
  least squares 573
  point estimates 574
  sum of squared errors 574
  unequal standard deviations
    571

Index for Data Sets

A

absorpt.dat 614
age18.dat 290, 307, 343, 344, 546
agebp.dat 167, 181, 189
arsenic.dat 122, 164, 174
assay.dat 479, 480

B

babywt.dat 497
ballbear.dat 482
bivgauss.dat 63
bivngaus.dat 66
bp.dat 326
bpweight.dat 477, 478

C

cabbage.dat 207, 208
cans.dat 200, 201
car.dat 80, 131, 170
car2.dat 97
car17.dat 337, 338
car20.dat 151, 152, 170
carbmon.dat 576, 577
cereal.dat 332
chamber.dat 433
chol.dat 215, 216
coil.dat 612
concrete.dat 464
contrast.dat 617
crystal.dat 118, 119, 148, 184,
    214, 361, 412, 434
crystal3.dat 447, 448

D

drugconc.dat 561, 562

E

eggshell.dat 443, 453
electric.dat 253, 254, 275, 280

ethnic.dat 414, 424
exam.dat 595

F

fat.dat 203

G

gifted.dat 512
gpa.dat 242, 243, 261, 266, 510,
    524, 535, 538, 545
grades.dat 100
grades26.dat 104, 105, 132, 181
gravity.dat 212
grocery.dat 287, 288

H

haze.dat 624

L

larch.dat 567, 568
light.dat 608

M

mammalwt.dat 367
minivan.dat 209
mouse.dat 335, 336
mpg.dat 489, 490

N

nickel.dat 491, 492

P

plant.dat 340, 341
plastic.dat 225, 230
premiums.dat 354
prison.dat 483, 484
process1.dat 126, 127
process2.dat 126, 127
profsal.dat 586, 587
pumpkin.dat 565, 566

R

ramus.dat 564

S

salaries.dat 486, 487
sales.dat 470, 471
serum.dat 622, 623
shelflif.dat 176, 177, 180
so2.dat 158, 188, 435, 581, 590
soyburgr.dat 592, 593
sulfuric.dat 461, 474

T

table161.dat 28
table164.dat 46
table323.dat 108, 193
table324.dat 109
table346.dat 131
table355.dat 158
table423.dat 225, 226, 227
table425.dat 229
table426.dat 231
table442.dat 238
table444.dat 246, 259
table445.dat 250
table541.dat 373, 374
table542.dat 379
table543.dat 386
table561.dat 399, 400
table631.dat 420, 421
table733.dat 516, 517
table742.dat 541
thermom.dat 430
track.dat 476
turkey.dat 495

U

usedcars.dat 407, 408, 411

Terminology

The following box summarizes terminology associated with straight line regression.

Population regression function, or simply, the **regression function:**

$$\mu_Y(x) = \beta_0 + \beta_1 x \qquad \text{for } a \leq x \leq b$$

Sample regression function:

$$\hat{\mu}_Y(x) = \hat{\beta}_0 + \hat{\beta}_1 x$$

Population regression model, or simply, the **regression model:**

$$Y_I = \beta_0 + \beta_1 X_I + E_I \qquad \text{for } I = 1, \ldots, N$$

Sample regression model:

$$y_i = \beta_0 + \beta_1 x_i + e_i \quad \text{for} \quad i = 1, \ldots, n$$

A randomly chosen Y value from the subpopulation determined by $X = x$:

$$Y(x)$$

Sample prediction function, or simply, **prediction function:**

$$\hat{Y}(x) = \hat{\beta}_0 + \hat{\beta}_1 x$$

Note: $\hat{\mu}_Y(x) = \hat{Y}(x).$
